Up and Running with AutoCAD 2019

Up and Running with AutoCAD 2019

2D Drafting and Design

Elliot J. Gindis
President - Vertical Technologies, Los Angeles, CA, United States

Robert C. Kaebisch
Construction Sciences - Architecture/Structural, Gateway Technical College
Sturtevant, WI, United States

Academic Press is an imprint of Elsevier
125 London Wall, London EC2Y 5AS, United Kingdom
525 B Street, Suite 1650, San Diego, CA 92101, United States
50 Hampshire Street, 5th Floor, Cambridge, MA 02139, United States
The Boulevard, Langford Lane, Kidlington, Oxford OX5 1GB, United Kingdom

Library of Congress Cataloging-in-Publication Data
A catalog record for this book is available from the Library of Congress

British Library Cataloguing-in-Publication Data
A catalogue record for this book is available from the British Library

ISBN: 978-0-12-816440-2

For information on all Academic Press publications visit our website at https://www.elsevier.com/books-and-journals

Publisher: Katey Birtcher
Senior Acquisition Editor: Steve Merken
Editorial Project Manager: Peter Jardim
Production Project Manager: Anitha Sivaraj
Designer: Victoria Pearson

Typeset by TNQ Technologies

Contents

Spotlight On: Interior Design

7. Blocks, Wblocks, Dynamic Blocks, Groups, and Purge

8. Polar, Rectangular, and Path Arrays

Spotlight On: Electrical Engineering

9. Basic Printing and Output

10. Advanced Output—Paper Space

Level 1
Answers to Review Questions

Level 2
Chapters 11–20

11. Advanced Linework

About the Authors

Elliot J. Gindis

Elliot J. Gindis started out using AutoCAD professionally at a New York City area civil engineering company in September 1996, moving on to consulting work shortly afterward. He has since drafted in a wide variety of fields ranging from all aspects of architecture and building design to electrical, mechanical, civil, structural, aerospace, and rail design. These assignments, including lengthy stays with IBM and Siemens Transportation Systems, totaled over 50 companies to date.

In 1999, Elliot began teaching part-time at the Pratt Institute of Design, followed by positions at Netcom Information Systems, RoboTECH CAD Solutions, and more recently at the New York Institute of Technology. In 2003, Elliot formed Vertical Technologies Consulting and Design an AutoCAD training firm that continues to train corporate clients nationwide in using and optimizing AutoCAD.

Elliot holds a bachelor's degree in aerospace engineering from Embry-Riddle Aeronautical University, and a master's degree in engineering management from Mercer University. As of 2016, he resides in the Los Angeles area and is a flight test engineer for the US Air Force. He also continues to be involved with AutoCAD education and CAD consulting. *Up and Running with AutoCAD 2019*, which carefully incorporates lessons learned from over 23 years of teaching and industry work, is his 14th overall textbook on the subject. His 2012 textbook has also been translated and is available in Spanish from Anaya Multimedia. He can be reached at Elliot.Gindis@gmail.com or via his website: www.verticaltechtraining.com.

Previous Textbooks by Elliot J. Gindis:
English

- *Operational AutoCAD 2008.* New York: Netcom Inc. (out of print)
- *Up and Running with AutoCAD 2009* [e-book only]. Waltham, MA: Elsevier, Inc.
- *Up and Running with AutoCAD 2010.* Waltham, MA: Elsevier, Inc.
- *Up and Running with AutoCAD 2011: 2D Drawing and Modeling.* Waltham, MA: Elsevier, Inc.
- *Up and Running with AutoCAD 2011: 2D and 3D Drawing and Modeling.* Waltham, MA: Elsevier, Inc.
- *Up and Running with AutoCAD 2012: 2D Drawing and Modeling.* Waltham, MA: Elsevier, Inc.
- *Up and Running with AutoCAD 2012: 2D and 3D Drawing and Modeling.* Waltham, MA: Elsevier, Inc.
- *Up and Running with AutoCAD 2013: 2D Drawing and Modeling.* Waltham, MA: Elsevier, Inc.
- *Up and Running with AutoCAD 2013: 2D and 3D Drawing and Modeling.* Waltham, MA: Elsevier, Inc.
- *Up and Running with AutoCAD 2014: 2D and 3D Drawing and Modeling.* Waltham, MA: Elsevier, Inc.
- *Up and Running with AutoCAD 2015: 2D and 3D Drawing and Modeling.* Waltham, MA: Elsevier, Inc.
- *Up and Running with AutoCAD 2016: 2D and 3D Drawing and Modeling.* Waltham, MA: Elsevier, Inc.
- *Up and Running with AutoCAD 2017: 2D and 3D Drawing and Modeling.* Waltham, MA: Elsevier, Inc.
- *Up and Running with AutoCAD 2018: 2D Drafting and Design.* Waltham, MA: Elsevier, Inc.

Spanish

- *AutoCAD 2012 - Dibujar y Modelar en 2D y 3D.* Madrid, Spain: Anaya Multimedia.

Robert C. Kaebisch

Robert C. Kaebisch started using AutoCAD in 1989 with R10 and continued using each subsequent release through college and throughout his career. Most of his drafting work has involved architecture, interior design, mechanical, electrical, plumbing, and other construction-related industries, from schematic designs through full construction documents.

After starting his architectural career in the mid-1990s, and spending several years in the field, Robert began teaching as a part-time adjunct instructor at ITT Technical Institute at the school of Design and Drafting in 2001. In 2004, he stepped away from teaching to focus back on architecture and obtained his architect's license in 2005. In 2008 he returned to teaching and became a full-time instructor in Gateway Technical College's Architectural-Structural Engineering Technician program. He continues to teach and be involved in architectural design and consulting, as well as AutoCAD and other software training.

Robert holds a bachelor's degree in architecture from the University of Wisconsin—Milwaukee's School of Architecture and Urban Planning, and a certificate in Project Management from the University of Wisconsin—Stout. He maintains his architect's license, technical college teaching license in WI, and national certifications in both AutoCAD and Autodesk Revit. He currently lives in the Milwaukee area and can be reached at Robert.Kaebisch@gmail.com.

Previous Textbooks by Robert C. Kaebisch:

- *Up and Running with AutoCAD 2018: 2D Drafting and Design.* Waltham, MA: Elsevier, Inc.

Preface

WHAT IS AUTOCAD?

AutoCAD is a drafting and design software package developed and marketed by Autodesk, Inc. As of 2018, it has been around for approximately 36 years - several lifetimes in the software industry. It has grown from modest beginnings to an industry standard, often imitated, sometimes exceeded, but never equaled. The basic premise of its design is simple and is the main reason for AutoCAD's success. Anything you can think of, you can draw quickly and easily. For many years, AutoCAD remained a superb 2D electronic drafting board, replacing the pencil and paper for an entire generation of technical professionals. In recent releases, its 3D capabilities finally matured, and AutoCAD is now also considered an excellent 3D visualization tool, especially for architecture and interior design.

The software has a rather steep learning curve to become an expert but a surprisingly easy one to just get started. Most important, it is well worth learning. This is truly global software that has been adopted by millions of architects, designers, and engineers worldwide. Over the years, Autodesk expanded this reach by introducing add-on packages that customize AutoCAD for industry-specific tasks, such as electrical, civil, and mechanical engineering. However, underneath all these add-ons is still plain AutoCAD. This software remains hugely popular. Learn it well, as it is still one of the best skills you can add to your resume.

ABOUT THIS BOOK

This book is not like most on the market. While many authors certainly view their particular text as unique and novel in its approach, I rarely reviewed one that was clear to a beginner student and distilled AutoCAD concepts down to basic, easy to understand explanations. The problem may be that many of the available books are written by either industry technical experts or teachers but rarely by someone who is actively both. One really needs to interact with the industry and the students, in equal measure, to bridge the gap between reality and the classroom.

After years of AutoCAD design work in the daytime and teaching nights and weekends, I set out to create a set of classroom notes that outlined, in an easy to understand manner, exactly how AutoCAD is used and applied, not theoretical musings or clinical descriptions of the commands. These notes eventually were expanded into this book that you now hold. The rationale was simple: I need this person to be up and running as soon as possible to do a job. How do we make this happen?

TEACHING METHODS

My teaching approach has its roots in a certain philosophy I developed while attending engineering school many years ago. While there, I had sometimes been frustrated with the complex presentation of what in retrospect amounted to rather simple topics. My favorite quote was, "Most ideas in engineering are not that hard to understand but often become so upon explanation." The moral of that quote was that concepts can usually be distilled to their essence and explained in an easy and straightforward manner. That is the job of a teacher: Not to blow away students with technical expertise but to use experience and top-level knowledge to sort out what is important and what is secondary and to explain the essentials in plain language.

Such is the approach to this AutoCAD book. I want everything here to be highly practical and easy to understand. There are few descriptions of procedures or commands that are rarely used in practice. If we talk about it, you will likely need it. The first thing you must learn is how to draw a line. You see this command on the first few pages of Chapter 1, not buried somewhere in later sections. It is essential to present the "core" of AutoCAD, essential knowledge common to just about any drafting situation, all of it meant to get you up and running quickly. This stripped down approach proved effective in the classroom and was carefully incorporated into this text.

TEXT ORGANIZATION

This book comes in three parts: Level 1, Level 2, and Level 3:

Level 1 Chapters 1–10 is meant to give you a wide breadth of knowledge on many topics, a sort of "mile wide" approach. These 10 chapters comprise, in my experience, the complete essential knowledge set of an intermediate user. You then can work on, if not necessarily set up and manage, moderate to complex drawings. If your CAD requirements are modest or if you are not required to draft full time, then this is where you stop.

Level 2 Chapters 11–20 is meant for advanced users who are CAD managers, full-time AutoCAD draftspersons, architects, or self-employed and must do everything themselves. The goal here is depth, as many features not deemed critically important in Level 1 are revisited to explore additional advanced options. Also introduced are advanced topics necessary to set up and manage complex drawings.

Level 3 (Chapters 21–30) is all about 3D. Solid knowledge of the previous two levels is highly recommended before starting these chapters. The 3D material covers all aspects of AutoCAD solid modeling including lights and rendering. For the 2018 and again for this 2019 edition the 3D chapters have been removed from the print and eBook version of the textbook, and instead are now offered online for free download.

Throughout this book, the following methods are used to present material:

- Explain the new concept or command and why it is important.
- Cover the command step by step (if needed), with your input and AutoCAD responses shown so you can follow and learn them.
- Give you a chance to apply just-learned knowledge to a real-life exercise, drawing, or model.
- Test yourself with end-of-chapter quizzes and drawing exercises that ask questions about the essential knowledge.

You will not see an extensive array of distracting "learning aids" in this text. You will, however, see some common features throughout, such as

Commands: These are presented in almost all cases in the form of a command matrix, such as the one shown here for a line. You can choose any of the methods for entering the command.	*Keyboard:* Type in **line** and press Enter *Cascading menus:* **Draw→Line** *Toolbar icon:* **Draw** toolbar *Ribbon:* **Home** tab**→Line**
Tips and tricks: These are seen mostly in the first few chapters and one is shown here. They are very specific, deliberate suggestions to smooth out the learning experience. Do take note.	**TIP 1:** The Esc (escape) key in the upper left-hand corner of your keyboard is your new best friend. It gets you out of just about any trouble you get yourself into. If something does not look right, just press the Esc key and repeat the command. Mine was worn out learning AutoCAD, so expect to use it often.
Step-by-step instructions: These are featured whenever practical and show you exactly how to execute the command, such as the example with line here. What you type in and what AutoCAD says are in the software's default font: `Courier New`. The rest of the steps are in the standard print font.	**Step 1.** Begin the line command via any of the previous methods. • AutoCAD says: Specify first point: **Step 2.** Using the mouse, left-click anywhere on the screen. • AutoCAD says: Specify next point or [Undo]: **Step 3.** Move the mouse elsewhere on the screen and left-click again. You can repeat Step 2 as many times as you wish. When you are done, click Enter or Esc.
Learning objectives and time for completion: Each chapter begins with this, which builds a "road map" for you to follow while progressing through the chapter, as well as sets expectations of what you will learn if you put in the time to go through the chapter. The time for completion is based on classroom teaching experience but is only an estimate. If you are learning AutoCAD in school, your instructor may choose to cover part of a chapter or more than one at a time.	In this chapter, we introduce AutoCAD and discuss the following: • Introduction and the basic commands • The Create Objects commands • The Edit and Modify Objects commands • The View Objects commands, etc. By the end of the chapter, you will ... Estimated time for completion of chapter: 3 hours.
Summary, review questions, exercises: Each chapter concludes with these. Be sure to not skip these pages and to review everything you learned.	**SUMMARY** **REVIEW QUESTIONS** **EXERCISES**

WHAT YOUR GOAL SHOULD BE

Just learning commands is not enough; you need to see the big picture and truly understand AutoCAD and how it functions for it to become effortless and transparent. The focus after all is on your design. AutoCAD is just one of the tools to realize it.

A good analogy is ice hockey. Professional players do not think about skating; to them, it is second nature. They are focused on strategy, scoring a goal, and getting by the defenders. This mentality should be yours as well. You must become proficient through study and practice, to the point where you are working with AutoCAD, not struggling against it. It then becomes "transparent" and you focus only on the design, to truly perform the best architecture or engineering work of which you are capable.

If you are in an instructor-led class, take good notes. If you are self-studying from this text, pay very close attention to every topic; nothing here is unimportant. Do not skip or cut corners, and complete every drawing assignment. Most important, you have to practice, daily if possible, as there is no substitute for sitting down and using the software. Not everyone these days has the opportunity to learn while working and getting paid; companies want ready-made experts and do not want to wait. If that is the case, you have to practice on your own in the evening or on weekends. Just taking a class or reading this book alone is not enough.

It may seem like a big mountain to climb right now, but it is completely doable. Once on top, you will find that AutoCAD is not the frustrating program it may have seemed in the early days but an intuitive software package that, with proficiency of use, becomes a natural extension of your mind when working on a new design. That, in the end, is the mark of successful software; it helps you do your job easier and faster. Feel free to contact myself or my co-author with questions or comments at Elliot.Gindis@gmail.com and Robert.Kaebisch@gmail.com. Also, be sure to visit my website at www.verticaltechtraining.com, where you will find not only the 3D chapters for download, but other extensive resource such as exercises, videos, articles and a discussion board to get your AutoCAD questions answered.

Acknowledgments

A textbook of this magnitude is rarely a product of only one person's effort. I thank all the early and ongoing reviewers of this text and Chris Ramirez of Vertical Technologies Consulting for research and ideas when most needed as well as using the text in his classroom. A big thank you also to Karen Miletsky (retired) formerly of Pratt Institute of Design, Russell and Titu Sarder at Netcom Information Technology, and everyone at the New York Institute of Technology, RoboTECH CAD Solutions, Future Media Concepts, and other premier training centers, colleges, and universities for their past and present support.

Extensive gratitude also goes to Joseph P. Hayton, Todd Green, Stephen R. Merken, Peter Jardim, Jeff Freeland, Kathleen Chaney, Gnomi Schrift Gouldin, Kiruthika Govindaraju, Maria Ines Cruz, and the rest of the team at Elsevier for believing in the project and for their invaluable support in getting the book out to market. Thank you also to Denis Cadu of Autodesk for all the support at the Autodesk Developers Network.

A big thank you to my new coauthor, Robert C. Kaebisch of Gateway Technical College, for assisting in bringing the 2018 edition to the next level and continuing his excellent work for this 2019 book.

Finally, I thank my friends and family, especially my parents, Boris and Tatyana Gindis, and my wonderful wife, Marina, for their patience and encouragement as I undertake the lengthy annual effort of bringing this textbook to life. This book is dedicated to the hundreds of students who have passed through my classrooms and made teaching the enjoyable adventure it has become.

Elliot J. Gindis
May 2018

First and foremost, I have to thank Elliot J. Gindis for asking me to join him on this journey and trusting me to help with the evolution of his creation. Thank you to the teachers who gave me the freedom to explore technical drafting, architecture, and all of their associated technologies: Mr. Peter Wilson, Mr. Paul Barry, and Mr. Rick Jules.

I could not do the work I do without the support of the Engineering Technology Wing faculty: Ray Koukari, Jr., Pat Hoppe, and Steve Whitmoyer. Another big thank you to the Gateway Technical College administration: Dr. Bryan D. Albrecht, Zina Haywood, Matt Janisin, and Dr. John Thibodeau for helping me grow as an instructor.

Finally, I thank my wife Cathi, my children, Alex and Emma, and my parents, Jerald and Sharon Kaebisch, Jr. for their continued support. This journey is proving to be a great learning experience, making me a better instructor and learner.

Robert C. Kaebisch
May 2018

Level 1. Chapters 1–10

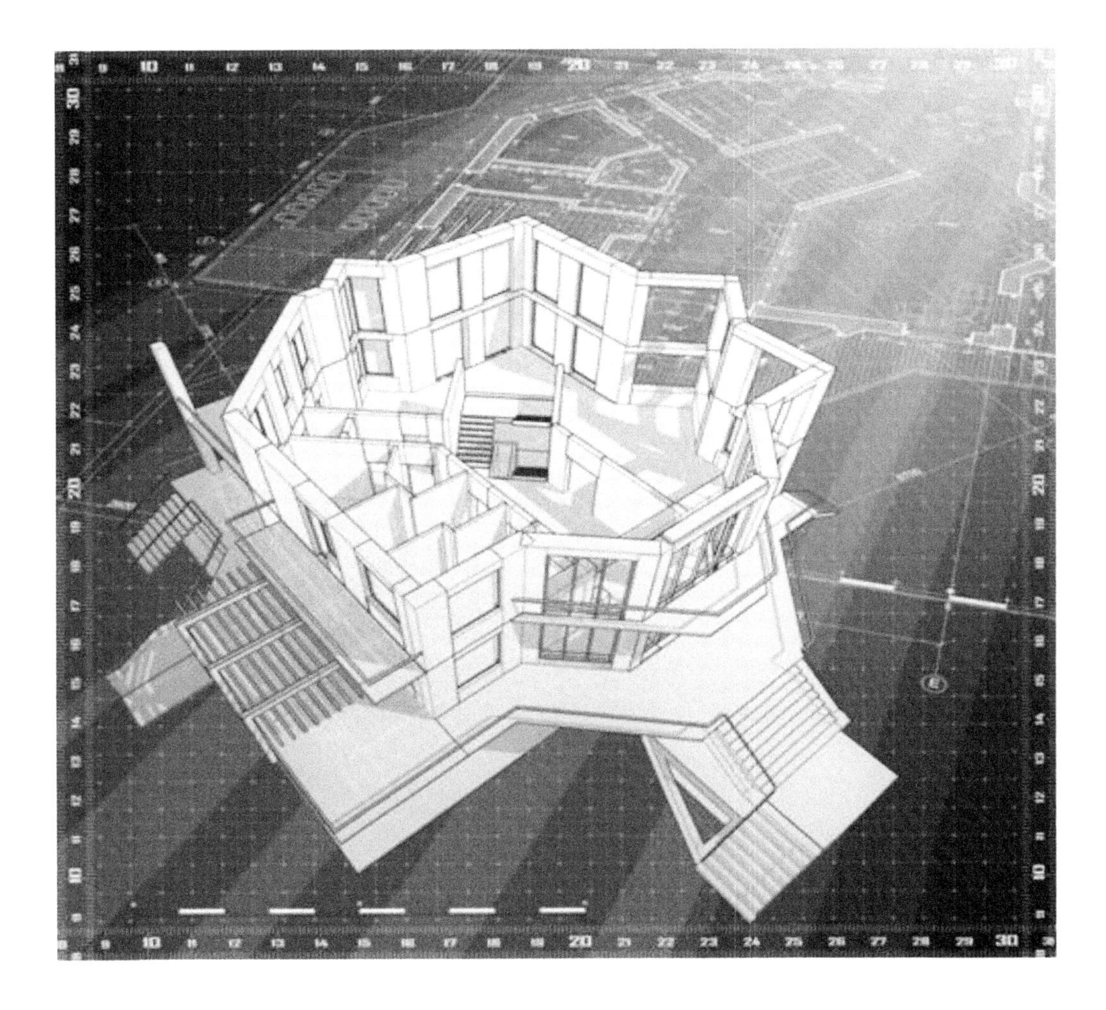

Level 1 is the very beginning of your studies. No prior knowledge of AutoCAD is assumed, only basic familiarity with computers and some technical aptitude. You are also at an advantage if you have hand drafted before, as many AutoCAD techniques flow from the old paper and pencil days, a fact alluded to later in the chapters.

We begin Chapter 1 by outlining the basic commands under Create Objects and Modify Objects followed by an introduction to the AutoCAD environment. We then introduce basic accuracy tools of Ortho and OSNAP.

Chapter 2 continues the basics by adding units and various data entry tools. These first two chapters are critically important, as success here ensures you will understand the rest and be able to function in the AutoCAD environment.

Chapter 3 continues on to layers, and then each succeeding chapter continues to deal with one or more major topics per chapter: text and mtext in Chapter 4, hatching in Chapter 5, and dimensioning in Chapter 6. In these six chapters, you are asked to not only practice what you learned but also apply the knowledge to a basic architectural floor plan. Chapter 7 introduces blocks and wblocks and Chapter 8 arrays. At this point, you are asked to draw another project, this time a mechanical device. Level 1 concludes with basic printing and output in Chapter 9 and finally advanced printing and output (Paper Space) in Chapter 10.

Be sure to dedicate as much time as possible to practicing what you learn; there really is no substitute.

Chapter 1

AutoCAD Fundamentals—Part I

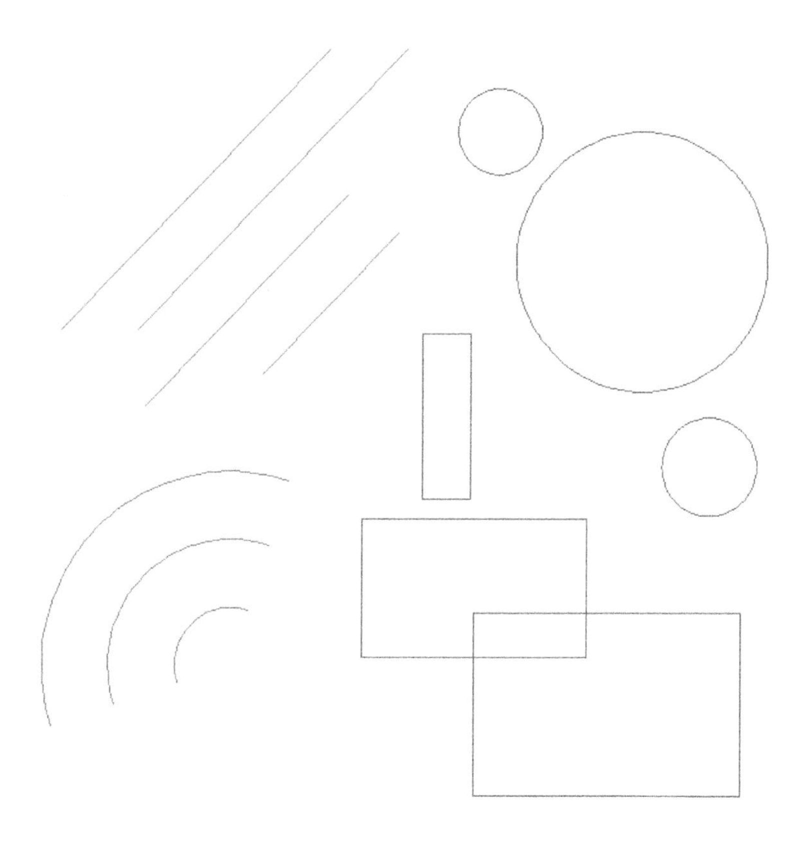

Learning Objectives

In this chapter, we introduce AutoCAD and discuss the following:

- Introduction and the basic commands
- The Create Objects commands
- The Edit/Modify Objects commands
- The View Objects commands
- The AutoCAD environment
- Interacting with AutoCAD
- Practicing the Create Objects commands
- Practicing the Edit/Modify Objects commands
- Selection methods—Window, Crossing, and Lasso
- Accuracy in drafting—Ortho (F8)
- Accuracy in drafting—OSNAPs (F3)

Up and Running with AutoCAD 2019. https://doi.org/10.1016/B978-0-12-816440-2.00001-X

By the end of this chapter, you will learn the essential basics of creating, modifying, and viewing objects; the AutoCAD environment; and accuracy in the form of straight lines and precise alignment of geometric objects via OSNAP points.

Estimated time for completion of this chapter: 3 hours.

1.1 INTRODUCTION AND BASIC COMMANDS

AutoCAD 2019 is a very complex program. If you are taking a class or reading this textbook, this is something you probably already know. The commands available to you, along with their submenus and various options, number in the thousands. So, how do you get a handle on them and begin using the software? Well, you have to realize two important facts.

First, you must understand that, on a typical workday, *95% of your AutoCAD drafting time is spent using only 5% of the available commands, over and over again.* So getting started is easy; you need to learn only a handful of key commands; and as you progress and build confidence, you can add depth to your knowledge by learning new ones.

Second, you must understand that even the most complex drawing is essentially *made up of only a few basic fundamental objects that appear over and over again* in various combinations on the screen. Once you learn how to create and edit them, you can draw surprisingly quickly. Understanding these facts is the key to learning the software. We are going to strip away the perceived complexities of AutoCAD and reduce it to its essential core. Let us go ahead now and develop the list of the basic commands.

For a moment, view AutoCAD as a fancy electronic hand-drafting board. In the old days of pencil, eraser, and T-square, what was the simplest thing that you could draft on a blank sheet of paper? That of course is a line. Let us make a list with the following header, "Create Objects," and below it add "line."

So, what other geometric objects can we draw? Think of basic building blocks, those that cannot be broken down any further. A *circle* qualifies and so does an *arc*. Because it is so common and useful, throw in a *rectangle* as well (even though you should note that it is a compound object, made up of four lines). Here is the final list of fundamental objects that we have just come up with:

Create Objects

- Line
- Circle
- Arc
- Rectangle

As surprising as it may sound, these four objects, in large quantities, make up the vast majority of a typical design, so already you have the basic tools. We will create these on the AutoCAD screen in a bit. For now, let us keep going and get the rest of the list down.

Now that you have the objects, what can you do with them? You can *erase* them, which is probably the most obvious. You can also *move* them around your screen and, in a similar manner, *copy* them. The objects can *rotate*, and you can also *scale* them up or down in size. With lines, if they are too long, you can *trim* them, and if they are too short, you can *extend* them. *Offset* is a sort of precise copy and one of the most useful commands in AutoCAD. *Mirror* is used, as the name implies, to make a mirror-image copy of an object. Finally, *fillet* is used to put a curve on two intersecting lines, among other things. We will learn a few more useful commands a bit later, but for now, under the header "Edit/Modify Objects," list the commands just mentioned:

Edit/Modify Objects

- Erase
- Move
- Copy

- Rotate
- Scale
- Trim
- Extend
- Offset
- Mirror
- Fillet

Once again, as surprising as it may sound, this short list represents almost the entire set of basic Edit/Modify Objects commands you need once you begin to draft. Start memorizing them.

To finish up, let us add several View Objects commands. With AutoCAD, unlike paper hand drafting, you do not always see your whole design in front of you. You may need to *zoom* in for a close-up or out to see the big picture. You may also need to *pan* around to view other parts of the drawing. With a wheeled mouse, virtually universal on computers these days, it is very easy to do both, as we soon see. To this list we add the *regen* command. It stands for *regenerate*, and it simply refreshes your screen, something you may find useful later. So here is the list for View Objects:

View Objects

- Zoom
- Pan
- Regen

So, this is it for now, just 17 commands making up the basic set. Here is what you need to do:

1. As mentioned before, memorize them so you always know what you have available.
2. Understand the basic idea, if not the details, behind each command. This should be easy to do because (except for maybe *offset* and *fillet*) the commands are intuitive and not cryptic in any way; *erase* means erase, whether it is AutoCAD, a marker on a whiteboard, or a pencil line.

We are ready now to start AutoCAD, discuss how to interact with the program, and try out all the commands.

1.2 THE AUTOCAD ENVIRONMENT

It is assumed that your computer, whether at home, school, work, or training class, is loaded with AutoCAD 2019. It is also assumed that AutoCAD starts up just fine (via the AutoCAD icon or Start menu) and everything is configured right. If not, ask your instructor, as there are just too many things that can go wrong on a particular PC or laptop, and it is beyond the scope of this book to cover these situations.

If all is well, start up AutoCAD. You should see the Start tab welcome screen depicted in Fig. 1.1. This is a change from older versions of AutoCAD, which defaulted to a blank drawing area when first opened. We do not spend too much time here, but note that on the bottom you have the option to switch to another screen called *Learn*. It contains a *What's New* section, some *Getting Started Videos*, and a collection of *Learning Tips* and *Online Resources*. Meanwhile the Create screen has some options to open drawings via *Get Started*, some *Recent Documents*, and *Notifications*. Take a quick tour of everything if you wish, and when you are ready to open a new blank file, press the "sheet of paper" icon at the very top left of the screen; the universal computer symbol for a new file. A Select template dialog box opens up, already set to open an acad.dwt file (Fig. 1.2). Press the Open button at the lower right, and a blank file opens (Fig. 1.3).

This is your basic out-of-the-box AutoCAD environment for the Drafting & Annotation workspace. From both the learning and teaching points of view, this screen layout (and the layout of the last few releases) is an improvement over older versions of AutoCAD, and Autodesk has done an admirable job in continuing to keep things clean and (relatively) simple.

With some previous versions of AutoCAD, a new palette, called *Design Feed*, (and sometimes other palettes) also appeared during start up. You may or may not see them with this release, but if you do, close any pop-up palette via the X at the upper left of the palette; we do not need any of them anytime soon and they just clutter up the screen.

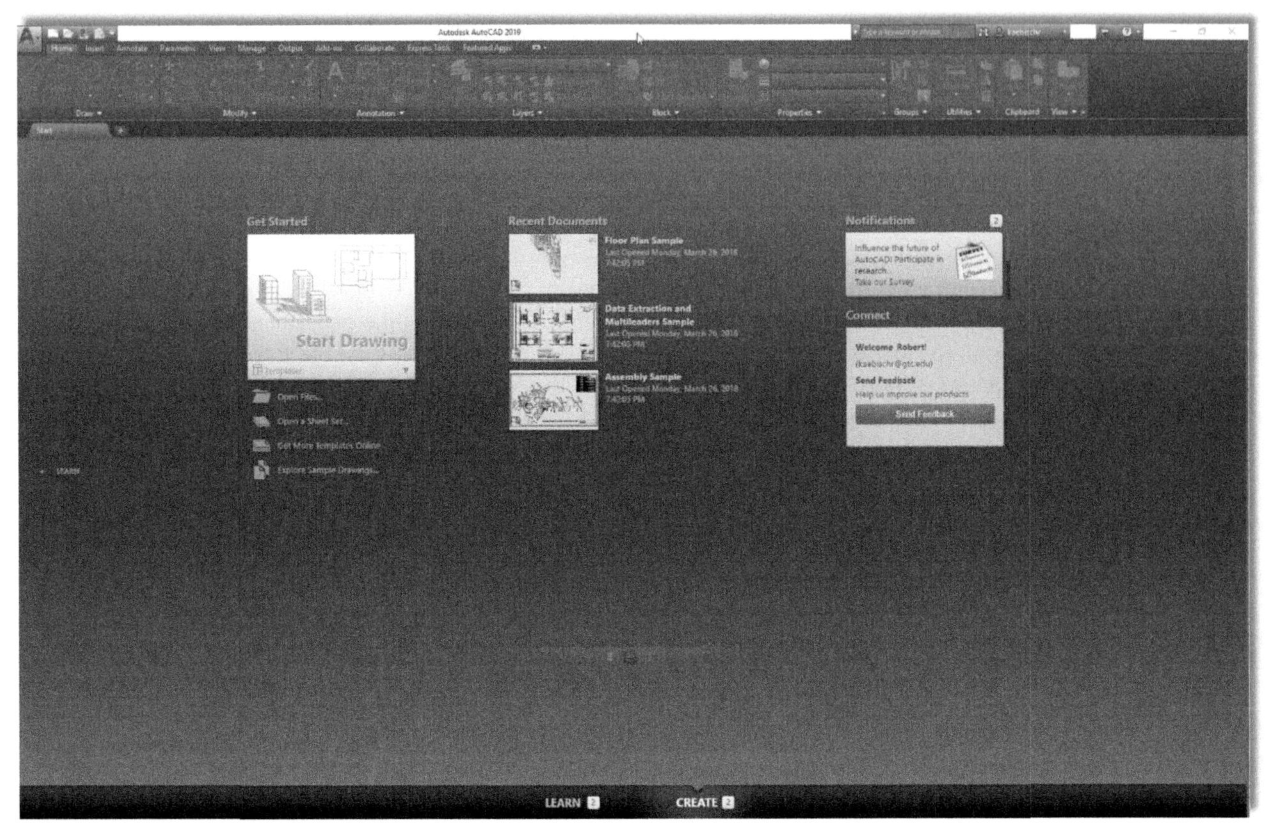

FIGURE 1.1 AutoCAD 2019—Start tab.

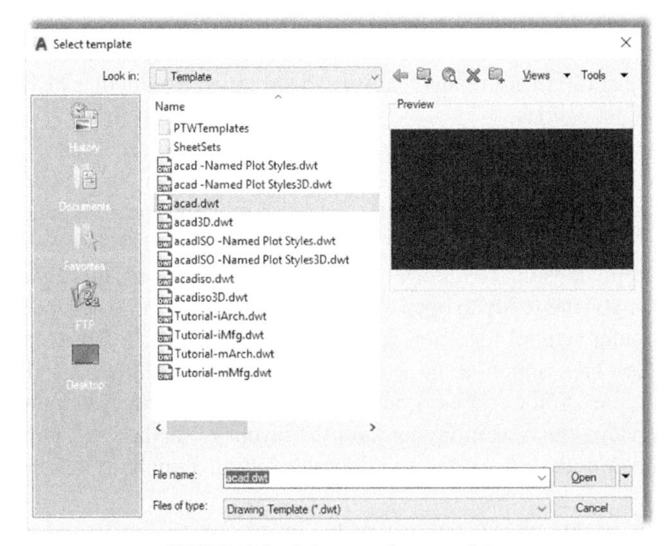

FIGURE 1.2 Select template—acad.dwt.

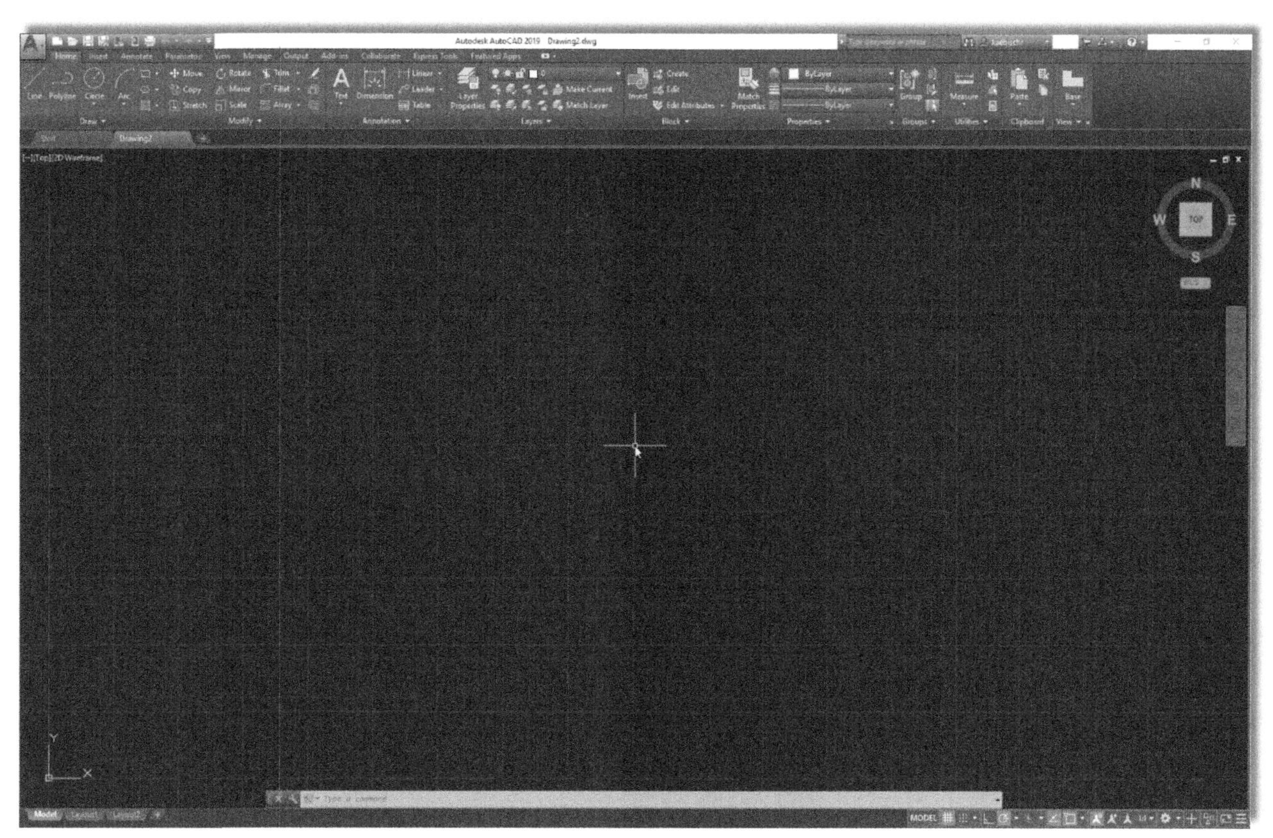

FIGURE 1.3 AutoCAD 2019 (running on Windows 10).

AutoCAD went through a major facelift in 2009 release, with some minor additions incorporated into every release ever since. If you caught a glimpse of earlier versions, you may have noticed toolbars present. They are still around, but what we have had for some time now, dominating the upper part of the screen, is called the *Ribbon*. After this many years, it is a well-established way of interacting with AutoCAD, and we discuss it in detail soon. Other screen layouts or workspaces are available to you; however, they are related to 3D and we discuss them in the 3D-downloadable online content. The Drafting & Annotation workspace you are in now is all you need to get started.

For clarity and ease of learning, we need to change the look of a few items on the screen. First, we lighten everything. By default AutoCAD 2019 opens up with a dark theme (it is easier on the eyes). We switch to the light theme and also change the work area color to white. This not only saves ink but also makes it a lot easier for you to see and understand the graphics on the printed page. We then turn the command line from its single-line configuration to a multiline one, so you can follow command sequences more easily. Finally, we add the cascading menu bar back in and make a few other minor tweaks to the drawing and construction aids at the bottom right, such as turning off the Grid (which you can do right away by pressing F7). You need not do all this; a lot of the steps are based on personal taste of how you want AutoCAD to look. However, it is necessary for the textbook and will make learning AutoCAD a bit easier for you initially as well. You may want to "go dark" later down the road however.

Let us now take a tour of the features you see on the screen. Along the way you will see descriptions of how to incorporate the previously mentioned changes. Your screen will then look like Fig. 1.4. Study all the screen elements carefully, so you know exactly what you are looking at as we progress further.

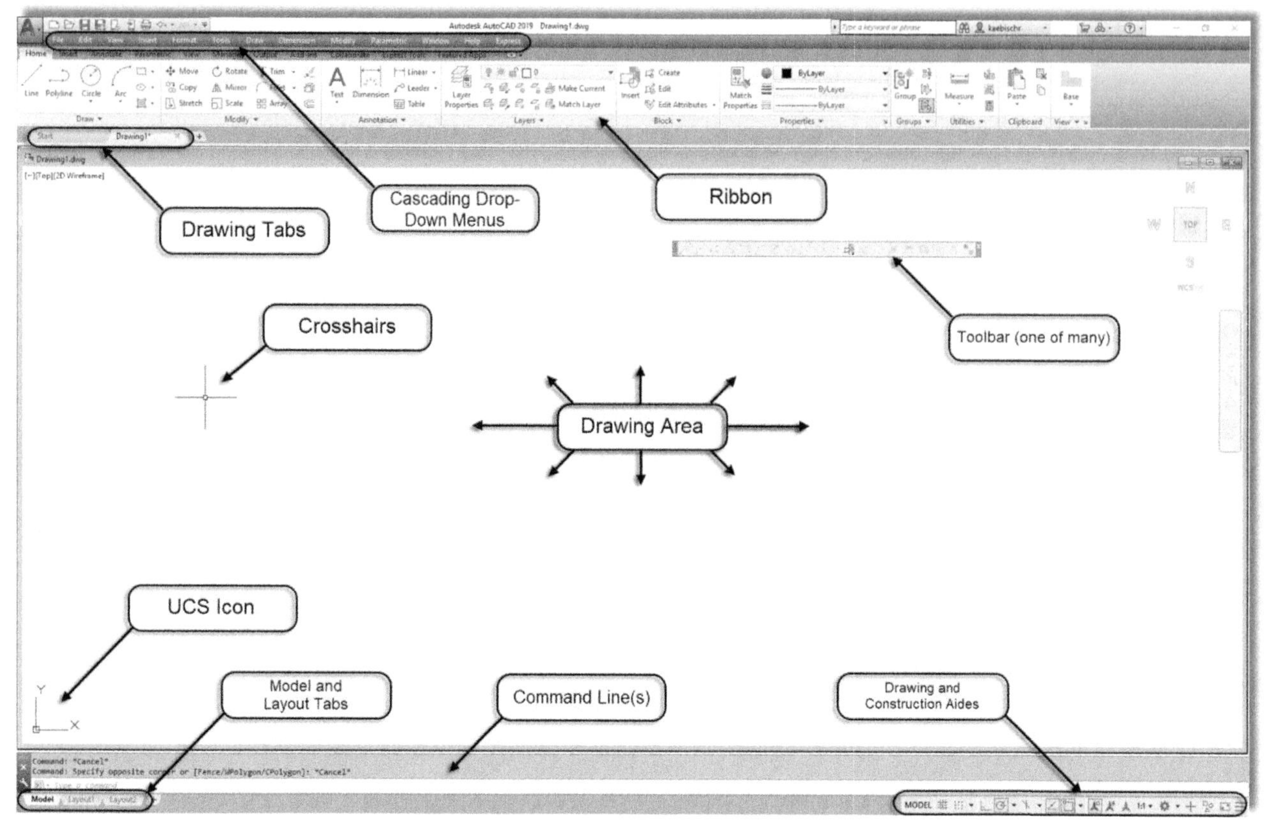

FIGURE 1.4 AutoCAD 2019 screen elements.

- **Drawing Area:** The drawing area takes up most of the screen and is colored a very dark gray in the default version of the environment. This is where you work and your design appears. Some users change it to all-black to further ease eyestrain, as less light radiates toward you. In this textbook, however (as in most textbooks), the color is always white to conserve ink and for clarity on a printed page. If you wish to change the color of the drawing area, you need to right-click into Options…, choose the Display tab, then Colors…. Finally, change the color from the drop-down menu on the right. We cover all this in more detail in Chapter 14, so you should ask your instructor for assistance in the meantime, if necessary.
- **Drawing Tabs:** One new change that was added back in the 2015 release is that the drawing area now has "tabs," visible at the upper left, which allow you to cycle through multiple drawings. As described earlier, AutoCAD 2019 defaults to the Start tab. You can close it while you are working on another drawing, but it will return if no other drawings are present.
- **Command Line(s):** Right below the drawing area are the command lines or, by default, just one floating line. It can be turned off and on via Ctrl+9. This is where the commands may be entered and also where AutoCAD tells you what it needs to continue. Always keep an eye on what appears here, as this is one of the main ways that AutoCAD communicates with you. Although we discuss the "heads-up" on-screen interaction method in Chapter 2, the command line remains very relevant and often used. It is colored gray and white by default but can be changed (in a manner similar to the drawing area) to all white if desired. We leave the color as default but dock it at the bottom and stretch it upward to include three lines for clarity. These are all minor details, so you can leave your command line completely "as is" if you wish.
- **UCS Icon:** This is a basic X-Y-Z (Z is not visible) grid symbol. It will be important later in advanced studies and 3D. It can be turned off via an icon on the Ribbon's View tab, but generally it is kept where it is by most users. The significance of this icon is great, and we study it in detail later on. For now, just observe that the Y axis is "up" and the X axis is "across."
- **Paper Space/Model Space Tabs:** These Model/Layout1/Layout2 tabs, similar to those used in Microsoft's Excel, indicate which drawing space you are in and are important in Chapter 10, when we cover Paper Space. Although you can click on them to see what happens, be sure to return to the Model tab to continue further.

- **Toolbar(s):** Toolbars contain *icons* that can be pressed to activate commands. They are an alternative to typing and the Ribbon, and most commands can be accessed this way. AutoCAD 2019 has over 50 of them. You may not have a toolbar present on your screen at the moment. If that is the case, do not worry, we activate a few toolbars shortly. Toolbars are generally falling out of favor among some users, who prefer the Ribbon, but which approach to use remains a personal choice. We always try to have the relevant toolbars open throughout the book for learning purposes.
- **Crosshairs:** Crosshairs are simply the mouse cursor and move around along with the movements of your mouse. They can be full size and span the entire screen or a small (flyspeck) size. You can change the size of the crosshairs if you wish, and full screen is recommended in some cases. For now, we leave it as is. The crosshairs have been empowered with a new intelligence a few releases ago. They now show, via a small icon, what you are doing while executing certain commands.
- **Drawing and Construction Aids:** These various settings assist you in drafting and modeling, and we introduce them as necessary. They were also redesigned some time ago. If you have seen earlier versions, the symbols were also present as written words, not just icons, but you no longer have this, they are all icons now. More importantly, however, some of them now contain selectable menus (look for the little black upside-down triangles), where you can set or choose even more options. This was one of the more significant interface changes from AutoCAD 2014 to 2015, a few years back, and gives you more power and convenience at your fingertips. The full menu of what is available can be seen by clicking the tab with the three short horizontal lines (Customization option, "hamburger menu") all the way on the right of the construction aids. If an option is checked, the corresponding button is visible in the horizontal menu. For all of the drawing aids, remember, if they are gray, they are *off*; if they are blue, they are *on*.
  - **Be sure to turn all them to "off" for now. We activate them as needed throughout the chapters.**
- **Ribbon:** It comprises a large panel that stretches across the screen and is populated with context-sensitive tabs. Each tab, in turn, contains many panels. The Ribbon can be turned off and on at will via the cascading menu's Tools→Palettes→Ribbon. We have more to say about the Ribbon in the next section.
- **Cascading Drop-Down Menus:** This is another way to access commands in AutoCAD. These menus, so named because they drop out like a waterfall, may be hidden initially, but you can easily make them visible via the down arrow at the very top left of the screen, just to the right of the Undo and Redo arrows. A lengthy menu appears. Select "Show Menu Bar" toward the bottom and the cascading menus appear as a band across the top of the screen, above the Ribbon. You should keep these menus in their spot from now on; they are referred to often. Why Autodesk designed them to be hidden by default is a mystery.

Review this bulleted list carefully; it contains a lot of useful "get to know AutoCAD" information. Once again, you do not have to change your environment exactly as suggested, but be sure to understand how to do it if necessary.

1.3 INTERACTING WITH AUTOCAD

OK, so you have the basic commands in hand and ideally a good understanding of what you are looking at on the AutoCAD screen. We are ready to try out the basic commands and eventually draft something. So, how do we interact with AutoCAD and tell it what we want drawn? Four primary ways (Methods 1–4) follow, roughly in the order they appeared over the years. There are also two outdated methods, called the *tablet* and the *screen side menu* (dating back to the very early days of AutoCAD), but we do not cover them.

Method 1: *Type in* the commands on the command line (AutoCAD v1.0–current).
Method 2: Select the commands from the *drop-down cascading menus* (AutoCAD v1.0–current).
Method 3: Use *toolbar* icons to activate the commands (AutoCAD 12/13–current).
Method 4: Use the *Ribbon* tabs, icons, and menus (AutoCAD 2009–current).

Details of each method, including the pros and cons, follow. Most commands are presented in all four primary ways, and you can experiment with each method to determine what you prefer. Eventually, you will settle on one particular way of interacting with AutoCAD or a hybrid of several.

Method 1. Type in the Commands on the Command Line

This was the original method of interacting with AutoCAD and, to this day, remains the most foolproof way to enter a command: good old-fashioned typing. AutoCAD is unique among leading CAD software in that it has retained this method while almost everything else moved to graphic icons, toolbars, and Ribbons. If you hate typing, this will probably not be your preferred choice.

However, do not discount keyboard entry entirely; AutoCAD has kept it for a reason. When the commands are abbreviated to one or two letters (Line = L, Arc = A, etc.), input can be incredibly fast. Just watch a professional typist for proof of the speed with which one can enter data via a keyboard. Other advantages to typing are that you no longer have toolbars or a Ribbon cluttering up precious screen space (there is never enough of it), and you no longer have to take your eyes off the design to find an icon; instead, the command is literally at your fingertips. The disadvantage is of course that you have to type.

To use this method, simply type in the desired command (spelling counts!) at the command line, as seen in Fig. 1.5, and press Enter. The sequence initiates and you can proceed. A number of shortcuts are built into AutoCAD (try using just the first letter or two of a command), and we learn how to make our own shortcuts in advanced chapters. This method is still preferred by many "legacy" users—a kind way to say they have been using AutoCAD forever. Note that AutoCAD has an AutoComplete (as well as an AutoCorrect) feature on the command line, which calls up a possible list of commands based on just the first letter or two. This can be useful as you start out your learning the software but can also be turned off via right-clicking on the command line, choosing Input Settings, and then checking off what you want or do not want.

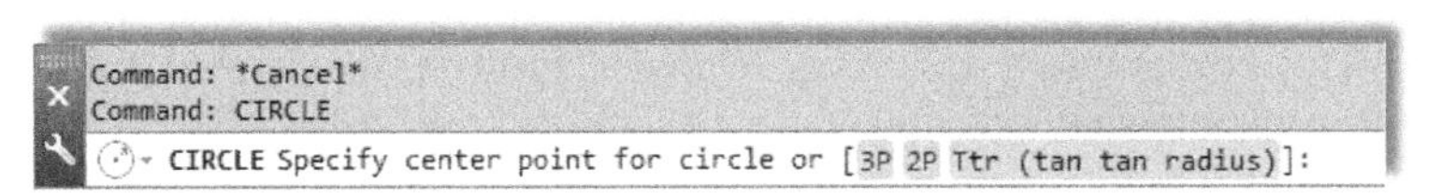

FIGURE 1.5 AutoCAD 2019 command line (CIRCLE typed in).

Method 2. Select the Commands From the Drop-Down Cascading Menus

This method has also been around since the beginning. It presents a way to access virtually every AutoCAD command, and indeed many students start out by checking out every one of them as a crash course on what is available—a fun but not very effective way of learning AutoCAD.

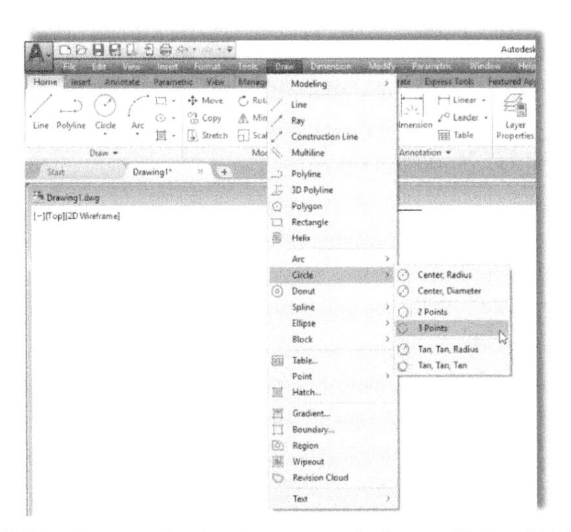

FIGURE 1.6 Cascading drop-down menus for Draw→Circle→3 Points.

Go ahead and examine the cascading menus; these are similar in basic arrangement to other software, and you should be able to navigate through them easily. We refer to them on occasion in the following format: Menu→ Command→Subcommand. So, for the sequence shown in Fig. 1.6, you would read Draw→Circle→3 Points.

Note that, in earlier versions of AutoCAD, the cascading menu commands featured underlined letters (or numbers). These were "hot keys" that you would have pressed to choose that command option while running through the menus. The visible underlining was discontinued in AutoCAD 2015, as part of an overall trend to move away from keyboard entry, but these hot keys still work—if you know which letters or numbers to press—and certainly expect to see plenty of very noticeable underlining in all earlier versions of AutoCAD. This book has kept some of the underlined hot keys visible in the command matrices (e.g., Line, Circle) as an example.

Method 3. Use Toolbar Icons to Activate the Commands

This method has been around since AutoCAD switched from DOS to Windows in the mid-1990s and was a favorite of a whole generation of users. Toolbars contain sets of icons (see Fig. 1.7 for an example), organized by categories (Draw toolbar, Modify toolbar, etc.). You press the icon you want, and a command is initiated. One disadvantage to toolbars, and the reason the Ribbon was developed, is that they take up a lot of space and, arguably, are not the most effective way of organizing commands on the screen.

FIGURE 1.7 AutoCAD 2019 Draw toolbar.

You can access all the toolbars at any time by simply selecting Tools→Toolbars→AutoCAD from the cascading menus. When you do that, a menu appears (Fig. 1.8), and you can simply check off the ones you want or do not want. You also can access the same menu by right-clicking on any of the toolbars themselves. Once you have a few up, you typically just dock them on top or off to the side. It is highly encouraged to bring up the *Standard*, *Draw*, and *Modify* toolbars for learning purposes. We do not require any other ones for now but will just bring them up as the need arises. Starting with Fig. 1.12, you will see those three toolbars docked neatly at the top of the drawing area.

Method 4. Use the Ribbon Tabs, Icons, and Menus

This is the most recently introduced method of interacting with AutoCAD and follows a new trend in software user interface design, such as that used by Microsoft and its Office 2013 (seen in Fig. 1.9 with Word). Notice how toolbars have been displaced by "tabbed" categories (Home, Insert, Page Layout, etc.), where information is grouped together by a common theme. So it goes with the new AutoCAD. The Ribbon was introduced with AutoCAD 2009, and it is here to stay. It is shown again by itself in Fig. 1.10.

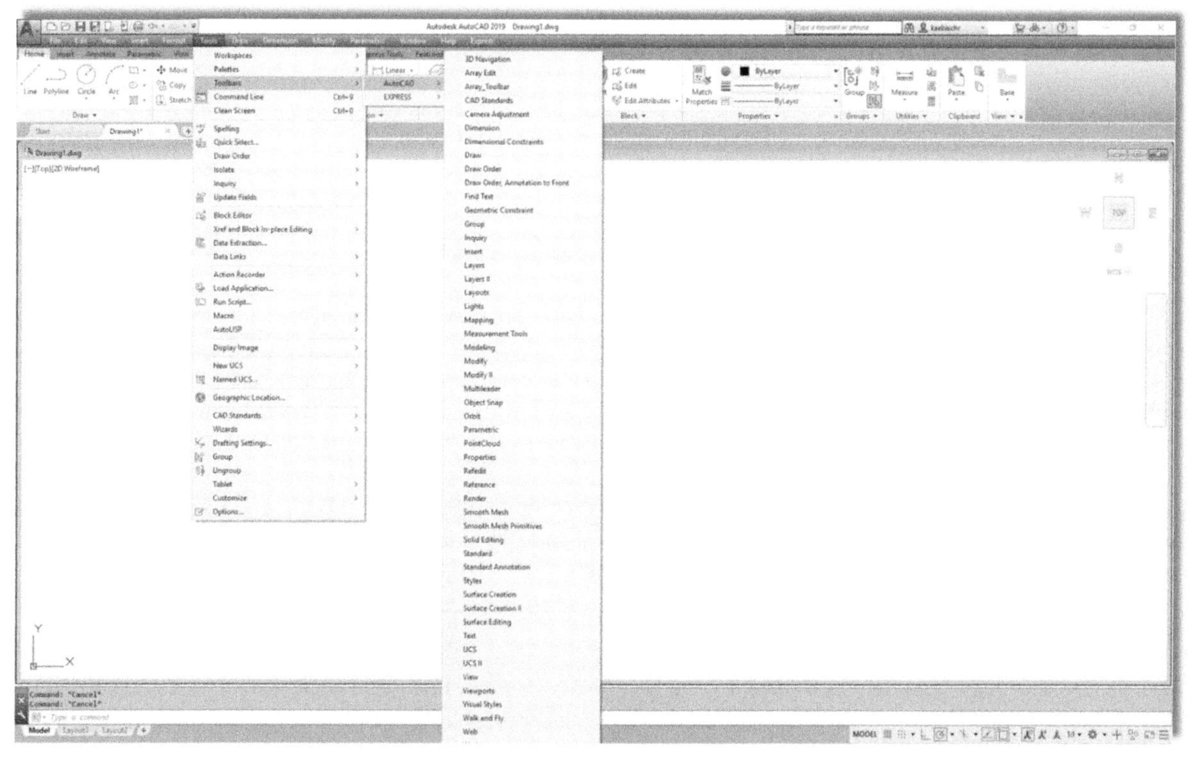

FIGURE 1.8 Toolbars menu.

FIGURE 1.9 MS Word Ribbon.

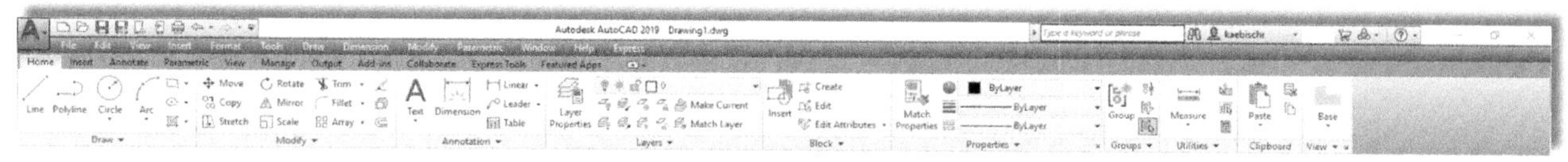

FIGURE 1.10 AutoCAD 2019 Ribbon.

Notice what we have here. A collection of tabs, indicating a subject category, is found at the top (Home, Insert, Annotate, etc.), and each tab reveals an extensive set of tools (Draw, Modify, Annotation, etc.). At the bottom of the Ribbon, additional options can be found by using the drop arrows. In this manner, the toolbars have been rearranged in what is, in principle, a more logical and space-saving manner.

Additionally, tool tips appear if you place your mouse over any particular tool for more than a second. Another second yields an even more detailed tool tip. In Fig. 1.11, you see the Home tab selected, followed by additional options via the drop arrow, and finally the mouse placed over the polygon command. A few moments of waiting reveal the full tool tip; pressing the icon activates the command.

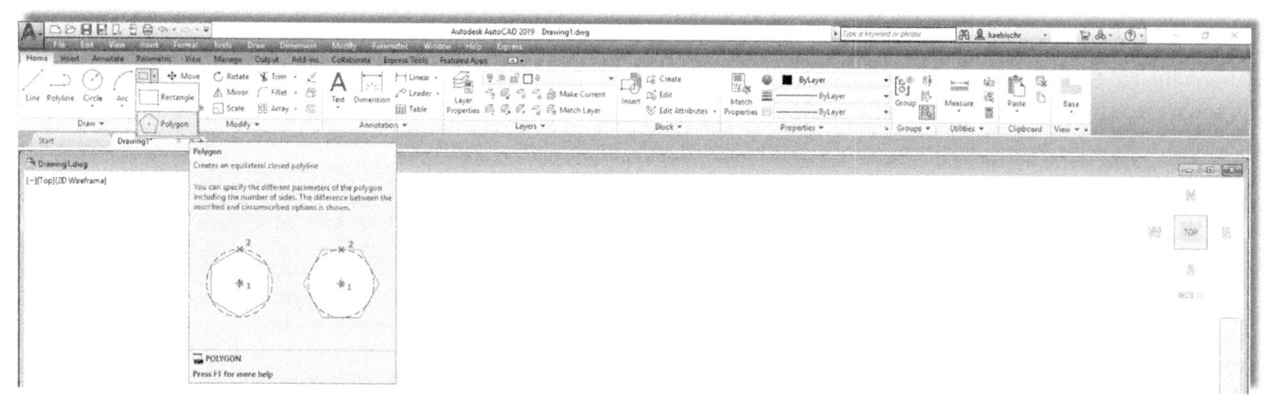

FIGURE 1.11 Polygon command and tool tip.

Familiarize yourself with the Ribbon by exploring it. It presents some layout advantages, and specialized Ribbons can be displayed via other workspaces. The disadvantage of this new method is that it is a relatively advanced tool that presents many advanced features right away and some confusion is liable to come up for a brand-new user. It is also not ideal for longtime users who type and some veterans in my update classes turn the feature off. The Ribbon is the single biggest change to AutoCAD's user interface and represents a jump forward in user/software interaction, but the ultimate decision to use it is up to you.

Before we try out our basic commands, let us begin a list of tips. These are a mix of good ideas, essential habits, and time-saving tricks passed along to you every once in a while (mostly in the first few chapters). Make a note of them as they are important. Here is the first, most urgently needed one.

TIP 1

The Esc (Escape) key in the upper left-hand corner of your keyboard is your new best friend while learning AutoCAD. It gets you out of just about any trouble you get yourself into. If something does not look right, just press the Esc key and repeat the command. Mine was worn out learning AutoCAD, so I expect you to use it often.

In the interest of trying out not just the Ribbon but also toolbars, cascading menus, and typing while learning the basics, bring up the three toolbars mentioned earlier, Standard, Draw, and Modify, shutting off all the rest. This is the setup you see in all the upcoming AutoCAD screen shots. Review everything learned thus far and proceed to the first commands.

1.4 PRACTICING THE CREATE OBJECTS COMMANDS

Let us try the commands now, one by one. Although it may not be all that exciting, it is extremely important that you memorize the sequence of prompts for each command, as these are really the ABCs of AutoCAD.

Remember: If a command is not working right or you see little blue squares (they are called *grips*, and we cover them very soon), just press Esc to get back to the Command: status line, at which point you can try it again. All four methods of command entry are presented: typing, cascading menus, toolbar icons, and the Ribbon. Alternate each method until you decide which one you prefer. It is perfectly OK to use a hybrid of methods.

Line

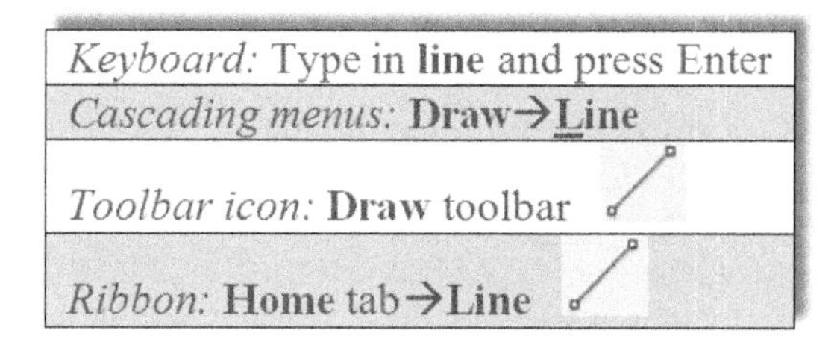

Keyboard: Type in **line** and press Enter
Cascading menus: **Draw→Line**
Toolbar icon: **Draw** toolbar
Ribbon: **Home** tab→**Line**

Step 1. Begin the line command via any of the preceding methods.

- AutoCAD says: `Specify first point:`

Step 2. Using the mouse, left-click anywhere on the screen.

- AutoCAD says: `Specify next point or [Undo]:`

Step 3. Move the mouse elsewhere on the screen and left-click again. You can repeat Step 2 as many times as you wish. When you are done, click Enter or Esc. You should have a bunch of lines on your screen, either separate or connected together, as shown in Fig. 1.12.

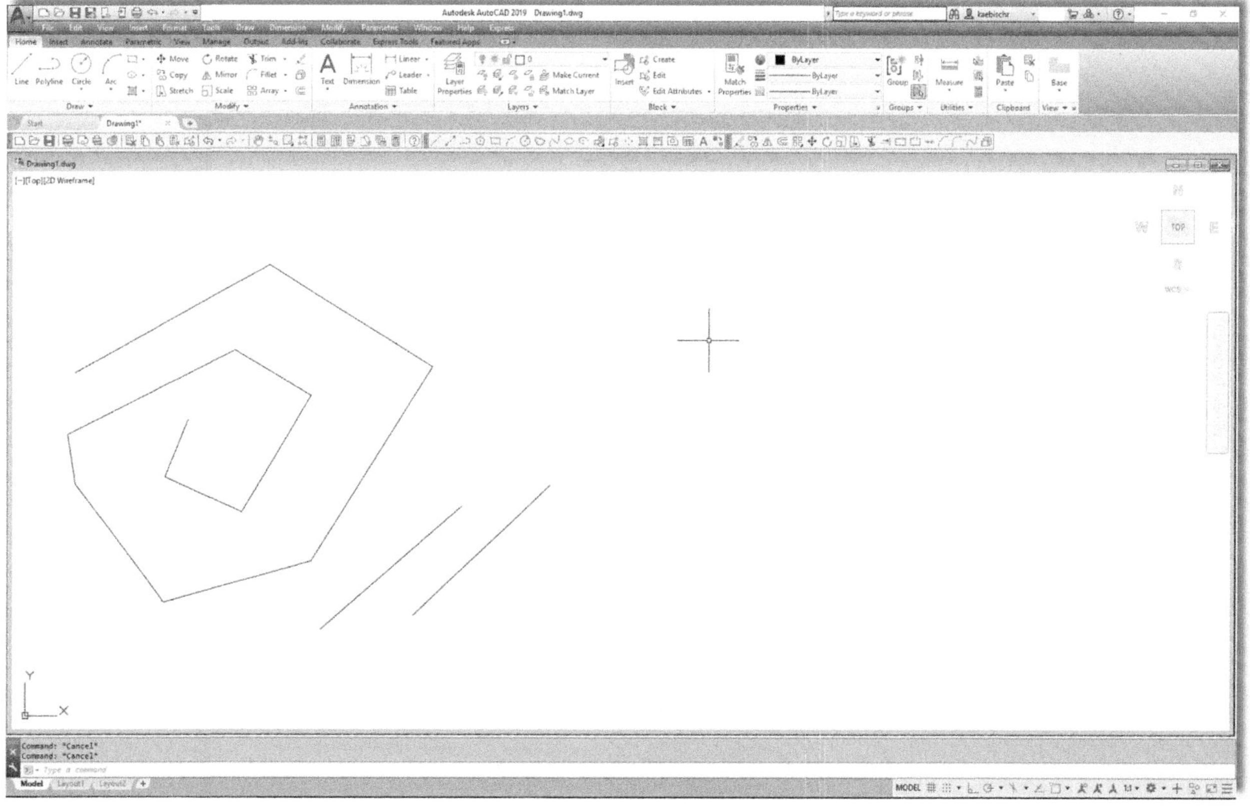

FIGURE 1.12 Lines drawn.

You need not worry about several things at this point. The first is accuracy; we introduce a tremendous amount of accuracy later in the learning process. The second is the options available in the brackets at the command line, such as `[Undo]`; we cover those as necessary. What is important is that you understand how you got those lines and how to do it again. In a similar manner, we move on to the other commands.

Circle

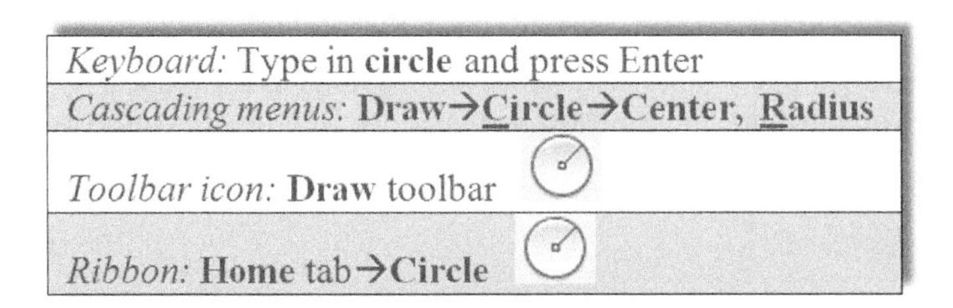

Keyboard: Type in **circle** and press Enter
Cascading menus: **Draw→Circle→Center, Radius**
Toolbar icon: **Draw** toolbar
Ribbon: **Home** tab→**Circle**

Step 1. Begin the circle command via any of the preceding methods.

- AutoCAD says: `Specify center point for circle or [3P/2P/Ttr (tan tan radius)]:`

Step 2. Using the mouse, left-click anywhere on the screen and move the mouse out away from that point.

- AutoCAD says: `Specify radius of circle or [Diameter] <1.9801>:`

Notice the circle that forms; it varies in size with the movement of your mouse. The value in the <> brackets will of course be different in your case or may not even be there at all if this is the first circle you just created.

Step 3. Left-click to finish the circle command. Repeat Steps 1 through 3 several times, and your screen should look like Fig. 1.13.

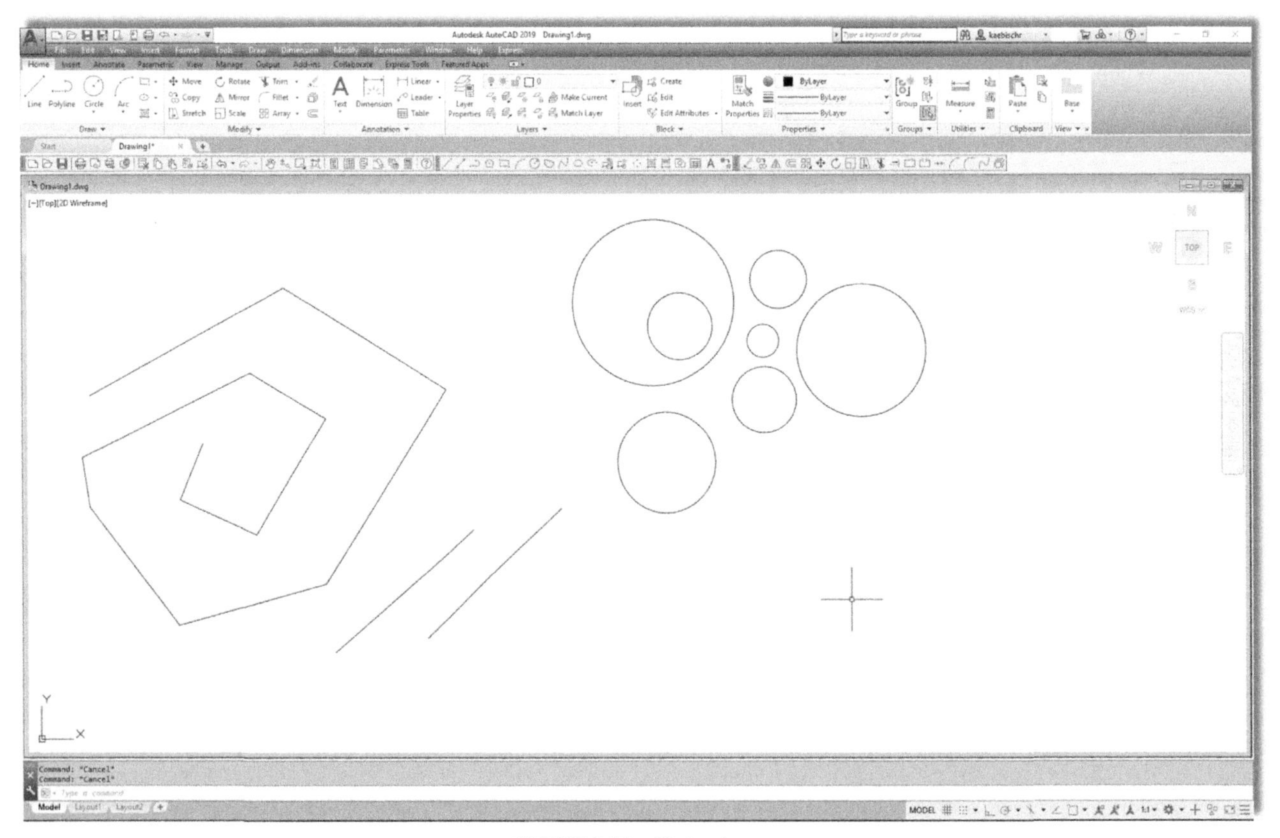

FIGURE 1.13 Circles drawn.

The method just used to create the circle was called *Center*, *Radius*, and you could specify an exact radius size if you wish, by just typing in a value after the first click (try it). As you may imagine, there are other ways to create circles—six ways to be precise, as seen with the Ribbon and cascading menus (Figs. 1.14 and 1.15).

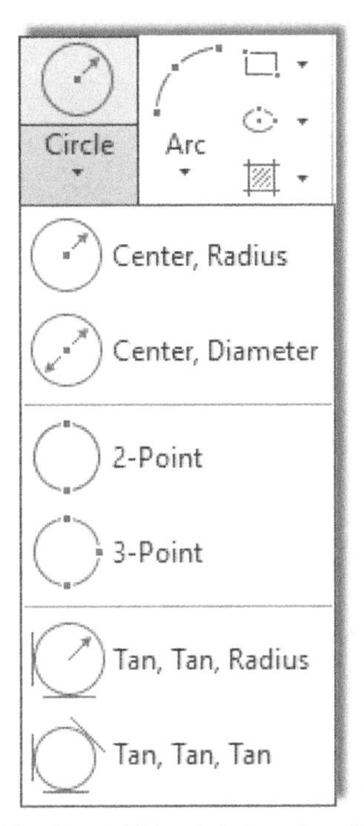

FIGURE 1.14 Additional circle options (Ribbon).

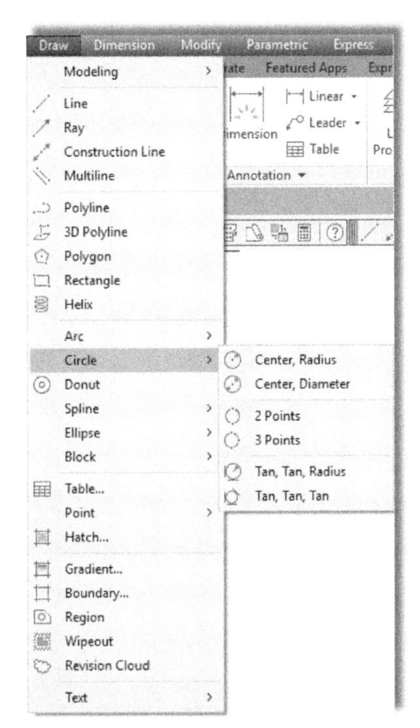

FIGURE 1.15 Additional circle options (Cascading menus).

Special attention should be paid to the Center, Diameter option. Often students are asked to create a circle of a certain diameter, and they inadvertently use radius. This is less of a problem when using the Ribbon or the cascading menus, but you need to watch out if typing or using toolbars, as you will need to press `d` for `[Diameter]` before entering a value; otherwise, guess what it will be. We focus much more on these bracketed options as the course progresses. *You get a chance to practice the other circle options as part of the end of chapter exercises.*

Arc

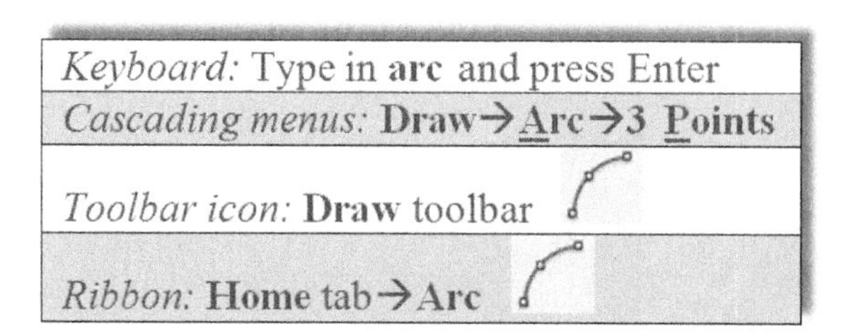

Keyboard: Type in **arc** and press Enter
Cascading menus: **Draw→Arc→3 Points**
Toolbar icon: **Draw** toolbar
Ribbon: **Home** tab→**Arc**

Step 1. Begin the arc command via any of the preceding methods.

- AutoCAD says: `Specify start point of arc or [Center]:`

Step 2. Left-click with the mouse anywhere on the screen. This is the first of three points necessary for the arc.

- AutoCAD says: `Specify second point of arc or [Center/End]:`

Step 3. Click somewhere else on the screen to place the second point.

- AutoCAD says: `Specify end point of arc:`

Step 4. Left-click a third (final) time, somewhere else on the screen, to finish the arc.

Practice this sequence several more times to fully understand the way AutoCAD places the arc. Your screen should look something like Fig. 1.16.

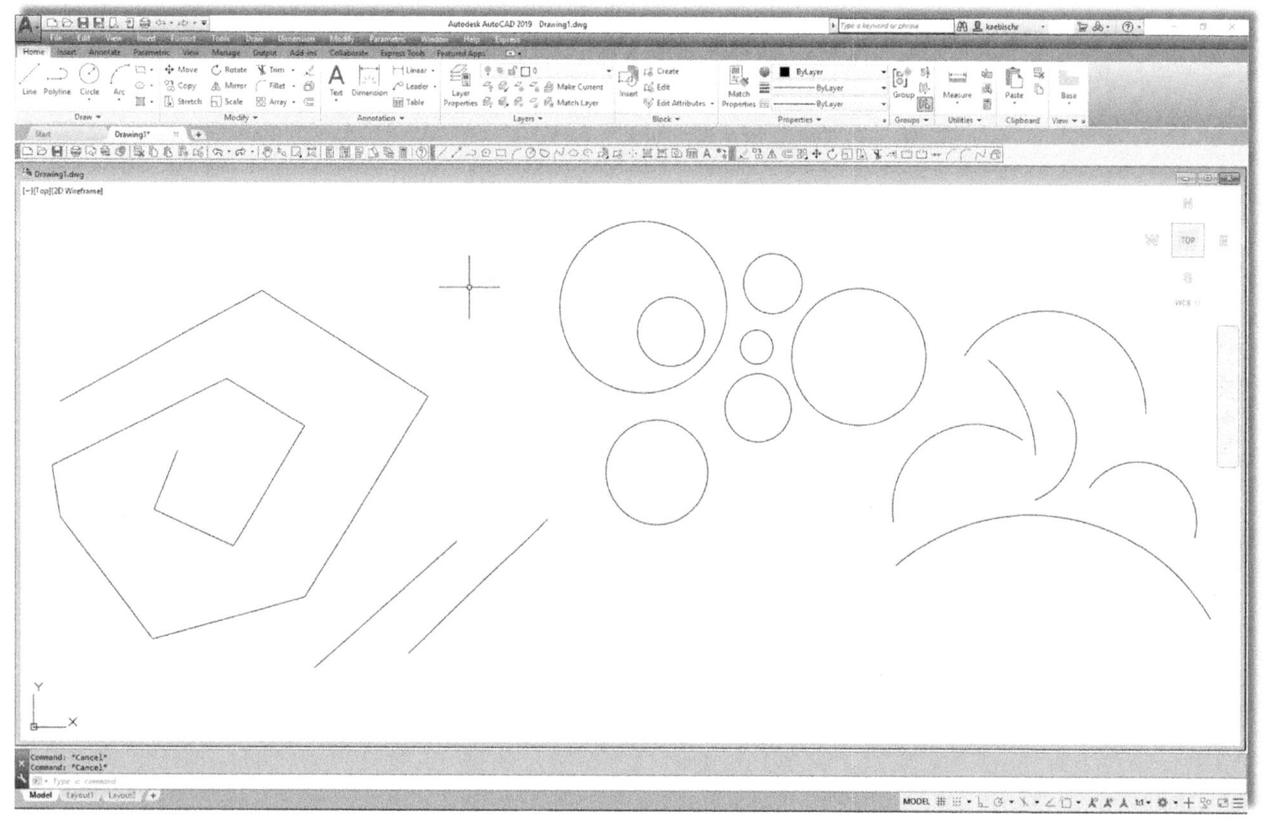

FIGURE 1.16 Arcs drawn.

The method just used to create the arcs is called *3 Point* and, without invoking other options, is a rather arbitrary (eyeball) method of creating them, which is just fine for some applications. Just as with circles, there are, of course, other ways to create arcs, 11 ways to be precise, as seen with the Ribbon and cascading menus (Figs. 1.17 and 1.18).

Not all of these options are used, and some you will probably never need, but it is worth going over them to know what is available. *You get a chance to practice the other arc options as part of the end of chapter exercises.*

Rectangle

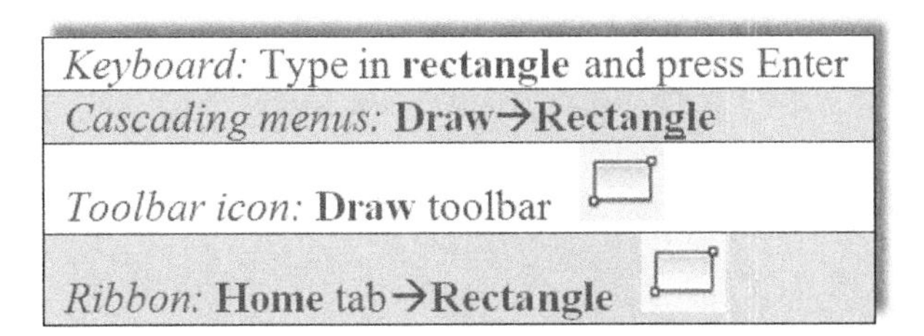

Keyboard: Type in **rectangle** and press Enter
Cascading menus: **Draw→Rectangle**
Toolbar icon: **Draw** toolbar
Ribbon: **Home** tab→**Rectangle**

Step 1. Begin the rectangle command via any of the preceding methods.
- AutoCAD says: `Specify first corner point or [Chamfer/Elevation/Fillet/Thickness/Width]:`

Step 2. Left-click, and move the mouse diagonally somewhere else on the screen.
- AutoCAD says: `Specify other corner point or [Area/Dimensions/Rotation]:`

Step 3. Left-click one more time to finish the command. Your screen should look more or less like Fig. 1.19.

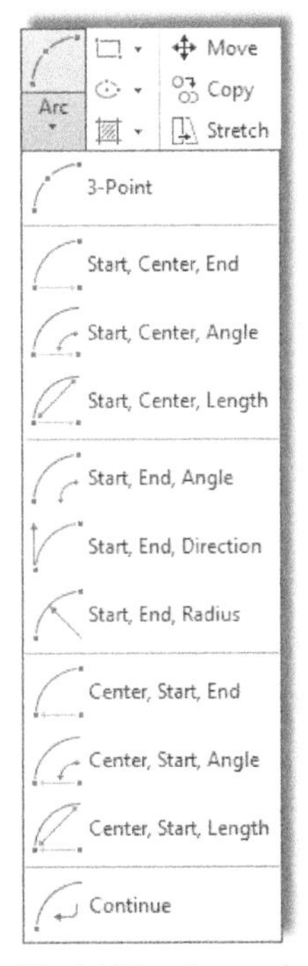

FIGURE 1.17 Additional arc options (Ribbon).

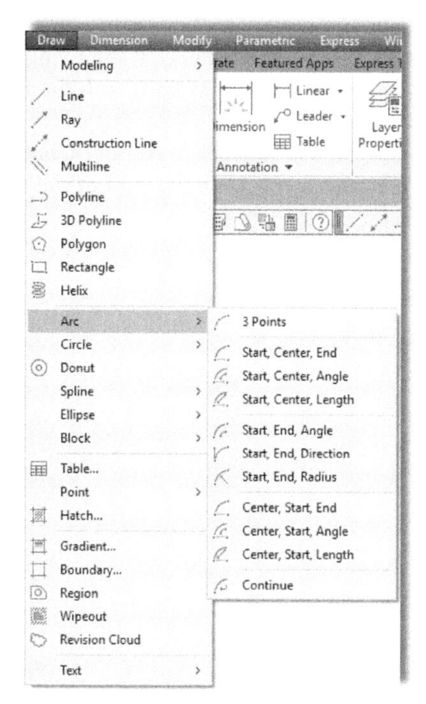

FIGURE 1.18 Additional arc options (Cascading menus).

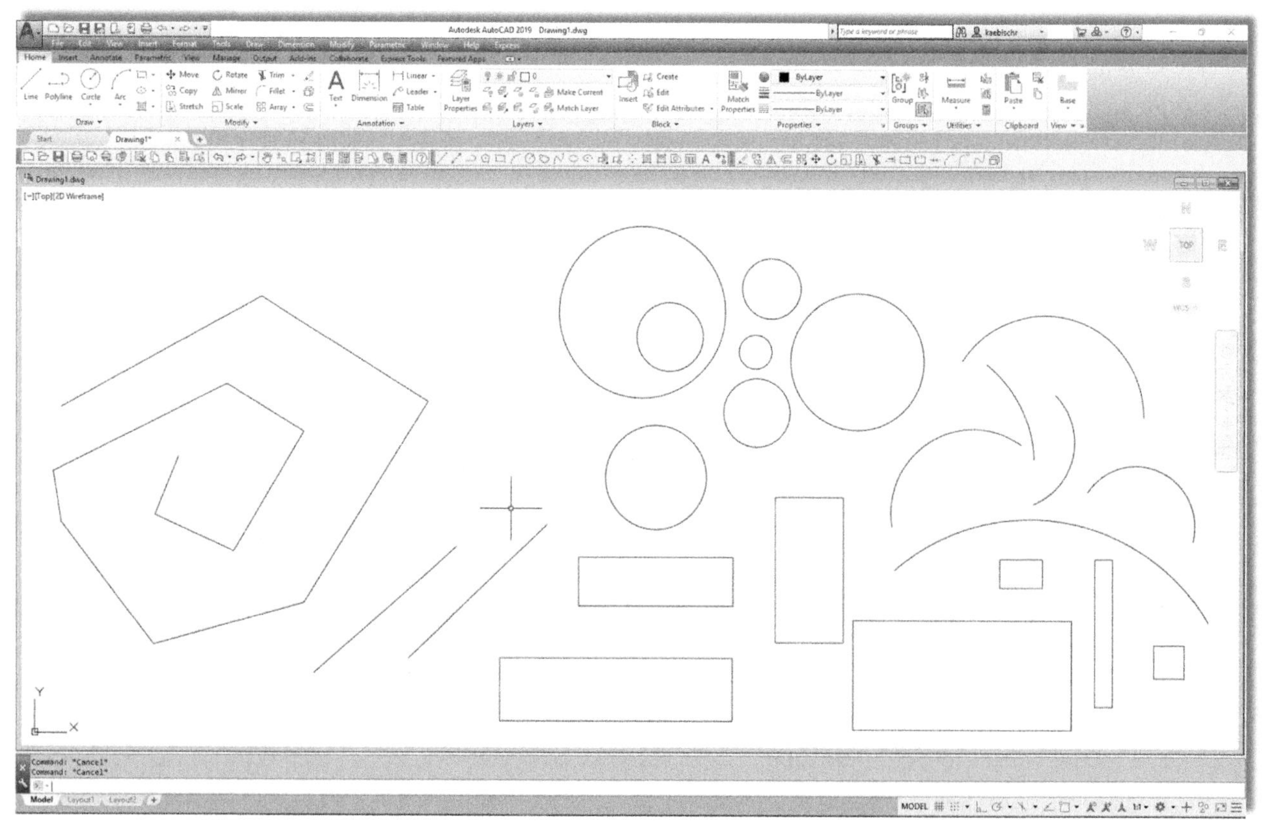

FIGURE 1.19 Rectangles drawn.

There are of course more precise ways to draw a rectangle. In Step 3, you can press `d` for `Dimensions` and follow the prompts to assign length, width, and a corner point (where you want it) to your rectangle. Try it out, but as with all basic shapes drawn so far, do not worry too much about sizing and accuracy, just be sure to understand and memorize the sequences for the basic command—that is what is important for now.

So, now you have created the four basic shapes used in AutoCAD drawings (and ended up with a lot of junk). Let us erase all of them and get back to a blank screen. To do this, we need to introduce Tip 2, which is quite easy to just type. Notice the use of the `e` abbreviation; we will see more of this as we progress.

TIP 2

To quickly erase everything on the screen, type in `e` for `erase`, press Enter, and type in `all`. Then press Enter twice. After the first Enter, all the objects are grayed out, indicating they were selected successfully for erasing. The second Enter finishes the command by deleting everything. Needless to say, this command sequence is useful only in the early stages of learning AutoCAD, when you do not intend to save what you are drawing. You may want to conveniently forget this particular tip when you begin to work on a company project. For now, it is very useful, but we will learn far more selective erase methods later.

So, let us put all of this to good use. Create the drawing shown in Fig. 1.20. Use the rectangle, line, circle, and arc commands. Do not get caught up trying to make everything line up perfectly, just do a quick sketch. Then, you can tell everyone that you learned how to draw a house in AutoCAD within the first half hour. Well, sort of.

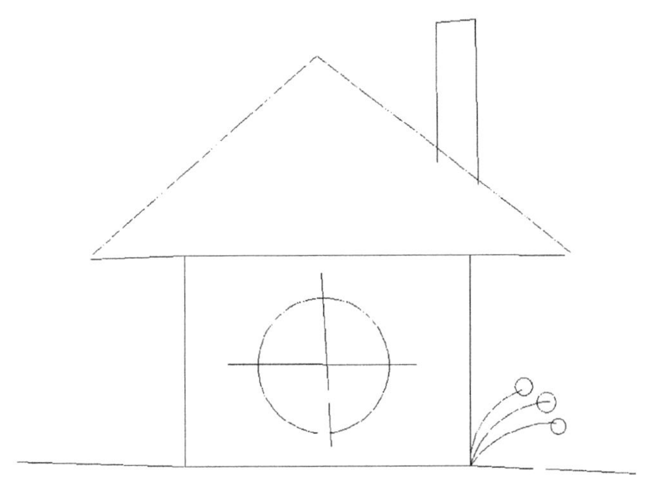

FIGURE 1.20 Simple house sketch.

At this point it is necessary to mention the zoom and pan commands. These are very easy if you have a mouse with a center wheel (as almost all computers do these days).

1.5 VIEW OBJECTS

Zoom

Put your finger on the mouse wheel and turn it back and forth without pressing down. The house you just created gets smaller (roll back) or larger (roll forward) on the screen. Remember though, the object remains the same; just your view is zooming in and out.

Pan

Put your finger on the mouse wheel, but this time press and hold. A hand symbol appears. Now move the mouse around while keeping the wheel depressed. You are able to pan around your drawing. Remember again, the drawing is not moving; you are moving around it.

Regen

This is the easiest command you will learn. Just type in regen and press Enter. The screen refreshes. One use for this is if you are panning over and AutoCAD does not let you go further, as if you hit a wall. Simply regen and proceed.

While the mouse method is easiest, there are other ways to zoom and pan. One alternative is using toolbars, and part of the Standard toolbar (which you should have on your screen) has some Zoom (magnifying glass) and Pan (hand) icons for this purpose, as seen in Fig. 1.21. Explore the icon options on your own, but one particular zoom option is critically important: Zoom to Extents. If you managed to pan your little house right off the screen and cannot find it, this comes in handy, and a full tip on Zoom to Extents is presented next.

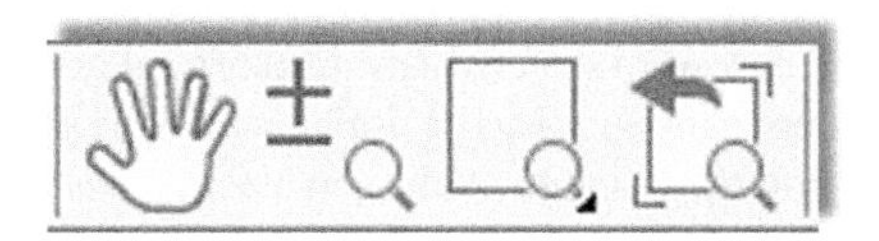

FIGURE 1.21 Zoom and Pan icons.

TIP 3

Type in z for `zoom`, press Enter, type in the letter `e`, and then press Enter again. This is called Zoom to Extents and makes AutoCAD display everything you sketched, filling up the available screen space. If you have a middle-button wheel on your mouse, then double-click it for the same effect. This technique is very important to fit everything on your screen.

What if you did something that you did not want to do and wished you could go back a step or two? Well, like most programs, AutoCAD has an undo command, which brings us to the fourth tip.

TIP 4

To perform the undo command in AutoCAD, just type in `u` and press Enter. Do not type the entire command—if you do that, more options appear, which we do not really need at this point, so a simple `u` suffices. You can do this as many times as you want to (even to the beginning of the drawing session). You can also use the familiar *Undo* and *Redo* arrows seen at the very top of the screen and also as part of the Standard toolbar. AutoCAD also has an *oops* command. This command just restores the last thing you erased, even if it was a while ago, without undoing any other steps. Keep it in mind; it can be useful as well. To try it, erase something, and then type in `oops` and press Enter.

Now that you have drawn your house, erase it and let us move on. We now have to go through all the Edit/Modify Objects commands, one by one. Each of them is critical to creating even the most basic drawings, so go through each carefully, paying close attention to the steps involved. Along the way, the occasional tip is added as the need arises.

Just to remind you: the Esc key gets you out of any mistakes you make and returns you to the basic command line. Be sure to perform each sequence several times to really memorize it. As you learn the first command, erase, you also are introduced to the *picking objects* selection method, which is similar throughout the rest of the commands. Also, keep a close eye on the command line, as it is where you and AutoCAD communicate for now, and it *is* a two-way conversation.

Finally, before you get started, here is another useful tip that students usually appreciate early on to save some time and effort.

TIP 5

Hate to keep having to type, pull a menu, or click on an icon each time you practice the same command? Well, no problem. Simply press the space bar or Enter, and this will repeat the last command you used, saving you a few minutes per day for that coffee break.

1.6 PRACTICING THE EDIT/MODIFY OBJECTS COMMANDS

Erase

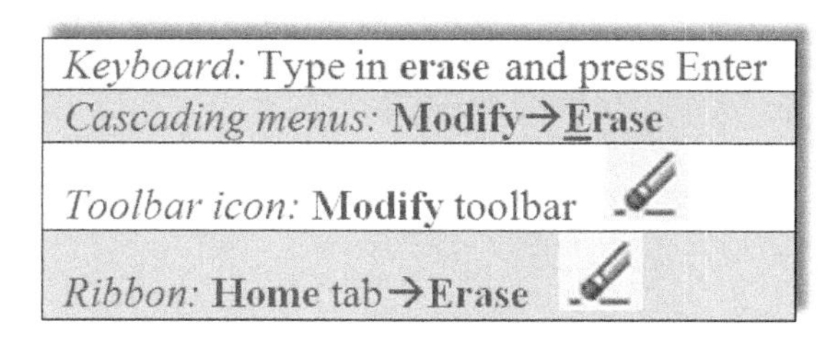

Keyboard: Type in **erase** and press Enter
Cascading menus: **Modify→Erase**
Toolbar icon: **Modify** toolbar
Ribbon: **Home** tab→**Erase**

Step 1. Draw a random sized rectangle on your screen. Begin the erase command via any of the preceding methods.
- AutoCAD says: `Select objects:`

Step 2. Select the rectangle by taking the mouse, positioning it over that object (not in the empty space), and left-clicking once.
- AutoCAD says: `Select objects: 1 found`. A red X appears over the crosshairs and the object becomes dashed. You saw this before when we erased everything in Tip 2. It means the object was selected. AutoCAD continues to ask you: `Select objects:` Watch out for this step. AutoCAD always asks this, in case you want to select more objects, so get used to it. You are done, however, so…

Step 3. Press Enter and the rectangle disappears. No figure is provided for this command as it is very straightforward and completely static.

Practice this several times, using the undo command to bring the rectangle back. You can of course also draw a few more rectangles and practice selecting more than one object, as this is the whole point of being asked to select again. As you click on each one, it becomes dashed. You can do this until you run out of objects to select; there is no limit. To deselect any objects, hold down the Shift key and click on them. They are removed from the selection set.

Move

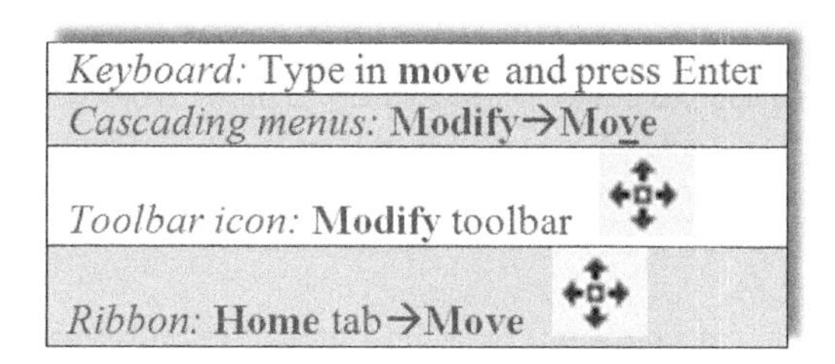

Keyboard: Type in **move** and press Enter
Cascading menus: **Modify→Move**
Toolbar icon: **Modify** toolbar
Ribbon: **Home** tab→**Move**

Step 1. Draw a random sized circle, and begin the move command via any of the preceding methods.
- AutoCAD says: `Select objects:`

Step 2. Select the circle by positioning the mouse directly over it (not the empty space) and left-clicking once.
- AutoCAD says: `Select objects: 1 found`. The circle becomes faded.
- AutoCAD then asks you again: `Select objects:`

Step 3. Unless you have more than one circle to move, you are done, so press Enter.
- AutoCAD says: `Specify base point or [Displacement] <Displacement>:`

Step 4. Left-click anywhere on or near the circle to "pick it up"; this is where you are moving it *from*.
- AutoCAD says: `Specify second point or < use first point as displacement>:`

Step 5. Move the mouse somewhere else on the screen and left-click to place the circle in the new location. This is where you are moving it *to*.

Undo what you just did and practice it again. Fig. 1.22 shows the move command in progress. Notice how the original circle fades in color slightly, and a dashed line follows your mouse until you indicate the new location for the circle.

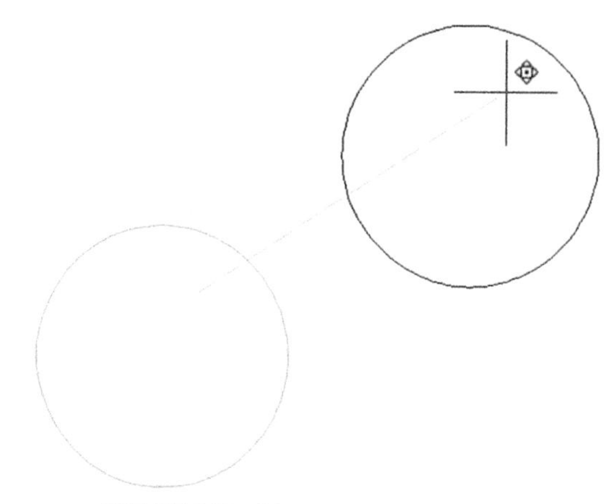

FIGURE 1.22 Move command in progress.

Copy

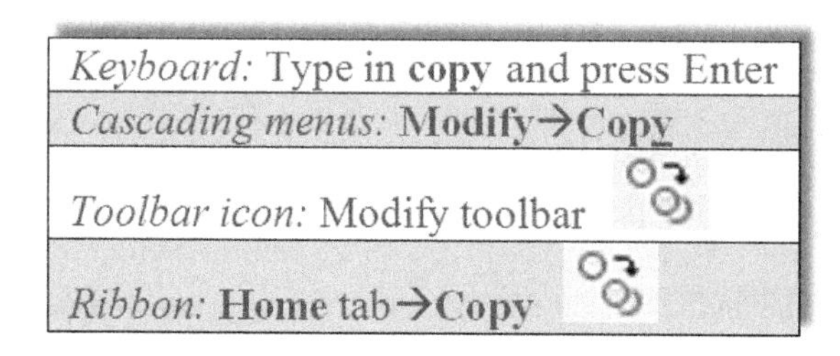

Keyboard: Type in **copy** and press Enter
Cascading menus: **Modify→Copy**
Toolbar icon: Modify toolbar
Ribbon: **Home** tab→**Copy**

Step 1. Leave the circle you just used on the screen. Begin the copy command via any of the preceding methods.

- AutoCAD says: `Select objects:`

Step 2. Select the circle by positioning the mouse over it and left-clicking once.

- AutoCAD says: `Select objects: 1 found`. The circle will become dashed, not faded.
- AutoCAD then asks you again: `Select objects:`

Step 3. Unless you have more than one circlet to copy, you are done, so press Enter.

- AutoCAD says: `Specify base point or [Displacement/mOde]<Displacement>:`

Step 4. Left-click anywhere on or near the circle to "pick it up"; this is where you are copying it *from*.

- AutoCAD says: `Specify second point or [Array] <use first point as displacement>:`

Step 5. Move the mouse somewhere else on the screen and left-click to copy the circle to its new location. This is where you are copying it *to*. Notice that a dashed copy of it remains in its original location until you complete the command.

Step 6. You can copy as many times as you want, pressing Enter or Esc to finish. Before the second click,

- AutoCAD says: `Specify second point or [Array/Exit/Undo] <Exit>:`

Undo what you just did and practice it again. Fig. 1.23 shows the copy command in progress. The dotted circle remains there as a "shadow," along with the dashed line attached to your mouse, until you finish the copy command.

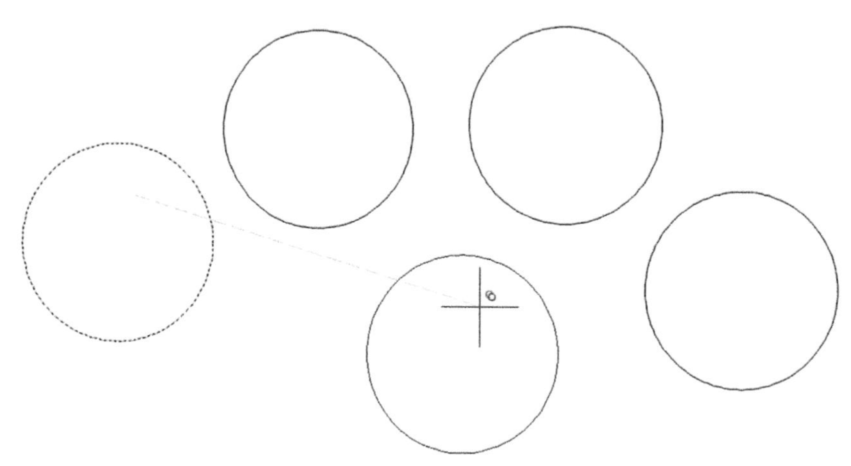

FIGURE 1.23 Copy command in progress.

Rotate

Keyboard: Type in **rotate** and press Enter
Cascading menus: **Modify→Rotate**
Toolbar icon: **Modify** toolbar
Ribbon: **Home** tab→**Rotate**

Step 1. For this command erase the circle and draw a rectangle. Begin the rotate command via any of the preceding methods.

- AutoCAD says: Current positive angle in UCS: ANGDIR = counterclockwise ANGBASE = 0, and on the next line: Select objects:

Step 2. Select the rectangle, remembering to press Enter again after the selection. So far, it is quite similar to erase, move, and copy.

- AutoCAD says: Specify base point: This means select the pivot point of the object's rotation (the point about which it rotates). If you do use a circle for your object, try to stay away from the center point, as your rotation efforts may be less than obvious.

Step 3. Click anywhere on or near the selected object.

- AutoCAD says: Specify rotation angle or [Copy/Reference] <0>:

Step 4. Move the mouse around in a wide circle and the rectangle rotates. Notice how the motion gets smoother as you move the mouse farther away. You can click anywhere for a random rotation angle or you can type in a specific numerical degree value.

Undo what you just did and practice it again. Fig. 1.24 shows the rotate command in progress. The dotted rectangle remains there as a "shadow," along with the dashed line attached to your mouse, until you rotate it into the new locations.

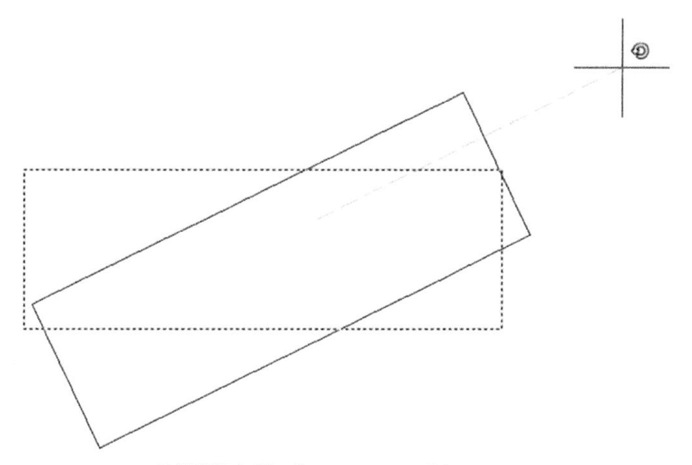

FIGURE 1.24 Rotate command in progress.

Scale

Keyboard: Type in **scale** and press Enter
Cascading menus: **Modify→Scale**
Toolbar icon: **Modify** toolbar
Ribbon: **Home** tab→**Scale**

Step 1. We reuse the previous rectangle to do scale. Begin the scale command via any of the preceding methods.
- AutoCAD says: `Select objects:`

Step 2. Select the rectangle as before, remembering to press Enter again after the first selection.
- AutoCAD says: `Specify base point:` This means select the point from which the scaling of the object (up or down) is to occur.

Step 3. Click somewhere on or near the rectangle or directly in the middle of it.
- AutoCAD says: `Specify scale factor or [Copy/Reference].`

Step 4. Move the mouse around the screen. The rectangle gets bigger or smaller. You can randomly scale it or enter a numerical value. For example, if you want it twice as big, enter `2`; half size will be `0.5`.

Undo what you just did and practice it again. Fig. 1.25 shows the scale command in progress. The dotted rectangle remains there as a "shadow," along with the dashed line attached to your mouse, until you finish the scale command.

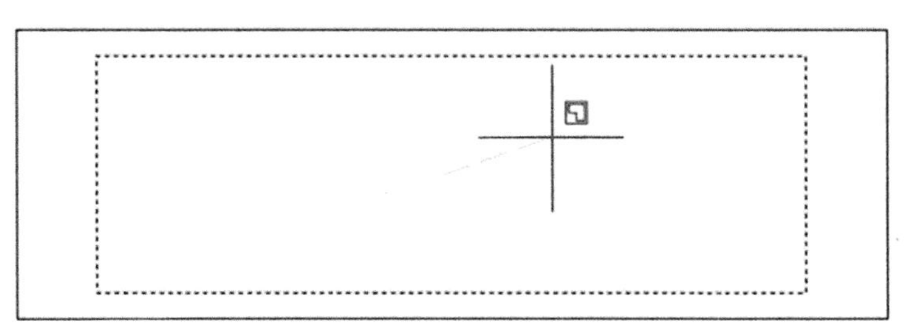

FIGURE 1.25 Scale command in progress.

Trim

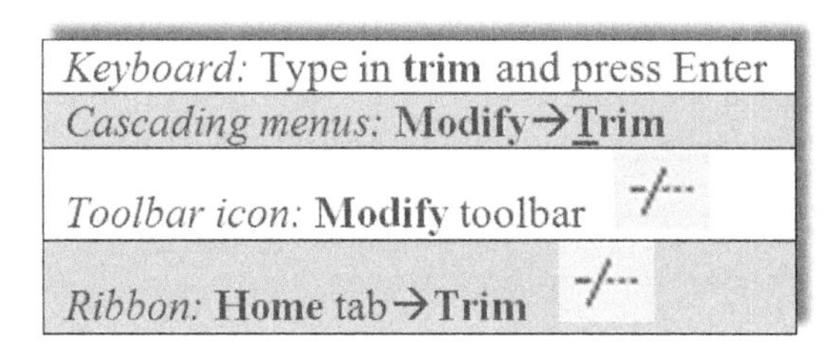

Keyboard: Type in **trim** and press Enter
Cascading menus: **Modify→Trim**
Toolbar icon: **Modify** toolbar
Ribbon: **Home** tab→**Trim**

To practice this command you first need to draw two intersecting lines, one horizontal and one vertical, similar to a plus sign. Once this is done, go ahead and perform the sequence that follows.

Step 1. Begin the trim command via any of the preceding methods.
- AutoCAD says:

```
Current settings: Projection = UCS, Edge = None
Select cutting edges...
Select objects or <select all>:
```

Step 2. Left-click on one of the lines. You can choose the vertical or horizontal, it does not matter. This is your *cutting edge*. The other line, the one you did not pick, will be trimmed at the intersection point of the two lines. The cutting-edge line becomes dashed.
- AutoCAD says: `Select objects or <select all>: 1 found`

Step 3. Press Enter.
- AutoCAD says: `Select object to trim or shift-select to extend or[Fence/Crossing/Project/Edge/eRase/Undo]:`

Step 4. Go ahead and pick anywhere on the line that you did *not* yet select. It turns gray and a red X appears over it. Click it once to trim it off.

You can repeat Step 4 as many times as you want, as long as there is something, but we are done, so press Enter. Practice this several times by pressing u then Enter and repeating Steps 1 through 4. Remember two things to avoid mistakes: Pick the cutting edge first, and do not forget to press Enter before picking the line to be chopped.

You can also do a trim between two or more lines, as shown in Fig. 1.26. This is something we need in the first major project, so practice both types of trims. In both cases, the lines selected as cutting edges are now dashed, and the lines about to be trimmed fade and have the cursor (small box) over them with a red X.

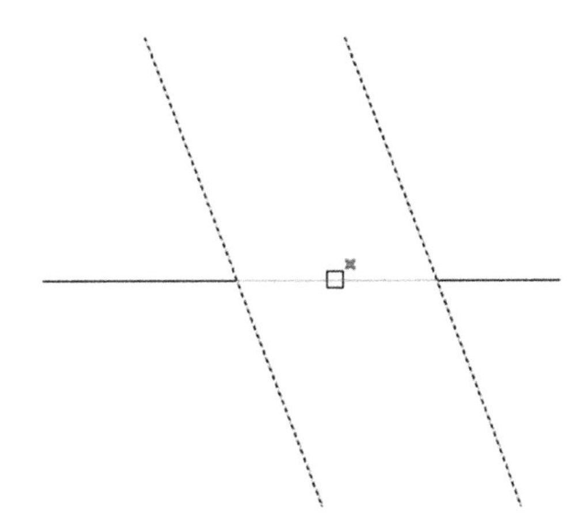

FIGURE 1.26 Trim command (with two lines) in progress.

Extend

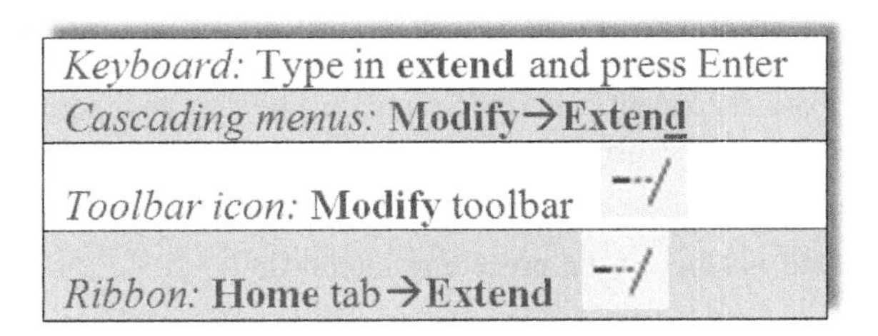

Keyboard: Type in **extend** and press Enter
Cascading menus: **Modify→Extend**
Toolbar icon: **Modify** toolbar
Ribbon: **Home** tab→**Extend**

To practice this command, we first need to draw two intersecting lines, just as for trim but then use the move command to relocate the vertical line directly to the right (or left) of the horizontal line, a short distance away, so we can extend the horizontal line into the vertical. You can see this in Fig. 1.27. Once this is done, go ahead and perform the extend command.

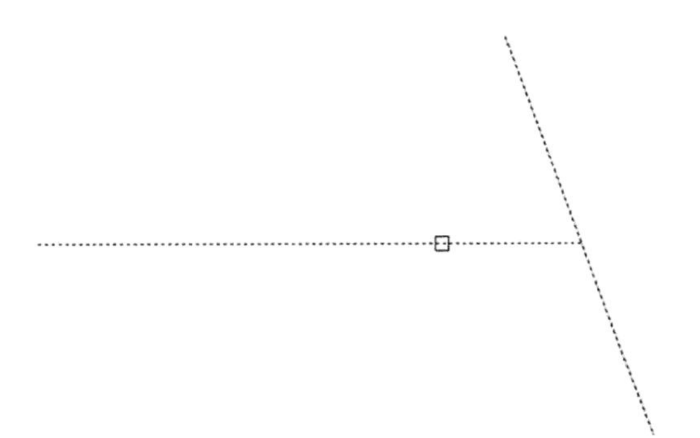

FIGURE 1.27 Extend command in progress.

Step 1. Begin the extend command via any of the preceding methods.

- AutoCAD says:

  Current settings: Projection = UCS, Edge = None
  Select boundary edges...
  Select objects or <select all>:

Step 2. At this point, left-click on the vertical line. This is the *target* into which you extend the horizontal line. It becomes dashed.

- AutoCAD says: Select objects: 1 found.

Step 3. Press Enter.

- AutoCAD says: Select object to extend or shift-select to trim or [Fence/Crossing/Project/Edge/Undo]:

Step 4. Pick the end of the horizontal line that is closest to the vertical line. The *arrow* is what extends into the vertical line target. It automatically extends to show you the result, but you have to click to make it permanent.

You can repeat Step 4 as many times as you want, but we are done, so press Enter. Practice this several times by pressing u then Enter and repeating Steps 1 through 4. In Fig. 1.27, the dashed vertical line is the "target" and the solid horizontal line is the "arrow." The cursor (box) is positioned to extend the arrow into the target by clicking once.

Offset

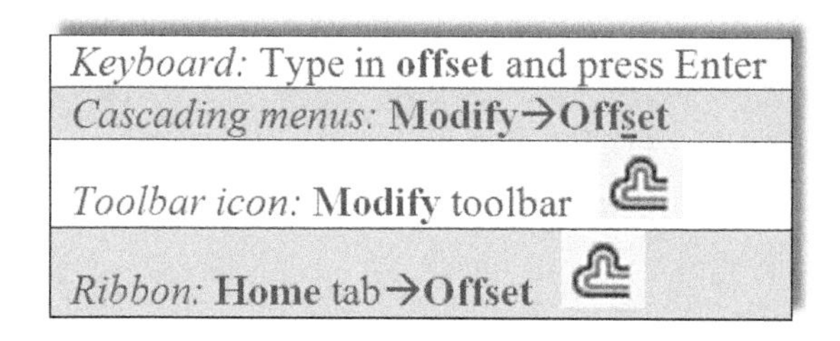

Keyboard: Type in **offset** and press Enter
Cascading menus: **Modify→Offset**
Toolbar icon: **Modify** toolbar
Ribbon: **Home** tab**→Offset**

This command is one of the more important ones you will learn. Its power lies in its simplicity. This command creates new lines by the directional parallel offset concept, where if you have a line on your screen, you can create a new one that is a certain distance away but parallel to the original. To illustrate it, first draw a random vertical or horizontal line anywhere on your screen.

Step 1. Begin the offset command via any of the preceding methods.

- AutoCAD says: Specify offset distance or [Through/Erase/Layer] <Through>:

Step 2. Enter an offset value (use a small number for now, 1 or 2). Then press Enter.

- AutoCAD says: Select object to offset or [Exit/Undo] <Exit>:

Step 3. Pick the line by left-clicking on it; the line becomes dashed.

- AutoCAD says: Specify point on side to offset or [Exit/Multiple/Undo] <Exit>:

Step 4. Pick a *direction* for the line to go, which is to one of the two sides of the original line. It does not matter exactly how far away from the original line you click, as the distance has already been specified. You can keep doing this over and over by selecting the new line then the direction (see Fig. 1.28). If you want to change the offset distance, you must repeat Steps 1 and 2.

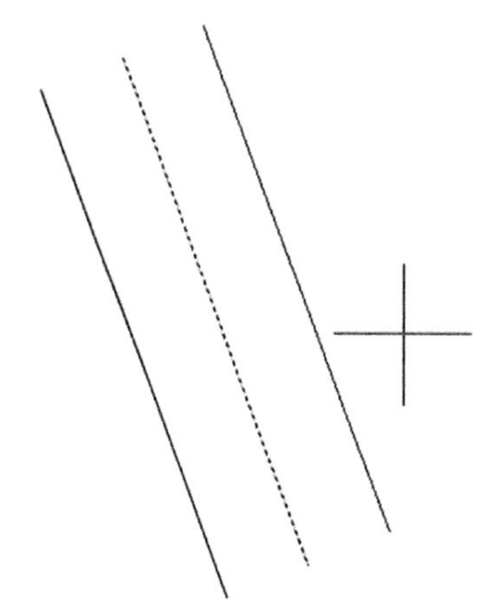

FIGURE 1.28 Offset command in progress.

The significance of the offset command may not be readily apparent. Many students, especially ones who may have done some hand drafting, come into learning AutoCAD with the preconceived notion that it is just a computerized version of pencil and paper. While this is true in some cases, in others, AutoCAD represents a radical departure. When you begin a new drawing on paper, you literally draw line after line. With AutoCAD, surprisingly little line drawing occurs. A few are drawn, and then the offset command is employed throughout. Learn this command well, you will need it.

Mirror

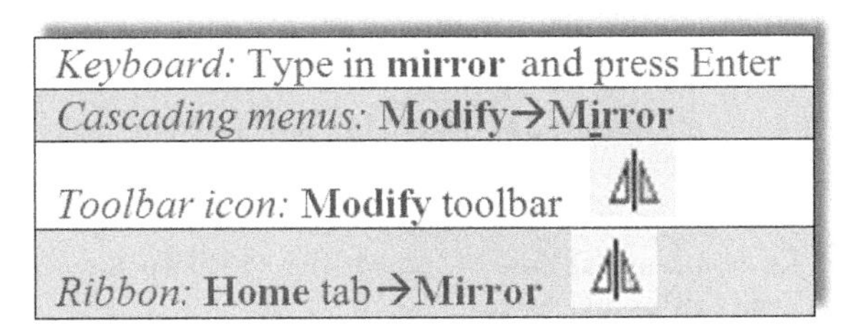

Keyboard: Type in **mirror** and press Enter
Cascading menus: **Modify→Mirror**
Toolbar icon: **Modify** toolbar
Ribbon: **Home** tab→**Mirror**

This command, much as the name implies, creates a mirror copy of an object over some plane. It is surprisingly useful because often just copying and rotating is not sufficient or takes too many steps. To learn this command, you have to think of a real-life mirror, with you standing in front of it. The original object is you, the plane is the door with the mirror attached (which can swing on a hinge), and the new object is your mirrored reflection. To practice this command, draw a triangle by joining three lines together, as seen in Fig. 1.29.

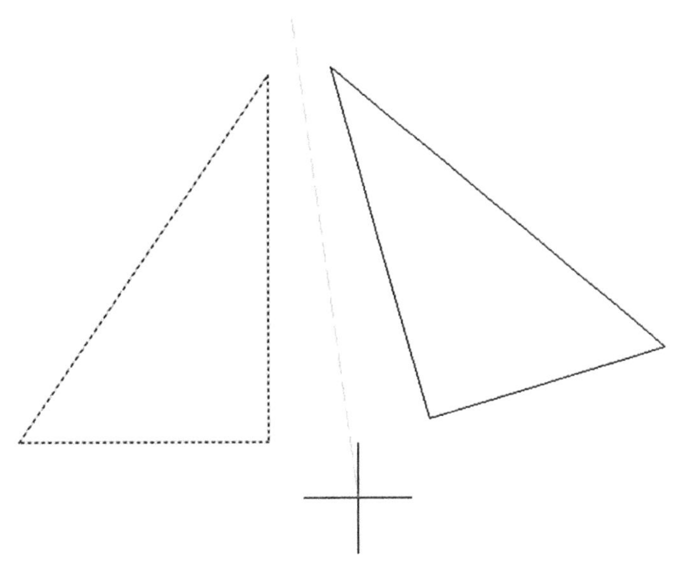

FIGURE 1.29 Mirror command in progress.

Step 1. Begin the mirror command via any of the preceding methods.
- AutoCAD says: `Select objects:`

Step 2. Select all three lines of the triangle and press Enter.
- AutoCAD says: `Specify first point of mirror line:`

Step 3. Click anywhere near the triangle and move the mouse around. You notice the new object appears to be anchored by one point. This is the swinging mirror reflection.
- AutoCAD says: `Specify second point of mirror line:`

Step 4. Click again near the object (make the second click follow an imaginary straight line down). You notice the new object disappear. Do not panic; rather check what the command line says.
- AutoCAD says: `Delete source objects? [Yes/No] <N>:`

Step 5. The response in Step 4 just asked you if you want to keep the original object. If you do (the most common response), then just press Enter and you are done. Fig. 1.29 illustrates Step 3.

Fillet

Keyboard: Type in **fillet** and press Enter
Cascading menus: **Modify→Fillet**
Toolbar icon: **Modify** toolbar
Ribbon: **Home** tab→**Mirror**

Fillets are all around us. This is not just an AutoCAD concept but rather an engineering description for a rounded edge on a corner of an object or at an intersection of two objects. These edges are added to avoid sharp corners for safety or other technical reasons (such as a weld). By definition, a fillet needs to have a radius greater than zero; but in AutoCAD, a fillet can have a radius of zero, allowing for some useful editing, as we see soon. To illustrate this command, first draw two perpendicular lines that intersect, as seen in the first box of Fig. 1.30.

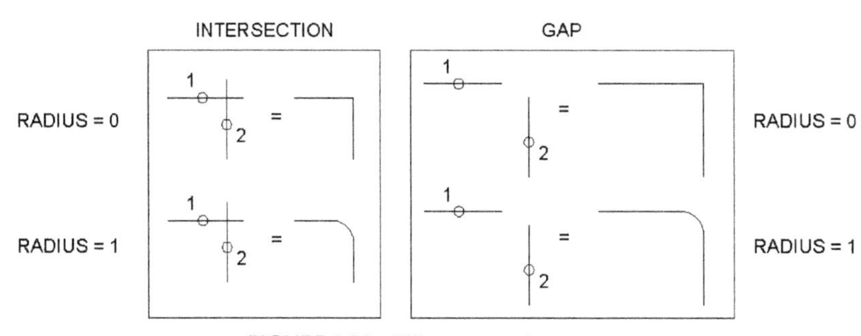

FIGURE 1.30 Fillet command summary.

- **Step 1.** Begin the fillet command via any of the preceding methods.
  - AutoCAD says: `Current settings: Mode = TRIM, Radius = 0.0000`
    `Select first object or [Undo/Polyline/Radius/Trim/Multiple]:`

  Step 2. Put a radius on the fillet. Type in `r` for *radius* and press Enter.
  - AutoCAD says: `Specify fillet radius <0.0000>:`

  Step 3. Enter a small value, perhaps 0.5 or 1. Press Enter.
  - AutoCAD says: `Select first object or [Undo/Polyline/Radius/Trim/Multiple]:`

  Step 4. Select the first (horizontal) line, somewhere near the intersection. It becomes faded, and a proposed fillet appears in dashed form. This is just a preview for you to get an idea of the final result.
  - AutoCAD says: `Select second object or shift-select to apply corner or [Radius]:`

  Step 5. As your mouse moves toward the second (vertical) line, you continue to see a preview of the resulting fillet (if the fillet is valid). This preview capability is a relatively new feature, first seen with AutoCAD 2012. Go ahead and click the second line somewhere near the intersection; a fillet of the radius you specified is added, the unneeded lines disappear and everything becomes solid.

These are the steps involved in adding a fillet with a radius. You can also do a fillet with a radius of zero, so repeat Steps 1 through 5, entering 0 for the radius value. This is very useful to speed up trimming and extending. Intersecting lines can be filleted to terminate at one point, and if they are some distance apart, they can be filleted to meet together. All this is possible with a radius of zero. In these cases, the preview feature shows where the lines intersect. Try out all the variations. Fig. 1.30 summarizes everything described so far. The 1 and 2 indicate a suggested order of clicking the lines for the desired effect.

Well, this is it for now. We covered the basic commands. These commands form the essentials of what we call *basic CAD theory*. All 2D computer-aided design programs need to be able to perform these functions to work, although other software may refer to them differently, for example using *displace* instead of *move*, but the capability still has to be there. Be sure to memorize these commands and practice them until they become second nature. They are truly the ABCs of basic computer drafting, their importance cannot be overstated, and you need to master them to move on.

1.7 SELECTION METHODS

Before we talk about the accuracy tools available to you, we need to discuss more advanced types of selection methods. The classic approaches are called *Window* and *Crossing*, and as of AutoCAD 2015 release, you also have the new *Lasso* tool.

You may have already seen the Window, Crossing and Lasso tools by accident: when you single-click randomly on the screen without selecting any objects and move your mouse to the right or left, a rectangular shape forms. It is either dashed (with a green fill) or solid (with a blue fill) and goes away once you click the mouse button again or press Esc. Alternatively, if you single-click randomly and keep *holding down* the left mouse button while moving the mouse you will get the Lasso. These all happen to be powerful selection tools. They are defined as follows:

- **Crossing:** A crossing appears when you single-click and move to the *left* of an imaginary vertical line (up or down) and is a *dashed* green-shaded rectangle. *Any object it touches or is inside of it gets selected.*
- **Window:** A window appears when you single-click and move to the *right* of an imaginary vertical line (up or down) and is a *solid* blue-shaded rectangle. *Any object falling completely inside it is selected.*

Fig. 1.31 shows what both the Crossing and the Window look like. It is not shown in the figure, but note that, depending on settings, your Crossing and Window may be filled with a color, green for crossing and blue for window. Practice selecting objects in conjunction with the erase command. Draw several random objects and start up the erase command. Then, instead of picking each one individually (or doing Erase, All), select them with either the Crossing or Window and make a note of how each functions. This is a very important and useful selection method.

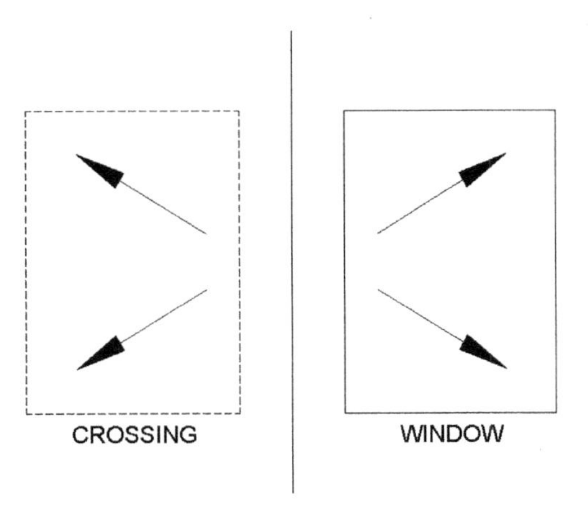

FIGURE 1.31 Crossing and Window.

Finally, let us discuss the new Lasso function. This is a technique borrowed from Photoshop and similar software, and it frees you from any geometric constraints whatsoever, allowing you to "free-hand" choose the selection area by "roping" around (or through) it. You may end up liking this new method the most. It can be applied both as a Window Lasso and a Crossing Lasso.

To try out the *Window Lasso*,

Step 1. Draw a set of circles, as seen in Fig. 1.32.

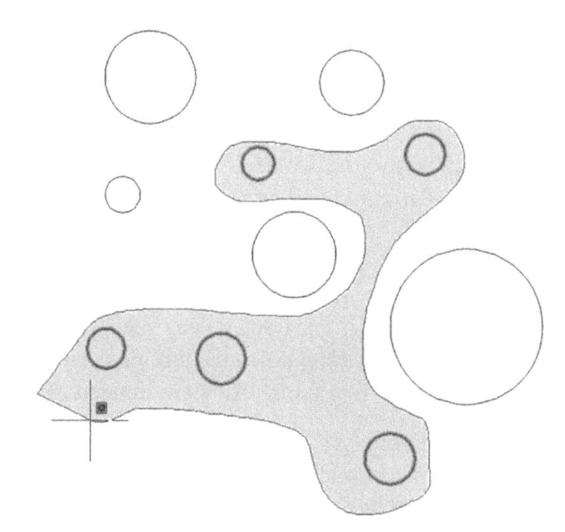

FIGURE 1.32 Window Lasso.

Step 2. Begin the erase command via any method you prefer.
Step 3. Do not type anything in, rather *press and hold down* the left mouse button, and drag it to the *right* (thus making it a *Window* Lasso) around several of the circles, as seen in Fig. 1.32.
Step 4. Continue to smoothly drag the mouse *around* the circles you wish to select, and right-click when done.

To try out the *Crossing Lasso*,

Step 1. Draw another set of circles, as seen in Fig. 1.33, or just undo your previous steps.

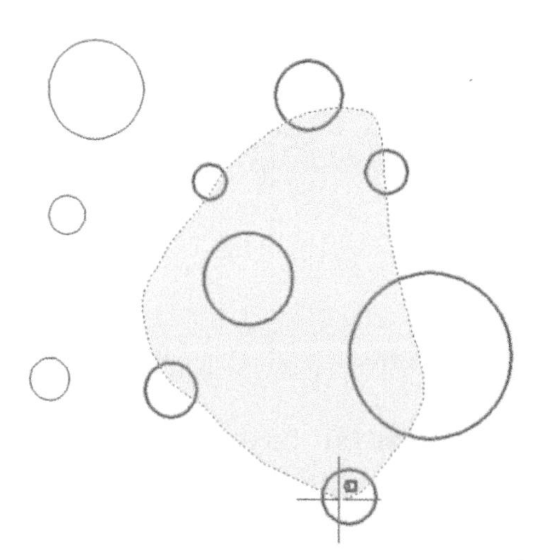

FIGURE 1.33 Crossing Lasso.

Step 2. Begin the erase command via any method you prefer.
Step 3. Do not type anything in, rather press *and hold down* the left mouse button, and drag it to the *left* (thus making it a *Crossing* Lasso) around several of the circles, as seen in Fig. 1.33.
Step 4. Continue to smoothly drag the mouse *through* the circles you wish to select, and right-click when done.

Note how, in both cases, this tool makes the selection process very smooth and free form, as opposed to being restricted to straight lines or rectangles. This may well end up being the primary approach to selecting objects in subsequent releases of AutoCAD, with all the other methods as a secondary backup.

1.8 DRAWING ACCURACY—PART 1

Ortho (F8)

So far, when you have created lines, they were not perfectly straight. Rather, you may have lined them up by eye. This of course is not accurate enough for most designs, so we need a better method. Ortho (which stands for *orthographic*) is a new concept we now introduce that allows you to draw perfectly straight vertical or horizontal lines. It is not so much a command as a condition you impose on your lines prior to starting to draw them. To turn on the Ortho feature, just press the F8 key at the top of your keyboard or click to depress the ORTHOMODE button at the bottom right of the screen in the drawing aids (looks like an "L" angle). Then, begin drawing lines. You see your mouse cursor is constrained to only vertical and horizontal motion. Ortho is usually not something you just set and forget about, as your design may call for a varied mix of straight and angled lines, so do not be shy about turning the feature on and off as often as necessary, sometimes right in the middle of drawing a line.

1.9 DRAWING ACCURACY—PART 2

OSNAPs

While drawing straight lines, we generally like those lines to connect to each other in a very precise way. So far we have been eyeballing their position relative to each other, but that is not accurate and is more reminiscent of sketching, not serious drafting. Lines (or for that matter all fundamental objects in AutoCAD) connect to each other in very distinct and specific locations (two lines are joined end to end, for example). Let us develop these points for a collection of objects.

ENDpoint

ENDpoints are the two locations at the ends of any line. Another object can be precisely attached to an endpoint. The symbol for it is a square, as illustrated in Fig. 1.34.

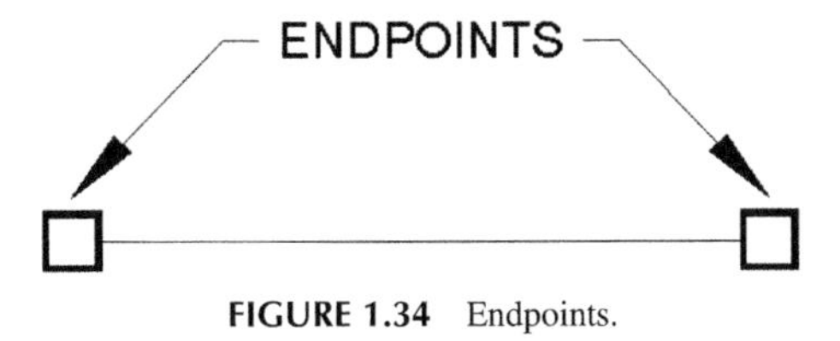

FIGURE 1.34 Endpoints.

MIDpoint

MIDpoint is the middle of any line. Another object can be precisely attached to a midpoint. The symbol for it is a triangle, as illustrated in Fig. 1.35.

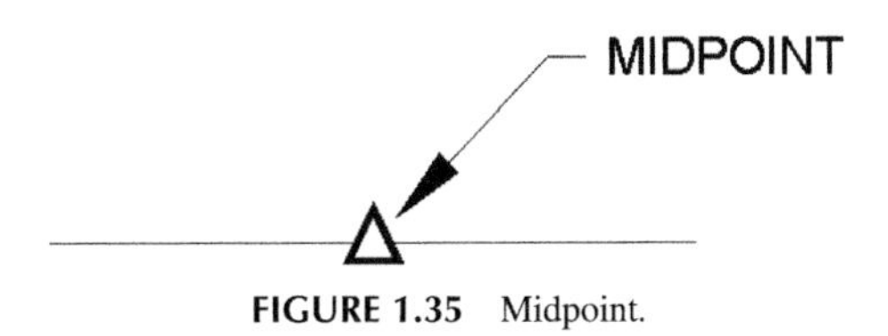

FIGURE 1.35 Midpoint.

CENter

CENter is the center of any circle. Another object can be precisely attached to a center point. The symbol for it is a circle, as illustrated in Fig. 1.36.

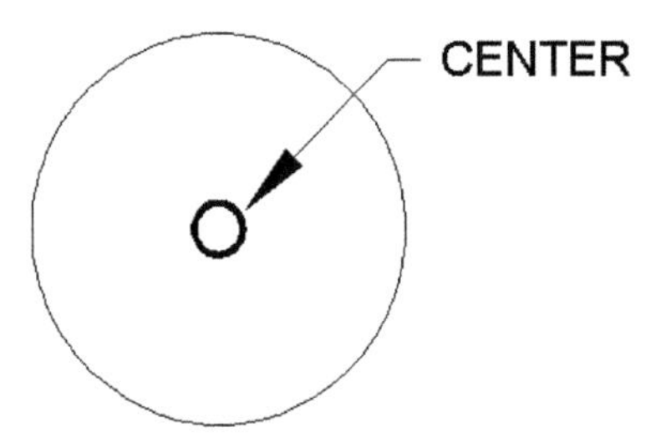

FIGURE 1.36 Center.

QUADrant

QUADrants are the four points at the North (90 degrees), West (180 degrees), South (270 degrees), and East (0°, 360 degrees) quadrants of any circle. Another object can be precisely attached to a quadrant point. The symbol for it is a diamond, as illustrated in Fig. 1.37.

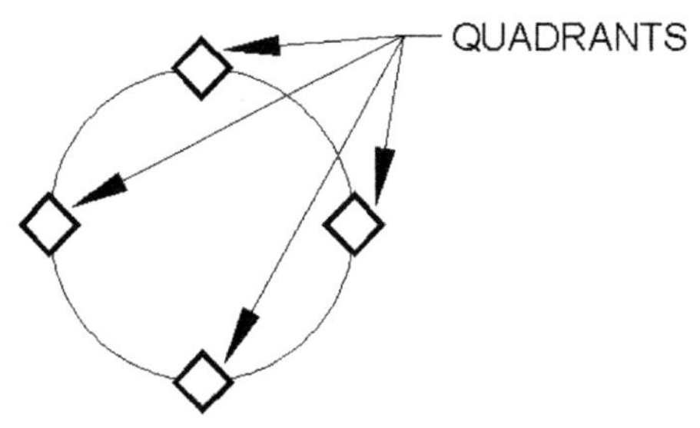

FIGURE 1.37 Quadrants.

INTersection

INTersection is a point located at the intersection of any two objects. You may begin a line or attach any object to that point. The symbol for it is an X, as illustrated in Fig. 1.38.

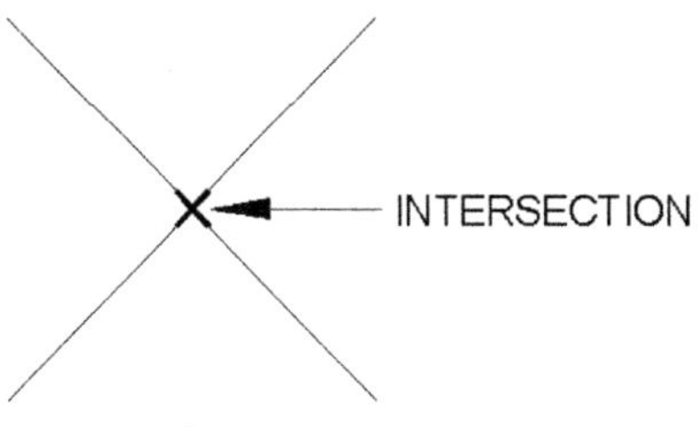

FIGURE 1.38 Quadrants.

PERPendicular

PERPendicular is an OSNAP point that allows you to begin or end a line perpendicular to another line. The symbol for it is the standard mathematical symbol for a perpendicular (90 degrees) angle, as illustrated in Fig. 1.39.

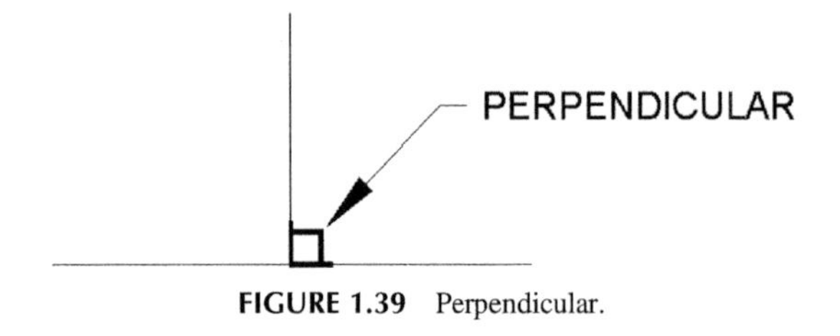

FIGURE 1.39 Perpendicular.

These are the six fundamental OSNAP points we mention for now. There are others of course, such as Geometric Center, Node, Extension, Insertion, Tangent, Nearest, Apparent Intersection, and Parallel (a total of 14), but those are used far less often. We touch on a few of them, such as Tangent, Nearest, Node, and Geometric Center (covered in upcoming chapters), but the remaining ones are not discussed.

So, how do you use the OSNAPs that we have developed so far? There are two methods. The first is to type in the OSNAP point as you need it (only the capitalized part needs to be typed in: END, MID, etc.). For example, draw a line. Then, begin drawing another line, but before clicking to place it, type in end and press Enter. AutoCAD adds the word of, at which point you need to move the cursor to roughly the end of the first line. The snap marker for endpoint (a square) appears, along with a tool tip stating what it is. Click to place the line; now your new line is precisely attached to the first line.

In the same manner, try out all the other snap points mentioned (draw a few circles to practice QUAD and CEN). In all cases, begin by typing in line, then one of the snap points, followed by clicking Enter. Next, bring the mouse cursor to the approximate location of that point and click.

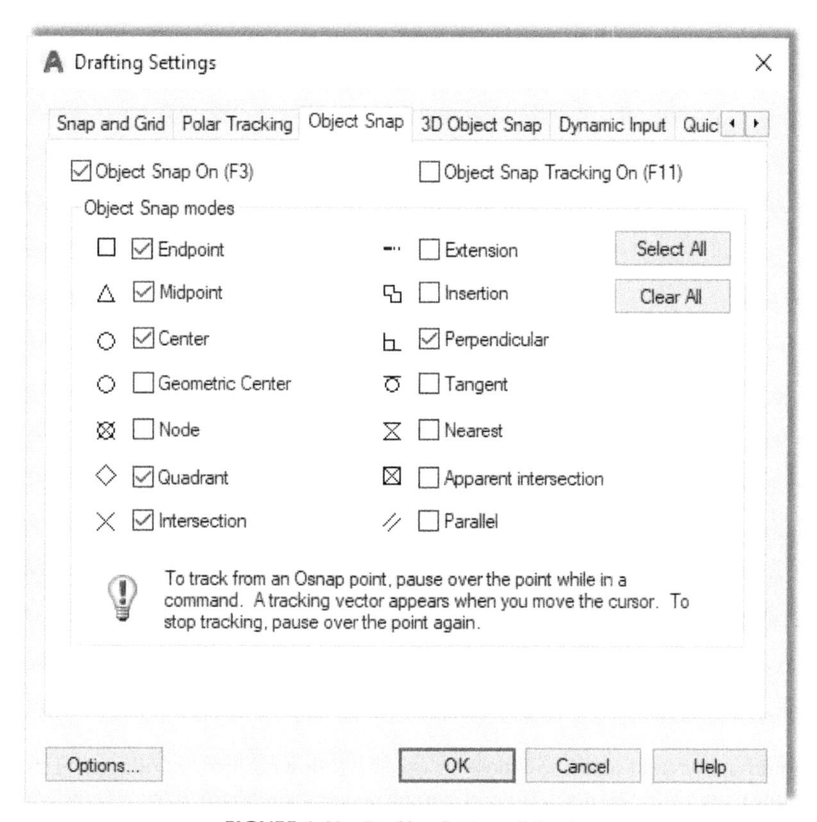

FIGURE 1.40 Drafting Settings dialog box.

1.10 OSNAP DRAFTING SETTINGS

As you may imagine, there is a faster way to use OSNAP. The idea is to have these settings always running in the background, so you need not type them in anymore. To set them up, you need to type in osnap or select Tools→Drafting Settings… from the cascading drop-down menu. The Drafting Setting dialog box appears (Fig. 1.40). First, make sure the Object Snap tab is selected, and then select the Clear All button. Finally, check off the six settings mentioned previously (Endpoint, Midpoint, Center, Quadrant, Intersection, Perpendicular), and press OK.

So now you have set the appropriate settings, but you still need to turn them on. For that, we need to mention another F key. If you press F3 or press the OSNAP button at the bottom right of the screen in the drawing aids (looks like a square), the OSNAP settings become active. Go ahead and do this, then begin the same line drawing exercise; see how much easier it is now as the snap settings appear on their own. This is the preferred way to draw, with them running in the background. Feel free to press F3 anytime you do not need them anymore; OSNAP is also not something that you may want all the time. Fig. 1.40 shows the Drafting Settings dialog box, with the discussed OSNAP points checked.

SUMMARY

You should understand and know how to use the following concepts and commands before moving on to Chapter 2*:*

- AutoCAD drawing environment
  - Drawing Area, Command Line, Cascading Menus, Toolbars, Ribbon
- Create Objects
  - Line
  - Circle
  - Arc
- Rectangle
- Edit/Modify Objects
  - Erase
  - Move
  - Copy
  - Rotate
  - Scale
  - Trim
  - Extend
  - Offset
  - Mirror
  - Fillet
- View Objects
  - Zoom
  - Pan
  - Regen
- Selection methods
  - Crossing
  - Window
  - Lasso
- Ortho (F8)
- OSNAP (F3)
  - ENDpoint
  - MIDpoint
  - CENter
  - QUADrant
  - INTersection
  - PERPendicular

Additionally, you should understand and know how to use the following tips introduced in this chapter:

TIP 1: The Esc (Escape) key in the upper left-hand corner of your keyboard is your new best friend while learning AutoCAD. It gets you out of just about any trouble you get yourself into. If something does not look right, just press the Esc key and repeat the command. Mine was worn out learning AutoCAD, so I expect you to use it often.

TIP 2: To quickly erase everything on the screen, type in `e` for `erase`, press Enter, and type in `all`. Then press Enter twice. After the first Enter, all the objects are grayed out, indicating they were selected successfully for erasing. The second Enter finishes the command by deleting everything. Needless to say, this command sequence is useful only in the early stages of learning AutoCAD, when you do not intend to save what you are drawing. You may want to conveniently forget this particular tip when you begin to work on a company project. For now, it is very useful, but we will learn far more selective erase methods later.

TIP 3: Type in `z` for `zoom`, press Enter, type in the letter `e`, and then press Enter again. This is called Zoom to Extents and makes AutoCAD display everything you sketched, filling up the available screen space. If you have a middle-button wheel on your mouse, then double-click it for the same effect. This technique is very important to fit everything on your screen.

TIP 4: To perform the undo command in AutoCAD, just type in `u` and press Enter. Do not type the entire command—if you do that, more options appear, which we do not really need at this point, so a simple `u` suffices. You can do this as many times as you want to (even to the beginning of the drawing session). You can also use the familiar Undo and Redo arrows seen at the very top of the screen and also as part of the Standard toolbar. AutoCAD also has an oops command. This command just restores the last thing you erased, even if it was a while ago, without undoing any other steps. Keep it in mind; it can be useful as well. To try it, erase something, then type in oops and press Enter.

TIP 5: Hate to keep having to type, pull a menu, or click an icon each time you practice the same command? Well, no problem. Simply press the space bar or Enter, and this repeats the last command you used.

REVIEW QUESTIONS

Answer the following questions based on what you learned in this chapter. You should have the basic commands memorized, so try to answer most of the questions without looking up anything:

1. List the four Create Objects commands covered.
2. List the 10 Edit/Modify Objects commands covered.
3. List the three View Objects commands covered.
4. What is a workspace?
5. What four main methods are used to issue a command in AutoCAD?
6. What is the difference between a selection Window, Crossing, and Lasso?
7. What is the name of the command that forces lines to be drawn perfectly straight horizontally and vertically? Which F key activates it?
8. What six main OSNAP points are covered? Draw the snap point's symbol next to your answer.
9. What command brings up the OSNAP dialog box?
10. Which F key turns the OSNAP feature on and off?
11. How do you terminate a command in progress in AutoCAD?
12. What sequence of commands quickly clears your screen?
13. What sequence of commands allows you to view your entire drawing?
14. How do you undo a command?
15. How do you repeat the last command you just used?

EXERCISES

1. Draw the following sets of lines on the left and practice the 2-Point, 3-Point, and Tan, Tan, Tan circle options. You may use the Ribbon, cascading menus, or typing. In all cases, use OSNAPs to connect the circle precisely to the drawn lines. The result you should get is shown on the right. (Difficulty level: Easy; Time to completion: <5 minutes.)

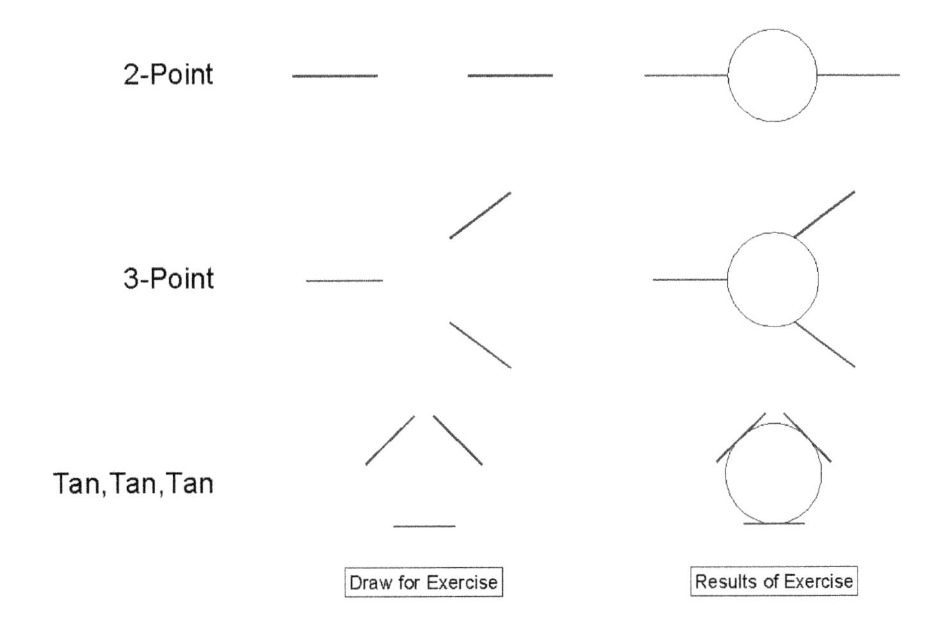

2. Draw an arc using the Start, Center, End option (a commonly used one) as follows. First, draw the three lines shown on the left, making use of the mirror command as needed, then draw the arc itself. Be sure to select from the bottom up, as shown, and use OSNAPs. The result you should get is shown on the right. Try the same exercise by picking from the top down, noticing the difference. (Difficulty level: Easy; Time to completion: <5 minutes.)

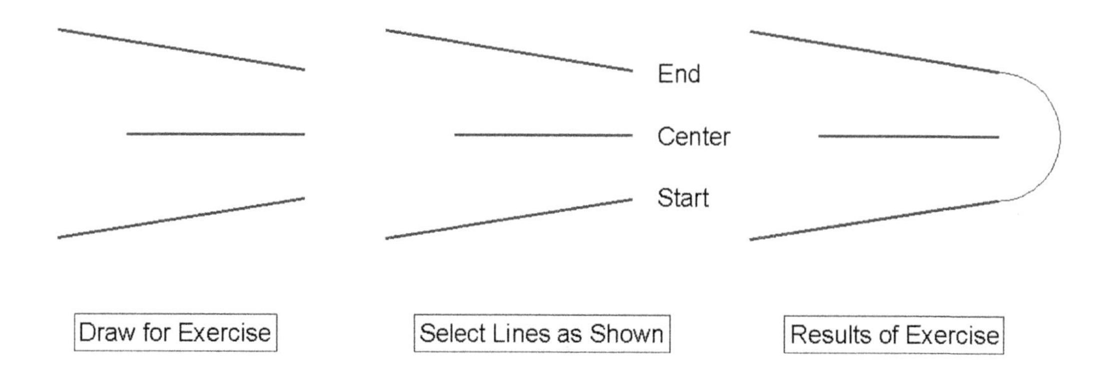

3. Draw a simple door using the 3-Point arc method. This requires some precise mouse clicking and we introduce an alternate method in Chapter 3, but for now this is a good exercise to try. First draw, copy, rotate, and position the lines as seen in Steps 1 through 4, then add the arc as shown in Step 5 (use Ortho and OSNAPs throughout). You may have to try it a few times to get it to look right. Finally, erase the bottom line to complete the door as seen in Step 6. (Difficulty level: Easy; Time to completion: <5 minutes.)

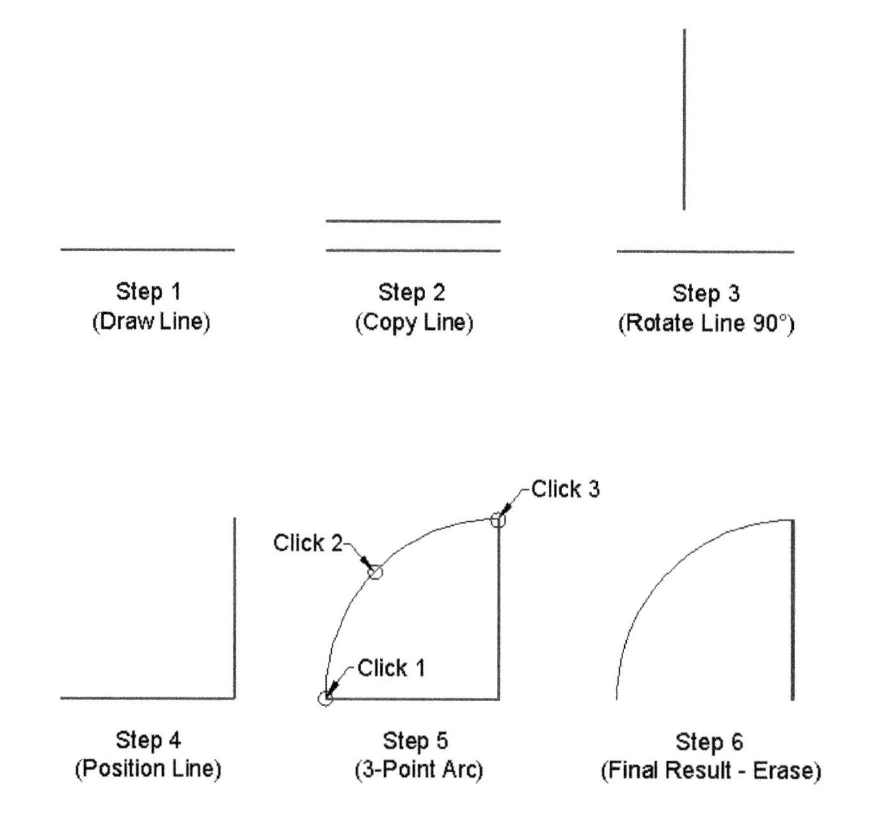

4. Draw the following sets of arbitrary electrical connection shapes. You need only the basic Create and Modify commands, including rectangle, circle, line, offset, fillet, and mirror, as well as OSNAPs and Ortho. Specific sizes are not important at this point, but make your drawing look similar, and use accuracy in connecting all the pieces. (Difficulty level: Easy; Time to completion: <10 minutes.)

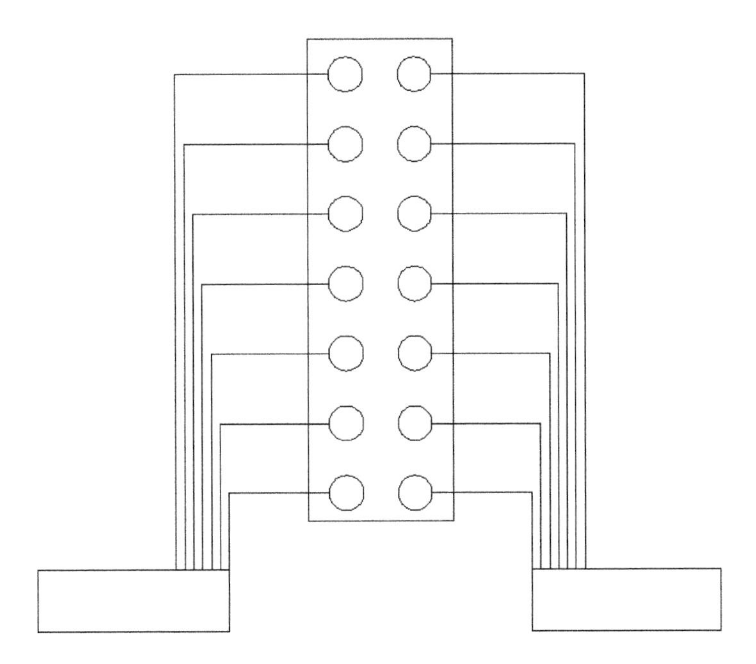

5. Draw the following house elevation using what you learned. You need only the basic Create and Modify commands. The overall sizing can be approximated; however, use Ortho (no crooked lines) and make sure all lines connect using OSNAPs. (Difficulty level: Intermediate; Time to completion: <15 minutes.)

6. Draw the following sets of random shapes. You need only the basic Create and Modify commands, but you use almost every one of them, including multiple cutting-edges trim. The overall sizing is completely arbitrary, and your design may not look exactly like this one, but be sure to draw all the elements you see here and use OSNAP precision throughout. (Difficulty level: Intermediate; Time to completion: <15 minutes.)

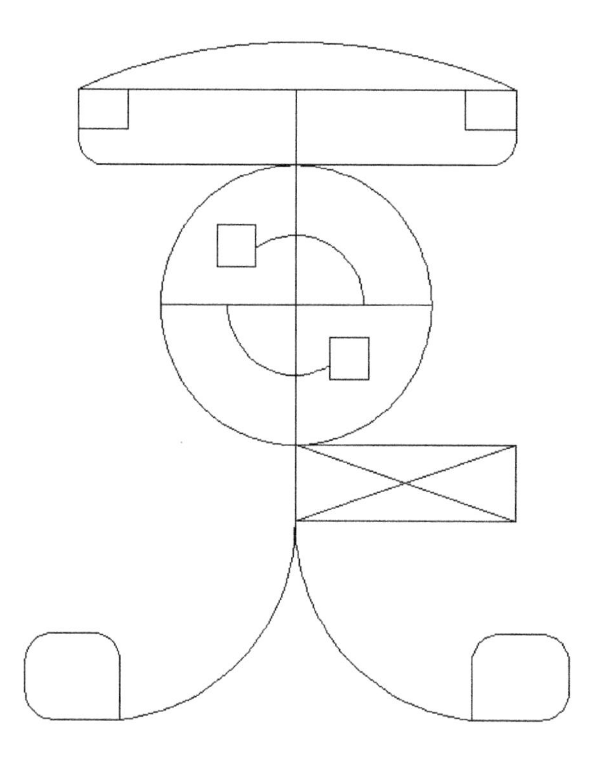

7. Draw the following side view of a car using what you learned. You need only the basic Create and Modify commands. The overall sizing (including multiple fillets needed on most corners) can be approximated; however, use accuracy and make sure all lines connect using OSNAPs. (Difficulty level: Intermediate; Time to completion: <20 minutes.)

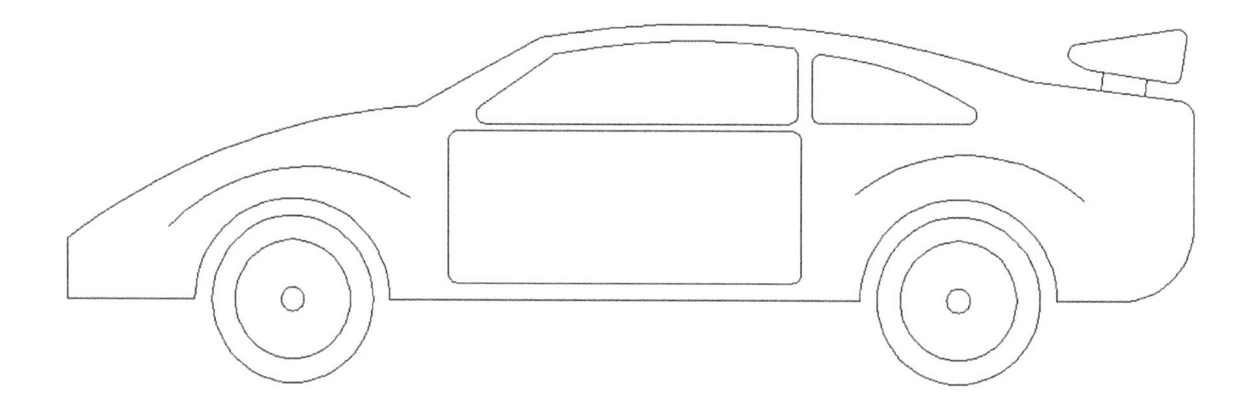

8. Draw the following elevation of a building using what you learned. You need only the basic Create and Modify commands. The overall sizing (windows, door, etc.) can be approximated; however, use accuracy and make sure all lines connect using OSNAPs. There is extensive use of mirroring to save time. (Difficulty level: Intermediate; Time to completion: <20 minutes.)

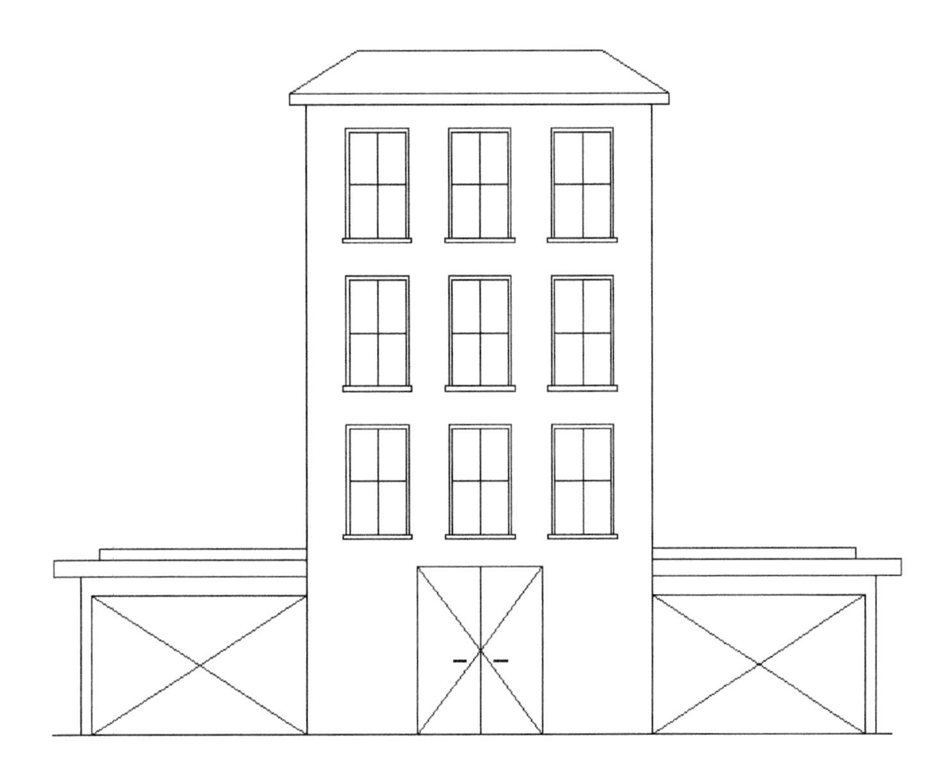

9. Logic diagrams are a familiar sight in electrical engineering; they are made up of components such as AND, OR, Not OR (NOR), Not AND (NAND), and others, as well as various inputs and outputs (I/O). They are used to construct signal flow diagrams by representing voltage flows as True (+V) and False (0 V). Truth tables (you may have seen them in junior high math classes) are derived from logic diagrams. These schematics lend themselves easily to AutoCAD. Draw the following diagram using basic Create/Modify tools. Extensive use is made of the copy command. (Difficulty level: Intermediate; Time to completion: <20 minutes.)

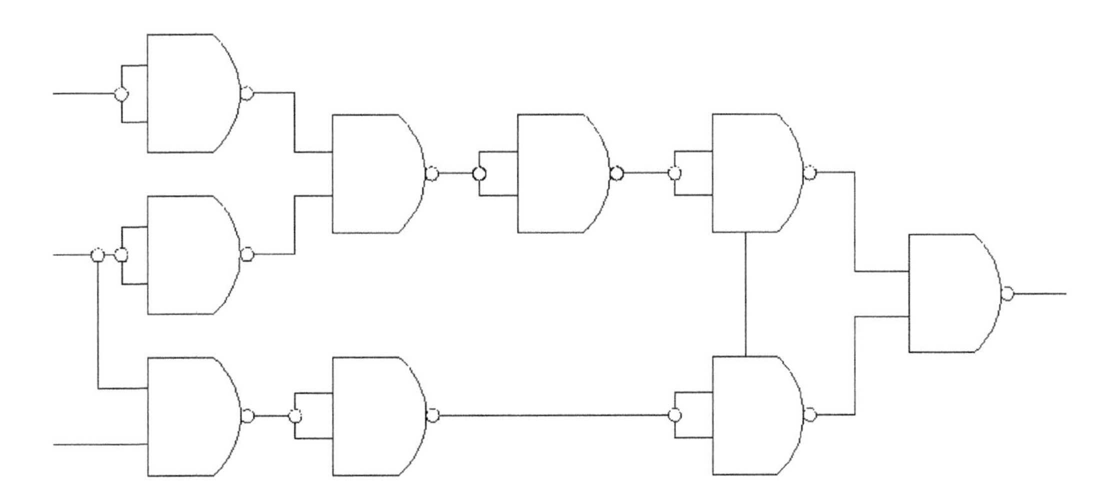

10. Draw the following set of pliers using just basic Draw/Modify commands. Extensive use is made of the copy and mirror commands, along with trim. The rest is just lines, arcs, and a few circles. (Difficulty level: Intermediate; Time to completion: <30 minutes.)

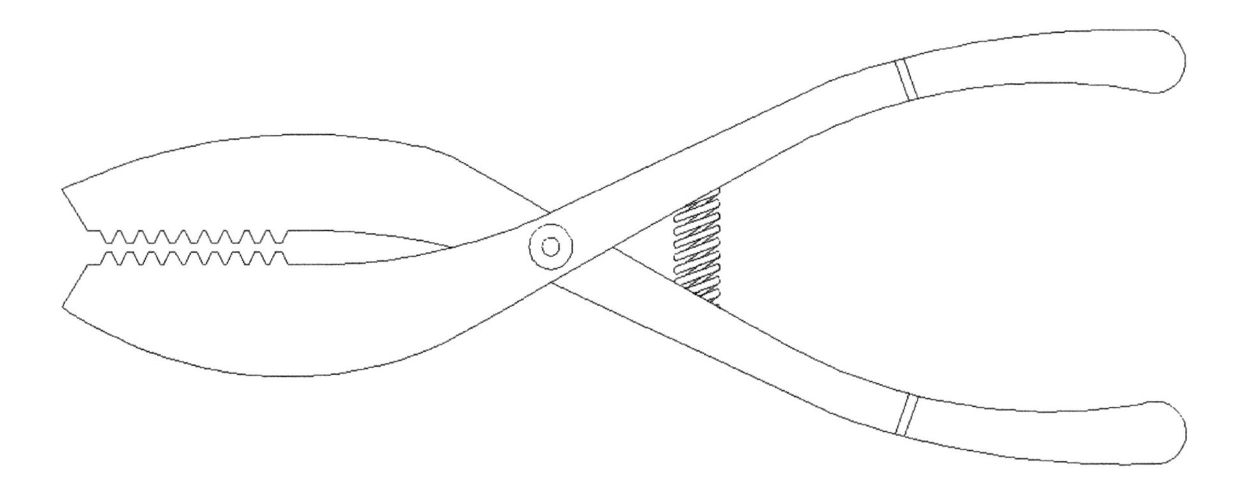

Chapter 2

AutoCAD Fundamentals—Part II

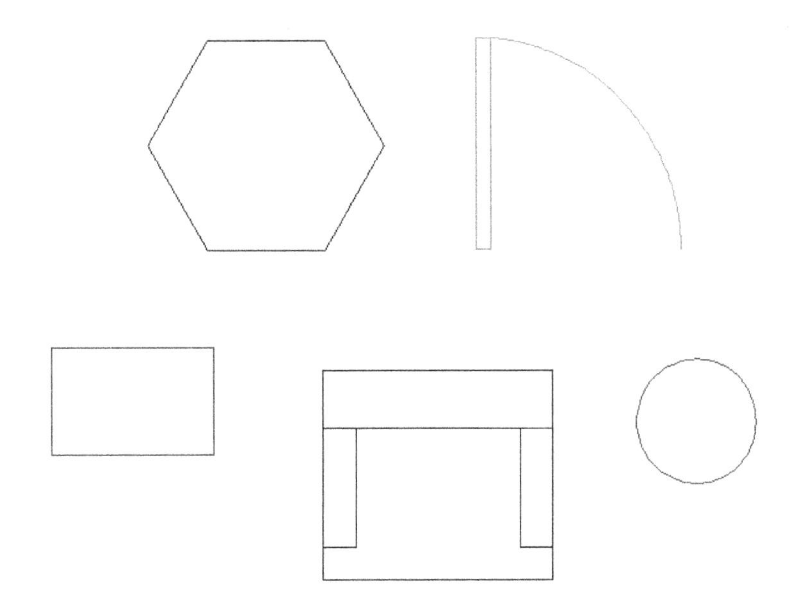

Learning Objectives

You learned a great deal in Chapter 1, AutoCAD Fundamentals—Part I, in fact the majority of what you need to know to get started on a design. This chapter serves as a "cleanup" and concludes the basic introduction to AutoCAD, as there are just a few more topics to get to. Here, we discuss the following:

- Using grips
- Setting units
- Using Snap and Grid
- Understanding the Cartesian coordinate system
- Geometric data entry
- Inquiry commands:
  - Area
  - Distance
  - List
  - ID
- Explode command
- Polygon command
- Ellipse command
- Chamfer command

Up and Running with AutoCAD 2019. https://doi.org/10.1016/B978-0-12-816440-2.00002-1

- Templates
- Setting limits
- Save
- Help files
- TANgent OSNAP
- Time

By the end of this chapter, you will have learned all the necessary basics to begin drafting your first project, including setting proper units, analyzing what you have done, and saving your work. After an introduction to layers in Chapter 3, you will begin a realistic architectural design. Estimated time for completion of this chapter: 2 hours.

2.1 GRIPS

We begin this chapter with grips, a natural progression from learning the OSNAP points. These are relatively advanced editing tools and chances are you have already made them appear by accident by clicking an object without selecting a command first. That is OK; just press Esc and they go away. But, what are these mysterious blue squares? Take a close look; after learning OSNAPs, these should look familiar.

Grips are control points, locations on objects where you can modify that object's size, shape, or location. They are quite similar to the "handles" that you find in Photoshop or Illustrator if you are familiar with those applications.

To try out the concept, first draw some lines, arcs, circles, and rectangles. Then, activate grips by selecting all of them (as seen in Fig. 2.1). The objects become dashed and the grips appear. When your mouse hovers over one of them, the grip turns pink. If you click on it, it becomes red, meaning it is active (or hot). With the grip activated, move the mouse around. The object changes in one of several ways. A line or circle can be moved around if the center grip is selected. A circle can be scaled up or down in size if the outer quadrants are selected. Arcs can be made larger or their length increased. Other grips, such as the endpoints of a line or corners of a rectangle, change (or stretch) the shape of that object if activated and moved. Try them all out.

There is of course much more to grips. Formally referred to as *multifunctional grips*, they perform other construction tasks via an available vertex menu that appears if you hold your mouse over them. We explore this in depth in a later chapter.

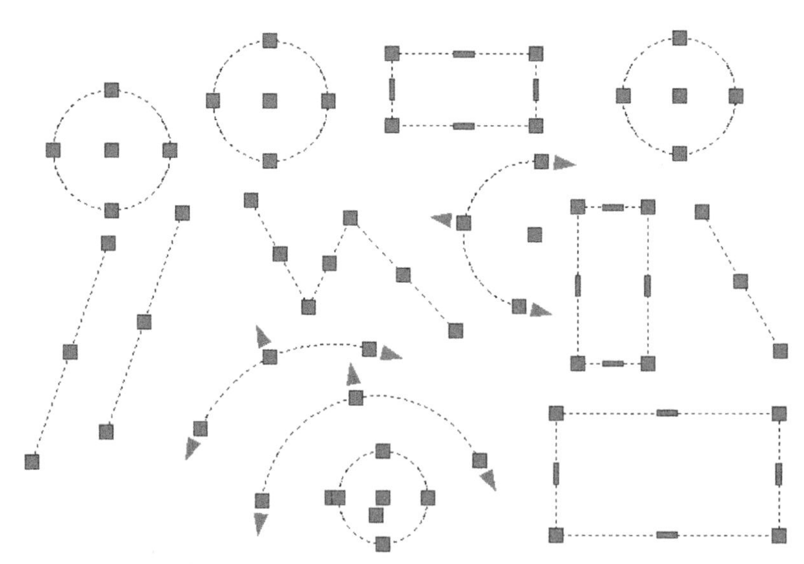

FIGURE 2.1 Grips on circles, rectangles, arcs, and lines.

2.2 UNITS AND SCALE

We have not yet talked about any drawing units, such as feet, inches, or centimeters. When you practiced the offset command, you entered a value, such as 1 or 2, but what did it mean? Well, as you inadvertently discovered, the default unit system in AutoCAD, until you change it, is a simple decimal system of the type we use in everyday life. However, it is unitless, meaning that the 1 or 2 you entered could have been anything—feet, inches, yards, meters, miles, or millimeters. Think about it: It is an easy and powerful idea, but it is up to you to set the frame of reference and stick to it. So, if you are designing a city and 1 is equal to a mile, then 2 is 2 miles, and 0.5 is half a mile. AutoCAD does not care; it gives you a system and lets you adapt it to any situation or design. Then, you just draw in real-life units or, as we say in CAD, 1-to-1 units. This is the golden rule of AutoCAD. *Always draw everything to real-life size. Do not randomly scale objects up or down.*

A decimal system of units is OK for many engineering applications, but what about architecture? Architects prefer their distances broken down to feet and inches because it is always easier to understand and visualize a distance of 22′–6″ than 270″. To switch to architectural units, type in `units` and press Enter, or select Format → Units... from the cascading menus, and the dialog box in Fig. 2.2 appears.

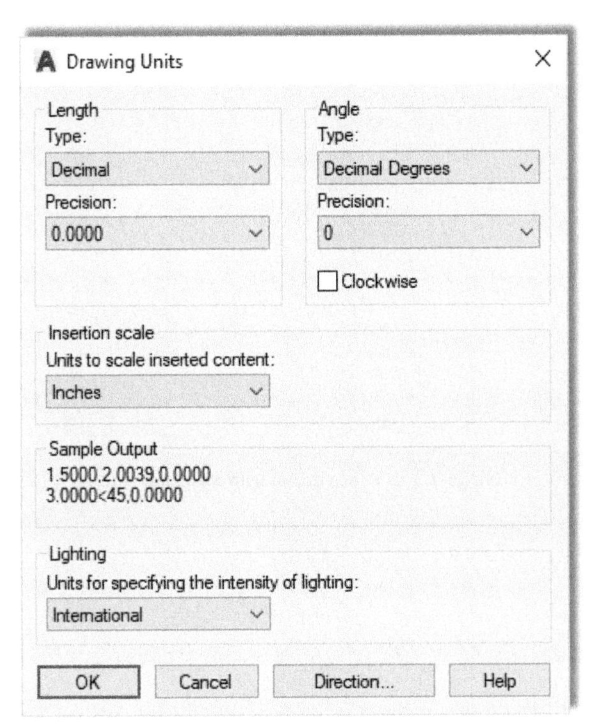

FIGURE 2.2 Drawing Units dialog box.

Simply select Architectural from the drop-down menu at the upper left and ignore the rest of the options in the dialog box; there is no need to change anything else at this point. Now, as you draw and enter in units, they are recognized as feet or inches. To enter in 5 feet (such as when using the offset command), type in 5′. To enter in 5 inches, there is no need for an inch sign, just type in the number 5; AutoCAD understands that an absence of a foot marker means inches. To mix feet and inches, type in 5′6, which is of course 5′–6″; there is no need for a hyphen either. If you need a fraction, such as 5 feet, 4 inches, and ¼, then type in 5′4.25. There is a way to enter actual fractions as well, but students make errors with this type of data entry on a regular basis, so we will not even discuss it. Just convert all fractions to decimal. So 3 inches and 13/16 will just be 3.8125. That is the easiest way to go.

2.3 SNAP AND GRID

These two concepts go hand in hand. The idea here is to set up a framework pattern on the screen that restricts the movement of the crosshairs to predetermined intervals. That way, drawing simple shapes is easy because you know that each jump of the mouse on the screen is equal to some preset unit. Snap is the feature that sets that interval, but you need the grid to make it visible. One without the other does not make sense, and usually they are set equal to each other. Snap and Grid are not used that often in the industry, but it is a great learning tool, and we do a drawing at the end of this chapter using them. It also helps knowing about this feature, as it can be inadvertently activated by pressing the Snap or Grid buttons at the bottom of the screen. Lately, having a Grid on has also been the default setting in AutoCAD.

To Set Snap

Type in `snap` and press Enter.

- AutoCAD says: `Specify snap spacing or [ON/OFF/Aspect/Legacy/Style/Type]<0'-0 1/2">:`

  Type in `1`, setting the snap to 1 inch.

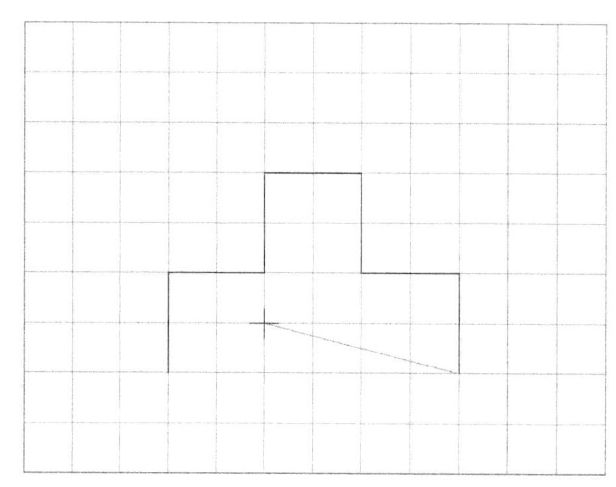

FIGURE 2.3 Figures drawn with Snap and Grid.

To Set Grid

Type in `grid` and press Enter.

- AutoCAD says: `Specify grid spacing(X) or [ON/OFF/Snap/Major/aDaptive/Limits/Follow/Aspect]<0'-0 1/2">:`

  Type in `1`, setting the grid to 1 inch.

Snap and Grid are now set, although in some cases you may have to zoom to extents (Tip 3) to see everything. In older versions of AutoCAD, the grid was hard-to-see dots. In AutoCAD 2012 and continuing with 2019, it became reminiscent of graph paper, as seen in Fig. 2.3. Try drawing a line; you notice right away that the cursor "jumps" according to the intervals you set. F9 turns Snap on and off, whereas F7 takes care of Grid, and you can always toggle back and forth using the drawing aids buttons at the bottom of the screen just to the right of the MODEL icon. Grid looks like a grid, and snap looks like a bunch of dots.

Fig. 2.3 is a screen shot of Snap and Grid in use while drawing a simple shape. You can, if desired, change the color of the grid's major and minor lines via the right-click Options…, Display tab, Colors… button. The entire Options dialog box is described later in the text, but you should explore this on your own as well.

2.4 CARTESIAN COORDINATE SYSTEM

AutoCAD, and really all CAD programs, rely on the concepts of axes and planes to orient their drawings or models in space. In 2D drafting, you are always drawing somewhere on AutoCAD's X–Y plane, which is made up of the intersections of the X and Y axes. When you switch to 3D, the Z axis is made visible to allow drawing into the third dimension. The X, Y, and Z axes are part of the Cartesian coordinate system, and we need to briefly review this bit of basic math and illustrate why it is helpful to know it. Fig. 2.4 illustrates the 2D coordinate system, made up of the X and Y axes and numbered as shown.

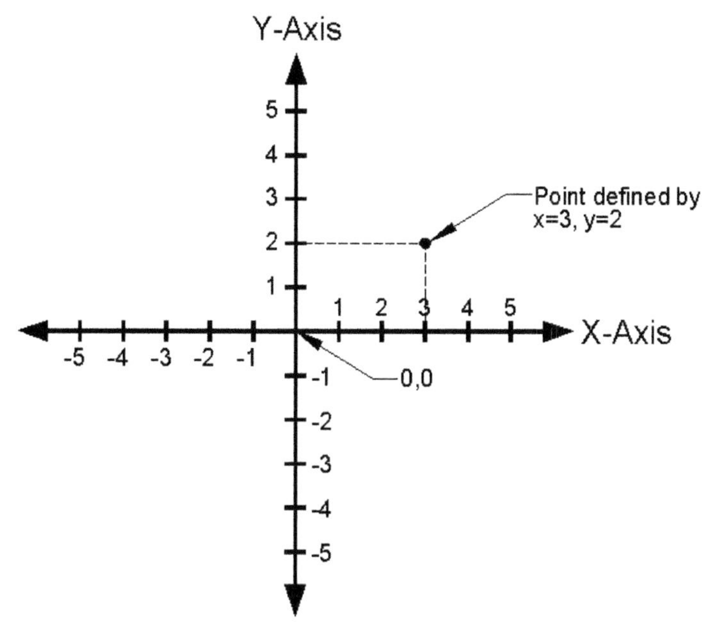

FIGURE 2.4 The Cartesian coordinate system.

This system is important not only because you may want to know where in space your drawing is but also you can create objects of a given size by entering their X and Y values. Moreover, we refer to this system often for a variety of other construction needs and distance entries, including inserting drawings into the point 0,0, so review and familiarize yourself with it again, as it is likely you may not have seen this in a while. Next, we introduce the various distance entry tools. As you go through Section 2.5, knowing the Cartesian coordinate system helps you understand what is going on.

2.5 GEOMETRIC DATA ENTRY

The ability to accept sizing information when creating basic objects is the "bread and butter" of any drafting or modeling software. Geometric data entry refers to the various techniques needed to convey to AutoCAD your design intent, which may involve everything from the length and angle of a line to the diameter of a circle, or the size of a rectangle.

AutoCAD has two precise methods to enter distances, angles, and anything else you may need. These techniques are the older "manual" method, where you enter data via the keyboard and it appears on your command line, and the newer "dynamic" method, where you still use the keyboard, but the data appear on your screen via what is known as a "heads-up" display. This type of display can of course be used with all commands, not just for data and sizing entry. Most students probably gravitate toward the newer method, and indeed it offers many advantages, not the least of which is faster data input. We consider this the primary method in this text but present the manual method as an optional alternative. Experience has shown that some students dislike the additional on-screen clutter that Dynamic Input adds, finding the manual method less distracting. Ultimately, it is up to you to decide what to use.

Dynamic Input

Erase anything that may be on your screen and turn off all drawing aids at the bottom of the screen, including Snap, Grid, and anything else you may have on. Now press the DYNMODE drawing aid, as seen in Fig. 2.5. Note that you may first need to add that to the drawing/construction aides menu via the Customization button (three horizontal bars at the very bottom right).

FIGURE 2.5 Dynamic Input turned on.

This type of input allows you to specify distances and angles directly on the screen. Begin drawing a line and you see an example of Dynamic Input in action. An arbitrary example is shown in Fig. 2.6. Your values of course may be different, but the basic display should be the same.

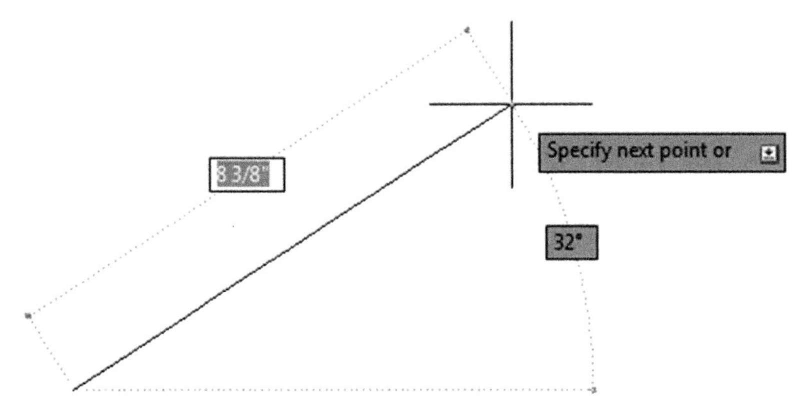

FIGURE 2.6 Drawing a line with Dynamic Input.

Notice what AutoCAD is allowing you to do. The 8 3/8″ value in Fig. 2.6 is surrounded by a box, meaning it is editable. All you have to do is enter a desired value, so enter a small value, such as 5, but do not press Enter yet. What if you wanted to change the angle of the line? Simply press the Tab key, tab over to the 32 degrees value, and enter something else, perhaps a 45. Finally, press Enter. Your line 5 units long at 45 degrees is done. The line command continues of course, and you can repeat this as much as needed, very quick and efficient.

This works with Ortho on as well. In that case, you enter only distance values, no angles, as Ortho forces the lines to all be 90 degrees. Be sure to "lead" the mouse (and in turn the lines being created) by slightly moving the mouse in the direction you want to go.

To use Dynamic Input to draw a rectangle (5″ × 3″ in this example), make sure DYNMODE is on, begin the command, and click anywhere to start the rectangle. You see fields in which to enter the values for the height and width (Fig. 2.7). As before, enter a new value in the first text field (5) and tab over to the next field, entering the other value (3). As you press Enter, the 5″ × 3″ rectangle is created. Note of course that the first value is the width and the second value is the height.

The usefulness of Dynamic Input extends out to just about any shape with which you may work. Fig. 2.8 shows DYNMODE with an arc (left) and a circle (right). As before, you can specify on-screen data needed to create that particular shape.

You may notice also that some of the "heads-up" menus show a downward pointing arrow. You can use the down arrow on your keyboard to drop down the menu and reveal other choices, which of course mimic what is available in the parentheses on the command line. Fig. 2.9 is one example with the offset command. You can cycle through these options to select what you need.

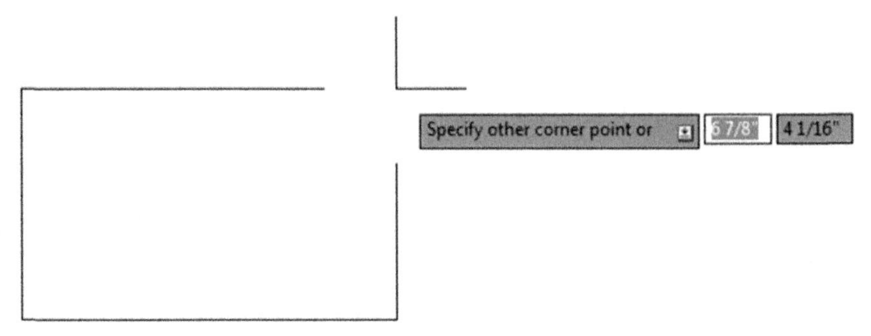

FIGURE 2.7 Drawing a rectangle with Dynamic Input.

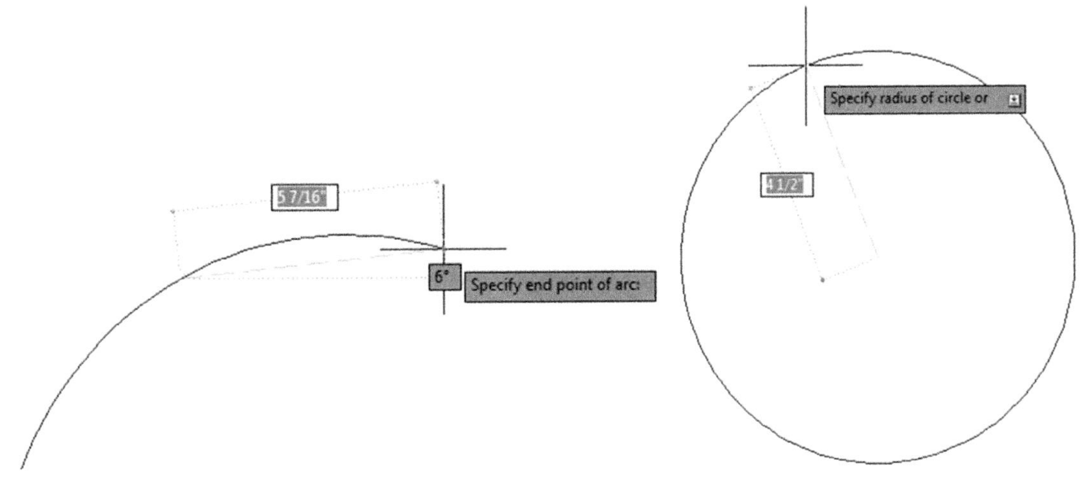

FIGURE 2.8 Drawing an arc (left) and a circle (right) with Dynamic Input.

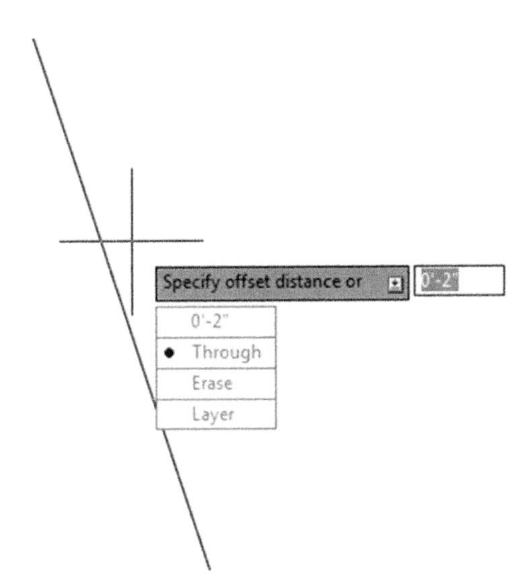

FIGURE 2.9 Dynamic Input menu, Offset command.

Manual Input

Using the manual (command line) method, also called Direct Distance Entry, you can achieve the same thing that can be done with the DYNMODE input. Let us run through a few scenarios and, along the way, introduce Absolute and Relative Entry. The simplest case just involves lines. Turn off the DYNMODE button and begin the line command, clicking where you want to start. Now, all you have to do is point the mouse in the direction you want the line to go (moving the mouse a bit in that direction), type a length value on the keyboard, press Enter, and the line appears. You can continue this "stitching" indefinitely. Note that we have not yet explored how to draw the lines at various precise angles, except for 90 degrees; for that, just turn on Ortho. If using Ortho, be sure to once again "lead" the linework as before, by moving your mouse in the direction you want to go, so AutoCAD understands your intent.

Whether done manually or via DYNMODE, this point-by-point stitching, called *Direct Distance Entry*, is a useful technique. One example of its use occurs after you do a basic room or building perimeter survey. This may just be a wall-by-wall sequential field measurement of a space. When you return to the office you can quickly enter the wall lengths into AutoCAD one at a time via this technique. Do take a few minutes to practice it.

Direct Distance Entry can be done manually one other way: *Absolute Coordinate Entry*. While not a common technique for day-to-day drafting, it really gets you to practice and understand the Cartesian coordinate system. What you do is enter

X and Y values (in pairs) one at a time. To practice this, begin the line command, then, instead of clicking anymore, type in the following values, pressing Enter after each one:

`0,0 0,2 2,2 2,4 4,4 4,2 6,2 6,0` then `c` for `close`

These are the X and Y values for the lines to follow; and if you typed correctly, then the shape seen in Fig. 2.10 should emerge on your screen one segment at a time.

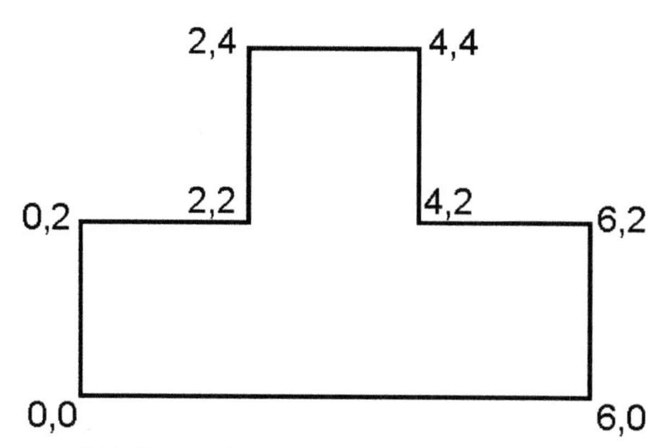

FIGURE 2.10 Shape created via Direct Distance Entry.

Another variation on Direct Distance Entry is called *Relative Data Entry*. This is useful for quickly and easily creating shapes, such as rectangles, anywhere on the screen. It also allows you to enter in various angles for lines besides the Ortho-induced 90 degrees. To do either you need to know two new symbols, @ and <.

The @ symbol frees you from the absolute 0,0 and can be interpreted as meaning "relative to." The last point is in the case of a rectangle; this means you can create it relative to wherever you are in space by simply starting the rectangle command and specifying an X and a Y value preceded by the @ symbol. So, for a 5″ × 3″ rectangle,

1. Start the rectangle command.
2. Click somewhere on your screen to anchor the first point.
3. Type in: @5,3.

The rectangle then appears.

Now that you know about the @ symbol and what it does, you can also draw lines at any angle, from any point, by employing the "less than" symbol (<). This is called Relative Distance Angle Entry. Any numerical value after this "alligator teeth" symbol is interpreted as an angle. The format is as follows: @Distance < Angle. Therefore, to create a line that is 10 units long at 45 degrees,

1. Start the line command.
2. Click somewhere on the screen to anchor the first point.
3. Type in: @10 < 45.

The line (10 units long, at 45 degrees) appears.

As we conclude Section 2.5, remember that, when it comes to creating a rectangle, the dynamic and basic manual methods are not the only ones available. A brief mention was made in Chapter 1, when the rectangle was first introduced, that you can also access length and width data as part of the command line suboptions, such as pressing d for Dimensions. You can vary that by specifying a total area and then either the length or width. It may be worth it to briefly review all this again to have the full set of options available to you. You need to create quite a few rectangles as you begin drafting. They are the basis of furniture, mechanical components, and even entire floor plans, as we will soon see. You should be well versed in all the ways to create them.

2.6 INQUIRY COMMANDS

Our next topic in AutoCAD fundamentals is the Inquiry commands set. These commands are there to give you information about the objects in front of you on the screen as well as other vital data, such as angles, distances, and areas. As such, it is important for you to learn them early on; the information these commands provide is especially valuable to a beginner student and will continue to be essential as you get more experienced. Inquiry commands were upgraded just a few releases ago, and they currently comprise the following:

- *Area* gives you the area of an object or any enclosed space.
- *Distance* gives you the distance between any two points.
- *List* gives you essential information on any object or entity.
- *ID* gives you the coordinates of a point.
- *Radius* gives you the radius of a circle or arc.
- *Angle* gives you the angle between two lines.
- *Volume* gives you the volume of a 3D object
- *Mass properties* lists various mass properties of a 3D solid.

We cover the first four inquiry commands in detail (area, distance, list, and ID) and include a brief mention of radius and angle. Volume and mass properties are not discussed here, as those are for 3D work. As before, all commands can be accessed through the usual four ways: typing, cascading menus, toolbars, and the Ribbon. The toolbar used is the Inquiry toolbar, as seen in Fig. 2.11, and should be brought up on screen using the methods described in Chapter 1.

FIGURE 2.11 Inquiry toolbar.

Area

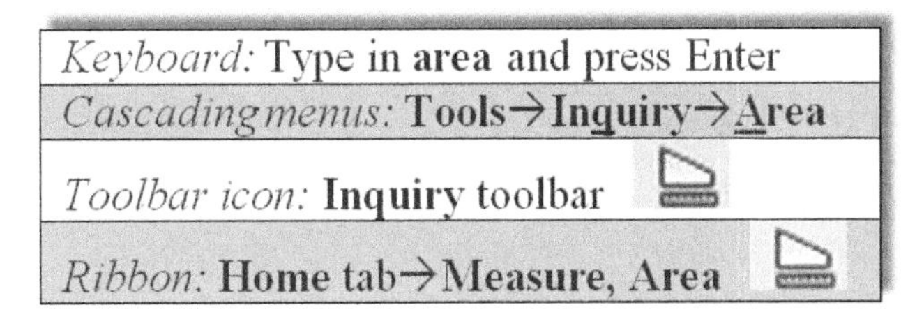

Keyboard: Type in **area** and press Enter
Cascading menus: **Tools→Inquiry→Area**
Toolbar icon: **Inquiry** toolbar
Ribbon: **Home** tab→**Measure, Area**

Area is a very important and useful Inquiry command. It is used often by architects to verify the area of a room or an entire floor plan. Area calculations are performed in two ways: either on an object or, more commonly, point by point (such as the area of a room). Draw a rectangle of any size. We go over both methods, although with a rectangle you would really use only the object method, saving the point-by-point approach for a more complex floor plan.

Point by Point

This is used when measuring an area defined by individual lines (not an object).

Step 1. Make sure your units are set to Architectural and you have a rectangle on your screen.

Step 2. Begin the area command via any of the preceding methods.

- AutoCAD says: `Specify first corner point or [Object/Add area/Subtract area] <Object>:`

Step 3. Activate your OSNAPs (F3) and click your way around the corners of the rectangle until the green shading covers the entire area to be measured (four clicks total).

- AutoCAD says:

```
Specify next point or [Arc/Length/Undo]:
Specify next point or [Arc/Length/Undo]:
Specify next point or [Arc/Length/Undo/Total] <Total>:
Specify next point or [Arc/Length/Undo/Total] <Total>:
```

Step 4. When done, press Enter. You get a total for area and perimeter. Mine said:

`Area = 67.9680, Perimeter = 34.2044`

Yours of course will likely say something different.

Object

This is used when you have an actual *object* such as a rectangle or circle (not a closed shape, defined by individual lines) to measure.

Step 1. Begin the area command via any of the preceding methods.

- AutoCAD says: `Specify first corner point or [Object/Add area/Subtract area/eXit] <Object>:`

Step 2. Press the letter `o` for `Object`.

- AutoCAD says: `Select objects:`

Step 3. Select the square. You should of course get the same area and perimeter measurements.

Distance

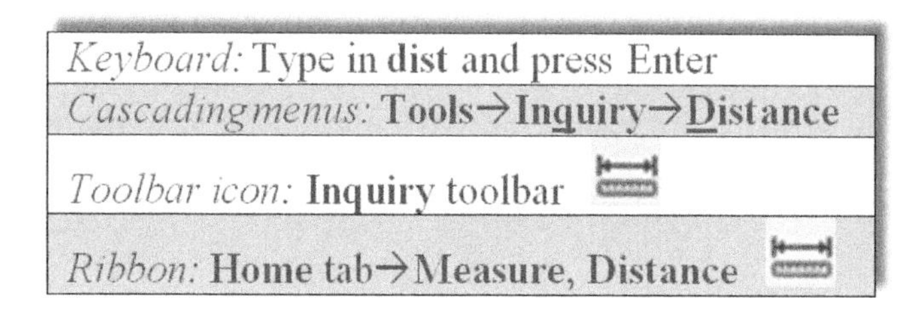

Keyboard: Type in **dist** and press Enter

Cascading menus: **Tools→Inquiry→Distance**

Toolbar icon: **Inquiry** toolbar

Ribbon: **Home tab→Measure, Distance**

The distance command gets you the distance between any two points. Although we practice on the corners of a rectangle, the command works on any two points, as selected by a mouse click, even an empty space.

Step 1. Make sure your units are set to Architectural, a rectangle is on your screen, and your OSNAPs are activated.

Step 2. Begin the distance command via any of the preceding methods. (Be sure to read the note after Step 4.)

- AutoCAD says: `Specify first point:`

Step 3. Select a corner of the rectangle.

- AutoCAD says: `Specify second point or [Multiple points]:`

Step 4. Select the other corner across the rectangle. AutoCAD then gives you some information. Mine said:

`Distance = 13.5521, Angle in XY Plane = 0, Angle from XY Plane = 0`

`Delta X = 13.5521, Delta Y = 0.0000, Delta Z = 0.0000`

Yours will likely say something different.

Note: There is an important difference in procedures if you use the icons or the Ribbon. In those cases, DYNMODE activates and you get a graphical representation of the distance, as seen in Fig. 2.12. If you type or use cascading menus, then you get the command line results seen in Steps 1 through 4.

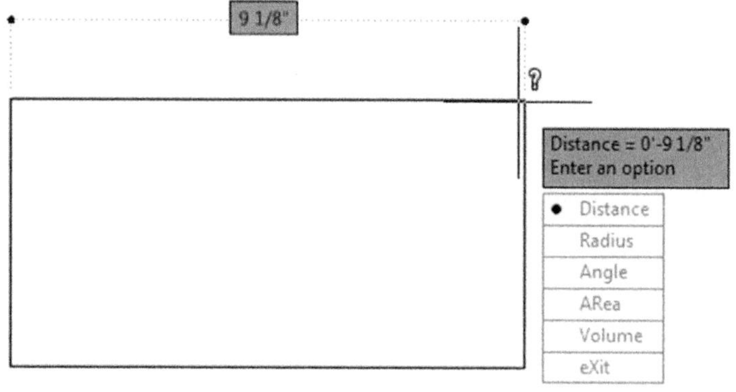

FIGURE 2.12 Distance command results via toolbar or Ribbon.

List

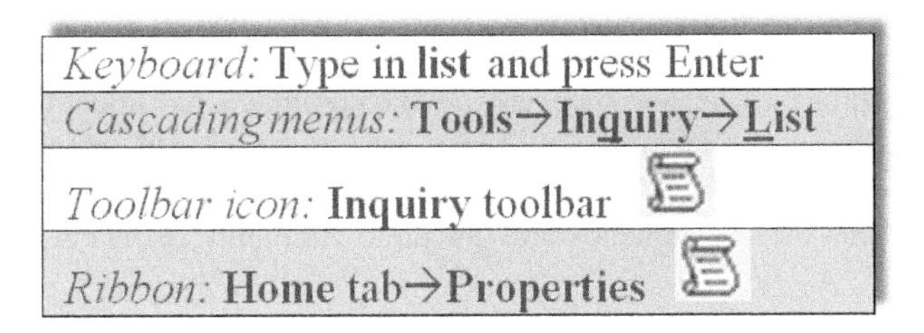

The list command gives you information about not just the geometric properties of the object but also its CAD properties, such as what layer it is on, which is the subject of the next chapter. It is an extremely useful command that clues you in on what you are looking at (whether a beginner or an expert, you will sometimes ask yourself this question).

Step 1. Make sure you have something on your screen to investigate. Draw a line, rectangle, or circle.

Step 2. Begin the list command via any of the preceding methods.

- AutoCAD says: `Select objects:`

Step 3. Select your object (it becomes highlighted) and press Enter. The command line history pops up above the command line. Look it over and close it with F2 (or just click in the command line) when done. What mine shows is a list about an arbitrary rectangle as shown in Fig. 2.13.

```
LIST
Select objects: Specify opposite corner: 1 found
Select objects:
                  LWPOLYLINE  Layer: "0"
                            Space: Model space
                   Handle = 20d
           Closed
    Constant width     0.0000
              area    64.9273
         perimeter    33.2829
          at point  X=  19.5530  Y=  15.2461  Z=   0.0000
          at point  X=  29.9492  Y=  15.2461  Z=   0.0000
          at point  X=  29.9492  Y=   9.0008  Z=   0.0000
          at point  X=  19.5530  Y=   9.0008  Z=   0.0000
```

FIGURE 2.13 List command AutoCAD text window.

ID

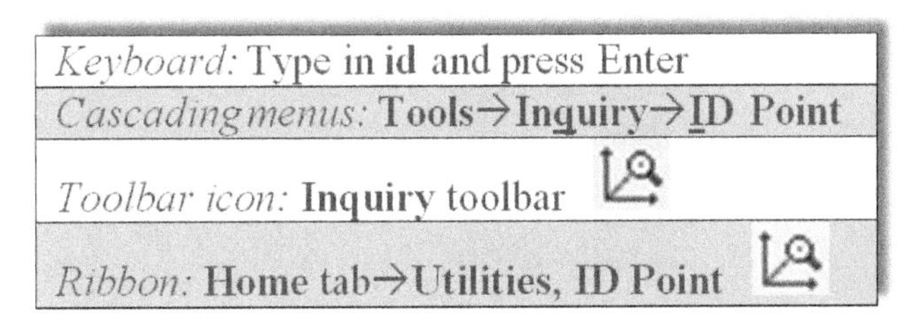

This command is the simplest of all. It tells you the coordinates of a point in space, such as the intersection of two lines. This is useful for inserting blocks (see Chapter 7) and just plain old figuring out the general location of an object, something that may be needed for Xrefs (Chapter 17).

Step 1. Make sure you have something on your screen to ID. Create a line or a rectangle.

Step 2. Begin the ID command via any of the preceding methods.

- AutoCAD says: `Specify point:`

Step 3. Using OSNAPs, select a point, such as the corner of a rectangle or end of a line. AutoCAD gives you the X, Y, and Z coordinates. Here is what I got; you of course will likely get different values:

`X = 11.1504 Y = 6.7753 Z = 0.0000`

Radius and Angle

You should now have a good feel for the Inquiry toolset, so go ahead and practice the remaining two commands on your own. Radius and angle cannot be typed in; you must use the toolbars, the cascading menus, or the Ribbon.

To practice radius, draw an arbitrary circle, and then activate the command and select it. A radius is displayed and you can just press Esc to get out. The result is shown in Fig. 2.14 (left). To get an angle reading, create two "alligator teeth" lines, activate the command, and select both lines. The angle is displayed, and again, you can press Esc to exit. The result is shown in Fig. 2.14 (right). Note that these commands are "dynamic" in nature, so DYN activates when you use them.

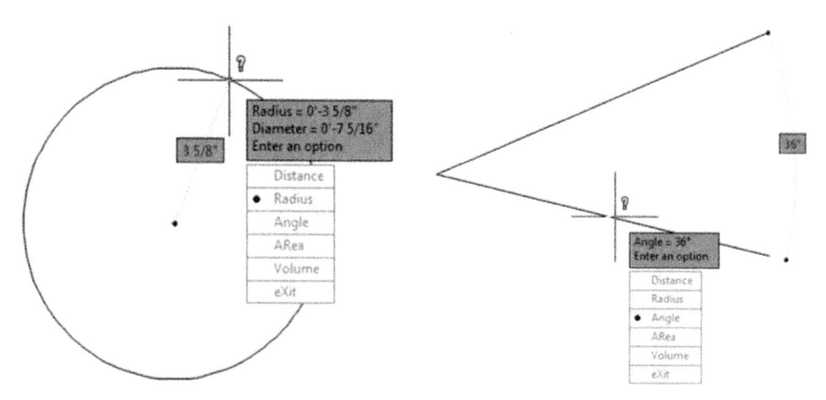

FIGURE 2.14 Radius (left) and Angle (right) measurements.

This concludes the section on Inquiry commands. We are now in the home stretch and will explore some commands and concepts that do not quite fit in neatly anywhere else but will prove to be very useful as you start a realistic project in the next chapter and, of course, everyday AutoCAD use.

2.7 ADDITIONAL DRAFTING COMMANDS

We are going to take a moment to explore four additional drafting commands: explode, polygon, ellipse, and chamfer. They are not mentioned in Chapter 1 because they are not part of the essential set of drafting commands. The need for lines and rectangles in a typical drawing is far greater than the need for ellipses or polygons. However, in the few cases you do need one of those shapes, there really is no substitute. We start off with the unusual explode command.

Explode

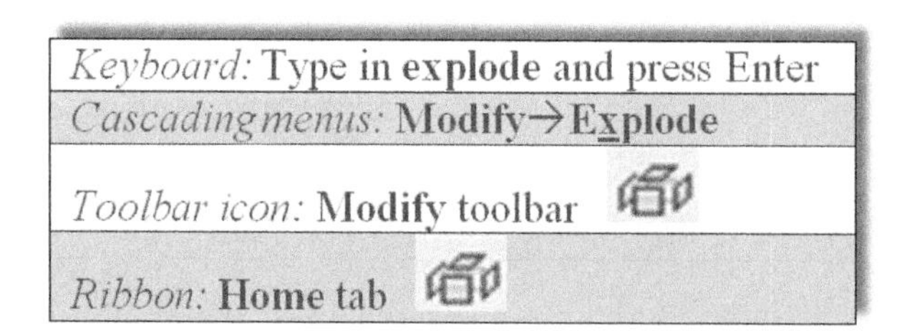

Keyboard: Type in **explode** and press Enter
Cascading menus: **Modify→Explode**
Toolbar icon: **Modify** toolbar
Ribbon: **Home** tab

Explode is used to make complex objects simpler. That means a rectangle becomes four separate lines, and other objects we have yet to study (blocks, text, dimensions, hatch, etc.) are simplified into their respective components. This command is mentioned in almost every chapter, as virtually all AutoCAD entities can be exploded—although many should never be! To try it out, we again need a rectangle of any size. Make sure you are using the actual rectangle command, not just four connected lines. Those are already in their simplest form, and explode does not work on them.

Step 1. Begin the explode command via any of the preceding methods.
- AutoCAD says: `Select objects:`

Step 2. Pick the rectangle and it becomes dashed.
- AutoCAD says: `1 found`

Step 3. Press Enter and the rectangle is exploded.

Click any of its lines and you notice that each line is independent, as seen in Fig. 2.15. Keep in mind that you cannot explode circles, arcs, and lines (previously mentioned), as those are already the simplest objects possible. There is a way to reverse explode, in addition to an immediate undo. It is called a *pedit* command (discussed in chapter: Advanced Linework), but that applies only to simple compound objects, such as a rectangle. If you explode a hatch pattern (do not do it) and save your work before using *undo*… that is it, no going back. This is why you see appropriate warnings when we get to those chapters.

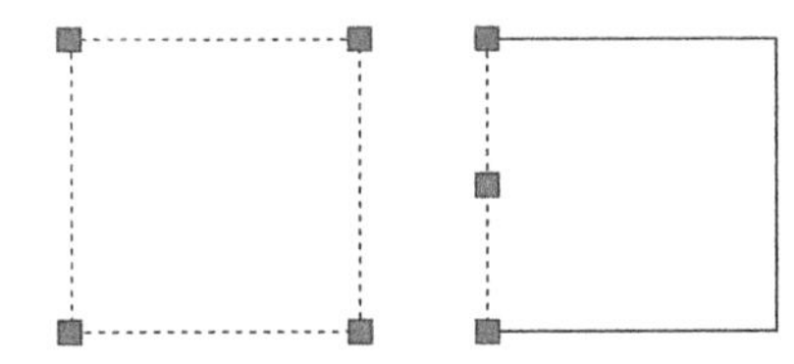

FIGURE 2.15 Unexploded rectangle (left) and exploded rectangle (right).

Polygon

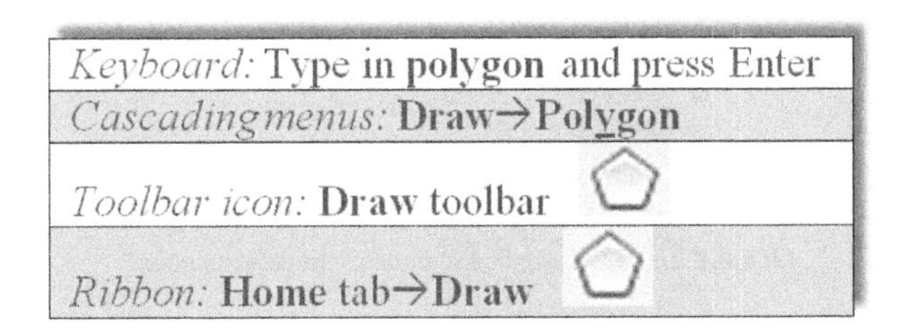

Keyboard: Type in **polygon** and press Enter
Cascading menus: **Draw→Polygon**
Toolbar icon: **Draw toolbar**
Ribbon: **Home tab→Draw**

The polygon command is a useful tool that we need in later chapters for a project. A polygon, strictly speaking, is any shape with three or more sides to it. Of course, in those cases when it has three sides (triangle), you can draw it with existing line techniques. Same thing with four sides (square or rectangle). However, what about five sides or more, such as a pentagon or hexagon? Here, a polygon command is needed, as it is a real chore to draw these shapes using the line command—just try figuring out all the correct angles.

Drawing polygons, however, presents a complication you do not have with simpler shapes. With a rectangle, you specify the height and width and that is it, but how do you do that with, say, a pentagon? Fig. 2.16 is a picture of one. What is the height? What is the width? It is not an easily measured shape.

The solution that drafters came up with a long time ago (many of AutoCAD's techniques are just adaptations of pencil-and-paper board-drafting methods used for decades) is to enclose the shape in a circle of a known radius or diameter. Of course, you can also enclose the circle inside the polygon.

With either approach, it is much easier to specify a size. You just say, "I need a pentagon that fits inside (or outside) a circle of 10 inches in diameter." This is where the terms *inscribed* and *circumscribed* come from. Fig. 2.17 is a diagram of two pentagons, one inscribed in a circle and one circumscribed about a circle. Note the tangencies. In one case, the points of the polygon touch the circle; in the other case, its sides touch it.

Let us go through the steps in making a polygon inscribed in a circle. Begin by drawing a circle with an arbitrary radius of 5. Note that we do not need this step to create a polygon, but for your first try, it is a great help in visualizing what you are doing. Then proceed as shown next.

Step 1. Begin the polygon command via any of the preceding methods.
- AutoCAD says: `Enter number of sides <4>:`

Step 2. Enter the number 5 for five sides (making it a pentagon), and press Enter.
- AutoCAD says: `Specify center of polygon or [Edge]:`

Step 3. With your OSNAPs on, click to select the center of the circle.

- AutoCAD says: `Enter an option [Inscribed in circle/Circumscribed about circle] <I>:`

Step 4. Now that you know what these terms mean, go ahead and press Enter to select `i` for `Inscribed in circle` (the default value). You can just as easily select c as well.

- AutoCAD says: `Specify radius of circle:`

Step 5. Enter the value `5`. A screen shot of the result is in Fig. 2.18.

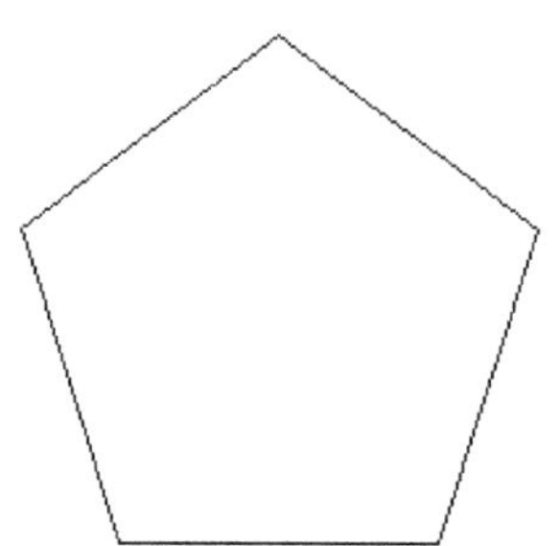

FIGURE 2.16 Basic pentagon.

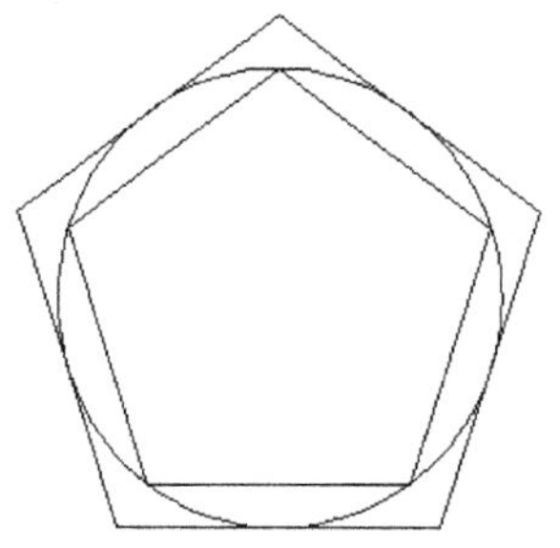

FIGURE 2.17 Inscribed and circumscribed pentagons.

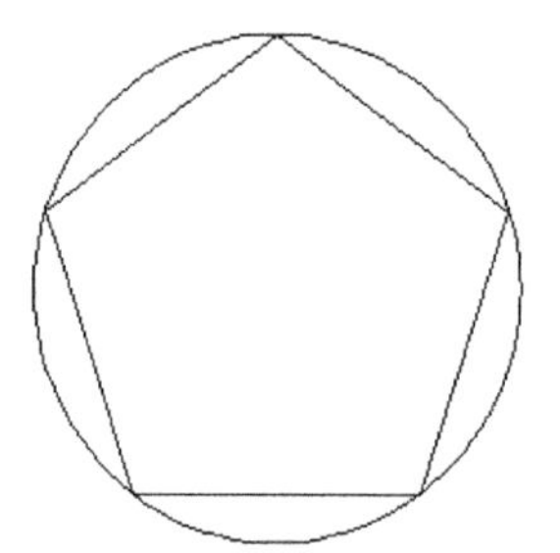

FIGURE 2.18 Results after Step 5.

You can now erase the circle, because it was just a guide. In the future you do not need it; just click to place the polygon anywhere on the screen. Go ahead and practice this command, but this time select a different number of sides (six or more) and also use the Circumscribed option. This new command leads us into the first challenge exercise.

Drawing Challenge—Star

Try this exercise based on what you learned. Draw a perfect five-sided star, such as the one shown in Fig. 2.19. Think about everything we talked about up to this point. Only four commands are needed to create the star. Take a few minutes to try to draw it. If you cannot get it to look right, the answer is at the end of this chapter, just before the summary.

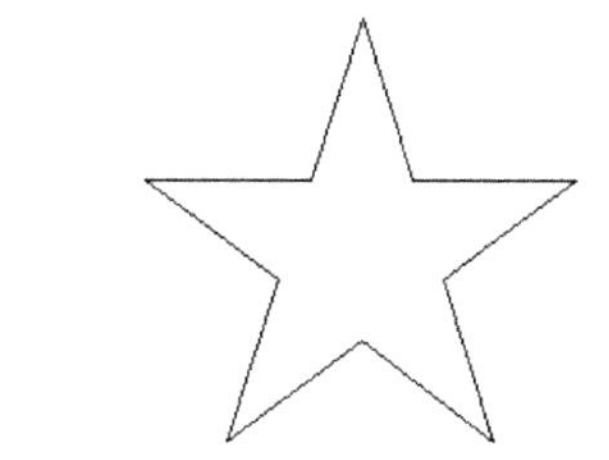

FIGURE 2.19 Drawing challenge—star.

Ellipse

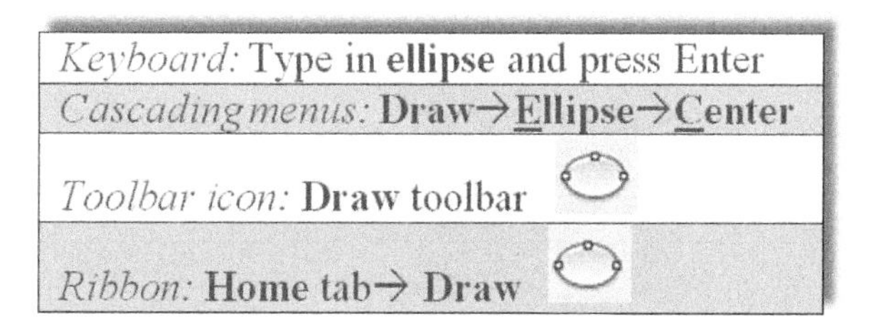

Keyboard: Type in **ellipse** and press Enter
Cascading menus: **Draw→Ellipse→Center**
Toolbar icon: **Draw** toolbar
Ribbon: **Home tab→ Draw**

The ellipse command is easy to use and is needed when we draw furniture in Chapter 4, Text, Mtext, Editing, and Style, so it is included here. The command can be executed in one of two ways: Center or Axis, End. You can also create a partial ellipse by defining just a section of it.

Step 1. Begin the ellipse command via any of the preceding methods.
- AutoCAD says: `Specify axis endpoint of ellipse or [Arc/Center]:`

Step 2. Click anywhere on the screen.
- AutoCAD says: `Specify other endpoint of axis:`

Step 3. With Ortho on (not required but good to have), click again somewhere off to the left or right of the first point. Then, move the mouse back toward the center as the ellipse is formed, as seen in Fig. 2.20.
- AutoCAD says: `Specify distance to other axis or [Rotation]:`

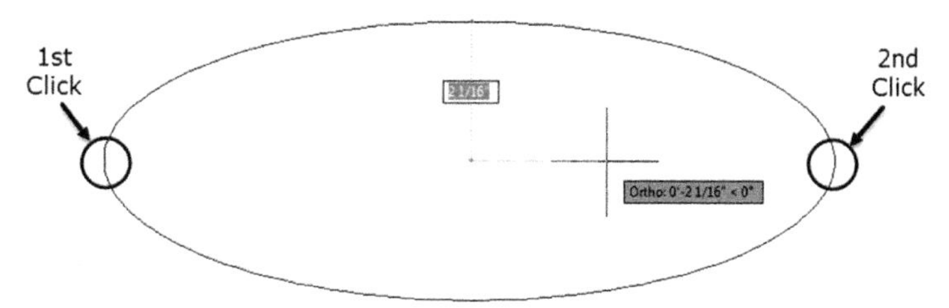

FIGURE 2.20 Ellipse after Step 3.

Step 4. Click anywhere again to complete the ellipse. You can also specify a rotation angle along the major axis (the line between clicks 1 and 2). This would be as if the ellipse were rotating away from (or toward) you. For a variation on this, try the Arc submenu of the ellipse, which is essentially similar in execution until the final step, when you trace how much of the ellipse you want to see, as shown in Fig. 2.21.

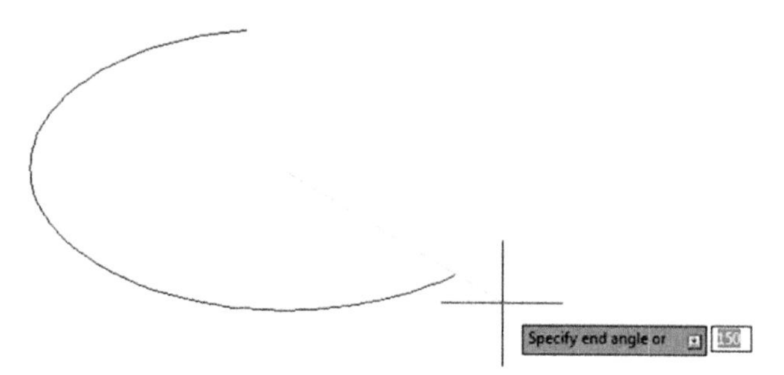

FIGURE 2.21 Ellipse, arc variation.

Chamfer

Keyboard: Type in **chamfer** and press Enter
Cascading menus: **Modify→Chamfer**
Toolbar icon: **Modify** toolbar
Ribbon: **Home tab→ Draw**

You may have already tried the chamfer command, as it sits right next to fillet in the cascading menus and the Modify toolbar. While normally not used as often as a fillet, it is still very much worth investigating. A chamfer is a beveled edge connecting two surfaces. Think of it as a fillet that is angled instead of round. Chamfers may have a 45 degrees corner, but this is not required; AutoCAD allows for asymmetrical chamfers.

To try out the command, draw two perpendicular lines, similar to the fillet exercise in Chapter 1 (reproduced in Fig. 2.22), and follow these steps given below:

Step 1. Begin the chamfer command via any of the preceding methods.
- AutoCAD says: `(TRIM mode) Current chamfer Dist1 = 0.0000, Dist2 = 0.0000`
`Select first line or [Undo/Polyline/Distance/Angle/Trim/mEthod/Multiple]:`

Step 2. No distance has been entered yet, and any attempt at chamfer results in a chamfer of zero. Press `d` for `Distance`.
- AutoCAD says: `Specify first chamfer distance <0.0000>:`

Step 3. Enter a small value, such as `1`.
- AutoCAD says: `Specify second chamfer distance <1.0000>:`

Step 4. Here the `1` is not needed again, so just press Enter.
- AutoCAD says: `Select first line or [Undo/Polyline/Distance/Angle/Trim/mEthod/Multiple]:`

Step 5. Select the first line.
- AutoCAD says: `Select second line or shift-select to apply corner or[Distance/Angle/Method]:`

Step 6. As you move the mouse toward the second line, you again see a preview of the completed command (similar to fillet). Select the second line and the chamfer is created as seen in Fig. 2.23. Zoom in if needed to see it.

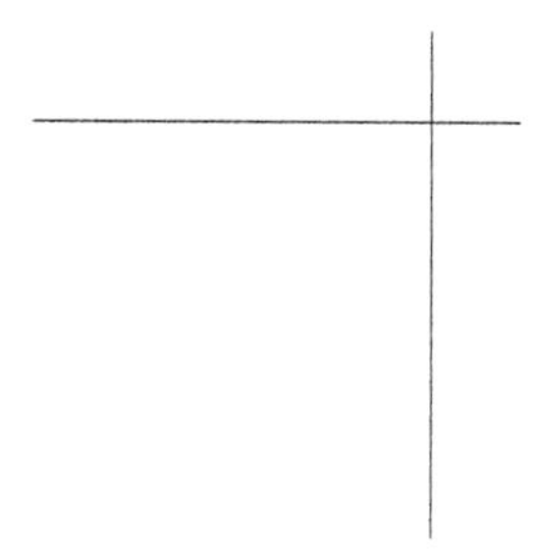

FIGURE 2.22 Chamfer setup.

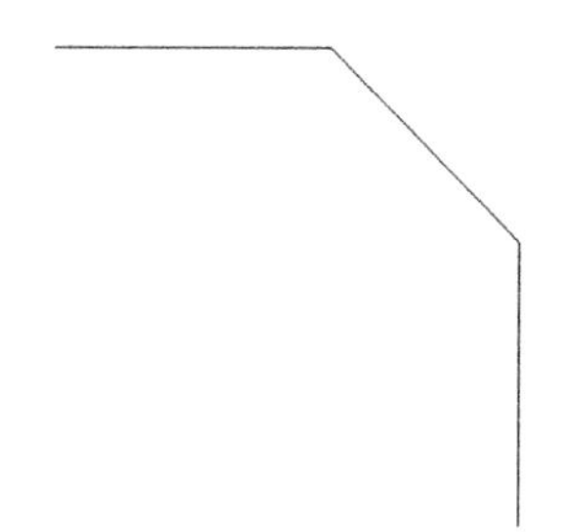

FIGURE 2.23 Chamfer executed.

If the chamfer fails, it is possible that you selected values too large for the available lines. Another way to create a chamfer is using the length of one line segment together with an angle, using the Angle option. Experiment with both.

Templates

Templates are something of a mystery for a beginner AutoCAD student. What exactly are they? While templates can get complicated in what they contain, they are really just files with some information embedded in them, with all the proper settings but no actual design work present. So it is a blank page with some intelligence built in and everything set up and ready for you to draw. These templates can be saved under a .dwt extension, and you can certainly make your own.

In this case, using the cascading menu File → New…, you default to the screen in Fig. 2.24. You should get a déjà vu with this. It is the same as Fig. 1.2, and it is how you opened a blank file when AutoCAD was first introduced in Chapter 1. Open up the highlighted acad.dwt and you have a blank file, and that is it. Except now you know a bit more about what it is. Note the other preset templates available. In Chapter 10 we come back here to grab an Architectural title block.

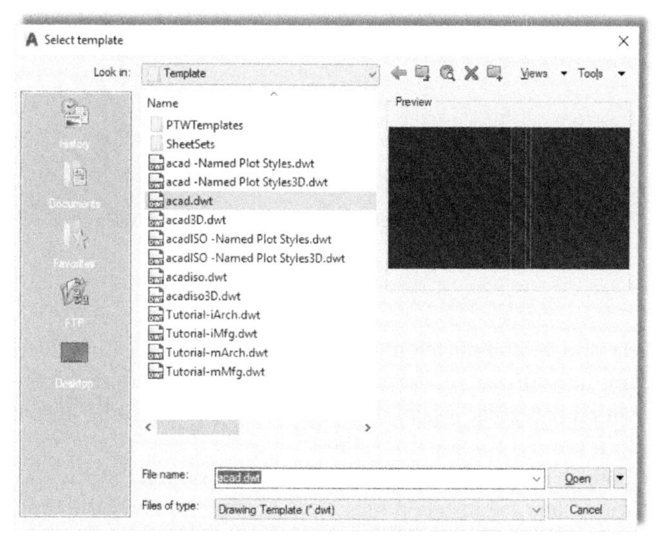

FIGURE 2.24 An acad.dwt template.

Limits

This is yet another topic that causes confusion. Limits are not technically necessary, but many textbooks persist in having you set them up. All they are is an indication of the size to which your drawing area is preset. Remember, however, that model space is limitless, so all limits refer to is the area you can work on without a regen. If you draw a shape that is way beyond your limits, you simply regenerate the screen via a Zoom, Extents (see Tip 3 in Chapter 1 for a reminder) and the limits move to just beyond this new shape, so setting them is rather redundant. If you would like to set them anyway, just type in `limits`, press Enter, and set the lower left corner to 0,0, then the upper right corner to perhaps 100′, 50′, or anything else you wish.

Save

Needless to say, you have to save your work often. While it is pretty rare these days, AutoCAD (like any software) can lock up or crash. If not, Windows will, there will be a power failure, you may spill coffee on your laptop, or an asteroid can hit you, so it is just a good habit to get into. Starting in the next chapter, you need to start saving your work. To do so, you can type in save and press Enter or press the graphic of a disk at the top of the screen. Another way is to drop down the arrow located next to the big red A at the upper left of the screen (Fig. 2.25) and choose Save or Save As. You then get the usual Save Drawing As dialog box from which you should know how to proceed.

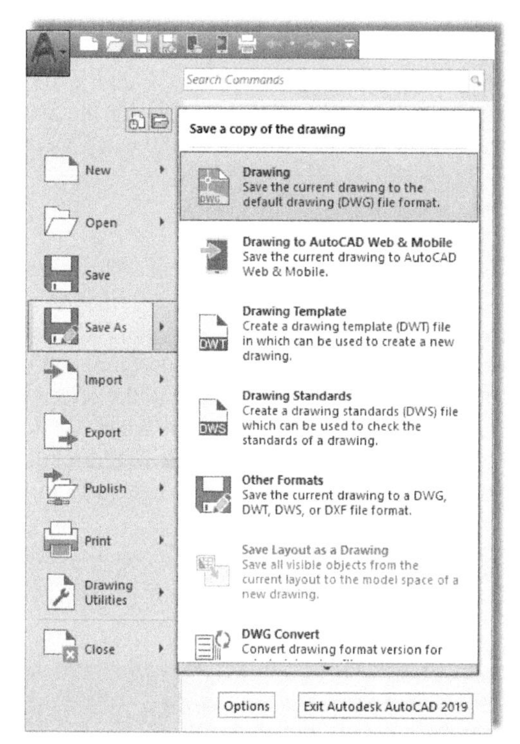

FIGURE 2.25 AutoCAD 2019 main menu with Save As selected.

Help Files

AutoCAD, like pretty much every software application, has extensive Help files, and indeed that is one way (though not recommended) to attempt to learn the software. A better application of Help is to explore it once you know how to use AutoCAD and need assistance with a particular aspect of a command. Tool tips that appear with every command can be considered to be part of the Help files.

AutoCAD's Help files have come a long way in recent releases. The original method to access them was generally to press F1 (same as for other software), and the installed files would come up. Early on, they became context sensitive and showed only relevant results based on what you were doing. You could also just search for what you wanted and access additional information, including examples. Autodesk even offered a printed Help files manual for those preferring paper.

These files, however, were part of AutoCAD, could not be updated by Autodesk between releases, and offered very little interactive content and access to other tools. All that began to change as internet access became virtually guaranteed as part of working on a computer. The monumental shift with Help files in recent years was to move them completely online and link them with additional tools and functionality. Today, with AutoCAD 2019, online content is accessed if the internet is available, and built-in files are not even offered with AutoCAD (though you can download them, as we discuss in a bit). As far as paper help manuals? I am not aware of them even being offered anymore, but you can probably print those downloads if you want to.

Help files today are part of a larger, interactive suite of online tools, applications, and resources, which we discuss shortly. To access just the regular Help files, you can still just press F1 (or the question mark at the very top right of your screen). You can do this with no commands active (resulting in Fig. 2.26) or in "context-sensitive" mode, while in the middle of a command (such as shown in Fig. 2.27 with "line").

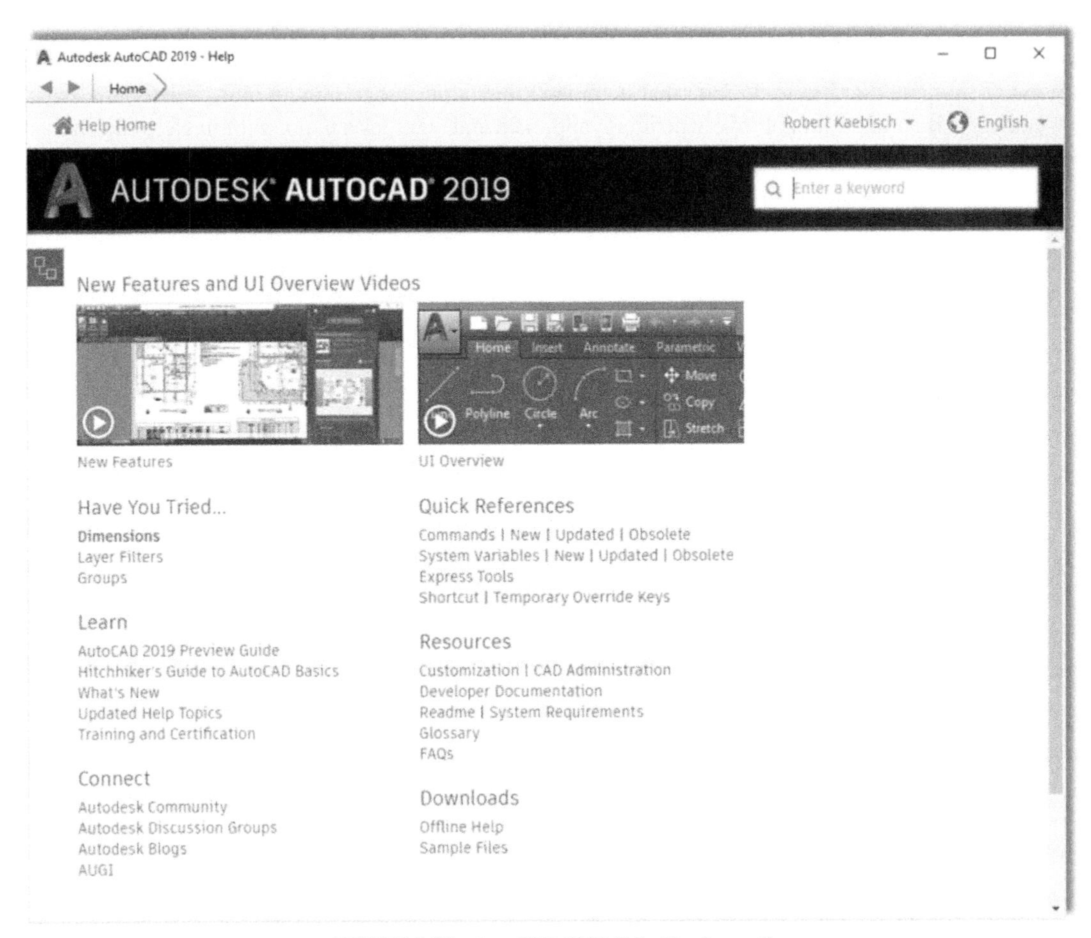

FIGURE 2.26 AutoCAD 2019 Help files (general).

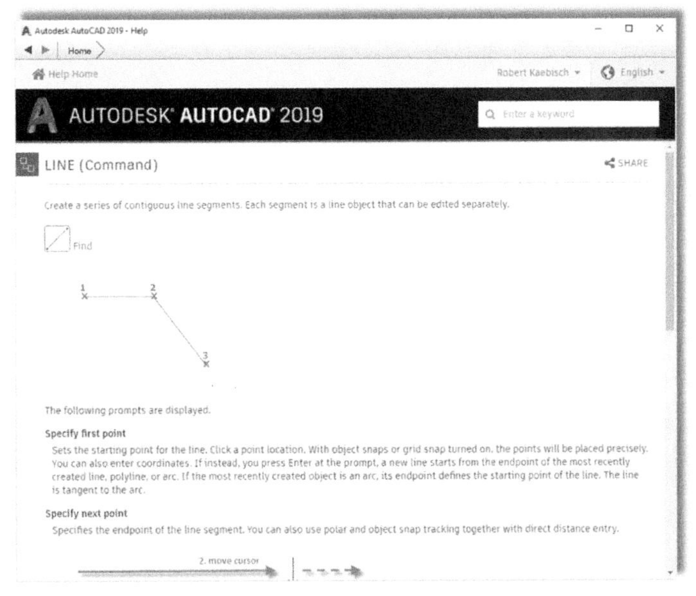

FIGURE 2.27 AutoCAD 2019 Help files (LINE Command).

Here, you can search for any other commands and find related tasks, references, and concepts. Note also how, on the left-hand side of the Help files, you can filter what you see based on your level of interest and responsibility. You can now access pretty much anything that has to do with that command, including programming tools, related tutorials, and even relevant blog entries. It is an interesting, if overwhelming, approach and signifies a major shift in Help file philosophy. You are now getting content-based, Google-like content to cover every angle, not just a cookie-cutter text file to explain the command.

As mentioned, there is far more to the Help files than just command explanations, and the extras can be found under the Help cascading menu, mostly in the Additional Resources flyout menu, as seen in Fig. 2.28.

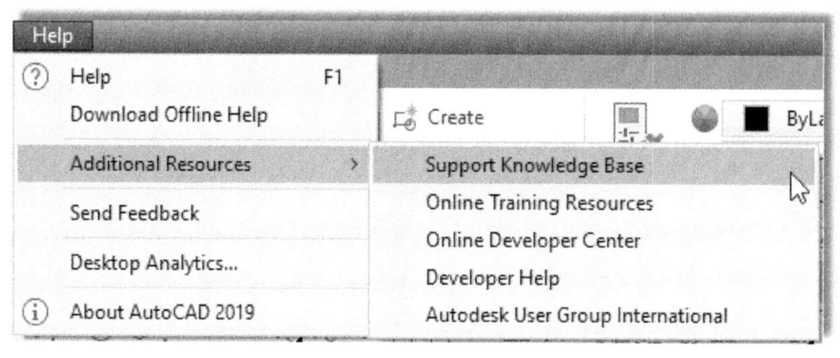

FIGURE 2.28 AutoCAD 2019 Help cascading menu.

You already have called up the regular command Help files via F1, so no need to select the first item from the top, but what about the rest of these choices? Let us go over a few of them. When you click the Support Knowledge Base you get the Autodesk web page called *Knowledge Network*, seen here in Fig. 2.29. Note that you may see a slightly different design, as web pages can certainly be updated long after this textbook comes out in print.

The web address, if you want to go there directly, bypassing AutoCAD is: https://knowledge.autodesk.com/support/autocad

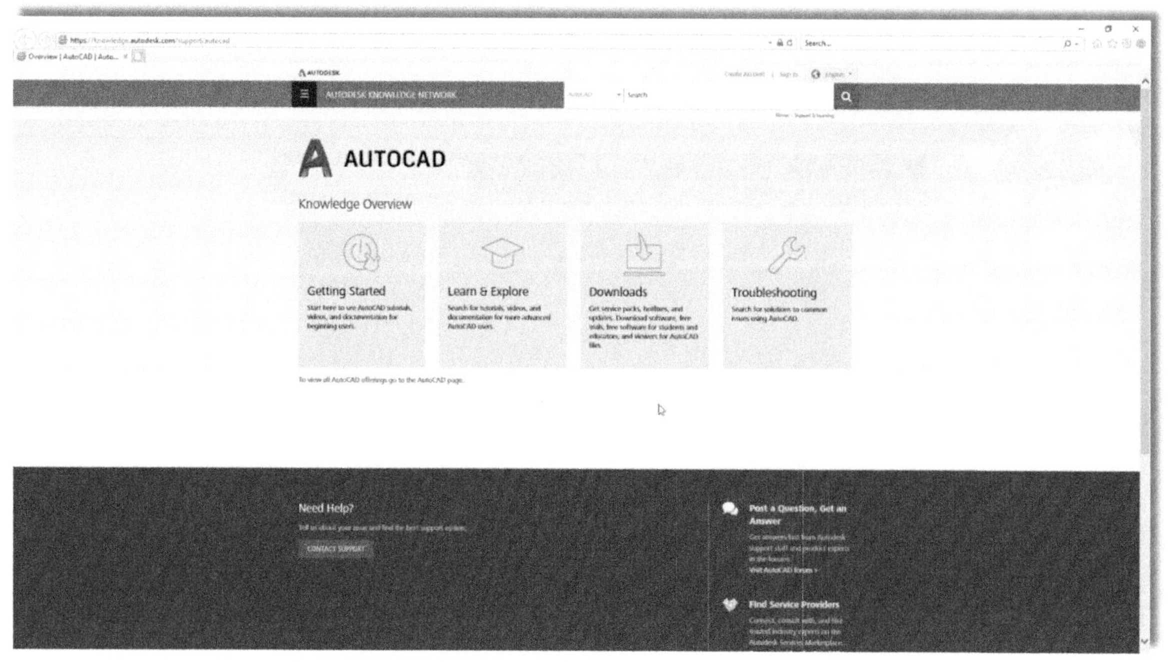

FIGURE 2.29 Autodesk's Knowledge Network web page—Overview tab.

Here, you find some troubleshooting guidance as well as Service Pack downloads and a vast array of instructional data. Click over to the other tabs, moving from left to right. Getting Started is a subset of the previous page and not directly accessible from AutoCAD itself. It contains another vast array of learning data. Moving to the right, the Learn & Explore tab correlates to the Help's Online Training Resources choice, seen in Fig. 2.28. It also contains, you guessed it, more learning data.

The next tab, Downloads, if accessed via clicking it, presents an array of fixes, templates, and other goodies. If you access the tab through AutoCAD's Download Offline Help choice from Fig. 2.28, the tab is a different one; and it gives you the chance to download AutoCAD 2019's Help documentation—all 176 MB of it as of this writing. Just choose your language and download away, ideally with a fast connection. When it all downloads, you get a self-extracting *.exe file, which needs to be installed. If you anticipate working offline a lot and generally want local access to the files, this may be a good idea. They are not regularly updated or interactive but do the job for most users.

There is some more useful stuff under the Troubleshooting tab and a bit more under Online Developer Center (going back to the cascading menu), though it's geared toward developers not a beginner student.

As we wind down our tour, a new feature of the Help files, first introduced with AutoCAD 2015, deserves mention. Take a look at Fig. 2.27 again. Do you see the word *Find* just to the right of the line icon? If you click on this while the Ribbon is active, it finds the relevant command for you and points a bobbing red arrow at it. The arrow can be seen all the way to the top left in the screen shot in Fig. 2.30 (minus the bobbing up and down, of course). This clever assistant works with any command contained in the Ribbon and as long as you have it up. Starting with AutoCAD 2016 and continuing with AutoCAD 2019, it now also works with application menus and the Status bar.

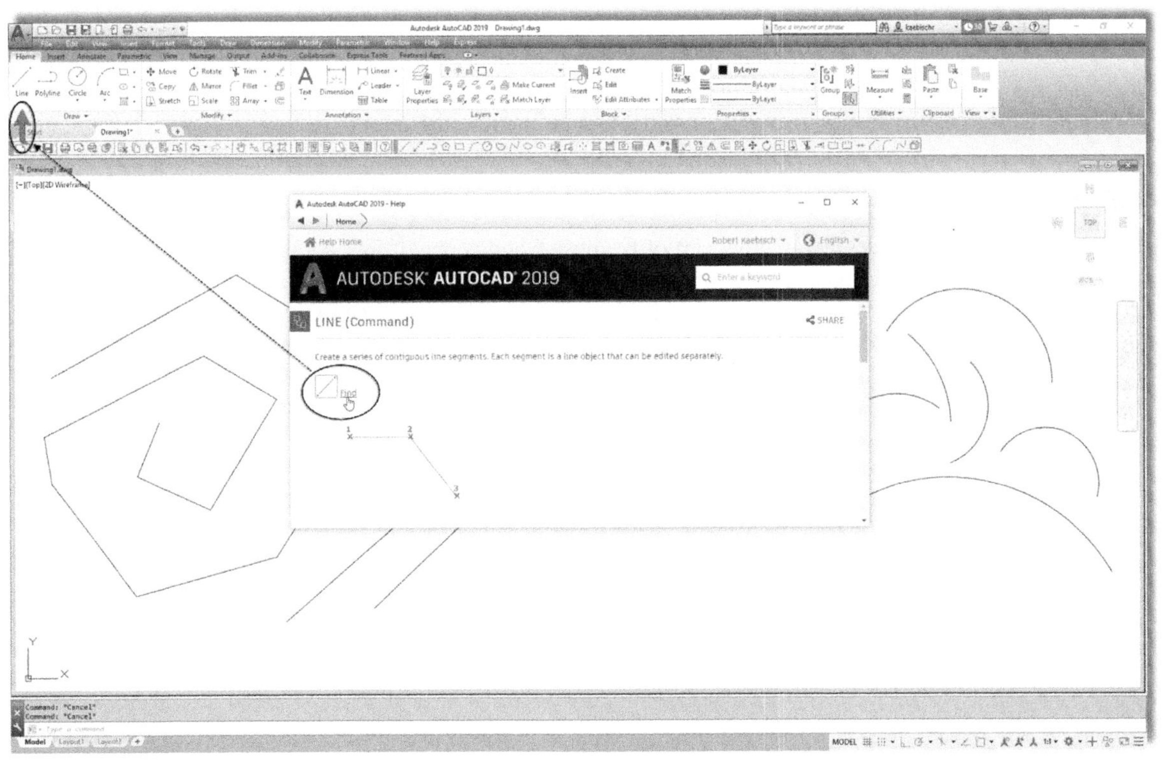

FIGURE 2.30 Find command tool.

Finally, we have the Autodesk Exchange (Fig. 2.31). It is not part of the Help files per se, but this is a good time to introduce it. You can access it at any time via a small X toward the upper right of the screen, to the right of the Autodesk 360 Sign In. The Exchange is also geared toward developers and "power users," with various tools and apps to customize, extend, or otherwise enhance AutoCAD. Although, as a beginning user, you probably will not need all this additional content, make a note of it for future reference. The Exchange's web address, for independent access to it is: https://apps.autodesk.com/ACD/en/Home/Index

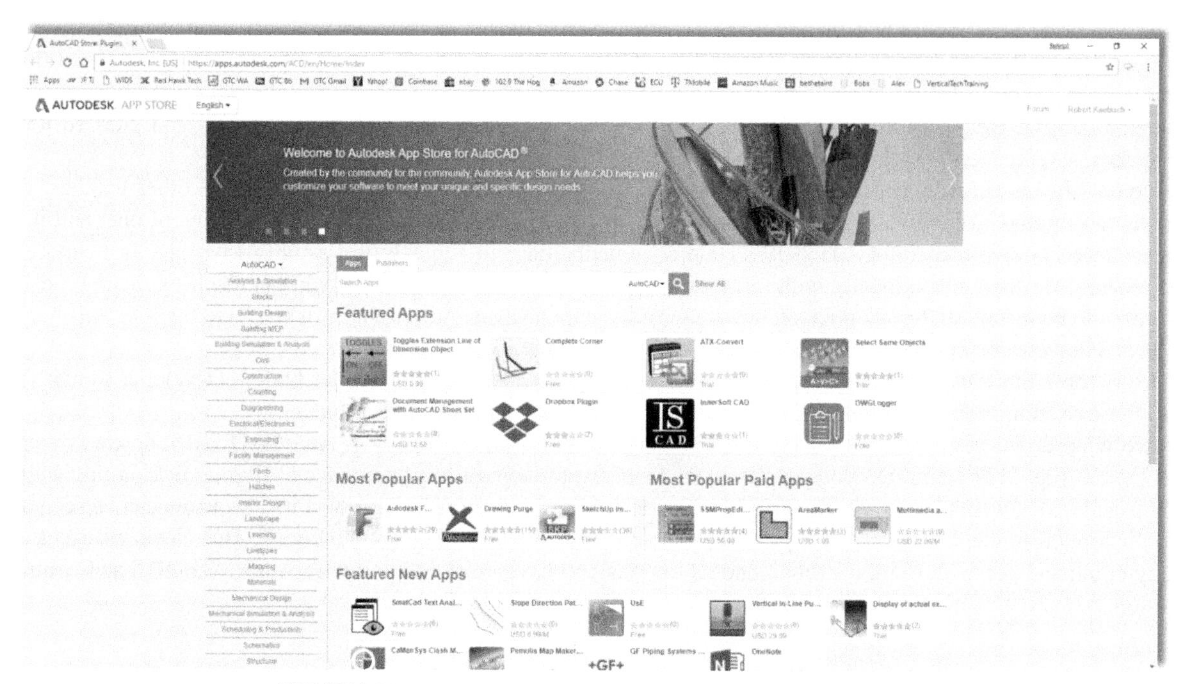

FIGURE 2.31 Autodesk Exchange App Store in Google Chrome browser.

As you can see, a tremendous amount of useful information is found in all these references. You will need to occasionally explore these files for answers, so get comfortable with them at the outset.

TANgent OSNAP

The Tangent OSNAP point is necessary to draw part of Exercise 6 and is one of four additional OSNAP points mentioned throughout upcoming chapters. Tangent works by allowing you to create a line that maintains perfect tangency to a circle or an arc. Because this OSNAP is used quite rarely, just type in tan when you need to create the tangency; you see the symbol shown in Fig. 2.32. Once started, create another tangency point on another circle or arc.

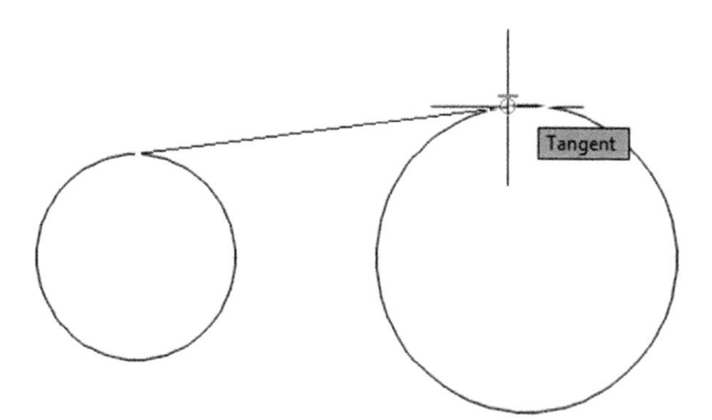

FIGURE 2.32 Using the Deferred Tangent OSNAP.

Time

This interesting command allows you to keep track of various time factors involved with a drawing, such as when it was created, last updated, or the total time spent on it. This could be useful for billing purposes when doing contract drafting or good when you are the boss and want to see how much an employee actually spent on a file. For those students just starting out, or if you are not anyone's boss yet, keep this command in mind, someone may be watching you! To use it with a drawing open, just type in time and press Enter. You will see the command line history pop up above the command line as in Fig. 2.33. Yours, of course, likely says something different.

```
Command:
Command: TIME
Current time:              Monday, March 26, 2018  8:55:05:000 AM
Times for this drawing:
  Created:                 Monday, March 26, 2018  8:54:58:000 AM
  Last updated:            Monday, March 26, 2018  8:54:58:000 AM
  Total editing time:      0 days 00:00:07:000
  Elapsed timer (on):      0 days 00:00:07:000
  Next automatic save in: <no modifications yet>
```

FIGURE 2.33 AutoCAD Text Window—Time command.

Review everything presented so far. In Chapter 3, Layers, Colors, Linetypes, and Properties, we introduce the layers concept and put all the preceding together to begin drawing a floor plan. Remember that learning AutoCAD is cumulative, and you need everything we have covered so far, no more and no less, to successfully deal with the upcoming material, so make sure you understand it all before moving on.

Drawing the Star

As promised, the technique of drawing the star is as follows. First, create a pentagon. Then, noticing that it is a five-point geometric shape, draw lines from endpoint to endpoint, connecting those vertices. Finally, erase the pentagon and use trim to erase the inner lines, as shown in the progression in Fig. 2.34.

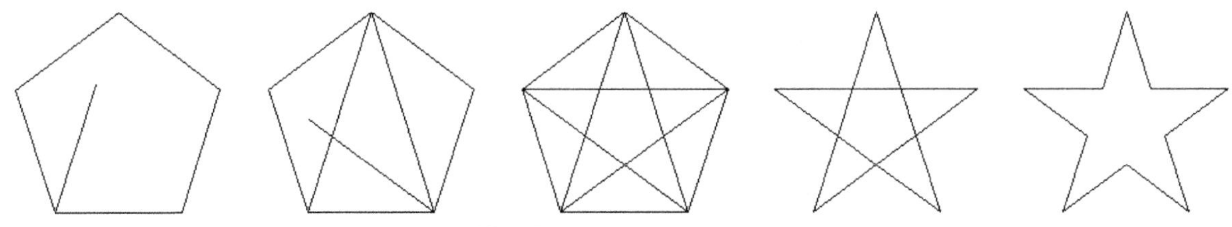

FIGURE 2.34 Drawing challenge.

The reason students do this simple exercise is to underscore a very important point. AutoCAD is used across a vast array of industries, each with its own symbols and unique objects. It would be impossible for the software to include all of them in its database. Instead, AutoCAD gives you basic tools and leaves it up to your creativity and skill to create the shape you need. So, when at a loss as to how to draw something you imagined in your mind, stop and think about the basics that you learned and how you can put them together into something new.

SUMMARY

You should understand and know how to use the following concepts and commands before moving on to Chapter 3*:*

- Using grips
  - Activating them, using Esc to get out
- Setting units
  - Architectural and Engineering (Decimal) units
- Using Snap and Grid
  - Setting both, turning them on and off

- Understanding the Cartesian coordinate system
  - X and Y axes, ordered pair (X,Y)
- Distance Entry tools
  - Direct Distance Entry
  - Relative Distance Entry
  - Dynamic Distance Entry
- Inquiry commands
  - Area
  - Dist
  - List
  - ID
  - Radius and angle
- Explode command
- Polygon command
  - Circumscribed and inscribed
- Ellipse command
- Chamfer command
- Templates
  - acad.dwt
- Setting limits
- Save
- Help files (F1)
- Tangent OSNAP
- Time

REVIEW QUESTIONS

Answer the following questions based on what you learned in this chapter. You should have the basic commands memorized, so try to answer most of the questions without looking up anything:

1. How do you activate grips on AutoCAD's geometry?
2. How do you make grips disappear?
3. What are grips used for?
4. What command brings up the Drawing Units dialog box?
5. The snap command is always used with what other command?
6. Correctly draw a Cartesian coordinate system grid. Label and number each axis appropriately, with values ranging from −5 to +5.
7. On the preceding Cartesian coordinate system grid, locate and label the following points: (3,2), (4,4), (−2,5), (−5,2), (−3,−1), (−2,−4), (1,−5), and (2,−2).
8. What are the two main types of geometric data entry?
9. Describe how Dynamic Input works when creating a line at an angle.
10. List the six inquiry commands covered.
11. What command simplifies a rectangle into four separate lines?
12. What is the difference between an inscribed and a circumscribed circle in the polygon command?
13. What is a chamfer as opposed to a fillet?
14. What template should you pick if you want a new file to open?
15. What is the idea behind limits?
16. Why should you save often?
17. What is the graphic symbol for save?
18. How do you get to the Help files in AutoCAD?

EXERCISES

1. Open a new file, set units to Architectural, and set both your snap and grid to 1″. Then, draw the following shape. Keep in mind the spacing between each dot is now 1″. You would be wise to use the mirror command to complete this faster. The linework in this example has been darkened slightly for clarity. (Difficulty level: Easy; Time to completion: <5 minutes.)

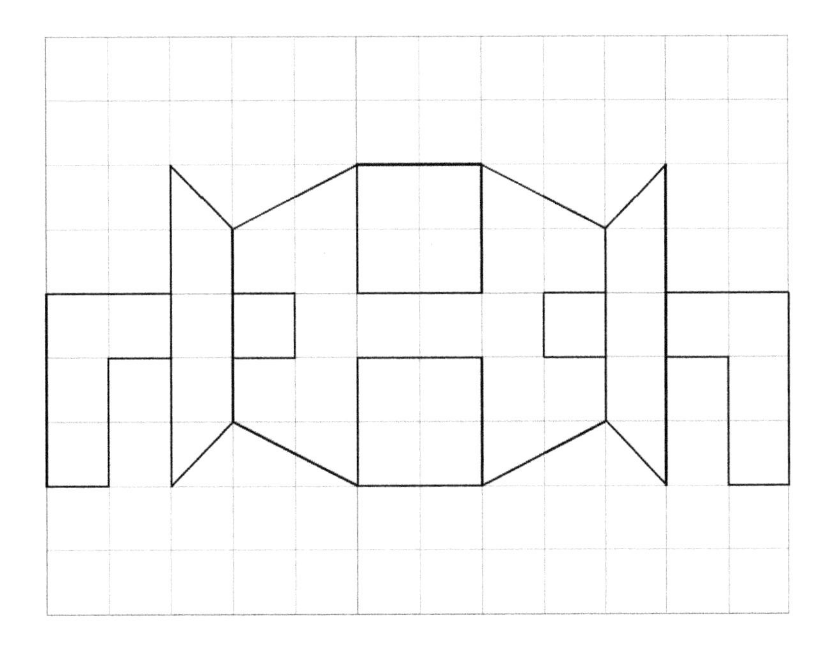

2. Open a new file and set units to Architectural. Then, draw the following figure using the DYNMODE drawing aid. When done, use the area command to find its area. Redraw the shape via the manual method. (Difficulty level: Easy; Time to completion: <5 minutes.)

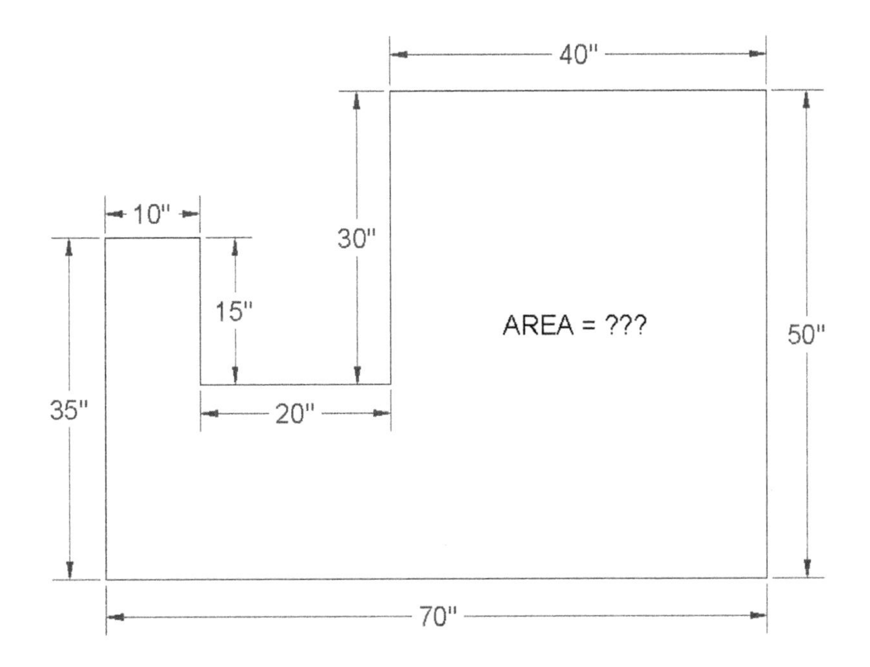

3. Open a new file and set units to Architectural. Draw the following shapes using the manual Relative Distance Entry (the @ sign) method. Repeat the exercise using the DYNMODE method. Repeat again using the rectangle command's length and width suboptions. Finally, use the area command (Object option) to find the area and perimeter of each shape. The sizes are 24″ × 36″, 17.5″ × 3″, 12″ × 12″, and 3″ × 40″. Remember that the first number is always the X; the second is the Y, as shown in the exercise figure. (Difficulty level: Easy; Time to completion: <10 minutes.)

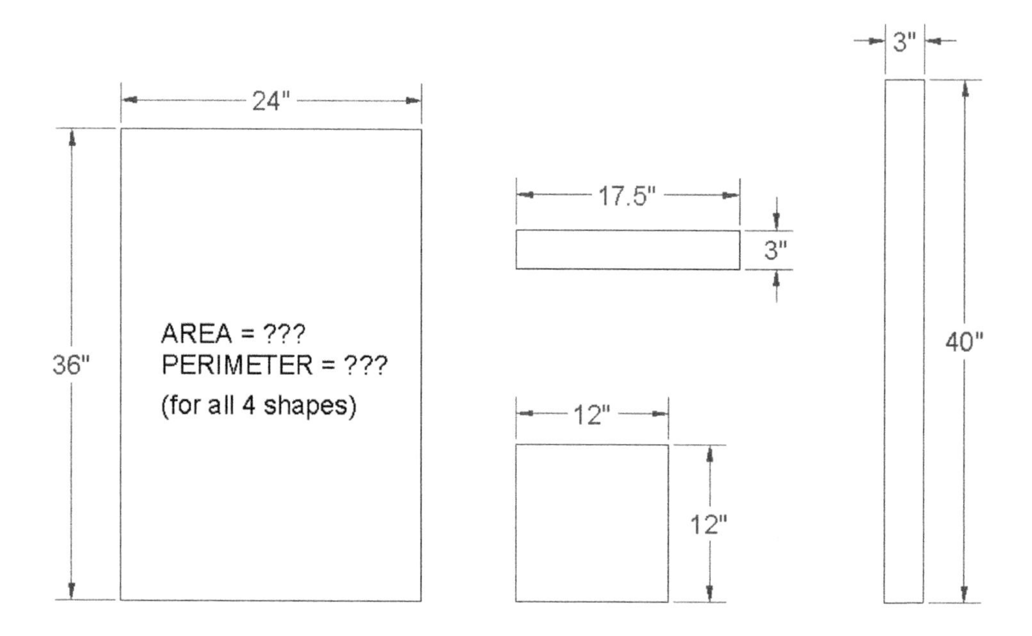

4. Open a new file and set units to Architectural. Draw the following figures using the basic circle, line, offset, trim, and fillet commands. Create the front view first, using circles and then project lines from the quadrants of the circles to create the side view. (Difficulty level: Easy; Time to completion: 5–7 minutes.)

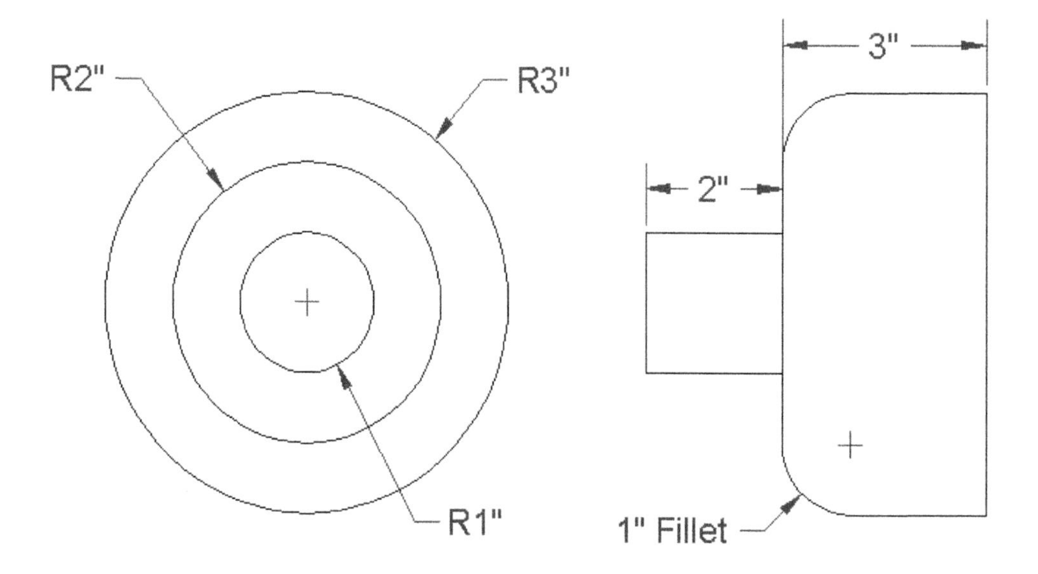

5. Open a new file and set units to Architectural. Draw the following bookshelf. Depending on your exact approach, you may need the rectangle, offset, explode, trim, rotate, copy, scale, and line commands. The spacing of the shelves and the size and rotation of the books can be anything you want. All dimensions are for reference only. (Difficulty level: Intermediate; Time to completion: <10 minutes.)

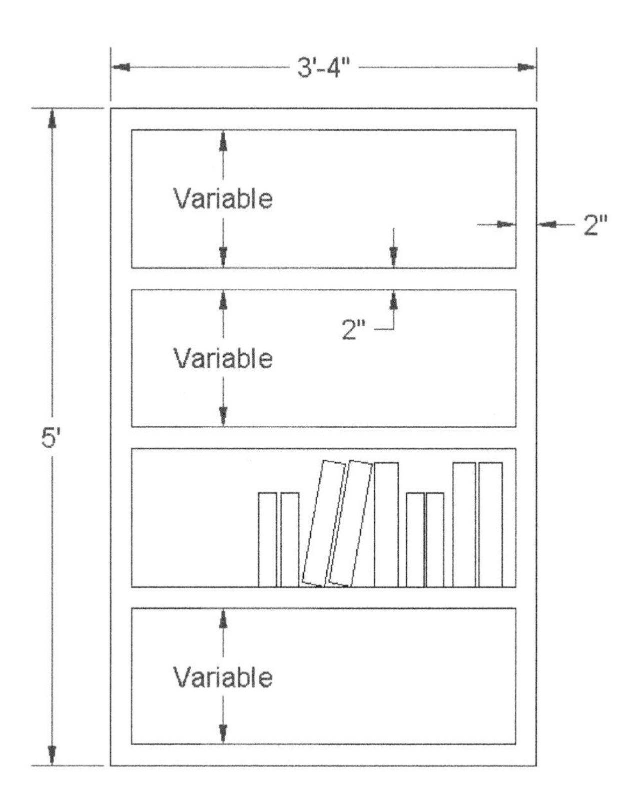

6. Electrical engineering schematics, because they generally comprise basic 2D linework, lend themselves easily to AutoCAD, and we have a few such layouts as exercises in this textbook. Open a new file and set units to Decimal. Draw the following basic capacitor, resistor, and diode schematic. The two dimensions are just for reference and to get you started, all sizing is otherwise approximated. (Difficulty level: Easy; Time to completion: <10 minutes.)

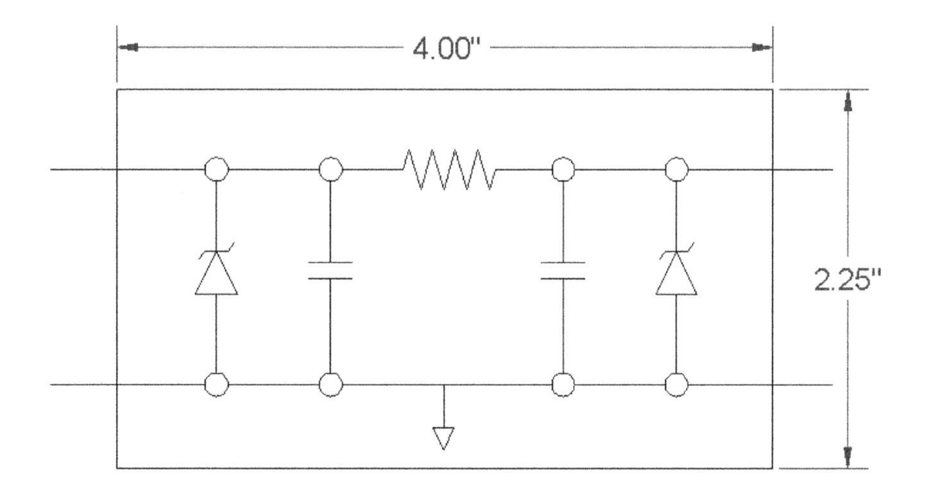

7. The remainder of the exercises in this chapter is mechanical in nature. Open a new file and set units to Decimal. Draw the following mechanical gasket. All dimensions and centerlines are for reference only; we have not yet covered how to create them. You need to use the Tangent OSNAP point. (Difficulty level: Intermediate; Time to completion: 10 minutes.)

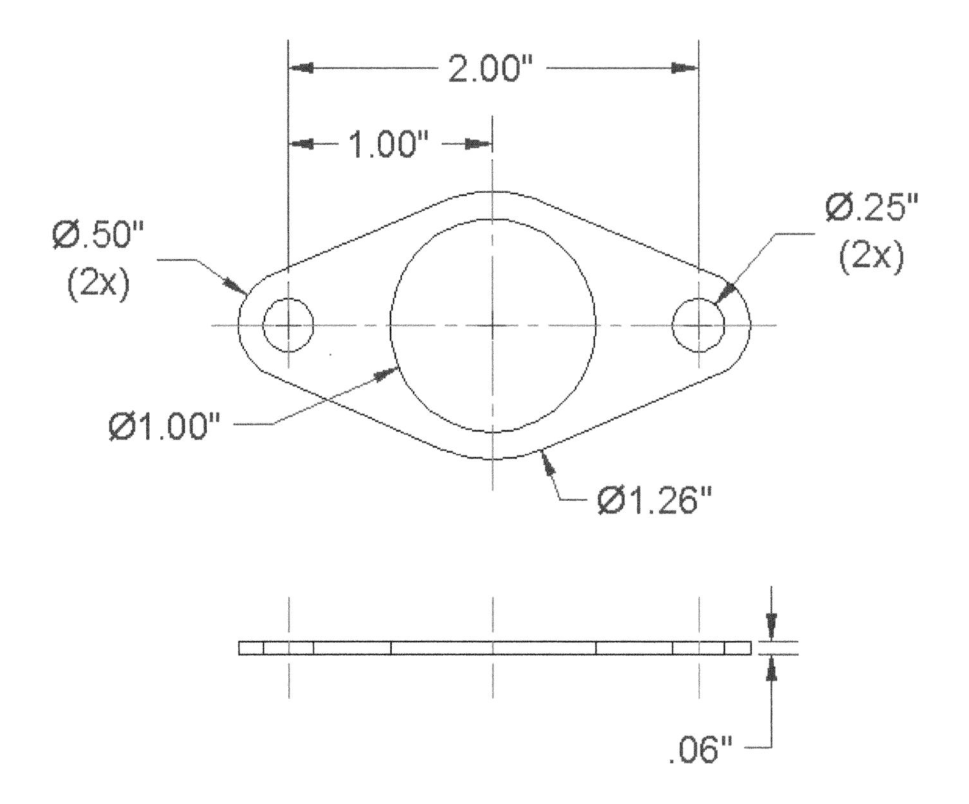

8. Open a new file and leave the units as Decimal. Draw the following simplified bolt diagram. You have all the information needed but may have to carefully project some parts of the side view. You may need the polygon, circle, line, trim, chamfer, and other commands. (Difficulty level: Intermediate; Time to completion: 10 minutes.)

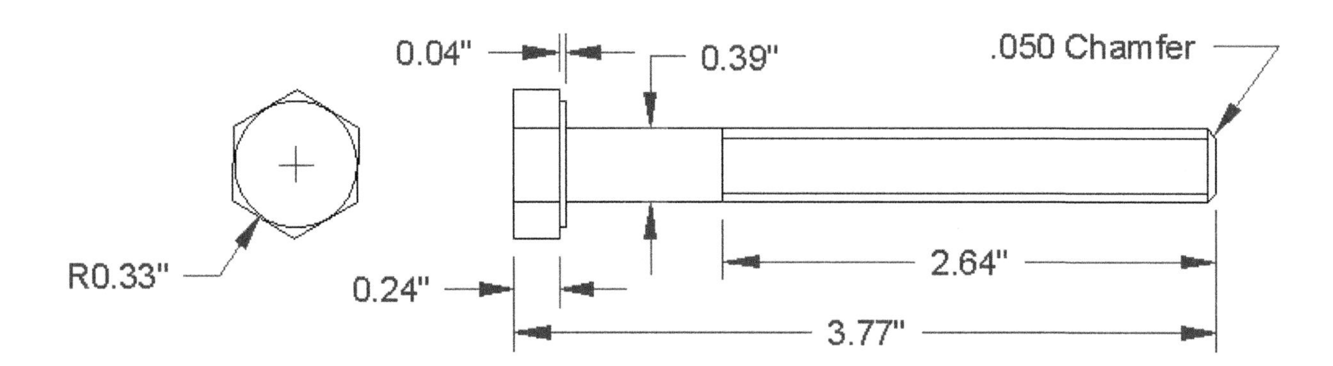

9. Open a new file and leave the units as Decimal. Draw the following mechanical bracket. You have all the information needed except for the bolts—you can improvise their sizing and position. You may need the rectangle, circle, line, trim, fillet, polygon, and other commands. (Difficulty level: Intermediate; Time to completion: 20 minutes.)

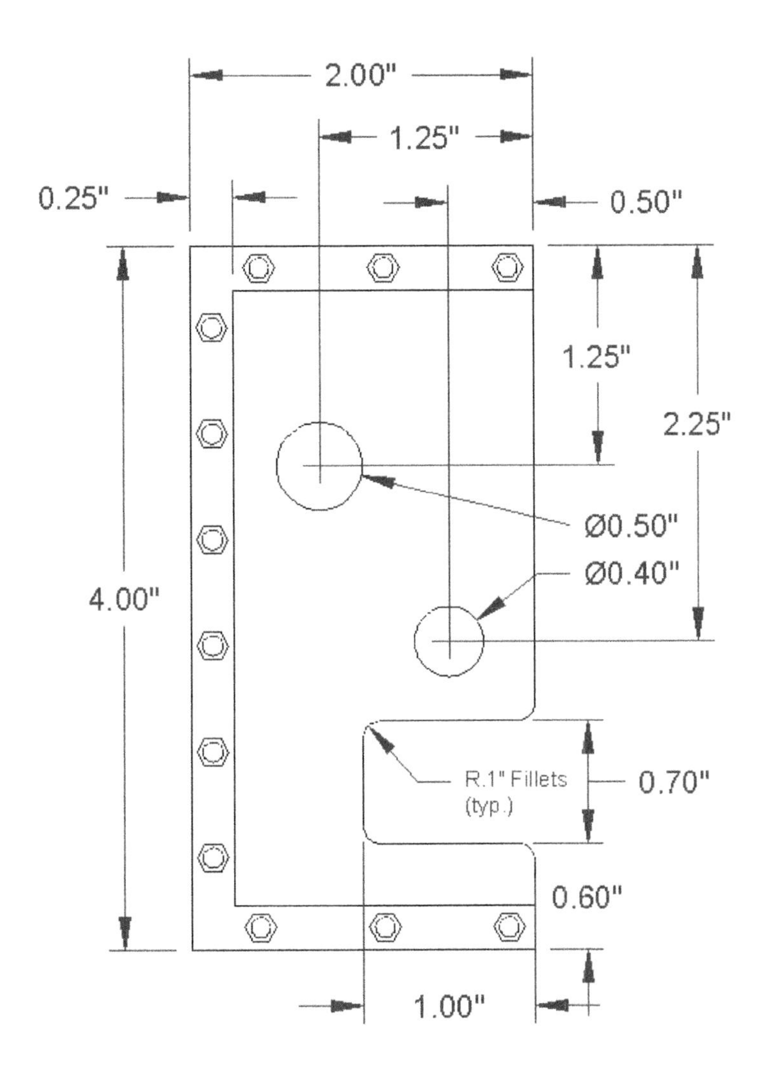

10. Open a new file and leave the units as Decimal. Draw the following mechanical bracket. You may need to use the rectangle, circle, fillet, offset, and other commands. The middle slot is made up of two circles, positioned as shown. Then, two lines are drawn using the Tangent OSNAP, with a trim to complete the slot. (Difficulty level: Intermediate; Time to completion: 30 minutes.)

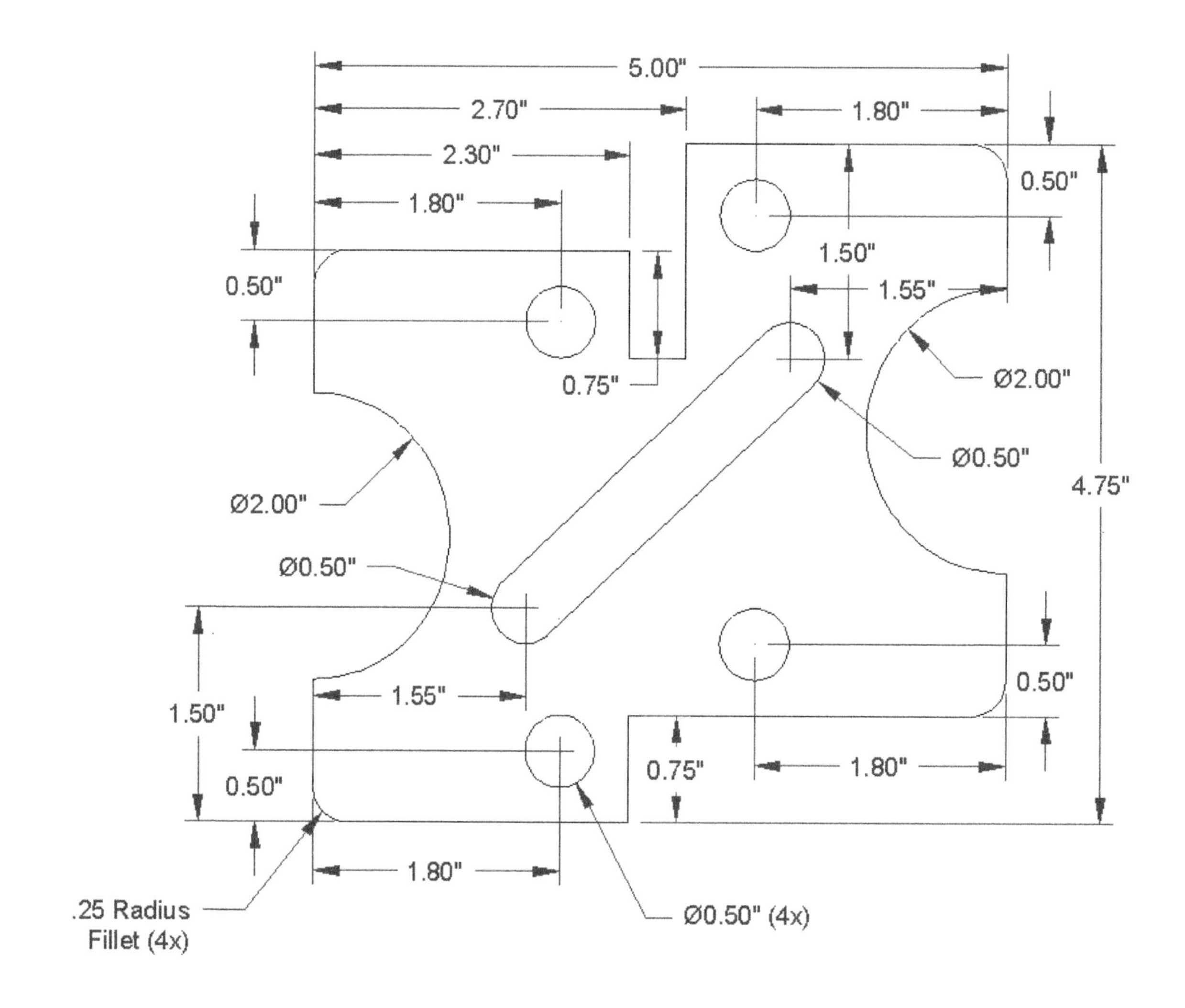

Spotlight On: Architecture

AutoCAD is the undisputed CAD software of choice among most architects, and indeed the software finds its widest application in architecture and related fields. Therefore, it makes sense to feature architecture in our first Spotlight On feature.

Architecture is the art and science of designing and constructing buildings and other physical structures for human shelter or use. It is a far-reaching field that encompasses not only the construction and layout of structures but also some civil engineering, landscape design, interior design, and quite a bit of project management.

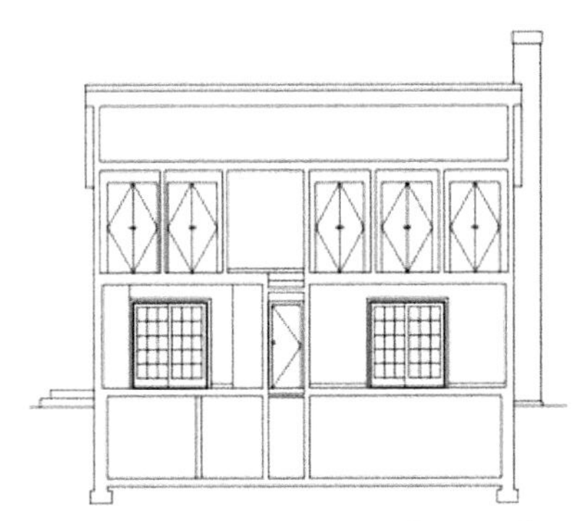

FIGURE 1 *Image source: Elliot Gindis.*

Architects are often described as a cross between engineers and artists. They not only need extensive knowledge of statics, strength of materials, construction methods, and codes but also artistic flair and well-developed esthetics. Although architects often work with specialists to complete projects, many are well versed in electrical, plumbing, HVAC, landscape, and lighting design. Add to that the necessary ability to manage projects and deal with clients as well as financial and legal matters, and one can see that architecture is a demanding and multifaceted profession.

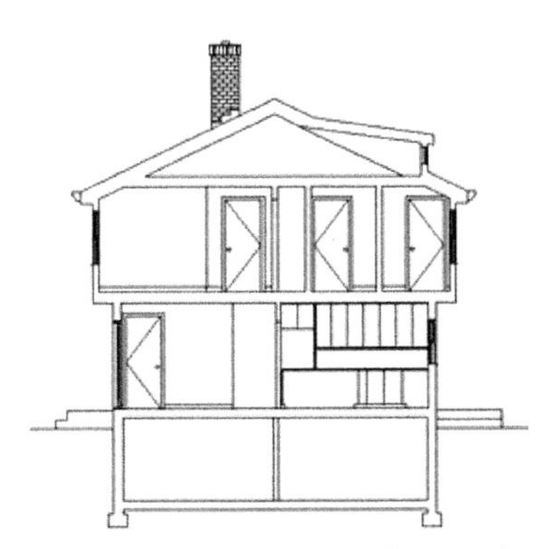

FIGURE 2 *Image source: Elliot Gindis.*

In the United States, requirements to become a licensed architect vary by state. Generally, a student needs to complete a 4- to 5-year Bachelor of Architecture program, followed by a Masters in Architecture (if desired). The student then has to gain a specified number of work experience hours, often while earning a relatively low salary, before being eligible to sit for the Architect Registration Examination, to be licensed in the state of his or her residence. While earnings for partners and senior architects can be substantial, it does take quite a while to get established and build a reputation and client base.

Virtually all architects today have switched to computer-aided drafting, and a group of CAD drafters is a common site in most offices. Drafting is time-consuming, so most senior architects focus on bringing in new business and creating top-level concepts and designs, leaving the details and drafting to project designers and CAD drafters (sometimes the same person). This works out well, as project designers then get an opportunity to learn by doing actual, useful work while "learning the ropes" of the business. Acquiring cutting-edge AutoCAD skills is therefore a great way to get your foot in the door of the profession.

Around 2015, the architecture profession finally started to rebound somewhat from the decline of 2008–13. This decline was tied to the housing market and recession of the previous 7 years. Because of lack of demand in new construction and, in turn, new design work, architects have seen many of their clients' cancel projects or postpone nonessential upgrade work. With an upturn in the economy and the rebound of the housing and real estate market, this should all continue to improve. Surveys done in 2012 and 2013 already indicated a designer shortfall for 2014 onward due to many architects leaving the profession, and as a result there seemed to be enough new ones entering to steadily increase the overall amount of new registered architects in those same years; a good sign indeed.

Despite the downturn, compensation for employed architects has remained steady or risen. The median annual wage for a licensed architect is $76,930 per year. The average salary for a graduate with a Bachelor of Architecture degree is about $63,000, and for a graduate with a Master of Architecture degree it was $74,000. Professional architects with 20 or more years of experience earn an average income of over $110,000.

So how do architects use AutoCAD, and what can you expect? A typical medium-sized project, such as a residential home, features a set of plans that may include

- T1 Title Sheet
- SP-1 Site Plan: The property surrounding the structure.
- A1.1 Demo Plan: What needs to be demolished to make room for the new design, if applicable.
- A1.2 Foundation Plan: The foundation, basement, and/or finished basement plan with mechanical systems
- A1.3-A1.4 Floor Plan(s): The actual design, which may have multiple floors.
- A1.5 Roof: Shows construction of roof beams, rafters, braces, joists, and other elements.
- A2.1-A2.2 Exterior Elevations: Shows exterior views, heights, and materials
- A3.1-A3.2 Building sections, wall sections, and details: "cuts" through the house to show details of construction.
- A4.1 Interior Elevations and Details: Shows cabinets and appliances and other interior elevation elements.
- A4.2 Bathroom Elevations: Shows bath, shower, sink, and other features.
- A5.1 Schedules: Lists of doors, windows, appliances, and other items, including sizing, quantity, etc.
- E1.1 Electrical Plan: The outlets, switches, panels for each floor level (may include switching, ceiling, and wall lighting)
- E1.2 Electrical Lighting Plan (reflected ceiling plan [RCP]): switching, wiring schematic, and ceiling/wall lighting
- GN-1 General Notes: Describes the overall job and any special instructions to the contractor (could be on other sheets).
- P1.1 Plumbing Riser Diagram: Describes flow of pipes and plumbing features.
- S1.1 Framing Plan: Describes the construction of the house frame.
- S2.1 Structural Details: Additional details of design; may run multiple pages.

Extensive use of AIA layering, or an in-house system based on it, is employed, as seen in Fig. 3. Figs. 4 and 5 show some architectural details. You should also expect to see extensive use of advanced Level 1 and Level 2 concepts, such as

- *Xrefs* (see Chapter 17): The main plan is externally referenced to the electrical and RCP plans.
- *Paper Space* (see Chapter 10): Each of the plans is a separate layout tab.
- *Attributes* (see Chapter 18): Found in the title block, door and window schedules, and other elements.

Architectural drafting is a demanding and skill-intensive endeavor. Many jobs are started with measurements of existing structures, if any exist. If the project calls for a brand-new building, not a renovation, then only a survey and site plan is needed and the design can get started right away. For commercial buildings, typically, a building core is laid out (which then becomes the Xref), and the architect then works on designing the internal details of the building. When the overall design is set, other systems are added, such as electrical, plumbing, HVAC, security, fire protection, mechanical, and much more. Corporate and commercial jobs are typically more involved than residential ones, and some of the previously mentioned subsystems are applicable mostly to those. Finally, extensive construction notes, details, and schedules need to be developed. The final document set may feature 20–30 sheets and sometimes much more.

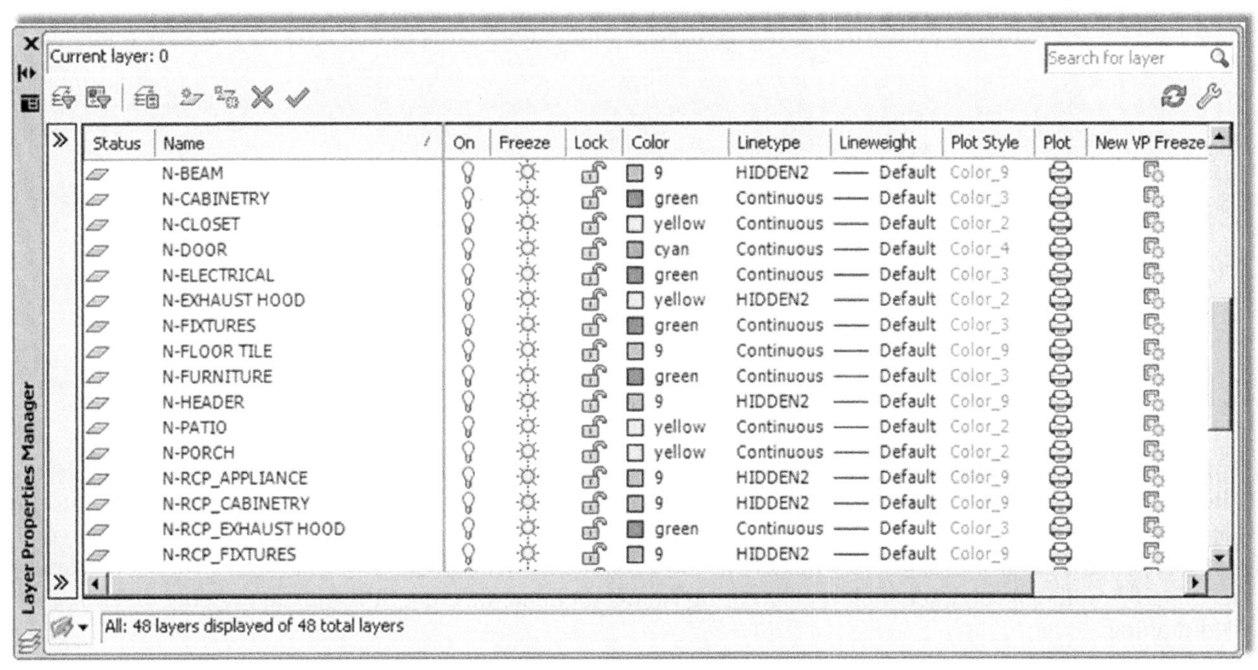

FIGURE 3 Typical AIA layers.

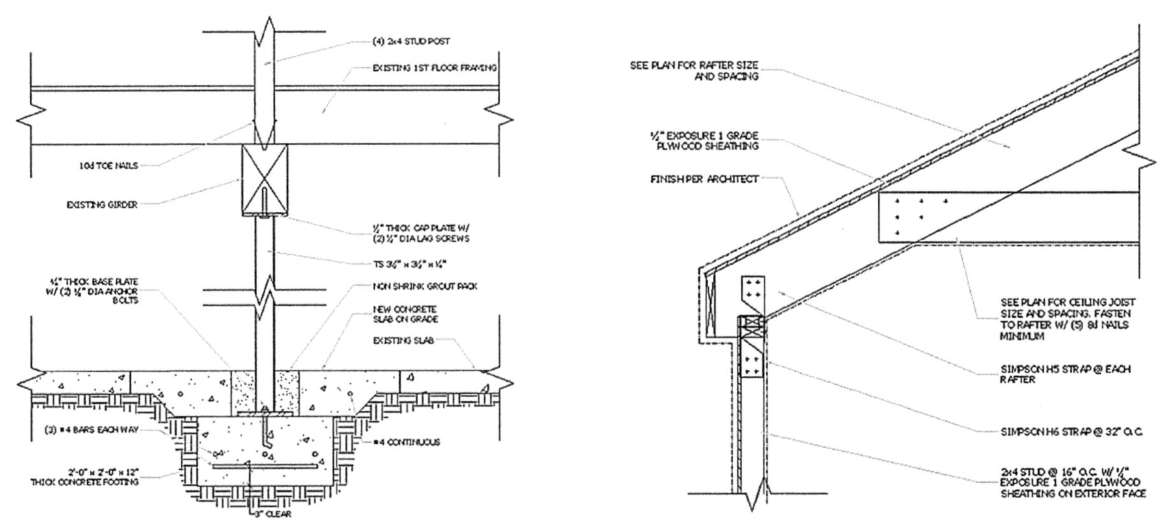

FIGURE 4 Architectural details. *Image source: RM Architect, PC, www.rma-nyc.com.*

Several factors make architectural drafting challenging:

- A complex building design may use virtually all of AutoCAD's tools and concepts; you need to be well versed in all of them. This means mastering every topic in Levels 1 and 2 of this book.
- There is a need for extreme accuracy, as the original design plans of the building structure are the anchor for the rest of the design. Mistakes made early on come back to haunt you.
- There is rarely a luxury of excess time. Architects bid on a job and must work within a set budget. A slow draftsperson who is not proficient (or efficient) takes too long to create the drawings and diminishes profits. Working quickly and accurately is a must. Rarely do you learn the basics and are you allowed to mature "on the job" anymore; it is just too expensive for a small architecture office.

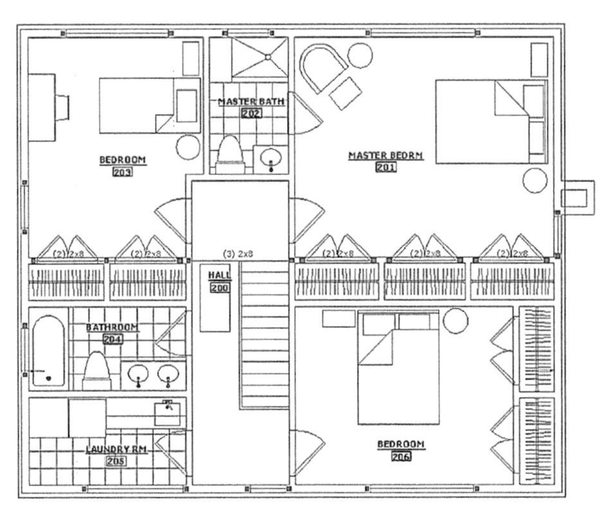

FIGURE 5 Residential architectural floor plan. *Image source: RM Architect, PC, www.rma-nyc.com.*

In spite of these challenges, or maybe because of them, architectural drafting is enjoyable and rewarding, especially if you are a new graduate and picking up knowledge in your profession along the way. Even if you just do nondesign drafting, the pay is good (salaries vary from $35,000 to $55,000 and up), and you acquire some frighteningly good AutoCAD skills very quickly in this fast-paced environment. Most of today's AutoCAD experts (including the authors) can trace their roots to using AutoCAD in architecture; it is the cutting edge of computer-aided drafting.

FIGURE 6 *Image source: www.archinect.com.*

Chapter 3

Layers, Colors, Linetypes, and Properties

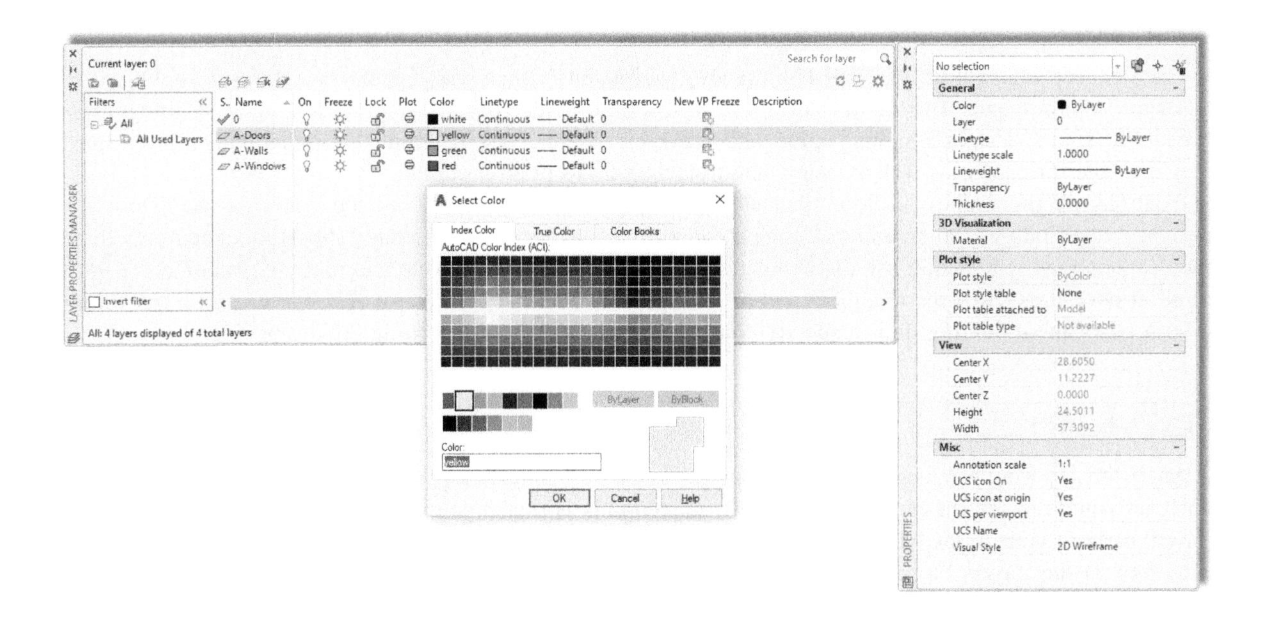

Learning Objectives

Layers are an essential concept and need to be introduced before a serious drawing is attempted. In this chapter, we introduce layers and discuss the following:

- What are layers?
- Why use them?
- Creating and deleting layers
- Making a layer current
- Assigning layer colors
- Index color
- True Color
- Color Books
- Layer Freeze/Thaw
- Layer On/Off
- Layer Lock/Unlock
- Loading and setting linetypes
- Properties palette
- Match Properties
- Layers toolbar

Up and Running with AutoCAD 2019. https://doi.org/10.1016/B978-0-12-816440-2.00003-3

Here, you learn all the remaining tools to begin creating a floor plan, which we start at the end of this chapter. Estimated time for completion of this chapter: 3 hours (lesson and project).

3.1 INTRODUCTION TO LAYERS

Layers are our first major topic after the previous chapters of basics. It is a very important concept in AutoCAD, and no designer or draftsperson would ever consider drawing anything more complex than a few lines without using them.

What Are Layers?

"Layers" is a concept that allows you to group drawn geometry in distinct and separate categories according to similar features or a common theme. This allows you to exercise control over your drawing by, among other things, applying properties to the layers, such as assigning colors and linetypes. You can also manipulate each individual layer, making it visible or invisible for clarity, as well as being able to lock them to prevent editing.

For example, all the walls of a floor plan will exist on the "Walls" layer, all the doors on the "Doors" layer, windows on the "Windows" layer, and so on. You can think of layers as transparency sheets in an anatomy textbook, with each sheet showing a unique and distinct category of data (skeleton, muscle, circulatory system, etc.), but when viewed all at once they show the complete picture of a person. Similarly, in AutoCAD, there can be any number of layers, with all of them holding their respective data and combining into the complete design.

Why Use Them?

As just stated, the key word here is *control*. If all your geometry were lumped together on one layer, as some beginners do, then whatever you do to one set of objects, you do to all; certainly not what you want. It is very important that you learn about layers early in your AutoCAD education and apply them right away to your first real drawing.

Let us look at the Layers Properties Manager. This is AutoCAD's Layers dialog box, where everything important related to layers happens. Open a new file and, if you decide to use toolbars, also bring up the Layer toolbar. We take a closer look at that toolbar in Section 3.3, but for now you need only the icon that calls up the Layers Property Manager.

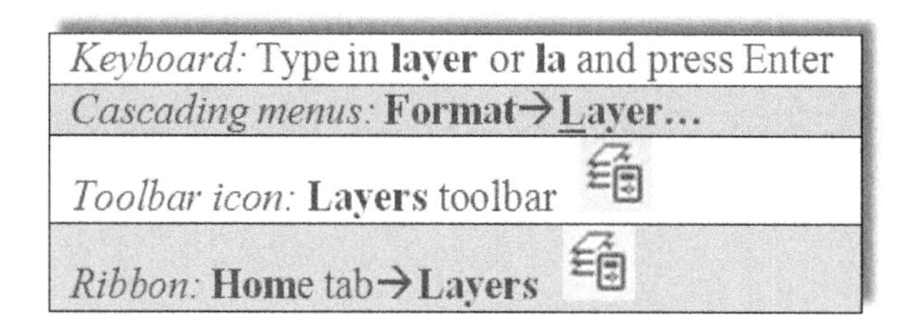

Keyboard: Type in **layer** or **la** and press Enter
Cascading menus: **Format→Layer...**
Toolbar icon: **Layers** toolbar
Ribbon: **Home** tab→**Layers**

Once you call up the Layers Property Manager, you will see Fig. 3.1.

Examining the dialog box from left to right, notice the filters taking up some room on the left. We do not talk about filters until Chapter 12, Advanced Layers, and you can minimize that whole area by clicking the ≪ arrows. To the right of the filters, notice the main category headers, such as "Status," "Name," "On," and "Freeze." Below the categories you find the only existing layer thus far, 0, and its properties. Notice that it is white with a "continuous" linetype. Above the categories are some action buttons for creating and deleting layers, which we cover next.

Creating and Deleting Layers

Assuming you are starting from a blank file, each new layer initially needs to be typed according to what you (or the company you work for) require.

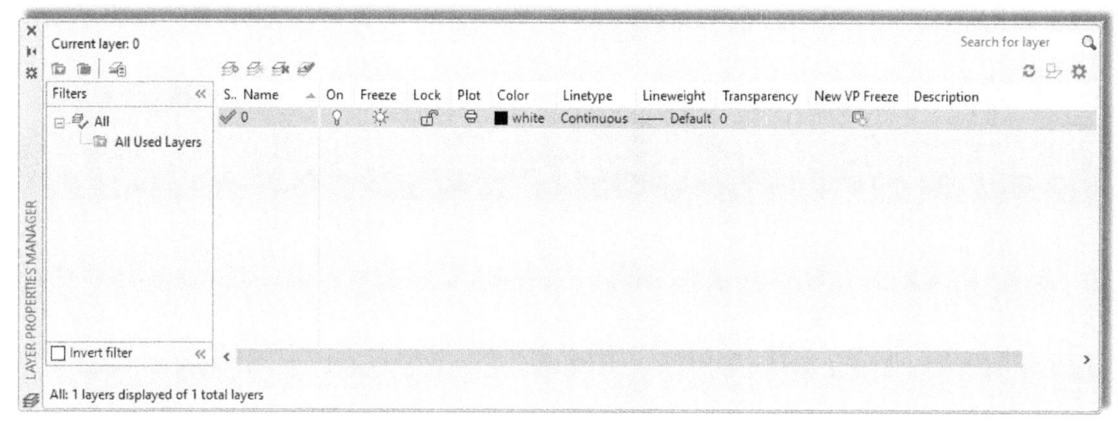

FIGURE 3.1 AutoCAD 2019 Layer Properties Manager.

To do that, you first need to click on the 0 layer to make sure it is highlighted with the blue band. Then, press the New Layer icon at the top of the layer box (two icons to the left of the red X), which looks like a sheet of paper with a sparkle next to it. You can also hold down the Alt key and press N. Either way, once you do that, a new band saying Layer 1 appears. Just type in your new layer's name followed by pressing Enter, and a new layer is created. Here are some helpful tips:

- If you notice that you made a spelling mistake after you already pressed Enter, you can just click back into the layer name once or twice until the cursor enters Edit mode and retype.
- Do not attempt to change the 0 layer's name. It is the default layer and cannot be deleted or renamed.
- If you want to delete a layer (perhaps an accidentally made one), simply highlight the layer, and click the red X icon, or press Alt-D. If the layer contains nothing, it will be deleted.

Following this procedure, create the following three layers:

- A-Doors
- A-Walls
- A-Windows

What you should see is shown in Fig. 3.2. The letter A in front of each layer simply means Architectural.

Making a Layer Current

The current layer is the active layer, on which you can draw at any given moment. You are always on some layer (even if you are not immediately sure which one exactly). So, logically, if you are about to draw a wall, you should be on the A-Walls layer and so on. To specify what layer to have as current, either click on the green check mark button, to the right of the red X, or double-click on the layer name itself. At the very top left of the Layers dialog box, it says Current layer: followed by the name of the layer. Check there if you are unsure what is what. For now, go ahead and make A-Walls the current layer.

Assigning Layer Colors

Assigning colors to layers is very important and should be done as you create them or soon after. The reasons to have color are twofold. On the most basic level, a colorful drawing is more pleasant to look at on the screen, but more important, it is easier to interpret and work with.

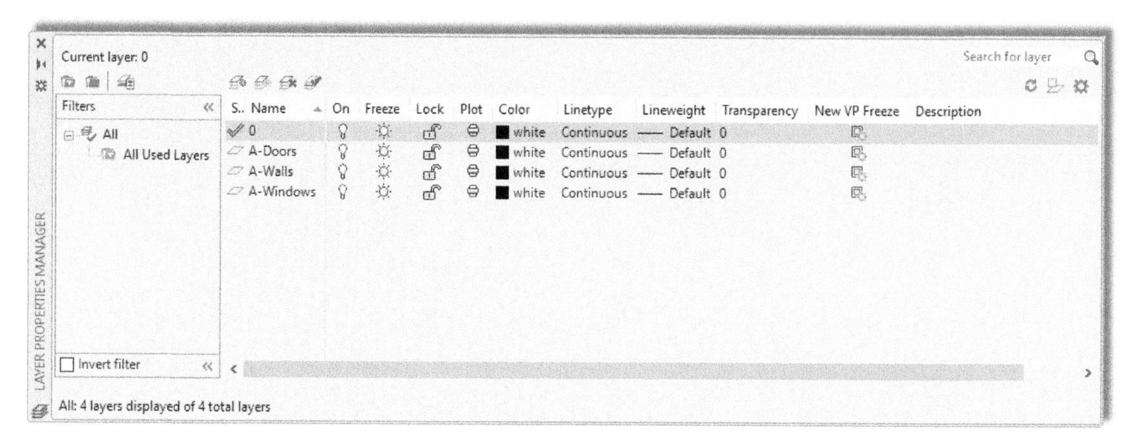

FIGURE 3.2 Layer Properties Manager with three new layers.

If, e.g., your walls are green and you keep this in mind, then your eye can quickly see and assess the design. Likewise, yellow doors are easy to spot. This *color separation* greatly aids in drafting. Imagine a scenario where your building floor plan is made up of all white lines; you would be unable to tell one line from another. As that drawing grew more complex, all the lines would melt into one monochrome mess. Finally, colors are assigned pen thicknesses (an advanced topic) and contribute to more professional-looking output. We address this in level 2.

To assign colors, let us turn our attention back to the Layers dialog box. Highlight, by clicking once, layer A-Walls. Now, follow the blue ribbon across until you see a box with the word *white* next to it. It will be under the Color column. Click on it, and the Select Color dialog box appears, as seen in Fig. 3.3.

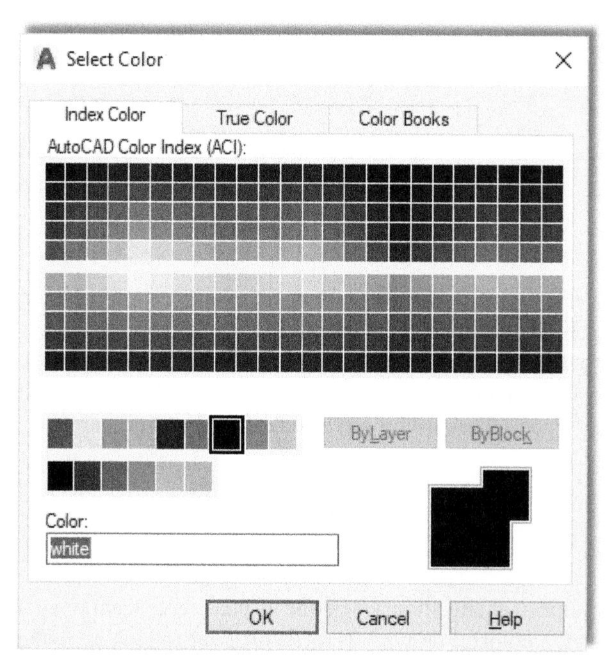

FIGURE 3.3 Select Color, Index Color.

Index Color

Let us take a tour of this Index Color tab. The two large 24 × 5 stripes of colors you immediately see at the top make up the AutoCAD Color Index, and it contains 240 colors. Do not be too quick in selecting one of those right away. Take a look below; there are two smaller color stripes. The rainbow-like stripes 1 through 9 are the ones in which we are most interested. If you take a close look, they are the *primary* colors and the 240 colors above them are just different shades of the primaries. You should always use up those nine colors first—they are very distinct from each other—and only then move on to the others. Also, this makes setting pen weights easier later down the road, as you need not scroll through dozens of colors to find the ones you used.

The colors from left to right are Red, Yellow, Green, Cyan (not Aqua), Blue, Magenta (not Maroon), and then White and two shades of Gray, 8 and 9. The color stripe below that contains just additional shades of gray, and we do not need them for now. So, finally, pick a color for the A-Walls layer (green is what many architects use for walls), click on it, then press Enter; you are taken back to the Layers dialog box to assign more colors to more layers. Yellow is recommended for A-Doors and red for A-Windows.

True Color

There is more to colors, of course. Go back to the Select Color dialog box. Now, click on the second tab, called True Color (Fig. 3.4). Here is where some experience in graphic design may come in handy.

The main color window you are looking at now contains over 16 million colors and should be familiar to you if you use Photoshop, Illustrator, Paint Shop, or any of those types of design programs. You can now move your mouse around, selecting colors based on the HSL method. HSL stands for Hue (the actual colors, moving across), Saturation (moving up or down), and Luminance (brighter or darker, moving the slider up or down next to the main window). You may also use the RGB (red, green, blue) model of color selection by using the drop-down selector at the upper right (Fig. 3.5).

Whichever color you settle on is shown by both color models. Designers rarely use these colors for 2D work (where you want objects to be bright and visible), but do pick these more subtle colors for realistic-looking 3D models.

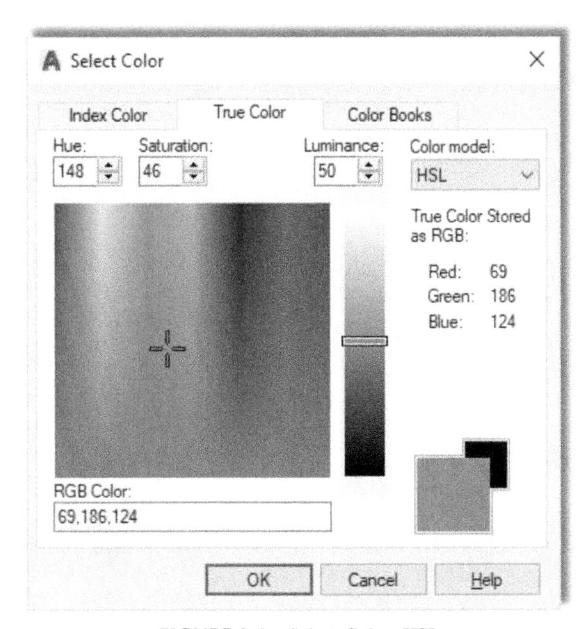

FIGURE 3.4 Select Color, HSL.

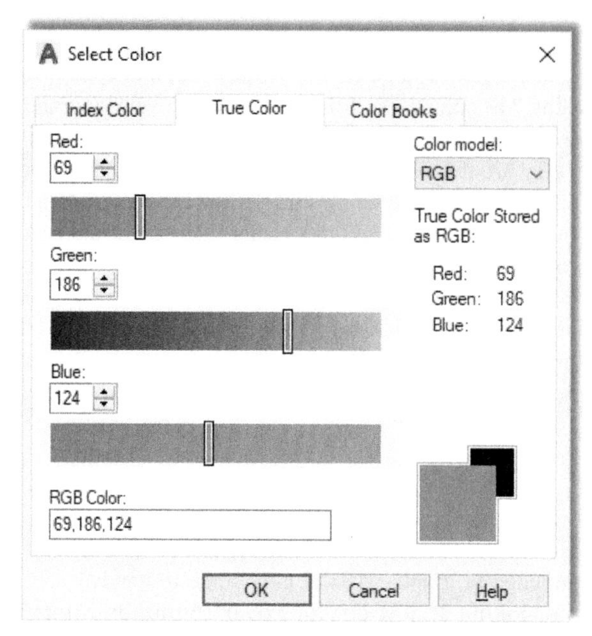

FIGURE 3.5 Select Color, RGB.

Color Books

Here is where things get even fancier. Color Books are Pantone colors. If you are an architect or an interior designer, you already know what these are. If you are shaking your head and have never heard of Pantone, do not worry about it, chances are you will not need to use them. Just to try, however, scroll through the palettes and select a color. Fig. 3.6 is a screen shot of the Color Books tab.

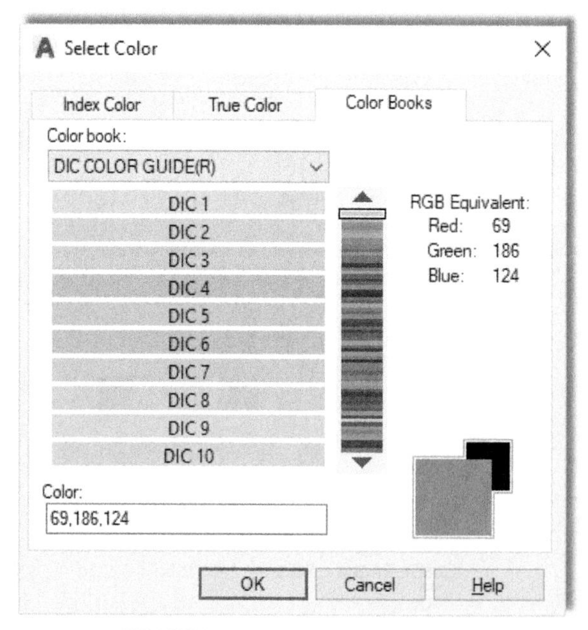

FIGURE 3.6 Select Color, Color Books.

Layer Freeze/Thaw and On/Off

Here, we come to a very important part of the Layers dialog box. Freeze and Thaw are opposites of the same command, one that makes a layer appear or disappear from view. The command freezes or thaws the layer for the entire drawing. Pick a layer, highlight it, follow the band across to the Freeze column and underneath you will find a yellow sun icon. If you click on it, that layer "freezes" and the icon becomes a blue snowflake. Clicking again undoes the action. So, what does freeze mean? It makes the layer (or more precisely all the objects on that layer) disappear from view, but remember, they are not deleted only invisible. You will use this feature routinely in drawings to control what is displayed. The On/Off feature represented by the light bulb (yellow to blue and back) is essentially the same as Freeze/Thaw, with some subtle (and minor) details, so we do not go into them. You can use either method; although stay with Freeze/Thaw just to keep things consistent.

Layer Lock/Unlock

This layer command, represented by the padlock, does exactly that; it locks the layer so you cannot edit or erase it. The layer remains visible but untouchable. The unlocked layer's icon is blue and looks "unlocked," whereas a locked layer's padlock turns yellow and changes to a "locked" position. To let you know which layers are locked, the layers fade somewhat, so you can tell them apart from unlocked geometry. You can control the level of this fading via a slider on the Ribbon. Go to the Home tab, and drop down the arrow in the Layers category. You see a slider that says Locked layer fading. Use it to make the locked layers darker or lighter to your liking.

Experiment with everything mentioned thus far to get a feel for layer controls. When done, make sure everything is reset (not locked or frozen) and click on the X in the upper corner of the layer box to close it.

3.2 INTRODUCTION TO LINETYPES

Linetypes are the different lines that come with AutoCAD. As a designer, architect, or engineer, you may need a variety of lines to convey different ideas in your design. Just as in hand drafting, where you create Dashed, Hidden, Phantom, and other linetypes to show cabinets, demolition work, or hidden geometry in part design, in AutoCAD you load the lines and use them as necessary.

The idea here is to load them all in the beginning, then assign the linetypes to either a layer or sometimes just a few items. You can also load them as needed, but it is recommended to get this out of the way so you have them at your fingertips.

The procedure to do so is as follows:

Step 1. Type in `linetype` and press Enter or choose Format → Linetype... from the cascading menu. The Linetype Manager (Fig. 3.7) appears. If it is full of linetypes, it means they have been loaded earlier. If not, and the Linetype Manager is empty, go on to Step 2.

Step 2. Press Load... The Load or Reload Linetypes box comes up (Fig. 3.8).

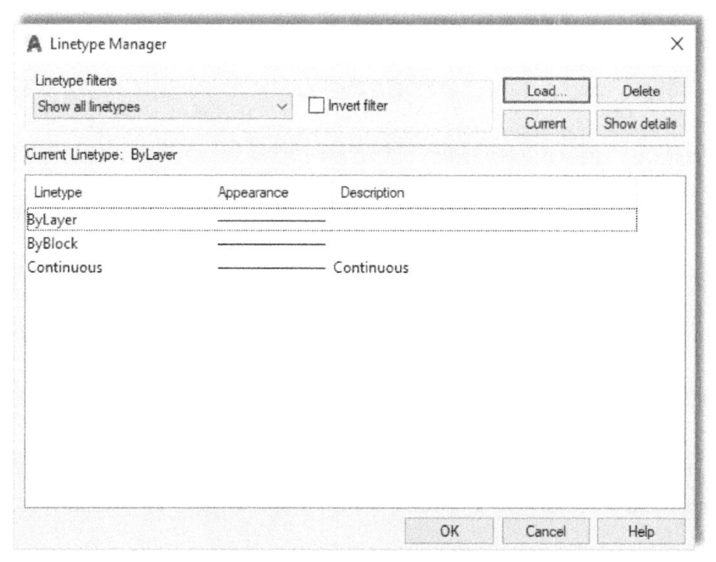

FIGURE 3.7 Linetype Manager.

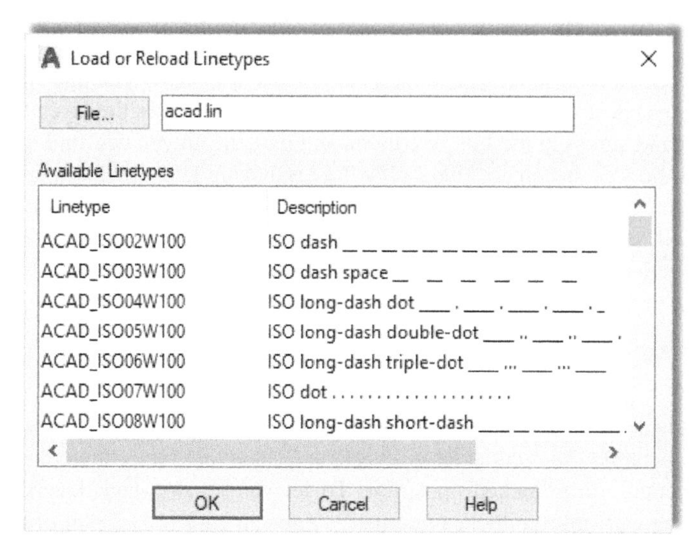

FIGURE 3.8 Load or Reload Linetypes.

Step 3. Position your mouse in the white empty space between the two columns (Linetype and Description) of the Load or Reload Linetypes box and right-click. The Select All/Clear All menu appears.

Step 4. Press Select All and every linetype in the left column turns blue (selected).

Step 5. Press OK (and possibly a Yes to All as well) and the dialog box disappears. All the linetypes are loaded now, so press OK again in the Linetype Manager, and you are done.

We can now use the loaded linetypes and assign a linetype to a layer. Go back to the Layers dialog box:

Step 1. Create a new layer called Hidden and assign a linetype to it (the "Hidden" linetype, naturally).

Step 2. Make sure the Hidden layer is highlighted with the blue bar, and move your mouse to the right until it matches up with the header category Linetype. It should say Continuous right now.

Step 3. Click on the word Continuous, and the Select Linetype (Fig. 3.9) dialog box comes up with all your linetypes preloaded. If we did not load them earlier, the box would be practically empty and we would have to do the loading at this point.

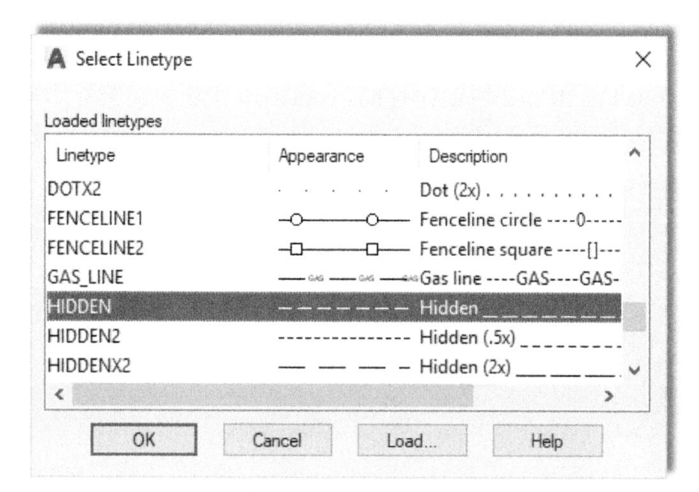

FIGURE 3.9 Select Linetype.

Step 4. So now, all we must do is scroll up and down the list to find the appropriate linetype (Hidden in this case), select it, and press OK.

Make a note of the important linetypes for future reference: Center, Dashed, Divide, Hidden, and Phantom are some of the key ones. Ignore the ones named ACAD_ISO… for the simple reason that their names are not descriptive enough to recognize at a glance. Also, make a note of why there are three types of the same linetype (such as DASHED, DASHED2, and DASHEDX2). By checking the appearance carefully, you see that the scale is different. Finally, let us get back to the Layers dialog box. If you set your Hidden layer to Current, then everything you draw is a dashed line. Go ahead and try it out.

We conclude with one final important note on linetypes. If you cannot see the linetype, and it appears as a solid line when it should be something else, chances are the scale is too small. Type in ltscale (linetype scale) and enter a larger number. This is mentioned again as a tip in Chapter 4, Text, Mtext, Editing, and Style.

3.3 INTRODUCTION TO PROPERTIES

Sometimes, you need to change objects from one layer to another (perhaps because of an error in originally placing it on the incorrect layer or a later decision to change to another layer). Also, you may need to change the color, linetype, or other properties of objects. AutoCAD has a collection of tools to change the properties of objects, with the three most common methods being the Properties (and QuickProperties) palette, the Match Properties tool, and the Layer toolbar. We explore each in order.

Properties Palette

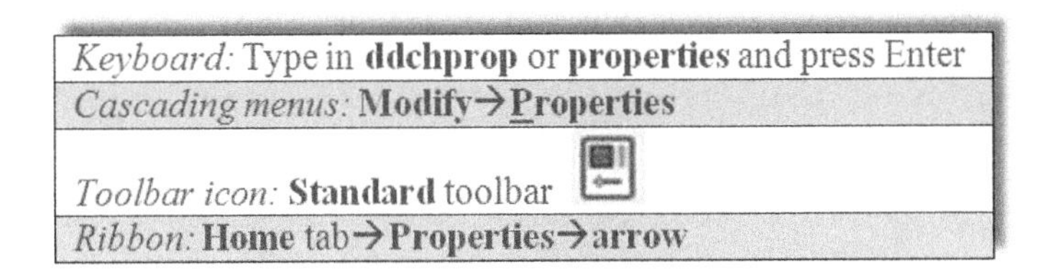

Keyboard: Type in **ddchprop** or **properties** and press Enter
Cascading menus: **Modify→Properties**
Toolbar icon: **Standard** toolbar
Ribbon: **Home** tab→**Properties→arrow**

Once you do any of these (also: Ctrl + 1), the Properties box shown in Fig. 3.10 appears.

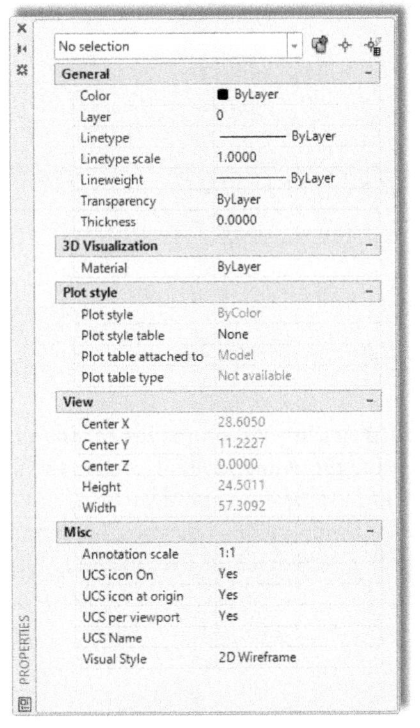

FIGURE 3.10 Properties palette.

Look at the palette in detail from top to bottom. There are many categories (General, View, etc.) and many features that can be modified in those categories (Color, Layer, Linetype, etc.). Also, there are two columns, a gray, one on the left and a mix of gray (inactive) and white (active) fields on the right. To use this palette, first draw several lines on your screen, each of which is on a different layer. Then, follow these steps:

Step 1. First, make sure you have selected one or more items to change. They become dashed. If you selected two items, only those properties common to both that can be changed are available.
Step 2. In the Properties box, find the category you want changed, color, for example.
Step 3. Click on the right-hand column directly across from that property.
Step 4. An arrow drops down. Click on it and view the choices within that category.
Step 5. Select the different property that you want the new objects to have, such as a new color.
Step 6. Close the Properties box (the X in the upper corner) and press Esc once or twice; the objects have that property.

The Properties toolbar is a quick, straightforward way to change many properties but there is another way, shown next. There is also a QuickProperties variation to the full Properties palette just described. This feature is accessible by just double-clicking on any of the objects in the drawing. The QuickProperties palette appears, as seen in Fig. 3.11. Notice a major difference here. The QuickProperties palette is much smaller and shows only the essential properties, such as Color, Layer, and Linetype. It omits various geometry data, which is probably OK, because you are unlikely to use them. Notice also that the QuickProperties palette is *context sensitive*, meaning that it varies slightly, depending on what you have selected, line, arc, and so forth. The process of using it, however, is the same as just described for the full Properties palette. Go ahead and try it out, then close it by just pressing Esc.

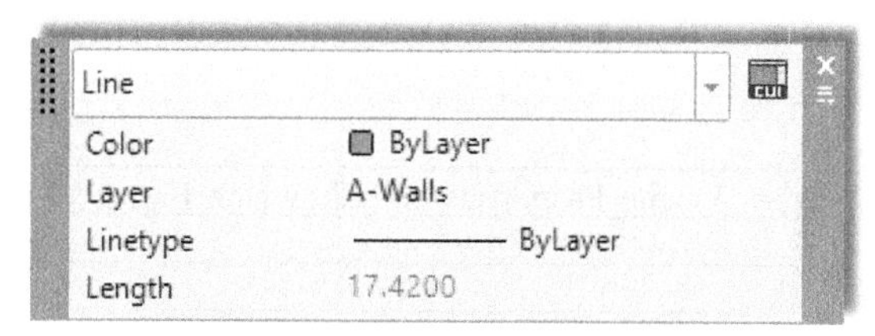

FIGURE 3.11 QuickProperties palette.

Match Properties

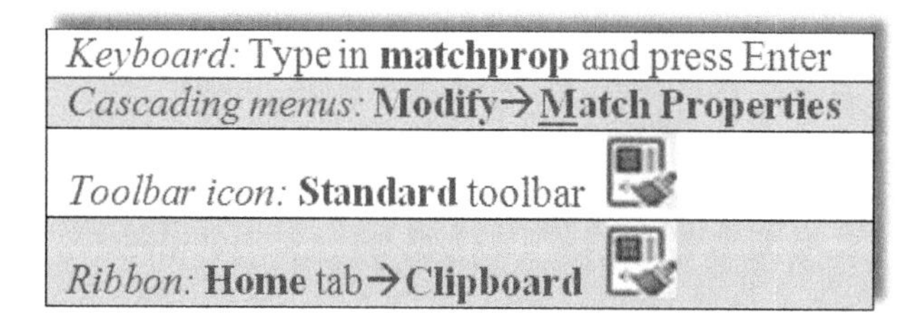

Keyboard: Type in **matchprop** and press Enter
Cascading menus: **Modify→Match Properties**
Toolbar icon: **Standard** toolbar
Ribbon: **Home** tab→**Clipboard**

This is a fantastic way to very quickly match up the properties of one object with the properties of another. The procedure is simple. Start the command via the preceding methods, select the first object whose properties you like, and then click on another object that you want to have those properties.

Match Properties matches almost anything: layers, colors, linetypes, linetype scales, text sizes, hatch patterns, and many more factors. Finally, we have the Layers toolbar, which has its own method of layer changes and deserves its own separate discussion.

Layers Toolbar

The Layers toolbar is shown in Fig. 3.12 with the three layers added. If you want to use the toolbars, then add it to your on-screen collection and use it throughout the remainder of this chapters.

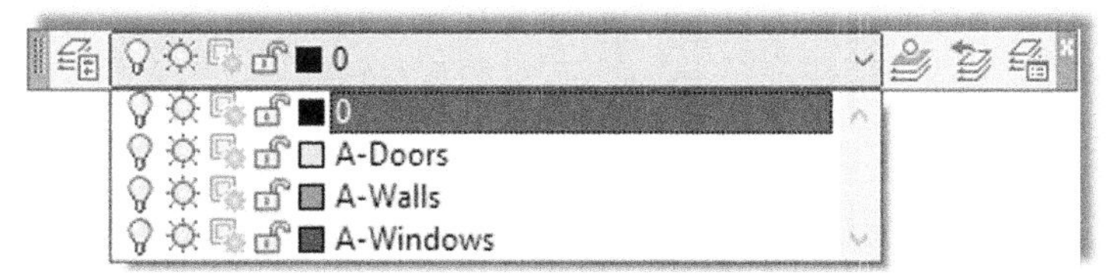

FIGURE 3.12 Layers toolbar.

To change an object's layer using this toolbar, simply select whatever object you would like to change to another layer (what you selected becomes dashed), and then drop down the Layer menu in the toolbar and pick the new layer on which you want the highlighted geometry to go, and it changes. You may have to press Esc to deselect the layer(s) afterward.

This toolbar is also useful for freezing, turning off, and locking layers. It is also a great visual reference to determine what layer is current, which of course is the layer being shown. The toolbar has some useful icons, although the only one we have covered so far was the one all the way to the left that brings up the Layer Properties Manager.

This is it for basic layers and associated topics. You should know the following:

- What layers are and why they are needed.
- How to open the Layers dialog box.
- How to enter a new layer, delete it, and change its name.
- How to make a layer current.
- How to assign a color to a layer and all the color options.
- How to Freeze/Thaw, turn Off/On, and Lock/Unlock the layers.
- How to select and set linetypes.
- How to use the Properties and QuickProperties palettes.
- How to use Match Properties.
- How to use the Layers toolbar.

We did not yet cover a few items, such as the previously mentioned Filters and Layer States, Plot, Lineweights, and a few other features of the Layers dialog box. We get to them in due time in advanced chapters. For now, it is time to put everything to good use and draw a full apartment floor plan. Review everything thus far and begin the assignment in Section 3.4.

3.4 IN-CLASS DRAWING PROJECT: FLOOR PLAN LAYOUT

This is your first major drawing assignment. You need to put together everything you have learned so far and apply it to a realistic drawing situation. Although this floor plan is very simple compared to major architectural or engineering projects created in the industry, it still features components of a professional drawing, just fewer of them, so you do not get overwhelmed right away. Follow all instructions carefully and proceed slowly. Speed comes later, when all the commands become second nature. Our goal is to draw the floor plan shown in Fig. 3.13.

This is a basic one-bedroom apartment. The kitchen is on the upper left, bedroom on the upper right, bathroom on the lower right, living room right in the middle, and entrance foyer in the lower left corner. There are two closets, six windows, and six doors (the kitchen has no door). The dimensions are only for you to look at and use, not draw in yet, as we do not cover them until Chapter 6. The numbers inside the rectangles are the door opening sizes. All walls are 5″ thick.

Outlined next is a complete discussion on how to start drawing the floor plan, intended for those students studying AutoCAD by themselves without an instructor. If you are taking a class, your instructor will assist you but should essentially follow the same format. Once you learn all the steps involved and build confidence, aim to draw it again in less than 60, then 30, and finally less than 10 minutes, starting from a blank screen. This may sound unreasonable now, but that is the required industry speed, and you will be able to get to that quite soon with some practice. An outline of steps with additional information (if needed) follows.

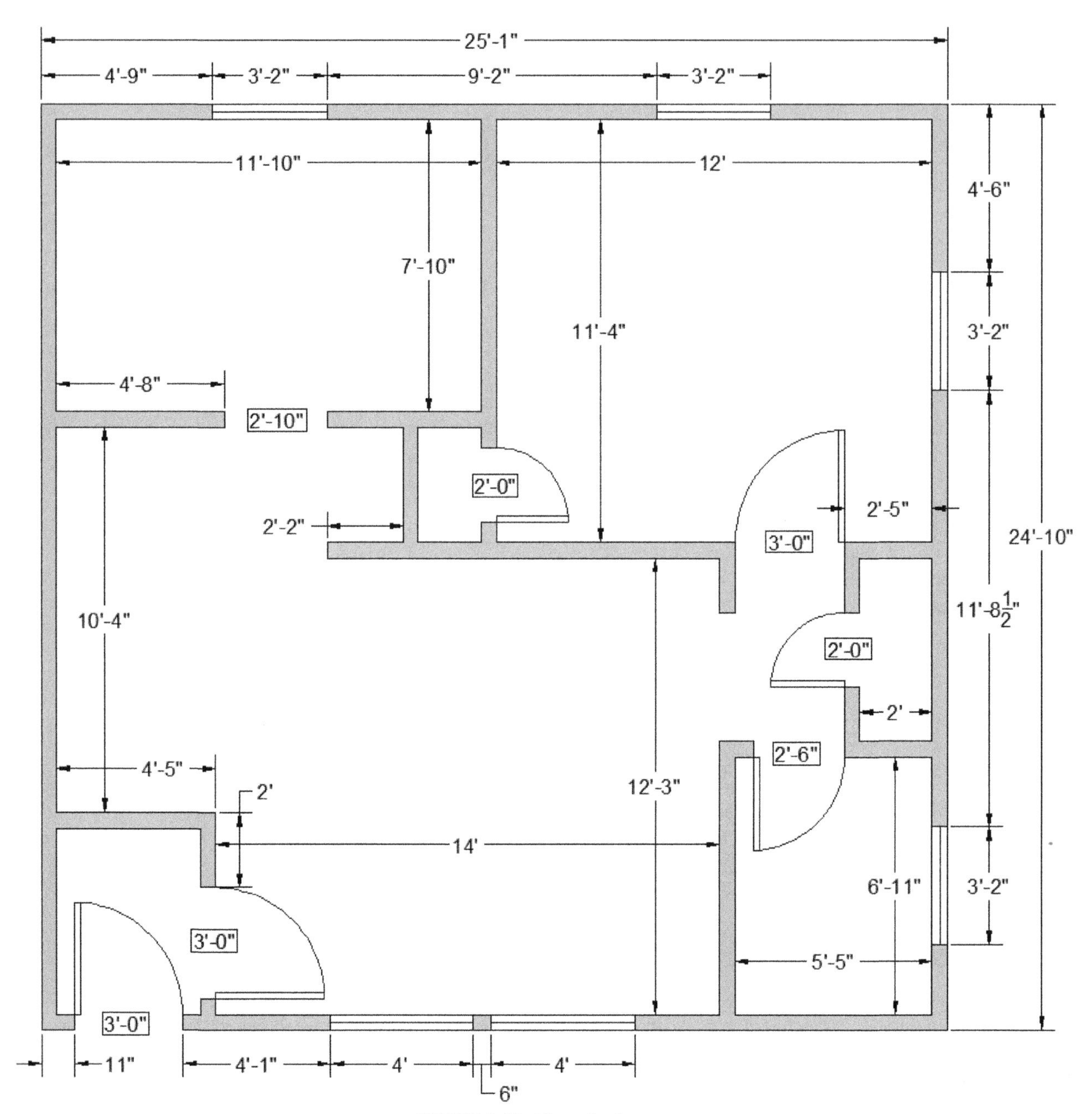

FIGURE 3.13 Floor plan layout.

Basic File Preparation

1. Open a brand-new AutoCAD file using the acad.dwt template. Save it as YourName_AptProject1.dwg. We start from the very beginning for practice purposes.
2. Set your units to Architectural using the units command.
3. Set up layers: A-Walls (green), A-Windows (red), and A-Doors (yellow).
4. Load all your linetypes (we need them later). If you were to save and close this file right now and use it later for another project, then it would be called a *template*, as first explained in Chapter 2.

Starting the Floor Plan

Now that the basic file is set up, how would you start this design from the very beginning? The exterior walls would be the first step, and there are a few ways to go about it. We could certainly draw line by line, but that is not too efficient. Think back to your set of tools learned thus far. The best approach is to draw a rectangle of given dimensions (25′−1″ × 24′−10″) and offset it 5″ to the inside. This first rectangle represents the outer walls and the offset one represents the inner walls.

5. Draw a rectangle that is 25′-1″ wide and 24′−10″ tall. Be careful and use the correct method of manual Relative Distance Entry or Dynamic Input, your choice. You can also just use the rectangle command, indicating the dimensions as allowed by one of the suboptions. If you do it manually, you start a rectangle, press Enter, left-click anywhere in the screen, then type in `@25′1,24′10` exactly as written. If you use DYN, start the rectangle and enter the values on screen, using the Tab key to hop between the first and second values. If you use the Rectangle Dimension option, start the rectangle, press d, and enter its length and width when prompted. Once again, make sure the units are Architectural or you will immediately run into problems.
6. Offset the rectangle 5″ to the inside to create the inner walls (offset command).
7. Explode both wall rectangles. This is needed to break apart the rectangles for use as the basis for the interior walls as well as doorways and windows. You should now have a total of eight individual lines, as seen in Fig. 3.14.

Drawing the Inner Wall Geometry

As mentioned in Chapter 1, one of the biggest misconceptions beginners have about drafting in AutoCAD is that they will spend lots of time drawing line after line. That expectation is natural; after all, in hand drafting, that is exactly what you would do—pick up the pencil and T-square and draw lines connecting them in intricate ways to create a design. Not so in AutoCAD. You still draw lines here and there, but a lot of the time you use the offset command to create new geometry, such as in this case. For example, we use the inner main walls as a basis for the room walls and offset as needed.

8. Offset the inner left vertical wall 11′−10″ to the right to create part of the kitchen wall. Then, right away, offset 5″ for the thickness of the wall. Repeat with the inner top horizontal wall (offset 7′−10″) to get the lower part of the kitchen. Offset 5″ below that line for thickness. Fig. 3.15 illustrates this. Note that, along the way, you will run into fractional lengths to offset (11′−8½″ for example). Do not forget that to enter fractions you just simply type in a decimal equivalent (11′8.5 in this case).

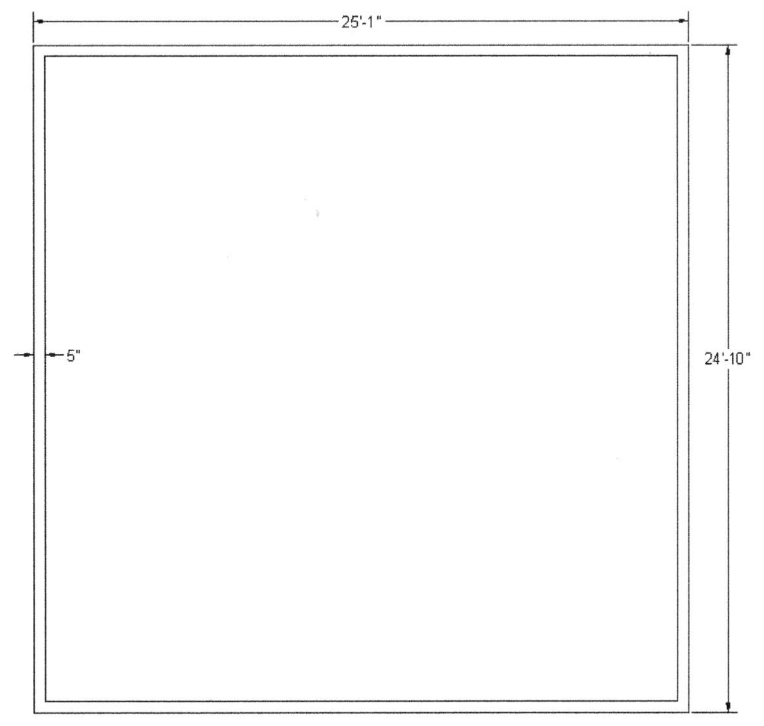

FIGURE 3.14 Floor plan, Step 7.

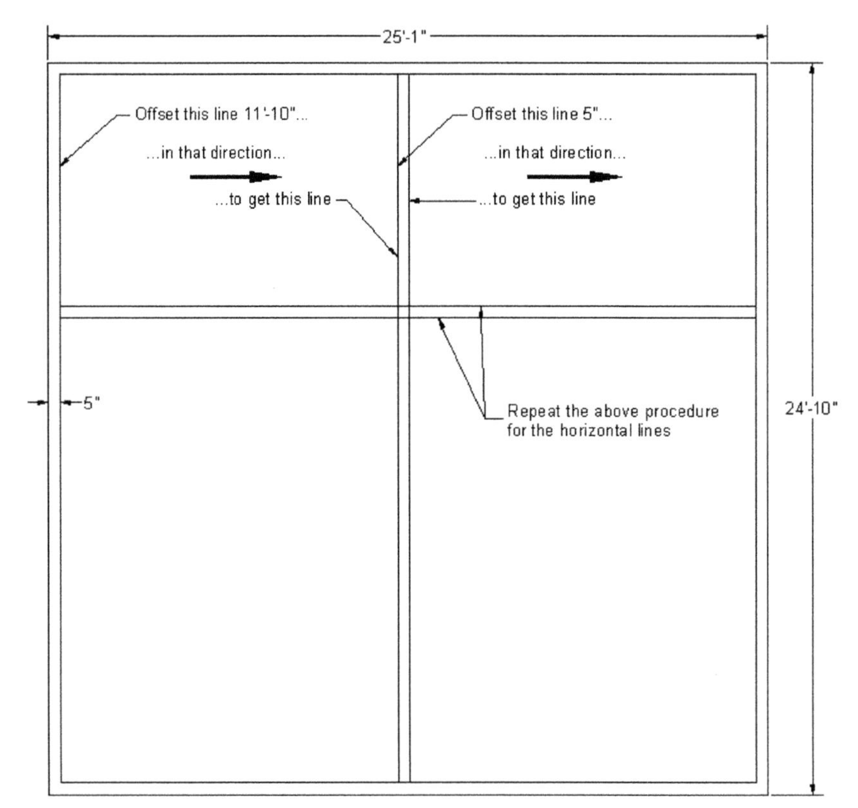

FIGURE 3.15 Floor plan, Step 8.

9. Fillet the two sets of walls to create a sharp corner. This is where the fillet command's advantage really shows. Follow it carefully, and select the two inner wall lines first, then the outer.
10. Create a 2′–10″ opening for the kitchen door. Once again use the offset command. Offset the inner wall 4′–8″ to the right as shown on the plan. Then, offset again 2′–10″ to frame the opening. Next trim it out by selecting all the lines using a crossing, press Enter, and click on the lines you do not need. You use this technique to create the windows in the same way.
11. Trim out the extra line where each inner wall meets the main wall. Fig. 3.16 shows you what you should see so far (be sure to ask your instructor if something is not working out).

If you understood the previous steps, then you should have a good idea of what is needed to draw the rest of the floor plan. It really is mostly similar. You of course must get creative on occasion, but just stop and think of the basic tools you have available first. If it seems overly complicated, you are doing it wrong. After the kitchen, go on to do the bathroom in the lower right corner and then the bedroom and entrance foyer.

Drawing the Doors and Windows

Create the openings for the doors and windows early in the process, but do the actual doors and windows last. Be sure to use the correct layers. Both a door and a window are shown in Fig. 3.17.

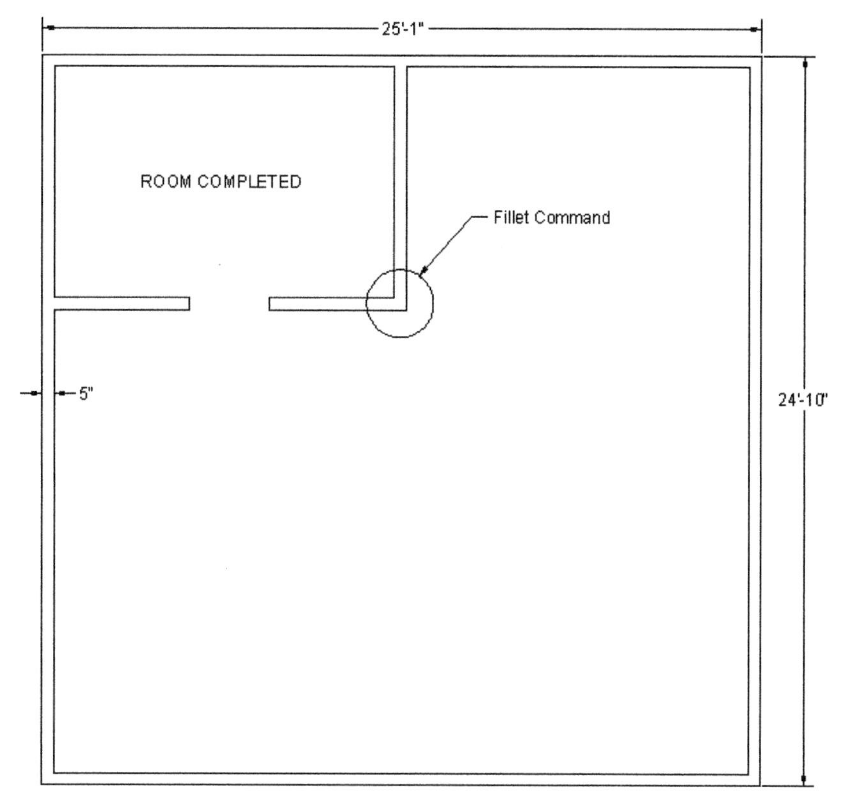

FIGURE 3.16 Floor plan, Step 11.

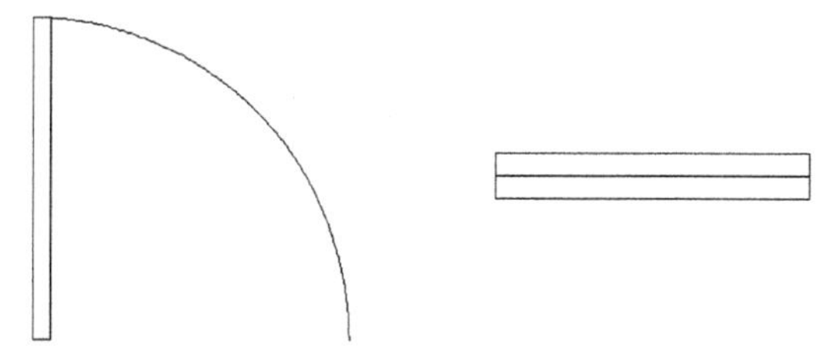

FIGURE 3.17 Door and Window symbols.

Here are some additional tips to help you along.

Doors

Once you have an opening in a wall of a certain size, adding the door involves creating a similarly sized rectangle, oriented in the proper direction. Then, you need to add the door swing to complete the door. You already know how to create rectangles in one of several ways from Chapter 1. You can even use what you did in Exercise 3 of Chapter 1. However, if you want the doors to look more authentic, you need to give them a thickness. We use a thickness of 2″ for all doors in this floor plan. Use DYN or any preferred method to create the door rectangle. A sample result is show in Fig. 3.18.

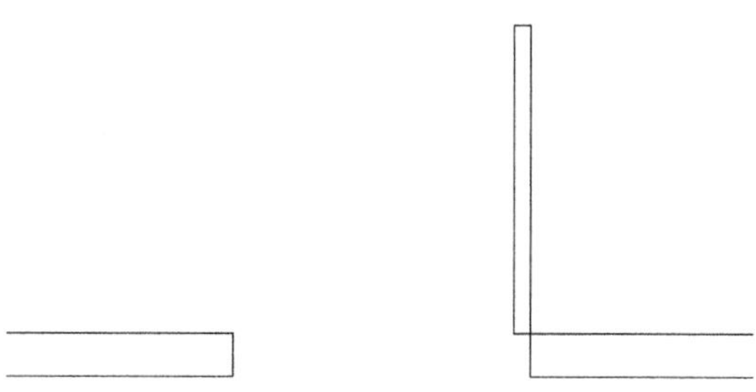

FIGURE 3.18 Creating the door symbol.

To create the door swing, you can carefully position a three-point arc (as in the previously mentioned in a Chapter 1 exercise) or, as a more accurate alternative, create a circle based at the hinge point and with a radius equal to the door width, as seen in Fig. 3.19. Then, just trim the unneeded three quarters of the circle by using the top of the door and the left wall as cutting edges.

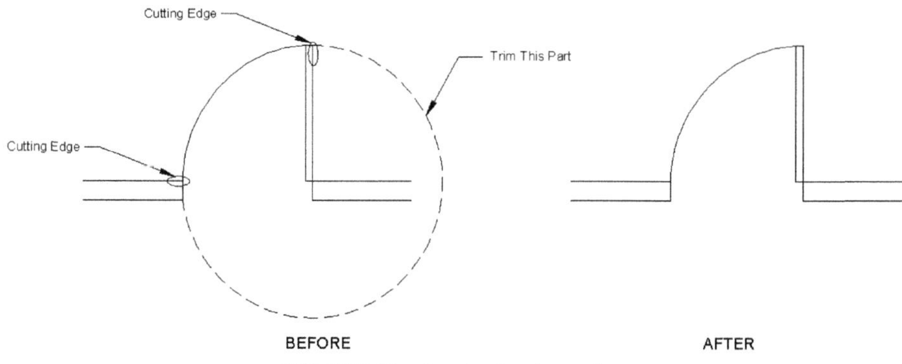

FIGURE 3.19 Creating the door swing.

Windows

To create windows, draw a rectangle of the proper size to fit into the opening and add a line through it. Then, position it into the window opening as seen in the four-step process of Fig. 3.20. These are of course very simplified windows; expect to see much more complex contraptions if you do architectural drafting. Finally, it bears repeating, for both the doors and the windows, be sure to be on the correct layer for each.

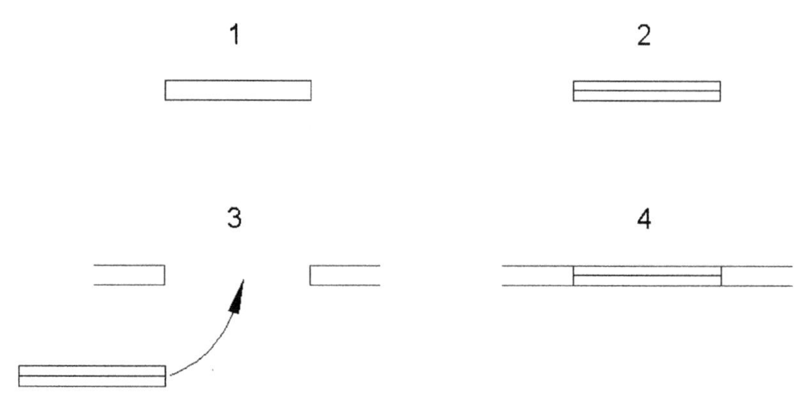

FIGURE 3.20 Creating a simple window.

SUMMARY

You should understand and know how to use the following concepts and commands before moving on to Chapter 4:

- Layers
  - Basic concept and purpose
  - Creating and deleting
  - Current layer
- Layer colors

- Setting and changing colors
  - Index color
  - True color
  - Color Books
- Layer Freeze/Thaw
- Layer On/Off
- Layer Lock/Unlock
- Linetypes
  - Loading and selecting
- Properties palette
- Match Properties
- Layers toolbar

REVIEW QUESTIONS

Answer the following based on what you learned in this chapter:

1. What is a layer?
2. What is the main reason for using layers in AutoCAD?
3. What typed command brings up the Layers dialog box?
4. List two ways to create a new layer.
5. How do you delete a layer?
6. How do you make a layer current?
7. What three Color tabs are available?
8. What does freezing a layer do?
9. What does locking a layer do?
10. What are some of the linetypes mentioned in the text?
11. What are three ways to bring up the Properties toolbar?
12. What two other ways to change properties or layers are discussed?

EXERCISES

1. **Step 1.** In a new file, load all linetypes, next open the Layers dialog box, and set up the following layers, colors, and linetypes. Then, lock all E-layers, make A-Walls-New current, and freeze A-Text. (Difficulty level: Easy; Time to completion: <5 minutes.)
   - A-Walls-Demo, Gray, Hidden
   - A-Walls-New, Green, Continuous
   - A-Windows-New, Red, Continuous
   - A-Doors-Demo, Gray, Hidden
   - A-Doors-New, Yellow, Continuous
   - A-Shelves, Blue, Hidden
   - A-Furniture, Magenta, Continuous
   - A-Text, Cyan, Continuous
   - A-Dims, Cyan, Continuous
   - A-Carpeting, Gray, Continuous
   - E-Outlets-Duplex, Red, Continuous
   - E-Switches, Red, Continuous
   - E-Switches-3 way, Red, Continuous

   Step 2. For practice: Lock all E-layers, make A-Walls-New current, and freeze A-Text. (Difficulty level: Easy; Time to completion: 1–2 minutes.)

2. In a new file, set units as Decimal; load all linetypes; set up layers M-Part, Hidden, and Center; and assign them proper linetypes and colors of your choosing. Then, draw the following mechanical object, including all hidden and centerlines. Align the front and side views as shown. NOTE: All text (CL stands for centerline) and dimensions are for your reference only; we learn how to create them in Chapter 4, Text, Mtext, Editing, and Style and Chapter 6, Dimensions, respectively, and you need not do either for now. (Difficulty level: Easy; Time to completion: 10–15 minutes.)

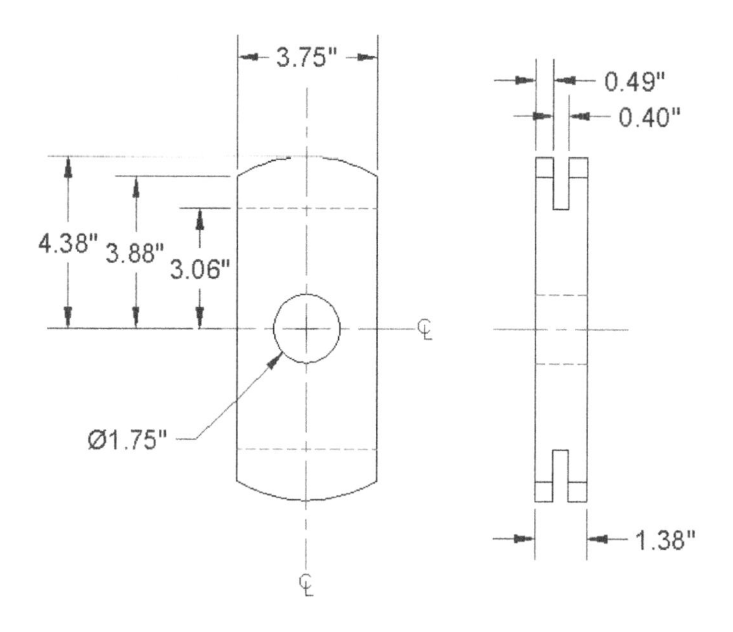

3. In a new file, set units as Decimal; load all linetypes; set up layers M-Part and Hidden, and assign them proper linetypes and colors of your choosing. Then, draw the following mechanical object, including all hidden lines. Align the front and side views as shown. NOTE: All dimensions are for your reference only. (Difficulty level: Easy; Time to completion: 10–15 minutes.)

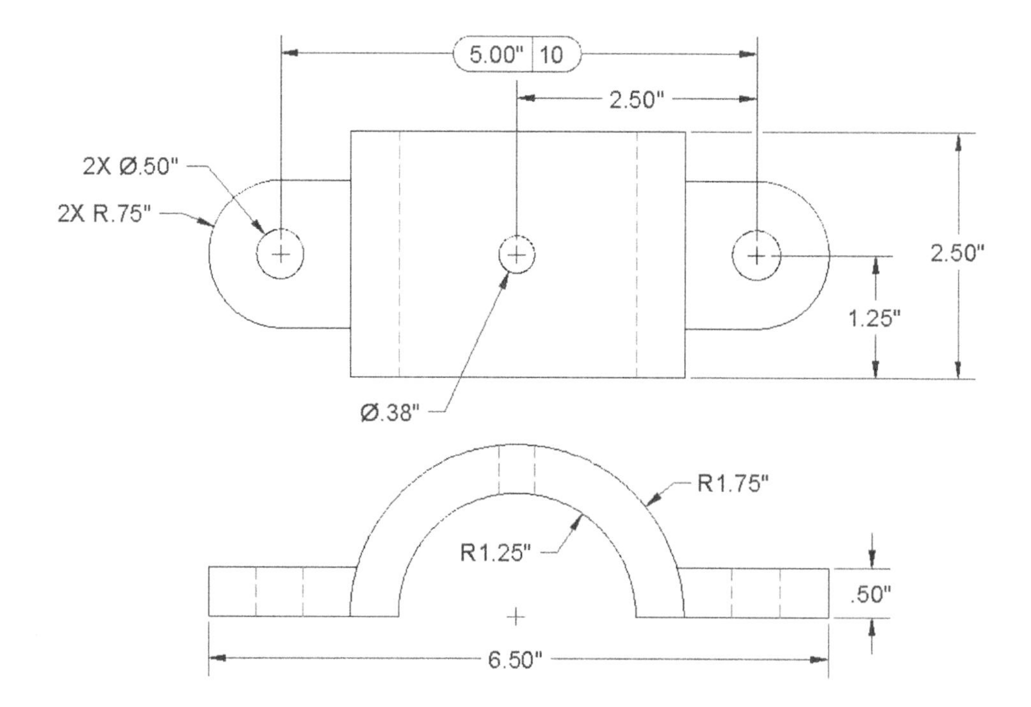

4. In a new file, set units as Decimal; load all linetypes; set up layers M-Part, Hidden, and Center; and assign them proper linetypes and colors of your choosing. Then, draw the following mechanical object, including all hidden and centerlines. NOTE: All dimensions are for your reference only. (Difficulty level: Easy; Time to completion: 15–20 minutes.)

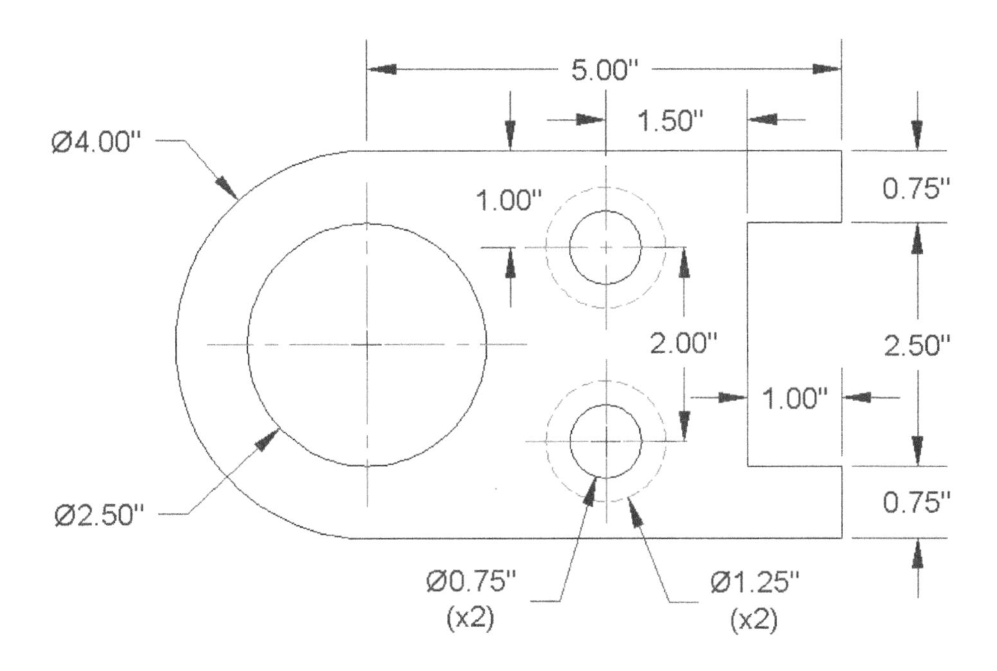

5. In a new file, set units as Decimal; load all linetypes; set up layers M-Part, Hidden, and Center; and assign them proper linetypes and colors of your choosing. Then, draw the following mechanical object, including all hidden and centerlines. NOTE: All dimensions are for your reference only. (Difficulty level: Easy; Time to completion: 15–20 minutes.)

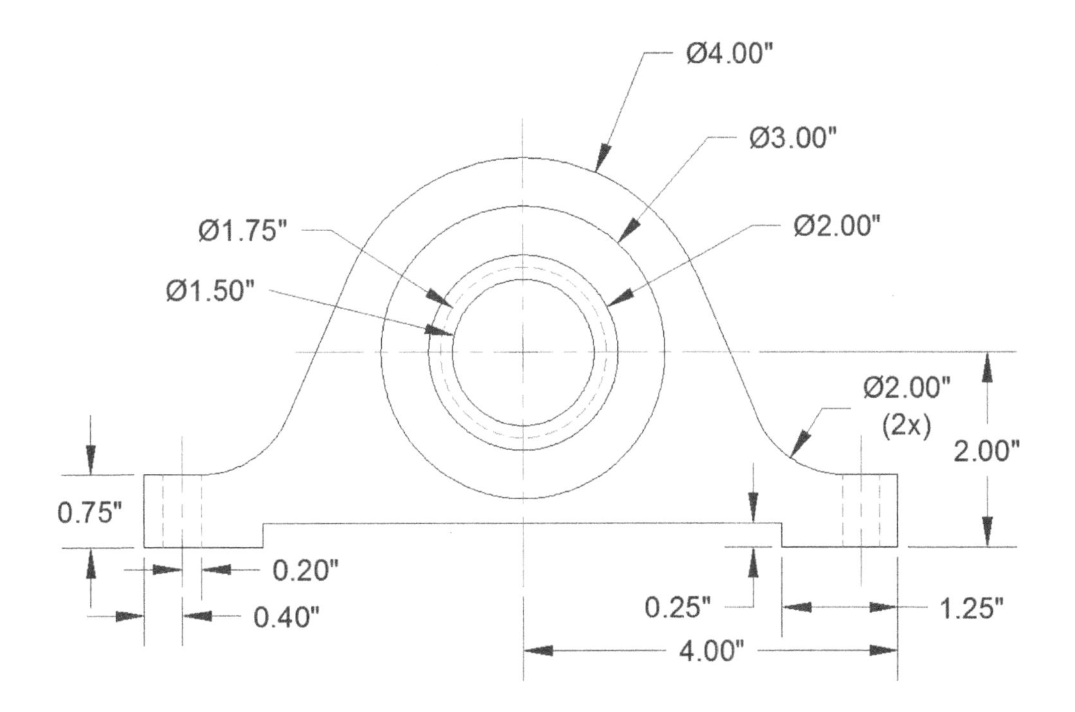

6. In a new file, set units as Decimal; load all linetypes; set up layers M-Part, Hidden, and Center; and assign them proper linetypes and colors of your choosing. Then, draw the following mechanical object, including all hidden and centerlines. Align the front and side views as shown. NOTE: All dimensions are for your reference only. (Difficulty level: Intermediate; Time to completion: 20–25 minutes.)

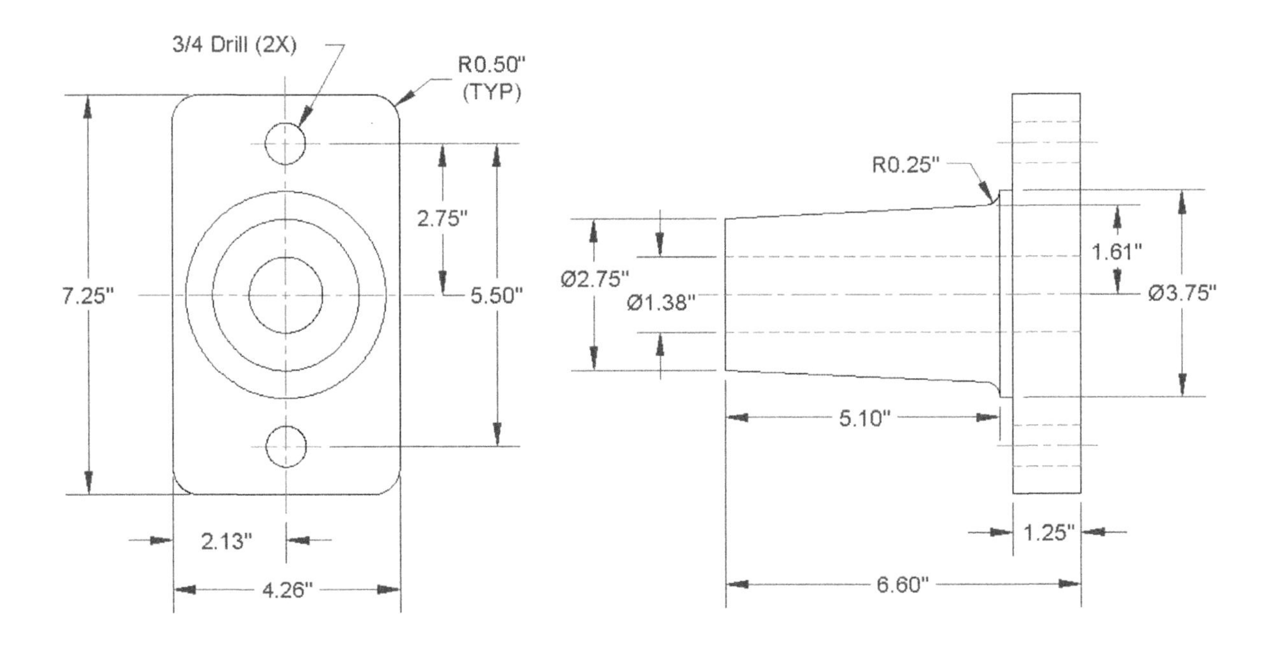

7. Often, in design work, you need to draw standard parts because no CAD files are available. Such is the case with the following electrical termination blocks, breakers, and thermostat equipment. You are provided minimal dimensions, and your job is not to design the part but simply to document its appearance so an engineer can insert it into a larger modular design. In a new file, leave units as Decimal and set up layer E-Part, assigning it any color you wish. Then, draw the electrical hardware, sizing everything relative to the overall provided dimensions and estimating most of the internal geometry. This is an excellent exercise in basic shape drafting. Be sure to use the copy command wisely to avoid significant duplication of drawing effort. (Difficulty level: Intermediate; Time to completion: 45 minutes.)

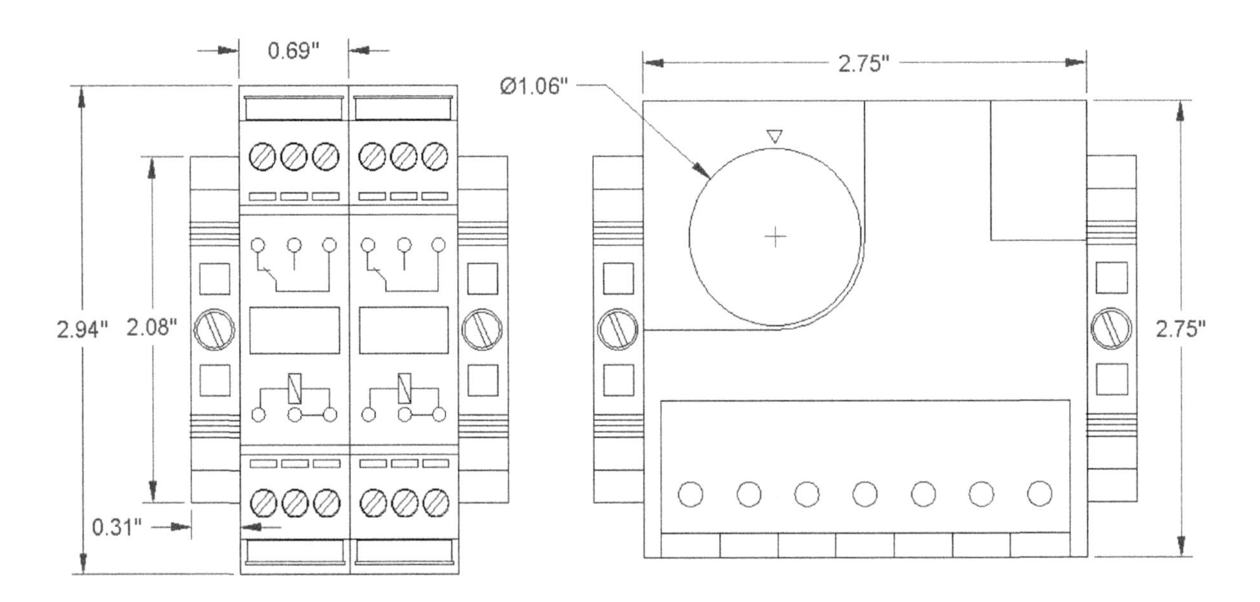

8. The next three exercises will all be Architectural in nature. In a new file, set up Architectural units, the correct layers (only A-Walls and A-Doors are needed), and draw the following floor plan. NOTE: Dimensions, text, and the solid hatch pattern are for clarity and construction purposes only. We learn them in upcoming chapters. (Difficulty level: Intermediate; Time to completion: 45—60 minutes.)

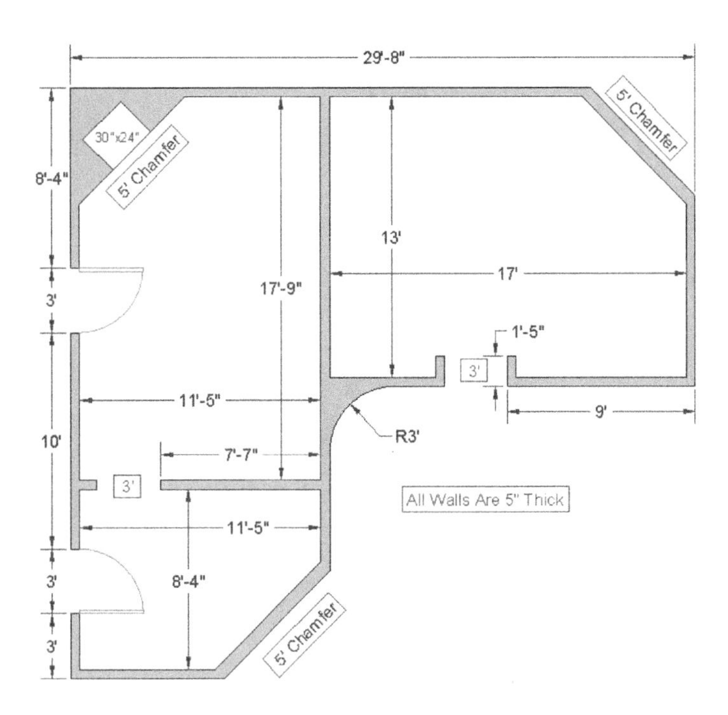

9. In a new file, set up Architectural units, the correct layers (A-Walls, A-Doors, and A-Windows), and draw the following floor plan. NOTE: Dimensions, text, and the solid hatch pattern are for clarity and construction purposes only; we learn them in upcoming chapters. (Difficulty level: Intermediate; Time to completion: 60—90 minutes.)

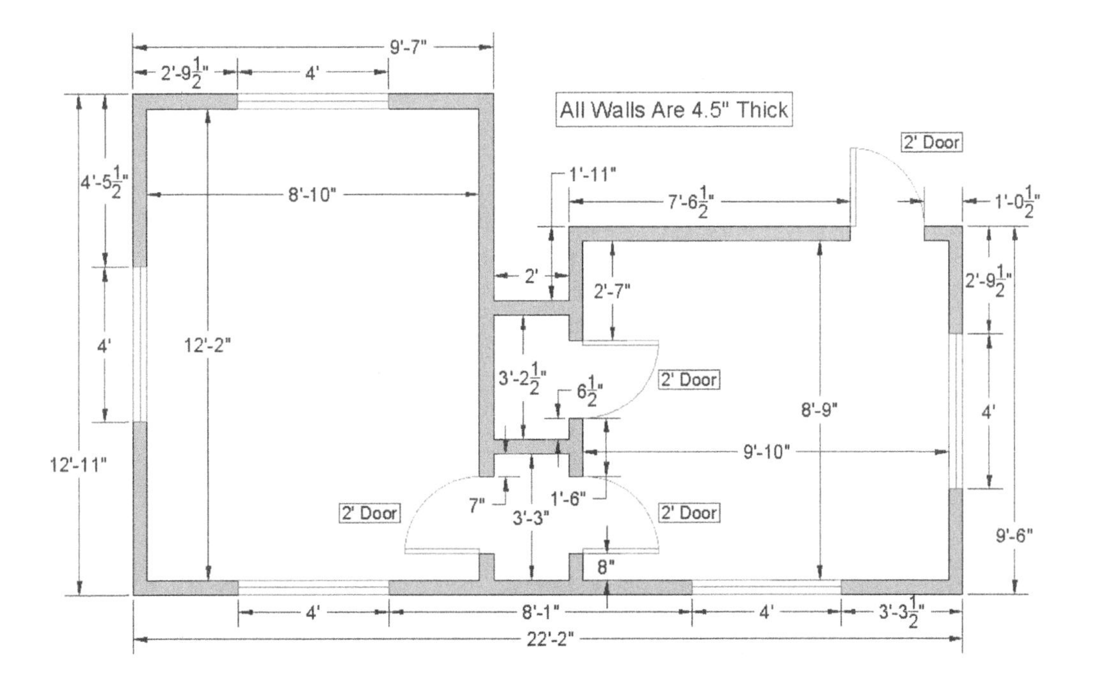

10. In a new file, set up Architectural units, the correct layers (A-Walls, A-Doors, A-Windows, and A-Furniture), and draw the following floor plan including the furniture. NOTE: Dimensions, text, and the solid hatch pattern are for clarity and construction purposes only; we learn them in upcoming chapters. Those dimensions not expressly given (such as with the furniture, kitchen, and bathroom elements and other minor details of the floor plan itself) can be assumed and free-drawn to appropriate size. (Difficulty level: Intermediate; Time to completion: 60–90 minutes.)

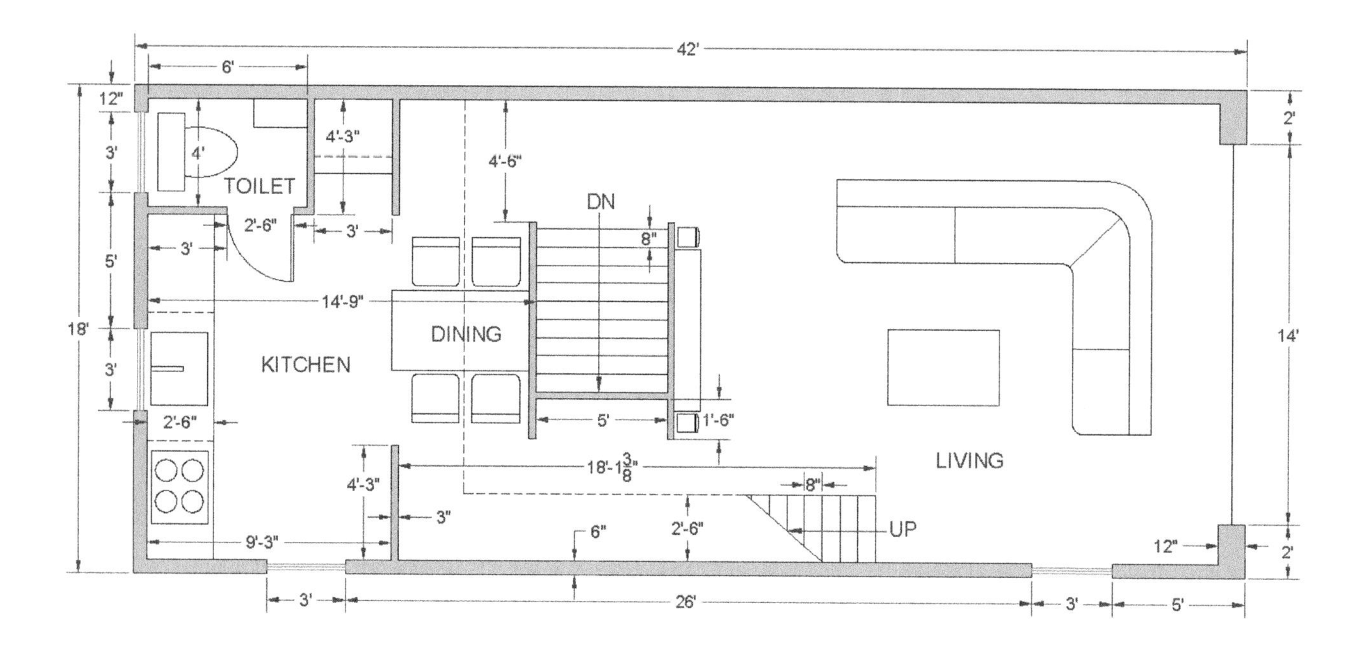

Chapter 4

Text, Mtext, Editing, and Style

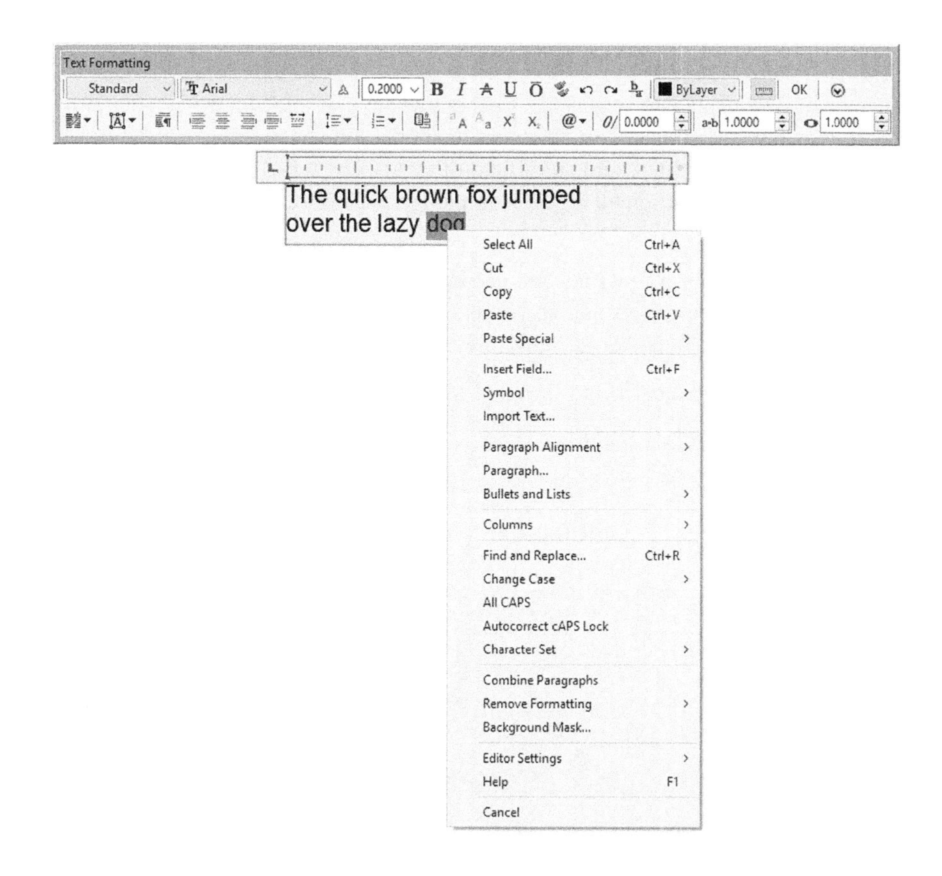

Learning Objectives

Text allows your design to communicate beyond what just a line drawing can. In this chapter, we introduce text, mtext, style, and editing and discuss the following topics:

- Text
- Properties and applications of text
- Editing all types of text
- Mtext
- Properties and applications of mtext
- Mtext formatting
- Mtext symbols
- Style
- Spell check
- Nearest OSNAP

At the end of this chapter you will be able to annotate your floor plan as well as add additional features, such as furniture and stairs.

Estimated time for completion of chapter: 2 hours (lesson and project).

Up and Running with AutoCAD 2019. https://doi.org/10.1016/B978-0-12-816440-2.00004-5

4.1 INTRODUCTION TO TEXT AND MTEXT

Adding text to your AutoCAD drawing is the next logical step once you learn how to create and edit basic designs. After all, the purpose of most drawings is to describe how to build something or show what a design looks like. Text goes a long way in assisting in this and, along with dimensions, is usually the next item to be added to a new design.

Text in AutoCAD comes in two versions: regular text (single-line text) and mtext (multiline text). The two share some overlaps and similarities, and we look at how to create and edit both types. We then look at how to choose and set fonts and conclude the chapter by adding text to the previously designed floor plan.

4.2 TEXT

This is your basic text creation command, and it creates a field anywhere you click on the screen, into which you can type whatever text you need. It does have a carriage return that goes to another line on pressing Enter, so you need to press Enter twice to get out of the field. While an unlimited number of lines of text can be typed, typically regular text is used for only a single line, as multiple lines of text are not joined together in paragraph form and cannot be formatted to any significant extent. Let us summarize:

- Text is used mostly when only one or a few lines are needed.
- The text cannot be formatted nor any effects added, beyond underlining and a few other minor effects.
- Multiple lines of text are not in paragraph form, rather they remain as individual lines.

Open a new file, bring up the Text toolbar (Fig. 4.1), and let us create a few sample lines.

FIGURE 4.1 Text toolbar.

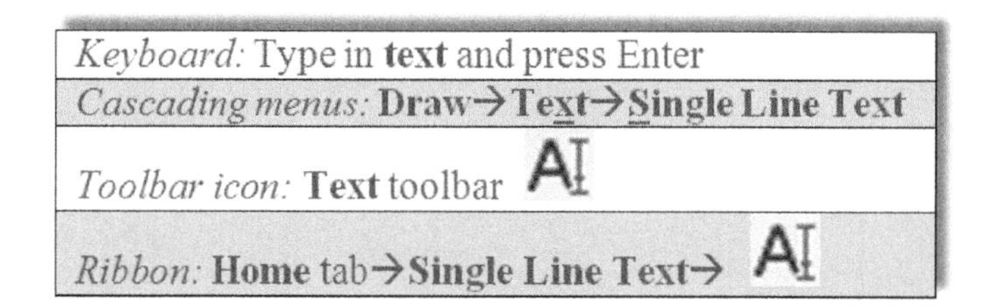

Keyboard: Type in **text** and press Enter
Cascading menus: **Draw→Text→Single Line Text**
Toolbar icon: **Text** toolbar
Ribbon: **Home** tab→**Single Line Text→**

Step 1. Begin the text command via any of the previous methods.
- AutoCAD says:

Current text style: "Standard" Text height: 0.2000
`Annotative: No Justify: Left`
`Specify start point of text or [Justify/Style]:`

Step 2. Left-click anywhere on the screen.
- AutoCAD says: `Specify height <0.2000>:`

Step 3. Enter a new height if desired, say 1.0, and press Enter.
- AutoCAD says: `Specify rotation angle of text <0>:`

You can rotate the text, but there really is no need to, so just type in `0` for now, and press Enter.

Step 4. A text field opens up with a blinking cursor. Go ahead and type something, pressing Enter once to go to the next line or twice to finish. Fig. 4.2 shows what the basic text looks like hanging out in the middle of a blank screen (just before you press Enter one last time). If the text is too large to see completely or too small to be even readable, just zoom in or out to fit it nicely on the screen. You can zoom and pan even while you are typing.

Sample text on three
lines with the word
"text" highlighted

FIGURE 4.2 Basic text.

The text is still rather simple; we have not yet addressed font, sizing, or any effects. Notice that, once you complete the command, the text is on three separate lines. You can edit, move, copy, or erase any of the lines without affecting the others. So how would you edit this text?

Editing Text

Editing, or changing, what the text says is very simple and follows the same procedure for both text and mtext, so we go over it right away for both cases. The easiest way to edit the text is to *double-click* on it, which opens both the text and mtext fields. This is such an overriding and best method, preferred by most students, that in this particular case you do not see an official command matrix detailing the other methods, just a brief mention of them.

A toolbar icon Edit… does exist on the Text toolbar for editing purposes. Finally, it is worth noting that an older typed-in command also exists, called ddedit. If you type that in, press Enter, and click on the text, you can edit it. This may come in handy with older versions of AutoCAD in cases of editing dimension values. A few releases ago, you could not edit them via a double-click. This was instead used for properties of the dimensions and not the text values, so ddedit came in handy.

Whichever method you use, when you are done editing, just press Enter again. Do not press Esc, as that cancels whatever editing you just did.

Here is a trick with regular text that you will not hear about much anymore. If you want to underline it, type in `%%u` just before the text string you want underlined. This old code dates back to DOS days, and it will automatically and immediately add an underline to your basic text. You can also experiment with %%c, %%p, and %%d. See what those three do. None of this is needed with mtext, as discussed next.

4.3 MTEXT

Mtext is short for *multiline* text. In other words, this is the command you use when you anticipate typing in a paragraph as opposed to one or two lines and, more important, when you require some advanced formatting and effects. As an added bonus, it also has a spell checker built in. Many designers use mtext for all their text needs, just in case formatting needs to be applied to even just one word. The only downsides are that it takes a few seconds longer to set up and the features make it a more complex command. The idea here is to define an area where your text will go. Once you do that, the new paragraph fits into that area. Let us summarize the mtext features and try the command:

- Used primarily for writing paragraphs
- Has extensive formatting and editing features
- Has an extensive symbol library and spell check
- Can accept extensive text importing

A few releases ago the mtext command was enhanced even further and has inherited more features from a dedicated word-processing program, such as MS Word, including auto bulleting, better auto stacking of fractions, and other useful tricks.

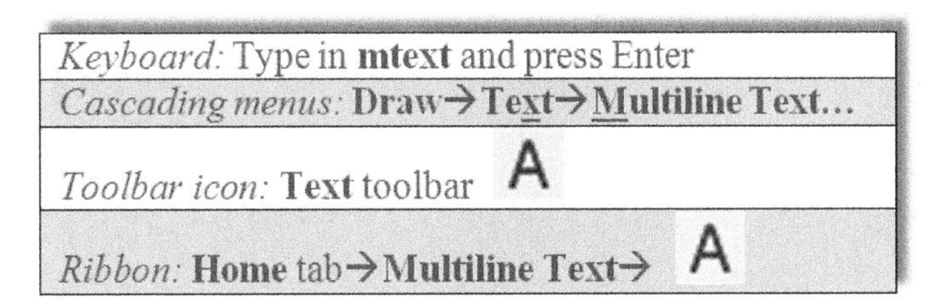

Keyboard: Type in **mtext** and press Enter
Cascading menus: **Draw→Text→Multiline Text…**
Toolbar icon: **Text** toolbar
Ribbon: **Home** tab→**Multiline Text→**

Step 1. Begin the mtext command via any of the previous methods.

- AutoCAD says: `Current text style: "Standard" Text height: 0.2000 Annotative: No`
  `Specify first corner`

Step 2. Left-click anywhere on the screen, noticing the abc next to the crosshairs. A rectangle with an arrow appears, as shown in Fig. 4.3. Continue to move your mouse down and across to the right, making the rectangle bigger. This is your text field; make it as large or small as you need it to be to hold all the text. As you do this,

- AutoCAD says: `Specify opposite corner or [Height/Justify/Line spacing/Rotation/Style/Width/Columns]:`

Step 3. Click again when you have defined the field.

Here, AutoCAD throws you a bit of a twist. If you are *not* using the Ribbon, you will see the text field with a toolset just above it (but not attached to it), as seen in Fig. 4.4, where you can type in your text, press Enter to go to the next line. To finish up, press OK in the upper right of the toolset or just click anywhere outside the field. We go over some of the tools in a moment, but what if you have the Ribbon up?

If you are using the Ribbon, the formatting duties are transferred completely to the Text Editor tab, and all you see below it is the text field, no toolset, as seen in Fig. 4.5 (although you can bring it back, via the Text Editor's Options category). The functionality is essentially the same, only the presentation differs. We cover mtext's extensive formatting tools in the next section.

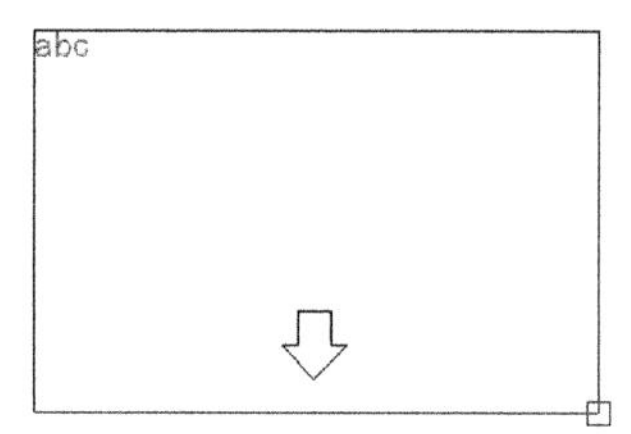

FIGURE 4.3 Mtext field.

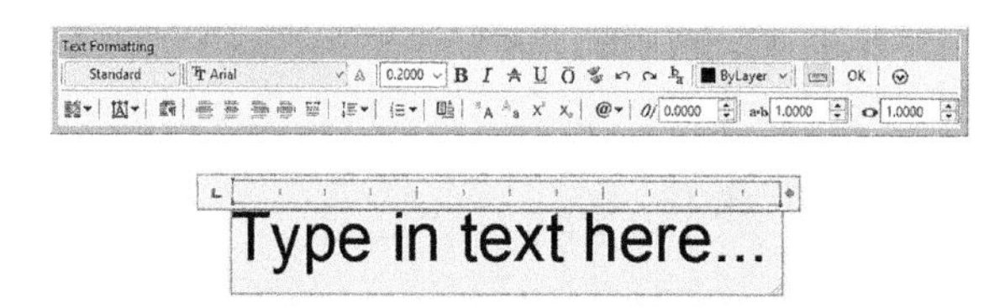

FIGURE 4.4 Text Formatting (no Ribbon).

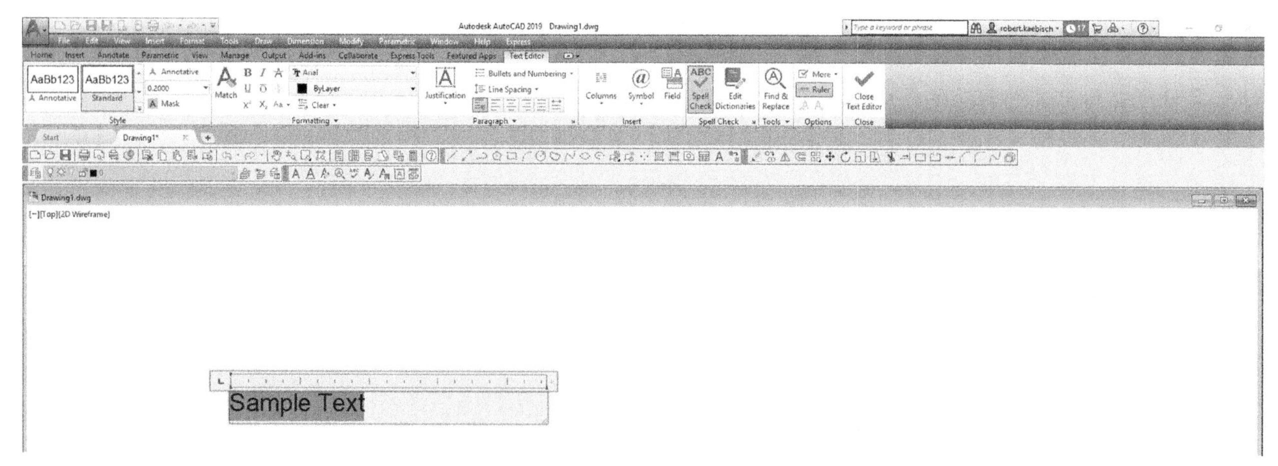

FIGURE 4.5 Text formatting via the Ribbon's Text Editor.

Formatting Mtext

First of all, notice something: If you click on mtext, it remains a paragraph, a big difference from a few lines of regular text. To edit mtext, as already mentioned, you do the same thing as for regular text, that is, double-click on it, use the icon, or type in ddedit.

Now, on to formatting. Mtext has a lot of additional features, and once you input the text, you can modify it significantly. Some intent was made here to mimic MS Word's text editing abilities, and although AutoCAD cannot really come close to a dedicated word-processing program, the available tools are still quite extensive. We take a look at the non-Ribbon toolset first and cover the tools again, this time using the Ribbon's Text Editor (in much less detail, as the tools essentially are the same). Looking at the Text Formatting toolbar from left to right, as shown in Fig. 4.6, we have the following groups:

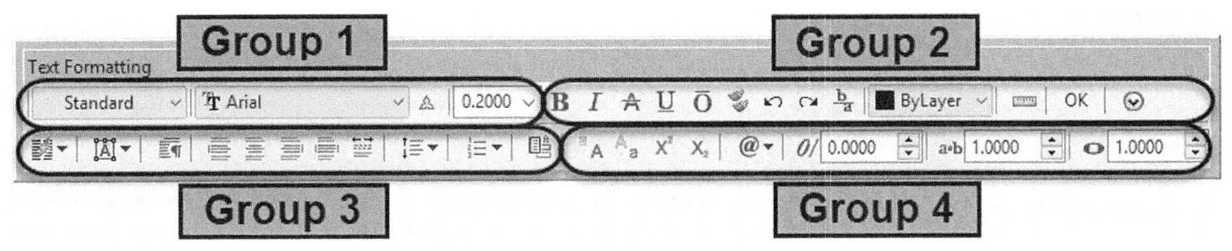

FIGURE 4.6 Text Formatting groups.

Group 1. This is where you set *name*, *font*, and *size*, although this is generally done using *style*, to be discussed later. You can also make the text annotative, to be covered in level 2.

Group 2. These are the standard *Bold*, *Italic*, *Strikeout*, *Underline*, *Overline*, and *Undo/Redo* buttons. This is also where you set *color*, although that is usually done by layer. The Ruler icon turns off the ruler grid; the OK button closes the field; and the down arrow brings up additional options, to be covered soon.

Group 3. Here you find a tool for making *columns* and some of the associated settings, *paragraph justification* tools, as well as tools to change *line spacing*, *numbering*, and *fields*.

Group 4. Here you find tools to change from *upper to lower case* (and vice versa), *superscript* and *subscript* (introduced in AutoCAD 2015), *oblique angle*, *tracking*, and *width factor tools*. The @ character is used to bring up additional symbols (to be discussed shortly).

You may be familiar with most of these if you ever typed a document in MS Word. Just as in Word, you need to highlight the text to which you want the changes to apply. Experiment with the buttons to see what they all do, and try to duplicate what you see in Fig. 4.7. Notice also how Arial is the default font in AutoCAD.

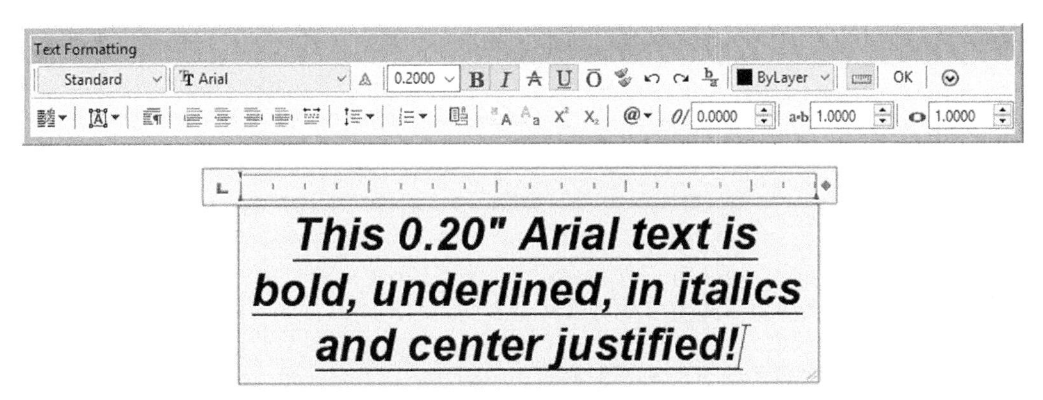

FIGURE 4.7 Mtext sample.

There is more to the mtext command than just the main toolbar detailed in Fig. 4.6, although the toolbar contains most of the often-used commands. Two additional menus are accessed by either right-clicking while inside the mtext box or pressing the @ symbol icon in Group 4. We cover this @ menu first; it introduces a vast array of available symbols.

Fig. 4.8 shows what this menu looks like. Examine it closely; it has interesting and useful options. In architecture and engineering, one typically encounters many industry- or trade-specific symbols. AutoCAD provides a significant database of them for your use. Browse through and try a few by clicking and making them appear in your text field.

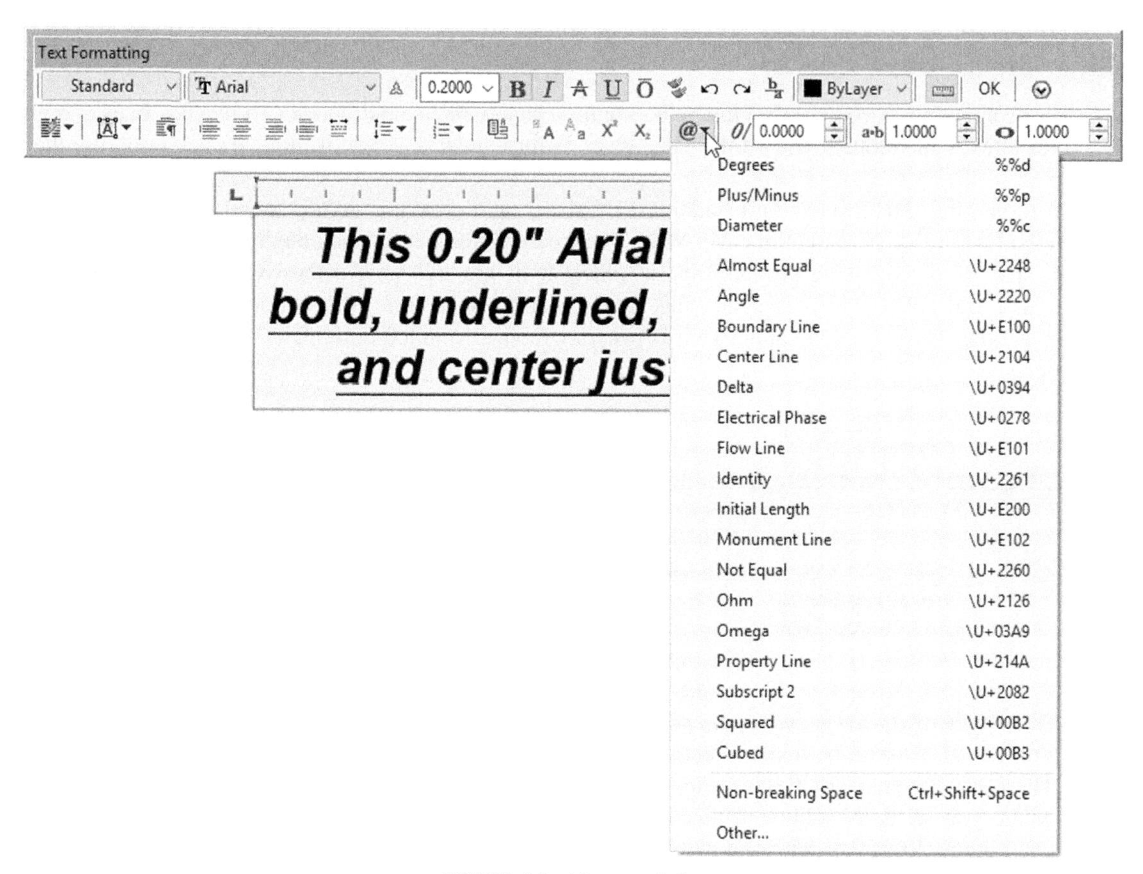

FIGURE 4.8 Mtext symbols.

In case none of those symbols is what you are looking for, there are more. At the bottom of the menu, click on Other… A *character map* appears, as shown in Fig. 4.9.

It is similar to the character map used by MS Word and functions the same way. This is basically a database of available letters (in English and several other languages) as well as symbols and characters. The choices change somewhat from font to font. To use this tool, simply click on a symbol you want (it will temporarily increase in size so you can examine it closer), and press the Select button. Then, minimize the Character Map box (or close it if you do not need it anymore), and right-click to paste the symbol into your mtext field.

Character maps are generally for symbols not found in the main list of the drop-down table (Fig. 4.8). If you see the symbol you want there first, use it. To introduce the final menu, right-click in the mtext field and you see the menu in Fig. 4.10.

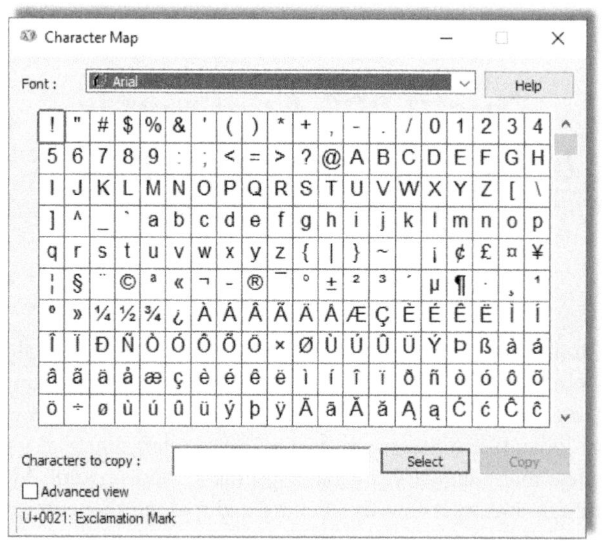

FIGURE 4.9 Character map.

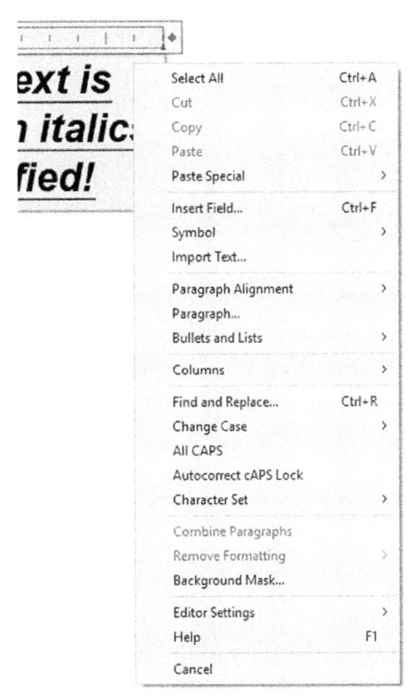

FIGURE 4.10 Mtext right-click menu.

Some of these menu choices duplicate what is already seen in the main mtext menu bar, and you can access all the symbols through here as well. Aside from some Cut and Paste choices, this menu just presents an alternate way of making option selections. As a final note, be aware that pressing Esc will exit you out of the Mtext window, and you need to save if you want to keep what you did. AutoCAD, of course, prompts you to do this.

Ribbon Text Editing

As mentioned before, Ribbon users are presented with essentially the same formatting tools but in a rearranged manner. Fig. 4.11 is the Ribbon's Text Editor, again with the various drop-down menus "thumbtacked" (by clicking on the tack symbol) on the drawing canvas.

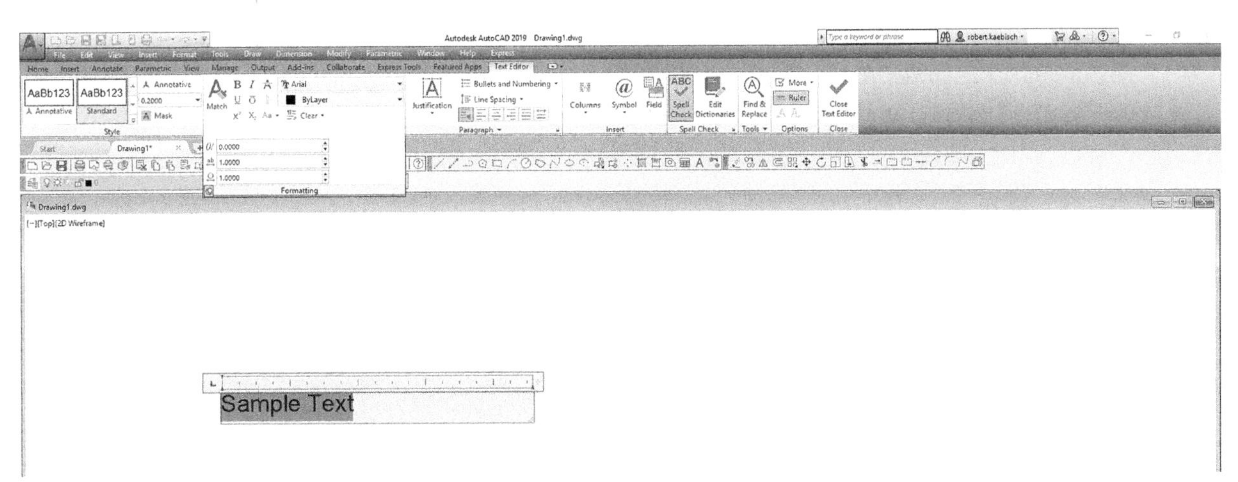

FIGURE 4.11 Ribbon Text Editor, thumbtacked.

The Ribbon's Text Editor categories are detailed next.

- *Style*: Annotation and size can be changed here.
- *Formatting*: Bold, italics, underline, overline, font, color, background mask (color), oblique, spacing, and width factor can be changed here.

- *Paragraph*: Justification, line spacing, bullets, and numbering can be changed here.
- *Insert*: Columns, symbols, and fields can be changed here.
- *Spell Check*: You can run the spell check and set dictionaries and settings here.
- *Tools*: Find and Replace, Import Text, and AutoCAPS are found here.
- *Options*: Here, you can make the Editor toolbar come back (AutoCAD never really gets rid of anything), add the ruler grid, and use Undo/Redo. You can also change the background of the Editor to opaque.
- *Close*: Closes the Editor, similar to the OK button.

Finally, before we move on to discussing style, a new feature needs to be mentioned that was just added recently. You can now put your mtext into a box frame, which was one of the top requests from past AutoCAD users. To do this, call up the Properties palette (first seen in Fig. 3.10). Then click your mtext one time to select it, which will make the Properties palette display four categories. Scroll to the bottom of the Text category and find Text frame on the left. Across from it will be a Yes/No drop-down menu. Select Yes and you are done. Your mtext will have an automatic frame around it, a very useful feature.

4.4 STYLE

The idea behind the style command is very straightforward. Pick a font, give it a name and a size, and use it throughout your drawing. Drawings typically use only one font throughout the main design and perhaps another, fancier one in the title block area for logos and other designations, so it makes sense to set one style and stick to it. You already may have changed the font while learning the mtext command, but it was only for *that* instance. You need to make sure that font is set globally, meaning for the entire drawing.

With AutoCAD 2019 and the previous 10 releases, the default font is Arial (the default size is 0.2), so you may want to stay with this popular font and not change it. If you need to, however, here is the procedure:

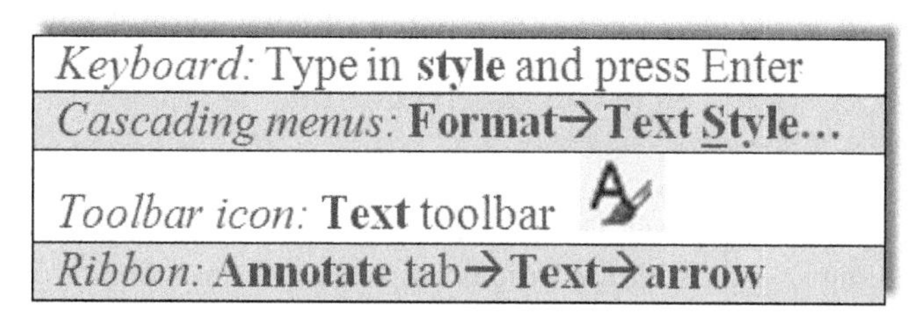

Keyboard: Type in **style** and press Enter
Cascading menus: **Format→Text Style...**
Toolbar icon: **Text** toolbar
Ribbon: **Annotate** tab→**Text→arrow**

Step 1. Begin the style command via any of the preceding methods. Whichever method you use, the Text Style dialog box of Fig. 4.12 appears. Taking a look at this dialog box, let us say Arial is not what we want here; we would like to set a Helvetica font that is 6″ high instead.

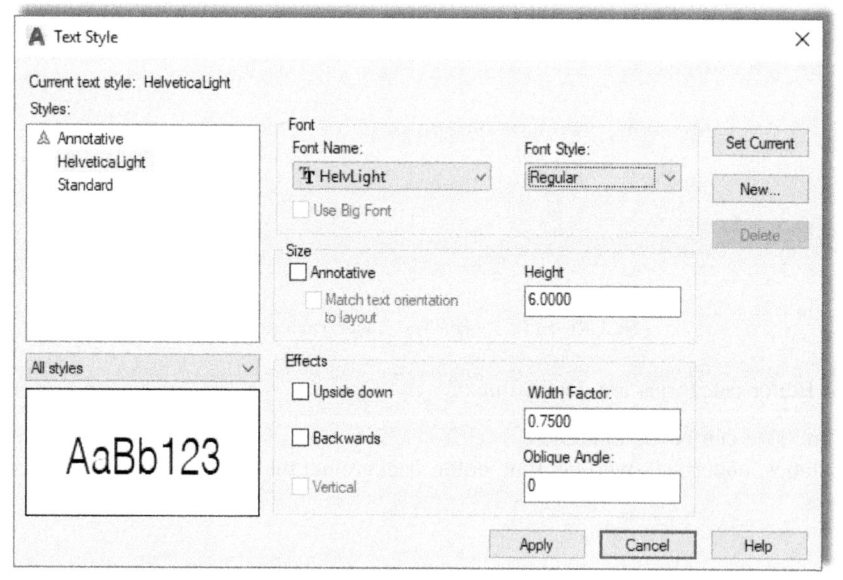

FIGURE 4.12 Text Style dialog box.

Step 2. Press the New... button and type in the name and size of the font: HelveticaLight 6.

Step 3. Pick HelvLight from the Font Name: drop-down menu.

Step 4. Highlight the 0.0000 in the Height field and type in 6 (no need for the inch symbol). Press Apply and Close.

All text that you type from now on will be HelvLight, 6″ font. To create another font just repeat these steps. To size your font up and down, you can also create new font styles, but in this case, it may be more practical to just use the scale command. (Note: If you leave the font height set to 0, the font can be used at any height when using the text or mtext commands.) The additional options in the style dialog box are not used that often, but review them just in case. A list follows. All effects can be previewed in real time in the Preview box in the lower right.

- *Upside down*: Flips the text upside down.
- *Backwards*: Flips the text backwards.
- *Vertical*: Stacks the text vertically.
- *Width Factor*: Widens the text if (>1), narrows text if (<1).
- *Annotative*: An advanced topic for Chapter 10.
- *Oblique Angle*: Leans the text to the right if positive (+) and to the left if negative (−).

4.5 SPELL CHECK

This section briefly describes AutoCAD's in-place, stand-alone spell checking abilities. Yes, AutoCAD does have spell check, and why not? For the price the software costs, it had better. It is very easy to use, so let us try it out on the misspelled phrase created via regular Text, as shown in Fig. 4.13.

The quick brown fox jmpded over the lazy dog

FIGURE 4.13 Misspelled word.

The first thing you notice is that the misspelled word (jumped) is underlined by a dotted line. If you right-click on the word while in editing mode, you get the in-place spell check with suggestions (Fig. 4.14). Simply select the word you want to correct, and AutoCAD obliges.

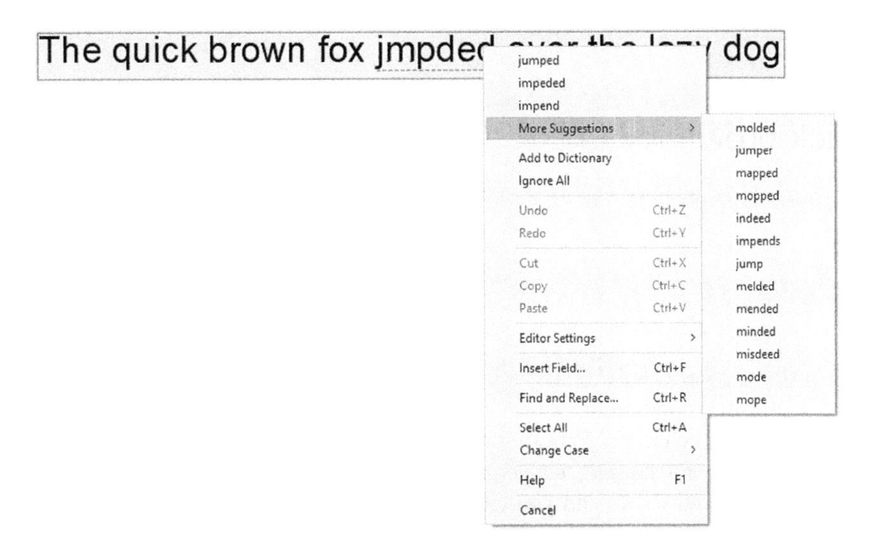

FIGURE 4.14 In-place Spell Check.

Now, what if you already have a significant amount of text (perhaps from an import) and need to spell check it? Then, you can use the stand-alone utility, as described next.

Keyboard: Type in **spell** and press Enter
Cascading menus: **Tools→Spelling**
Toolbar icon: **Text** toolbar
Ribbon: **Annotate** tab→**Check Spelling**

Using any of these methods, the Check Spelling dialog box appears, as seen in Fig. 4.15.

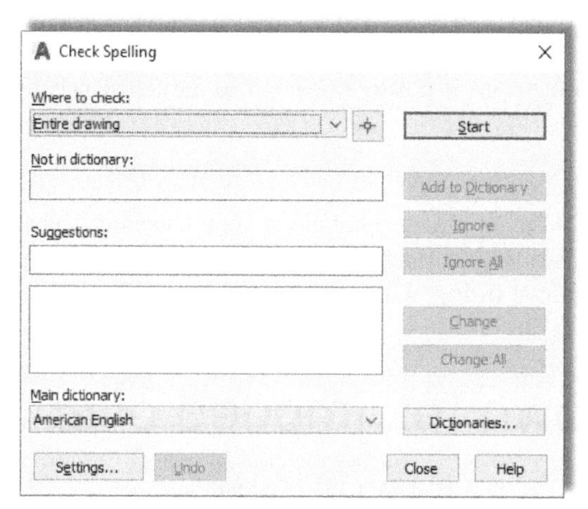

FIGURE 4.15 Check Spelling.

Using the drop-down menu of the Check Spelling dialog box, you can select specific paragraphs or lines of text or just press Start to initiate spell checker for all the text in the drawing. In any case, Check Spelling appears, as seen in Fig. 4.16.

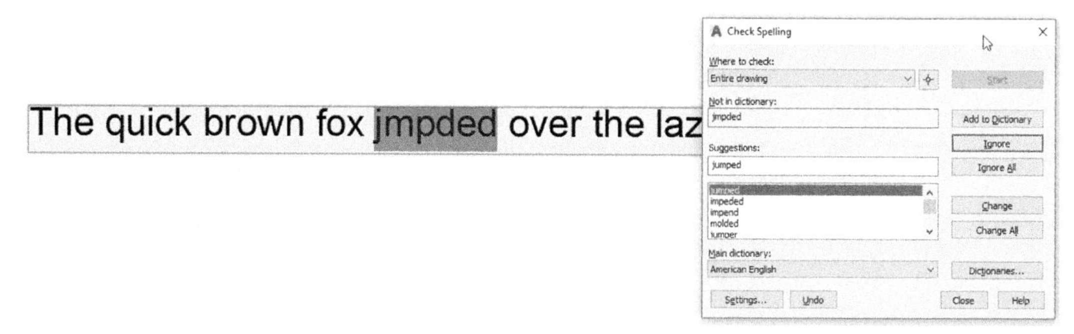

FIGURE 4.16 Check Spelling initiated.

Much as with the MS Word spell checker, you can run through some options, such as ignoring the word, changing it, or adding it to the dictionary so it is not flagged again (good for names and acronyms). Even a variety of languages are available. When done, spell check tells you, and you can close it.

4.6 IN-CLASS DRAWING PROJECT: ADDING TEXT AND FURNITURE TO FLOOR PLAN LAYOUT

Let us now apply what you learned to our floor plan. Shown in Fig. 4.17 is the same floor plan you worked on in Chapter 3, with the addition of text in the rooms, closet shelving (the dashed lines), new furniture, and appliances.

Here are two general tips before you get started:

- Make your own layers and colors for each set of data, being sure the names are logical. For example, call kitchen appliances something along the lines of A-Appliances and so on.
- None of the furniture is drawn to scale, so approximate all the sizes. However, draft everything carefully, connecting all lines with OSNAP and using Ortho for straight lines. Although you in theory make blocks out of this furniture, we have not covered that topic yet, so this is just a basic drafting exercise. Most shapes are based on rectangles, with some fillets, arcs, and circles. All were done in the simplest manner possible, and you should be able to create them with basic techniques learned in previous chapters. Some hints are listed next.

Some additional specific drawing tips:

- All kitchen furniture on the left side of the kitchen is formed using basic rectangles, drawn to no particular size. At the top left is a refrigerator with a microwave on top. Next to it is a countertop and a sink. Next to that is the range top with four burners. Fig. 4.18 is a suggested procedure for creating the range top. In Chapter 6 we cover some new ways to create center marks, just introduced in AutoCAD 2017, but here is the classic approach.

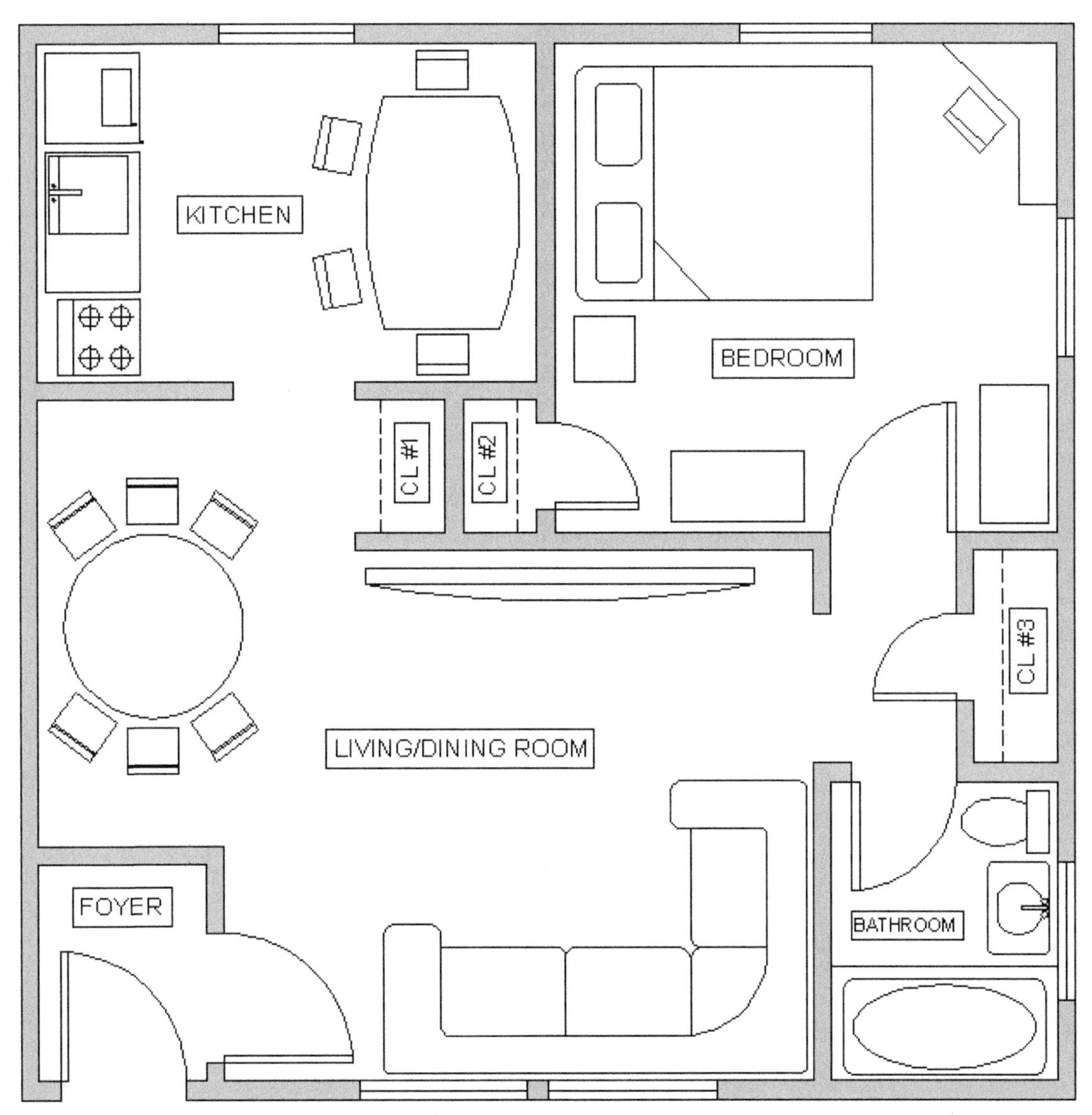

FIGURE 4.17 Floor Plan layout.

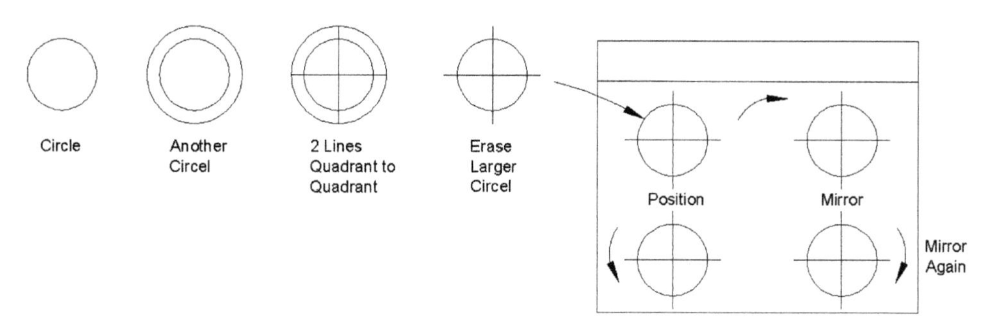

FIGURE 4.18 Constructing the range top.

- To create the kitchen table you could use a rectangle with a small straight guideline sticking out from the midpoint of the left side. That could be the anchor for the second point of the arc. Then, a mirror followed by an "explode" and "erase" finishes it off, as seen in Fig. 4.19.

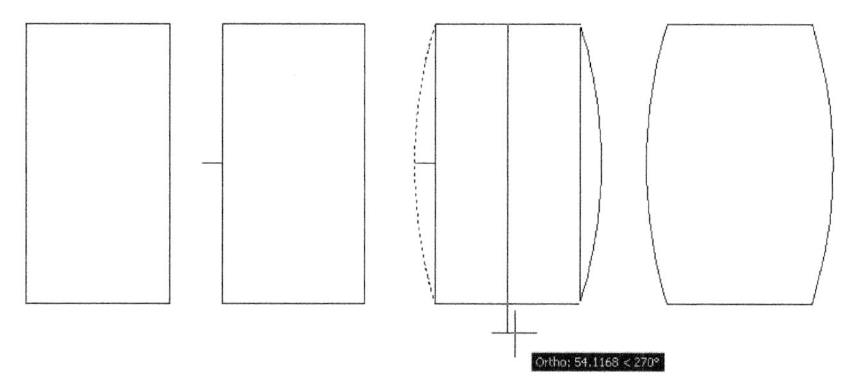

FIGURE 4.19 Constructing the table.

- Draw rectangles around each of the room text fields; we need this later for hatch.
- The bathtub and toilet are done using the basic ellipse command, introduced in Chapter 2.
- The corner desk in the bedroom, the pulled-back bed cover, and all closet shelves are constructed using lines and a new OSNAP point, the second (of three) that was not initially introduced in Chapter 1; a description follows.

NEArest OSNAP

You may occasionally need a new OSNAP point called *NEArest*. It begins or ends shapes (usually new lines) along random spots of other geometry (usually other lines). In a sense, NEArest is the exact opposite of the precise end, mid, and other points you learned. NEArest simply shadows a line or any object and allows you to begin a new object anywhere along that perimeter. We do not generally leave NEArest running along with the other OSNAP points, as it makes using them more difficult; rather, just type in `nea` and press Enter or use the toolbar as needed. An hourglass figure appears; just click on wherever you want to start a line. Fig. 4.20 is a screen shot of the NEArest OSNAP in action.

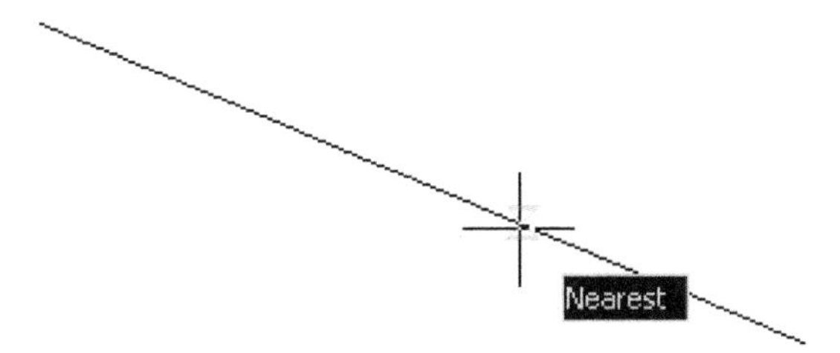

FIGURE 4.20 Nearest OSNAP.

Finally, we have two more tips. Tip 6 has to do with ltscale, something briefly mentioned in Chapter 3. Tip 7 shows you how to get rid of the UCS icon, something many students ask about.

TIP 6
If you draw your hidden lines for the closet shelving and they look solid not dashed, then the problem may be incorrect linetype scaling, a common error to watch out for. Type in ltscale, press Enter, and input a new value (a larger one because your floor plan is bigger than the default original space of 8.5″ × 11″). Do this a few times until you get it to look just right.

TIP 7
Want to get rid of that pesky Universal Coordinate System icon hanging out in the drawing area? It is somewhere near the lower left-hand corner of your screen, looks like a big L and has a Y and an X on its tips. To make it go away, type in ucsicon, press Enter, then type in off. Of course, if it does not bother you and you actually like it there, then ignore this whole tip.

SUMMARY

You should understand and know how to use the following concepts and commands before moving on to Chapter 5:

- Text
  - Creating basic text
- Mtext
  - Creating basic mtext
  - Mtext formatting options
    - Bold
    - Underline
    - Italics
    - Justification
    - Symbols
    - Other
- Editing text
  - Double-click
  - ddedit
- Style
  - Font name
  - Font type
  - Font size
  - Width factor
  - Oblique angle

- Spell check
- Nearest OSNAP

REVIEW QUESTIONS

Answer the following based on what you learned in this chapter:

1. What are the two types of text in AutoCAD? State some properties of each, and describe when you would likely use each type.
2. What is the easiest method to edit text in AutoCAD? What is another, typed method?
3. What feature allows you to add extensive symbols to your mtext?
4. What are the three essential features to input when using Style?
5. What is the width factor?
6. What is an oblique angle?
7. How do you initiate spell check?
8. What new OSNAP is introduced in this chapter?
9. How do you get rid of the UCS icon?
10. What can you do if you cannot see the dashes in a Hidden linetype?

EXERCISES

1. Creating tables of data, such as a Bill of Materials, is an important part of many designs. AutoCAD has a table creation tool (similar to MS Word), which we cover in Level 2, but for this exercise, you need to create one from scratch with basic linework and the offset command, populating it with text data, as shown. Open a blank file and create a new style called Helvetica Light 0.2; choose font: HelvLight, height: 0.2, and width factor: 0.95. Then, create the table shown, using the suggested sizing. Finally, fill in the data, using your choice of text or mtext commands. Position and size the text/mtext as needed. (Difficulty level: Easy; Time to completion: 20 minutes.)

10" (overall width); 5" (overall height); .5" (title row height); .4" (row height)

LIST OF MAIN PARTS				
ITEM	QTY	DESCRIPTION	SIZE DWG	DWG. #
1	1	SKID ASSEMBLY	A1	83006
2	1	STRAINER ASSEMBLY	A1	83008
3	1	GENERATOR ASSEMBLY	A1	59470-001
4	1	GENERATOR ASSEMBLY	D1	59470-002
5	2	POWER SUPPLY (OIL IMMERSED)	B1	58717-001
6	1	POWER SUPPLY (OIL IMMERSED)	B1	58717-002
7	1	LOCAL CONTROL PANEL ASSEMBLY	D1	83017-001
8	2	LOCAL CONTROL PANEL ASSEMBLY	A1	88017-002
9	1	INSTRUMENT AIR ASSEMBLY	A1	83022
10	1	AIR PUMP	A1	86584
1"	1"	5"	1"	2"

2. This next exercise is similar to the previous one. Using a combination of text and mtext, create the following table of tooling part numbers based on the starter dimensions given, including the text height. Exact text positioning is not specified and can be approximated. Use text styles similar to the previous example, but leave the width factor as 1.0 and adjust the height to the value shown. Create the fractions via mtext or by hand. (Difficulty level: Easy; Time to completion: 20 minutes.)

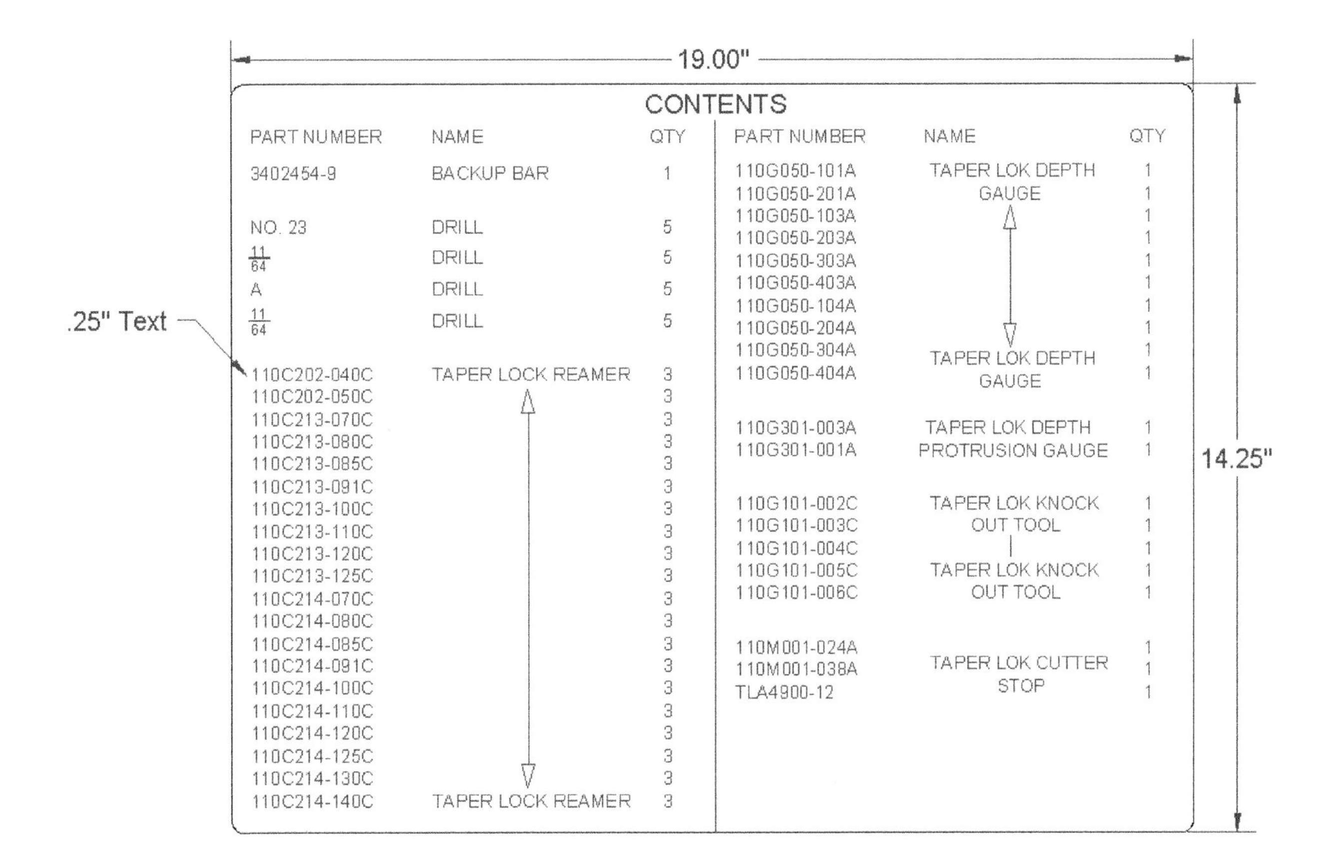

3. As seen with Exercise 7 of Chapter 3, Layers, Colors, Linetypes, and Properties (electrical pieces), you may be asked to draft something based on a picture and use arbitrary, not strict, design dimensions. You need to be fast and efficient while maintaining strict accuracy. A good grasp of the basics is essential, and you must know exactly how you will draft something in your mind, just moments before your fingers do it via mouse and keyboard. Such is the nature of this next exercise—to build up this accuracy and speed. Draw the following set of P&ID (piping and instrumentation diagram) symbols. They are not to scale and are approximated, but you have all of the necessary tools at your disposal. Before drafting each shape, recite to yourself the approach you will use, then execute it. Finally, add the text next to the respective symbols. Use the Arial font of appropriate size. (Difficulty level: Intermediate; Time to completion: 35–45 minutes.)

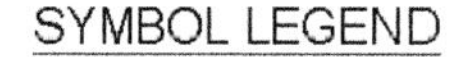

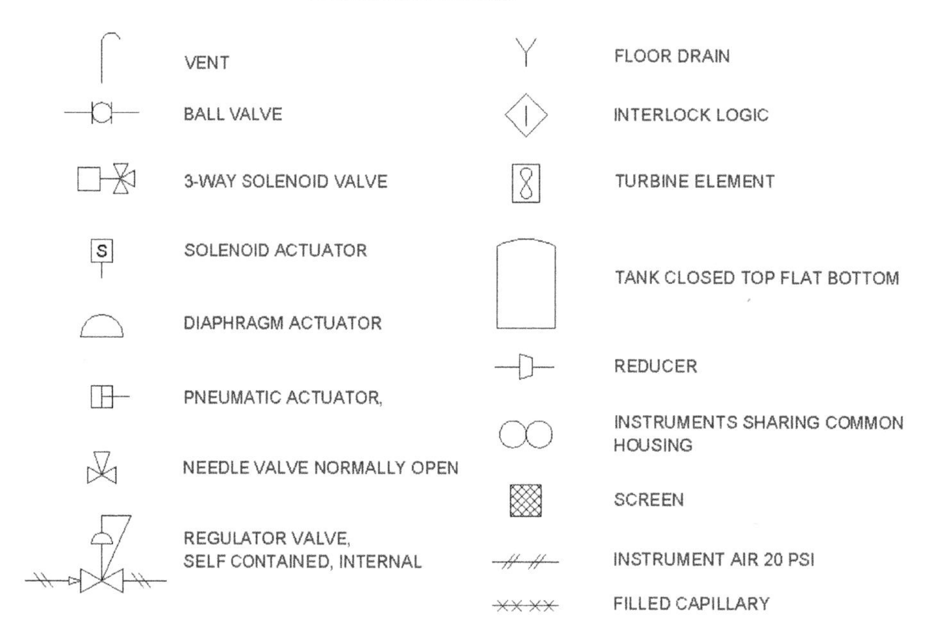

4. This exercise shows a simple P&ID schematic using some additional industry symbols. Draft it based on the reference dimensions given, with all internal pieces and text sizing estimated in relation to those. Use accuracy, connecting all lines with OSNAPs. (Difficulty level: Easy/Intermediate; Time to completion: 20–30 minutes.)

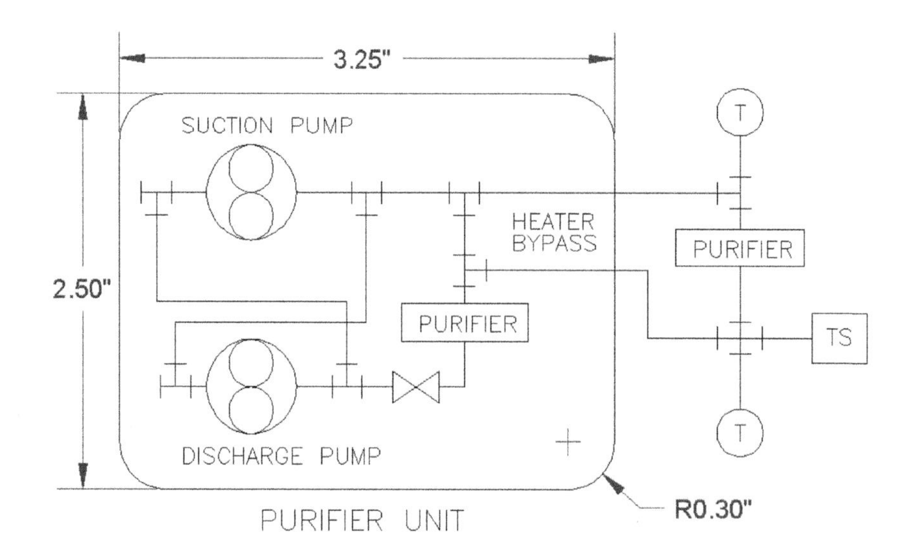

5. While we are on the subject of P&IDs, let us create a simplified design of a pipe. Open a new file and set up Architectural units, creating proper layers, colors, and linetypes (Hidden and Center), as shown. Then, draw the following pipe design. Be careful in how you approach the hidden lines, noting that they are on the same plane as the solid lines. This is commonly seen on mechanical drawings. (Difficulty level: Intermediate; Time to completion: 20–30 minutes.)

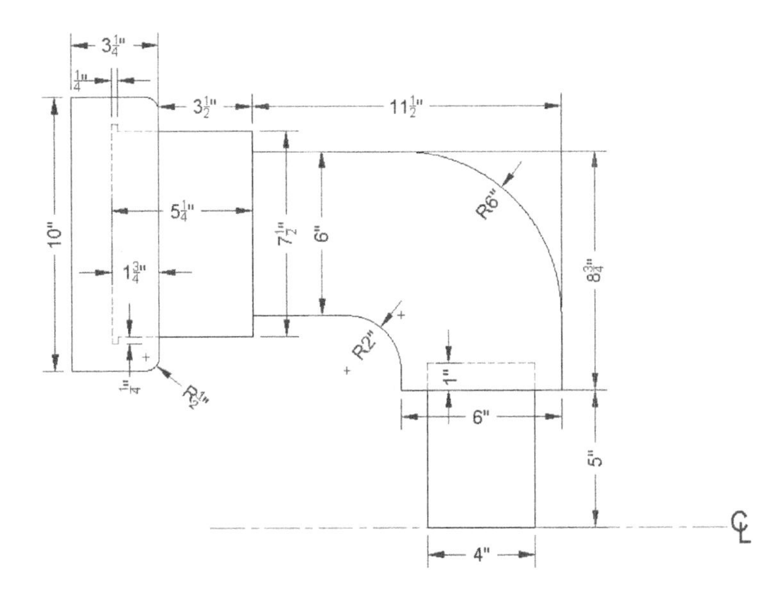

6. For this drawing, we return to basic bracket design, although this one is a bit more intricate. Set up the appropriate layers and linetypes and create the following mechanical part. (Difficulty level: Intermediate; Time to completion: 20–30 minutes.)

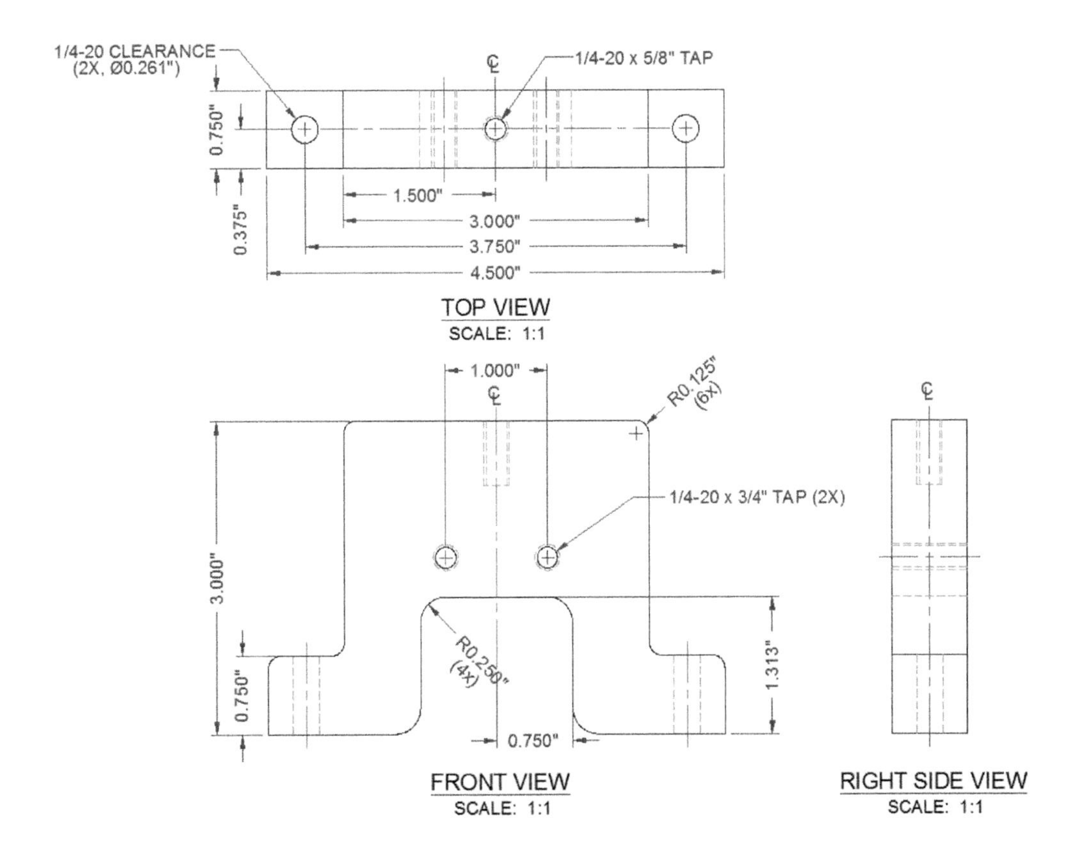

7. This exercise involves a floor plan. Similar to previous activities, open a new file, set up A-Walls and A-Doors layers and colors, and leave your units as Architectural. Then, complete the following basic floor plan. All walls are 6″ thick. All the doors are 3′ wide and have no thickness; you may give them thickness if you like. (Difficulty level: Intermediate; Time to completion: 45–60 minutes.)

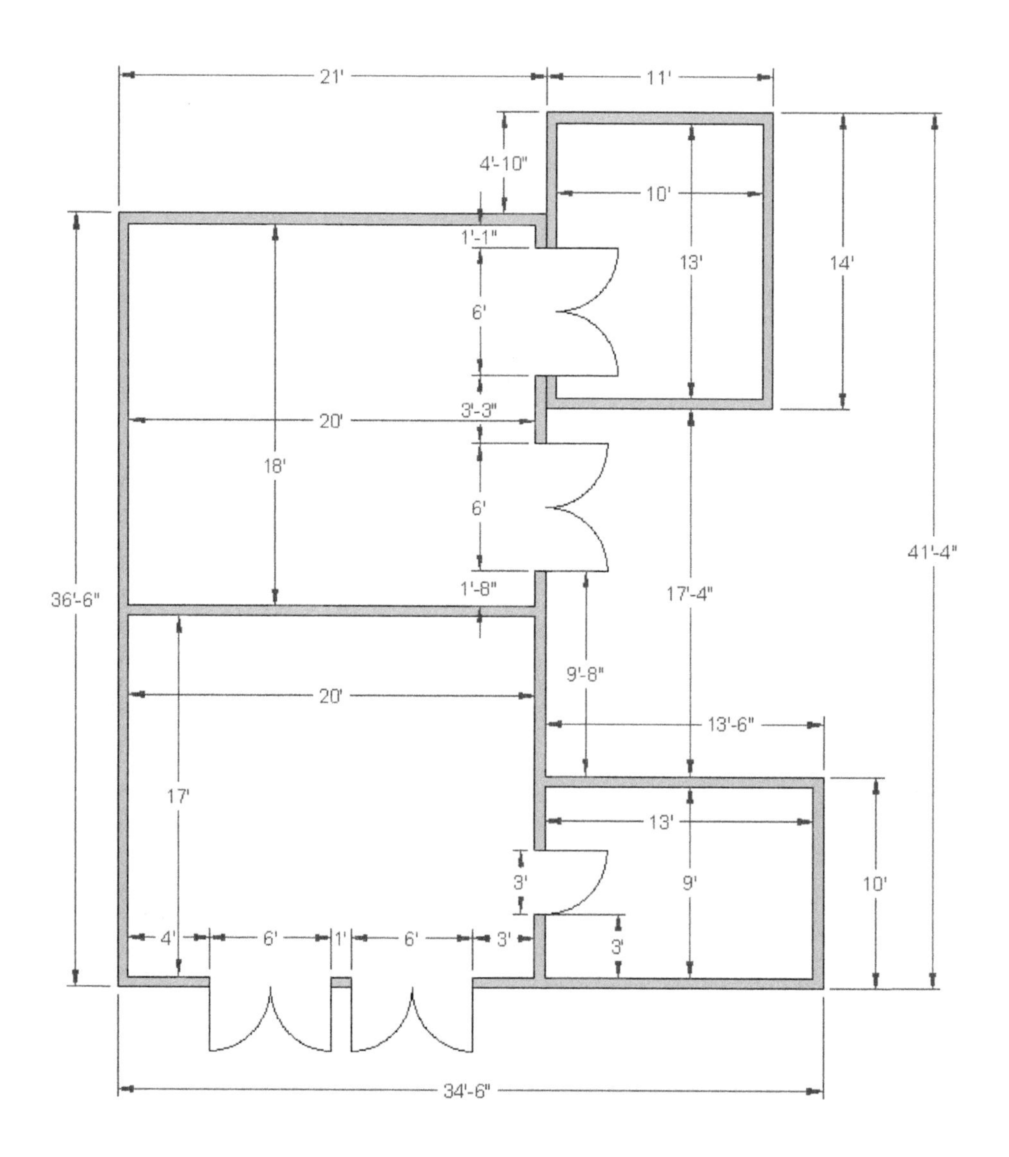

8. This exercise is a similar, if slightly more involved, floor plan. As before, open a new file; set up A-Walls, A-Doors, and A-Furniture layers and colors; and leave your units as Architectural. Then, complete the following floor plan layout. All walls are either 7.5″ or 5″ thick. All the doors are either 3′ or 4′ wide and have a 2″ thickness. The furniture sizing can be approximated. Dimensions and shading are for clarity only. (Difficulty level: Intermediate; Time to completion: 45–60 minutes.)

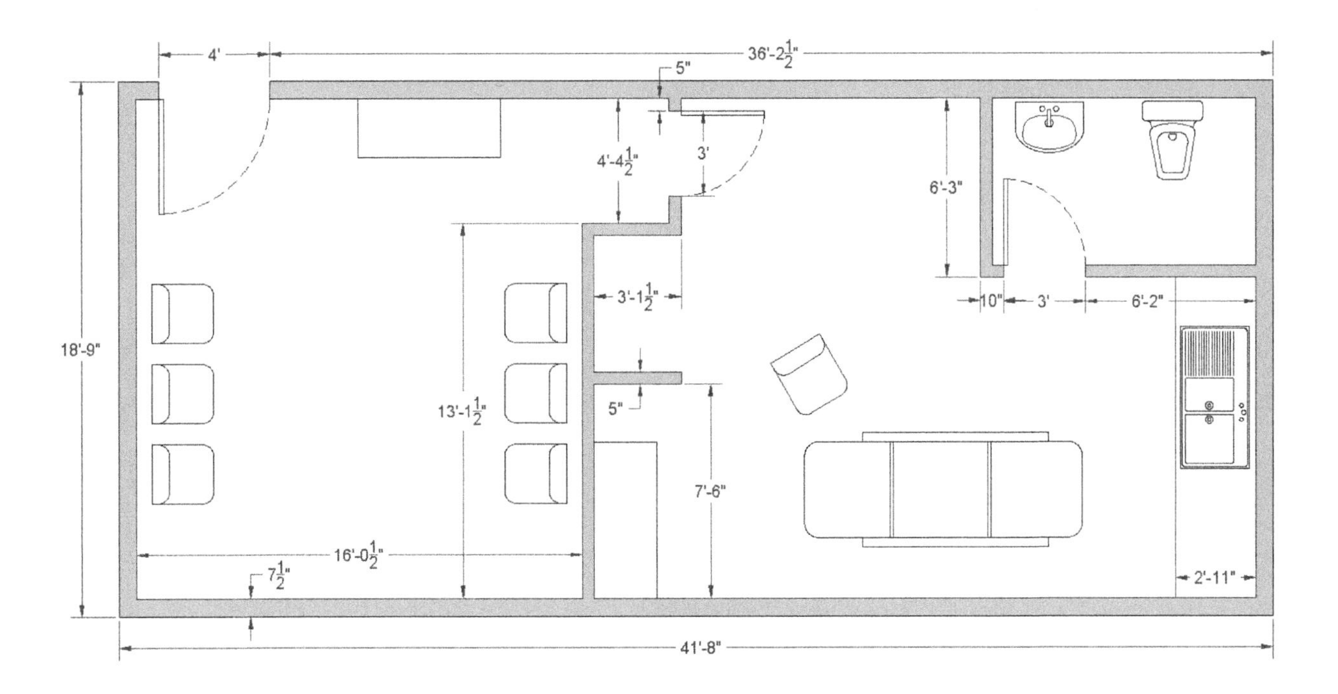

9. Open a new file and set up layers of your choosing. Then, create the following electrical power strip according to the critical dimensions shown. Where no dimensions are given, assume an appropriate value. Add text where shown. Dimensions and some hatching are shown just for clarity and for construction purposes. (Difficulty level: Easy; Time to completion: 20–30 minutes.)

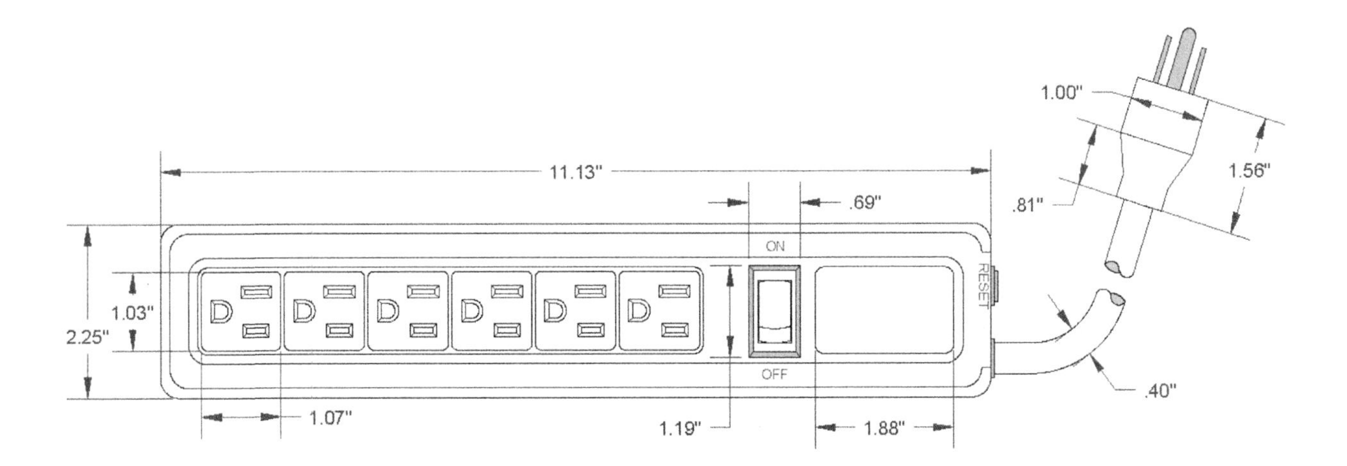

10. This exercise is another mechanical bracket. In a fashion similar to Exercise 6, it is projected in top, front, and side views, which is very typical for mechanical design drawings. Open a new file, set up your units as Decimal, and draw the shapes as shown. The dimensions and shading are for your use and clarity; we cover both shortly. (Difficulty level: Intermediate/Advanced; Time to completion: 60 minutes.)

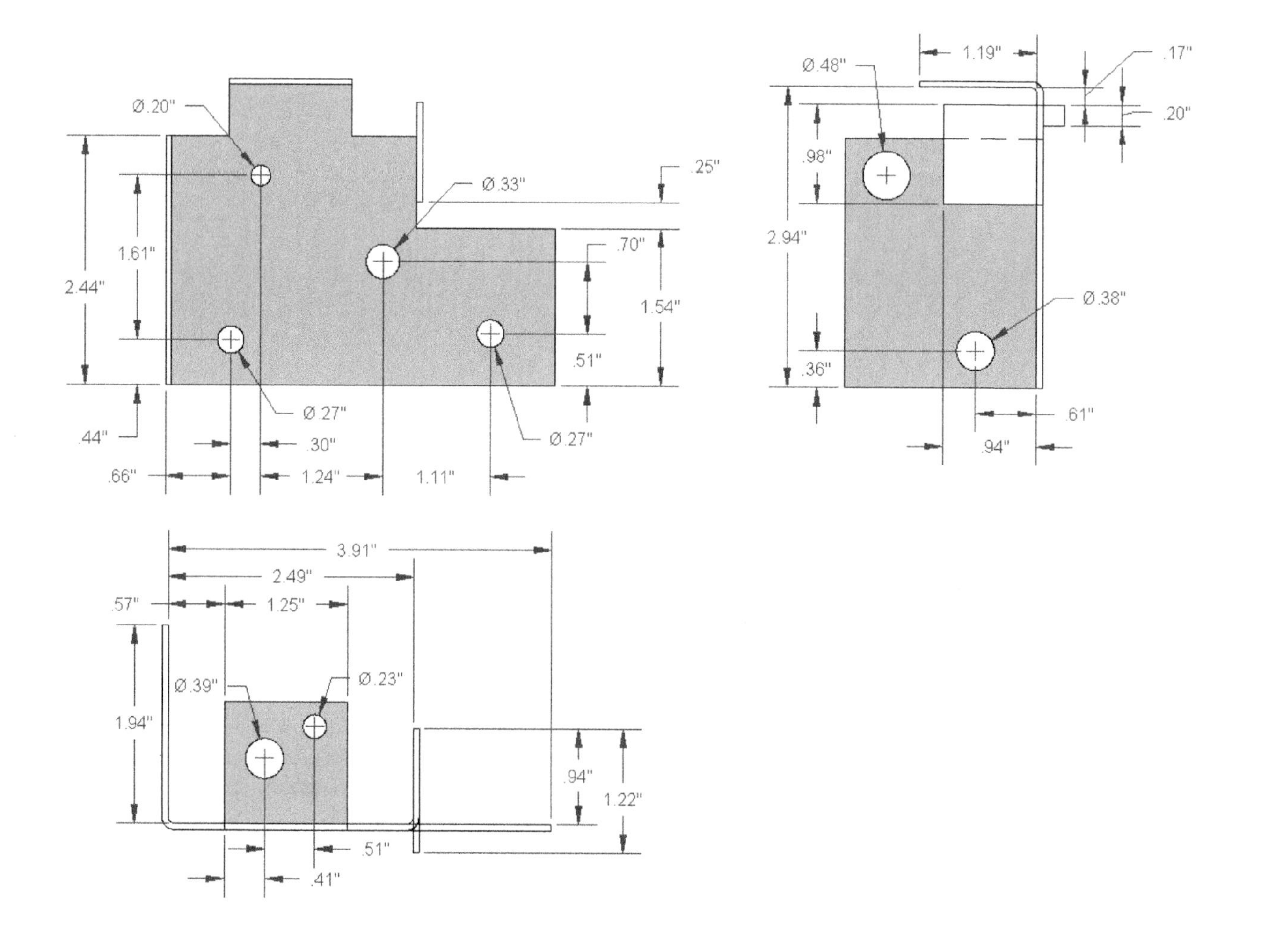

Spotlight On: Mechanical Engineering

Mechanical engineering is the science of designing and manufacturing mechanical systems or devices. It is one of the oldest and most widespread engineering specialties and is often the profession one has in mind when thinking of a "typical" engineer. Mechanical engineering is also by far the most diverse of the engineering specialties, and "mechies" can be found working in fields as wide-ranging as aerospace, automotive, naval, rail, power, infrastructure, robotics, manufacturing, consumer products, and much more. It is truly a profession that keeps our society moving.

FIGURE 1 *Image source: Chrysler.*

Mechanical engineers need to master a wide variety of engineering sciences, structures, materials, vibrations, machine design, control systems, electrical engineering, and more. They also need to be well versed in business practices, cost analysis, and risk assessment, as design and production of just about anything is closely tied to these. Mechanical engineers have been the driving force behind development of high-end software tools, such as finite element analysis for structural work.

FIGURE 2 *Image source: Mercruiser.*

Education for mechanical engineers starts out similar to education for the other engineering disciplines. In the United States, all engineers typically have to attend a 4-year ABET-accredited school for their entry-level degree, a Bachelor of Science. While there, all students go through a somewhat similar program in their first 2 years, regardless of future specialization. Classes taken include extensive math, physics, and some chemistry courses, followed by statics, dynamics, mechanics, thermodynamics, fluid dynamics, and material science. In their final 2 years, engineers specialize by taking courses relevant to their chosen field. For mechanical engineering, this includes classes on structures, controls, vibration analysis, thermodynamics, and machine design, among others.

On graduation, mechanical engineers can immediately enter the workforce or go on to graduate school. Although not required, some engineers choose to pursue a Professional Engineer (P.E.) license. The process first involves passing a Fundamentals of Engineering (F.E.) exam, followed by several years of work experience under a registered P.E., and finally sitting for the P.E. exam itself. Mechanical engineering is also an excellent undergraduate degree for entry into law, medicine, or business school. Just some of the careers you can pursue with an engineering bachelor's followed by a JD, MD, or an MBA include patent attorney, biomedical engineer, and manager or CEO of an engineering firm. If you can handle the educational requirements, the mechanical engineering undergraduate degree may truly be one of the most useful ones in existence.

Mechanical engineers can generally expect starting salaries (with a bachelor's degree) in the $62,000 per year range, which is in the middle of the pack among engineering specialties, behind computer, petroleum, and chemical engineering. This, of course, depends highly on market demand and location. A master's degree is highly desirable and required for many management spots. The median salary is around $80,000, with the top 10% making over $120,000. The job outlook for mechanical engineers is excellent, as the profession is so diversified that there is always a strong demand in some of the sectors, even if others are in a downturn.

So, how do mechanical engineers use AutoCAD and what can you expect? Industry wide, AutoCAD enjoys significant amount of use, even as its share shrinks as companies switch to low-cost 3D solutions such as SolidWorks and Inventor. In truly high-end design (aerospace, automotive, and naval, for example), the dominant software applications continue to be CATIA, NX, and Pro/Engineer (Creo); and AutoCAD does not compete with these. However, this still leaves many applications where AutoCAD is appropriate, such as small part design, schematics, and overall layout of systems. Many companies also do not manufacture but rather assemble modular pieces into new products; here, too, AutoCAD is the appropriate low-cost solution. Let us take a look at a few examples.

The acronym P&ID stands for piping and instrumentation diagrams, a major mechanical engineering field, commonly referred to as *process control*. Engineers use the principles of fluid flow, material science, and hydrodynamics to design piping and valves that regulate fluid flow. This can be anything from a chemical plant to an electrochlorination system to waste treatment. Usually, systems are assembled out of standard industry parts, so AutoCAD is the appropriate tool to put them together into a design. Fig. 3 shows a typical piping valve profile diagram.

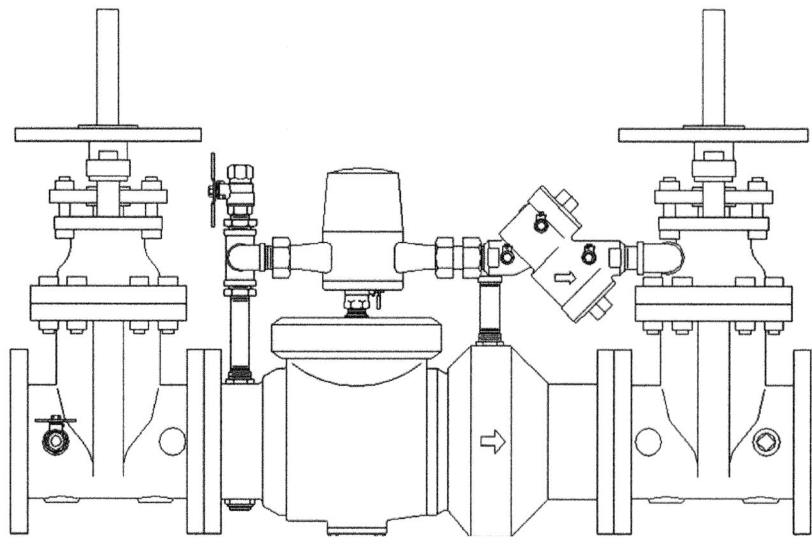

FIGURE 3 Piping valves. *Image source: Elliot Gindis.*

Small device assemblies are another example of mechanical design. If the parts are standard and no manufacturing is necessary, only assembly drawings, then AutoCAD can be used to depict the ideas. Fig. 4 shows an AutoCAD drawing of an instrument tuning peg assembly.

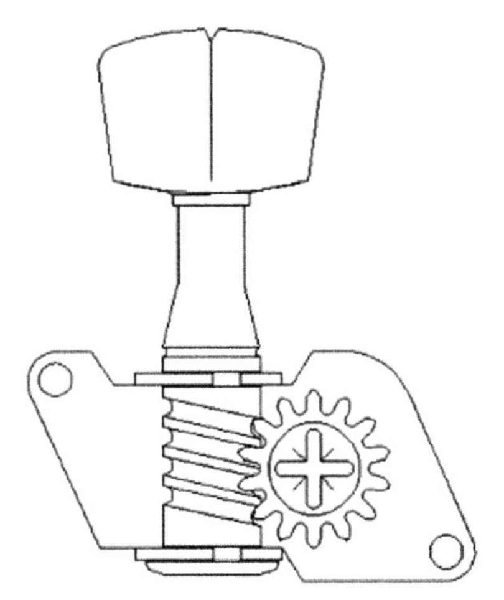

FIGURE 4 Tuning peg assembly. *Image source: Elliot Gindis.*

Chapter 5

Hatch Patterns

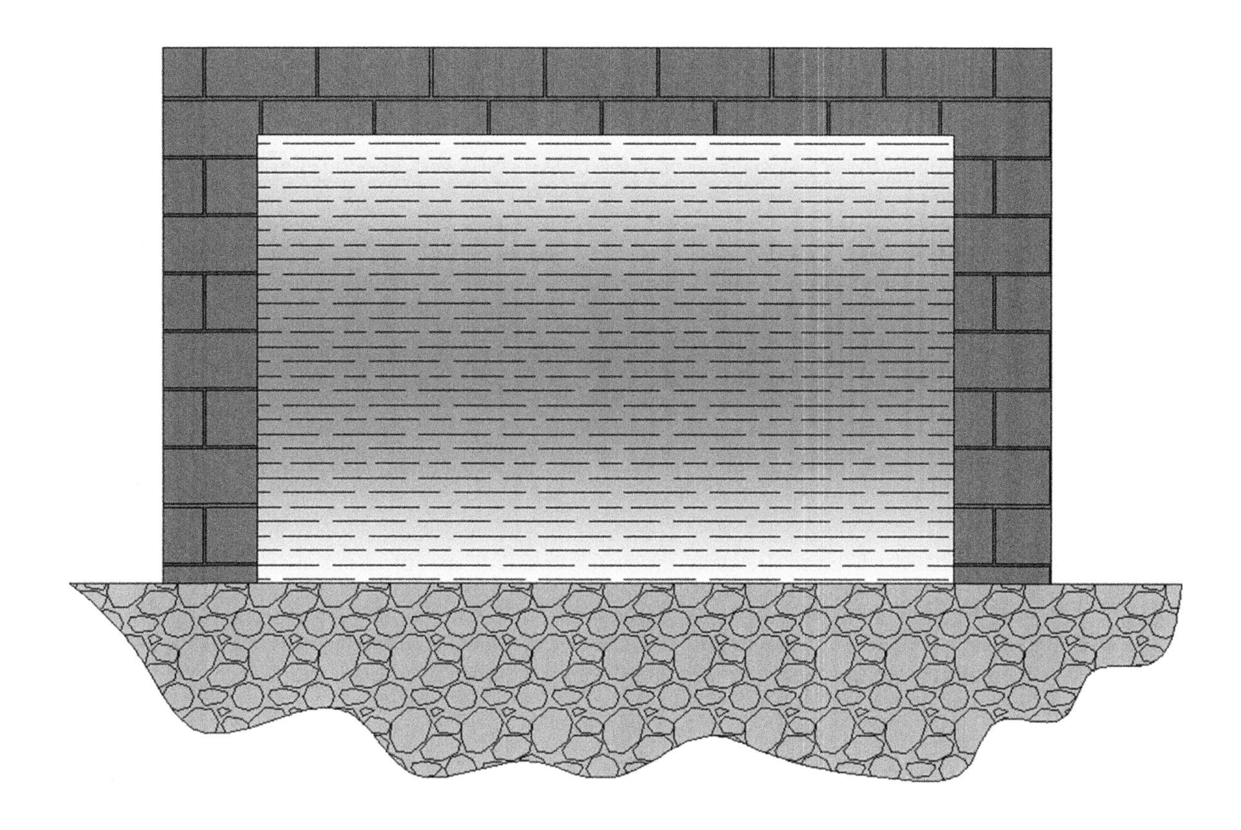

Learning Objectives

Hatch patterns allow you to indicate a variety of surfaces, from wood to concrete, on your design. In engineering, they are also used to indicate cross-section surfaces. In this chapter, we introduce boundary hatch and discuss the following:

- Hatch fundamentals
- Picking patterns
- Picking area or objects to hatch
- Adjusting scale and orientation
- Working with hatch patterns
- Gap tolerance
- Gradients and solid fills
- Superhatch

At the end of this chapter, you will be able to add hatch patterns to your floor plan.
Estimated time for completion of this chapter: 2 hours (lesson and project).

Up and Running with AutoCAD 2019. https://doi.org/10.1016/B978-0-12-816440-2.00005-7

5.1 INTRODUCTION TO HATCH

This is a chapter on the concept of boundary hatch or just *hatch* for short. This is simply a tool for creating patterns and adding fills to your design. These are not merely decorative; rather, they serve an important function of indicating types of materials, ground and floor coverings, and cross sections. As such, this topic is quite important and is the next logical concept to master in AutoCAD.

We cover hatch patterns in two sections. The first is with the Ribbon turned off, which makes the hatch command default to the classic dialog box seen in Fig. 5.1. It is easier for a beginner to visualize hatch concepts via this method. It was also the only way to access hatch up until AutoCAD 2011, good to know if you are using an older version of AutoCAD. Starting with AutoCAD 2011 and continuing through 2019, hatch can still be accessed the old way if the Ribbon is not on or, if it is, then via the Ribbon only. The Ribbon essentially duplicates the hatch dialog box, with maybe a few more bells and whistles; we cover this approach as well.

Hatch patterns are not an AutoCAD invention. In the days of hand drafting, there was also a need to visually indicate what sort of material one was looking at. Architects and engineers created easy-to-understand repeating patterns that somewhat, if not closely, resembled the actual material or at least gave you a very good idea of what it was. Just a few of the uses of hatch patterns follow:

Architecture:

- *Brick*: An important and popular pattern to render the exterior and (sometimes) interior of buildings. AutoCAD has numerous brick patterns available.
- *Herringbone*, *parquet*: Important for flooring designations.
- *Honeycomb*: Insulation designation.

Civil engineering:

- *Concrete*, *sand*, *clay*, *earth*, *gravel*: All used in designating surfaces in civil and site plan design.

Mechanical engineering:

- *Various ANSI diagonal patterns*: Used to designate the visible inside of an object cut in cross section.

With AutoCAD, creating these patterns is easy, as they are all predrawn and saved in a library, which is called up anytime you use the command. Often a designer can purchase or download additional patterns, if the basic ones are inadequate. You can even make your own and save them for future use, although this is a tedious process and only briefly covered in Appendix D.

5.2 HATCH PROCEDURES

At the most basic level, AutoCAD's hatch command requires only four steps:

Step 1. Pick the hatch pattern you want to use.
Step 2. Indicate where you want the pattern to go.
Step 3. Fine-tune the pattern by adjusting scale and angle (if necessary).
Step 4. Preview the pattern and accept it if OK.

Memorize the four steps and their sequences; this will help you when looking at the Hatch dialog box.

Before we get into the details, it goes without saying that you first need something to hatch. The command will not work on a blank screen or a bunch of unconnected lines. You need a closed area (although later on we violate that rule), and the easiest way to do this is to form a circle or rectangle. Open a new file and draw a square, sized 10 × 10″. Follow the steps carefully as described next. All advanced hatch functions flow from these basics.

Step 1. Pick the Hatch Pattern You Want to Use

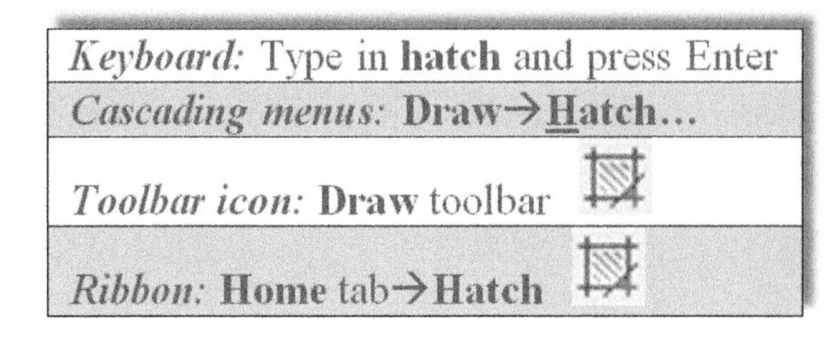

Keyboard: Type in **hatch** and press Enter
Cascading menus: **Draw→Hatch...**
Toolbar icon: **Draw** toolbar
Ribbon: **Home** tab→**Hatch**

With the Ribbon turned *off* (Tools → Palettes → Ribbon), begin the hatch command via the keyboard, cascading menus, or icons. The Hatch and Gradient dialog box appears (Fig. 5.1). At the bottom right of the dialog box is an arrow that expands or collapses additional options. Make sure it is expanded, as seen in the figure.

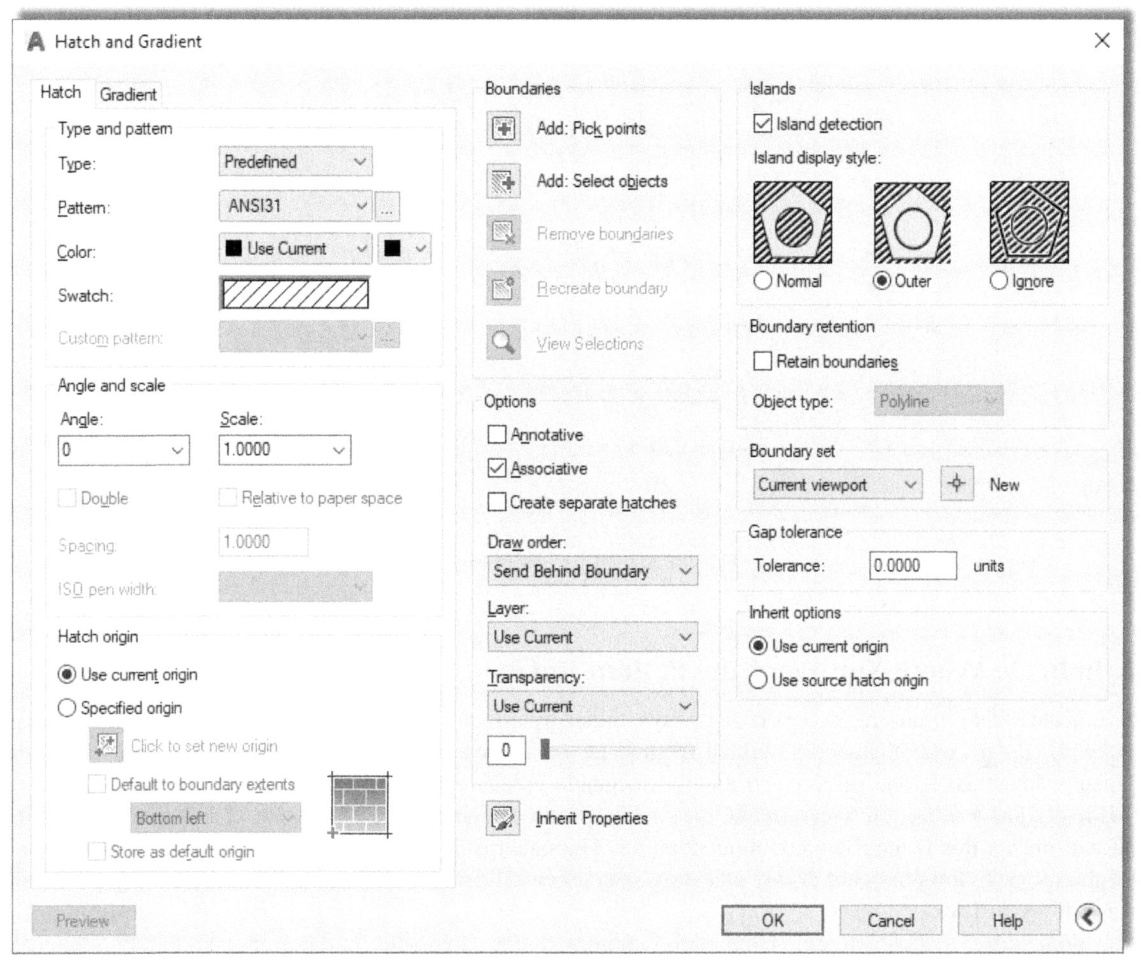

FIGURE 5.1 Hatch and Gradient dialog box.

Now select a pattern. The choices are listed in the upper left of the dialog box under the heading *Type and pattern*. Below that are four fields, called *Type:*, *Pattern:*, *Color:*,and *Swatch:*. Leave Type as Predefined and also ignore Color for now. For the actual pattern selection, if you use Pattern, then select it by name; if you pick Swatch, patterns are selected *visually*, a much better option. Click on the diagonal lines to the right of Swatch: (or the little box with the three dots to the right of Pattern), and a new box, called the *Hatch Pattern Palette*, appears (Fig. 5.2).

Take a look through the tabs: ANSI, ISO, Other Predefined, and Custom. We have a use for some of the ANSI patterns, not so much for the ISO ones, and Custom is where the custom defined patterns go if you are inclined to create them. We are most interested in the Other Predefined tab. Go ahead and select it; you will see what is shown in Fig. 5.3.

Scroll up and down the patterns. You will notice some of the ones mentioned at the start of this chapter. Go ahead and pick one that you like, preferably one that is distinct and stands out; AR-HBONE is a good choice and is used in this example. When you click on it, the pattern is highlighted blue. Go ahead and click on OK. We are done with Step 1, and the new pattern choice is reflected in the Swatch area.

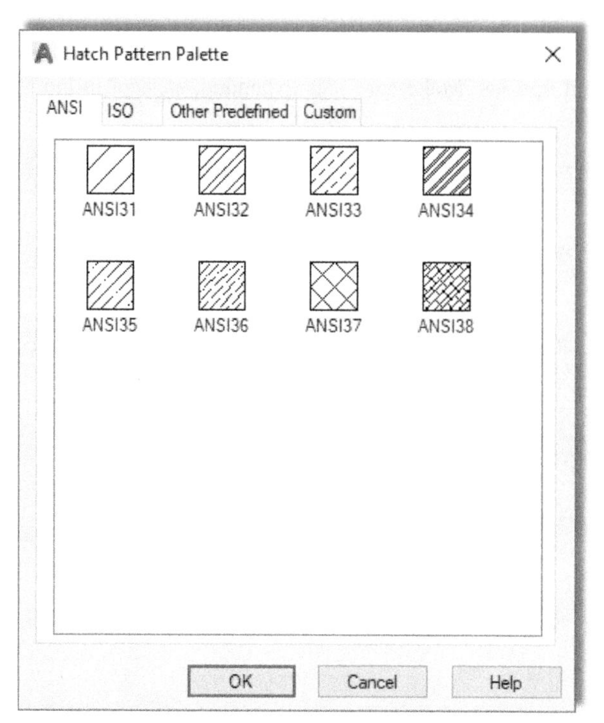

FIGURE 5.2 Hatch Pattern Palette.

Step 2. Indicate Where You Want the Pattern to Go

You can indicate where to put the pattern in two ways, either by *directly picking* the object that will contain the pattern or by *picking a point inside* that object. Which method to use is easy to determine. If the object is actually made of joined-together lines and is one piece (such as your rectangle or circle), then it can be picked directly. If not and the "object" is really just a collection of connected lines defining an area, then the best way is to pick a point in the middle of that area; and indeed this is the more common situation. The pattern then behaves like a bucket of paint spilled in the middle of the room. It flows outward evenly and stops only when it hits a wall. Once again, no holes or gaps are allowed yet; that is addressed later.

In the Boundaries section of the Hatch and Gradient dialog box (top middle), notice the two choices just discussed—Add: Pick points and Add: Select objects. Choose either one to try. The AutoCAD procedure for both is outlined next.

Add: Pick Points

If you choose this option, AutoCAD says: `Pick internal point or [Select objects/remove Boundaries]:`

Click somewhere *inside* the object, and AutoCAD says:

```
Selecting everything...
Selecting everything visible...
Analyzing the selected data...
Analyzing internal islands...
```

Press Enter and you return to the Hatch dialog box with Step 2 completed. Notice that a preview of the hatch is shown to you as you complete this step.

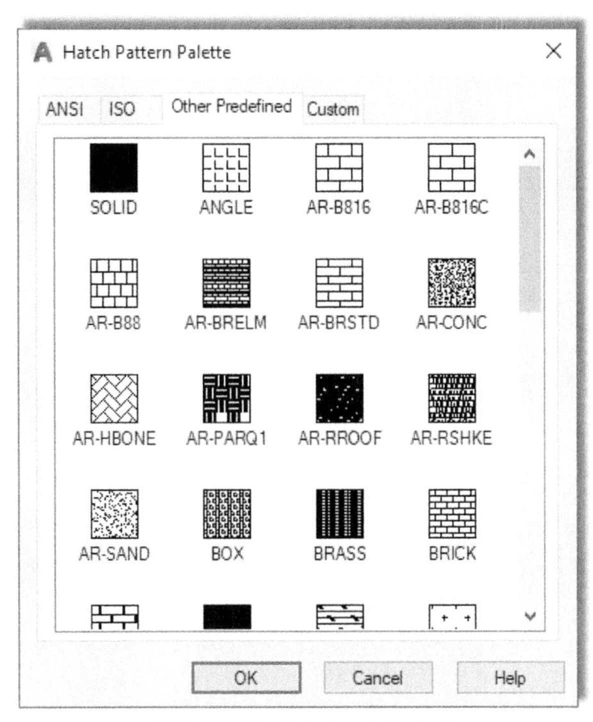

FIGURE 5.3 Other Predefined tab.

Add: Select Objects

Alternatively, if you pick this option, AutoCAD says: `Select objects or [picK internal point/remove Boundaries]:`

Click on the object itself (not the empty space) and press Enter. You return to the Hatch dialog box with Step 2 completed (Fig. 5.4) and the Preview button active.

Successfully picking a boundary can sometimes be a tricky business. The existence of gaps, no matter how small, can be a source of occasional frustration to designers; and until relatively recently, it was difficult to tell where those gaps were hiding to fix them. As of AutoCAD 2010, red circles now appear where the gaps are, aiding you in locating them, but this all can be avoided in the first place by not doing sloppy drafting (i.e., not connecting lines together properly). It is a sign of advanced AutoCAD skills when a complex area hatches right the first time.

In cases where boundary selection is not successful, complex areas can often be broken down to smaller ones by using lines to divide the area into smaller pieces and the nonproblematic portions hatched first, although this is a last resort. A few releases ago, AutoCAD added the gap tolerance command, which ignores gaps up to a preset limit. This is a great tool, which we cover shortly, along with the entire gap issue in general, but it is somewhat of a Band-Aid and still does not address the underlying sloppy drafting but merely allows you to get away with it.

Remember, you get only the level of accuracy that you put in. It is essential to master the fundamentals of drafting early on. The hatch command is an "early warning" to students. If they are having problems using it smoothly, they need to go back and refine their basic drafting (linework and accuracy) skills.

Step 3. Fine-Tune the Pattern by Adjusting Scale and Angle (If Necessary)

You are almost done and, at this point, could probably just press OK and finish the hatch pattern; however, press Preview instead. It is found at the bottom left of the dialog box and is generally a good habit to get into. What you see is your hatch pattern with the 10″ × 10″ square border in dashed lines (Fig. 5.5).

AutoCAD will also say what to do next:

```
Pick or press Esc to return to dialog or<Right-click to accept hatch>
```

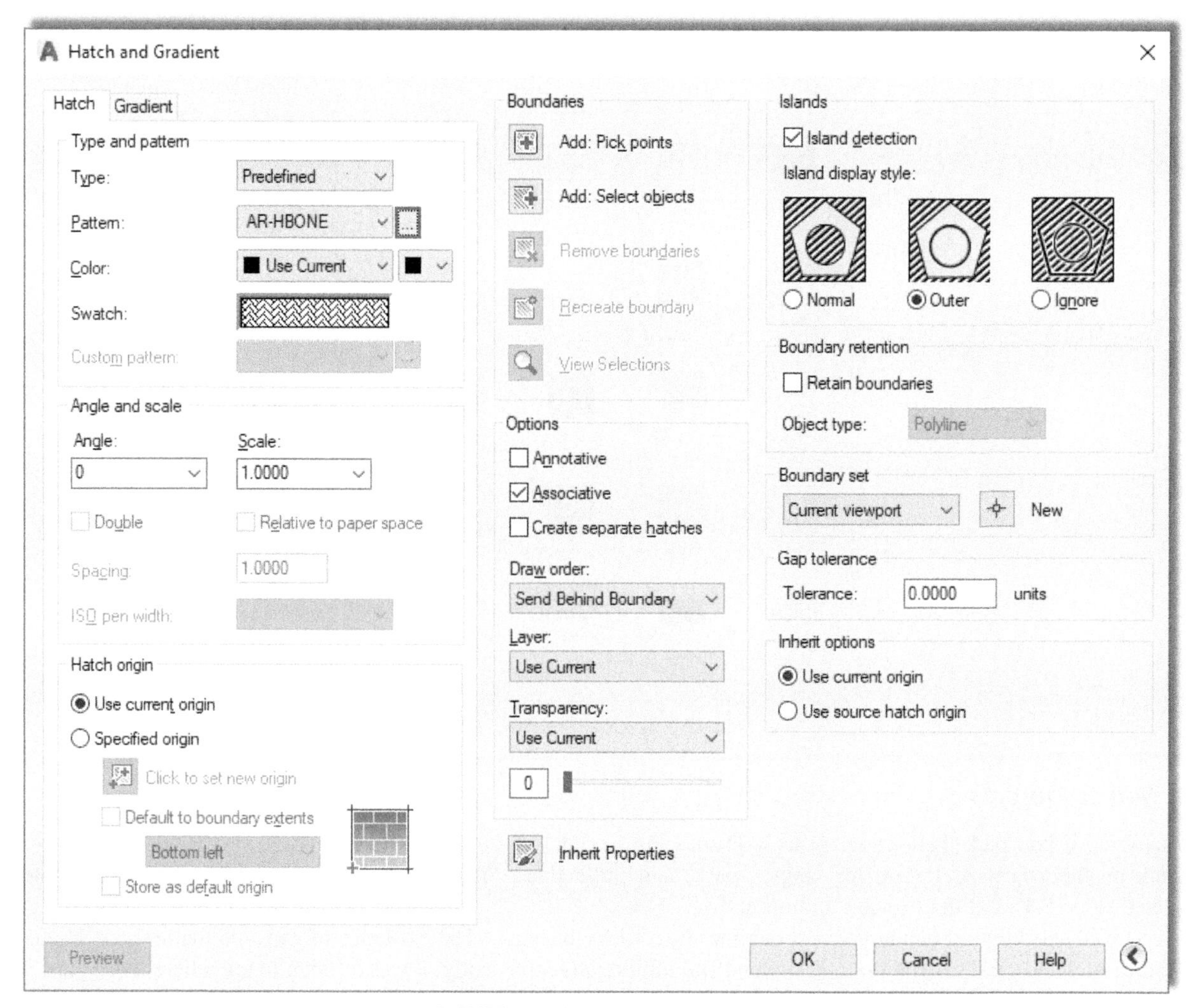

FIGURE 5.4 Steps 1 and 2 completed.

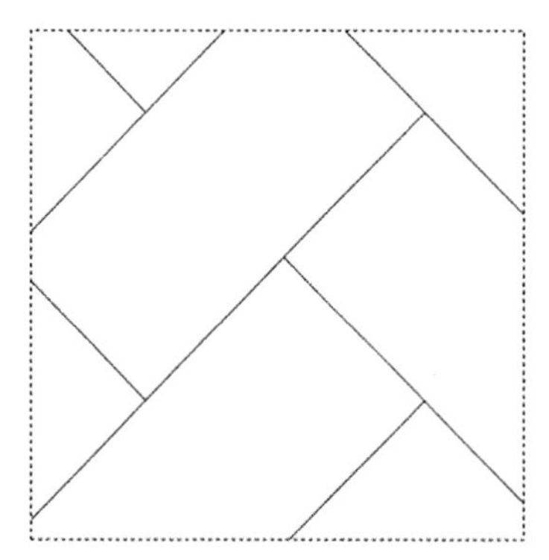

FIGURE 5.5 Hatch preview.

Here, you would go ahead and right-click or press Enter. Sometimes, however, the pattern is too big or too small. You can usually tell it is too big by either not being able to see it or seeing a small part of it. If it is too small, the problem is worse because you may not be able to finish the pattern at all. In those cases, nothing can be seen and AutoCAD tells you the following:

```
Hatch spacing too dense, or dash size too small. Pick or press Esc to return to dialog or<Right-click to accept hatch>:
```

For learning purposes, let us practice changing the scale as the final step in our hatch creation. To do this, press Esc once and you return to the dialog box. Under Angle and scale (Fig. 5.6), change the scale to a smaller number, such as 0.25, by typing it in. In some cases where the scale is way off, you may need to do this several times until the hatch pattern is just right. In the same manner, you may adjust the angle of the entire pattern if desired.

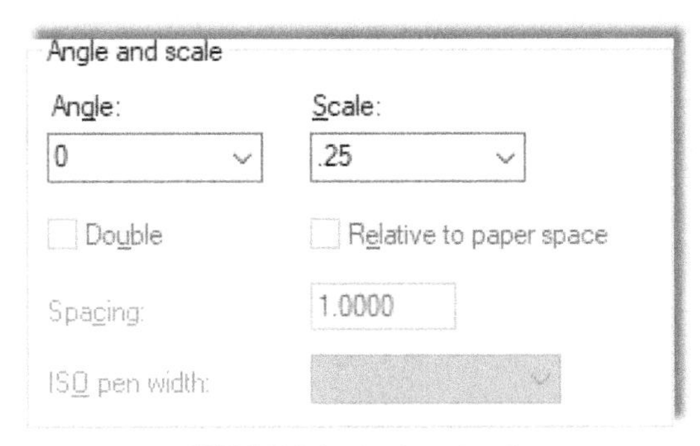

FIGURE 5.6 Angle and scale.

Step 4. Preview the Pattern and Accept It If OK

Finally, you are done. Do one more preview and press OK. If you followed the previous steps and used a rectangle, the AR-HBONE pattern, and the scale suggested, then you should get Fig. 5.7 or something reasonably close.

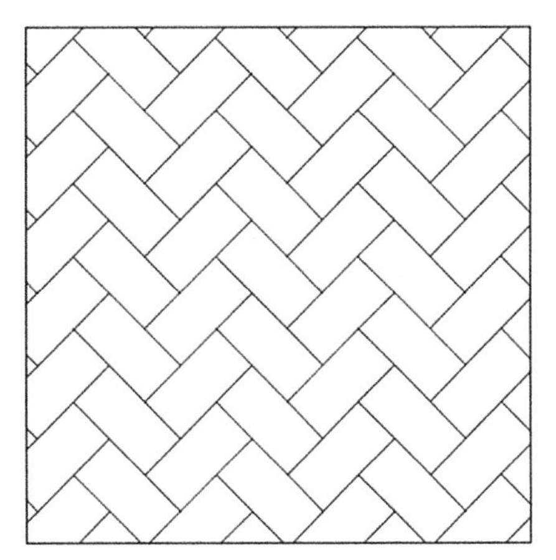

FIGURE 5.7 Final hatch pattern.

5.3 WORKING WITH HATCH PATTERNS

In this section, we mention some of the additional tools and concepts that pertain to hatch patterns. First of all, the most basic question at this point is: How do you edit a hatch pattern after you have created it? It is very easy: To edit a hatch pattern, simply double-click on it. Make sure you get one of the lines of the pattern, not just empty space. You will see a grip and the Quick Properties palette (Fig. 5.8); here, you can change whatever you need to. You can, of course, use the full Properties palette if needed as well (recall how from chapter: Layers, Colors, Linetypes, and Properties). This is different from older releases of AutoCAD, where you were taken back to the Hatch dialog box. The thinking was that the Hatch dialog box was too big and covered up your work, so a sleeker Quick Properties (or even Properties) palette was a better option. If you are using the Ribbon, to be discussed shortly, you need to click only once to edit the pattern.

Exploding Hatch Patterns

This may be one of the worst things you can do. Never explode hatch patterns; depending on the size, they turn into hundreds of pieces that cannot be put back together again in the same editable form. All the reasons designers give for exploding hatch patterns (making them fit, trimming them, moving the containment border, etc.) can be addressed with additional hatch tools.

Hatch Pattern Layers and Colors

Hatch patterns should be on a generic A-Hatch layer or whatever layer name best describes what the pattern represents (A-Carpet, M-Cross-Sec, C-Gravel, etc.). Choose any color you want for the hatch patterns, though you should not go with anything too bright, as it may overwhelm the design. Shades of gray are common for cross sections in mechanical design, as are various shades of rusty orange for brick. As a side note, hatch patterns can be easily erased by the usual erasing methods. The patterns can also be moved, copied, rotated, and mirrored, although there is rarely a reason to do so.

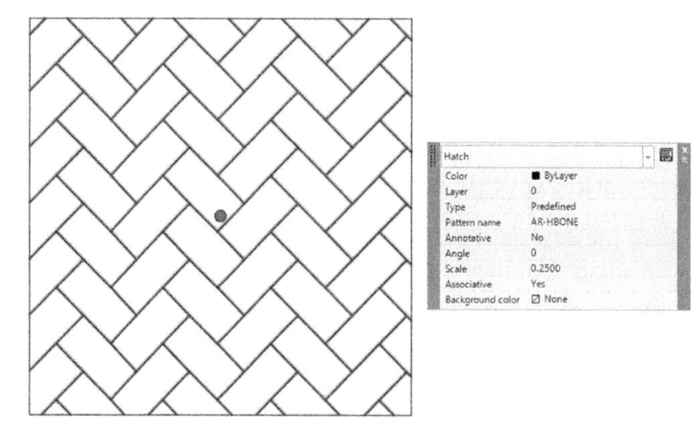

FIGURE 5.8 Editing the hatch pattern via the Quick Properties palette (no Ribbon).

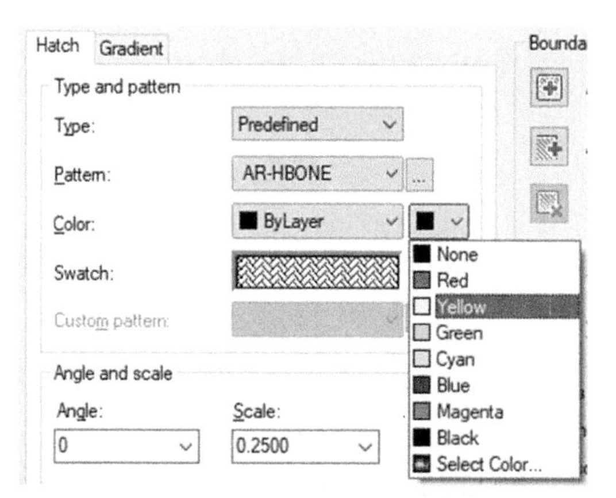

FIGURE 5.9 Hatch (Bylayer) and background (yellow) color selection.

As of AutoCAD 2011, you can assign a color directly to a hatch pattern regardless of what layer and associated color, it is on. You can also assign a background color (solid, not a gradient). To do this when first creating the hatch, you can select both the hatch color and the background color (Fig. 5.9) from two drop-down menus at the top left of the Hatch dialog box.

The same steps can be performed with the Ribbon, as shown a bit later. One sample result of assigning a hatch and background color is shown in Fig. 5.10.

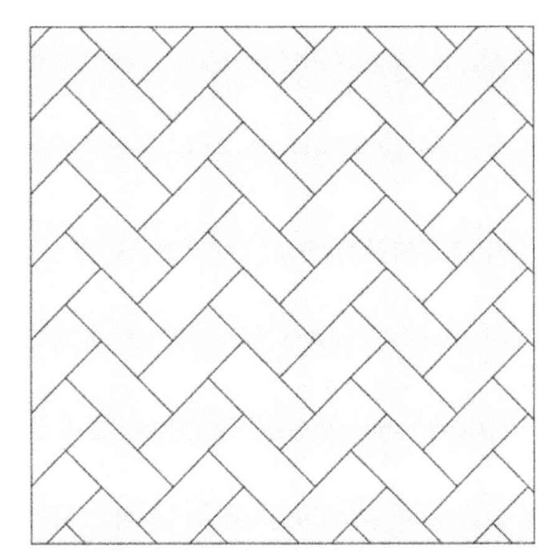

FIGURE 5.10 Hatch with a background color assigned.

Advanced Hatch Topics

Here, we summarize some additional advanced options available for working with hatches, including a more detailed discussion concerning the gap tolerance feature. You should explore these on your own as well; this is an important part of learning AutoCAD and greatly enhances your confidence with the software.

Hatch Origin

This allows you to begin hatch patterns from a precise location (origin) as opposed to a random fill-in. AutoCAD chooses the origin on the pattern itself, and you get to choose to what point on the object that origin is aligned. Try it out.

Options

- *Annotative*: This will not be covered until Level 2.
- *Associative*: Keep this checked. This simply means that, if you move the border, the hatch moves with it, very useful in case of future updates.
- *Create separate hatches*: If there are multiple separate areas to hatch, this option keeps them separate; sometimes useful, sometimes not.
- *Draw order*: This is exactly what it sounds like; it gives you options on how you want to stack the hatch and its boundary. Check out the options in the drop-down menu.
- *Inherit properties*: This option lets you borrow the features of another hatch pattern and use them in a new pattern you are creating, saving you some effort and time in picking out a pattern from scratch.

Islands

This gives AutoCAD guidance on how to interpret boundaries inside boundaries (e.g., a chair in the middle of your floor). It simply tells AutoCAD to ignore them or, if not, how to deal with them. Try drawing another rectangle inside your first one and running through the options represented by the pentagon shapes.

Gap Tolerance

This is the basic hatching aid that was mentioned earlier. Create a new rectangle with values close to what you did earlier, and then create a gap in one of the sides, such as what is seen in Fig. 5.11. To do this, you can use a trim command or the new break command (which we officially introduce in chapter: Advanced Design and File Management Tools). Then, if you try to fill this shape with a hatch pattern, you get what you see in Fig. 5.11. Note that the red circles may or may not be present. If they are, they can be removed via the regen command.

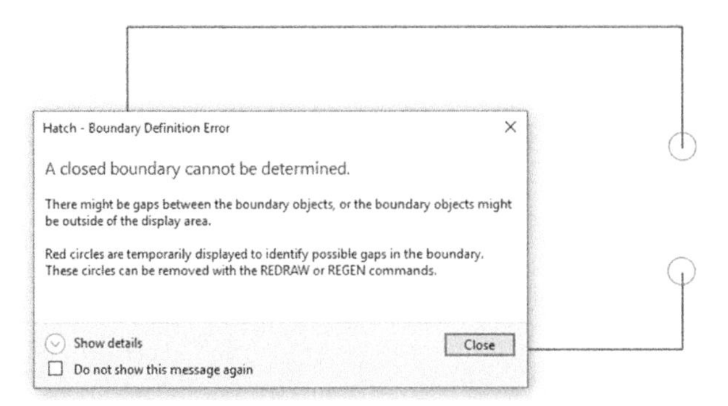

FIGURE 5.11 Boundary not closed alert.

Close the alert and press Esc to continue. Then, returning to the main dialog box, set a value for the gap tolerance (you can use 10 for starters) and rehatch the area. You get another warning, which can be eliminated from future occurrence by checking off *Always perform my current choice*, as seen in Fig. 5.12.

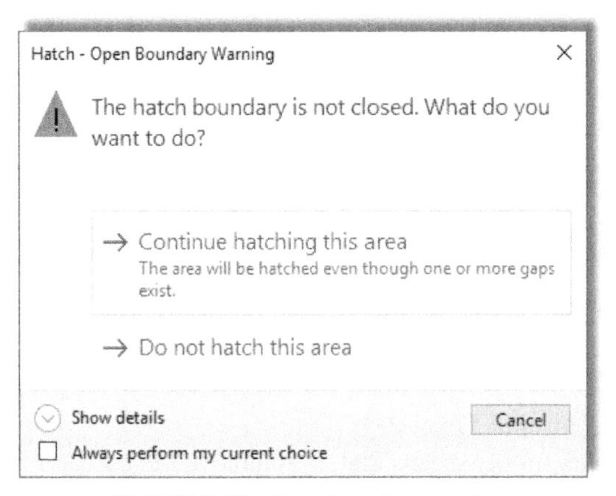

FIGURE 5.12 Open boundary warning.

Select *Continue hatching this area* and AutoCAD temporarily closes the gap to continue the hatch command. After successful completion, the gap shows again but the hatch remains, as seen in Fig. 5.13. Some designers have used the command to ignore door openings when hatching carpeting in floor plans. You can try this as well on your floor plan later.

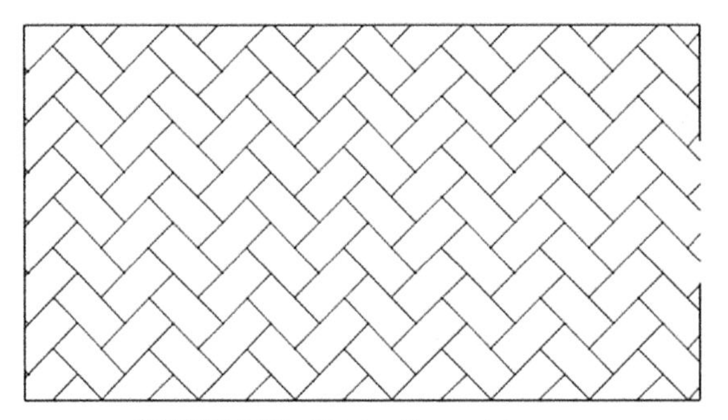

FIGURE 5.13 Hatch with gap tolerance.

5.4 GRADIENT AND SOLID FILL

At the top left of the Hatch and Gradient dialog box, click on the Gradient tab. The left side of the box changes, whereas the right-hand side stays essentially the same, as shown in Fig. 5.14.

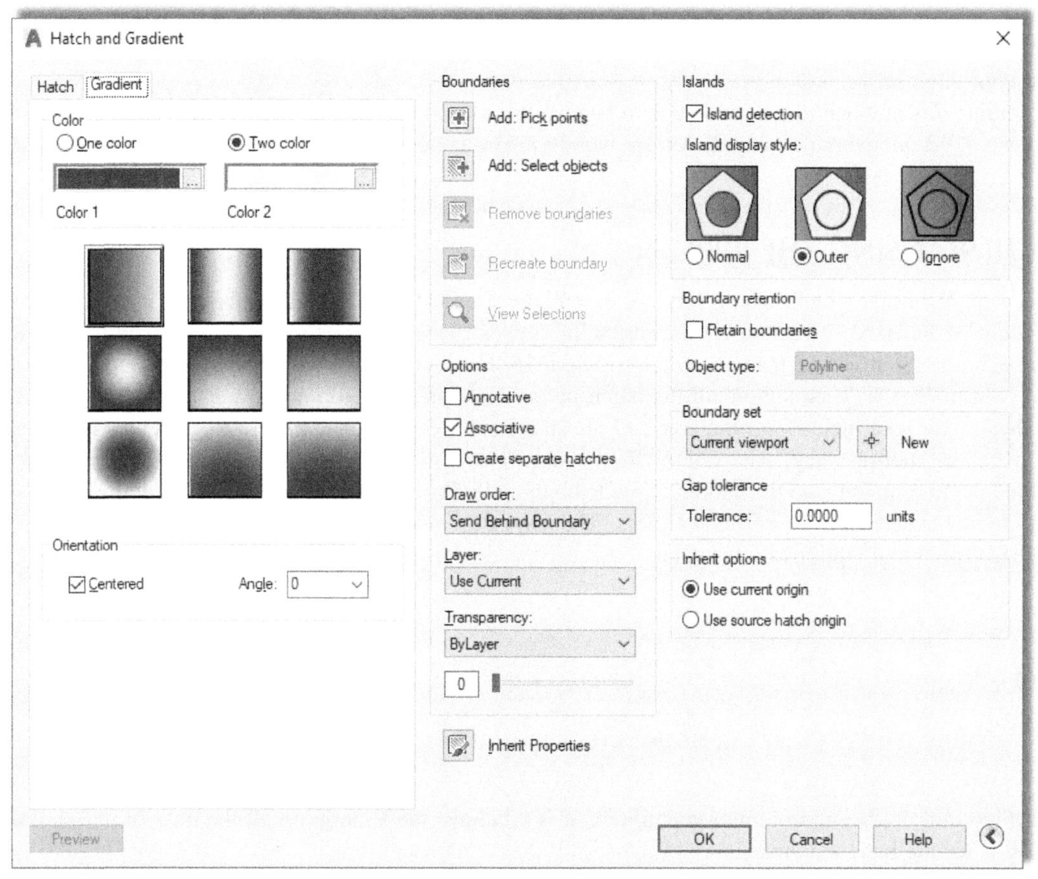

FIGURE 5.14 Gradient tab.

The gradient option is just a solid fill with some flair. You can choose the desired pattern, in one or two colors, and adjust the Shade, Tint, Orientation, and Angle of the gradient. The rest of the steps are similar: Pick point or Select object, Preview, and done.

Shown in Fig. 5.15 is a possible use for gradient in rendering a brick wall. A brick pattern is created and a brick-colored fill tops it. Note that you may need to use Tools → Draw Order → Bring to Front from the cascading menu to position the brick above the gradient fill (or the gradient fill behind the brick, depending on what you pick first). Try it.

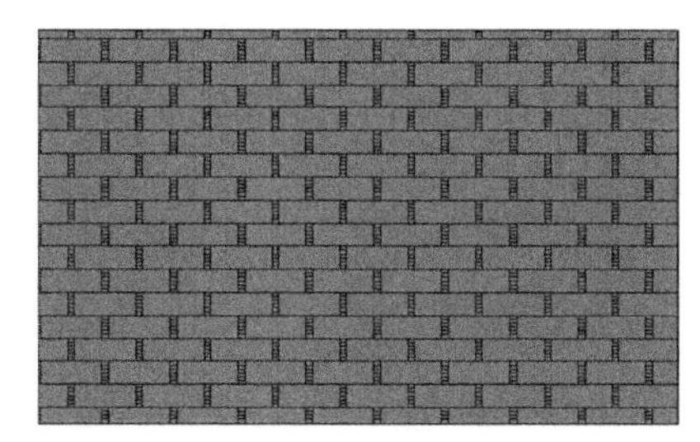

FIGURE 5.15 Brick wall created with Hatch and Gradient.

Solid Fill

This fill is the first choice (black box, top left) in the Other Predefined tab and deserves special mention as our final topic. This solid fill is similar to a gradient but with no fancy color mixing, just one color. It is very handy as a fill for walls, some types of cross sections, or any location where you want a dark area.

It becomes even more useful when paired up with a lineweight feature called *screening*, which is discussed in detail in Chapter 19, Advanced Output and Pen Settings. It has to do with being able to fade out (or screen) the intensity of any color, typically down to 30%, and is perfect for indicating "area of work" on key plans, some types of carpeting, and even pavement on civil engineering drawings.

A way to mimic this now without screening is to change the solid fill's color to Gray 9; the soft gray haze then seen is distinct and visible, yet not overwhelming. If you are wondering how the walls were filled in on the apartment drawing, this is how.

5.5 HATCHING USING THE RIBBON

The previous steps using the Hatch dialog box are essentially duplicated using the Ribbon. Bring it back by typing in Ribbon (or Tools → Palettes → Ribbon via the cascading menus). Draw a shape to hatch and begin the command in the same manner as before; you see the Ribbon switch to Fig. 5.16. Here, you still have to pick the pattern by dropping down the menu, as seen in Fig. 5.17. Note how all the hatch patterns are present and you do not have tabs anymore.

You can then hover your mouse over the area and see an instant preview of the pattern to get an idea of what you have. When you move the mouse away, the pattern disappears. Selection of boundaries is all the way on the left, under Boundaries, and the fine-tuning is in the middle of the Ribbon under Properties. Here, you can select the Scale, Angle (with instant rotation via a slider), and Background Colors. The rest of the features, such as Gap Tolerance, Origin, and others, are found under Origin and Options on the right.

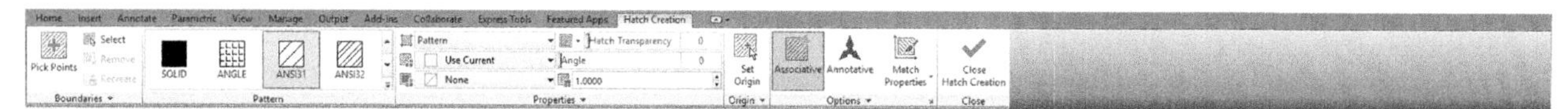

FIGURE 5.16 Ribbon—Hatch Creation.

Which method you use is entirely up to you, although if you have the Ribbon up all the time, it makes more sense to continue using it for the hatch command.

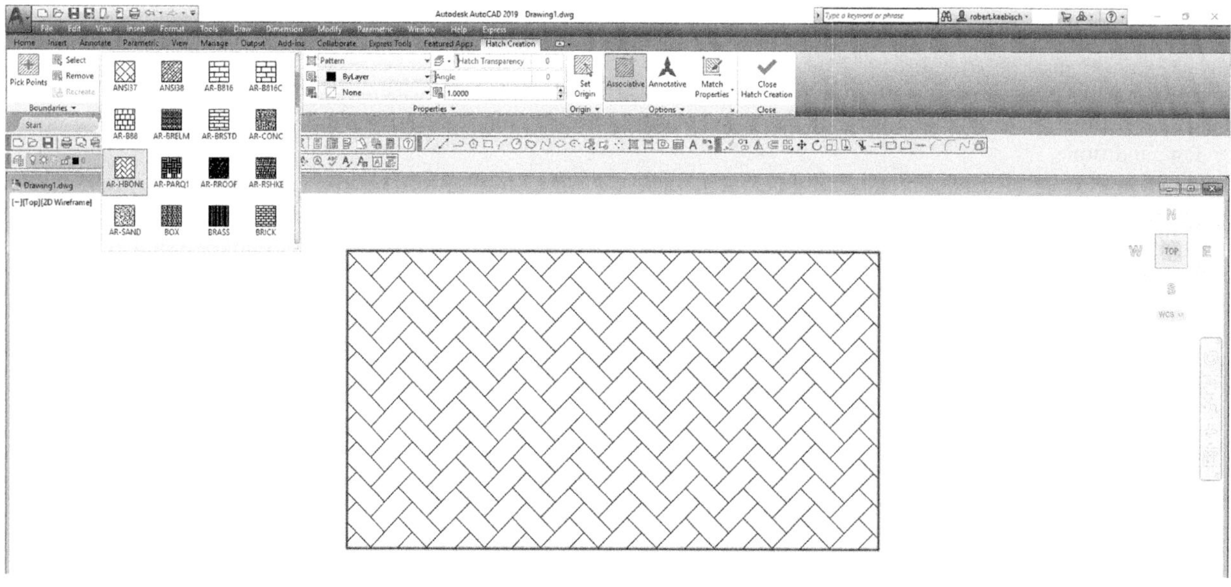

FIGURE 5.17 Selecting patterns.

5.6 SUPERHATCH

Superhatch is an interesting variation of the hatch command. The basic idea is that instead of using a premade pattern, you select a source image of your own, and AutoCAD duplicates whatever you chose into a pattern. Over several years the command evolved to allow you to choose non-image items as well such as Blocks and Xrefs, but the most common application is still patterning images.

When using this command, be careful with the sizing of the original image. For example, if you use a photograph that is about 1 MB in size (a commonly encountered size for a high-resolution photo), then your superhatch effort creates 1000 of these—well, you can see how the drawing file size will balloon. It may also take a while to superhatch a large area with large source images, and it's a great way to bring even a powerful workstation to a crawl.

With that in mind though, let us take a look at the command. Create a rectangular sample area to superhatch and also prepare a good candidate for the source image. In this textbook example we will pattern a photo of some flowers, but obviously you would pick something that serves a purpose in the context of your design. You generally use a superhatch only if needed to fulfill a certain hatch need that is not addressed by more conventional patterns.

Step 1. Type in `superhatch` and press Enter. You will see the dialog box shown in Fig. 5.18.

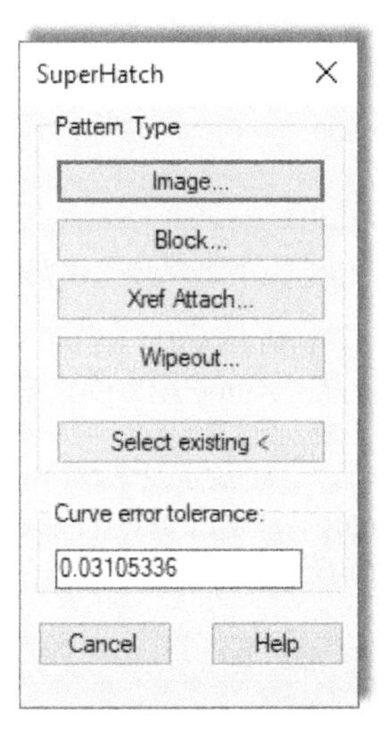

FIGURE 5.18 Superhatch.

Step 2. Click on Image… and browse around to find your chosen source image. Double-click it and you will see the dialog box in Fig. 5.19.

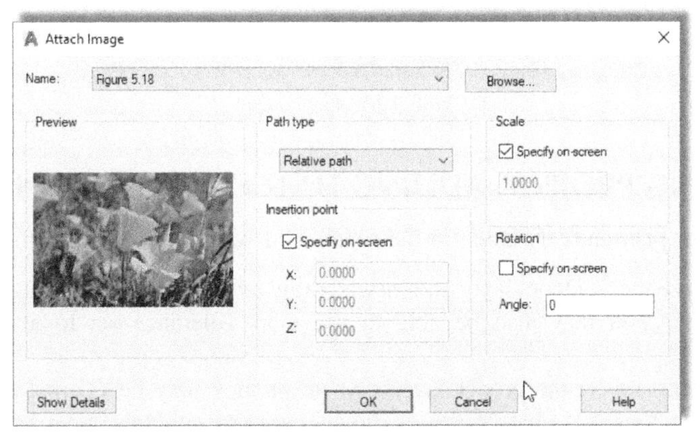

FIGURE 5.19 Attach Image.

Step 3. Click OK and position the image (using ENDpoint) at the lower left of the rectangle. AutoCAD will say: `Base image size: Width: 1, Height: 0.662338, Inches. Specify scale factor <1>:` Your sizing may of course be different. Go ahead and adjust the scale factor by either typing in a value or just dragging the mouse out. Make the image relatively small compared to the overall area (Fig. 5.20). Click when done and one copy of the image will appear where you placed it (Fig. 5.21).

FIGURE 5.20 Scale Factor.

FIGURE 5.21 Source image positioning.

Step 4. AutoCAD will ask: `Is the placement of this IMAGE acceptable? [Yes/No] <Yes>:`
If you like where it is, press Enter and AutoCAD will say: `Selecting visible objects for boundary detection...Done.`

Step 5. AutoCAD will say: `Specify an option [Advanced options] <Internal point>:` Click anywhere in the rectangular area and press Enter. AutoCAD will say: `Preparing hatch objects for display...Done.` The superhatch should then form a pattern (Fig. 5.22).

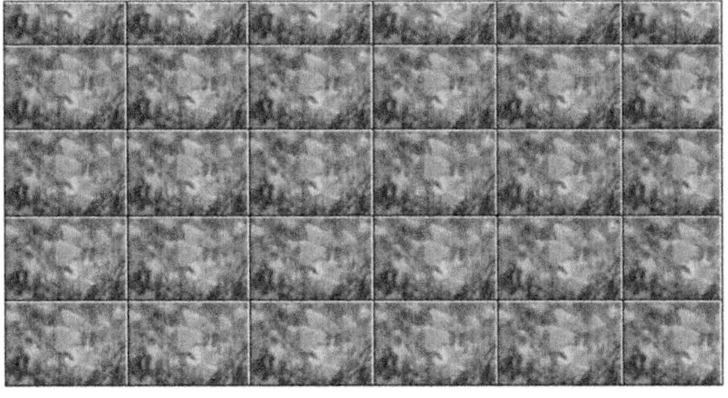

FIGURE 5.22 Final output.

5.7 IN-CLASS DRAWING PROJECT: ADDING HATCH TO FLOOR PLAN LAYOUT

We continue work on the floor plan and add hatch to the rooms and walls. Some tips follow:

- To add hatch to rooms, freeze the A-Doors layer and instead draw *temporary* lines closing off the door entrances. After hatching, remove the lines. You may also be able to use Gap Tolerance set to at least 3 feet, but that is not recommended.
- Make sure the hatch patterns are on the proper layers, whatever they may be (Carpeting or just Hatch_1, Hatch_2; you decide). You do not have to pick the hatch patterns shown, but make sure they are reasonable and at the proper scale.

To fill in the apartment walls, freeze everything except the A-Walls and the Hatch layer. Then, use the solid fill, color Gray 9, as described earlier. If you created everything properly, there should be no gaps or breaks in the wall islands. Check to make sure the windows are drawn correctly, meaning the wall has to have an actual closed gap, into which the window is inserted. Fig. 5.23 shows what everything should look like prior to wall hatching (A) and right after (B). One possible approach to hatching the apartment is shown in Fig. 5.24.

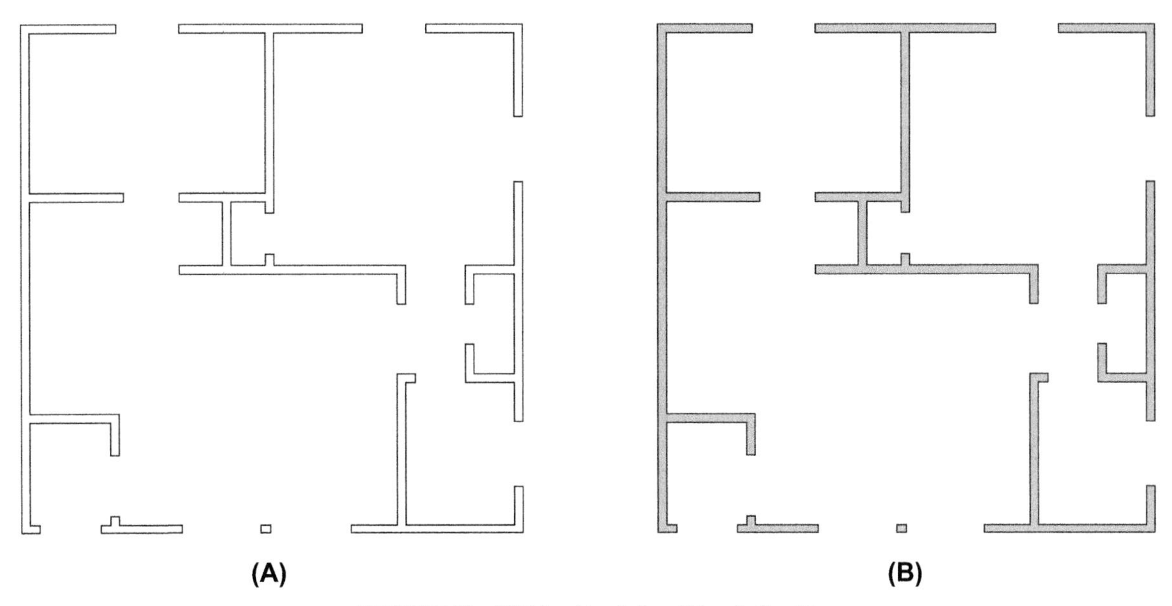

FIGURE 5.23 Wall hatching before (A) and after (B).

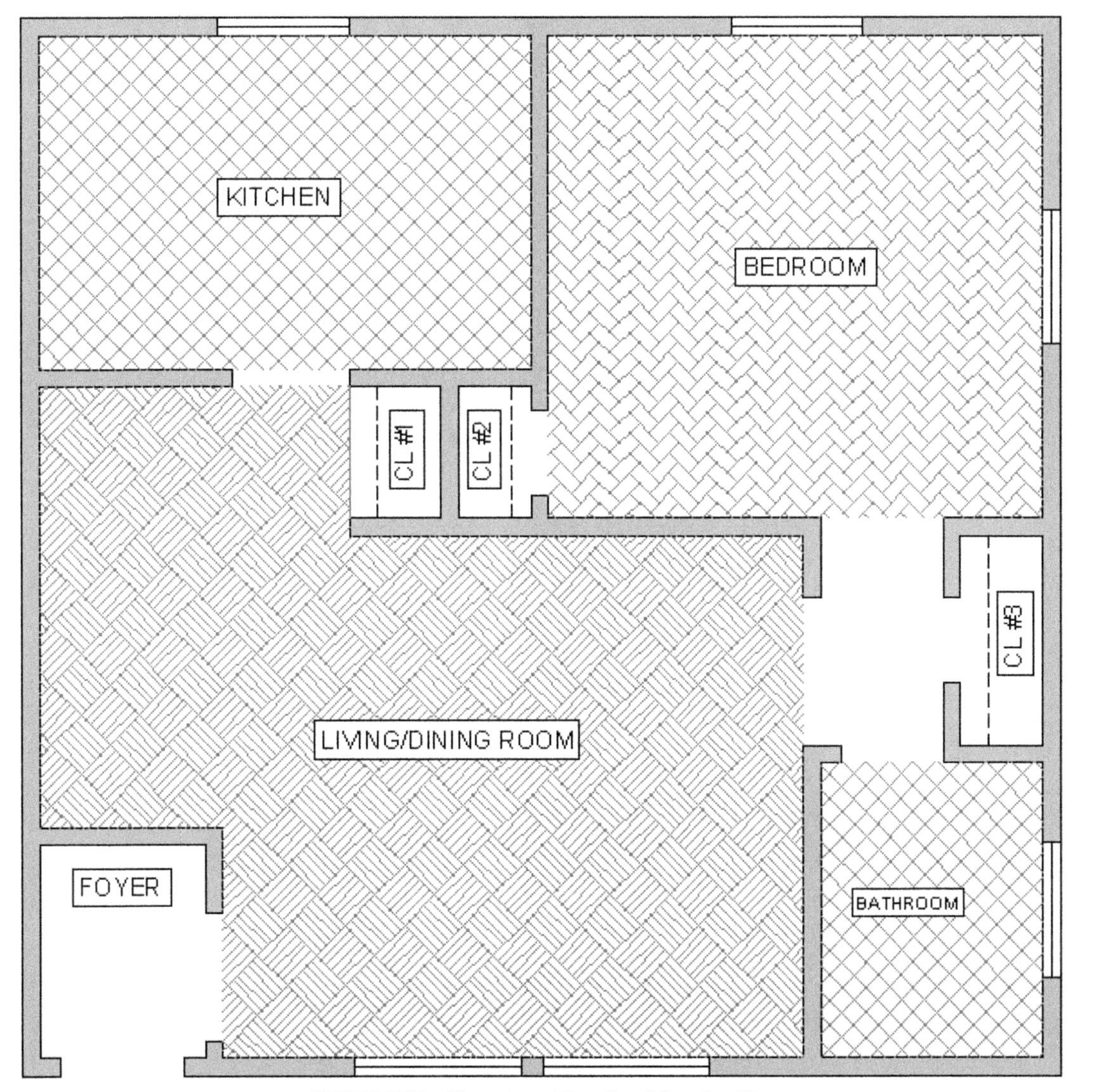

FIGURE 5.24 Floor plan with wall and floor hatching.

SUMMARY

You should understand and know how to use the following concepts and commands before moving on to Chapter 6:

- Hatch
  - Picking a pattern
  - Selecting area or object
  - Scale and orientation adjustments
  - Preview
- Editing a hatch
  - Double-click
- Additional options
  - Islands
  - Inherit properties
  - Associative
  - Draw order
  - Gap tolerance
- Gradients and fills

REVIEW QUESTIONS

Answer the following based on what you learned in this chapter:

1. List the four steps in creating a basic hatch pattern.
2. What two methods are used to select where to put the hatch pattern?
3. What commonly occurring problem can prevent a hatch being created?
4. What is the name of the tool recently added to AutoCAD to address this problem?
5. How do you edit an existing hatch pattern?
6. What is the effect of exploding hatch patterns? Is it recommended?
7. What especially useful pattern to fill in walls is mentioned in the chapter?

EXERCISES

1. In a new file, create the following 10″ rectangle, offset 2″ to get the smaller ones, and fill them all in with the shown hatch patterns. Be sure to create a layer for the rectangles, and a separate layer for the hatch patterns. Do try to duplicate what is shown, including hatch patterns, scale, and angles as closely as possible. (Difficulty level: Easy; Time to completion: <5 minutes.)

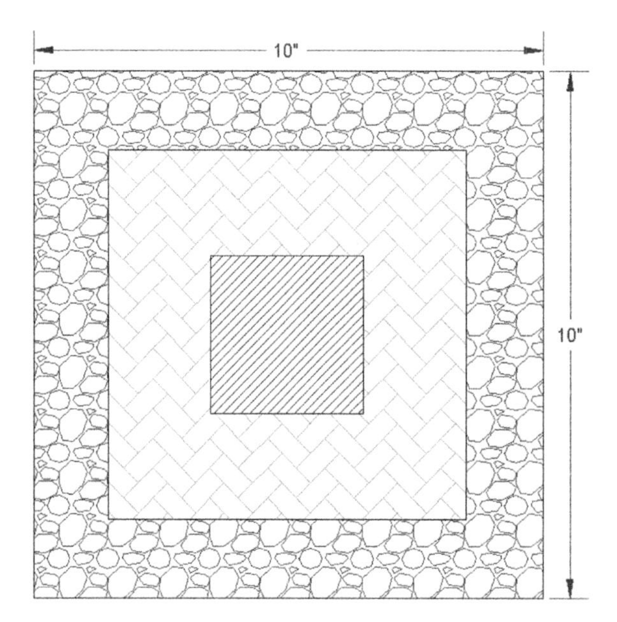

2. In a new file, create the following shapes and hatch patterns. Some overall sizing is shown, but you have to improvise the rest via the offset and line commands. Be sure to create a layer for background linework, and a separate layer for the hatch patterns. Do try to duplicate what is shown, including hatch patterns, scale, and angles as closely as possible. (Difficulty level: Easy; Time to completion: <5 minutes.)

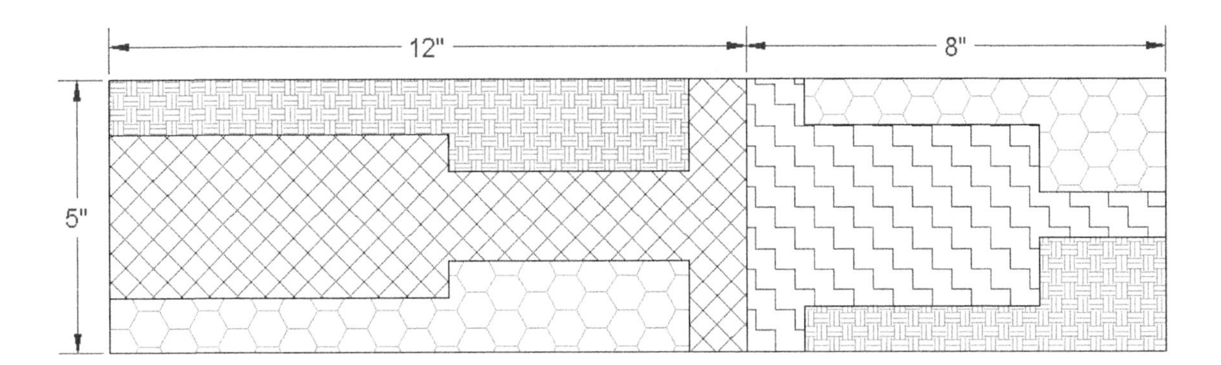

3. Open up (or redraw) Exercise 5 from Chapter 1 and add the following hatch patterns. Be sure to create a layer for the background linework, and a separate layer for the hatch patterns. Do try to duplicate what is shown, including hatch patterns, scale, and angles as closely as possible. (Difficulty level: Intermediate; Time to completion: 10 minutes.)

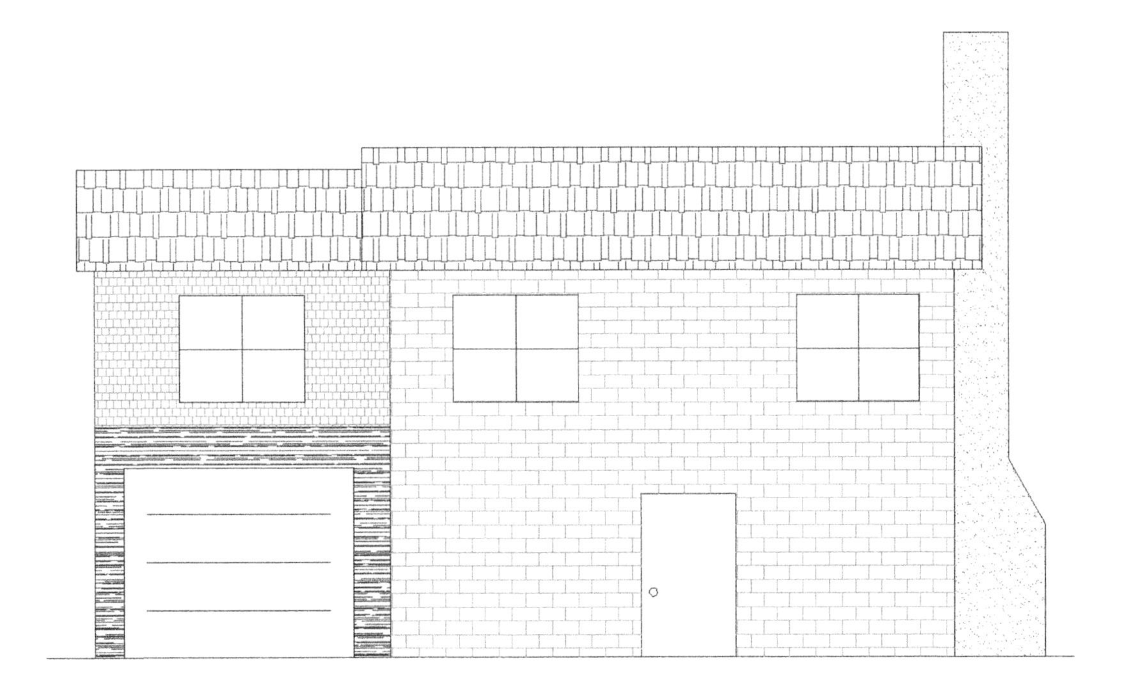

4. Open a new file and set units to Architectural. Create the following brick wall design, a slightly modified version of what is seen on the cover page of this chapter. Note that some of the backgrounds are gradients. Sizing and scale is up to you, but do try to duplicate what you see. (Difficulty level: Easy; Time to completion: <10 minutes.)

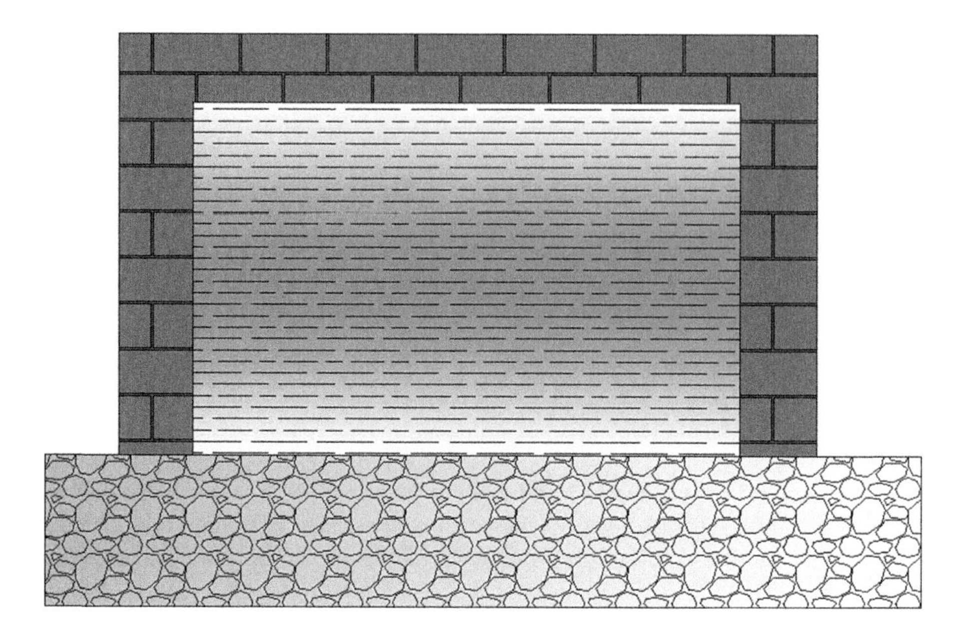

5. Open a new file, create the appropriate layers, and set units to Architectural. Create the following stair layout and hatch patterns. Any sizes and dimensions that are not given can be estimated, as can the scale of the patterns. Add in text where needed. (Difficulty level: Intermediate; Time to completion: 20–30 minutes.)

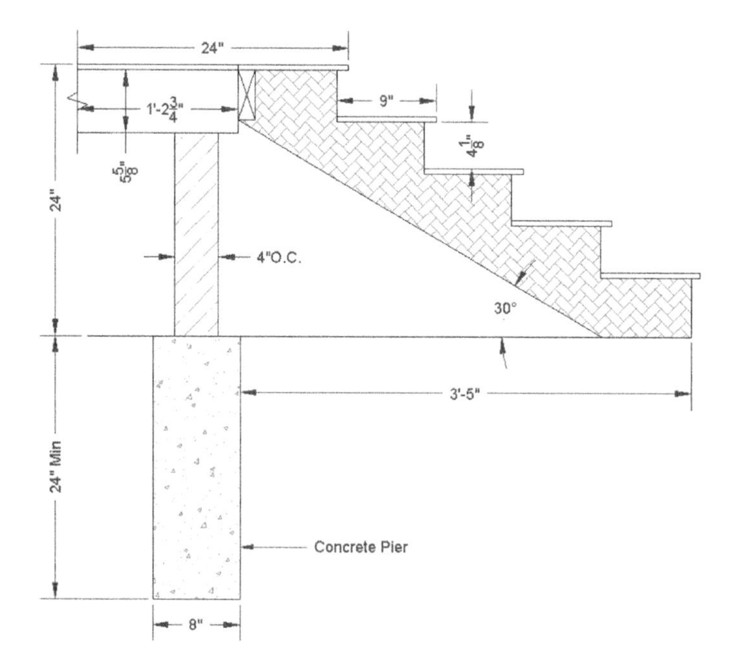

6. Open a new file and set units to Architectural. Create the following layout of a fictional set of cross-sectioned cylinders. Dimensions are provided, but sizing is of secondary importance to creating the hatch and gradient patterns. Notice how gradients are used to depict curvature inside the pipe cross sections, a good trick to add realism to 2D drawings. (Difficulty level: Intermediate; Time to completion: 20 minutes.)

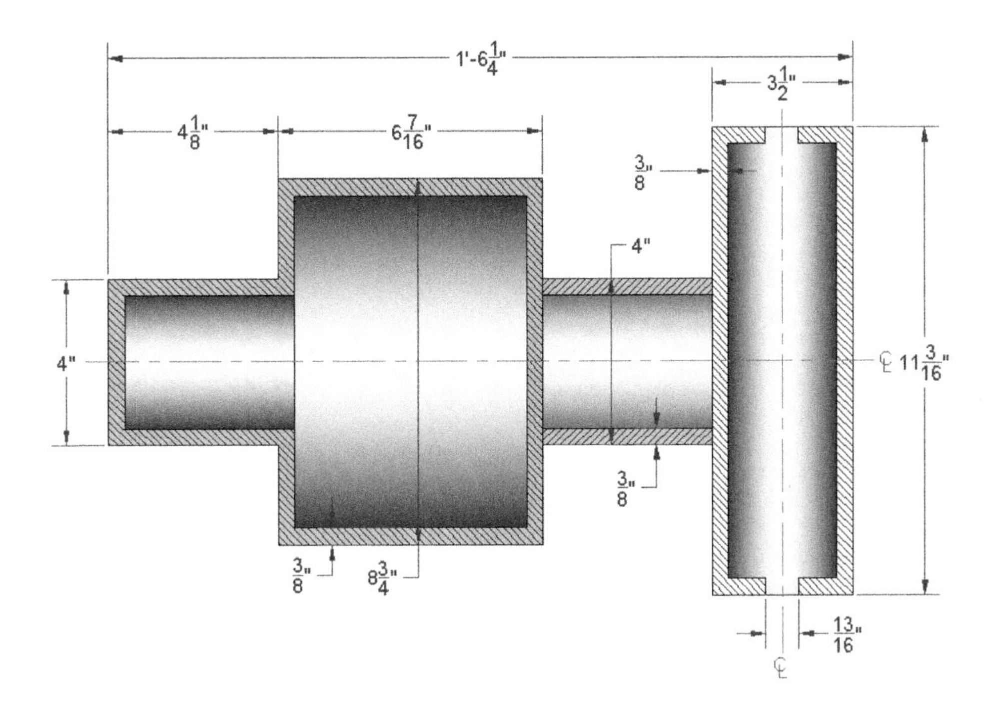

7. Create the following basic road signs using the starter dimensions given. All other sizing is to be approximated off of them. Add solid hatch fill as shown. (Difficulty level: Easy; Time to completion: 15 minutes.)

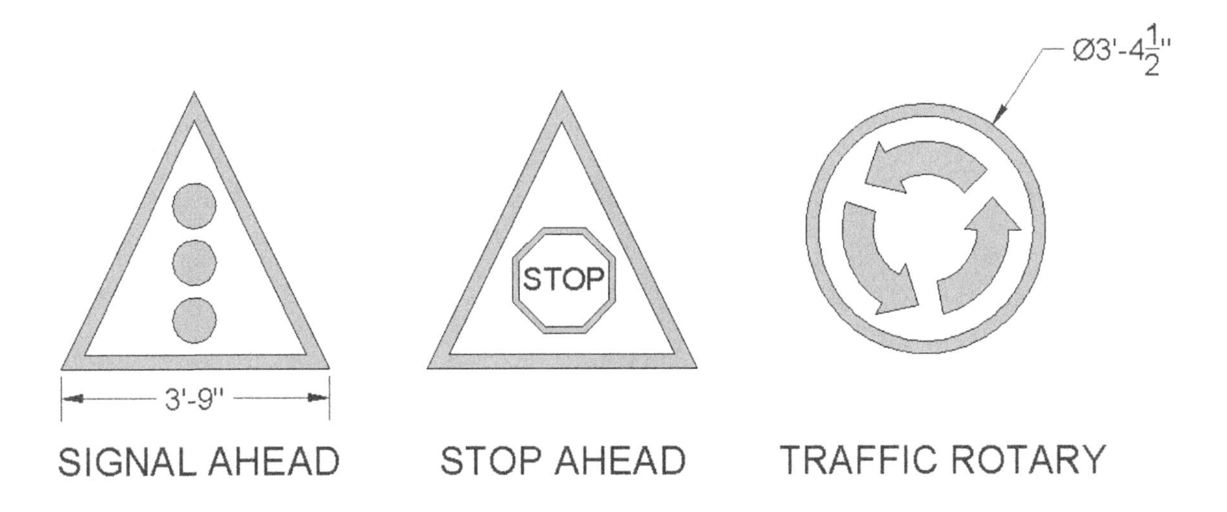

8. This exercise features some additional (slightly more complex) road signs. Create them using the starter dimensions given. All other sizing is to be approximated off of them. Add solid hatch fill as shown. (Difficulty level: Easy; Time to completion: 25 minutes.)

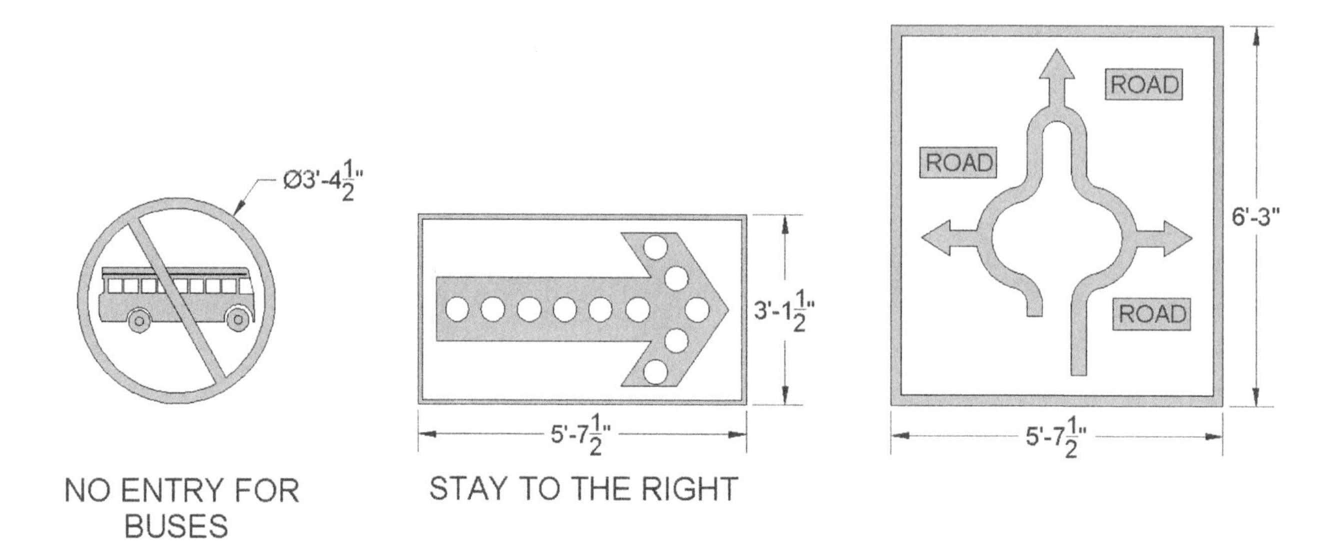

9. Create the following architectural detail. Many (but not all) of the dimensions are given. Use them as a basis to size the rest. Then, approximate the hatch patterns as shown. The squiggly vertical one is not a hatch pattern but a linetype, called *Batting*, often used to show insulation. Adjust its ltscale until it looks like what is shown. The rest of the hatch patterns should be familiar. (Difficulty level: Easy; Time to completion: 25–35 minutes.)

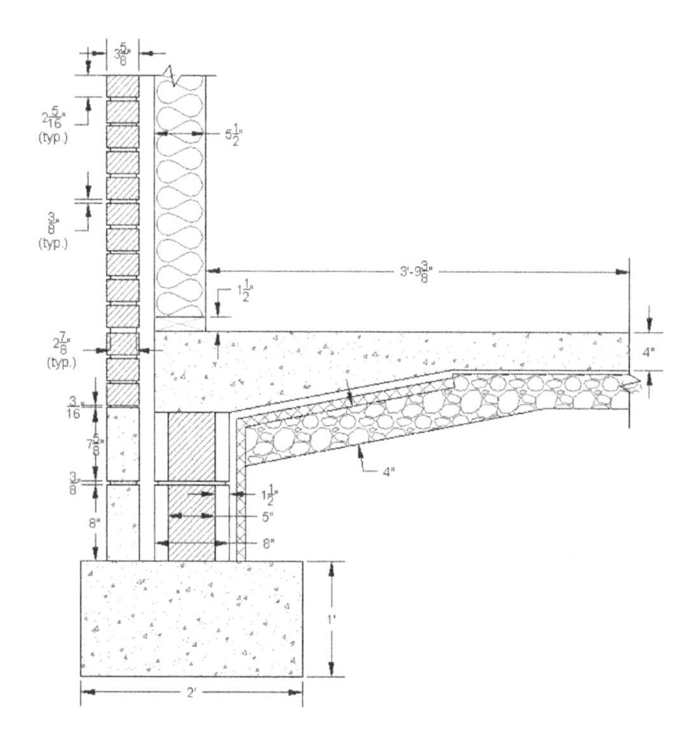

10. For the final exercise of this chapter, set up your file with the appropriate layers and text settings and draft the following detail of the roof and support beams. All necessary dimensions are found in either the main detail or the two supporting details. Any dimensions not given can be assumed. Then, approximate the hatch patterns as shown. The 6 and 12 above the roof indicates its slope and may be used to calculate the angle via some basic trigonometry, if needed. (Difficulty level: Moderate; Time to completion: 30–40 minutes.)

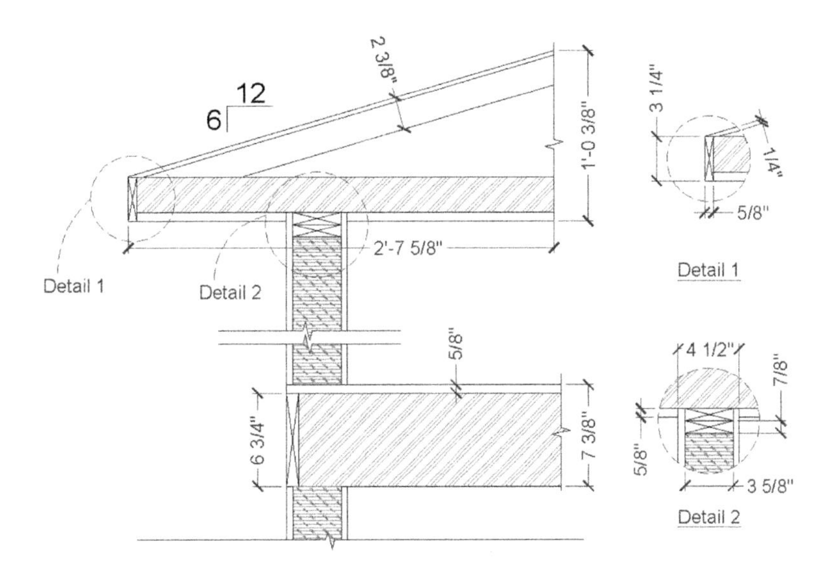

Chapter 6

Dimensions

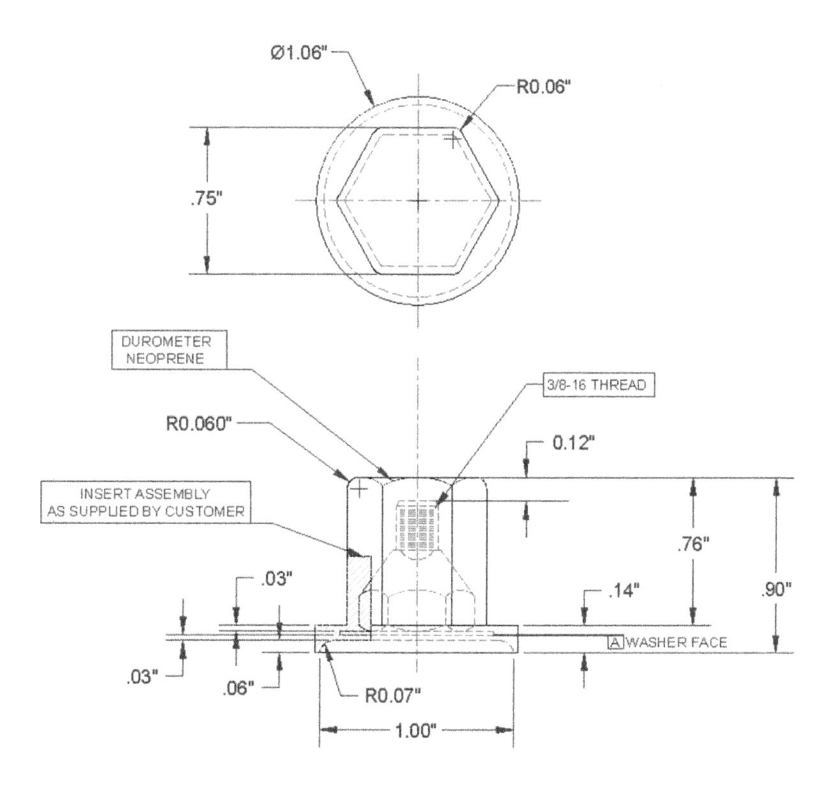

Learning Objectives

In this chapter, we introduce the first parts of AutoCAD's extensive dimensioning capabilities and discuss the following:

- Linear (horizontal and vertical) dimensions
- Aligned dimensions
- Diameter and radius dimensions
- Angular dimensions
- Continuous and baseline dimensions
- Leader and multileader
- Arc length dimensions
- Jogged and jogged linear dimensions
- Ordinate dimensions
- ddedit
- ddim
- Dimension units
- Dimension font
- Dimension arrowheads
- Dimension overall size
- Center marks and centerlines

At the end of this chapter, you will be able to add dimensions to your floor plan.
Estimated time for completion of this chapter: 3 hours (lesson and project).

Up and Running with AutoCAD 2019. https://doi.org/10.1016/B978-0-12-816440-2.00006-9

6.1 INTRODUCTION TO DIMENSIONS

Dimensioning in AutoCAD is a major topic, one so extensive that it is split into two separate discussions. In this chapter, we address the basics of what dimensions are, which ones are available, and how to use and edit them. In Chapter 13, Advanced Dimensions, we look at extensive customization and additional options and explore some relatively new additions to the dimensioning family: parametric dimensioning and the concept of constraints and dimension-driven design.

So, what are dimensions? They are simply visible measurements for purposes of conveying information to the audience who will be looking at your design. It is how you describe the size of the design and where it is in relation to everything else. Dimensions can be natural, meaning they display the actual value, or forced, when they display an altered value, such as when you dimension an object with a break line.

6.2 TYPES OF DIMENSIONS

The first step in learning dimensions is to know what is available, so you can use the appropriate type in any situation. The primary and secondary dimensions available to you are shown next. Commit them to memory, as you may need most of them on a complex project. AutoCAD 2019 features some rather fast methods for entering basic dimensions. We need to go over these but also mention other ways for those using slightly older releases.

Primary dimensions are the dimensions used most often by the typical architectural or engineering designer:

- Linear (horizontal and vertical)
- Aligned
- Diameter
- Radius
- Angular
- Continuous
- Baseline
- Leader

Secondary dimensions are used less often or are quite specialized but still need to be learned:

- Arc length
- Jogged
- Jogged linear
- Ordinate

Let us first take the primary list then the secondary and discuss the dimensions one by one to understand how each is used. In each case, you need to draw the appropriate shape to practice with, and then add the new dimension(s) by following the step-by-step instructions.

As before, we use typing, toolbars, cascading menus, and the Ribbon. Keep in mind however, there are numerous ways to input dimensions into AutoCAD. The preferred methods for AutoCAD 2019 are the Ribbon and command-line typing. As it happens, both approaches lead to the same revised DIM command, which gives you a brand-new "one-touch" method of dimension creation and placement. For users of older versions of AutoCAD (that is, 2015 and older), the closest you have for this is the Quick Dimension (or QDIM). You can also rely on the "legacy" linear methods of selecting placement points for the new dimensions. The number of different methods can be a bit overwhelming, but you will quickly gravitate toward one over another. All methods are detailed exhaustively with the first linear dimension, and less so for the other ones, as the approach is quite similar.

If you are a toolbar user, go ahead and open up the Dimension toolbar (Fig. 6.1), adding it to those already on your screen.

FIGURE 6.1 Dimension toolbar.

Note an important point if typing dimensions: You need to type in `dim` and press Enter. This gets you into the dimension submenu, and you are able to operate with making new dimensions (and related features) but not with any other commands. Simply press Esc to get out of this.

Linear Dimensions

These are random dimensions that are strictly horizontal or vertical (Fig. 6.2). We describe a general approach next, applicable to either the horizontal or vertical dimension, so be sure to practice both of them. Before you begin, create a generic square or rectangle as seen in Fig. 6.2.

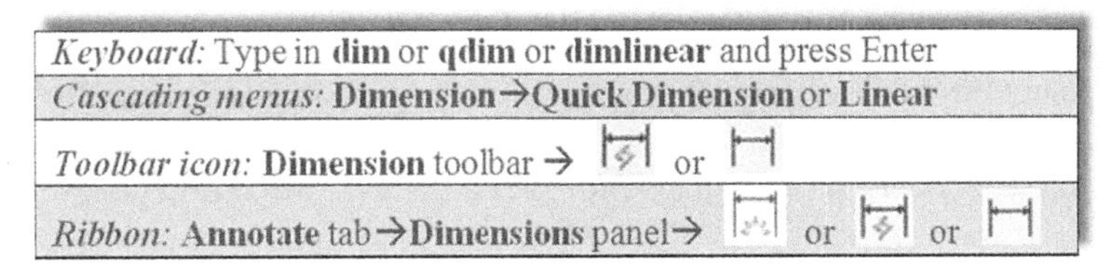
Keyboard: Type in **dim** or **qdim** or **dimlinear** and press Enter
Cascading menus: **Dimension→Quick Dimension** or **Linear**
Toolbar icon: **Dimension** toolbar → or
Ribbon: **Annotate** tab→**Dimensions** panel→ or or

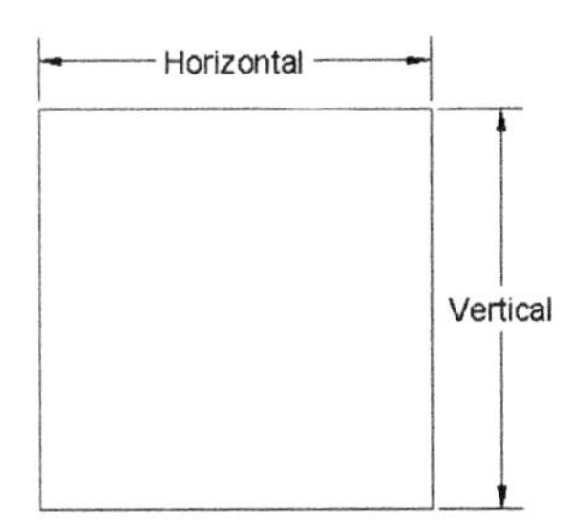

FIGURE 6.2 Linear (horizontal and vertical) dimensions.

For Ribbon or Typing Input (Dim, Primary Method)

Step 1. Press the first Ribbon icon in the command matrix (with the sunburst pattern), or type in `dim` and press Enter.
- AutoCAD says: `Select objects or specify first extension line origin or [Angular/Baseline/Continue/Ordinate/aliGn/Distribute/Layer/Undo]:`

Step 2. Hover the mouse over any vertical or horizontal line making up the square shape. It becomes dashed and a horizontal or vertical dimension appears in "preview" mode, shifting around as you move the mouse.
- AutoCAD says: `Select line to specify extension lines origin:`

Step 3. Click once and drag the dimension out to its appropriate location, just outside the square.
- AutoCAD says: `Specify dimension line location or second line for angle [Mtext/Text/text aNgle/Undo]:`

Step 4. Click one last time to fix the dimension in place.
- AutoCAD says: `Select extension line origin as baseline or [Continue]:`

You are done at this point and can move your mouse to another line that you may wish to dimension. This "one-touch" method of entry was introduced 3 years ago with AutoCAD 2016, somewhat speeding up dimensioning, and it continues with AutoCAD 2019.

For Typing, Cascading Menus, Toolbar, or Ribbon Input (Quick Dim, Secondary Method)

Step 1. Type in `qdim` and press Enter, select Dimension → Quick Dimension from the cascading menus, or press the appropriate toolbar or Ribbon icon (with the basic dimension and lightning bolt).
- AutoCAD says: `Associative dimension priority = Endpoint`
  `Select geometry to dimension:`

Step 2. Pick any of the lines of your square, with the entire square becoming dashed.
- AutoCAD informs you: `1 found.`

Go ahead and press Enter.

Step 3. Drag the dimension out to its appropriate location, just to the outside of the square. You will not yet see the actual dimension, only the extension lines, as you move the mouse up and down.
- AutoCAD says: `Specify dimension line position, or [Continuous/Staggered/Baseline/Ordinate/Radius/Diameter/datumPoint/Edit/seTtings] <Continuous>:`

Step 4. Click one last time to fix the dimension in place. You are done at this point and can repeat the command for another location.

For Typing, Cascading Menus, Toolbar, or Ribbon Input (Linear Dim, "Legacy" Method)

Step 1. Type in dimlinear and press Enter, select Dimension → Linear from the cascading menus, or press the appropriate toolbar or Ribbon icon (with a basic horizontal dimension).

- AutoCAD says: Specify first extension line origin or <select object>:

Step 2. With the ENDpoint OSNAP select one of the corners of the square.

- AutoCAD says: Specify second extension line origin:

Step 3. With the ENDpoint OSNAP select any other nearby corner of the square—vertically or horizontally. A dimension appears; go ahead and drag it to its appropriate location.

- AutoCAD says: Specify dimension line location or [Mtext/Text/Angle/Horizontal/Vertical/Rotated]:

Step 4. Click one last time to fix the dimension in place. The dimension appears and the command line shows the same value you already saw on the screen (e.g., Dimension text = 2.8422).

To summarize, as you just saw, there are essentially three ways to create virtually any dimension. The first is via the relatively new "one-touch" method. This can be done only via the Ribbon and command-line input. This method allows you to begin the dimensioning command and "touch" any geometry you want to dimension, with the appropriate dimension appearing automatically. Second is the qdim approach. It has been around for a while and can be accomplished via all four methods of input. It is a bit more cumbersome than Dim, but not by much, giving you the dimensions of some, but not all, geometry selected. It is mostly limited to linear and radius/diameter dimensions. The final version is the much older, but still effective, Linear command. It can also be executed via all four input methods. It requires two points, not a line, to create a linear dimension and takes a bit more time to do.

For users of AutoCAD 2015 and older releases, you can also type in Dim:, press Enter, and type in hor for a horizontal or vert for a vertical dimension, then proceed to place it point by point following the previous Linear description. This method of command input is not available from AutoCAD 2016 on, but that does not matter much, as the more advanced Dim command can handle point-by-point dimension entry just as easily as the "one-touch" approach if you happen to pick corners instead of lines.

In general, linear dimensions are by far the most common in drawings and are also the basis for the continuous and baseline dimensions, to be covered later on. Be sure to run through both the horizontal and vertical linear dimensions, with all methods shown, before settling on one. Next is the aligned dimension, which gives the true distance of a slanted surface. We describe it using the brand new one-touch approach, but the command matrix shows all options.

Aligned Dimension

This dimension measures a slanted line or object, as seen in Fig. 6.3. To practice it, you first need to create a small- to medium-sized square and chop off any corner, as also seen in Fig. 6.3. Only the first Ribbon/typing approach is described in detail, but experiment with the other approaches (listed in the command matrix) if you prefer them.

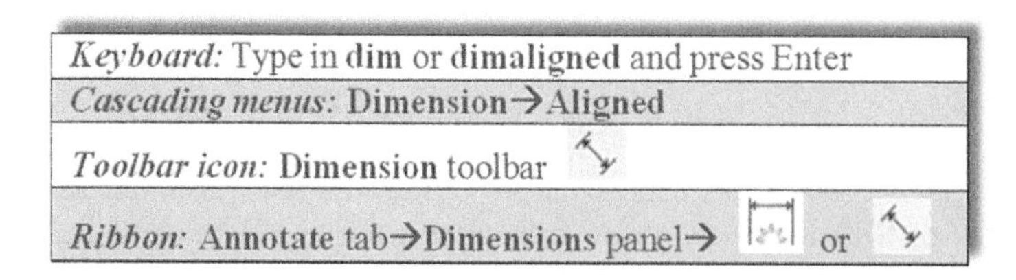

Keyboard: Type in **dim** or **dimaligned** and press Enter
Cascading menus: **Dimension→Aligned**
Toolbar icon: **Dimension** toolbar
Ribbon: **Annotate** tab→**Dimensions** panel→ or

Step 1. Press the first Ribbon icon in the command matrix (with the sunburst pattern) or type in dim and press Enter.

- AutoCAD says: Select objects or specify first extension line origin or [Angular/Baseline/Continue/Ordinate/aliGn/Distribute/Layer/Undo]:

Step 2. Hover your mouse over the slanted line of the square shape. It becomes dashed and an aligned dimension appears in preview mode, shifting around as you move the mouse.

- AutoCAD says: Select line to specify extension lines origin:

Step 3. Click once and drag the dimension out to its appropriate location, just outside the square.

- AutoCAD says: `Specify dimension line location or second line for angle [Mtext/Text/text aNgle/Undo]:`

Step 4. Click one last time to fix the dimension in place.

- AutoCAD says: `Select extension line origin as baseline or [Continue]:`

You are done at this point and can move your mouse to another line that you may wish to dimension.

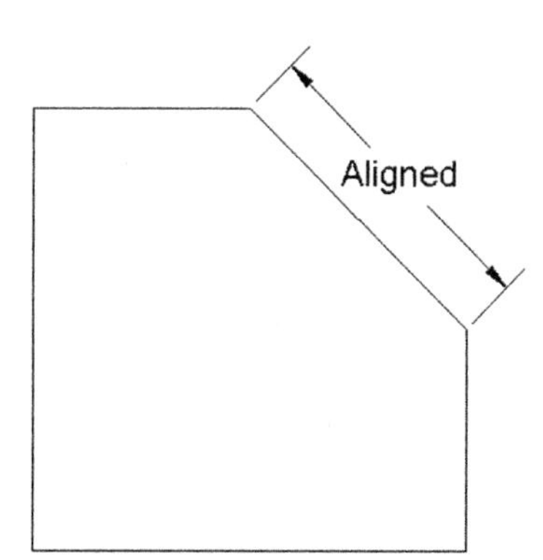

FIGURE 6.3 Aligned dimensions.

The next two dimensions, diameter and radius, are essentially similar, although we cover them separately for clarity. Note the primary differences: the symbol indicating diameter (Ø) and the symbol for radius (R) prior to the value. Although you can set this up, by default there is no line crossing the circle halfway for radius or all the way across for diameter, so you have to read the values carefully and look for the appropriate symbol to know what you are looking at.

Also, note the location of the values. They are positioned at roughly 10, 2, 4, and 8 o'clock relative to the circle. Do not just put them anywhere; hand-drafting rules still apply in CAD, and designers have typically left them in these positions for clarity, consistency, and style.

Diameter Dimension

This dimension measures the diameter of a circle or an arc, as seen in Fig. 6.4. To practice it, you first need to create a small- to medium-sized circle, as also seen in Fig. 6.4. Only the first Ribbon/typing approach is described in detail, but experiment with the other approaches (listed in the command matrix) if you prefer them.

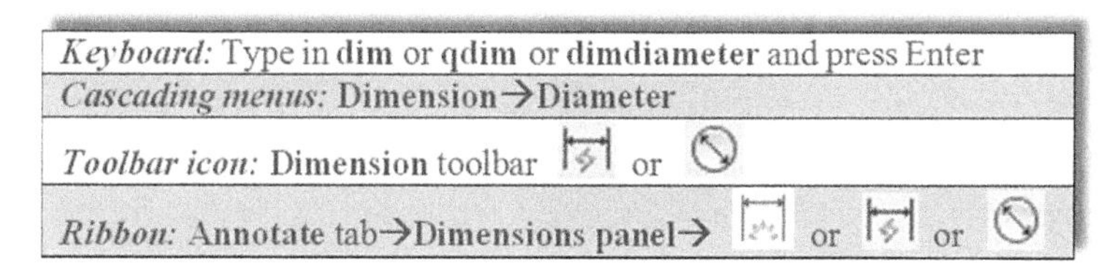

Keyboard: Type in **dim** or **qdim** or **dimdiameter** and press Enter
Cascading menus: **Dimension→Diameter**
Toolbar icon: **Dimension** toolbar or
Ribbon: **Annotate** tab→**Dimensions panel**→ or or

Step 1. Press the first Ribbon icon in the command matrix (with the sunburst pattern) or type in `dim` and press Enter.

- AutoCAD says: `Select objects or specify first extension line origin or [Angular/Baseline/Continue/Ordinate/aliGn/Distribute/Layer/Undo]:`

Step 2. Hover your mouse over any part of the circle. It becomes dashed and a diameter dimension (along with a center mark) appears in preview mode, shifting around as you move the mouse.

- AutoCAD says: `Select circle to specify diameter or [Radius/Jogged/Center mark/Angular]:`

Step 3. Position your mouse somewhere to the upper right of the circle (rarely inside, usually outside) at 2 o'clock or at any of the other accepted positions.

- AutoCAD says: `Specify diameter dimension location or [Radius/Mtext/Text/text aNgle/Undo]:`

Step 4. Click one last time to fix the dimension in place. You are done at this point and can move your mouse to anything else that you may wish to dimension. You have a value with a diameter symbol (e.g., ø2.25).

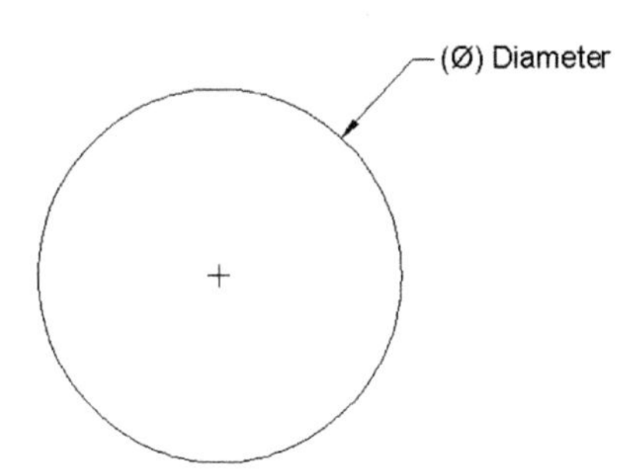

FIGURE 6.4 Diameter dimension.

Radius Dimension

This is a dimension that measures the radius of a circle or an arc, as seen in Fig. 6.5. To practice it, you first need to create a small- to medium-sized circle (or retain the one you already have), as also seen in Fig. 6.5. Only the first Ribbon/typing approach is described in detail, but experiment with the other approaches (listed in the command matrix) if you prefer them.

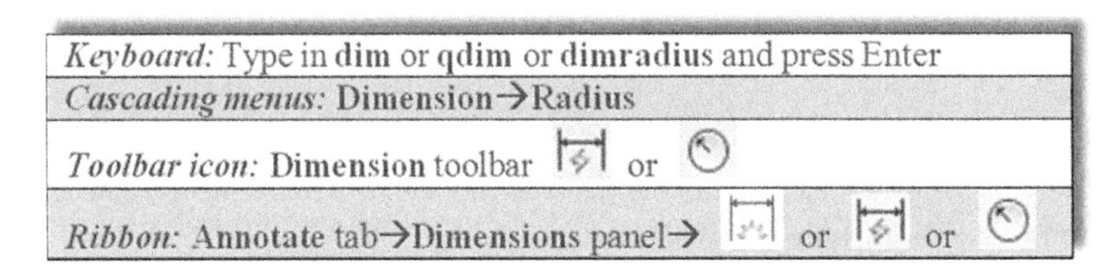

Keyboard: Type in **dim** or **qdim** or **dimradius** and press Enter
Cascading menus: **Dimension→Radius**
Toolbar icon: **Dimension** toolbar or
Ribbon: **Annotate** tab→**Dimensions** panel→ or or

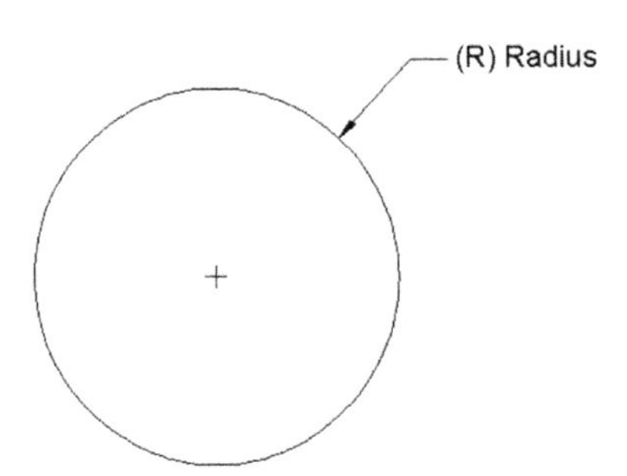

FIGURE 6.5 Radius dimension.

Step 1. Press the first Ribbon icon in the command matrix (with the sunburst pattern) or type in dim and press Enter.
- AutoCAD says: `Select objects or specify first extension line origin or [Angular/Baseline/Continue/Ordinate/aliGn/Distribute/Layer/Undo]:`

Step 2. Hover your mouse over any part of the circle. It becomes dashed and a radius dimension (along with a center mark) appears in preview mode, shifting around as you move the mouse. *Note an important point:* If AutoCAD shows the diameter dimension (left over from the last attempt) instead of the expected radius, then simply press `r` for `Radius`, as seen in the menu that follows, and AutoCAD switches to a radius instantly. The same goes for the other way around, if you are expecting a diameter and get a radius, press d when you get to this point.
- AutoCAD says: `Select circle to specify diameter or [Radius/Jogged/Center mark/Angular]:`

Step 3. Position your mouse somewhere to the upper right of the circle (rarely inside, usually outside) at 2 o'clock or at any of the other accepted positions.
- AutoCAD says: `Specify diameter dimension location or [Radius/Mtext/Text/text aNgle/Undo]:`

Step 4. Click one last time to fix the dimension in place. You are done at this point and can move your mouse to anything else that you may wish to dimension. You have a value with an R (e.g., R1.25).

Angular Dimension

This dimension measures the angle between two lines or objects, as seen in Fig. 6.6. To practice it, you first need to create two lines at some random angle to each other, as also seen in Fig. 6.6. For this particular dimension command, using the older toolbar icon or the equivalent Ribbon one is easier than the one-touch newer method, so we detail that one only, but experiment with the other approaches (listed in the command matrix) if you prefer them.

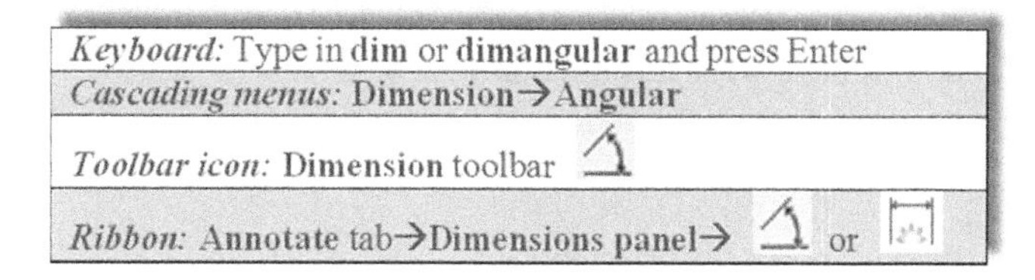

Keyboard: Type in dim or dimangular and press Enter
Cascading menus: Dimension→Angular
Toolbar icon: Dimension toolbar
Ribbon: Annotate tab→Dimensions panel→ or

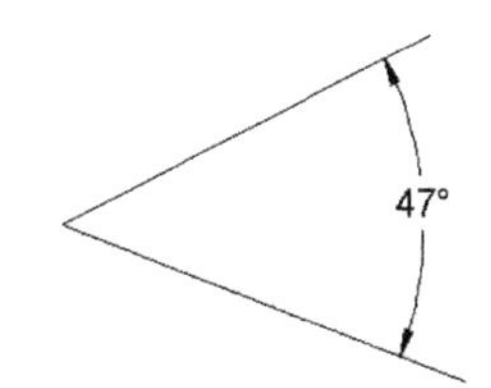

FIGURE 6.6 Angular dimensions.

Step 1. Begin the angular dimension command via the Ribbon's Angular icon.
- AutoCAD says: `_dimangular Select arc, circle, line, or <specify vertex>:`

Step 2. Pick the first line (either one).
- AutoCAD says: `Select second line:`

Step 3. Pick the second line, and an angular dimension appears, attached to every move of your mouse.
- AutoCAD says: `Specify dimension arc line location or [Mtext/Text/Angle/Quadrant]:`

Step 4. Drag the mouse out and move it around, selecting the best position for the new dimension and click when you find a spot you like. Be careful, as the supplement of the degree value shows if you move behind the lines, which of course may be desirable in some cases. Upon that final click, AutoCAD gives you a value on the command line that matches what is on the screen (e.g., Dimension text = 47).

Note that, if you do use the new one-touch approach, the procedure is more or less the same as in all the previous dimensions, but AutoCAD does not recognize that you are asking for an angle value right away when you touch the first line and attempts to give you an angular reading. Only when you touch the second line does the software "get it" and gives an angular dimension. So, as stated earlier, in this particular case, the new approach is actually a bit more cumbersome.

Angular completes the basic set of new fundamental dimensions. Next, we have continuous and baseline, which are not really new, as we will see, but rather an extension (or automation) of the linear dimension learned earlier. We then conclude with a leader, not so much a true dimension but a useful member of that family.

Continuous Dimensions

These are continuous *strings* of dimensions, as seen in Fig. 6.7.

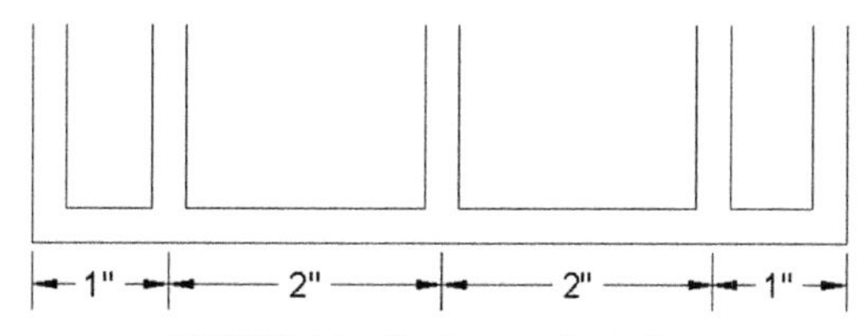

FIGURE 6.7 Continuous dimensions.

A short explanation is in order, as continuous and baseline dimensions sometimes give new students initial problems. Continuous dimensions are nothing more than a string of familiar horizontal or vertical ones. The idea here is to create one of those two types of linear dimensions, and then start up the continuous dimension where you just left off and let AutoCAD quickly fill them in as you pick contact points. It is really the same as using linear dimensions over and over again but faster and more accurate because it is partially automated.

The easiest way to practice continuous dimensions is to draw a set of squares attached to each other. These represent a simplified view of building and room walls. Draw one random-sized square and copy it, endpoint (lower left corner) to endpoint (lower right corner), until you have what is shown in Fig. 6.8.

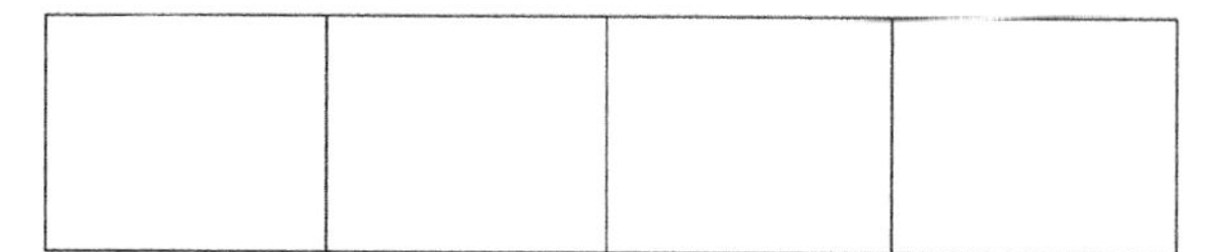

FIGURE 6.8 Setup to practice continuous dimensions.

The next step is to draw one horizontal linear dimension, as you already have done (it could have easily been a vertical line, too, if the rectangles were stacked vertically). Locate it about where you would like the entire string to go, as seen in Fig. 6.9.

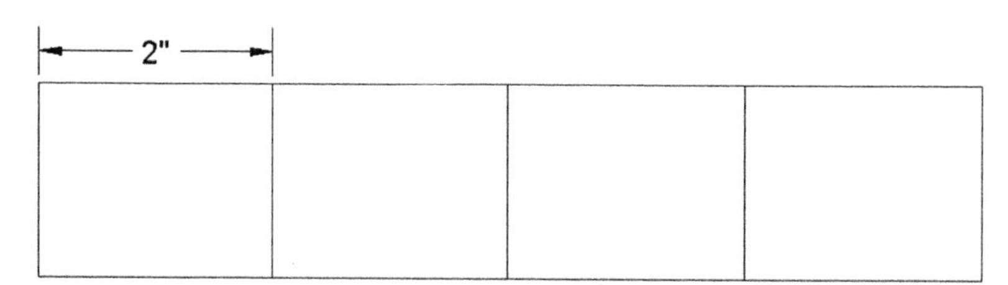

FIGURE 6.9 First step to create continuous dimensions.

This is the essential first step in creating both continuous and baseline dimensions. Note that you *can* create an initial linear dimension then do something else for a while. AutoCAD remembers (at least in most cases) the last place you left off and starts the new string from there. This changes a bit once you create another similar dimension somewhere else. Then, AutoCAD does not know to which one you are referring, so you have to slightly modify your approach and pick the dimension that will be the start of the continuous string. In the instructions to follow, we assume you began to create the string right after making an initial linear dimension.

Keyboard: Type in **dim** or **dimcontinuous** and press Enter
Cascading menus: **Dimension→Continue**
Toolbar icon: **Dimension** toolbar
Ribbon: **Annotate** tab→**Dimensions** panel→

Step 1. Once you have the squares and one linear dimension, as seen in Fig. 6.9, start up the continuous command via any of the preceding methods.

- AutoCAD says: `Specify a second extension line origin or [Undo/Select]<Select>:`

Step 2. A new dimension appears. Pick the *next* point (corner) along the string of rectangles and click on it (always using OSNAP points, no eyeballing).

- AutoCAD says: `Dimension text = 2.0`

Your value, of course, may be different.

Step 3. You can continue this until you run out of objects to dimension; at that point, just press Esc. Your result is shown in Fig. 6.10. Notice how all the values are neatly spaced in a row.

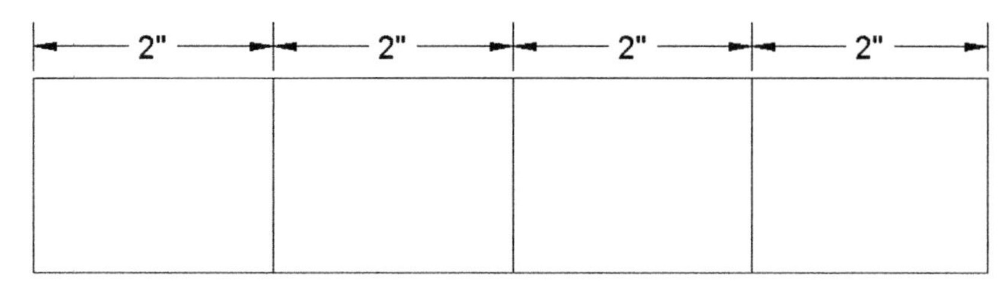

FIGURE 6.10 Continuous dimensions result.

Occasionally (and for a number of reasons), AutoCAD may not automatically know what the source for the continuous string may have been and may ask you to `Select continued dimension:`. Just pick the linear dimension you wish to serve as the "source" and continue with Step 2 as before.

Baseline Dimensions

These are continuous *stacks* of dimensions, as seen in Fig. 6.11.

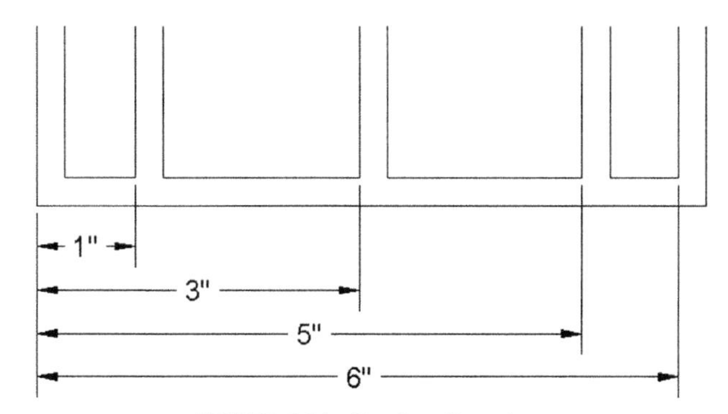

FIGURE 6.11 Baseline dimensions.

The baseline dimension, as mentioned before, is very similar in principle to the continuous dimension. The goal is to make a neat stack of evenly spaced dimensions that all start at one point (the base in baseline). To begin, erase the previous continuous dimensions (leaving the squares) and once again draw one linear (horizontal) dimension, as seen before in Fig. 6.9.

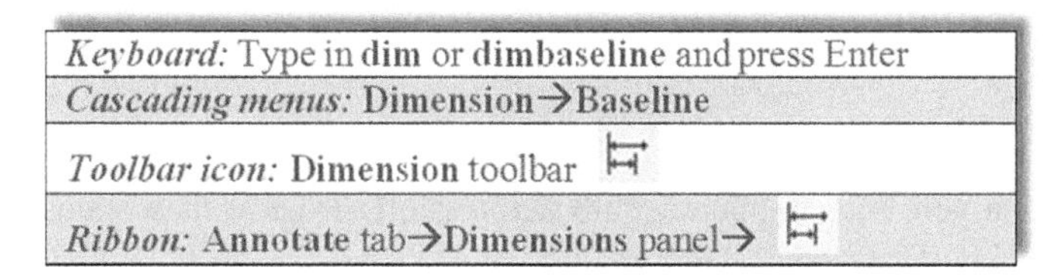

Keyboard: Type in **dim** or **dimbaseline** and press Enter
Cascading menus: **Dimension→Baseline**
Toolbar icon: **Dimension** toolbar
Ribbon: **Annotate** tab→**Dimensions** panel→

Step 1. Once you have the squares and one linear dimension, start up the baseline command via any of the preceding methods.

- AutoCAD says: `Specify a second extension line origin or [Undo/Select]<Select>:`

Step 2. A new dimension appears. Pick the *next* point (corner) along the string of rectangles and click on it (always using OSNAP points, no eyeballing).

- AutoCAD says: `Dimension text = 4.0`

Your value, of course, may be different.

Step 3. You can continue this until you run out of objects to dimension; at that point, just press Esc. Your result is shown in Fig. 6.12.

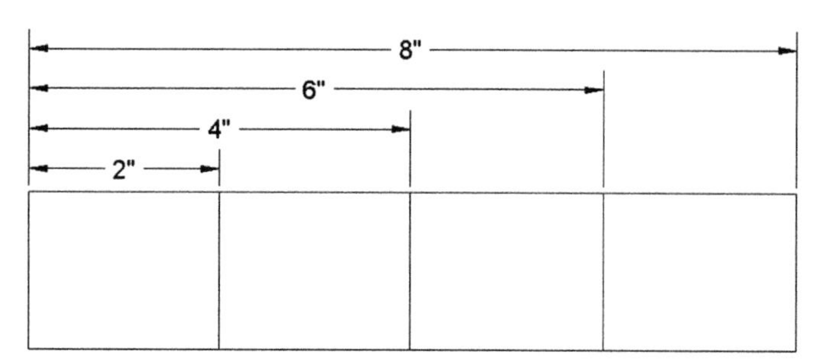

FIGURE 6.12 Baseline dimensions result.

Occasionally, just like with the continuous dimension, AutoCAD may not automatically know what the source for the baseline string may have been and ask you to Select base dimension:. Just pick the linear dimension you wish to serve as the "source" and continue with Step 2 as before.

Leader and Multileader

This is an *arrow and label* combination pointing at something (Fig. 6.13).

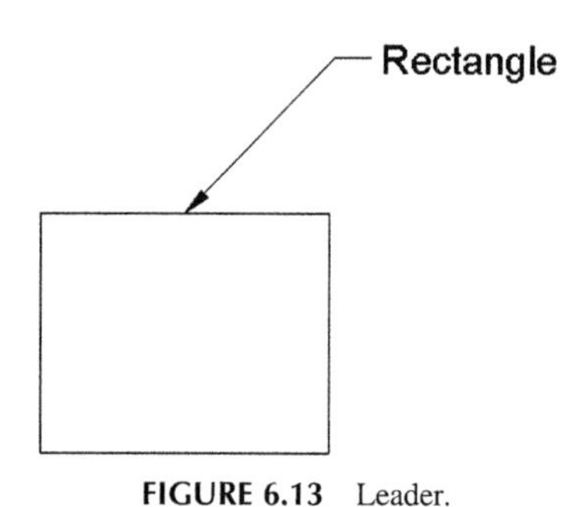

FIGURE 6.13 Leader.

While a leader is not a true dimension by definition, it is still a very common and necessary member of the dimension family. The values shown by the leader can be not only numerical but also text comments from the designer, concerning the part or object to which it is pointing.

Leaders are so important that AutoCAD has given them a major rework in recent releases. We go over the two main options in increasing order of complexity, starting with the basic leader command, followed by the more feature-rich multileader, with its Add, Remove, Align, and Collect options.

Leader

Step 1. Draw a small box on your screen, making sure Ortho is off. Type in the command leader and press Enter.
- AutoCAD says: `Specify leader start point:`

Step 2. Click with OSNAP precision on your shape, perhaps a midpoint on the top side, as seen in Fig. 6.13.
- AutoCAD says: `Specify next point:`

Step 3. Move your mouse at a 45 degrees angle away from the first point and click again at a reasonable distance away. You create the *leader line and arrowhead* seen in Fig. 6.14.
- AutoCAD says: `Specify next point or [Annotation/Format/Undo] <Annotation>:`

Step 4. Turn Ortho back on and draw a short line to the right, clicking when done; this is your *horizontal landing*, as seen in Fig. 6.15.
- AutoCAD says: `Specify next point or [Annotation/Format/Undo] <Annotation>:`

Step 5. You are done, so press Enter.
- AutoCAD says: `Enter first line of annotation text or <options>:`

Type something in, pressing Enter if you wish to do another line, and Enter again if you are done, as seen in Fig. 6.16.

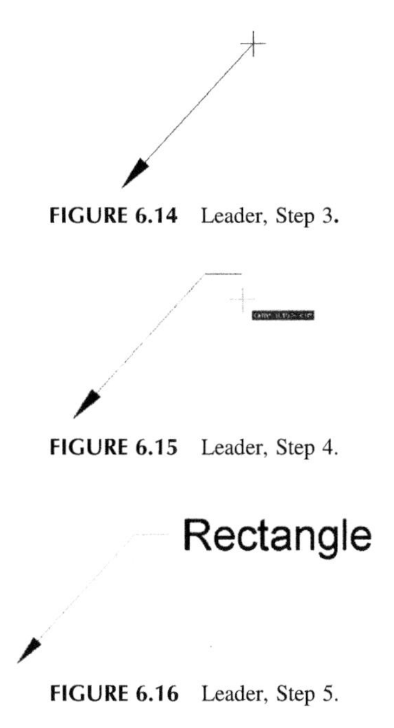

FIGURE 6.14 Leader, Step 3.

FIGURE 6.15 Leader, Step 4.

FIGURE 6.16 Leader, Step 5.

Multileader

This method is relatively new and is meant to add more flexibility and usefulness (and inadvertently some complexity) to the leader command. To work with this, you need to bring up the Multileader toolbar, shown in Fig. 6.17.

FIGURE 6.17 Multileader toolbar.

There is something new to the Multileader that you did not have with the basic leader. You can set up a Multileader style, so all leaders have the exact look you want. In our sample case, we give our Multileader a bubble in which to add text. Press the very last button on the right (mleader with a paint brush) and the dialog box shown in Fig. 6.18 appears.

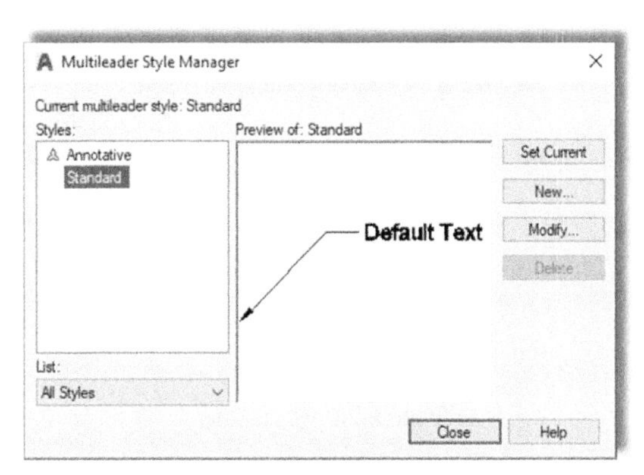

FIGURE 6.18 Multileader Style Manager.

Press New… and give the Multileader a name, such as Sample Style, and press Continue. You are taken to the Modify Multileader Style dialog box (Fig. 6.19). Examine the three tabs, Leader Format, Leader Structure, and Content, carefully. Most of the options are reasonably self-explanatory, and you will see some of them again later on. Under the Content tab, select Multileader type: as Block and Source block: as Detail Callout. Finally, press OK, Set Current, and Close. Now, let us try out our new Multileader style.

Keyboard: Type in **mleader**
Cascading menus: **Dimension→Multileader**
Toolbar icon: Multileader toolbar
Ribbon: **Annotate** tab→**Leaders** panel→

Step 1. Start the Multileader command via any of the preceding methods.

- AutoCAD says: `Specify leader arrowhead location or [leader Landing first/Content first/Options] <Options>:`

Step 2. Click anywhere and extend your mouse (Ortho off) at 45 degrees to the right.

- AutoCAD says: `Specify leader landing location:`

Step 3. Click again to place the Multileader and bubble. You see something like Fig. 6.20.

- AutoCAD says: `Enter view number <VIEWNUMBER>:`

Step 4. Enter in some value.

- AutoCAD says: `Enter sheet number<SHEETNUMBER>:`

Step 5. Enter in another value.

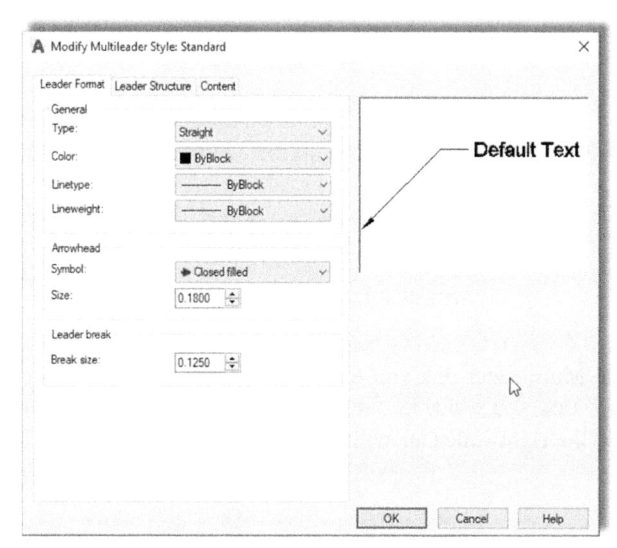

FIGURE 6.19 Modify Multileader Style.

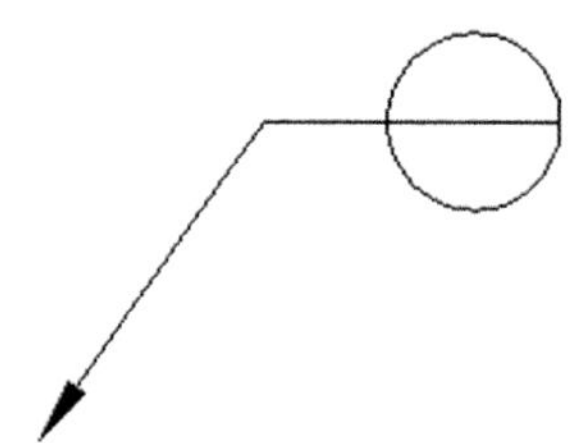

FIGURE 6.20 Multileader.

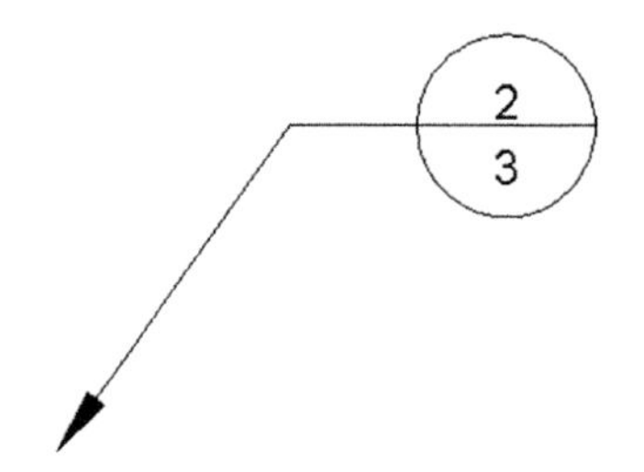

FIGURE 6.21 Multileader with block.

Once you do this, you see the final result (Fig. 6.21), the appearance of which can, of course, be fine-tuned. Now that you have one leader on the screen, it is easy to experiment with other interesting options. Move across the toolbar, starting first with Add Leader (Fig. 6.22) and get rid of it via Remove Leader. Then, after adding it back in, create a new set of leaders and line them up using Align Multileaders (Fig. 6.23), which aligns the leaders with one that you pick as the primary. Finally, Collect Multileaders (Fig. 6.24) combines them into one string.

Secondary Dimensions

It was mentioned that some dimensions are secondary, meaning they are not used as often. We briefly discuss them here, and you may want to also go over them in detail on your own, especially if you feel one or more of them may be extremely useful to your particular field. They are:

- Arc length
- Jogged
- Jogged linear
- Ordinate

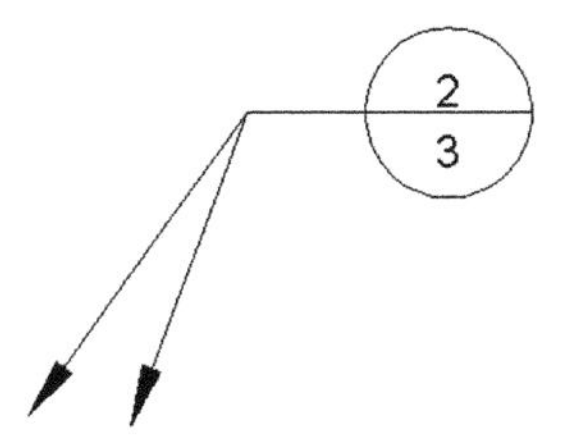

FIGURE 6.22 Add Leader.

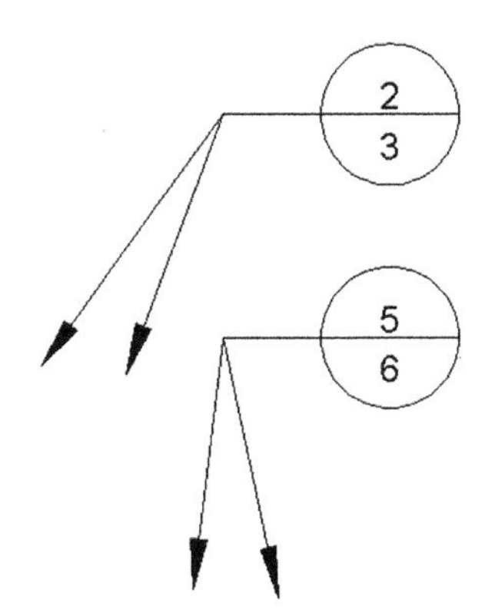

FIGURE 6.23 Align Multileaders.

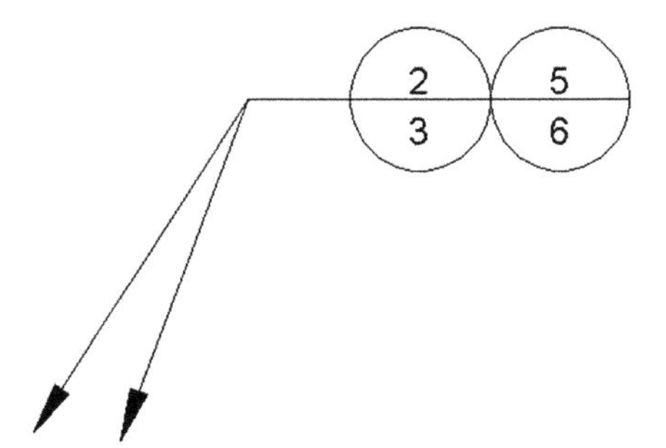

FIGURE 6.24 Collect Multileaders.

Arc length, as you may have guessed, measures the length of an arc. Simply select the toolbar icon and click on the arc you want to measure, as seen in Fig. 6.25. You can place the dimension value outside or inside the arc itself.

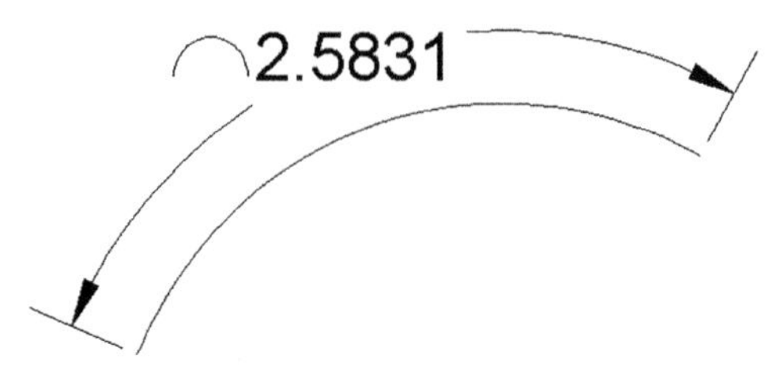

FIGURE 6.25 Arc length dimension.

Jogged dimensions are useful when the radius or diameter dimension's center is off the sheet of paper and cannot be shown directly, so a "break line" of sorts is used. You simply select the arc or circle then the center location override, and a jogged dimension appears (Fig. 6.26).

Jogged linear dimensions are useful for representing break lines in a linear format. Jogged linear is not a dimension type as much as a modification to a linear (horizontal, vertical, or aligned) dimension. To try it out, create a horizontal dimension, as you have done numerous times before, and press the Jog linear icon on the Dim toolbar or Ribbon Dimensions panel (it is also available in the Dimension cascading menus). Then, click the dimension line of the linear dimension you wish to modify and one more click to position the jog. If you do not select a location for it, AutoCAD will do it for you. The final result is seen in Fig. 6.27.

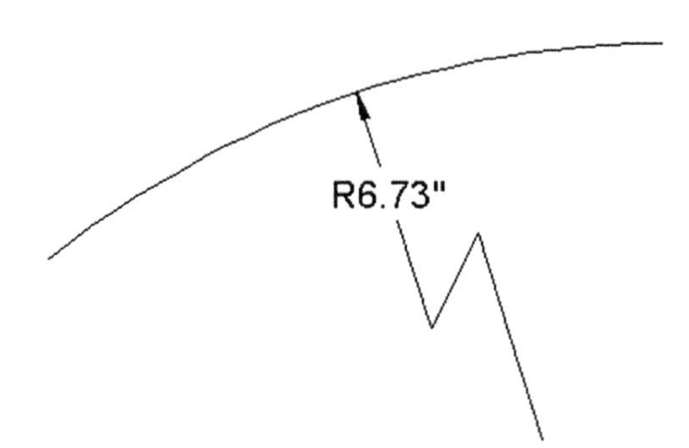

FIGURE 6.26 Jogged dimension.

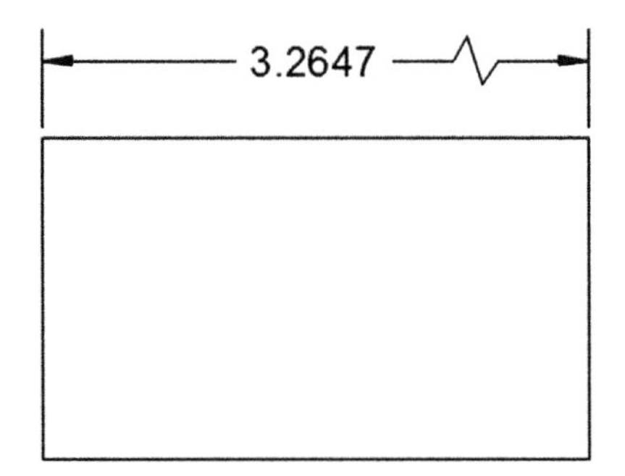

FIGURE 6.27 Jogged linear dimension.

Ordinate dimensions (Fig. 6.28) measure horizontal or vertical distances from a datum point (0,0 in this case). They are used in manufacturing to prevent accumulation of errors that can occur using continuous measurements. Draw the shapes shown in Fig. 6.28, positioning the lower left corner of the rectangle at 0,0, and begin the ordinate dimension. Then, click major points along the way in either horizontal or vertical directions to get precise values at that location from the origin.

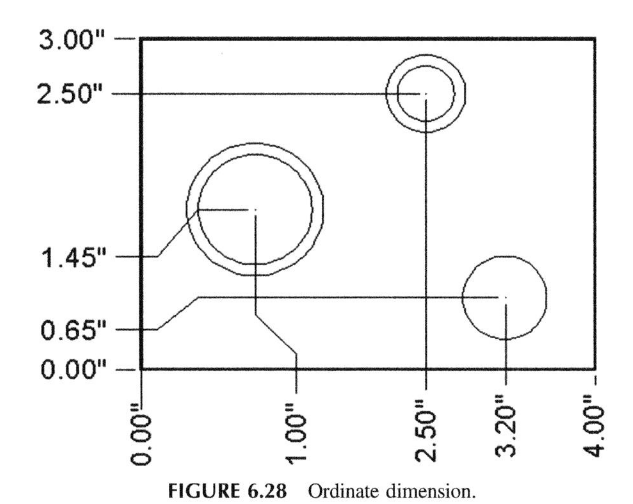

FIGURE 6.28 Ordinate dimension.

There are other dimension options we have not yet explored. We come back to them in Level 2. For now, we must move on and learn how to do editing and customizing.

6.3 EDITING DIMENSIONS

You should now be quite familiar with the types of dimensions available and how to apply them to basic shapes. The next step is to be able to edit them or change their values, if needed. This is a very simple and short topic, but before we start, note something very important. You generally should not be changing dimension values manually, but rather seek to find out why your dimension is wrong. Perhaps you did not attach it to the geometry correctly, or the geometry itself is flawed. Try to get to the root of the problem and change dimensions manually only as a last resort!

You already learned that you can edit text and mtext by double-clicking on it. The text in the dimensions is essentially editable mtext, so go ahead and double-click any part of the dimension to edit it. Recall also, in Chapter 4, Text, Mtext, Editing, and Style, we covered ddedit and mentioned it was also useful for editing purposes. You can use that command as well, and indeed with AutoCAD versions prior to 2013, you needed to use ddedit, as double-clicking on the dimension brings up the Properties palette. This "issue" was resolved for later releases.

What happens next depends on whether you have the Ribbon on your screen or not. If you do not, then you see the basic mtext editor, as in Fig. 6.29. You can just edit the value and press OK. If you do have the Ribbon active, then the editing field appears but not the rest of the mtext editor, with the Ribbon's Text Editor tab highlighted instead.

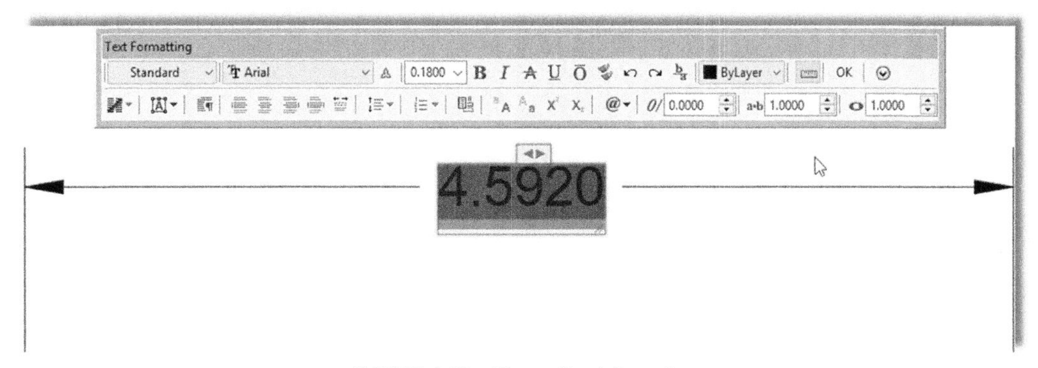

FIGURE 6.29 Mtext edit of dimensions.

Here is a good tip. If you want to reset your forced dimension value back to its natural value and you forgot what that was, just type in <>. These two "alligator teeth" brackets restore the natural value. Type them in instead of the forced value and click on OK. Try it yourself.

6.4 CUSTOMIZING DIMENSIONS

We have one more topic to cover on our way to proficiency in basic dimensioning. It is customization; and as mentioned at the start of this chapter, it is an extensive subject, necessitating it being split into the fundamentals here and advanced customization in Chapter 13. The goal here is to learn what experience has shown to be the most important four customization tools. This allows you to be productive in most of the drafting situations you are likely to encounter as a beginner. Later, you will add to this knowledge by learning tools used by CAD administrators and senior designers.

So, what customization is necessary at this level? We need to learn how to

- Change the *units* of the dimensions.
- Change the *font* of the dimensions.
- Change the *arrowheads* of the dimensions.
- Change the *fit* (size) of the dimensions.

To be able to change any of these, we need to introduce a new tool, the Dimension Style Manager dialog box, or *dimstyle* for short. There, you find a button to bring up the Modify Dimension Style dialog box. It is a "one-stop-shopping" dialog box for dimension modification that you will eventually need to learn backward and forward.

Dimstyle

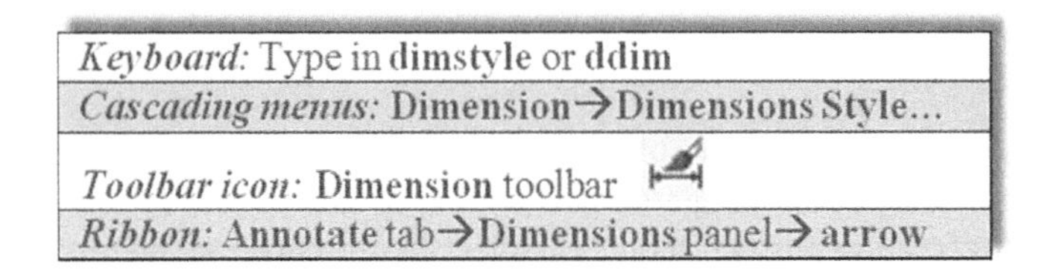

Keyboard: Type in dimstyle or ddim
Cascading menus: Dimension→Dimensions Style...
Toolbar icon: Dimension toolbar
Ribbon: Annotate tab→Dimensions panel→ arrow

Using any of the preceding methods, bring up the Dimension Style Manager, as seen in Fig. 6.30.

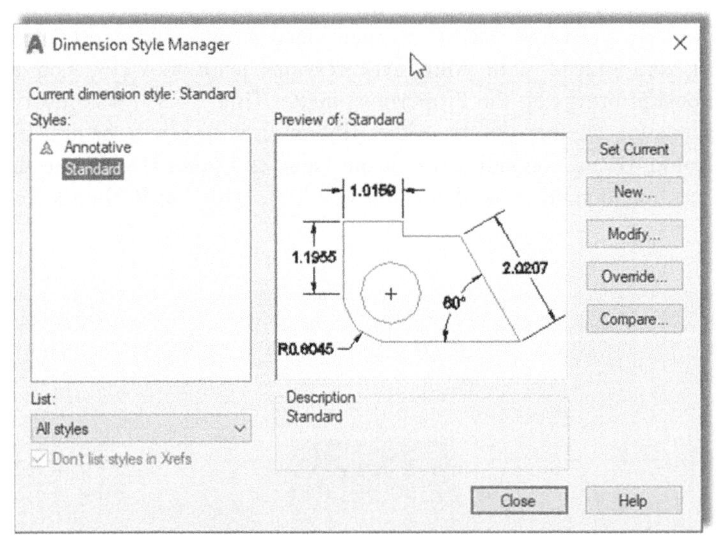

FIGURE 6.30 Dimension Style Manager.

The Dimension Style Manager dialog box has some options for creating a new dimension style or just modifying an existing one. For us at this point, it really does not matter which we pick, so just press Modify....

You then are taken to the Modify Dimension Style dialog box, shown in Fig. 6.31. Note that you may not necessarily see the Lines tab as the open tab, as seen in Fig. 6.31. If you do not, just click on it to have the same thing on your screen as what is in the text.

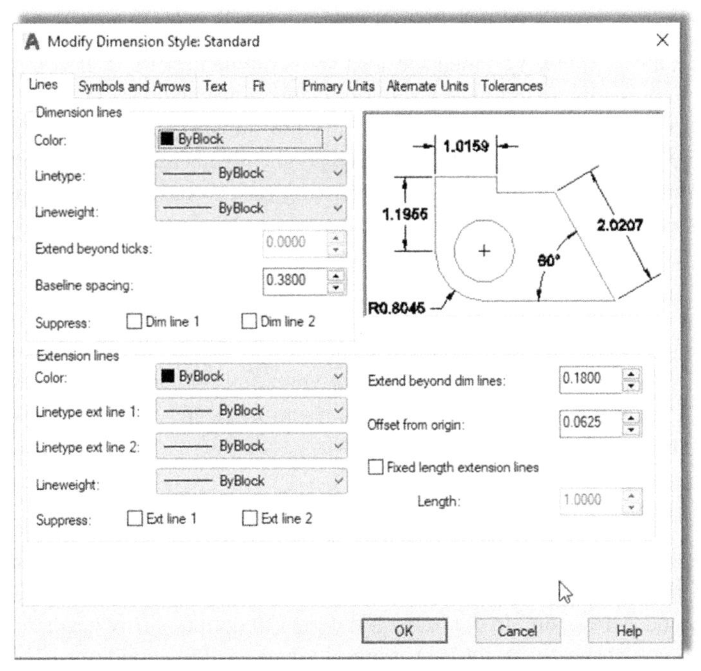

FIGURE 6.31 Modify Dimension Styles (Lines tab).

This rather imposing dialog box contains tools to significantly modify any dimension. We, in fact, do so in Chapter 13, Advanced Dimensions, as we explore nearly every feature. For now, we just want to focus on learning how to modify the handful of items shown in the bulleted list at the start of Section 6.4.

Step 1. Change the Units of the Dimensions

Pick the Primary Units tab and simply select the drop-down menu at the very top left, where it says Unit format:. The default value is Decimal; change it to Architectural and select the appropriate Precision in the menu just below it (the default value is fine). That is all we need from that tab for now. Notice how the preview window on the upper right reflects your choice of units, as seen in Fig. 6.32.

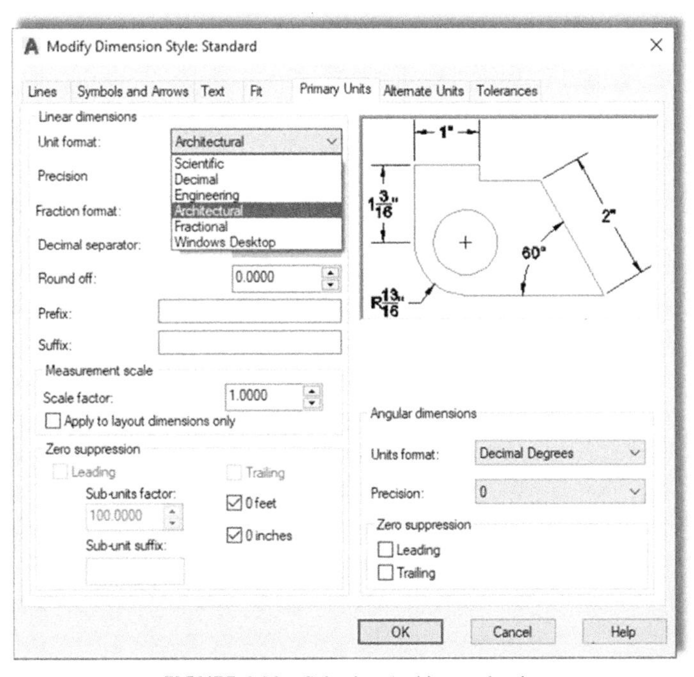

FIGURE 6.32 Selecting Architectural units.

Step 2. Change the Font of the Dimensions

To do this you need to click on the Text tab. You then see Text style: on the upper left. If you just opened a new AutoCAD file to practice dimensions, you probably do not have a font set and get only Standard as your choice if you click the down arrow. Fortunately, Standard is now a more attractive Arial font, not the stick-figure Simplex it used to be, so changing fonts is optional but still useful to know. For example, the HelvLight font is an alternative discussed in Chapter 4.

Something else to keep in mind is that, in a real working drawing, you would likely set your text style before you work on any dimensions (an important point, take note). Then, you would just select the font from the list. The idea here is to match up your regular font used in text and mtext with the font used in dimensioning, another important point. In general, you should stick to one font in a given drawing and just vary the size as needed. The title block is exempt from this, as that may have special fonts, logos, and the like.

AutoCAD, of course, anticipates that you may not have set a font style when you first set up dimensions and allows you to do this from the Text tab by bringing up the Style box when you click on the button just to the right of the Text style menu (with the three dots, …). Go ahead and set a different font if you wish, HelvLight perhaps with a Text Height of 0.25, and the new setting appears, as seen in Fig. 6.33.

Step 3. Change the Arrowheads of the Dimensions

The default for all dimensioning is the standard arrowhead. You can easily change that to any other type, including the Architectural tick, popular with architects. You can even create a custom arrowhead (not something we try to do here).

Click over to the Symbols and Arrows tab. In the upper left corner under the Arrowheads section, you see First:. That is the first arrowhead type, and if you change it, AutoCAD assumes you want the second one to be the same and changes it as well. Simply click the down arrow and select the Architectural tick. Leave everything else the same. The result is shown in Fig. 6.34.

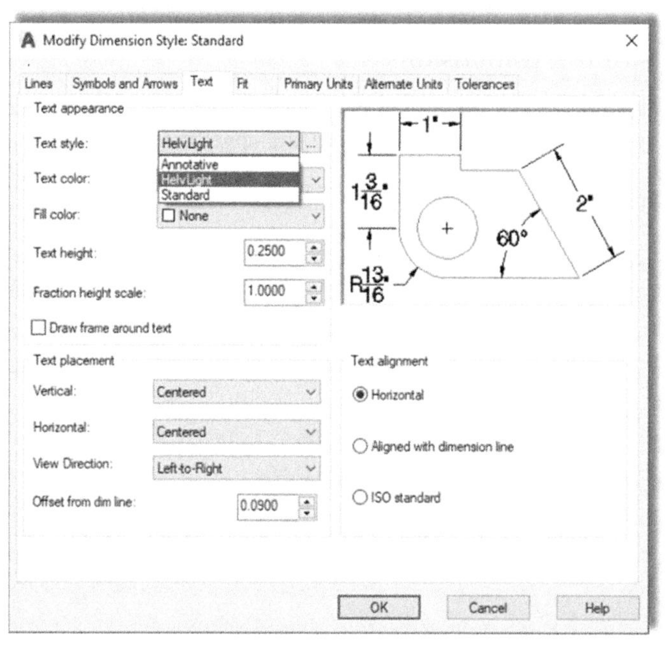

FIGURE 6.33 Setting a new font.

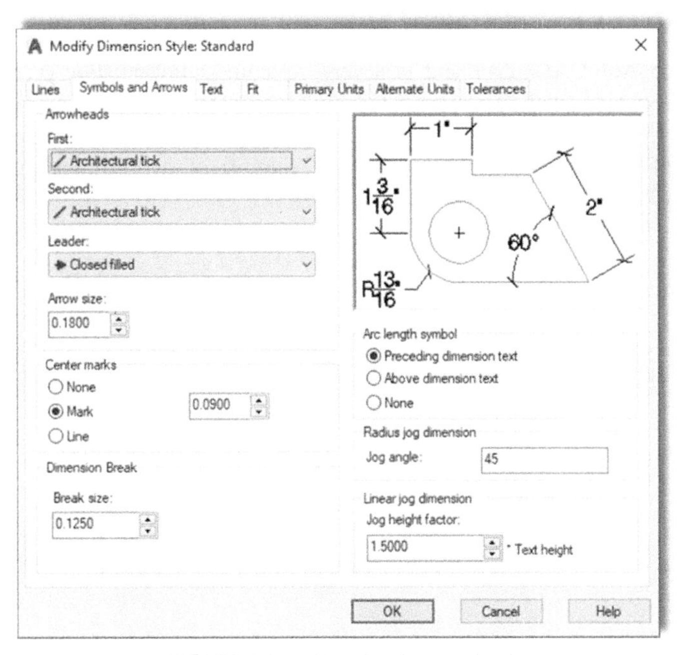

FIGURE 6.34 Changing the arrowheads.

Step 4. Change the Fit (Size) of the Dimensions

This function is quite important. Located in the Fit tab, just underneath the preview window, it features only a few lines and is shown in Fig. 6.35.

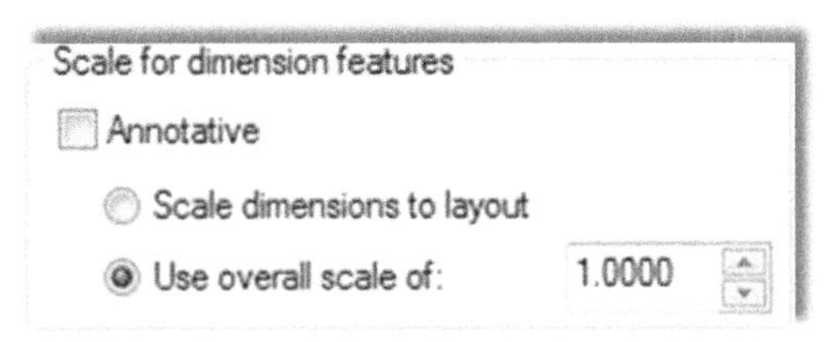

FIGURE 6.35 Scale for dimension features.

The idea here is to increase the size of the dimensions proportionately, so everything scales up evenly and smoothly in step with the size and scale of your overall drawing. This concept, however, opens up a number of questions. Chief among them: What value do you enter there and why? We are not ready to discuss this yet, but we do in Chapter 10, Advanced Output—Paper Space (on paper space). So, for now, do not enter in anything but be aware of how to do it when needed. When you get to dimensioning the floor plan in the next section, enter 15 in that space.

This is it for now; just click on OK and AutoCAD returns you to the Dimension Style Manager, where you can just press Close. Try creating a dimension and see what it looks like now. Fig. 6.36 shows a set of horizontal dimensions before the changes we just went through and another one afterward. The difference is obvious. For now, this is all you need to make professional-looking dimensions. Review and memorize everything you learned.

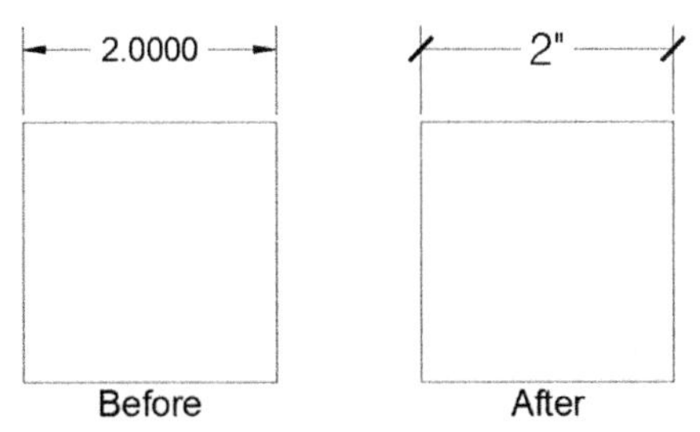

FIGURE 6.36 Scale for size of dimension.

6.5 CENTER MARKS AND CENTERLINES

This feature was introduced in AutoCAD 2017 and is a very welcome addition. For years creating center marks and especially centerlines was a somewhat tedious process that involved a few steps. A simple example of creating circle centerlines is shown in Fig. 4.18 back in Chapter 4. Good drafting practice requires these centerlines to be even and properly centered, with use of the correct linetype. A typical example of two circular and one rectangular part with centerlines is shown in Fig. 6.37.

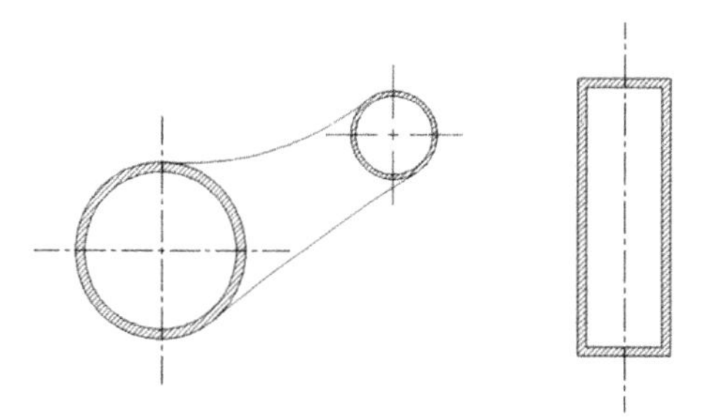

FIGURE 6.37 Centerlines in use.

These center marks and centerlines can now be created quickly and easily with just a few clicks—though what you will get will be default centerlines that use the Center2 linetype, and of a fixed size relative to the part you are annotating. If you need something different, outside those parameters, including an actual center mark, things do get slightly more involved. The command is still a big time-saver and definitely speeds up the process.

To try this out, first create a few circles, then for some variety, a rectangle and a pair of parallel lines, as seen in Fig. 6.38. Next, locate the new buttons on the Ribbon, under the Annotate tab, Centerline panel, as seen in Fig. 6.39.

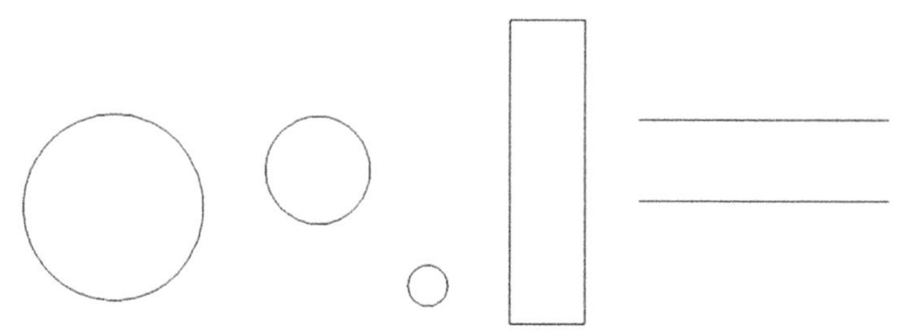

FIGURE 6.38 Basic shapes for center mark and centerlines.

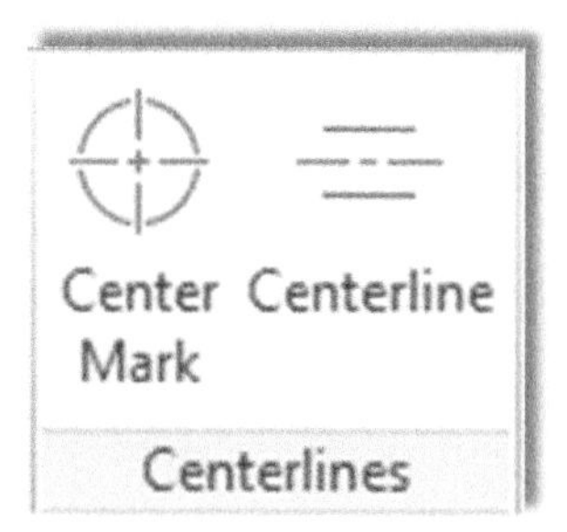

FIGURE 6.39 Centerlines panel.

Go ahead and click the Center Mark button. AutoCAD says: Select circle or arc to add center mark: at which point, just click each of the circles. In the same manner select the Centerline button and select the parallel lines (both the rectangle and free drawn). The result is shown in Fig. 6.40.

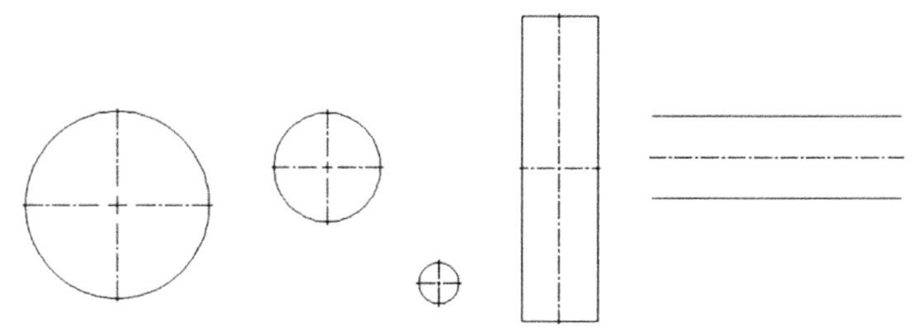

FIGURE 6.40 Completed centerlines.

These centerlines are editable via grips, and of course, additional options are available to customize them further. In Fig. 6.41 a double-click of the centerline brought up the options palette, and two of the extensions have been changed to a new length and color.

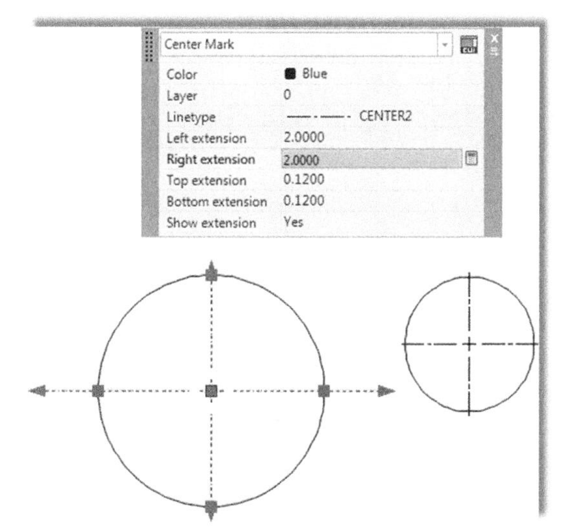

FIGURE 6.41 Manual modification of centerlines.

Some additional commands to control center mark and center line appearance include:

- CENTEREXE: This controls the length of extension line overshoot (the amount of line outside of the shape). Determining what looks right may require some trial and error. Setting a value here will then apply it to all future centerlines. The ones already created remain the same until manually updated.
- CENTERMARKEXE: This command creates true center marks, with no extension lines (basically a small "plus" sign inside the circle). It is of the On/Off variety, either it is a center mark or a centerline. Setting a value here applies it to all future marks. The ones already created remain the same until manually updated.
- CENTERLTYPE, CENTERLAYER, and CENTERLTSCALE: These specify the linetype, layer, and linetype scale, respectively. Enter the desired type, layer, and scale when promoted. The settings apply globally to all future center marks or centerlines. The first two items can also be entered locally on a case-by-case basis via the properties palette seen in Fig. 6.41.

There are a few more of these commands, and the full listing can be found in the Help files, but these are the essential ones need to customize this useful new feature of AutoCAD. Some of these commands hint at what are called *system variables*. We mention them again in Chapter 15, Advanced Design and File Management Tools.

6.6 IN-CLASS DRAWING PROJECT: ADDING DIMENSIONS TO FLOOR PLAN LAYOUT

Let us now apply what we learned to the floor plan. Open the file and freeze all the layers except the walls and windows. If you really want to get fancy, put the wall solid hatch on its own (visible) layer and freeze the regular floor hatch layer so the carpeting and floors do not show. Next, create a new layer for the dimensions, A-Dim. Finally, set up the dimensions: Use Architectural units and Arial 6″ font, leave the arrowheads as they are, and change the scale under the Fit tab to 15. Dimension the floor plan any way you want; what is shown in Fig. 6.42 is a guide but not the only way to do it. Note that the outer dimensions are continuous.

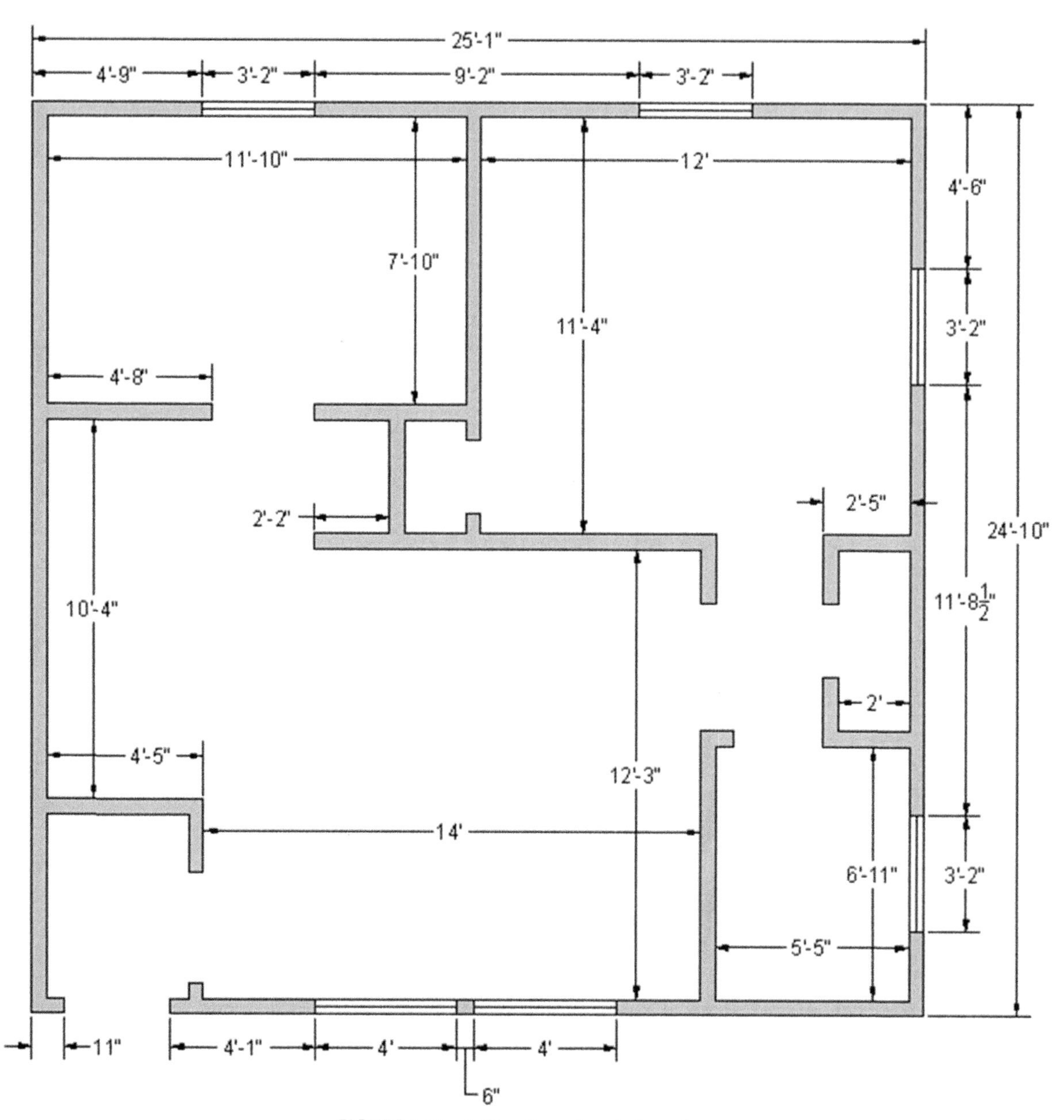

FIGURE 6.42 Adding dimensions to floor plan.

SUMMARY

You should understand and know how to use the following concepts and commands before moving on to Chapter 7:

- Dimensions
  - Linear (horizontal and vertical)
  - Aligned
  - Diameter
  - Radius
  - Angular
  - Continuous
  - Baseline
- Leader and multileader
- Arc length
- Jogged
- Jogged linear
- Ordinate
- Editing dimension values

- ddedit
- The effect of <>
- ddim command
- Dimension units
- Dimension font
- Dimension arrowheads
- Dimension overall size
- Center marks and centerlines

REVIEW QUESTIONS

Answer the following based on what you learned in this chapter:

1. List the 11 types of dimensions discussed.
2. What command is best for editing dimension values?
3. What is the difference between natural and forced dimensions?
4. How do you restore a natural dimension value when you have forgotten what it was?
5. List the four items that needed to be customized in our dimensions.
6. What command brings up the Dimension Style Manager?
7. What type of arrowhead do architects usually prefer?

EXERCISES

1. Retrieve Exercises 2, 3, 4, 5, and 6 from Chapter 3, Layers, Colors, Linetypes, and Properties, then set up and add all shown dimensions. (Difficulty level: Easy; Time to completion: 10 minutes/exercise.)
2. Retrieve Exercises 5, 6, 7, 8, and 9 from Chapter 4, Text, Mtext, Editing, and Style, then set up and add all shown dimensions. (Difficulty level: Easy; Time to completion: 10–15 minutes/exercise.)
3. Retrieve Exercises 5, 6, 9, and 10 from Chapter 5, Hatch Patterns, then set up and add all shown dimensions. (Difficulty level: Easy; Time to completion: 10–15 minutes/exercise.)
4. In a new file, set up the appropriate layer, colors, and linetypes. Then, draw and fully dimension the following mechanical part. Select Decimal as the dimstyle, with a precision of 0.00 and inch symbols (") for a suffix. (Difficulty level: Easy/Moderate; Time to completion: 35–45 minutes.)

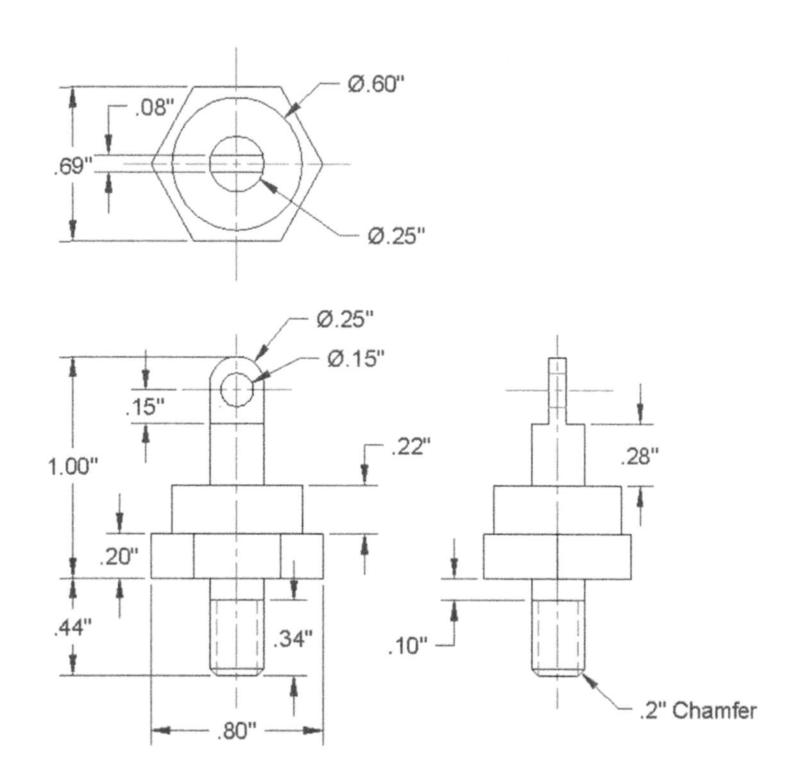

5. In a new file, set up the appropriate layer, colors, and linetypes. Then, draw and fully dimension the following mechanical part. Select Decimal as the dimstyle, with a precision of 0.00 and inch symbols (") for a suffix. Make careful use of mirror to minimize redundant work. All sizing is based on the smallest two circles and their position from an arbitrary centerline. (Difficulty level: Moderate; Time to completion: 35–45 minutes.)

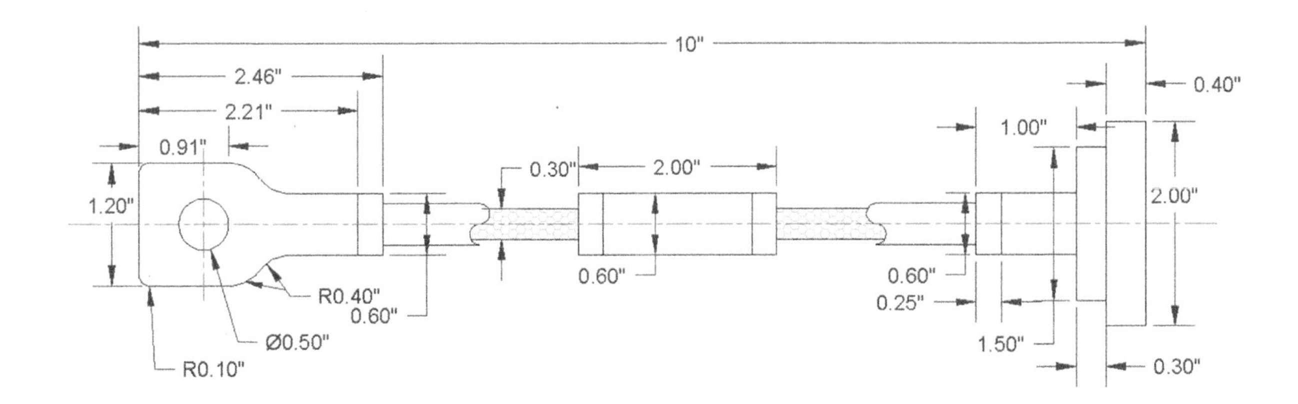

6. In a new file, set up the appropriate layer, colors, and linetypes. Then, draw and fully dimension the following mechanical part. Select Decimal as the dimstyle, with a precision of 0.00 and inch symbols (") for a suffix. (Difficulty level: Moderate; Time to completion: 35–45 minutes.)

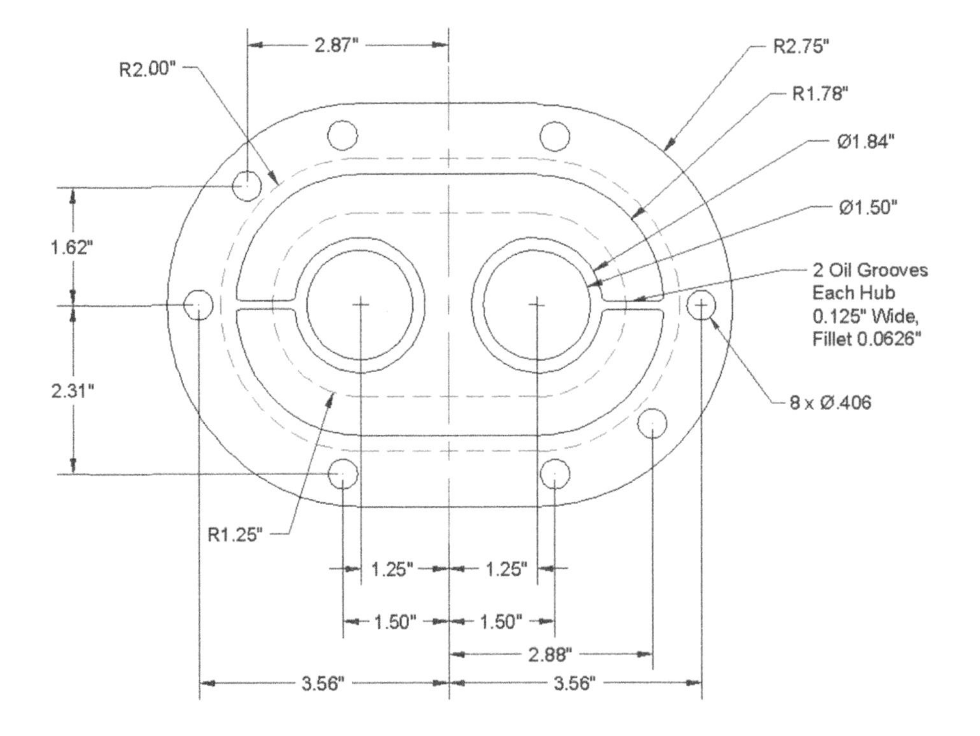

7. In a new file, set up the appropriate layer, colors, and linetypes. Then, draw and fully dimension the following mechanical part. Select Decimal as the dimstyle, with a precision of 0.00 and inch symbols (") for a suffix. (Difficulty level: Moderate; Time to completion: 35–45 minutes.)

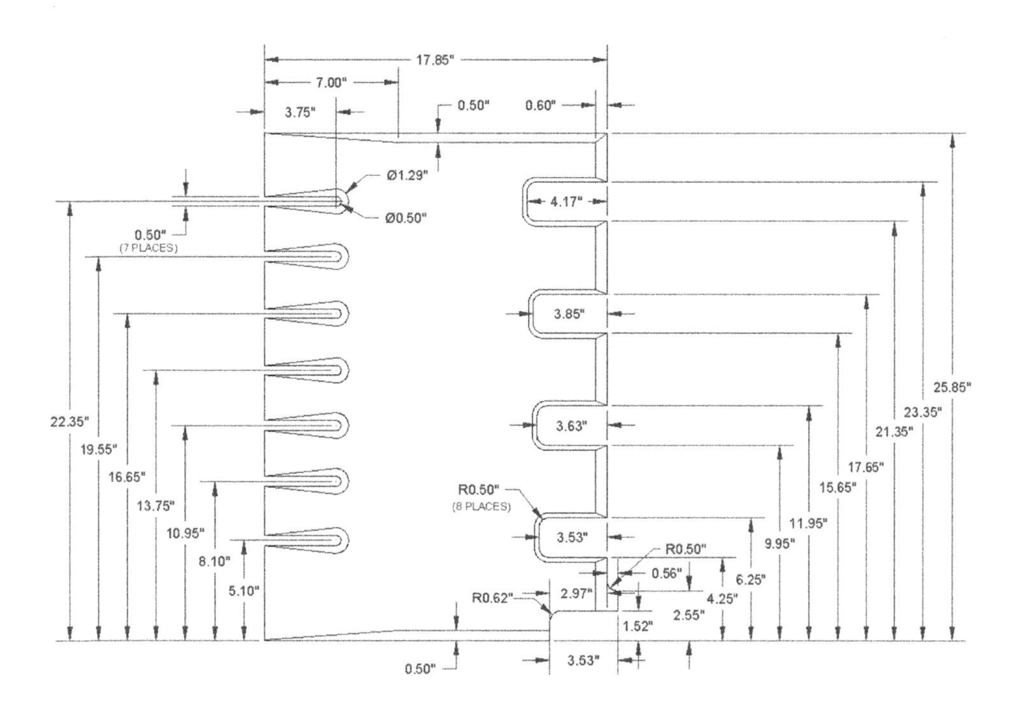

8. In a new file, set up the appropriate layer, colors, and linetypes. Then, draw and fully dimension the slightly simplified version of the bolt assembly seen on the opening of this chapter. Select Decimal as the dimstyle, with a precision of 0.00 and inch symbols (") for a suffix. Note that no dimensions are available for some internal parts of the bolt; you have to improvise. (Difficulty level: Moderate/Advanced; Time to completion: 60 minutes.)

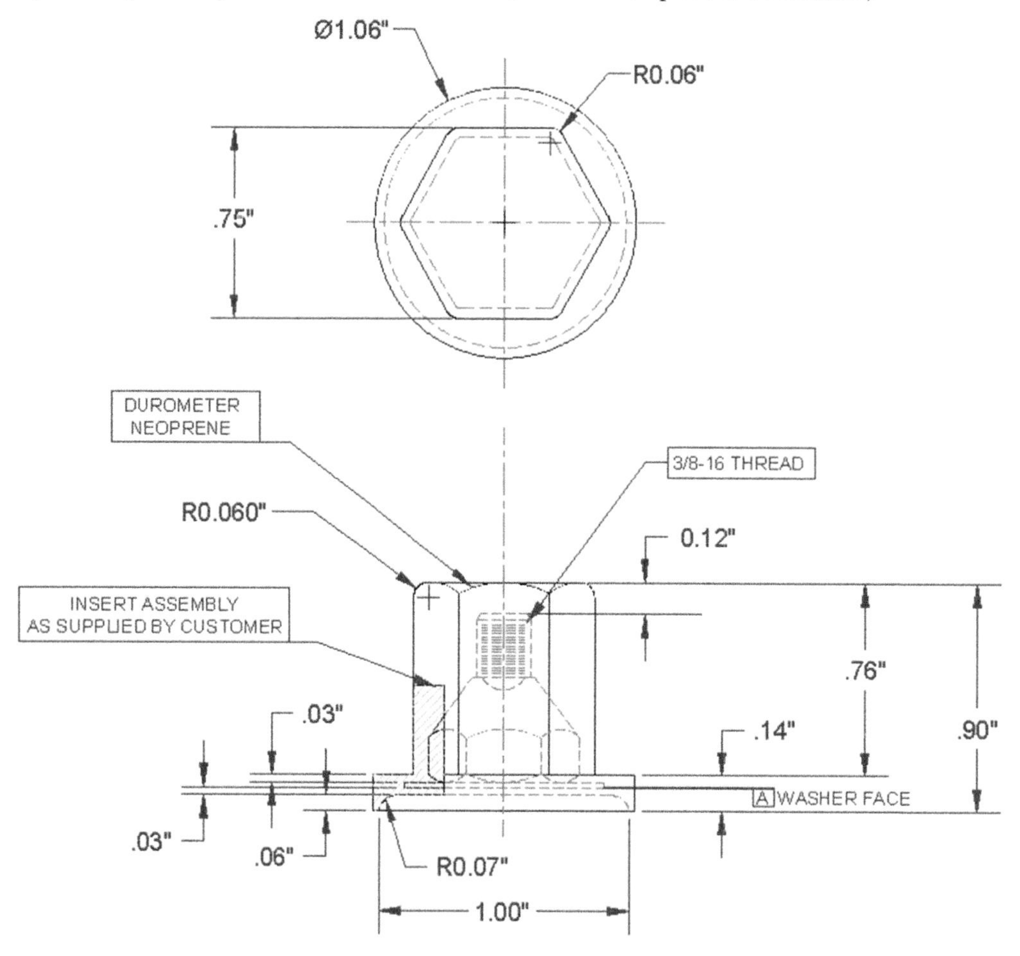

9. In a new file, set up the appropriate layer and colors. Then, draw and fully dimension (including text) the following architectural detail. Any dimensions not expressly given can be assumed. The hatch borders were created using arcs that were later erased. (Difficulty level: Moderate/Advanced; Time to completion: 60 minutes.)

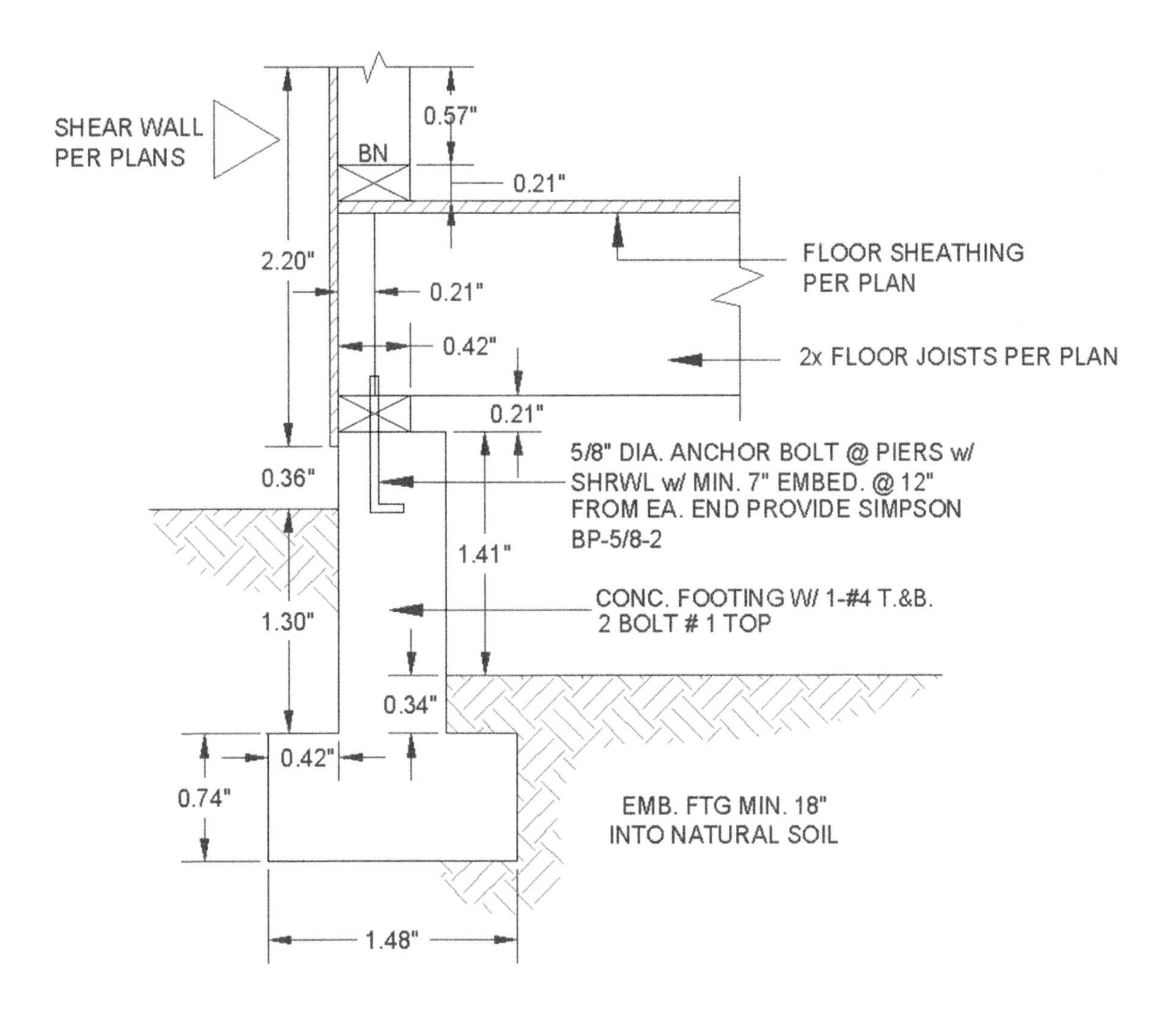

10. The final exercise of this chapter is a basic floor plan to keep your drawing skills sharp. It also features some electrical symbols, which you should get used to seeing, as they are standard in any electrical layout. Create one copy of each and duplicate them throughout. We cover blocks for this purpose in the following chapter. The numbers inside the rectangles are the doorway dimensions. A few secondary dimensions are not given due to space limitations; you need to make a best estimate. Use proper layers and add text and dimensions as shown. (Difficulty level: Moderate/Advanced; Time to completion: 90 minutes.)

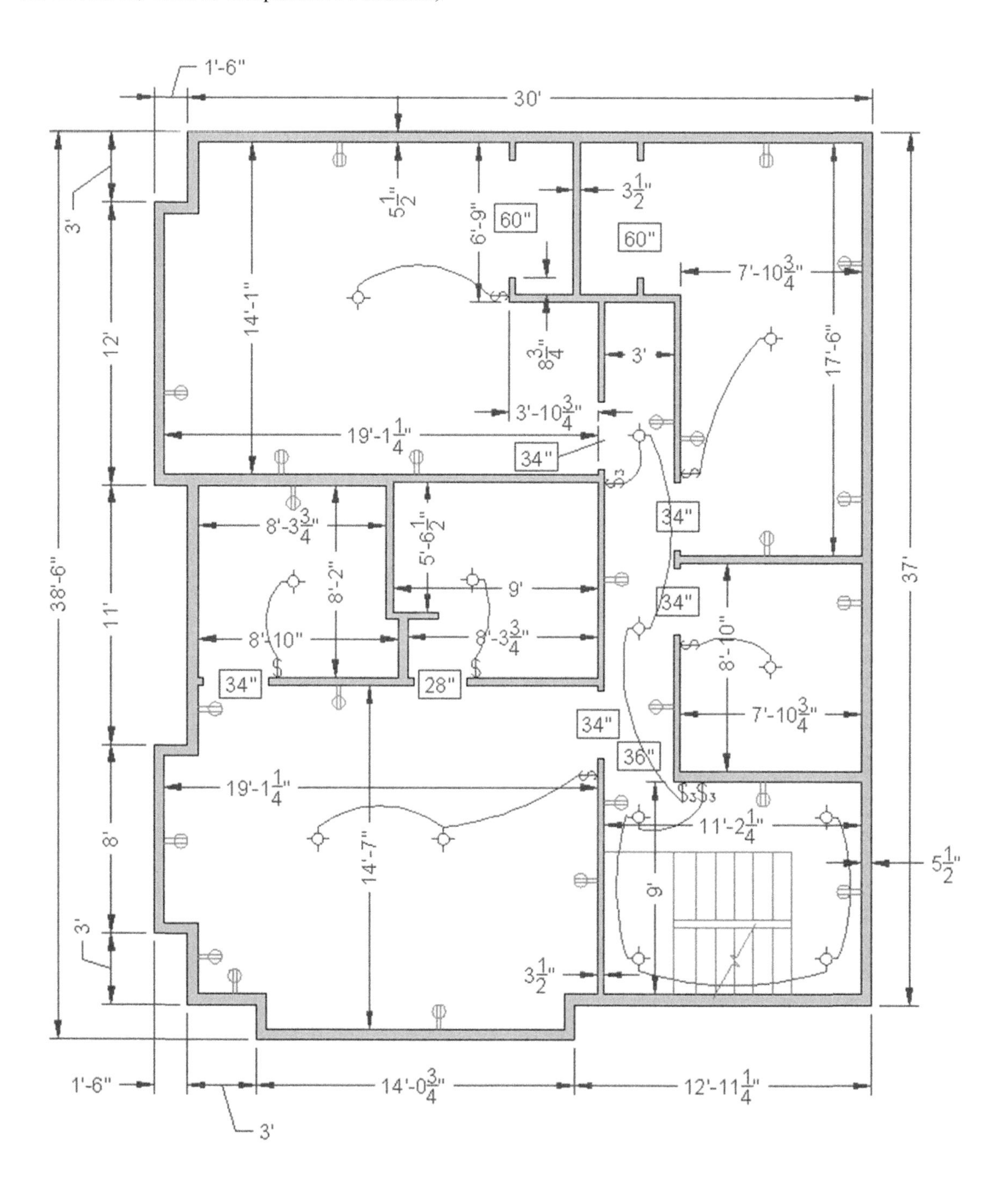

Spotlight On: Interior Design

Interior design is a major branch of architecture that concerns itself with interior living spaces. The goal is to create a personal, work, or public space that is both functional and esthetically attractive by the creative use of colors, lighting, and materials. Interior design is not to be confused with interior decorating. Interior designers are often well versed in the principles of architecture, engineering, and materials and must generally hold at least a 4-year degree. They can be called on to create challenging spaces that are tailored to certain ecological or handicapped-accessible needs. It is a creative profession, but one that still requires analytical skills.

FIGURE 1 *Image source: www.coldcoldlake.com.*

Education for interior designers (in the United States) involves a 2-year associate's or 4-year baccalaureate degree in interior design. Masters degrees (MS, MA, MFA, and recently the MID) in interior design are also available, although this advanced degree is not as common. Many professionals pursue advanced degrees in related subjects, such as industrial design, fine art, or education. Doctoral programs in interior design are increasing in number at various institutions of higher education. It is worth noting that some of the more notable interior designers held no formal education, though this is the exception not the rule.

Following formal training, graduates usually enter a 1- to 3-year apprenticeship to gain experience before taking a national licensing exam or joining a professional association. The National Council for Interior Design Qualification administers the licensing exam. To be eligible to take the exam, applicants must have at least 6 years of combined education and experience in interior design, of which at least 2 years constitute postsecondary education in design. Once candidates have passed the qualifying exam, they are granted the title of Certified, Registered, or Licensed Interior Designer, depending on the state. Some states require continuing education units to maintain the license.

Interior designer earnings vary, based on employer, number of years of experience, and the reputation of the individual. For residential projects, self-employed interior designers usually earn a per-hour fee plus a percentage of the total cost of furniture, lighting, artwork, and other design elements. For commercial projects, they may charge per-hour fees or a flat fee for the whole project. The median annual pay for interior designers is about $48,000, with the top 10% earning over $75,000.

So, how do interior designers use AutoCAD and what can you expect? Very often, as far as AutoCAD goes, interior design is tied in with the general architecture design of a building. Therefore, interior designers work side by side with architects on their CAD files. The layering convention often follows AIA standards. You find layers such as A-Furniture, A-Carpeting, and the like. Designers also work with blocks and attributes and make extensive use of custom hatch patterns and a wide array of colors (such as Pantones).

Interior designers often work in 3D, as the whole idea is often to present and sell the design concept, such as the rendering shown in Fig. 2. In these cases, they have to master 3D AutoCAD and additional rendering software. With 3D work, AutoCAD is not the only choice, and one can find Rhino, Form Z, and other applications running side by side with AutoCAD or even bypassing AutoCAD completely.

FIGURE 2 *Image source: www.revolutionp.com.*

Shown in Fig. 3 is a 2D plan view of a residential interior. It makes use of solid shading and custom hatch patterns to give the client a good idea of the final outcome of the design effort.

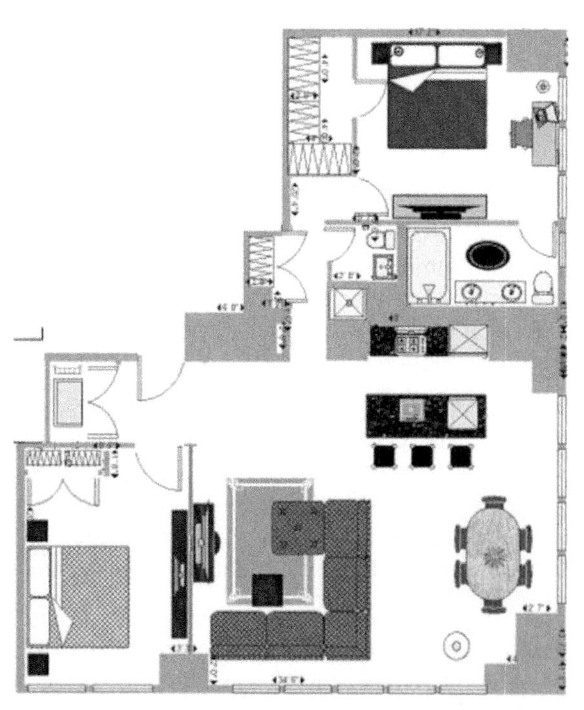

FIGURE 3 2D plan view of a residential interior. *Image source: www.homein.design.com.*

Shown in Fig. 4 is a 3D photorealistic rendering of another design. Although AutoCAD can be used to easily create the underlying architecture, additional rendering tools are usually used for this level of sophistication and realism.

FIGURE 4 3D photorealistic rendering of a design. *Image source: www.spazio3d.com.*

Chapter 7

Blocks, Wblocks, Dynamic Blocks, Groups, and Purge

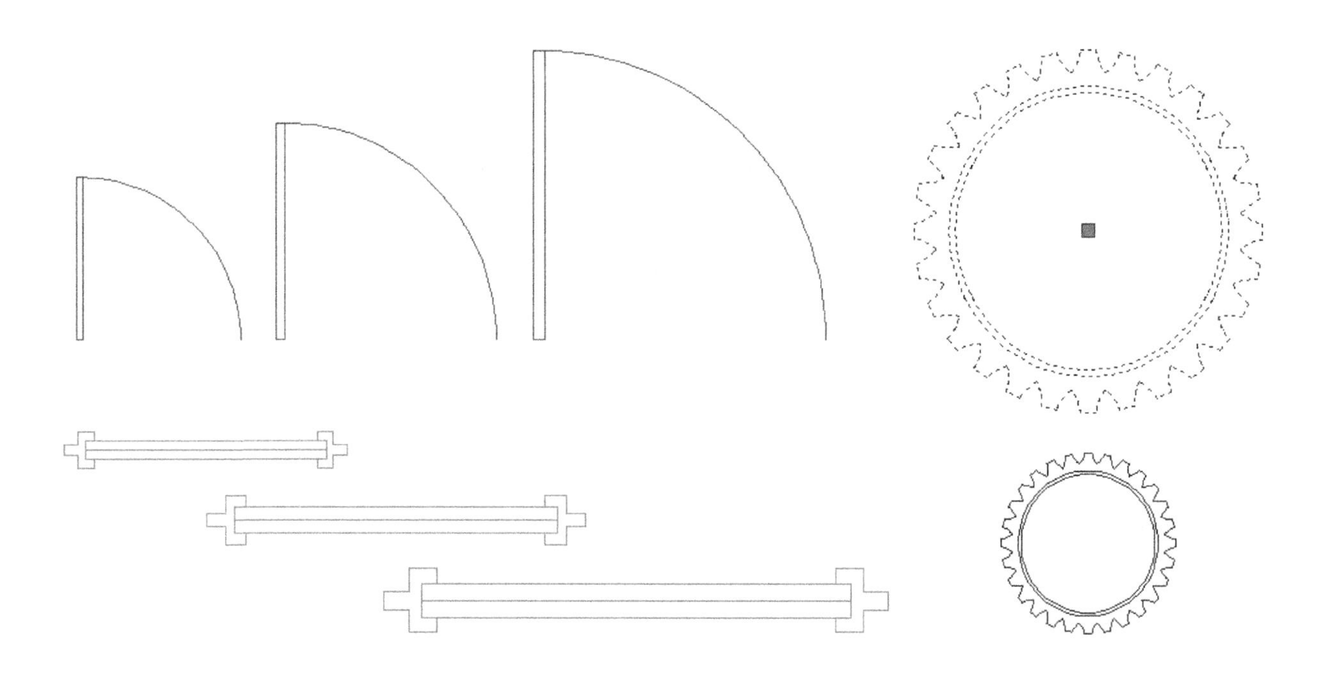

Learning Objectives

In this chapter, we introduce the concept of blocks and discuss the following:

- Creating and working with blocks
- Retain, convert, and delete options
- Redefining blocks
- Inserting blocks
- Purge
- Creating and working with wblocks
- Creating and working with dynamic blocks
- Creating and working with groups

At the end of this chapter, you will be able to create blocks, wblocks, dynamic blocks, and groups as well as insert symbol libraries and purge your drawing.

Estimated time for completion of this chapter: 1–2 hours.

Up and Running with AutoCAD 2019. https://doi.org/10.1016/B978-0-12-816440-2.00007-0

7.1 INTRODUCTION TO BLOCKS

This chapter is about some very useful AutoCAD concepts that greatly simplify your workload. Blocks, wblocks, dynamic blocks, and groups are all members of the same family. They exist around the idea that objects can and should be grouped together, if possible, to allow for easier handling and storage for future reuse.

Let us formally define a block. It is a collection of objects that are grouped and held together under some identifying name. These objects can then be copied, moved, erased, and just about anything else, all as one unit. Think of blocks as electronic "glue" that binds the objects they contain to each other.

There are many benefits to this. If you have an office table with eight chairs around it, and you need to move them all around many times while optimizing a room layout, it is a lot easier to click on a block once than attempt to select all nine pieces using a Window or a Crossing or one at a time. Imagine an engineering example involving a simple gear with all its intricately designed teeth. Dozens if not hundreds of lines and arcs may make up the gear. It would be not just wise but mandatory to form them all into a single block.

There is however another, more compelling, reason for blocks. That is the concept of *reuse* and *symbol libraries*. Blocks can be saved and reused for future drawings that may need the same exact object. That way you never have to draw the same thing twice, an AutoCAD Golden Rule. A collection of these objects cataloged in some orderly fashion becomes a symbol library, an indispensable part of just about any designer's toolset. Likely candidates for these libraries include doors, windows, furniture, bolts, standard parts, fasteners, and really just about anything that can be used in more than one design or drawing.

Difference Between Blocks and Wblocks

Ideally, you are sold on the importance of blocks as we go through the specifics. The differences between blocks and wblocks (write blocks) are blurred; they are alike in many ways. The key difference is that blocks are *internal* to the file you are working on and wblocks are *external*. This means that blocks are saved inside whatever file you are currently working on. As such, their main use is for grouping things together and reuse in that particular AutoCAD drawing. While they can certainly be moved between files, via the clipboard Copy/Paste tool, this is not really what they are intended for.

Wblocks are intended for creating symbol libraries. Wblocks are saved in the current drawing *and* to an external file, as an independent, stand-alone AutoCAD drawing, which you can open and work on if needed. Both types of blocks can, of course, be easily inserted back into drawings, and we cover that procedure in detail.

Creating a Block

To create a block you first need something from which to make a block. Once you draw something, only three things need to be done:

1. Select the object(s).
2. Give the block a name.
3. Select a base point.

Let us try this out. Draw a chair, as shown in Fig. 7.1. It is made up of three rectangles and three lines. Use accuracy tools.

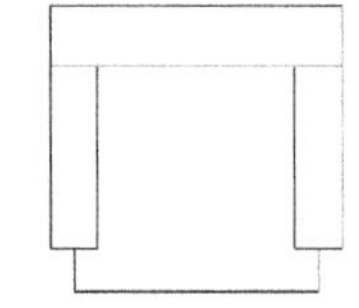

FIGURE 7.1 Basic chair symbol.

We need to now group these lines all together into one object.

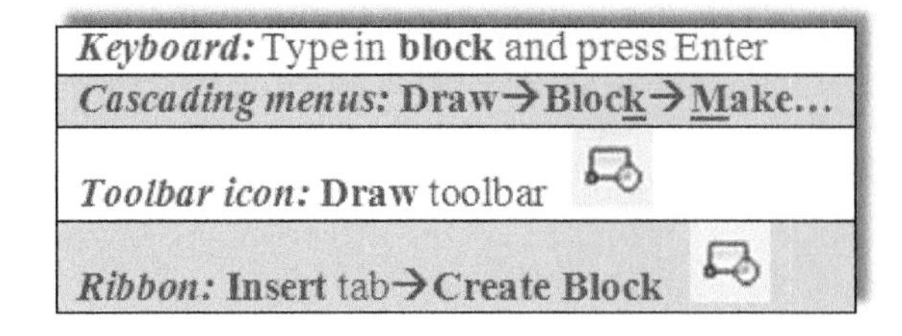

Keyboard: Type in **block** and press Enter
Cascading menus: **Draw→Block→Make...**
Toolbar icon: **Draw** toolbar
Ribbon: **Insert** tab→**Create Block**

Start the block command via any of the preceding methods. The Block Definition dialog box appears, as seen in Fig. 7.2.

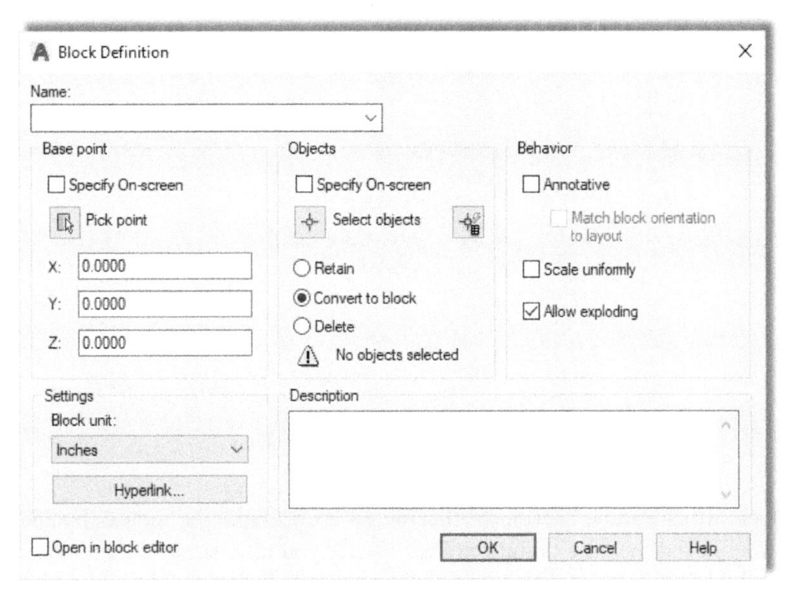

FIGURE 7.2 Block Definition dialog box.

Step 1. Press the Select objects button in the middle column. Select the entire chair and press Enter.
Step 2. Enter a name for this block in the top left field under Name.
Step 3. Pick a base point with the Pick point button at the upper left. Select a corner or midpoint somewhere on the object.

The significance of this becomes more apparent when we insert the object back into the file.

There are a few other items in the dialog box under Settings, Behavior, and Description, of course, but they are not usually needed (although feel free to enter a description). Annotative is covered in advanced chapters. The one setting of importance, however, is the Retain, Convert to block, and Delete choices under the Select objects button in the middle column. They are explained next and appear with almost similar wording in wblock as well.

Retain

This means that the block is created and stored in memory, but the original object (the chair you are looking at) remains in separate pieces, useful if you wish to make several blocks of the chair, one after the other, each slightly different yet based on the same original.

Convert to Block

This option is what you are likely to use the most. It creates the block and stores it in memory and also makes a block out of the original object. Use it when you are sure; one block is all you need to make out of this one object and you have a need for it right away.

Delete

Here, the block is created and stored in memory, but the original is deleted from the drawing. This is useful when you are making a block for future use but have no need for it now.

The common theme with all three of these choices is that, whichever way you use it, the block is created and saved to memory. The only real choice is what to do with the original. Finally, you are ready to click on OK. The Block Definition dialog box should look like Fig. 7.3. Notice the small preview window in the upper middle of the dialog box.

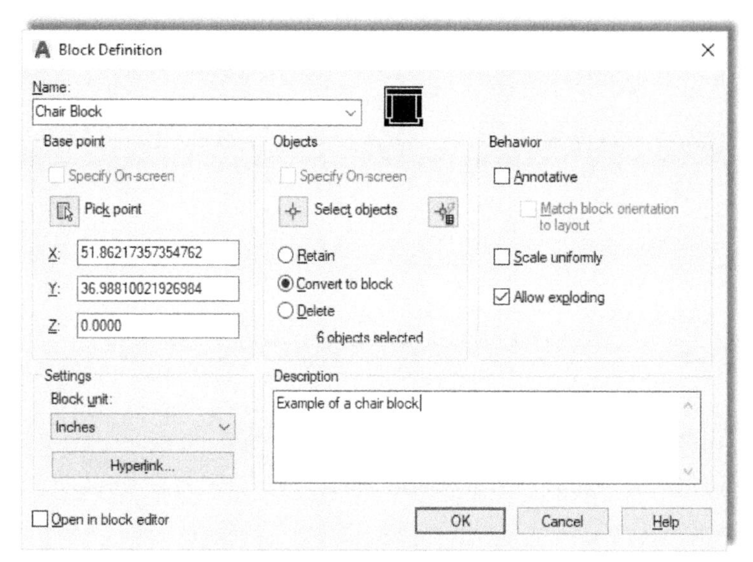

FIGURE 7.3 Completed block definition.

The chair is now a block, which can be proven by clicking on it once with the mouse. Notice that only one blue grip point appears (wherever you picked the base point; see Fig. 7.4). If you now erase it, the entire chair will disappear. Go ahead and erase it; we need to do this to introduce two new commands before we discuss wblocks.

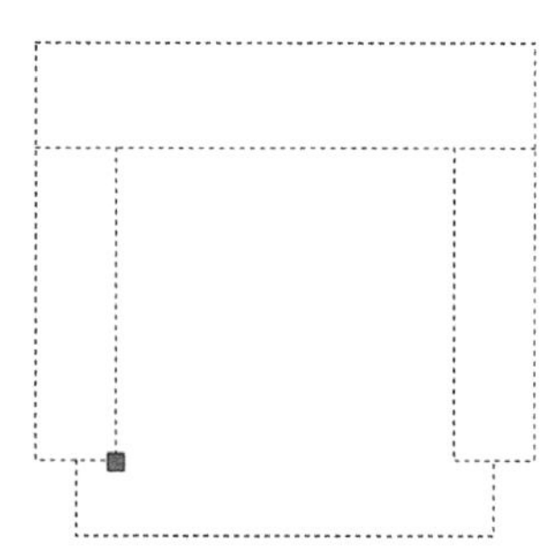

FIGURE 7.4 Chair as a block.

7.2 INSERT

As mentioned before, the process of making the block adds it into the memory of the file you have opened. Then, you can insert it back into the drawing when needed.

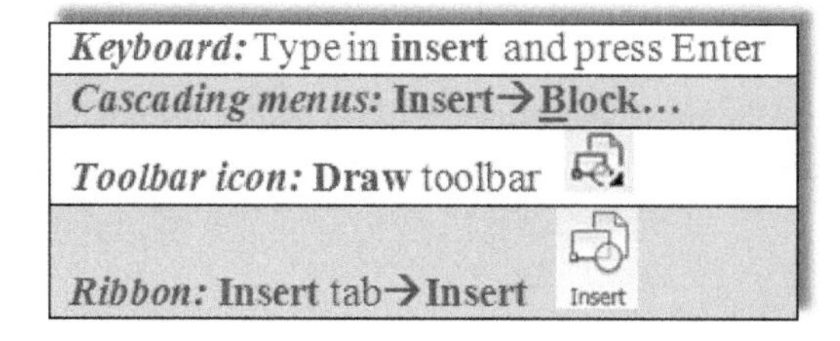

Keyboard: Type in **insert** and press Enter
Cascading menus: **Insert→Block...**
Toolbar icon: **Draw** toolbar
Ribbon: **Insert** tab→**Insert**

Start up the insert command via any of the preceding methods and the Insert dialog box appears, as seen in Fig. 7.5. Notice the chair is cued up and ready to go, as it is the only block created so far. A large preview of the chair appears on the right, increased in size over AutoCAD 2016s Insert dialog box preview.

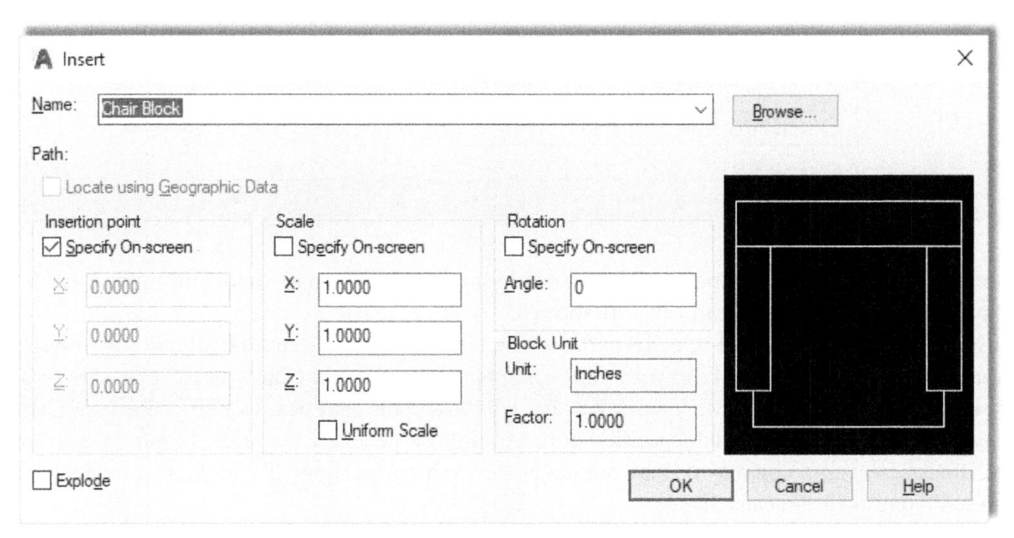

FIGURE 7.5 Insert dialog box.

Look over the various options, such as Insertion point, Scale, Rotation, and Block Unit, but at this point, you need not do anything extra. Go ahead and click on OK; the chair appears attached at the insertion point to the mouse. The point of attachment was specified when you made the block. Click anywhere on the screen and the block becomes part of the drawing. This process can be repeated as often as necessary as long as the block remains in memory. Note that you can always specify a precise point of insertion for the block by typing in the coordinates (e.g., 0,0,0).

There is something important to remember concerning redefining blocks. If a block is saved under a certain name and another block is created and saved under the same name, the first block is redefined, a sort of overwriting process. Sometimes, this is bad, as you are destroying the old block. Sometimes, however, this is desirable, such as when you are looking to update the appearance of the block. When there are many blocks in the drawing under the same name, all are updated in this manner.

7.3 PURGE

Between introduction of the block and wblock concepts, it is a good time to take an intermission of sorts and discuss a command called *purge*. Purge is a very useful tool to "clean up" your drawing of unwanted and often unseen geometry and data. It is especially useful when it comes to blocks and wblocks, hence its inclusion here.

Conceptually, here is the idea. A complex AutoCAD drawing often grows in size and becomes weighed down, similar to an iceberg floating in the ocean. Much like the mountain of ice, mostly hidden under the surface, what you see on screen may be only a small fraction of what actually is present in the file.

The reason becomes apparent when you think about the drafting process for a moment. If you create 50 layers for anticipated future use but end up needing only 40 of them, the other 10 are uselessly hanging out in your layer dialog box. Same thing with unused fonts or linetypes (remember loading *all* of them but using only a few). Layers, linetypes, and fonts really take up little space, but with blocks and wblocks, it is a whole different story.

Wblocks are miniature drawings all to themselves, and while many are just simple doors or window symbols, others can be quite large, such as entire furniture sets or appliances in an architectural layout. Worse yet, these blocks rarely come alone; they bring friends, lots of them. Complex multistory sprawling building layouts feature not dozens but *hundreds* of blocks. They all take up room in a file, ballooning it up to a rather huge size.

If all these blocks are needed, then fine; this is a necessary evil. But if not—say, a type of chair was deleted from the specs—they need to be removed. Erasing them all is easy enough, but are you done? Are they all gone for good? Unfortunately, no. They may not be visible, but they are still there weighing down your file, slowing down computer performance and just about everything else. This is where purge comes into play.

Purge gets rid of items that are not actually used in the drawing file and permanently deletes them. You can use the command to selectively purge items or purge everything all at once. The two main points to remember are

- Purge does not get rid of objects that are visible and present in the drawing—it is not an erase command.
- Purge is permanent (unless you immediately undo it), so be careful of what you purge. If you anticipate eventually needing something, do not purge it out.

Let us give the command a try.

Keyboard: Type in **purge** and press Enter
Cascading menus: **File→Drawing Utilities→Purge...**
Toolbar icon: **none**
Ribbon: **none**

First of all, erase the chair from your screen. Then, start up the purge command via one of the two methods shown in the command matrix, and the dialog box in Fig. 7.6 appears.

As you can see from looking at the list of items in the dialog box, quite a few items are eligible for purging. Some were already mentioned, such as Layers and Linetypes, whereas others you may not have heard of, such as Materials and Shapes. Either way, what is important is whether or not the item has a plus sign next to it. If it does, you can click on the plus sign and expand the folder to see what specific items are in those general categories. In our case, you see the chair block under the Blocks category. You may or may not have a few other plus signs. If you do, then whatever may be there will just get purged along with the chair block.

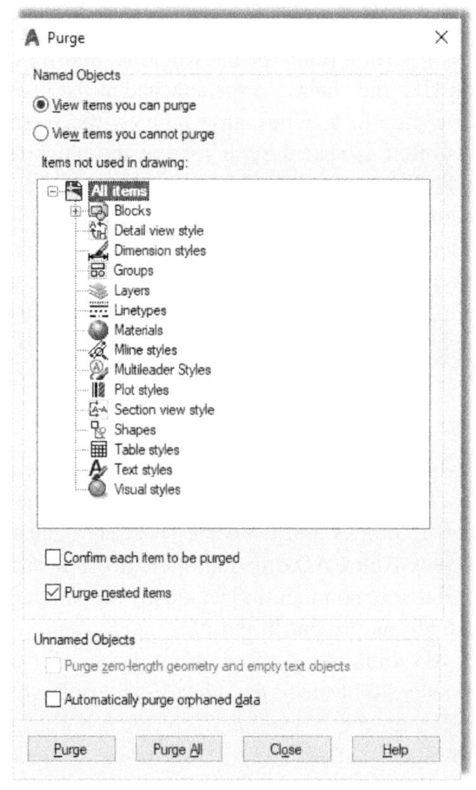

FIGURE 7.6 Purge dialog box.

The actual process of purging is easy. You can either select each item one at a time (usually done only when you want to selectively purge) or you can click on Purge All at the bottom. Be sure the Purge nested items box is checked and the one above it is not, as seen in Fig. 7.6. When all the items are purged, there are no more plus signs and the buttons gray out as well. Press Close after you are done and that is it.

Experienced users purge often and keep little useless junk in the file unless absolutely necessary. Purging is certainly something that should be done prior to saving the file for the day, emailing it, or archiving it for storage and record keeping. Be careful though, and do not get rid of needed items. If in doubt, do not purge the items.

7.4 WBLOCKS

Pretty much everything said about blocks applies to wblocks as well, and so this section is a bit of a review. The central idea to keep in mind is that a wblock is external, and as such, you need to tell AutoCAD where to put the wblock as you make it. Let us try this out. Clear your screen and make sure all items from the previous exercise are purged. Now go ahead and draw another furniture object, perhaps a sofa this time.

You can start up this command via the Ribbon's Insert tab or just by typing in `wblock` and pressing Enter. Once you do that, the dialog box in Fig. 7.7 appears. For the screen shot in this figure, a specific path has not been selected, but you can do that by clicking on the button with the three dots, …, to the right of File name and path and browsing around until you find where you want the file to go. Be sure to also change the name of the file at the end of the string, so it does not just say "new block." Then, select the sofa and pick one of the three options to Retain, Convert, or Delete from drawing the original. When done, press on OK and the wblock is created and dropped off where you indicated in the path.

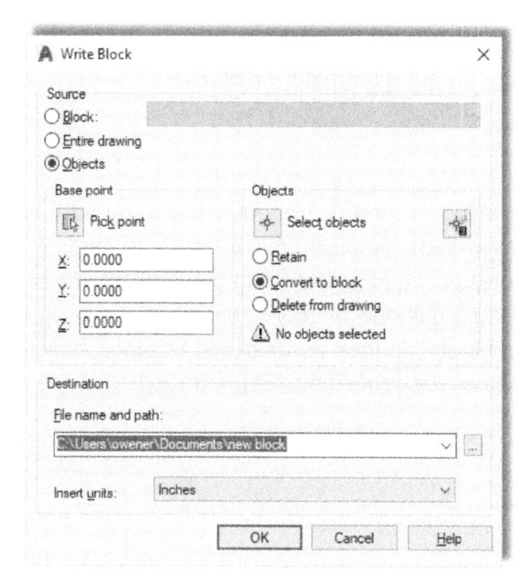

FIGURE 7.7 Write Block dialog box.

It is important to review again the three options you have for dealing with the original block that is in front of you. The last of these three is worded slightly differently. Here they all are again:

Retain

This means that the block is created and stored in memory, but the original object (the chair you are looking at) remains in separate pieces, useful if you wish to make several blocks of the chair, one after the other, each slightly different yet based on the same original.

Convert to Block

This option is what you are likely to use the most. It creates the block and stores it in memory and also makes a block out of the original object. Use it when you are sure; one block is all you need to make out of this one object and you have a need for it right away.

Delete from drawing

Here, the block is created and stored in memory, but the original is deleted from the drawing. This is useful when you are making a block for future use but have no need for it now.

Inserting Wblocks

Go ahead and insert your sofa back in using the insert command. The procedure is essentially the same, except that now you have to browse for the file. The Browse… button is just to the right of the Name: field. Look around for your block and when you find it, click on it and press OK. It inserts just as with the previous regular block command.

7.5 DYNAMIC BLOCKS

Dynamic blocks were introduced back in AutoCAD 2006. Like many new features, they are not a radically new concept. Rather, they are an evolution of the regular block and wblock. The idea here is to be able to modify the block in response to different design conditions. A simple example is of a door. If you insert a block of a 3′ door, opened at 45 degrees, and realize that you need a 3′–6″ door, shown in a closed position, you can change that in place without redefining the block or inserting another one. You can even go a step further and, using the visibility option, change the block to something else; in other words, not just its orientation and size, but what it is completely, akin to having many blocks in one.

So what we are essentially doing here is adding a level of intelligence and automation to the standard block and wblock concept. We use this for scaling, stretching, mirroring, and otherwise modifying the block's sizing and orientation. We also use this intelligence to add menus to a block, and it allows you to select many different versions of the block as dictated by your design needs. AutoCAD provides options for almost every conceivable design intent, so let us take a look at this from the basics on up by first considering very simple parameter changes.

Actions and Parameters

To create a generic dynamic block, you need to modify a regular block by adding parameters and actions. Three things make up a dynamic block: the *geometry* itself, at least one *parameter*, and at least one *action* associated with that parameter. A parameter defines the features of the geometry, and the action defines the modification you would like to act on that parameter. The result is a block that has one or more special grip points, each defined with a certain modifying property (e.g., stretch, rotate, scale). The block can then be modified in place as needed. For a simple example, recreate or retrieve our existing chair block, and follow the procedure outlined next.

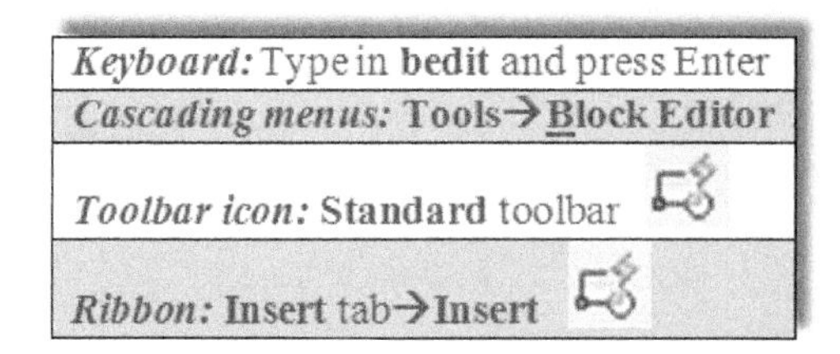

Keyboard: Type in **bedit** and press Enter
Cascading menus: **Tools→Block Editor**
Toolbar icon: **Standard** toolbar
Ribbon: **Insert** tab→**Insert**

Regardless of which method you use to start the command (you can also just double-click on the block itself), you see the dialog box in Fig. 7.8. Select the block you want to work with, which in this case is the only one available.

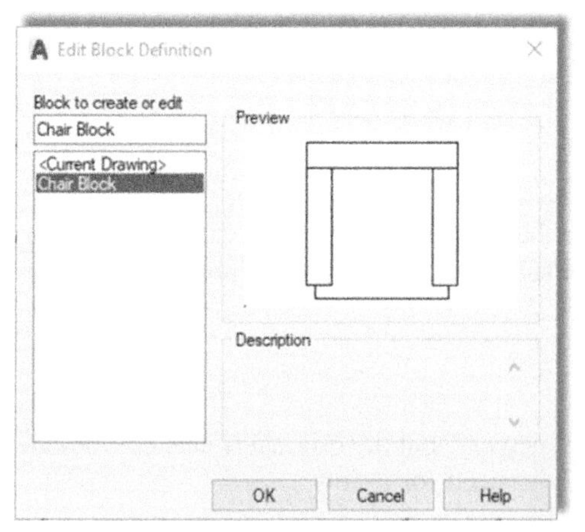

FIGURE 7.8 Edit Block Definition.

After you select Chair Block, you are taken to the main editing screen, called a *Block Editor* (recognizable, unless changed via Options, by its medium gray background), shown in Fig. 7.9. This is where the settings are applied.

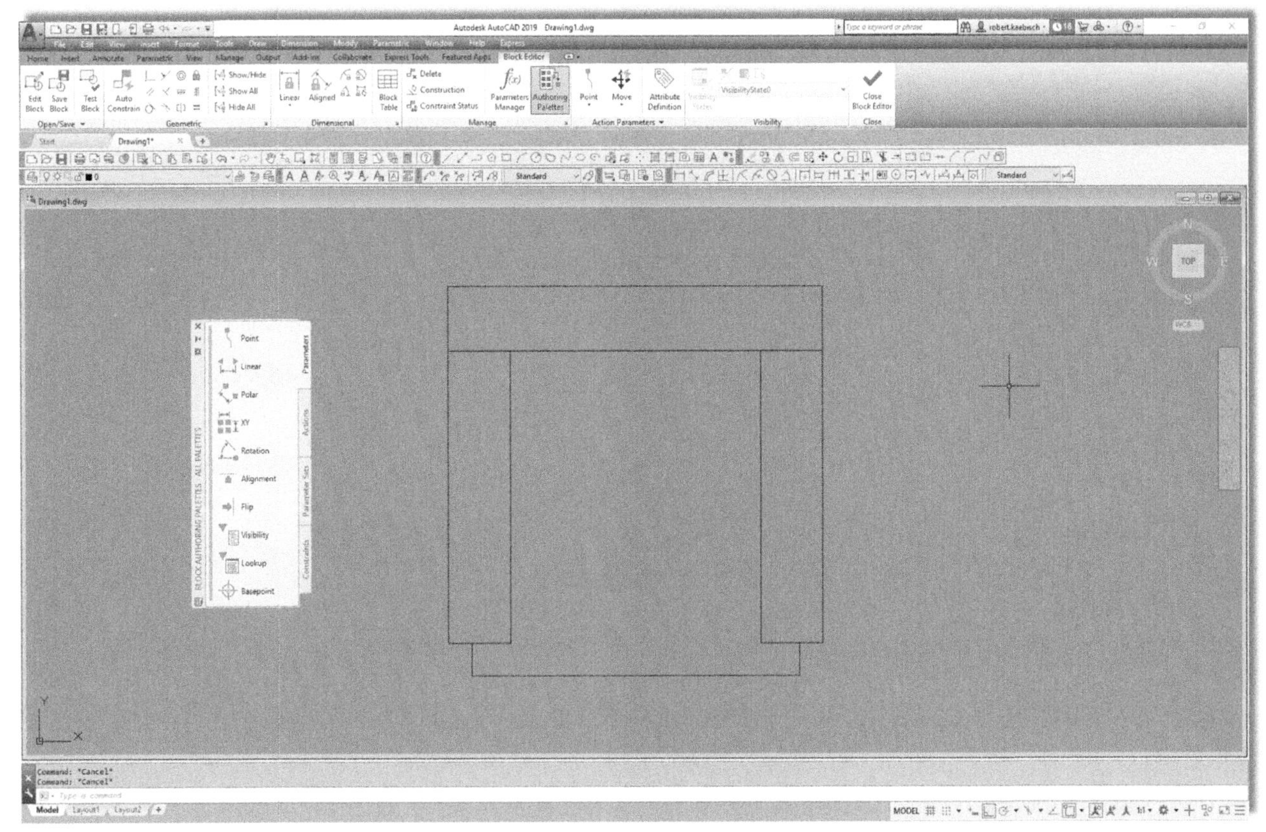

FIGURE 7.9 Dynamic Block Editor.

A Block Authoring palette also appears. Sets of parameters and actions are available from this palette, and the block can be set up to have any of those. When done, press Close Block Editor (top right of screen) to return to the regular drawing space. So what can we do here? Well, let us add two actions and parameters to the chair that give it some scaling and rotational abilities.

Step 1. Open the Parameter tab on the palette and pick the Linear parameter. That, after all, is the way scale works, by linearly scaling up an object. First, pick the upper left corner start point and then pick the upper right corner endpoint of the chair (like adding a horizontal dimension). The result is shown in Fig. 7.10. Note that a yellow exclamation point is seen as a reminder that no action has yet been added to this new parameter.

Step 2. While you are there, add a rotation by running through the required Base Point (lower left), Radius (any value), and Angle (30 degrees) prompts. The result is shown in Fig. 7.11.

Step 3. Now, you can add actions associated with those parameters. Pick the Actions tab and pick Scale. Then, select the Parameter and the Objects (the chair), as prompted. Repeat the same steps for the rotation. The result is shown in Fig. 7.12. Notice the new (faded) scale and rotate symbols, and the disappearance of the exclamation point.

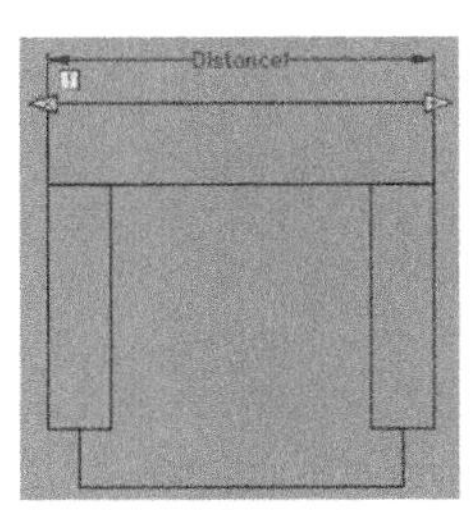

FIGURE 7.10 Dynamic Block, Step 1.

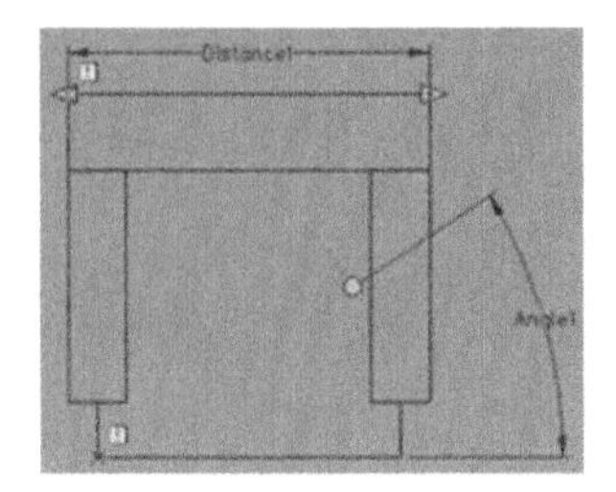

FIGURE 7.11 Dynamic Block, Step 2.

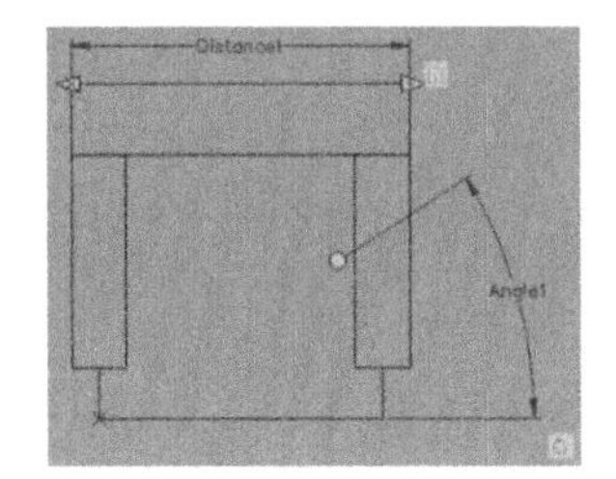

FIGURE 7.12 Dynamic Block, Step 3.

Close the Block Editor (button at upper right of the Ribbon) and you are prompted to save your work, as seen in Fig. 7.13.

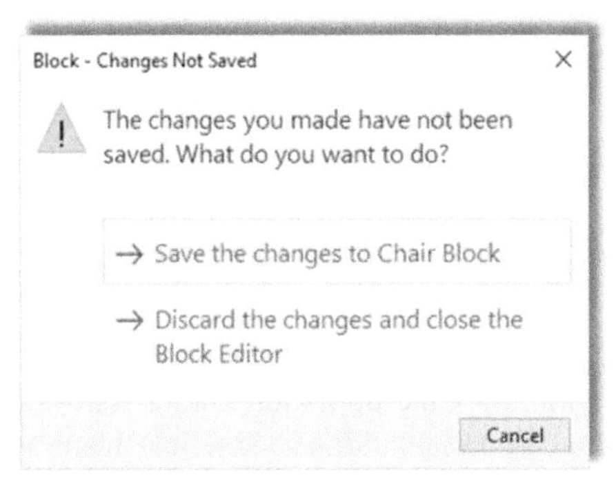

FIGURE 7.13 Save changes prompt.

Finally, you are back where you started. Click on the new block and notice the dynamic grips, as seen in Fig. 7.14. You can use these to change the size and rotation of the block, as seen in Fig. 7.15 with scaling.

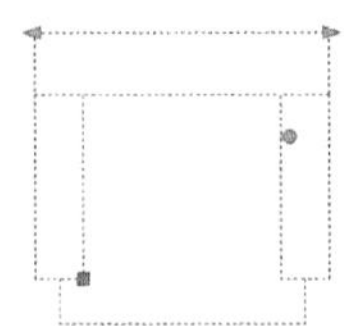

FIGURE 7.14 Dynamic Block grips.

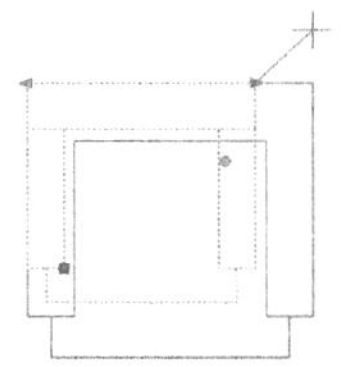

FIGURE 7.15 Dynamic Block modification.

This procedure features the absolute basics of creating a dynamic block. Let us try this again, with a door this time, and discuss a few more features. Create the door shown in Fig. 7.16, and call the block Door_36. Be sure to add in the door swing, the hatching, and the door jams. The dimensions are only for your reference.

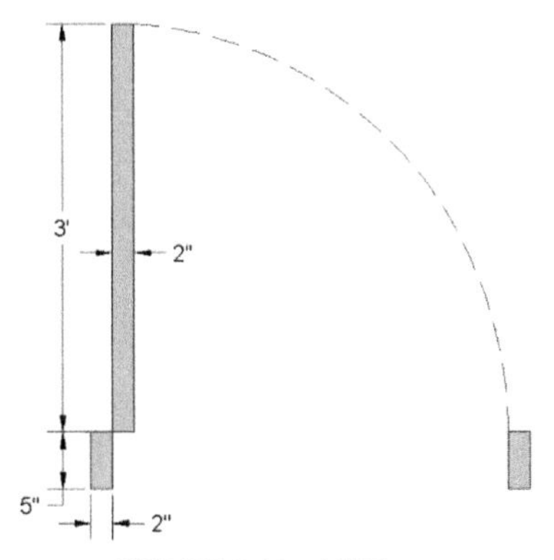

FIGURE 7.16 36″ Door.

Go into dynamic block edit mode and let us add the Flip action. This allows you to reverse the opening of the door after it has been inserted into place.

Step 1. Add a horizontal reference line across the bottom from the left side of the door to the far right end of the door swing.

Step 2. Add the Flip parameter using the midpoint of the reference line and tracing upward using Ortho. Position the Flip state1 somewhere near the line. You can rename this text string to Flip Left-Right or something similarly descriptive if you wish, using the change properties palette.

Step 3. Add the flip action by selecting the icon, then the parameter, and then all the objects. Fig. 7.17 shows what you should have on your screen. You can go ahead and erase the horizontal guideline at this point.

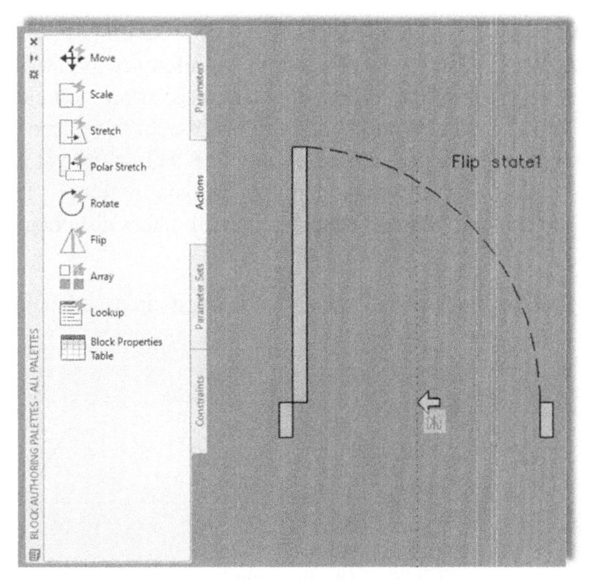

FIGURE 7.17 Flip parameter and action.

Step 4. As a final step, press the Test Block button at the upper left of the Ribbon. This takes you into an intermediate area (still in block edit mode) where you can test out what you have done. In our case here, click on the door and you see a blue arrow pointing to the left at the bottom. Click on that and the door's swing flips around (Fig. 7.18).

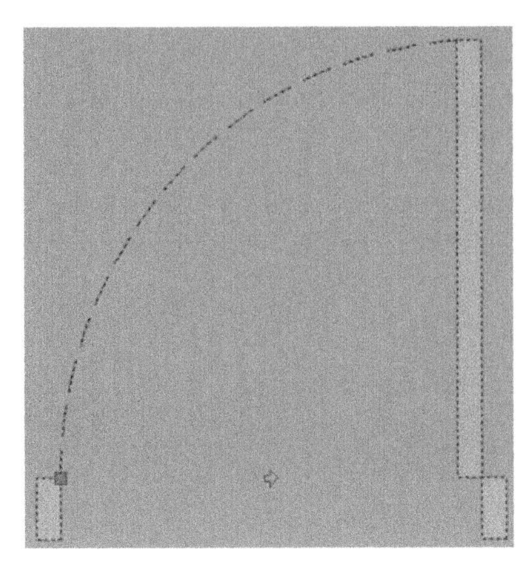

FIGURE 7.18 Flip executed.

You can also set up the flip so the door opens "downward" and not "inward," as it does now, of course. Your reference point would then be the midpoints of the two door jamb blocks. Try that as an exercise.

Another useful feature you can try is called *Value Set.* Add a linear parameter and action to the door (exactly like Distance 1 in Fig. 7.10). Then, while Distance 1 is selected, go to the Properties palette, Value Set category, and for the Dist type on the left side, select List on the right. Just below that for Dist value list, enter a few incremental values greater than 36″ (the current door size), such as 48″, 60″, and 72″, into the Add Distance Value dialog box. After you complete this step and go to test the door by clicking on the scale grip, you see a series of tick marks appear and you can increase the sizing only incrementally, as opposed to arbitrarily. There is quite a bit more to dynamic blocks, of course, and you can combine many features to create new ones.

Visibility

The final item to cover with dynamic blocks is a very useful feature called *Visibility*. This option is a favorite among quite a few designers because it allows you to combine virtually unlimited variations of a block into one and then simply select what you want via a drop-down menu. It can take a while to set up the numerous blocks and set the visibility states, but the payoff in the end is significant.

Here is the basic idea. You make a block by combining a number of other blocks (or even nonblock geometry by itself) by putting them one on top of the other, usually with a common reference point. You then create named visibility states—with each state representing a version of the block you wish to see when you click that name. You then set all but the relevant block as invisible. The process gets repeated until you process all of the geometry in such a manner. The result is one large block that contains numerous sets of geometry, but only one set is visible at a time. To pick which is the visible one, you select from a drop-down menu, which appears when you click on the block. There, you see all of your visibility state names, as you click each one the relevant geometry appears and all others disappear. Let us give this a try with a very simple example.

Step 1. Create five shapes and make a block out of each—a square, a circle, a hexagon, a triangle, and an ellipse—as seen in Fig. 7.19.

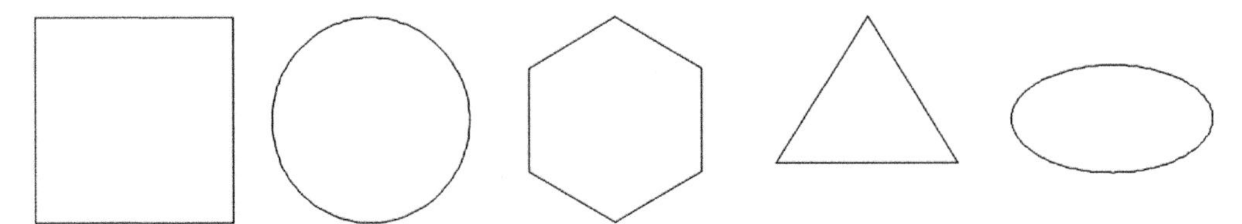

FIGURE 7.19 Basic block shapes.

Step 2. Put all the shapes on top of each other, centering them at the top. Make a block out of all of them calling it Main_Block. Then go into the dynamic block–editing mode via bedit or just double-clicking on Main_Block. You should see what is shown in Fig. 7.20.

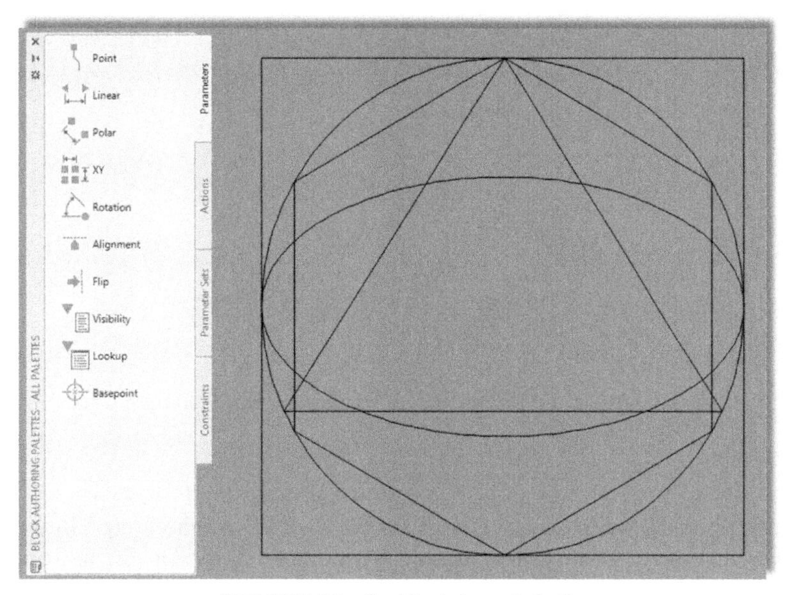

FIGURE 7.20 Combined shapes in bedit.

Step 3. Now add the visibility parameter (highlighted with a black rectangle in Fig. 7.21) by clicking on the Parameters tab. AutoCAD asks where to locate it. Select any convenient spot next to the shapes, as seen in Fig. 7.21.

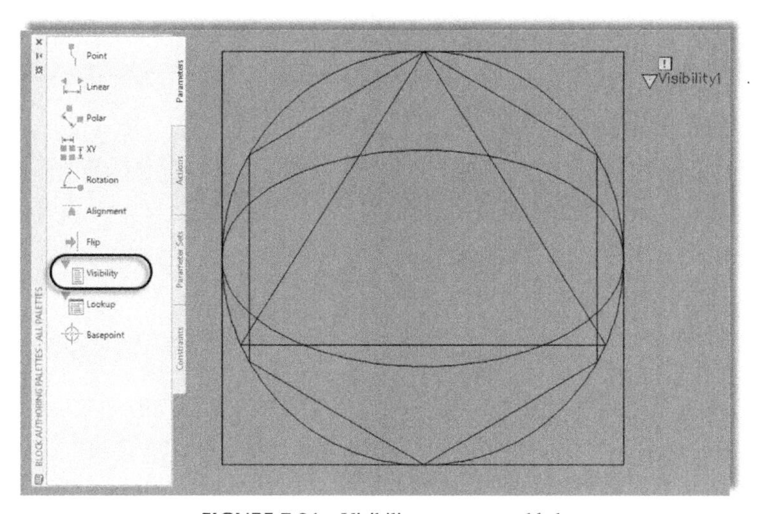

FIGURE 7.21 Visibility parameter added.

Step 4. Select the Visibility States button in the Ribbon's Block Editor tab, Visibility panel (Fig. 7.22). The Visibility States dialog box appears. Rename the existing state to Square and add the other four via the New... button (leaving any options as default in the dialog box). The result is seen in Fig. 7.23. Press OK.

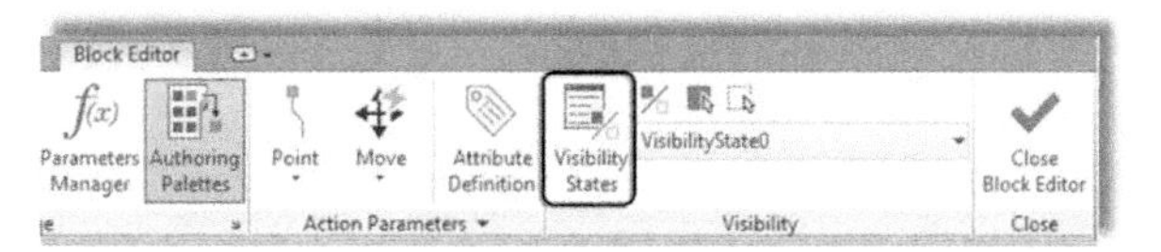

FIGURE 7.22 Visibility States.

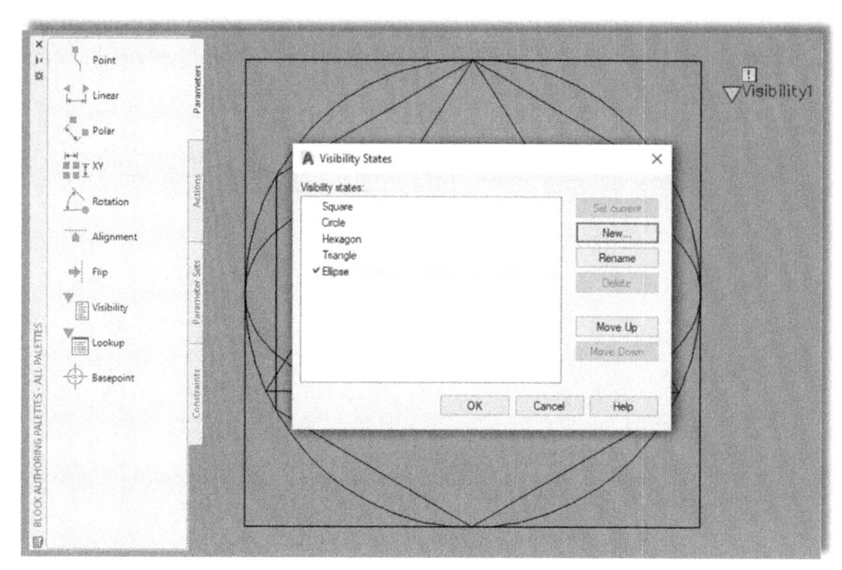

FIGURE 7.23 Visibility States completed.

Step 5. You now need to actually set the invisibility parameter of each visibility state. Go up to the Ribbon and find the drop-down menu just to the right of the Visibility States button (it can be seen in Fig. 7.22). Drop it down and double-click on the Square state. It is now the current visibility state, and it is here that you need to make all the other shapes invisible. To do that, press the button just above the drop-down menu called *Make Invisible* (it looks like a faded square with an arrow and can also be seen in Fig. 7.22.). AutoCAD says Select objects to hide:. Go ahead and pick *all* of the shapes *except* for the square. The selected shapes all fade to a gray color, as seen in Fig. 7.24.

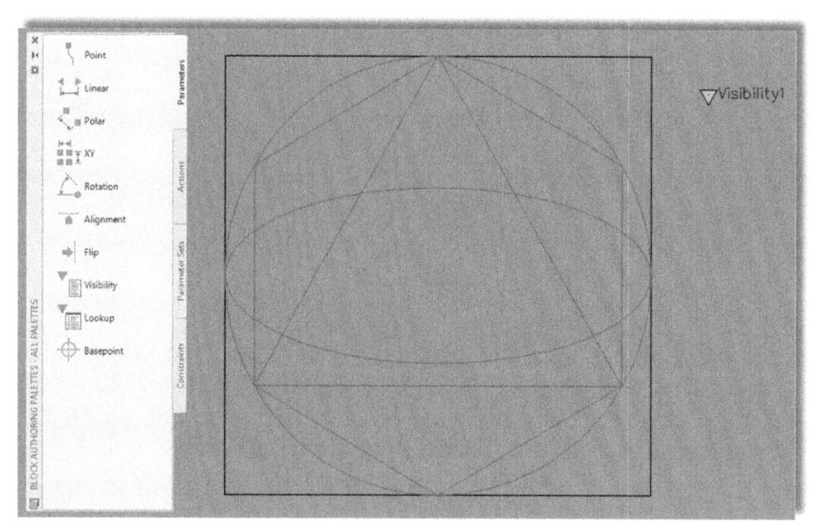

FIGURE 7.24 Square visibility state.

Step 6. Repeat this process for the other shapes by going again to the drop-down menu and double-clicking on the next shape on the list (the circle). All the shapes reappear, and now you have to use the Make Invisible button to select *all* of the shapes *except* the circle. This is why it was remarked that it may take a while if you have several dozen or more blocks embedded in one. The effort is well worth it, however.

Step 7. When you completed setting all of the visibility states, test them out via the Test Block button all the way on the left of the Ribbon's Block Editor tab. A new tab opens; go ahead and click on the block. Fig. 7.25 shows the first visibility state (square) with the drop-down menu revealing the rest of the choices. Click through each one to test each visibility state. When done, exit the test tab and save the dynamic block. It is now ready for use.

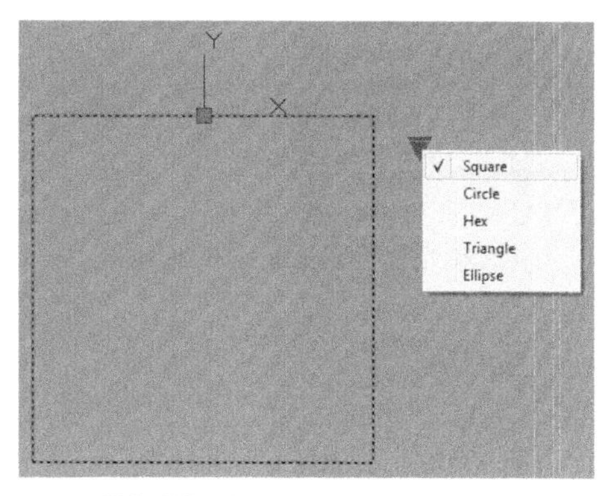

FIGURE 7.25 Visibility state menu.

This of course was a very basic example, but the process for more complex, realistic parts is identical. Keep them as blocks, stack them one on top of the other, create and set visibility states, save, and use. Here is a real-life example from the world of electrical (power) engineering. Fig. 7.26 is a top view of an actual relay switch commonly found at power substations.

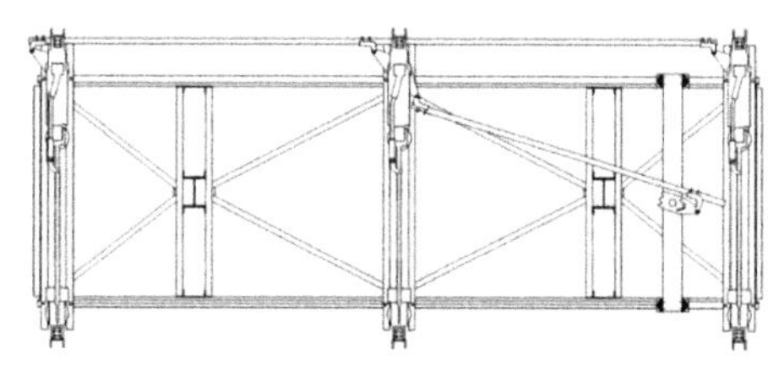

FIGURE 7.26 Relay switch.

As it happened in this case, a number of switches were at the disposal of the electrical engineer, all of them essentially the same in their basic function but very different in how they looked. It made sense to create one main block of them and just select what was needed for a particular job. That way the designer need not keep track of numerous blocks in a library. Such was the approach to drafting the switch in Fig. 7.26, as seen in the next image (Fig. 7.27). Notice the menu of available visibility state alternatives, with "Disconnect – Vertical Break SS20220 (B-42)" as the currently visible one.

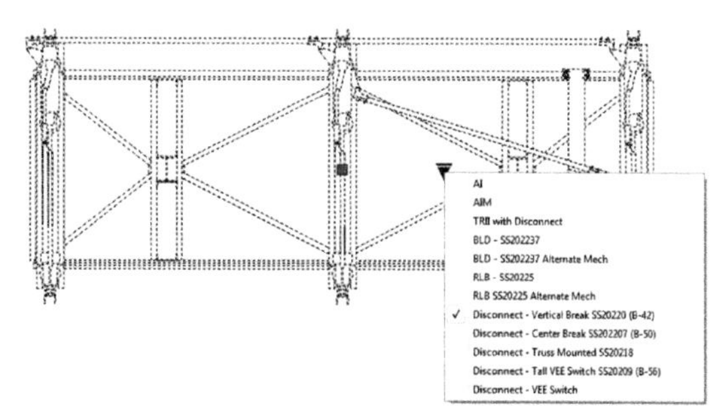

FIGURE 7.27 Available visibility states.

Figs. 7.28 and 7.29 show two other versions of the same switch. As you can see, they are complex blocks and quite different from one another but all sourced from the same file. Be sure to practice and get well-versed in setting up these types of dynamic blocks, visibility states are very useful in actual design work!

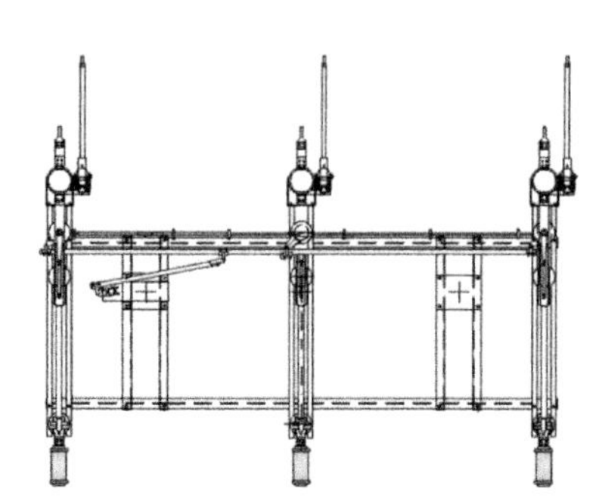

FIGURE 7.28 Alternate visibility state 1.

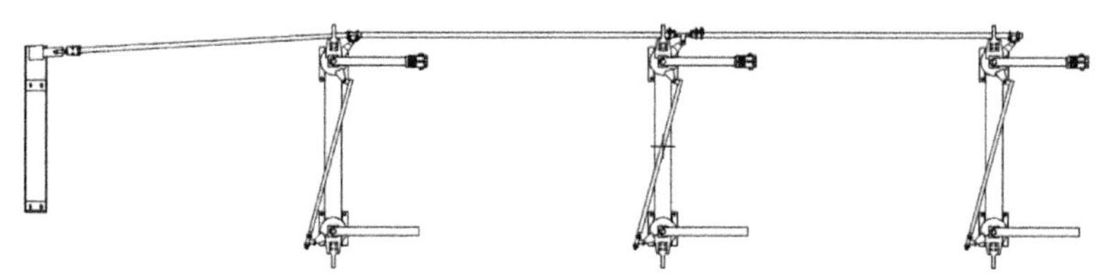

FIGURE 7.29 Alternate visibility state 2.

So, how do dynamic blocks fit into the big picture? Are they used as often as they could be? The results are mixed. Clearly, they are a big step forward for blocks and the built-in intelligence is very useful. However, one potential problem for long-time users is that, if they already have an extensive library of nondynamic blocks in use, there may not be any real incentive to rework them into dynamic blocks. This is an effort that sees more payoff if done from the beginning, but because dynamic blocks are relative latecomers (having appeared only in 2006), some may not want to upgrade their blocks and long-standing libraries if they have been in use and working fine for all these years. It is worth the effort however, as dynamic blocks, by allowing multiple variations on a single block, actually reduce the sizes and complexities of these libraries.

The bottom line for you as a student, however, is that it is a good skill to pick up, so try to explore further this interesting topic on your own. Chances are that, in the future, these will be the main types of blocks you will run into out there in the industry.

7.6 GROUPS

A group is a useful, but often overlooked, command from the block family. It is, perhaps, the easiest way to collect objects together and presents some advantages (but also some disadvantages) over the familiar block, as shown next.

Similar to a block, a group can

- Be given a name.
- Bind a collection of objects together.
- Be moved around, copied, and rotated.

Advantages of a group are as follows:

- A group can have objects added and subtracted from it.
- A group can have individual objects (and their properties) edited.

Disadvantage of a group are as follows:

- A group cannot be transferred between drawings (for "local" use only).
- An update to one group does not update other groups with the same name.

The group command went through some changes in the 2012 release of AutoCAD. You can now use the command line, Ribbon, or the Group toolbar (Fig. 7.30) to create and manage the groups, but the classic Object Grouping dialog box, which has been around forever, is still available as one of the options and is discussed later.

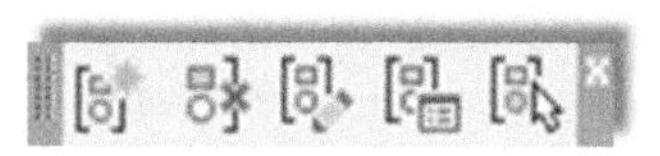

FIGURE 7.30 Group toolbar.

To try this command, first draw a random collection of objects, as seen in Fig. 7.31, then follow the steps shown.

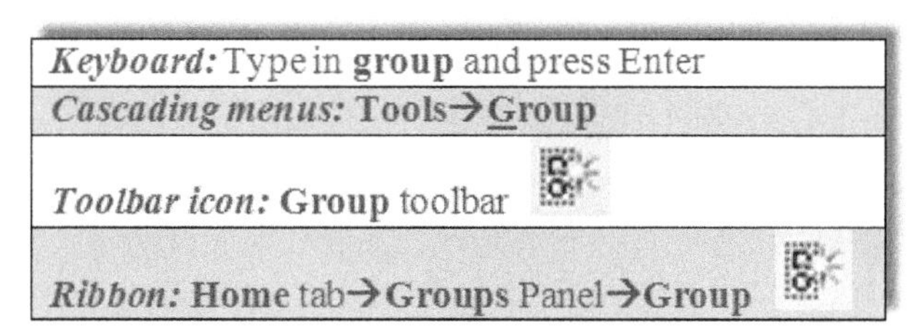

Keyboard: Type in **group** and press Enter
Cascading menus: **Tools→Group**
Toolbar icon: **Group** toolbar
Ribbon: **Home** tab→**Groups** Panel→**Group**

Step 1. Begin the group command via any of the preceding methods.
- AutoCAD says: `Select objects or [Name/Description]:`

Step 2. Press `n` for `Name`.
- AutoCAD says: `Enter a Group Name or [?]:`

Step 3. Type in a descriptive name: Sample_Group in this case. Note that spaces in the name are not allowed.
- AutoCAD says: `Select objects or [Name/Description]:`

Step 4. Select all the objects.
- AutoCAD says: `Group "SAMPLE_GROUP" has been created.`

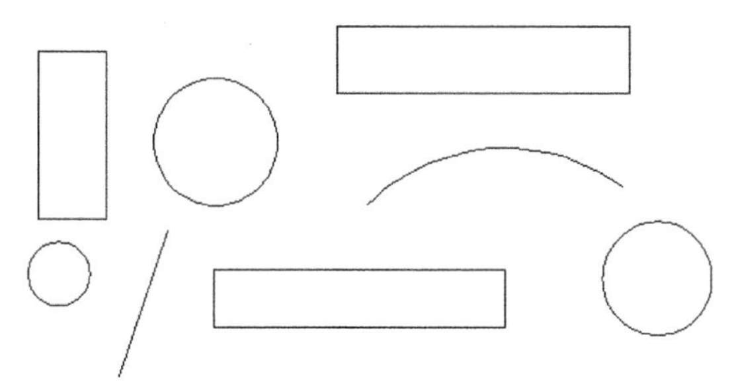

FIGURE 7.31 Random objects for grouping.

The group is now completed, and you can see results by clicking on any object (Fig. 7.32). Note how there is a rectangle around the objects and a single grip in the middle. You can now manipulate all the objects together as one unit.

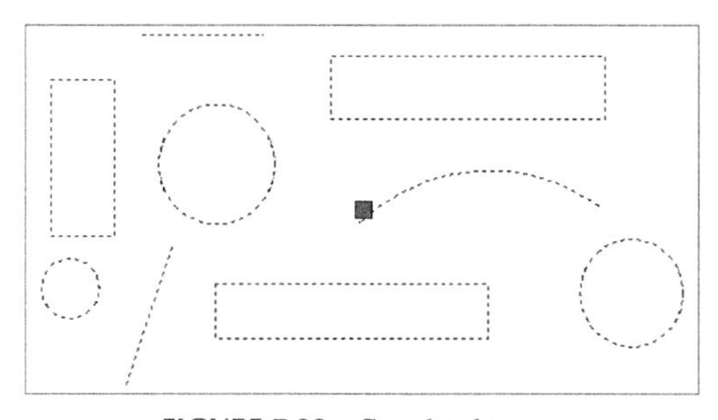

FIGURE 7.32 Completed group.

Let us take a look now at some of the features of a group and some additional functionality. One property you can take advantage of right away, and the main reason to use a group, is the ability to alter the properties of the objects in the group (or the shapes of the objects themselves). Double-click on any of the shapes and a palette called *QuickProperties* appears, as seen in Fig. 7.33.

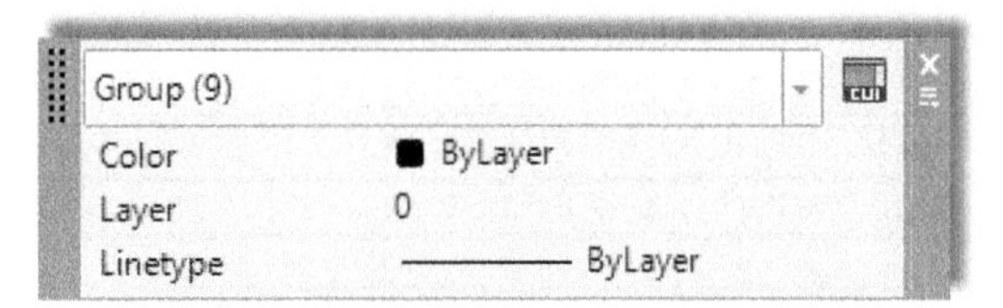

FIGURE 7.33 Group QuickProperties.

Drop down the top menu and you see a breakdown of the group, object by object. There are three rectangles (polyline), two lines, one arc, and three circles, for a total of nine objects. Select Circle (3) in the menu and the palette shows only the properties of the circles. Go ahead and change the color to red and observe the result (Fig. 7.34).

You can also work with individual objects in the group by clicking on the very last icon on the Group toolbar, called *Group Selection On/Off*. You can do the same by typing in pickstyle and entering 0 when prompted. Pickstyle is a system variable, and more is said about system variables in Chapter 15, Advanced Design and File Management Tools. For now, go ahead and modify any of the shapes, and when done, press the same icon or type in the same system variable (entering 1 this time) to get back to the grouping. Fig. 7.35 shows the editing in progress.

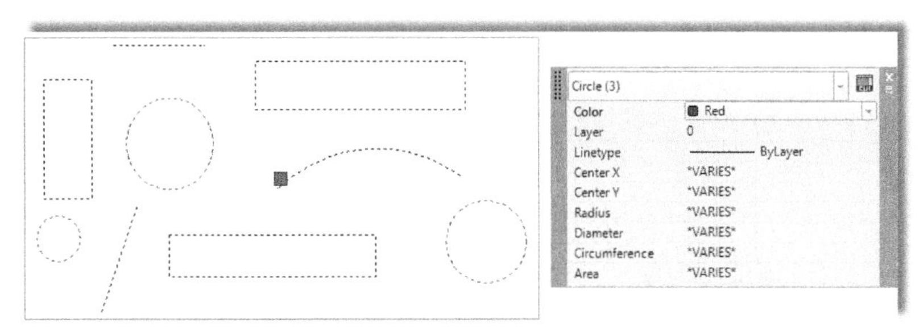

FIGURE 7.34 Color change to *circles* in group.

Take a look at the other toolbar icon. The second one from left, called *Ungroup*, as you would expect, removes the grouping from the objects. Note that exploding the group does *not* delete the group itself but rather just explodes the objects inside of it—only the rectangles in this case.

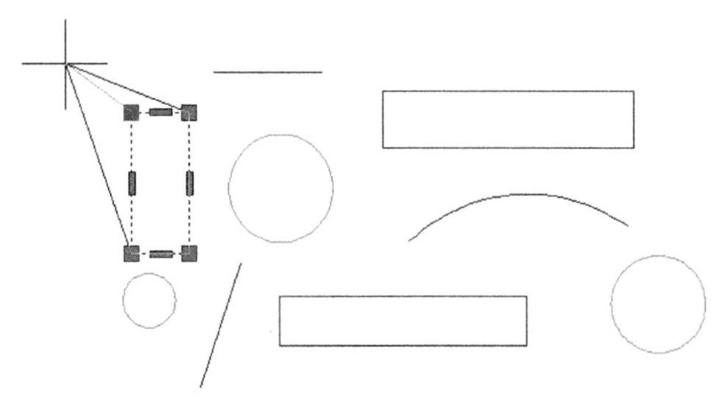

FIGURE 7.35 Group object editing.

The next icon to the right (with the ±), called *Group Edit*, allows you to add an object to or subtract an object from the group. Go ahead and try this on your own by deleting a rectangle via the Remove objects option and adding a new object (a polygon in this example) via the Add objects option. An example of the result is shown in Fig. 7.36.

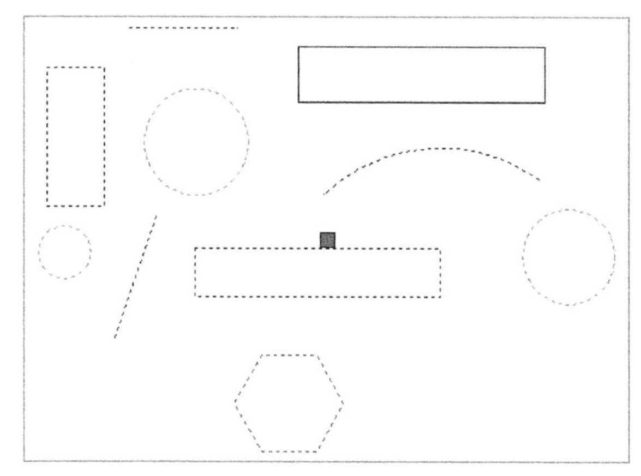

FIGURE 7.36 Object added (polygon) and another subtracted (rectangle).

Finally, it is necessary to mention the legacy Object Grouping dialog box (Fig. 7.37). It serves two purposes here. One is to provide you an alternate way of managing your groups. The second is to familiarize you with the process if you end up using an older version of AutoCAD, prior to the 2012 release, where this was the only way to create and manage groups.

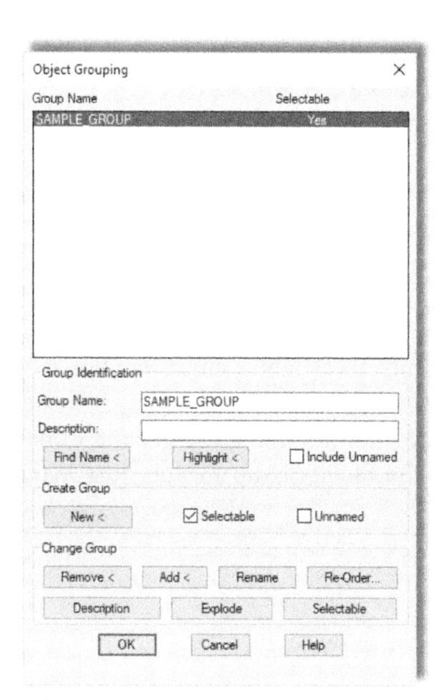

FIGURE 7.37 Object Grouping dialog box.

The Object Grouping dialog box can be called up by typing in `classicgroup` and pressing Enter. Note that, in this example, SAMPLE_GROUP is already created and is highlighted in the dialog box to reveal a multitude of options that should generally be familiar to you, based on the preceding discussion.

From this dialog box, you can create new groups, remove and add objects, and execute many other functions. Take some time to go over this dialog box on your own to complete your tour of the group command.

SUMMARY

You should understand and know how to use the following concepts and commands before moving on to Chapter 8.

- Block
  - Select objects
  - Name the block
  - Select a base point

- Wblock
  - Select objects
  - Select path and enter block name
  - Select a base point
- Options
  - Retain
  - Convert to block
  - Delete
- Insert command
- Purge command
- Dynamic blocks
  - Parameters
  - Actions
  - Visibility
- Groups

REVIEW QUESTIONS

Answer the following based on what you learned in this chapter:

1. What are blocks and why are they useful?
2. What is the difference between blocks and wblocks?
3. What are the essential steps in making a block? A wblock?
4. How do you bring the block or wblock back into the drawing?
5. Explain the purpose of purge.
6. Explain the basic idea behind dynamic blocks. What steps are involved? What is visibility?
7. What are the advantages and disadvantages of using groups?

EXERCISES

1. In a new file, create the following: architectural table, chair, keyboard, phone, and computer screen, and make a block out of the individual pieces as well as the overall "assembly." Nested blocks, such as these, are quite common in AutoCAD design. Insert the block back into the file. Only general dimensions are given, so you have to improvise for the individual pieces. Also, naming the blocks is up to you, just make them descriptive. (Difficulty level: Easy; Time to completion: 15–20 minutes.)

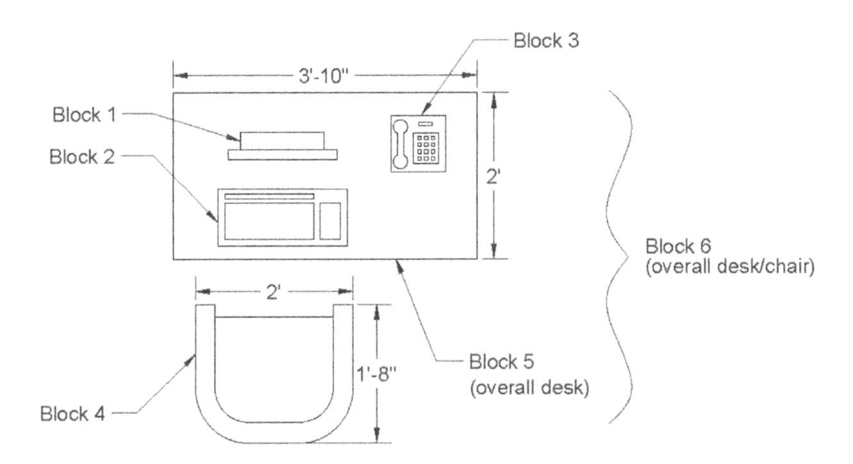

2. In a new file, create the following set of bathroom toilet and sink fixtures, doors, and other items. The overall dimensions are provided, but you need to create the details on your own. Be sure to use the appropriate layers. Create basic blocks out of the fixtures, and copy as needed. (Difficulty level: Easy/Moderate; Time to completion: 30–45 minutes.)

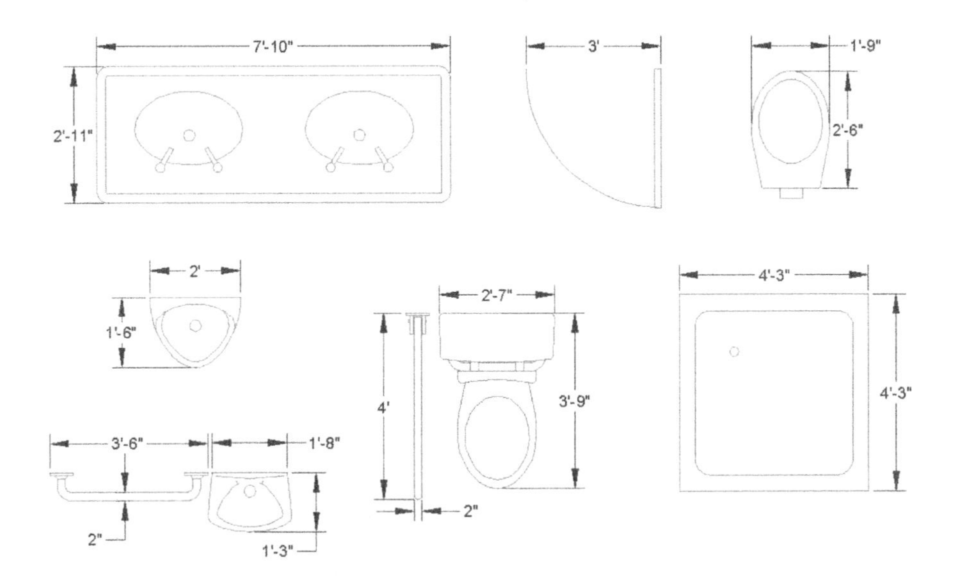

3. In a new file, create the floor plan depicted next. Insert the blocks from Exercise 2 into the floor plan as shown, adding the wall hatching as a last step. (Difficulty level: Moderate; Time to completion: 20–30 minutes.)

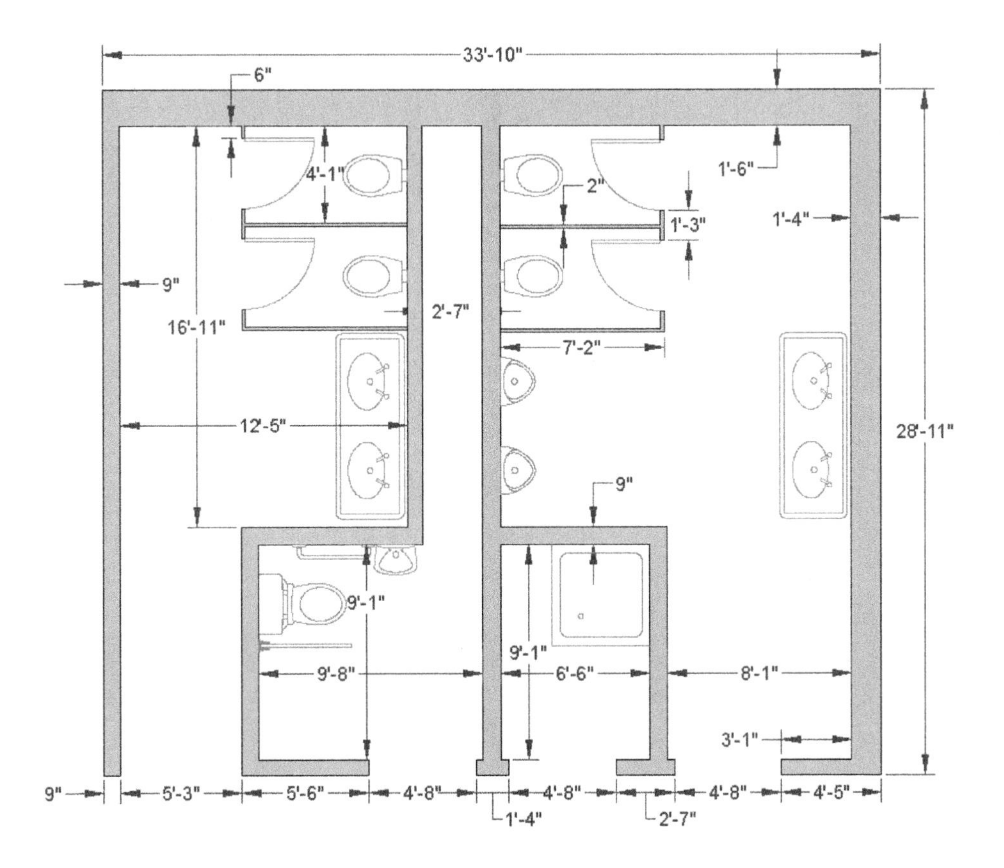

4. In a new file, create the following set of doors and make a wblock out of each one (Door_24in, Door_30in, Door_36in). Drop them all into a premade folder called Doors. Insert each wblock back into the file. When all three doors are inserted, explode them all and purge the file. Then, create dynamic blocks out of all three. Add the linear parameter followed by the scale action. (Difficulty level: Easy; Time to completion: 5–10 min per block set.)

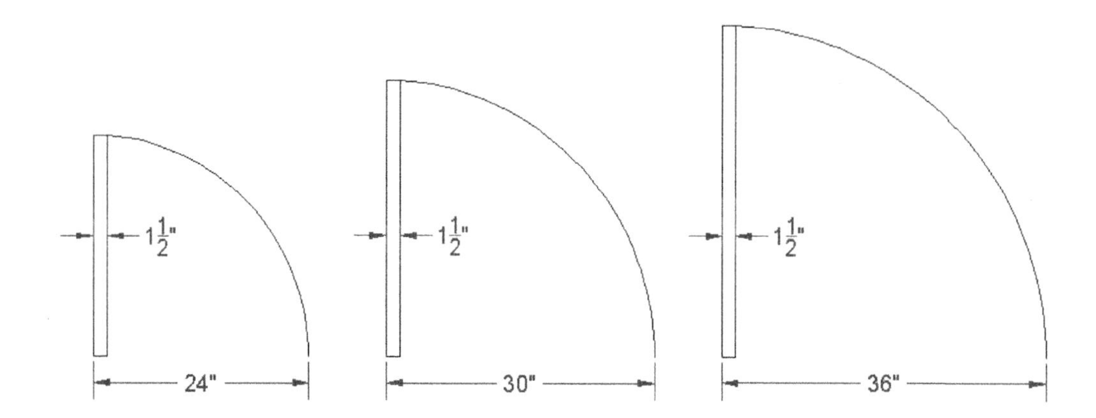

5. Create the following dynamic block double door. The basic dimensions are 2″ × 24″, with the door handle sized off of that. Add in shading, arbitrary-sized door jams, and the hidden line door swing. Create a variety of parameters and actions, including Flip, Rotate, Scale, and others as you see fit. (Difficulty level: Easy; Time to completion: 15–20 min.)

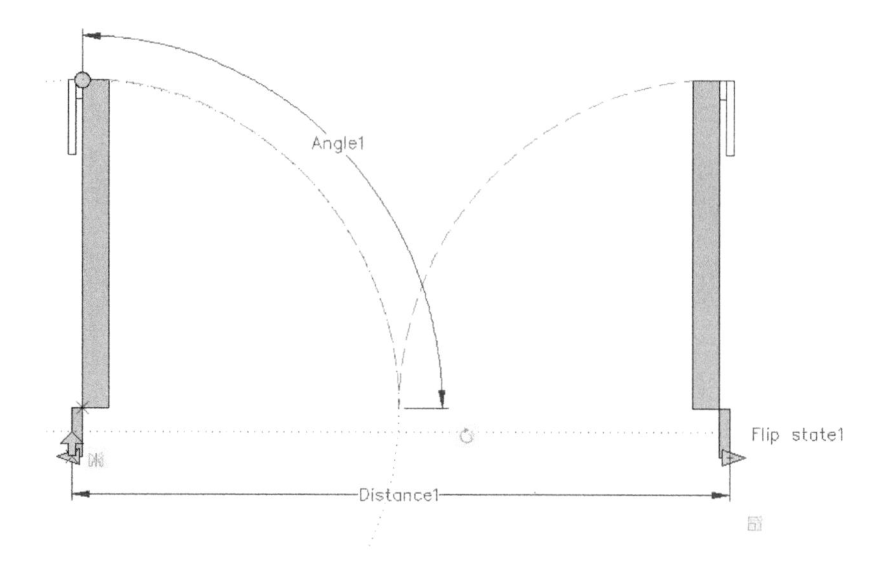

6. Electrical symbols are all usually blocks and electrical designers/draftspersons have an extensive library at their fingertips. You have already drawn a few electrical symbols in the previous chapter; now let us turn our attention to a few more. Draw the following symbol library and create a block out of each symbol. While you are drawing these symbols, try to memorize them as well. You will see them again in almost any architecture-related field. (Difficulty level: Moderate; Time to completion: 30–45 minutes.)

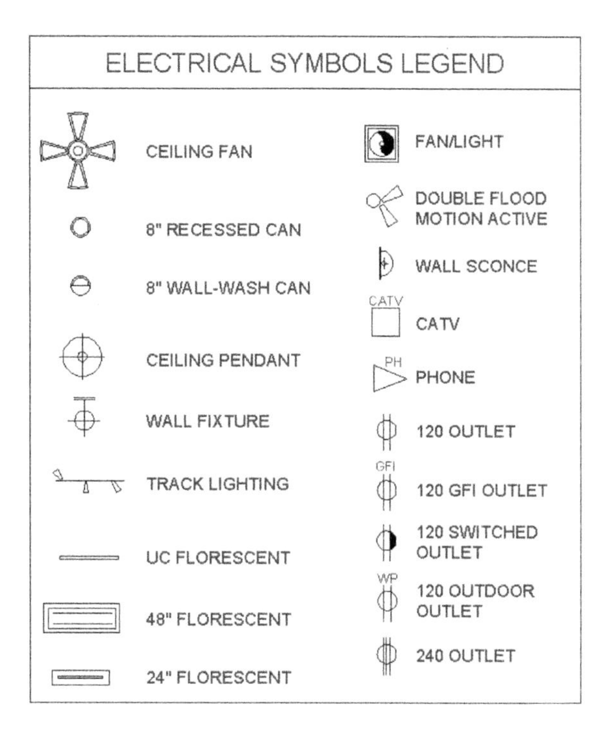

7. This exercise is another example of electrical schematics. Draft all the linework and add text as shown. Like most schematics, the drawing is not to any scale, but accuracy must still be maintained as far as connecting lines and other shapes, as well as text placement. Create blocks whenever practical and make use of the copy and mirror commands to expedite the drafting. (Difficulty level: Moderate; Time to completion: 30–45 minutes.)

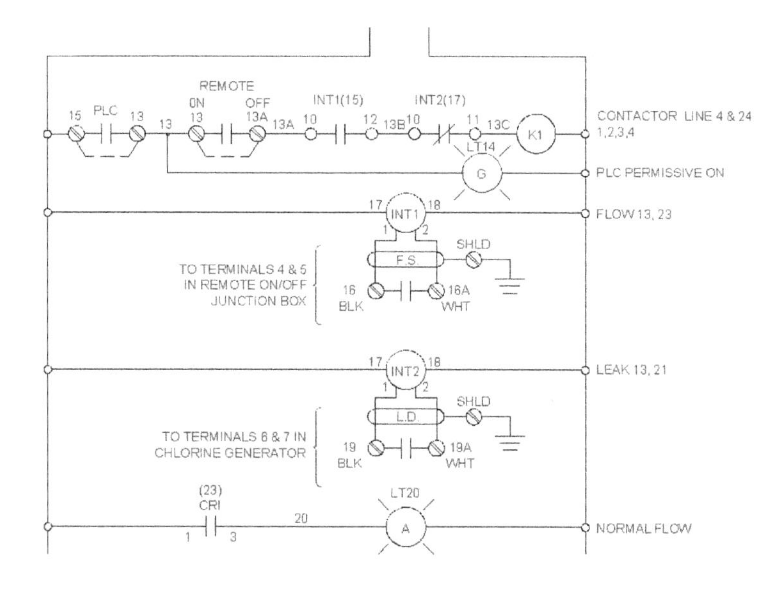

8. For this exercise, we go back to mechanical drafting while incorporating elements of what we just covered. Set your units to Mechanical; set up all appropriate layers, fonts, linetypes, and mechanical dimensions (with 0.000 accuracy); and create the following mechanical fixture, including the two section views. The symbols used for sections are typical in the industry. Be sure to make a dynamic block out of them with the Flip action, so they can easily be reversed to point in the other direction if needed. For additional practice, add in all hatching and dimensions/text. (Difficulty level: Moderate; Time to completion: 45–60 min.)

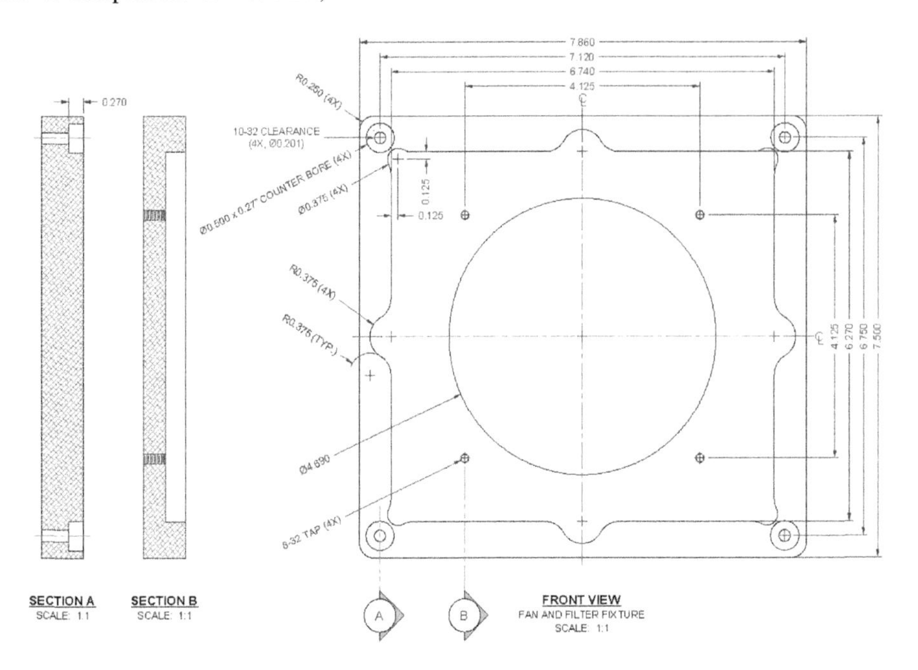

9. Valves are an important part of P&IDs. In a new file, set up all appropriate layers, fonts, linetypes, and mechanical dimensions (with 0.00 accuracy and a ″ suffix), and then, draw the valve shown. Be sure to use the mirror command wisely to avoid duplication of drawing effort. The hex nuts are not specifically detailed out; you have to improvise. Make a block out of each hex nut with the Flip action added. (Difficulty level: Intermediate/Advanced; Time to completion: 45–60 min.)

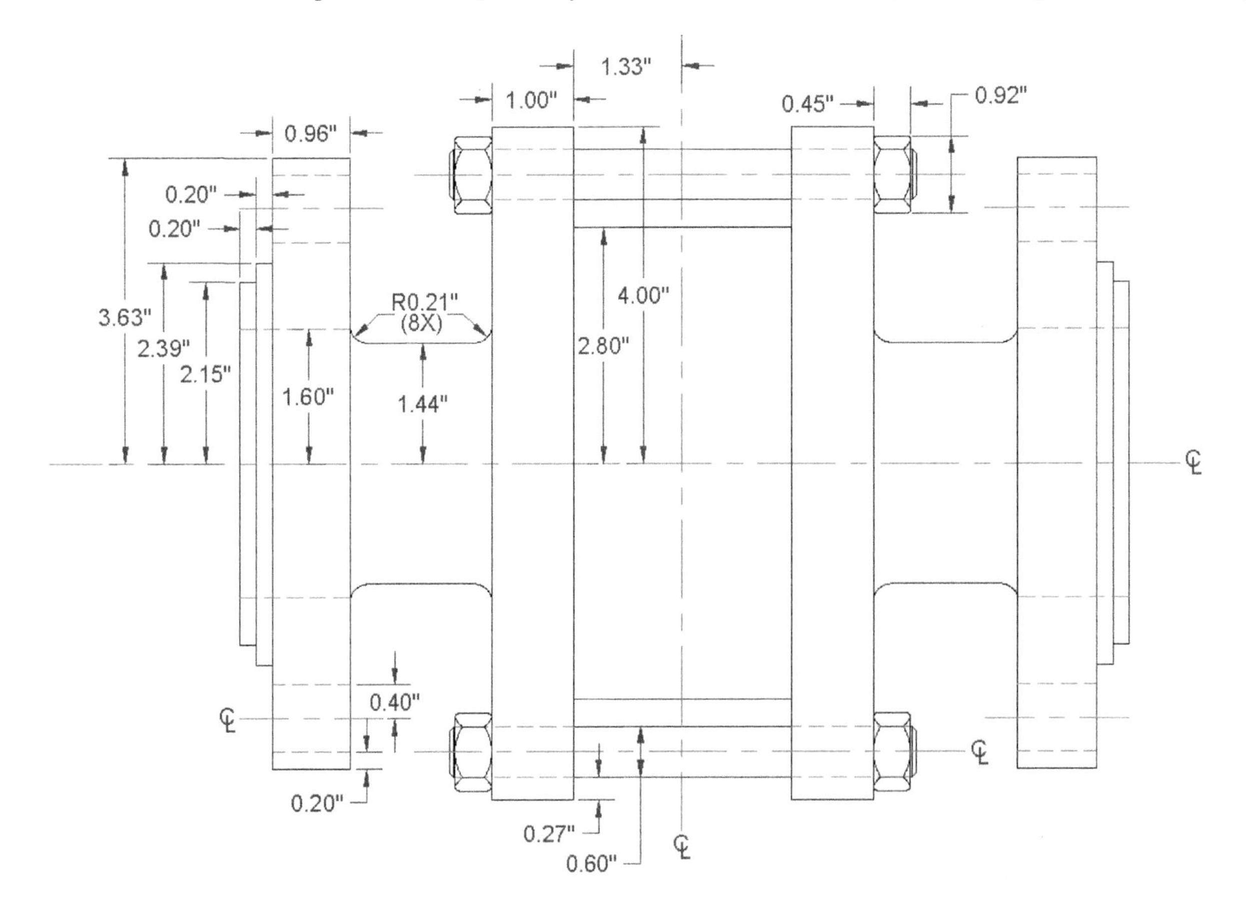

10. The final exercise of this chapter focuses on the basic drafting of an architectural detail. Set up the proper layers and draft what you see, based on the given primary dimensions. Where none are shown, you have to do a reasonable estimate. Note how lines and arcs were used to create borders for some of the hatches, then deleted. This is a common technique that we try again with plines and splines in Chapter 11, Advanced Linework to achieve a smoother effect. (Difficulty level: Moderate; Time to completion: 60–90 min.)

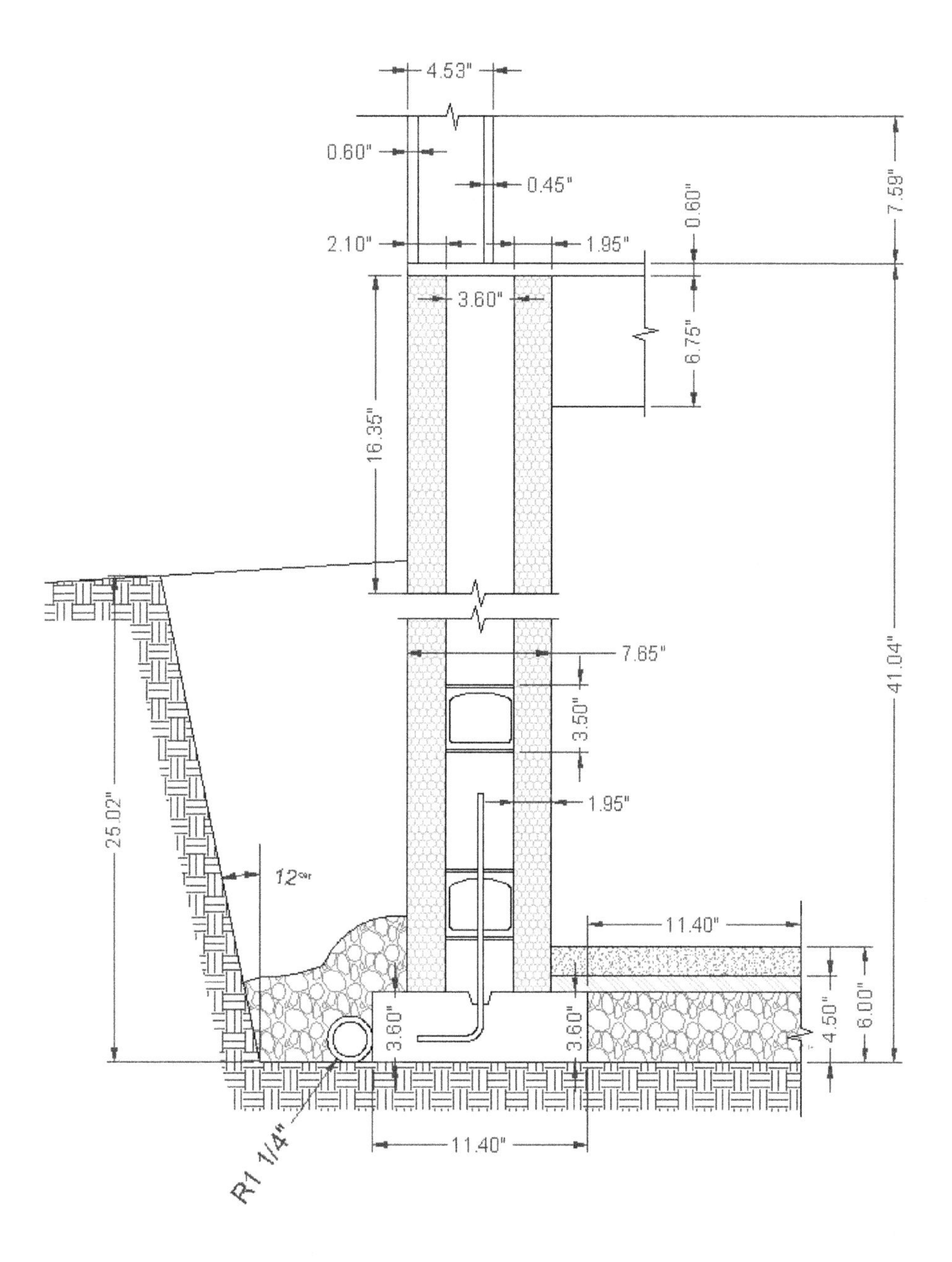

Chapter 8

Polar, Rectangular, and Path Arrays

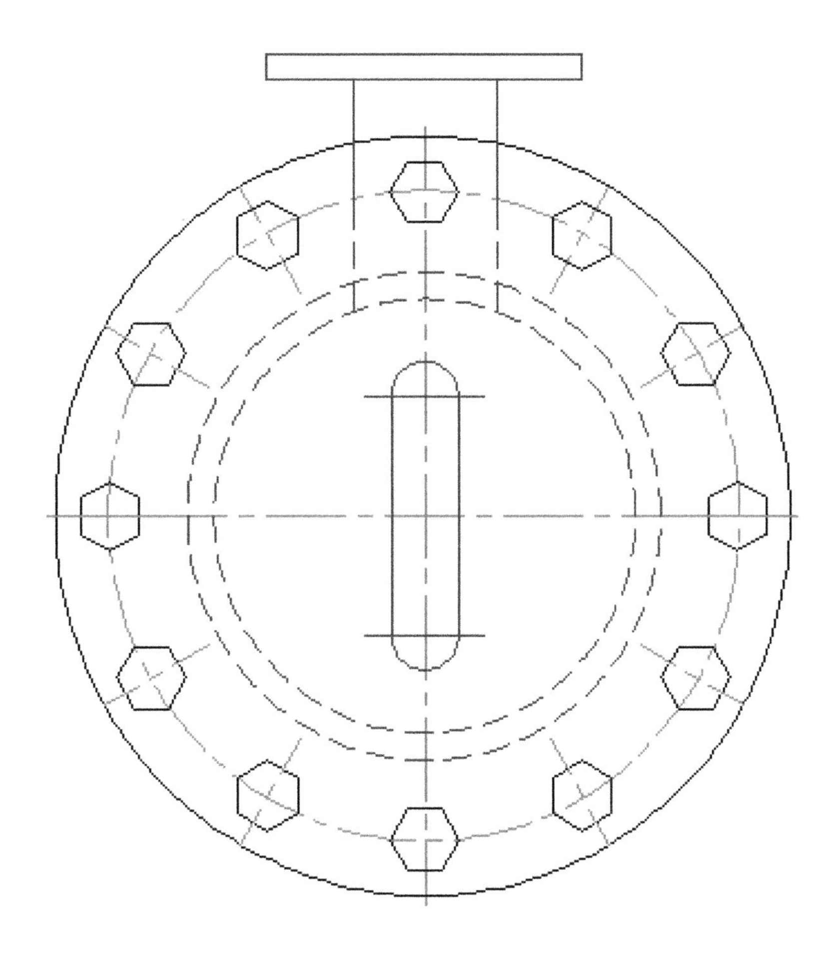

Learning Objectives

In this chapter, we introduce the concept of an array and discuss the following:

- Creating a polar array
- Object, center, quantity, and degrees
- Additional operations with polar arrays
- The legacy polar array
- Creating a rectangular array
- Object, rows and columns, and distances
- Additional operations with rectangular arrays
- Legacy rectangular array
- Creating a path array
- Additional operations with path arrays

Up and Running with AutoCAD 2019. https://doi.org/10.1016/B978-0-12-816440-2.00008-2

At the end of this chapter you will be able to create polar, rectangular, and path arrays. You then start a new in-class mechanical project.

Estimated time for completion of this chapter (lesson and project): 3 hours.

Arrays are very useful tools in AutoCAD to create patterns of objects. These patterns can be of a circular, linear, or path type, hence the polar array, rectangular array, and path array discussions that follow. In a strict sense, these tools are not something entirely new but merely automate what you can do (albeit tediously) by commands learned in Chapter 1, such as Move, Copy and Rotate, and they are huge time-savers.

The array command underwent a major rework in AutoCAD 2012, with some additional minor enhancements for 2013 and 2014. If you are using this textbook to refresh your skills and have used older versions of AutoCAD, you are in for a surprise. Not only has an entirely new type of array been added (path array) but also the tool is now command-line activated, with the Ribbon assisting with modifications, or almost entirely Ribbon driven. For better or for worse, the familiar Array dialog box is gone (though it can still be recalled), and a whole new set of possibilities is opened up.

This chapter introduces the various arrays by assuming that the student is new to AutoCAD and has never used or seen the older, legacy version of the command. It is, however, referred to and described, just in case you end up using a pre-2012 AutoCAD at some point in the future.

8.1 POLAR ARRAY

A polar array is a collection of objects around some common point arranged in a circle (or part thereof). An example in architecture is a group of chairs arranged around a circular table. The idea is to draw the table, then one chair, and after positioning the chair exactly where it will be in relation to the table, use a polar array to copy it around the table to create, let us say, 10 of them, as seen in Fig. 8.1 on the left. The array command copies and rotates them into position and spaces them out evenly, a task that would have taken a while to do one chair at a time.

In mechanical engineering, an example may be the teeth of a sprocket gear. You draw the circle representing the wheel of the gear; put in one tooth properly drawn, detailed, and centered on top; and then array to get the full gear. Fig. 8.1 shows illustrations of what was just discussed. There are, of course, many other examples.

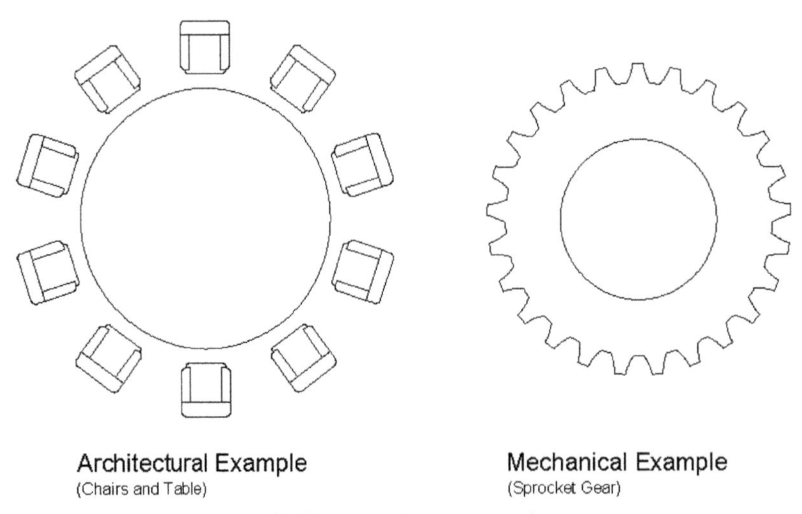

FIGURE 8.1 Polar array examples.

Steps in Creating a Polar Array

First of all, we need to create something to array. So go ahead and draw a circle of any size to represent the table just discussed, and then create a convincing-looking chair. You may want to make a block out of it after you are done for convenience.

Next, position the chair at the top of the table, as seen in Fig. 8.2. Be sure to line it up perfectly on the vertical axis, or else the array will be skewed off center; and move it back a bit from the table with Ortho on. What does AutoCAD need to know to do the array? The following four items are critical to a polar array; although note that only the first two items are explicitly asked for by AutoCAD as you create the array:

- It needs to know *what object(s)* to array (that would be the chair, not the table).
- It needs to know *the center point* of the array (center of the table).
- It needs to know *how many objects* to create (let us stick to 10 or so).
- It needs to know if the pattern goes all the way around (i.e., is it 360 degrees or less).

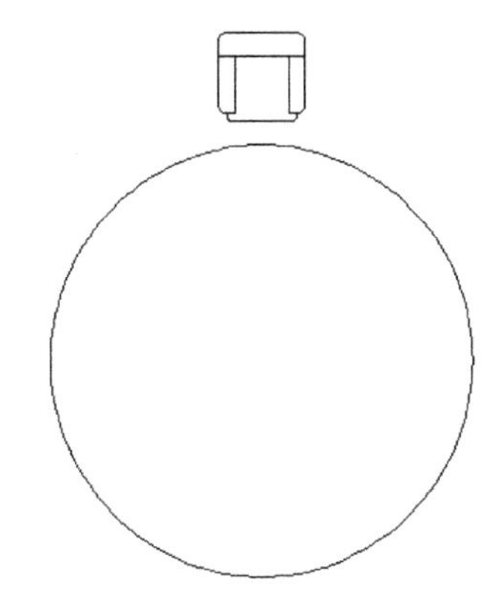

FIGURE 8.2 Chair and table, initial setup.

It is useful to memorize these steps and realize what you need to do before even starting up the array command; that way you have a clear idea of what to do and what data to enter as you get going. Once you have the table and chair drawn, follow the steps shown to create the polar array.

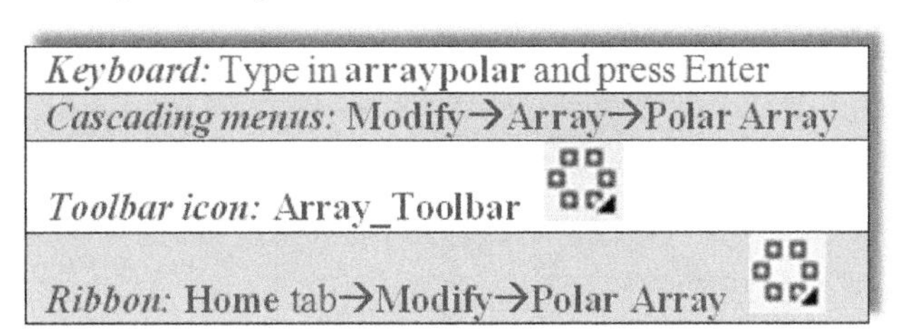

Keyboard: Type in **arraypolar** and press Enter
Cascading menus: Modify→Array→Polar Array
Toolbar icon: Array_Toolbar
Ribbon: Home tab→Modify→Polar Array

Step 1. Start up the array command via any of the preceding methods (it is recommended to have the Ribbon turned on).
- AutoCAD says: `Select objects:`

Step 2. Select the chair block (or the chair geometry if not using a block)
- AutoCAD says: `1 found`

Step 3. Press Enter.
- AutoCAD says: `Type = Polar Associative = Yes`
  `Specify center point of array or [Base point/Axis of rotation]:`

Step 4. You successfully selected the *object to array.* You now need to specify the *center point of the array.* Simply select the center of the large circle (the table) using the CENter OSNAP. AutoCAD creates a set of chairs around the table.
- AutoCAD says: `Enter elect grip to edit array or [ASsociative/Base point/Items/Angle between/Fill angle/ROWs/Levels/ROTate items/eXit]<eXit>:`

Step 5. Note a few things at this point. The array is automatically associative (look for that button to be active in the Ribbon). This means all the pieces are linked and not separate, a very desirable new property of arrays. Also note the details shown in the Ribbon. Here, you can modify the number of items (chairs) and the angle to fill under the Items tab. Make sure you have 10 chairs for this exercise.

Step 6. Your array is essentially done, so just press Enter to accept. If you wish, experiment with some of the variables such as Fill and Between. Also note how the chairs are rotated to face the table via the Rotate Items button that was automatically set (under Properties tab). You can, of course, turn that feature off. The completed polar array is shown in Fig. 8.3.

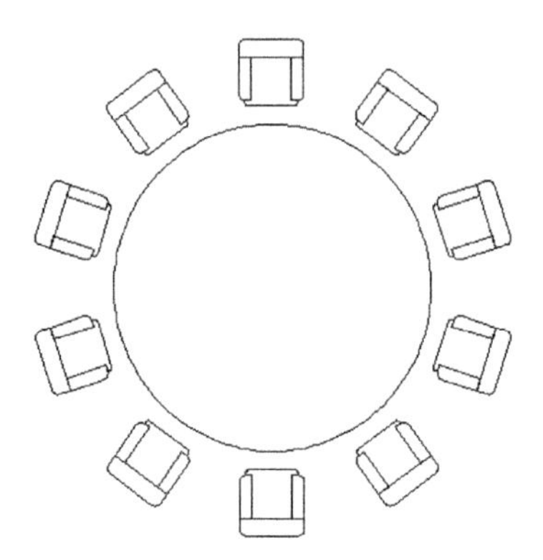

FIGURE 8.3 Polar array completed.

Additional Operations with Polar Array

Your new array behaves like a block (assuming Associative is set), a major difference first introduced in AutoCAD 2012, prior to which the new pieces of the array were not connected in any way. Click on one of the chairs with your mouse and it becomes selected, with several varieties of grips visible. If you hover your mouse over the chair grip, you see a menu of additional options. Finally, a new Ribbon Array tab appears. All of this is seen in Fig. 8.4. Let us take a closer look at the Ribbon and the Grip menu.

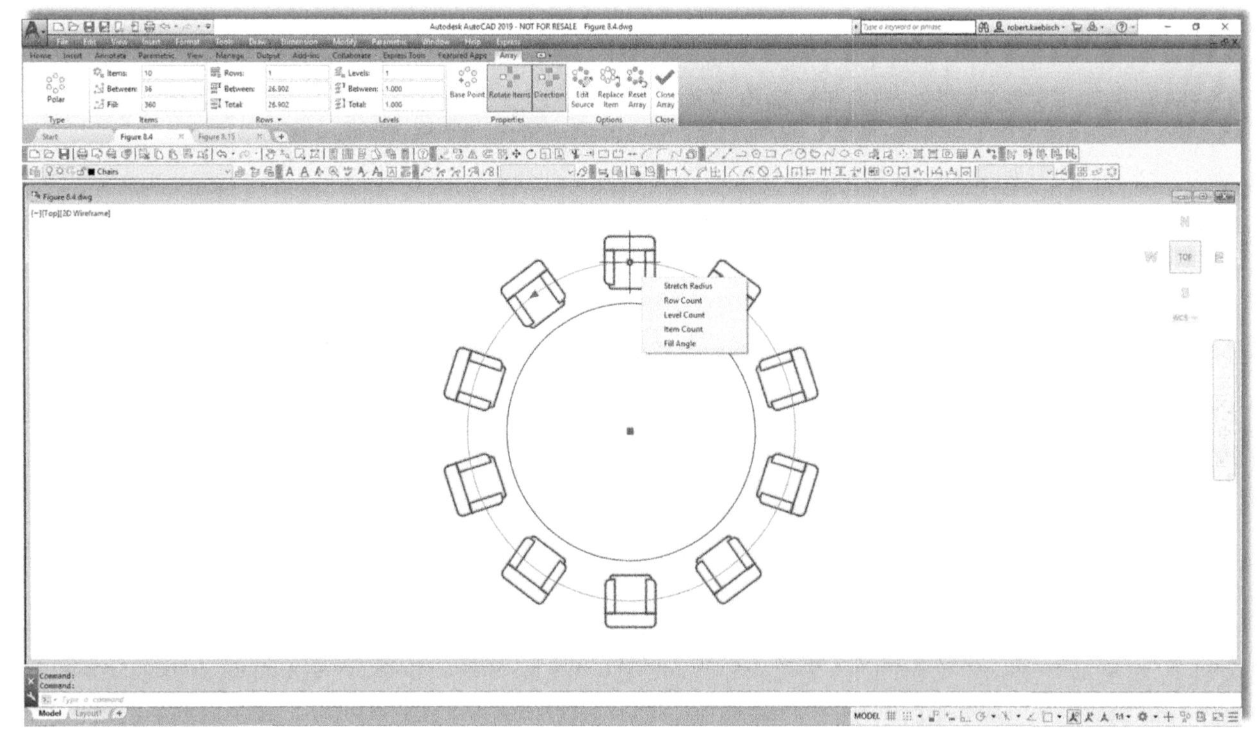

FIGURE 8.4 Polar array editing.

The Polar Array Ribbon has three areas of interest, the Items, Properties, and the Options/Close category, as seen in Fig. 8.5. The Items category we already mentioned is relatively self-explanatory. Top to bottom, the first value (10) is the number of items in the array, and you can change it for an instant update. Below that is the angle between all the objects (36 degrees), and finally, below that, the total angle filled (360 degrees, or a full circle). This should make sense as $36 \times 10 = 360$.

The Properties category's first button allows you to select a different base point for the rotation, useful if you did not quite get the first one right. The second button rotates the objects to face the rotation point as opposed to them all facing in one direction. The last button changes the direction of the array, from clockwise to counterclockwise, or vice versa. The last two buttons may be activated already on your sample table/chair array.

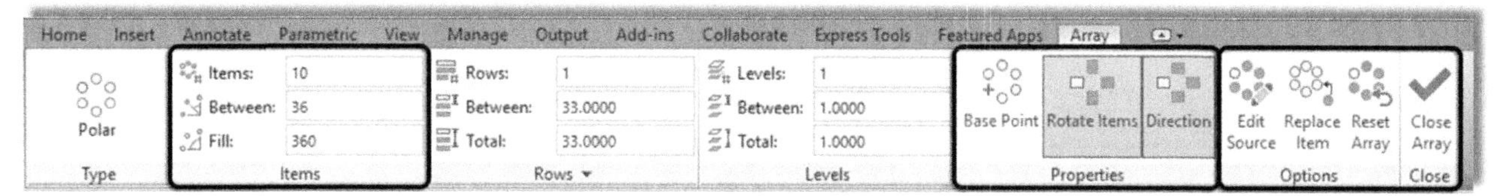

FIGURE 8.5 Polar array tab: Items, Properties, and Options/Close.

The Options category's tools are a bit more involved. They are therefore described in more detail next:

- *Edit Source*: This option allows you to edit and modify the source object. These changes are then reflected in the overall array. If you forget which object was the original source, it does not matter; you can pick any of the chairs and work on it. Execute the command, say "Yes" if you see the Array editing State dialog box, select the object, modify it, and when done, type in arrayclose, or click on Save Changes on the Ribbon at the far right. Fig. 8.6 shows Edit Source in progress.

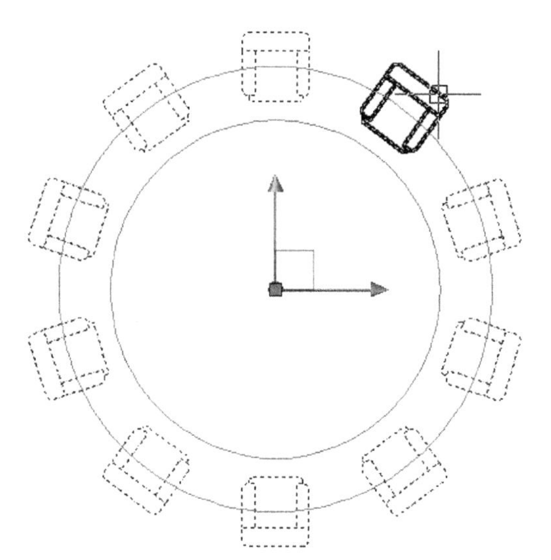

FIGURE 8.6 Edit Source.

- *Replace Item*: If editing the source is too time-consuming, then just replace it with another design via this option. AutoCAD asks you to select the replacement object and a base point. Use the Key Point option to pick a point on the replacement object. Finally, select an item (or several items) in the array to replace. Fig. 8.7 shows Replace Item in action as squares replace the chair symbol.

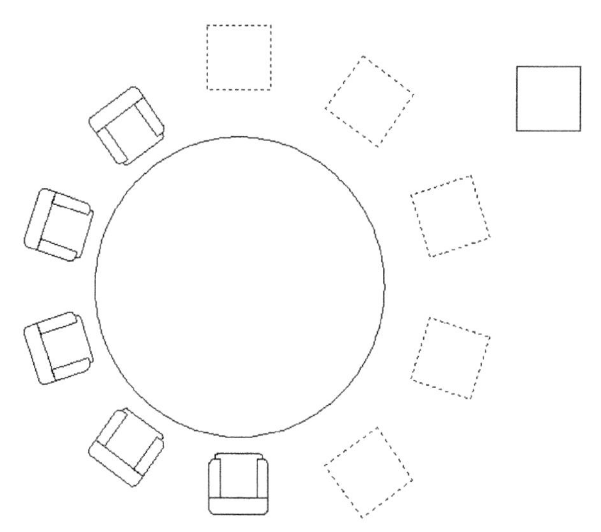

FIGURE 8.7 Replace Item in progress.

- *Reset Array*: This is an easy one. Autodesk figured you will mess up a few arrays while learning this tool (speaking from experience, of course) and included a Reset button. Its function is apparent immediately on use, though it is limited to just restoring erased items and some overrides.

Next, we have the menu seen when the mouse is held over the chair grip. Run through all the options as you read the descriptions. It contains the following tools:

- *Stretch Radius*: This option simply expands the radius of the overall array. The effect is the same as if you just activate the grip and move the mouse. Stretch Radius in action is seen in Fig. 8.8.
- *Row Count*: This option adds another set of row(s) of items in front or behind the main row, as seen in Fig. 8.9.
- *Level Count*: This option adds 3D levels and its discussion is not be included in this 2D-only chapter.
- *Item Count*: This option adds or deletes items (similar to the Ribbon function), or it can just be used to tell you how many items there are in the array, also a useful feature.
- *Fill Angle*: This option (also via the triangle grip) changes the included angle, currently set to 360 degrees. An example is shown in Fig. 8.10, where the fill angle has been rolled back to 270 degrees.

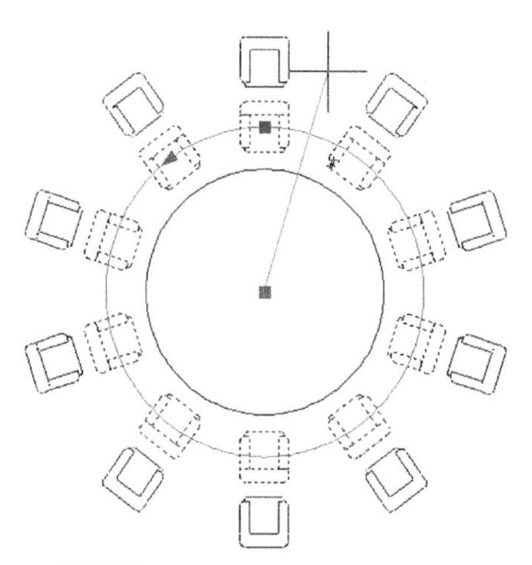

FIGURE 8.8 Stretch Radius in progress.

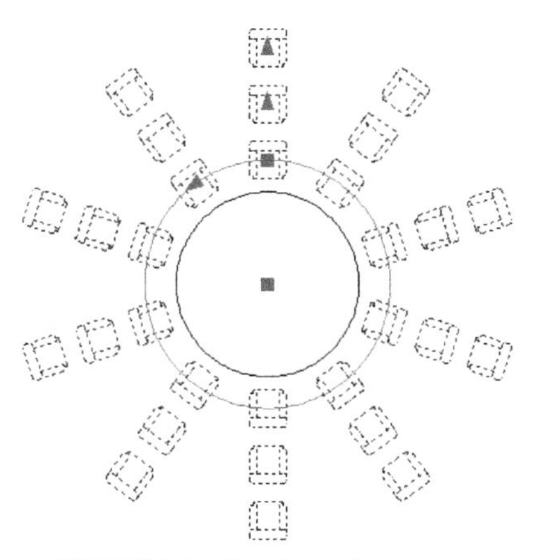

FIGURE 8.9 Row Count in progress.

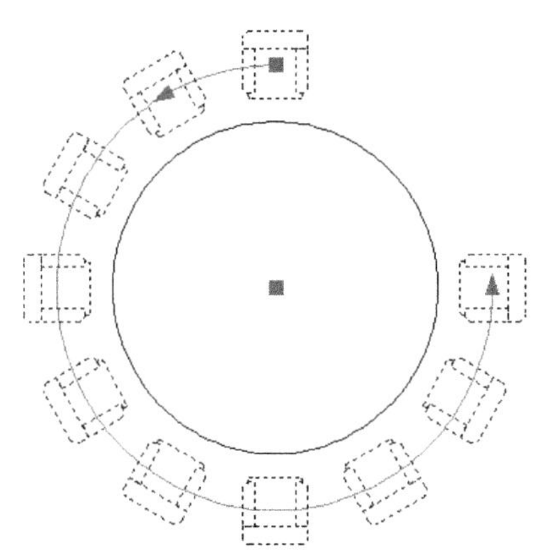

FIGURE 8.10 Fill Angle in progress.

Go over all these features until you are comfortable working with them. Be aware that you also can explode the array, which makes it behave like the "legacy" array, where each piece is separate. A short discussion of the legacy tools follows.

Legacy Polar Array (Pre-AutoCAD 2012)

Because the array command was so heavily redesigned for the 2012 release of AutoCAD, it makes sense to briefly mention the legacy polar array tool encountered via a dialog box prior to that release. If you have AutoCAD 2011 or any of the older versions still in common use, then the polar array can be accessed by typing in `array` and pressing Enter, or using the usual toolbar, cascading menus, or Ribbon (if available) methods. For the latest version of AutoCAD, you can still access this dialog box by typing in `arrayclassic`.

What you then see is the classic Array dialog box (Fig. 8.11, taken from AutoCAD 2011). You first need to make sure you have selected the Polar Array radio button at the very top. Then, run through the same steps outlined in the beginning of this chapter and shown here again:

- Select the object to array (button at top right).
- Select the center point of the array (button center top right).
- Enter how many objects to create (in the middle).
- Enter how many degrees to fill (just below the previous entry).

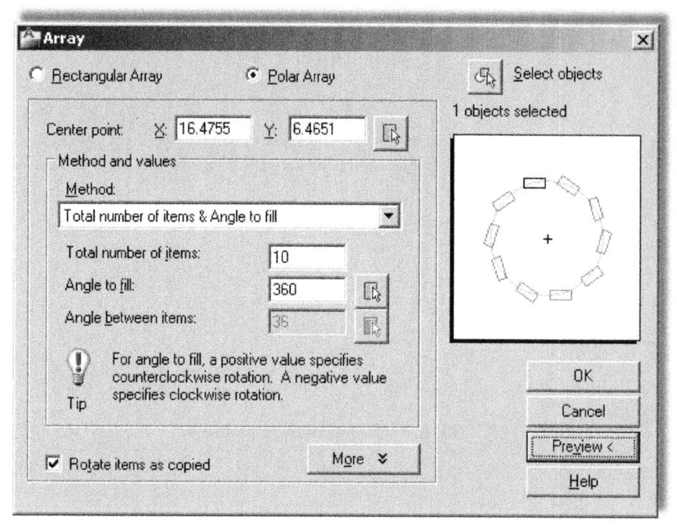

FIGURE 8.11 Legacy Polar Array dialog box (AutoCAD 2011).

Preview the array via the Preview button and press OK to finish off the array. Note that the entire array is not a block, just a collection of objects; and to redo the whole thing, you need to erase all the copies, leaving just the table and the original chair, and then repeat everything. That is why the Preview button is so important. Do not be so quick to press OK. This array, of course, has none of the editing options introduced in AutoCAD 2012 and continued with AutoCAD 2018.

8.2 RECTANGULAR ARRAY

The idea behind a rectangular array is not fundamentally different from a polar one. We are looking to replicate objects, but this time in rows (horizontal) and columns (vertical). In 3D, you can also do levels, which we do not address here. An example, as shown in Fig. 8.12, may be columns of a warehouse. If they are spread out evenly, then this is the perfect tool to create them. You simply draw one I-beam column, make a block out of it, and array it up and across the appropriate area.

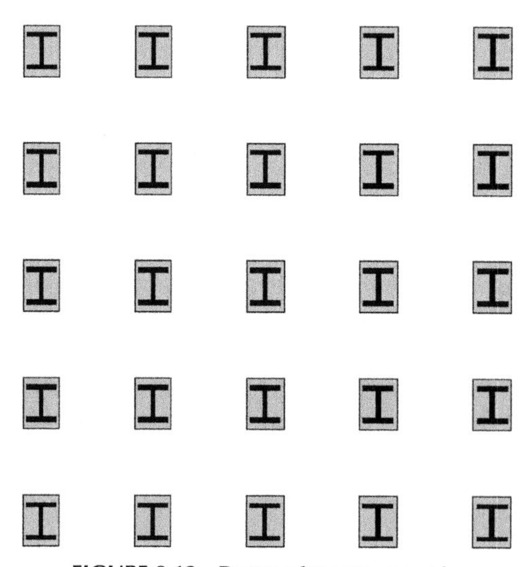

FIGURE 8.12 Rectangular array example.

Steps in Creating a Rectangular Array

The first thing to do is to draw a convincing-looking I-beam. Because we need actual distances, you cannot just randomly draw any object; use known sizing this time around. With one of the columns in Fig. 8.12 as a model, create a 20″ wide by 26″ tall rectangle. Inside of it, put an I-beam drawn with basic linework and a solid hatch fill, and shade the background a light gray (Fig. 8.13). Finally, make a block out of it. That sets up the column for an array.

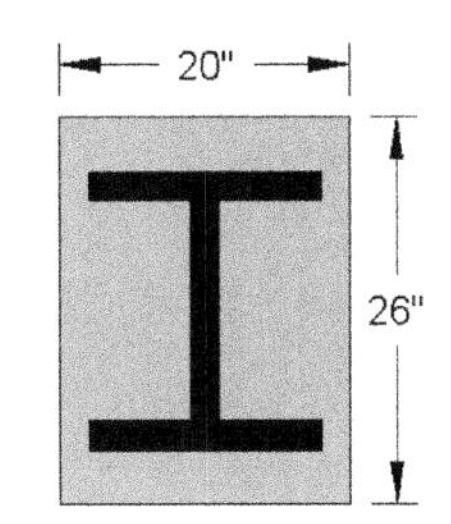

FIGURE 8.13 I-beam for column.

Now, what does AutoCAD need to know to create a rectangular array? The following three items are critical to a rectangular array; although, note that only the first item is explicitly asked for by AutoCAD as you create the array:

- It needs to know *what object(s)* to array (the I-beam column).
- It needs to know *how many rows* and *columns* to create (rows = across, columns = up and down).
- It needs to know the *distance between the rows and the columns* (from centerline to centerline).

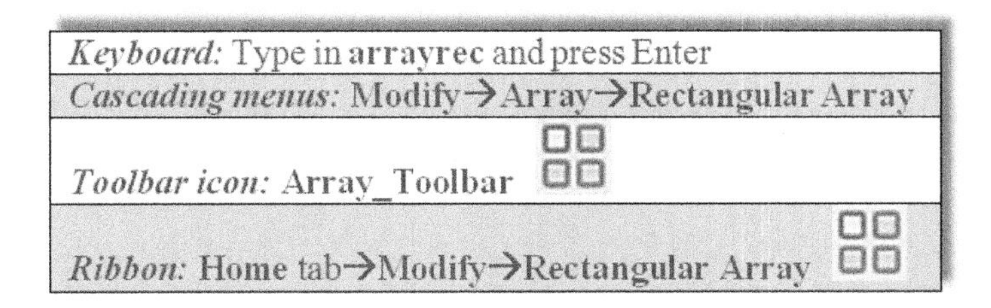

Keyboard: Type in arrayrec and press Enter
Cascading menus: Modify→Array→Rectangular Array
Toolbar icon: Array_Toolbar
Ribbon: Home tab→Modify→Rectangular Array

Step 1. Start up the array command via any of the preceding methods (it is recommended to have the Ribbon turned on).

- AutoCAD says: `Select objects:`

Step 2. Select the I-beam column block.

- AutoCAD says: `1 found`

Step 3. Press Enter. A small array of items are created, with grips visible.

- AutoCAD says: `Type = Rectangular Associative = Yes`

  `Select grip to edit array or [ASsociative/Base point/COUnt/Spacing/COLumns/Rows/Levels/eXit]<eXit>:`

Step 4. You have now successfully created the basic rectangular array. You still need to define the exact number of rows and columns, however, as well as the distance between them, as this current array is rather arbitrary. You can do this either via command line, dynamically (using grips), or via the Ribbon. Fig. 8.14 shows how you can pull the items diagonally in a dynamic fashion using the center grip. You can also do this via the outer arrow grips. This dynamic method is not really appropriate for setting distances between the columns, which are random values for now, but it does make creating the pattern very easy.

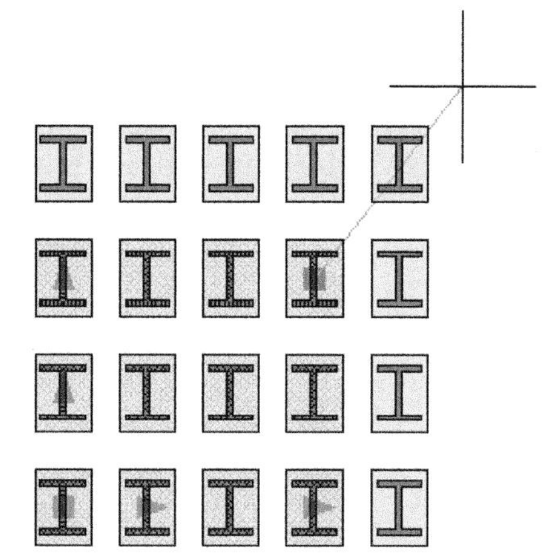

FIGURE 8.14 Creating a rectangular array dynamically.

Step 5. If you have the Ribbon up, then you just enter values. We want to create a 5 × 5 array, so start by entering those values into the Ribbon's Columns: and Rows: fields. For distances between the columns and rows use the value 72.

Step 6. If using the command line, you just have to follow the prompts for Columns, Rows, and Spacing. Regardless of which you start with (Rows or Columns), AutoCAD asks you for the spacing as needed. Press Enter when done. Your rectangular array should look like Fig. 8.12.

Additional Operations With Rectangular Array

After a lengthy discussion on additional operations with the polar array, you will find these options with the rectangular array to be quite familiar. Click on the new array of I-beam column, and notice a set of grips, a menu, and an Array tab in the Ribbon, all similar to the earlier polar array (Fig. 8.15).

The Rectangular Array Ribbon also has three areas of interest: the Columns, Rows, and Option/Closes categories, as seen in Fig. 8.16. The Columns category allows you to adjust the number of columns, the distance between them, and the total distance. The Rows category is the exact same thing but for rows. Finally, in the Options category, we already discussed the functionality of Edit Source, Replace Item, and Reset Array. They are all exactly the same as with the polar array.

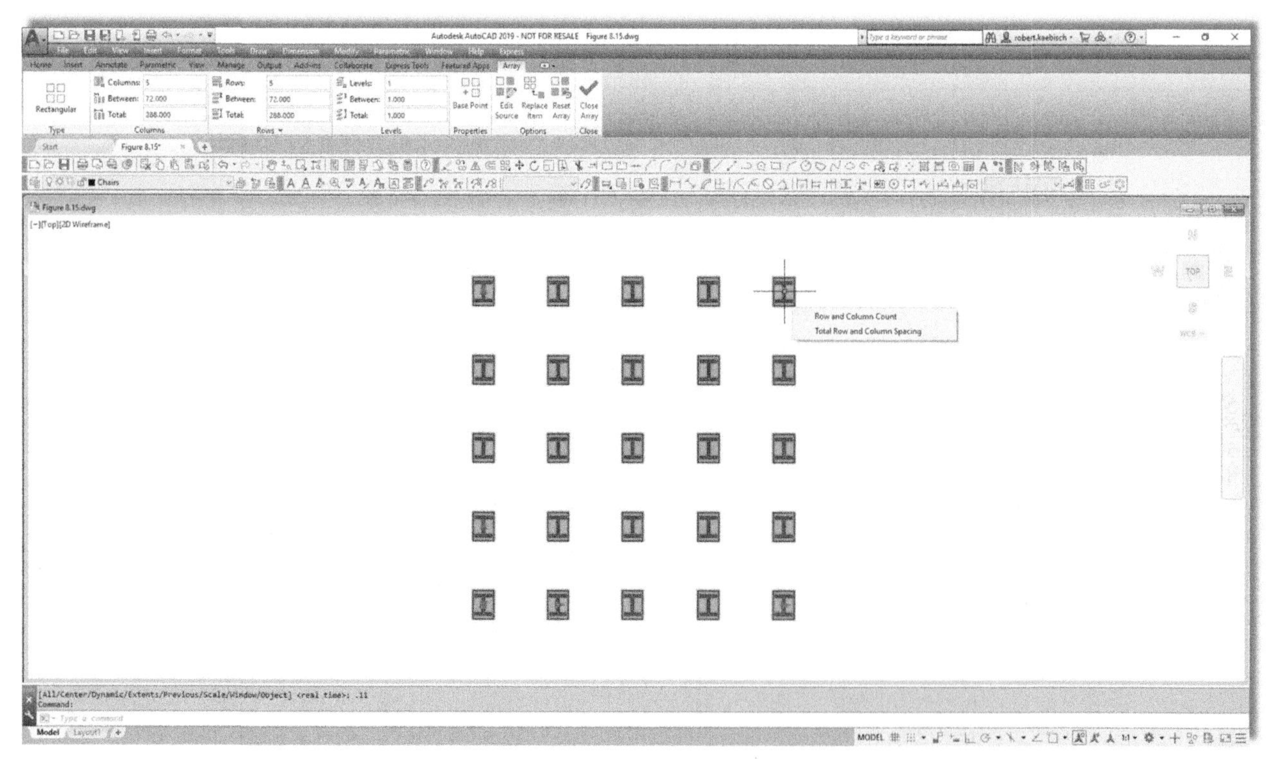

FIGURE 8.15 Rectangular array editing.

Next, we have the menu seen when the mouse is held over the top right I-beam column grip. The two choices are Row and Column Count and Total Row and Column Spacing. Their functions should be relatively easy to see by just trying them out. If you activate the first choice and move the mouse, the number of rows and columns increases, but the spacing remains the same. If you activate the second choice, the spacing increases but not the number of rows and columns.

Finally, try out the other grips on the array, specifically the ones in the lower left corner. You find that the square grip moves the entire array and the inner triangles increase the spacing. The outer triangle grips (toward the edges of the array), however, are used to increase the count. Experiment with all the options.

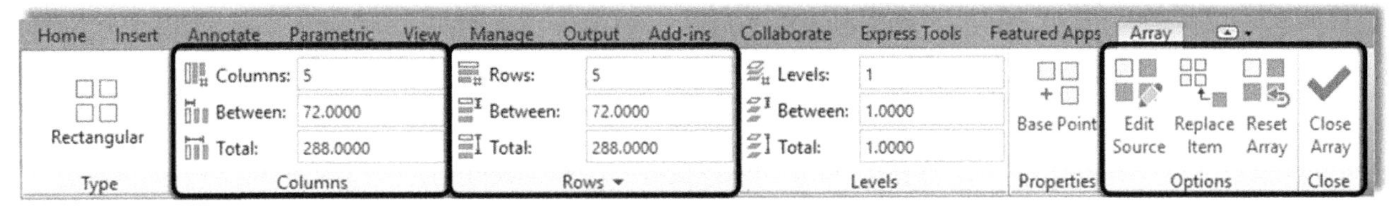

FIGURE 8.16 Rectangular array tab: Columns, Rows, and Options/Close.

Legacy Rectangular Array (Pre-AutoCAD 2012)

The legacy rectangular array is accessed via the same methods as the legacy polar array. Do not forget that, for the latest version of AutoCAD, you can still access this dialog box by typing in arrayclassic. When the dialog box appears, make sure the Rectangular array radio button is selected and you see what is shown in Fig. 8.17 (also taken from AutoCAD 2011). The procedure remains basically the same. You must select the object and then enter the number of rows and columns, followed by the distances between those rows and columns. Do a preview and press OK.

Note again that the entire array is not a block, just a collection of objects; and to redo the whole thing, you need to erase all the copies, leaving just one I-beam column and repeat everything. This array, of course, has none of the editing options introduced in AutoCAD 2012 either.

In general, polar arrays are used more often than the rectangular ones, but both are important to know.

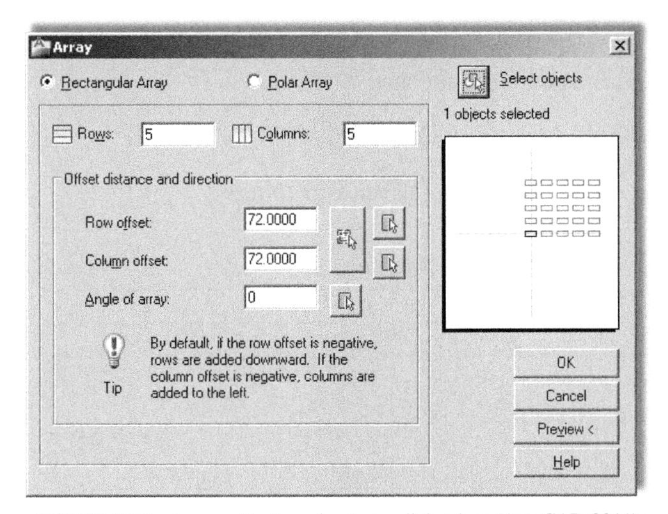

FIGURE 8.17 Legacy Rectangular Array dialog box (AutoCAD 2011).

8.3 PATH ARRAY

The path array is the (relatively) new kid on the block. Requested by users, this feature was finally added in AutoCAD 2012 along with the general redesign of the array command. The idea is simple, and it is surprising that this feature was not added earlier. A path array creates a pattern along any *predetermined pathway*, of any shape, as opposed to a strictly circular or linear one. This path can be an arc, a pline, or a spline. Because plines and splines have not yet been discussed (a Chapter 11, Advanced Linework topic), we focus on arcs as the pathways. A path array can also be created in 3D, but we do not get into that right now. An example of a path array in progress is shown in Fig. 8.18.

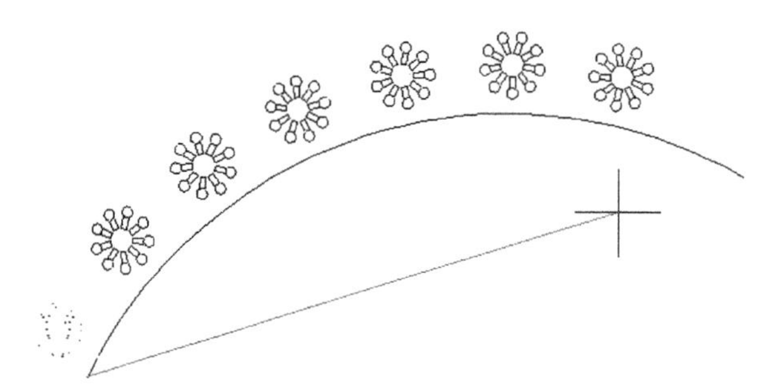

FIGURE 8.18 Path Array in progress.

Steps in Creating a Path Array

Boiled down to its essence, the path array just needs an *object to array* and a *path to follow*. Much like the previous two arrays, the objects can be spread out dynamically with mouse movements (as seen being done in Fig. 8.19) or with an input of a specific value. All other options flow from this basic approach. Let us give it a try. Create an arc and some random object to array; a rectangle will do, as seen in Fig. 8.19.

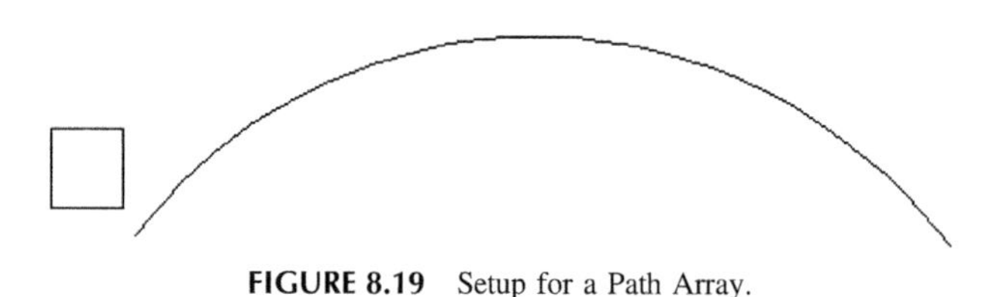

FIGURE 8.19 Setup for a Path Array.

So, what does AutoCAD need to know to do the array?

- It needs to know *what object(s)* to array (the rectangle).
- It needs to know *the path* (the arc).
- It needs to know *how many objects* to create (let us stick to 10 or so).

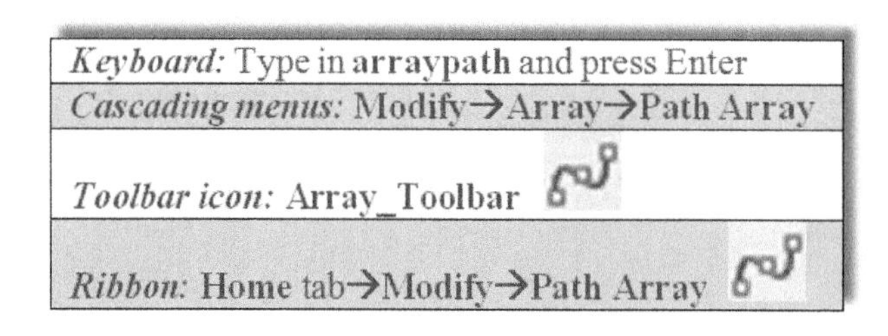

Keyboard: Type in **arraypath** and press Enter
Cascading menus: Modify→Array→Path Array
Toolbar icon: Array_Toolbar
Ribbon: Home tab→Modify→Path Array

Step 1. Start up the array command via any of the preceding methods.

- AutoCAD says: `Select objects:`

Step 2. Select the rectangle.

- AutoCAD says: `1 found`

Step 3. Press Enter.

- AutoCAD says: `Type = Path Associative = Yes`

`Select path curve:`

Step 4. You successfully selected the *object to array*. Now, select the *path* by clicking on the arc. A number of copies of the rectangle appear along the arc.

- AutoCAD says: `Select grip to edit array or [ASsociative/Method/Base point/Tangent direction/Items/Rows/Levels/Align items/Z direction/eXit]<eXit>:`

Step 5. You have essentially created the basic path array, but you can do a tremendous amount of modification to this basic layout. The easiest way to modify it is via the Ribbon. Under the Items: tab you have the most basic of the "mods," such as total number of items and the distance between them. Also note how you can alter the path array via dynamic movement of the grips. Be sure to keep the array associative, as with the previous examples. The completed array is seen in Fig. 8.20.

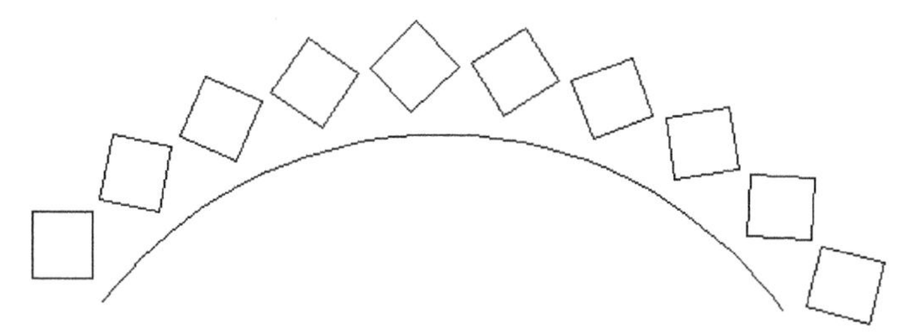

FIGURE 8.20 Path Array completed.

Additional Operations With Path Array

Just as with the previous two arrays, the path array has its own Ribbon tab, which is revealed by selecting one of the elements in the array, as well as its own menu, revealed as you hover the mouse over the grip, as seen in Fig. 8.21.

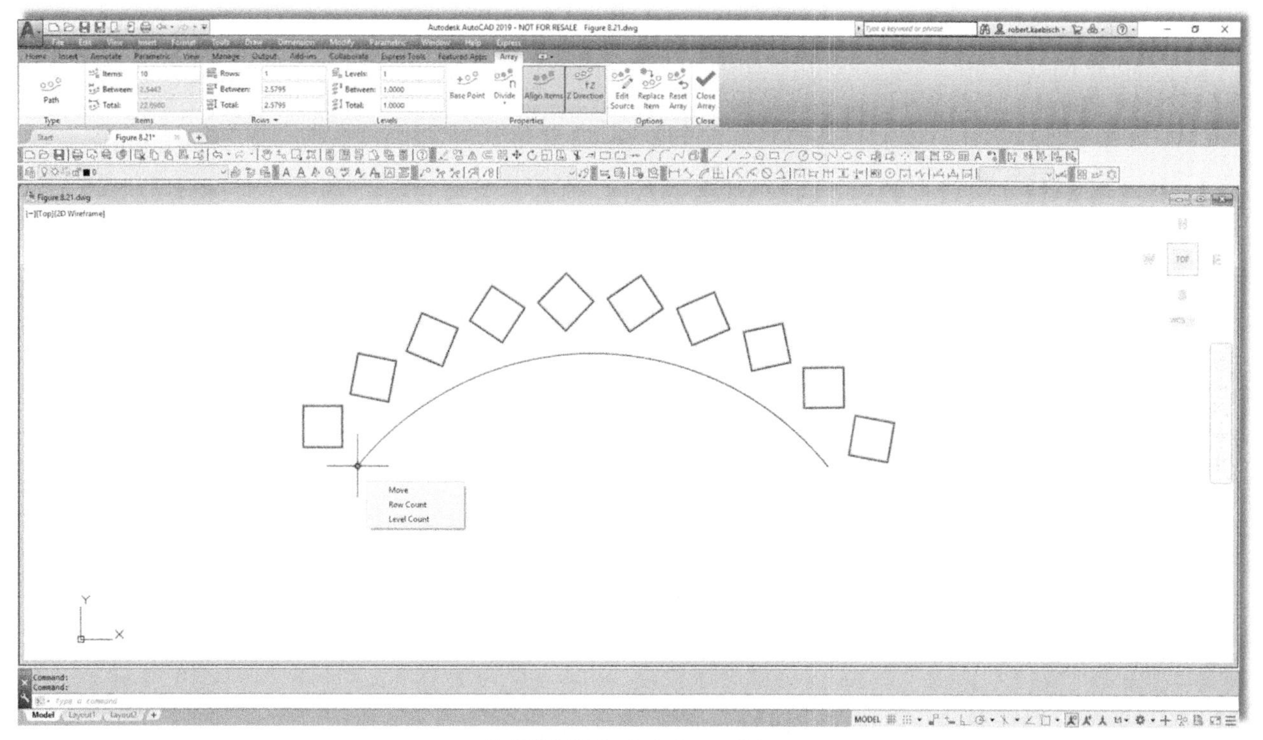

FIGURE 8.21 Path Array editing.

The Items category panel is of some interest and at this point should be somewhat familiar. In the Items panel of the Ribbon, you can vary the *Items* number when the *Divide* option is current in the Properties panel. Alternately, you can vary the *Between* distance when the Measure option is current in the properties panel. AutoCAD does the math for you in the Total category (Fig. 8.22). Under the Properties category, experiment with redefining the base point—this shifts your pattern around. Also check out Align Items—this makes all your elements face the same direction versus angling with the curve. Edit Source, Replace Item, and Reset Array all have the same meanings as with the previous polar and rectangular arrays. Spend a few moments understanding the cause and effect of these options.

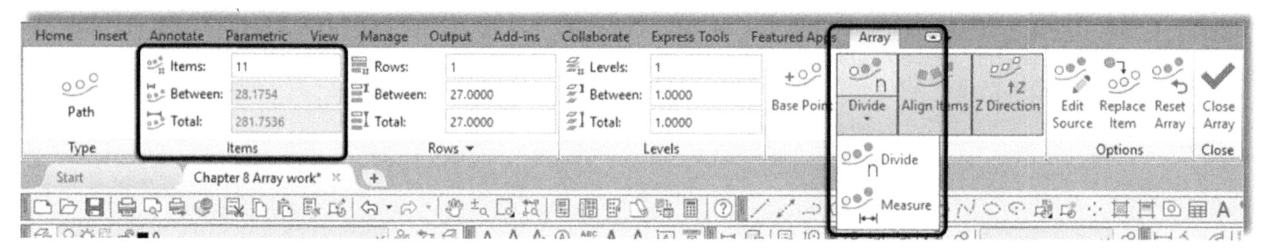

FIGURE 8.22 Path Array Ribbon editing—Divide/Items and Measure/Between Properties options.

Finally, let us take a look at the grip menu (Fig. 8.21). It also should be more familiar after your experiences with the previous two arrays. Move moves the array away from its originating path and triggers a warning along the way (Fig. 8.23). You may have seen the warning with earlier arrays as well, if you tried to move them around. Simply press Continue and carry on.

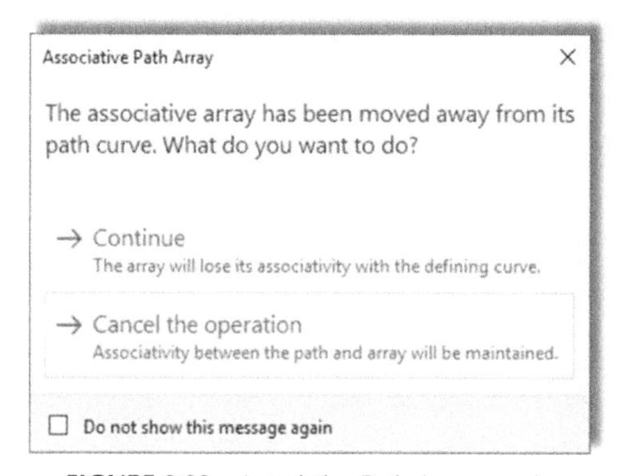

FIGURE 8.23 Associative Path Array warning.

Row Count creates additional rows of the pattern, as seen in Fig. 8.24. This option also repeats what is available with the other arrays. Level Count is not covered here, as it is a 3D function.

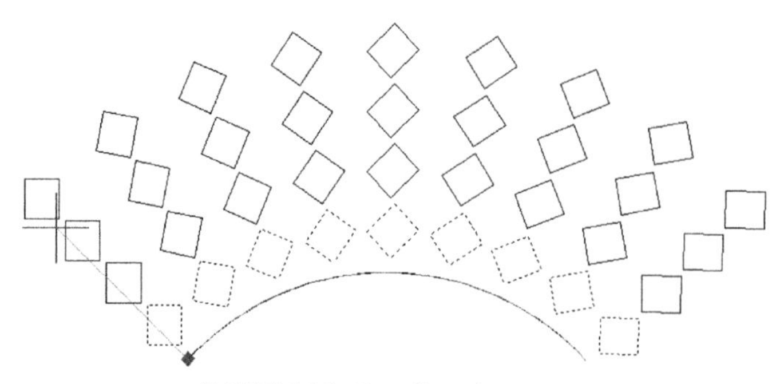

FIGURE 8.24 Row Count in progress.

8.4 IN-CLASS DRAWING PROJECT: MECHANICAL DEVICE

Let us now put what you learned to use by drawing a mechanical gadget (a suspension strut of sorts), created years ago for a class. Although what it is exactly is still debatable, the gadget proved to be an excellent drawing example and features quite a bit of what we covered in this and preceding chapters as well as a few minor nuances of AutoCAD drafting that, once learned, raise your skill levels considerably. The trick is to make your drawing look exactly like the example. If you approximate and cut corners, you miss out on some of the important concepts that give a more professional look to your drawings.

A full-page view of the completed device is found toward the end of the chapter, see Fig. 8.34. Take a good look at it and decide on a strategy to draw it. Over the course of the next few pages, you will see a step-by-step process to actually do it. Follow the steps carefully, and do not worry about the thick border; i.e., a polyline, to be covered in Level 2. The title block is otherwise just a simple collection of lines and text.

Step 1. For starters open a new file, give it a name (Save As...), and set up your layers. We need, at a minimum,
M-Part, Color: Green (for most of the parts of the device).
M-Text, Color: Cyan.
M-Dims, Color: Cyan (for the dimensions).
M-Hidden, Color: Yellow, Linetype: Hidden (for all hidden lines).
M-Center, Color: Red, Linetype: Center (for the main centerlines).
M-Hatch, Color: 9 (a gray for cross-section cutaways).
You may also want additional layers for the spring and the title block. You can decide on the exact layers and their colors; the preceding is just a guide. Also, the images in this chapter remain black and white, for clarity on a white page.

Step 2. Set up all the necessary items for smooth drafting, such as the Text Style (Arial, 0.20), the Dim Style, and Units (either Architectural or Mechanical is fine), and get rid of the UCS icon if you do not like it.

Step 3. To begin the drawing itself, we need to start with the hex bolts. Go ahead and draw one of them according to the information in Fig. 8.25. Be very careful, all sizes are diameters not radiuses; you must tell AutoCAD this by pressing d for Diameter while making a circle. This has tripped up many students in the past. Also, do not forget to make layer M-Part current. If the bolt is anything but green, you missed this step. The dimensions are only for your reference for now; there is no need to draw them.

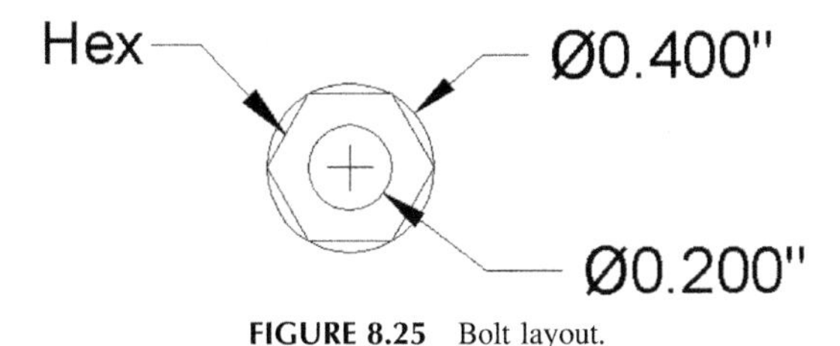

FIGURE 8.25 Bolt layout.

Step 4. Now make a block out of the bolt, calling it Bolt-Top View.

Step 5. Complete the rest of the top view by drawing the remaining circles and positioning the bolt at the top, as shown in Fig. 8.26. Use accuracy (OSNAPs), although the bolt can be any distance from the top of the circle. Be careful to use diameter, as indicated.

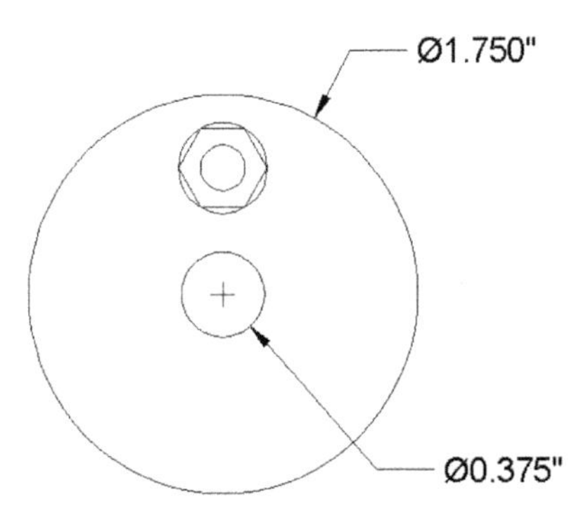

FIGURE 8.26 Bolt array.

Step 6. Array the bolt around the top view four times, as shown in Fig. 8.27.

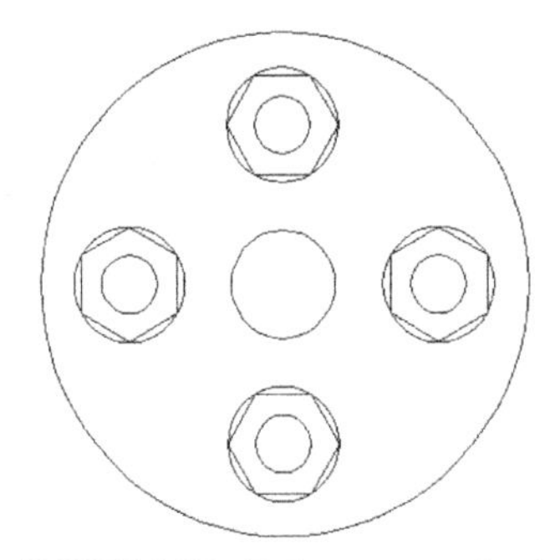

FIGURE 8.27 Bolt array completed.

Step 7. You now have to project the main body of the part based on this top view. The best way is by drawing two horizontal guidelines, of any length, starting from the top and bottom quadrants of the big circle; then, draw an arbitrary vertical line to mark the beginning of the part on the right, as shown in Fig. 8.28.

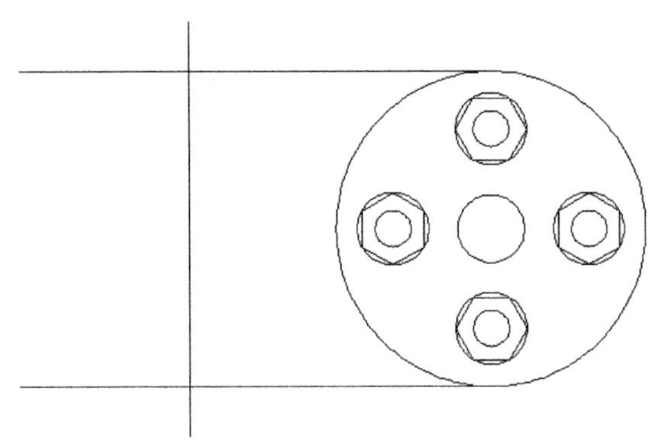

FIGURE 8.28 Projection lines.

Step 8. Trim or fillet the main vertical guideline and continue offsetting, based on the given geometry, to "sketch out" the basic outline of the shape, as seen in Fig. 8.29. Do not yet add dimensions; they are for construction only.

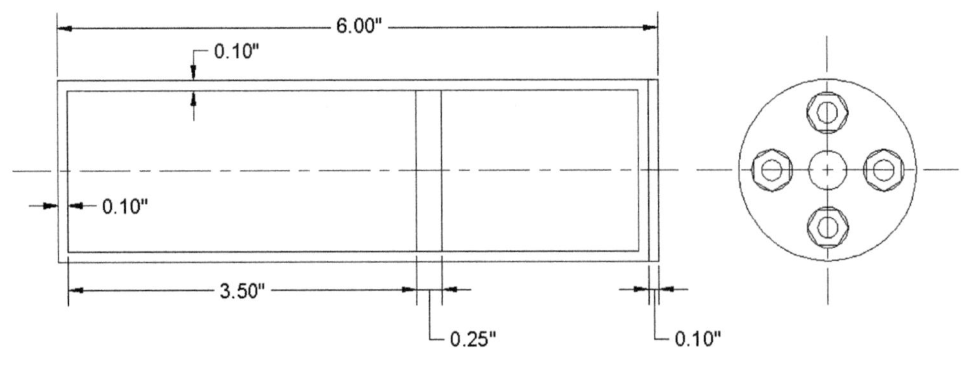

FIGURE 8.29 Body of part.

Step 9. Continue adding the basic geometry to the side view, as well as hidden and centerlines, as shown in Fig. 8.30. Set LTSCALE to 0.4.

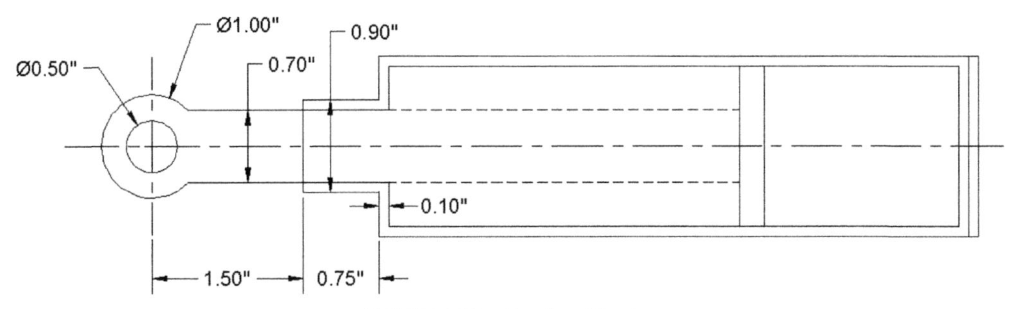

FIGURE 8.30 Creating side view.

Step 10. Add the bolt projections, as seen in Fig. 8.31(the depth into the side view can be approximated). Finally, add hatch patterns, using the Angle option to reverse the hatch directions, which is a common technique in mechanical design to differentiate the parts from each other in a cross section.

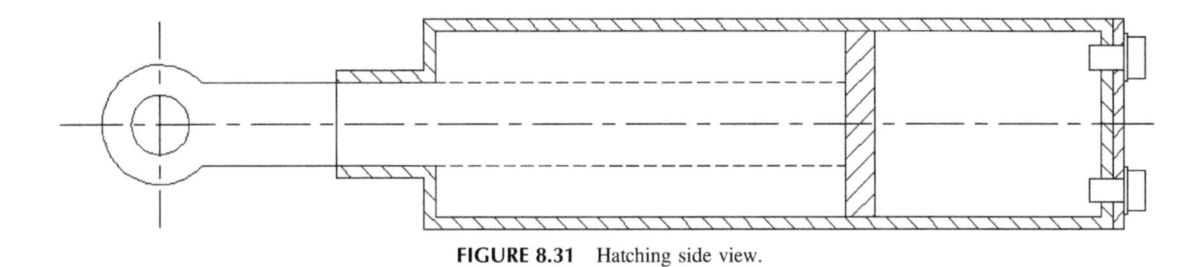

FIGURE 8.31 Hatching side view.

Step 11. Now, how do you do the spring? Well, this is where the Circle/TTR comes in. Although we have not practiced this version of the circle command, it is very simple to do. Start it up and select the TTR option. Then, pick the vertical and the horizontal lines where the circle needs to go (seen in Fig. 8.32) and specify a radius (0.10″). One way of proceeding is to draw a straight arbitrary line down from the center of that circle, rotate the line 15 degrees about that center point, then offset 0.1″ in each direction, erase the original line, extend the two new ones to the bottom (if necessary), and repeat the process, as seen in Fig. 8.32. Once a set is done, you can mirror the rest and trim carefully to create the impression of a coiled spring.

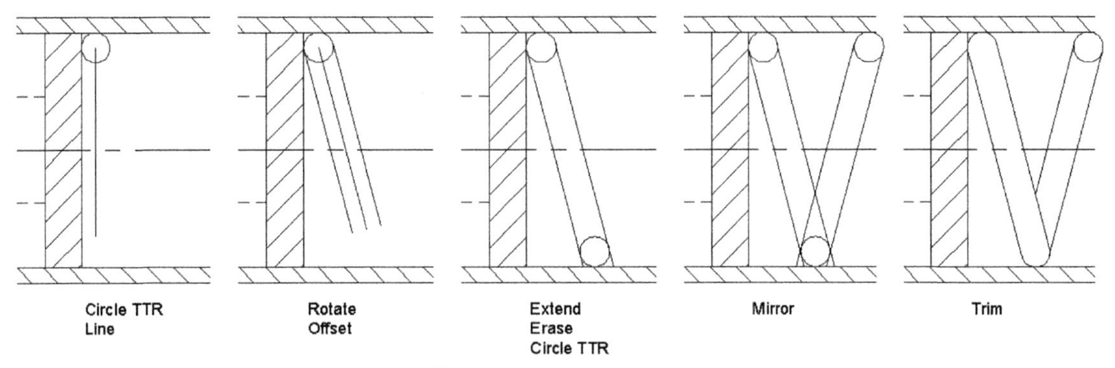

FIGURE 8.32 Drawing the spring.

Step 12. Finally, add dimensions (Fig. 8.33) and a title block (Fig. 8.34) by use of rectangles. The title block and its contents are typical of a mechanical engineering drawing.

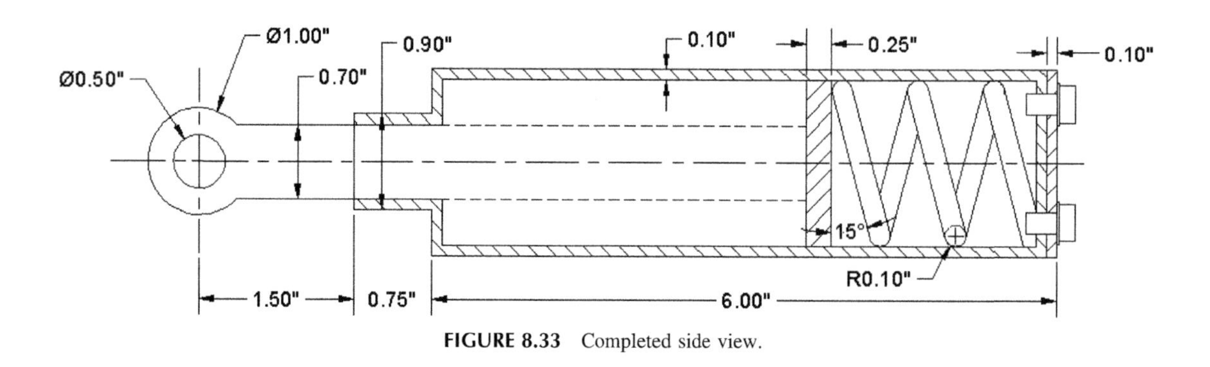

FIGURE 8.33 Completed side view.

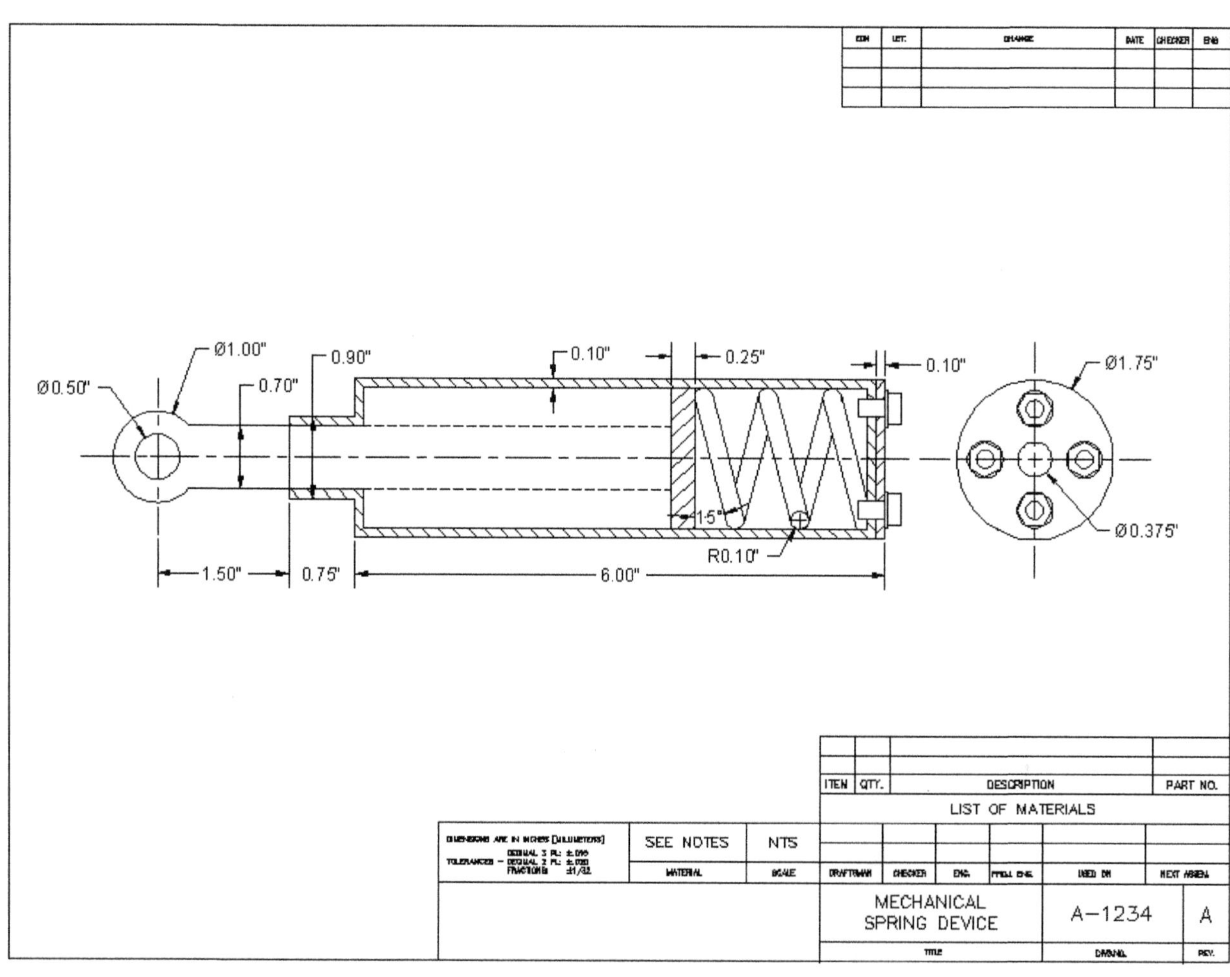

FIGURE 8.34 Mechanical project—final.

SUMMARY

You should understand and know how to use the following concepts and commands before moving on to Chapter 9:

- Polar array
  - Select object
  - Select center point of array
  - Number of objects to create
  - Angle to fill
  - Additional operations and options
- Rectangular array
  - Select object
  - Number of rows and columns
  - Distance between rows and columns
  - Additional operations and options
- Path array
  - Select object
  - Select path
  - Additional operations and options

REVIEW QUESTIONS

Answer the following based on what you learned in this chapter:

1. List the three types of arrays available to you.
2. What steps are needed to create a polar array?
3. To array chairs around a table, where is the center point of the array?
4. What steps are needed to create a rectangular array?
5. What steps are needed to create a path array?

EXERCISES

1. In a new file, draw the following five row by five column rectangular array. The chairs are 20″ by 20″ and the offset between them is 50″. The array is also at a 45 degrees angle. (Difficulty level: Easy; Time to completion: <10 minutes.)

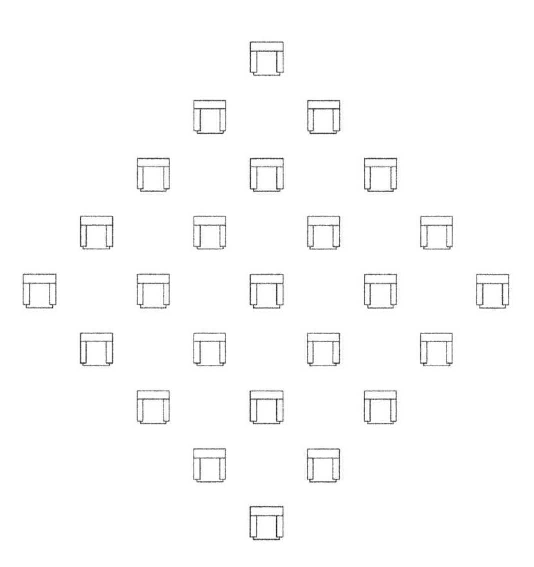

2. In a new file, draw the following seven rows by six column rectangular array, including the centerline grid and the grid lettering and numbering. The I-beams and the offset between them can be of any appropriate size, but approximate everything to look like the image shown. (Difficulty level: Easy; Time to completion: 10–15 minutes.)

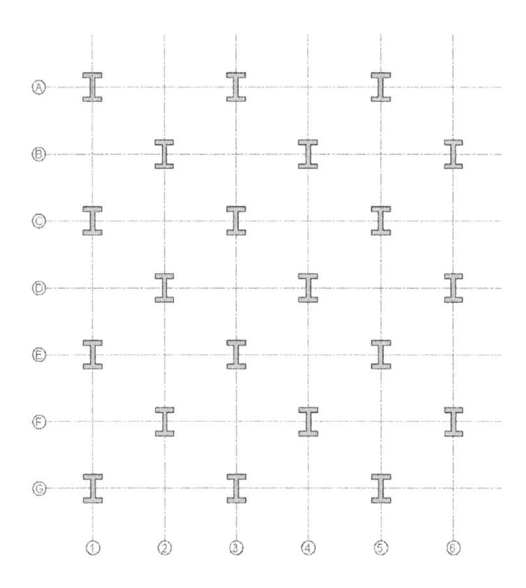

3. In a new file, draw the following set of path arrays with multiple rows. The two paths are simple arcs, and the symbols used are scaled up and shown next to the patterns. (Difficulty level: Easy; Time to completion: <15 minutes.)

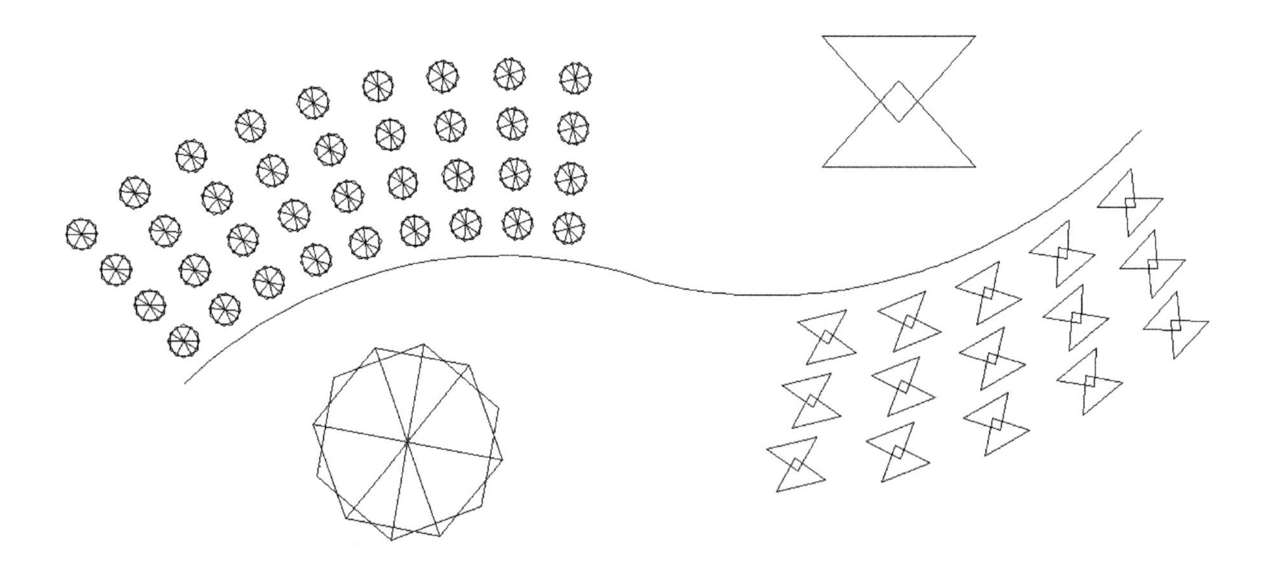

4. In a new file, draw the following set of path arrays with multiple rows. The three paths are simple arcs, and the symbols used are basic top view sketches of a flower pot. (Difficulty level: Easy; Time to completion: <15 minutes.)

5. In a new file, draw the following *polar* array, made up of a table, chairs, and computers. No sizing is given and all elements can be approximated. Place each element on the appropriate layer, and make a block out of it prior to using it in the array. (Difficulty level: Moderate; Time to completion: <20 minutes.)

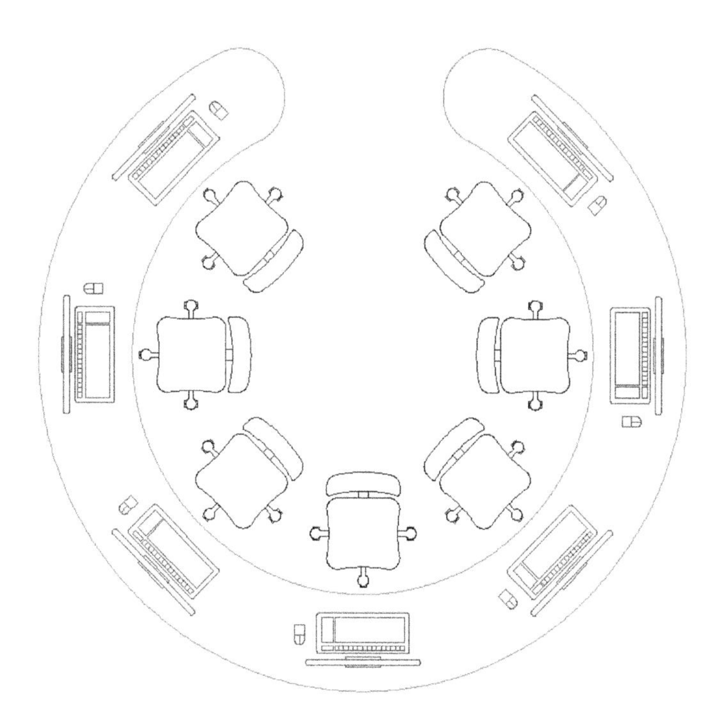

6. In a new file, draw this slightly altered version of the image found at the opening of this chapter, shown here with all the necessary dimensions. For practice, add the dimensioning as well. (Difficulty level: Easy; Time to completion: 15–20 minutes.)

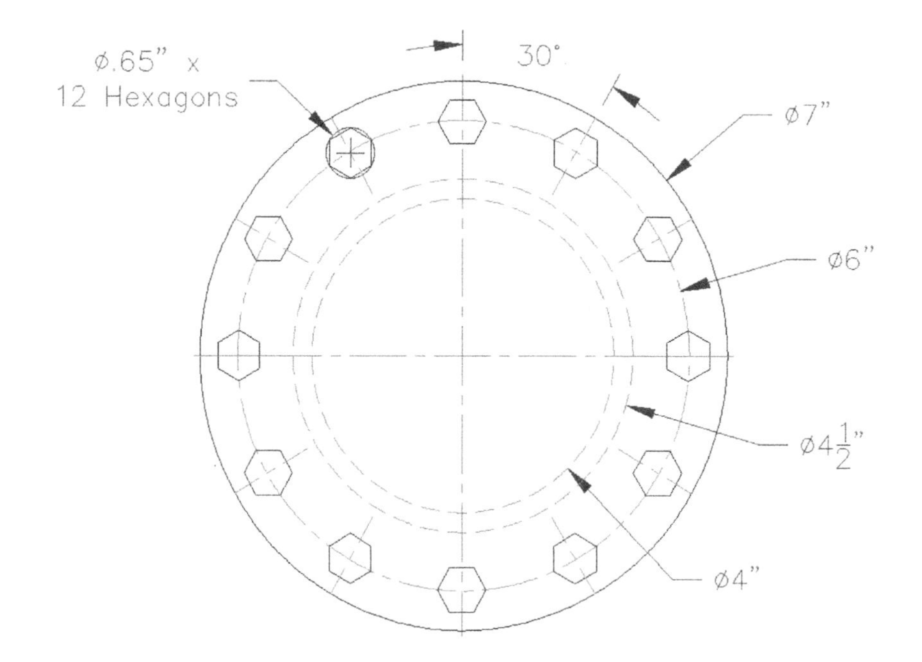

7. In a new file, draw the following gear array. Begin with circles of the indicated diameter. Then, focus on creating the first tooth, as seen in the detail. Make sure it is centered and aligned. Finally, array the tooth (24 teeth are used in the example) and clean up the design with trimming. The final result follows. (Difficulty level: Moderate; Time to completion: <20 minutes.)

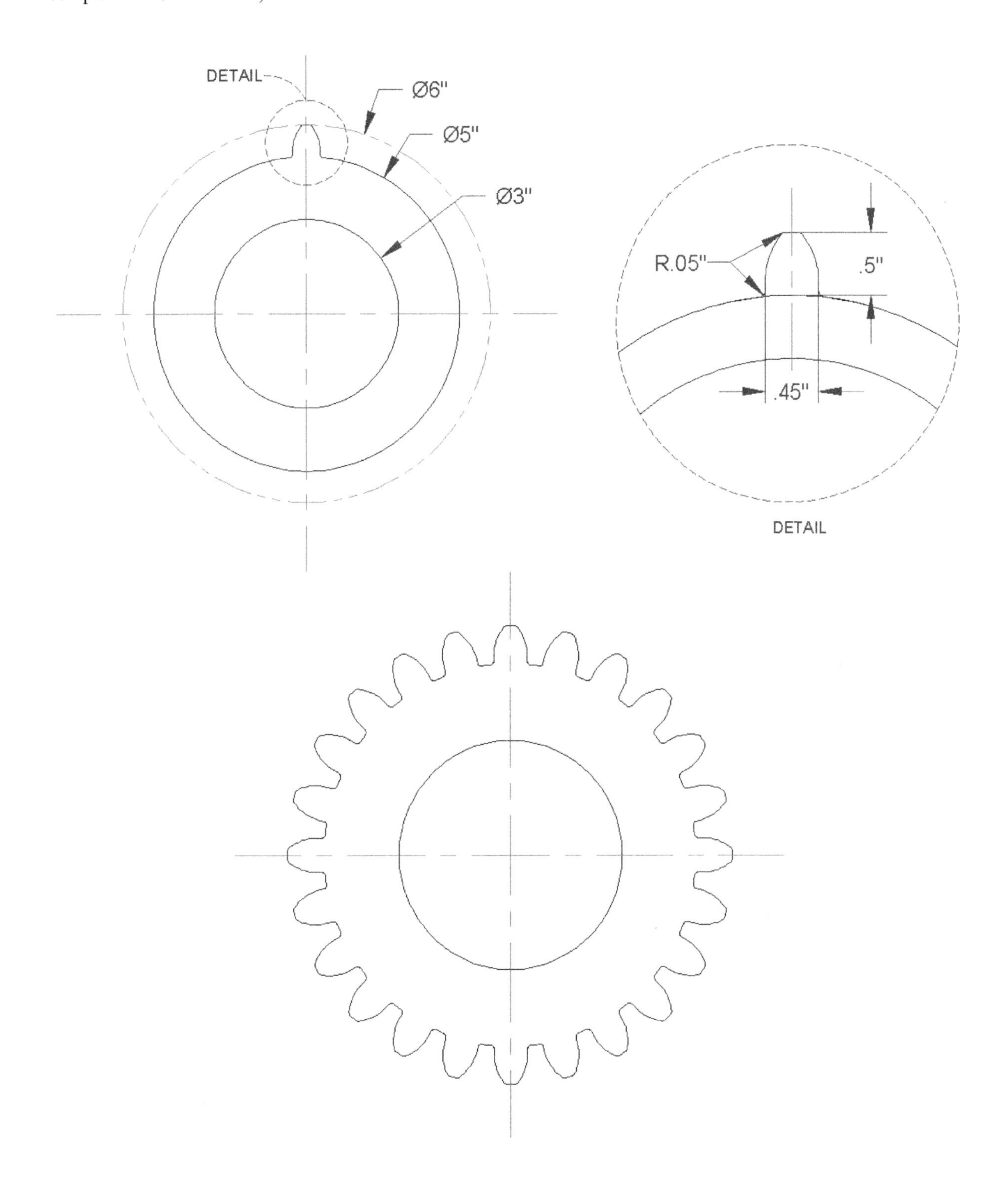

8. In a new file, draw the following pulley and its section view using polar array. All necessary diameters and widths are given. Load the center and phantom linetypes and set units and dims to Decimal (accuracy 0.00, with a ″ suffix). (Difficulty level: Moderate; Time to completion: 30–45 minutes.)

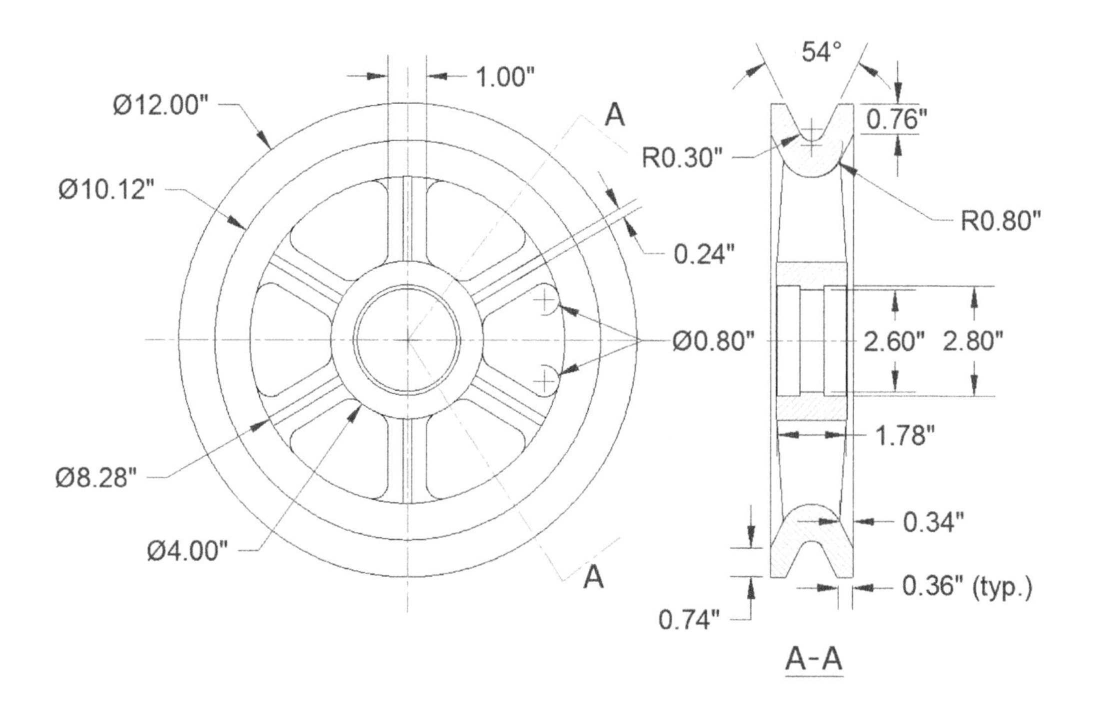

9. In a new file, draw the following fan and its side view using polar array. All necessary diameters and widths are given. Load the center linetype and set units and dims to Decimal (accuracy 0.00, with a ″ suffix). (Difficulty level: Moderate; Time to completion: 30–45 minutes.)

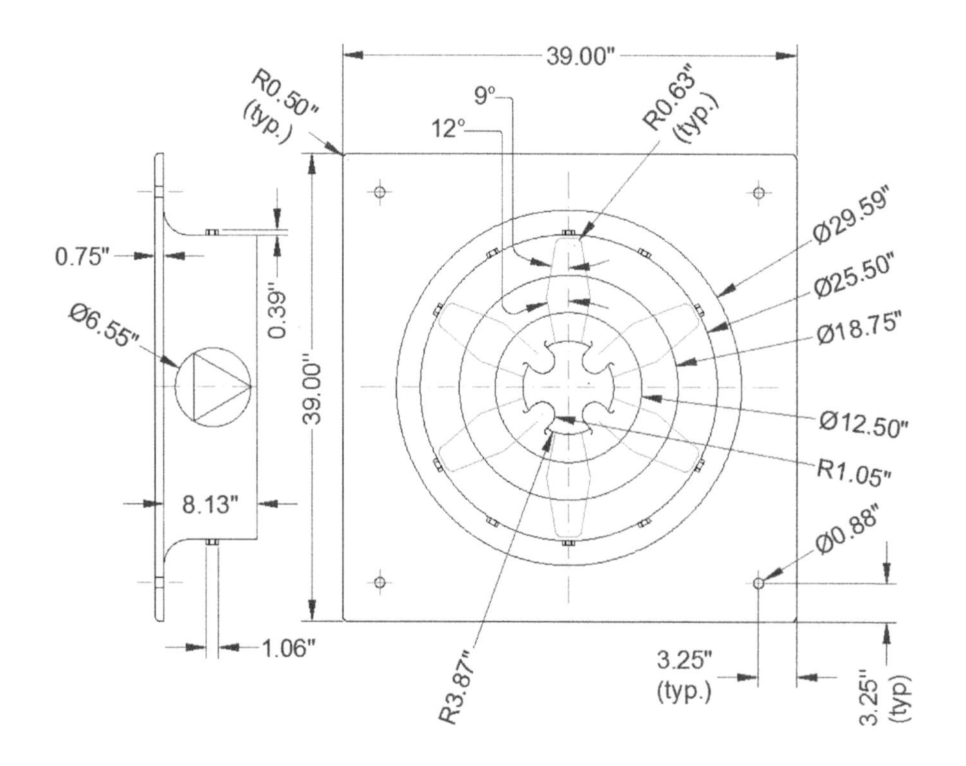

10. In a new file, draw the following mechanical device utilizing several polar arrays. All necessary diameters are given. Load the Center2 and Hidden2 linetype and set units and dims to Decimal (accuracy 0.00, with a ″ suffix). The details are nothing more than scaled up copies of the circled areas; be sure to draft them as well. (Difficulty level: Moderate/Advanced; Time to completion: 60−90 minutes.)

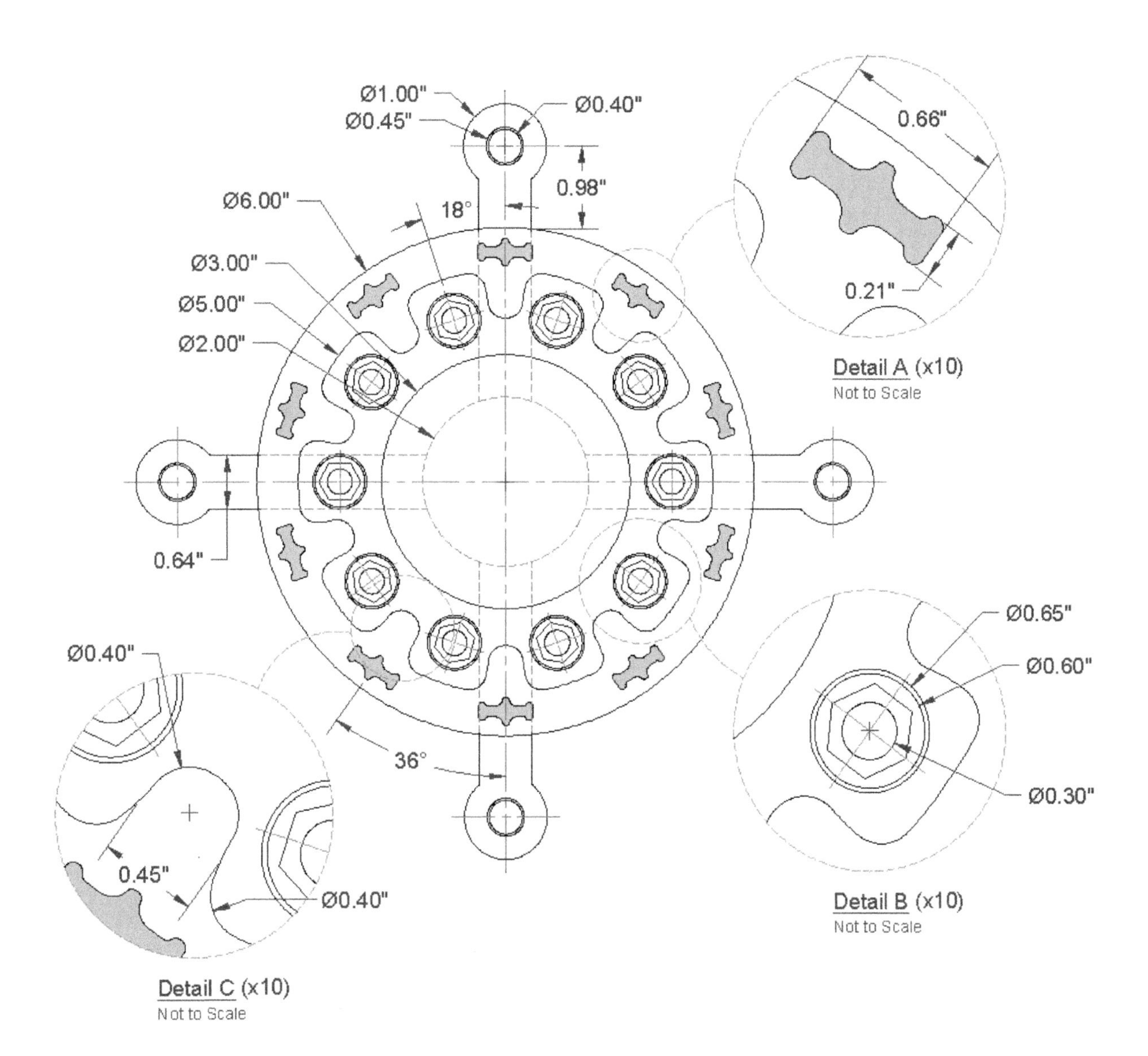

Spotlight On: Electrical Engineering

Electrical engineering is the branch of engineering that deals with electricity and its applications. This can be on a large scale, such as power generation and transmission or small scale, such as the design of chips and consumer electronics, sometimes referred to as *electronics engineering*. The profession appeared in the mid-1800s, as major advancements in the understanding of electrical phenomena were put forward by Ohm, Tesla, Faraday, Maxwell, and others. Today, electrical engineering is a wide-ranging profession vital to our society. Few human-made devices are without any electrical components, and electrical engineering figures prominently in virtually every product or device you see before you. Electrical and electronic engineers held about 315,900 jobs in 2014, making this the second largest branch of the US engineering community (behind software engineering).

Just some of electrical engineering's specialties include power, controls, electronics, microelectronics, signal processing, telecommunication, and computer engineering. Electrical engineers need a mastery of a wide variety of engineering sciences, such as basic and advanced circuit theory, electromagnetics, embedded systems, signal processing, controls, solid state physics, and computer science and programming, among others.

Education for electrical engineers starts out similar to that for the other engineering disciplines. In the United States, all engineers typically have to attend a 4-year ABET-accredited school for their entry-level degree, a Bachelor of Science. While there, all students go through a somewhat similar program in their first 2 years, regardless of future specialization. Classes taken include extensive math, physics, and some chemistry courses, followed by statics, dynamics, mechanics, thermodynamics, fluid dynamics, and material science. In their final 2 years, engineers specialize by taking courses relevant to their chosen field. For electrical engineering, this includes courses in most or all of the sciences just listed.

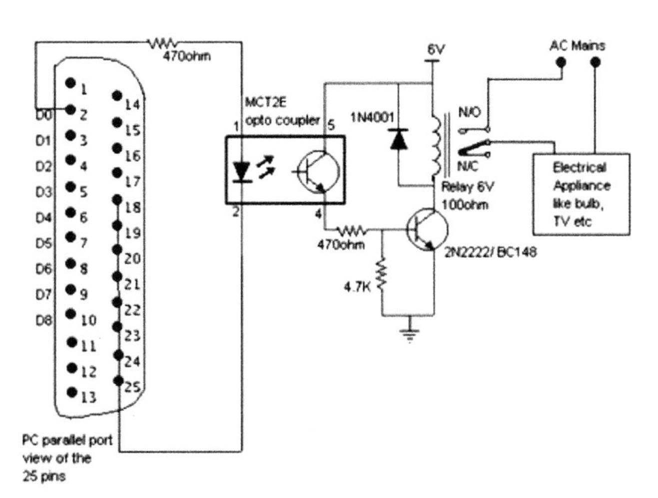

FIGURE 1 Electrical schematic 1. *Image source: Elliot Gindis.*

On graduation, electrical engineers can immediately enter the workforce or go on to graduate school. Though not required, some engineers choose to pursue a Professional Engineer (P.E.) license. The process first involves passing a Fundamentals of Engineering exam, followed by several years of work experience under a registered P.E., and finally sitting for the P.E. exam itself.

Currently, electrical engineers receive starting salaries averaging $63,000 with a bachelor's degree and $73,000 with a master's degree. The median pay is around $88,000. This, of course, depends highly on market demand and location. A master's degree is highly desirable and required for many management spots. The job outlook for electrical engineers was projected as good, with a 6% growth rate per year for the next 5 years. The profession is extremely diversified; and by selecting a specialization carefully, an engineer can join one of many growing fields, such as controls, nanoengineering, robotics, and other such career paths.

So how do electrical engineers use AutoCAD and what can you expect? Industry wide, AutoCAD enjoys a significant amount of use for the simple reason that most electrical work is 2D in nature and plays into AutoCAD's strength. AutoCAD layering is far simpler in electrical engineering than in architecture. There is no AIA standard to follow, only an internal company standard. As always, the layer names should be clear as to what they contain. The overall standard for much of electrical engineering work is provided by the IEEE, which may include CAD standards as well.

In electrical engineering, you will not likely need advanced concepts such as Paper Space and Xrefs, but often you need to build attributes that hold information on connectors and wires. You also need to have a library of standard electrical symbols handy, so very little needs to be drawn from scratch every time. The key in electrical engineering drafting is accuracy (lines straight and connected), as the actual CAD work is quite simple. A related field, network engineering, is similar. Here, you may use plines or splines that indicate cables going into routers and switches. Shown in Figs. 1 and 2 are examples of electrical schematics. Notice the extensive use of symbols for capacitors, diodes, transformers, and switches.

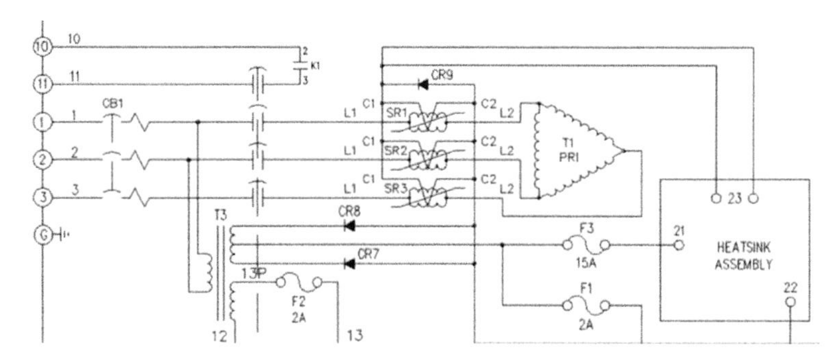

FIGURE 2 Electrical schematic 2. *Image source: Elliot Gindis.*

Fig. 3 is an example of some racked equipment in a network/electrical engineering drawing. There is nothing complex in this image, just careful drafting and a lot of copying. You will see this example again when wipeout is discussed. Fig. 4 shows some power transmission equipment from a substation.

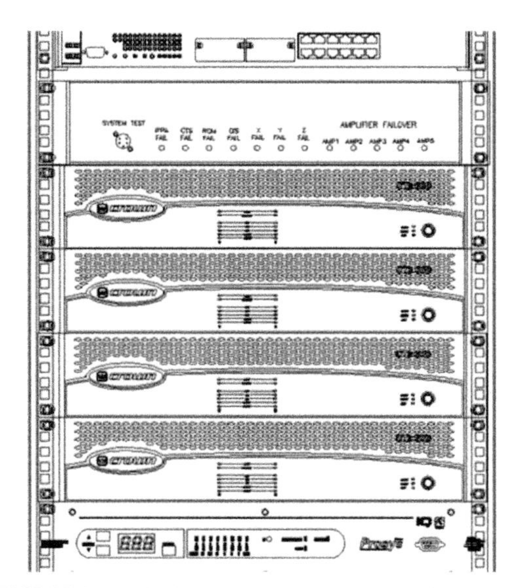

FIGURE 3 Racked equipment. *Image source: Elliot Gindis.*

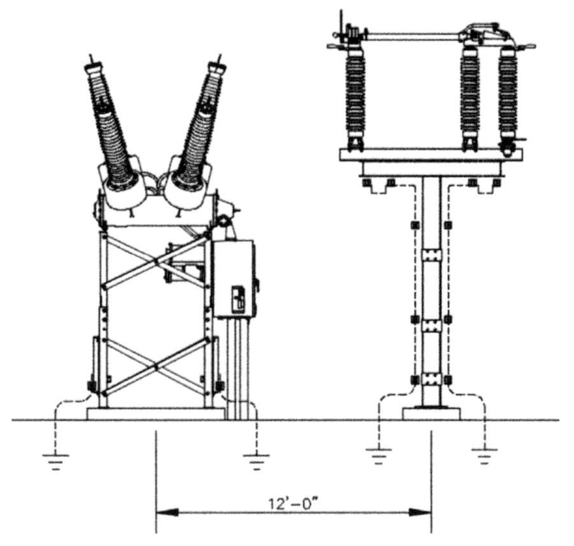

FIGURE 4 Substation power equipment. *Image source: Elliot Gindis.*

Chapter 9

Basic Printing and Output

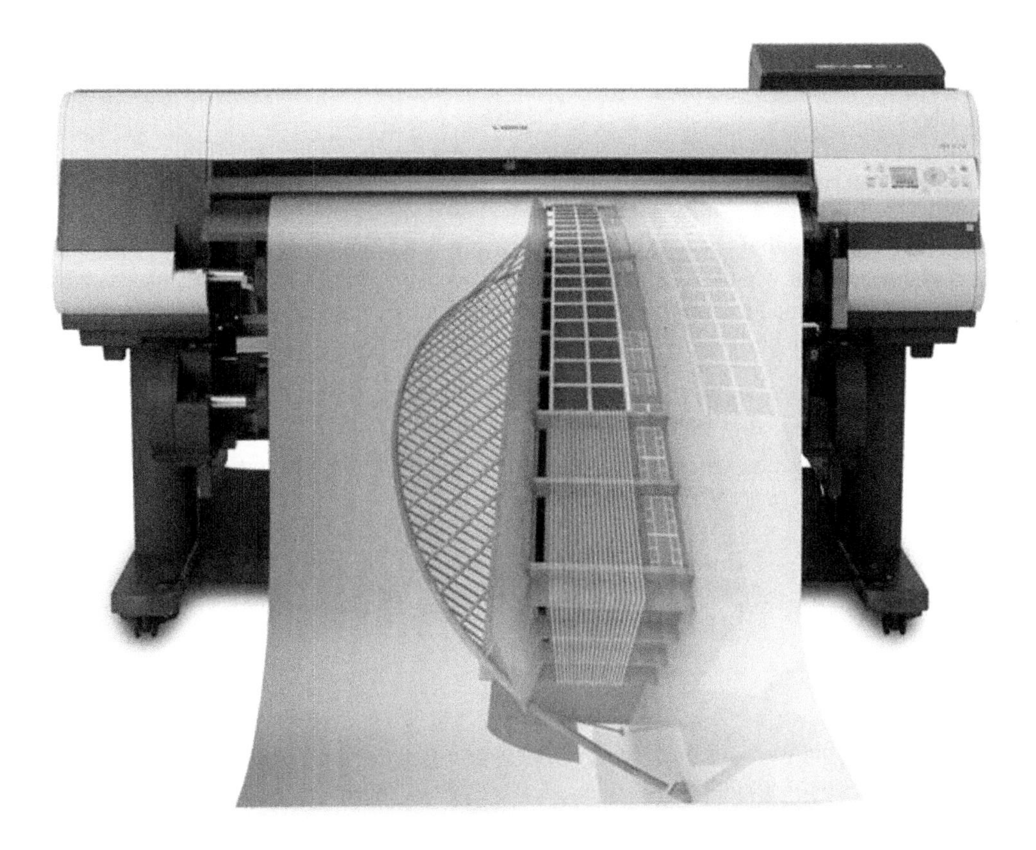

Learning Objectives

In this chapter we introduce the Plot dialog box and the Page Setup Manager and go through all the associated options to produce a print or a plot, including the following:

- What *printer* or *plotter* to use
- What *paper size* to use
- What *area* to plot
- At what *scale* to plot
- What *pen settings* to use
- What *orientation* to use
- What *offset* (if any) to use
- The Page Setup Manager

On the completion of this chapter you will be able to recognize and set up all the features needed to print or plot any type of drawing from any AutoCAD workstation.

Estimated time for completion of this chapter: 1 hour.

Up and Running with AutoCAD 2019. https://doi.org/10.1016/B978-0-12-816440-2.00009-4

9.1 INTRODUCTION TO PRINTING AND PLOTTING

Designers and their clients usually want something tangible to look at and use. Because of this, most design work in AutoCAD must be created with the ultimate goal of displaying it on a piece of paper, not just on screen. Indeed, the fundamental immediate "product" of most design work is paper output. This output is then used to manufacture, fabricate, build, or assemble the design. It also may be needed for pre- and postdesign purposes, such as sales or archiving. These needs influence the evolution of a drawing from the outset, and a skilled AutoCAD designer selects paper size, output scales, and lineweights early in the process so the inevitable output proceeds smoothly.

Professional-looking output in AutoCAD is no accident and comes with its own set of challenges. Although you may be able to walk up to an expertly set up drawing and just select File → Plot → OK, such smooth results come from only a thorough understanding of what is involved. More important, if the requirements call for a different output of the same drawing (e.g., for an informal check of a small section of the drawing vs. a final full-size submission to the client), you must know what to change and by how much.

AutoCAD allows for a wide array of output, each method with its own tricks and details. Output spans this chapter as well as part of Chapters 10 and 19. Our goal right now is only to familiarize you with the essentials and what AutoCAD needs to know to print or plot a drawing.

Take note, as you go through this chapter, that although some specifics, such as type of printer or plotter, may be unique to your particular office, home, or school, most of what you need to know stays constant from situation to situation and from one release of AutoCAD to another (even if the locations of the buttons and menus change). Therefore, it is important that you memorize and understand the essential underlying concepts, not just "which buttons to push."

9.2 THE ESSENTIALS

You need to give AutoCAD the following information when setting up a drawing for printing or plotting:

- What *printer* or *plotter* to use.
- What *paper size* to use.
- What *area* to plot.
- At what *scale* to plot.
- What *pen settings* to use.
- What *orientation* to use.
- What *offset* (if any) to use.

There are, of course, some other subtleties, but we focus on these essential ones for now. Note that we do not look at the actual Plot dialog box until later, as you first need to understand what the preceding means.

What Printer or Plotter to Use

This is where you select what output device to use. In most architecture offices, engineering companies, and schools, you have a plotter (color or black and white) for C-, D-, and E-sized drawings (paper sizes are discussed next), and a LaserJet printer (color or black and white) for A- and B-sized drawings. Your typical AutoCAD station usually is networked to access both, as requirements may change daily. You need to select which device is getting the drawing. (AutoCAD does not care—it will print a floor plan of a sports stadium on a postcard if you let it.) This choice, of course, is tied in with the scale and the ultimate destination of the printed drawing (is it a check plot or for a client?). We discuss this in more detail later.

What Paper Size to Use

You would think that paper sizing is a cut-and-dried, straightforward matter (somewhere a printing company manager is laughing), but this is certainly not the case. Paper comes in a bewildering array of international sizes and standards in both inches and metric (millimeter) sizing. Some common standards include ANSI, ISO, JIS, and Arch. For the purposes of this discussion, we need to simplify matters and present all paper as existing in the following basic sizes as viewed in Landscape mode, meaning the first (larger) value is the horizontal size:

- A size: 11″ × 8.5″ sheet of paper (letter).
- B size: 17″ × 11″ sheet of paper (ledger).

- C size: 22″ × 17″ sheet of paper.
- D size: 36″ × 24″ sheet of paper.
- E size: 48″ × 36″ sheet of paper.

Generally speaking, these are the maximum paper dimensions. Some standards list essentially the same sizes but with the border accounted for and removed, so a D-size sheet is 34″ × 22″ and so on. Note that legal (14″ × 8.5″) size paper, while certainly can be used, is not common in engineering or architecture.

As a side note, we can now define precisely the difference between printing and plotting.

- *Printing* usually refers to the output of drawings on A- or B-sized paper, which is easily done on most office desktop size printers. Most home printers can do only A size, but this of course varies with the size (and cost) of the printer.
- *Plotting* usually refers to output of drawings on C-, D-, or E-sized paper, which must be done on plotters, as a desktop printer generally cannot accept paper of this size.

Memorize the paper sizes presented here, as it is pretty much the industry standard as far as AutoCAD design is concerned. You will hear these sizes asked for and referred to, come printing time; and if there is any deviation from these sizes, rest assured it will not be anything radical, as printers and plotters can accept only so much flexibility, at least among the models that small companies and schools can afford.

What Area to Plot

What "area to plot" can be taken quite literally, as in: What do you want to see on the printed paper? The essential choices really boil down to two. Do you want to see the entire drawing (usually) or parts thereof (sometimes)? Your choices in that regard include the following.

Extents

This option prints everything visible in the AutoCAD file. However, this means everything visible to AutoCAD (and not necessarily to you); so if you have a floor plan with some stray lines that you did not notice located some distance away, AutoCAD shrinks the important floor plan to make room for those unimportant stray lines. This is exactly the same as typing in Zoom, then Extents (from Chapter 1). So, it is a very good idea to zoom to extents prior to printing. Remember this acronym: What you see is what you get (WYSIWYG). If you can see it on the screen after zooming, it will print unless prior arrangements are made (freezing the layer, etc.) or the geometry is on the Defpoints layer. For a carefully drawn layout, Extents is the only choice when you want to see the entire drawing.

Window

As you may have guessed, this is for those situations where you want to see just a piece of the drawing. You simply select that choice, click on the Window button that appears to the right of it, and draw a selection window around what you want to see.

Display and Limits

The remaining two choices are of little use in most situations. Limits plots the entire drawing from 0,0 to the preset limits, something most users do not set anyway. Display plots the view in the current viewport in the Model tab or the current Paper Space view in a Layout tab, something we have not covered yet. One can generally use Window in those situations, but Display is fine as well if you want only the current viewport printed.

At What Scale to Plot

Of all options in Plot, this one causes the most confusion to a beginner AutoCAD user, for the simple reason that scale is closely tied in with a very important topic: Paper Space. The essential idea here is that while, yes, you can assign a scale from the Plot dialog box, in reality this is not usually done for most drawings, except simple and small mechanical pieces, such as bolts and brackets. What is done instead is that a sheet with a title block is set up in Paper Space and the design (in model space) is displayed and scaled through viewports (much more on this in Chapter 10), so, in the Plot dialog box, you can select 1:1 (one-to-one) and a properly set up drawing prints to scale.

What if you do not care about scale and just want a print to look over? Here, the pendulum swings in the other direction and things become very easy. Just check off the Fit to Paper box above the scale and the drawing fits on whatever paper you are using, with no regard to scale whatsoever. While this technique is not for final prints of to-scale architectural layouts, it is used very often for nonscaled work, such as electrical wiring schematics, network design, conceptual layouts, and of course check plots.

What Pen Settings to Use

This is a topic that has an entire chapter all to itself (see Chapter 19), and we just do a brief overview here. Proper pen settings are a major factor in giving your drawings a professional look. The concept is simple, usually addressed in hand drafting, yet often overlooked in the computerized world of AutoCAD. It deals with the fact that not all lineworks in a drawing are of the same size width or thickness. Primary features, such as walls on a floor plan, should appear darker on a print, whereas secondary features, furniture, for example, should appear slightly lighter. This was the reason for the existence of 2H, 4H, and other pencil leads, for those students old enough to remember, and the reason today for the existence of pen settings.

With AutoCAD, different colors represent the different pencil leads, and a strict standard needs to be followed as to what drawing features have what colors assigned to them. In this manner, drawings come out with different shades of linework, something noticed by the eye; and as a result, the drawings are much more readable and professional looking. These settings are created, customized, adjusted, and saved under the ctb extension, for all to use according to company standards set by management.

For now, select acad.ctb as your ctb file; the full set of customization tools are presented in Chapter 19.

What Orientation to Use

This is simply a question of whether you want the drawing to be oriented Portrait or Landscape; in other words, do you want the longer side of the sheet of paper oriented up/down or across? This should be familiar to just about every computer user who has printed anything from a word processor or other software.

One twist here is that you can check off a box to plot upside down. A useful application for this option appears when you are printing many sets of similar-looking (but different) drawings. If you want to know which drawing is coming out of the plotter without waiting until it plots completely, send it in backwards. That way, the title block (with the identifying drawing number) comes out first and the question is instantly resolved.

What Offset to Use

This refers to whether or not you need to shift your drawing around to account for a border and stapling area on the left (or elsewhere). If you check off Center the plot, it will be automatically centered on the paper, and in most cases, this is what you would do.

Miscellaneous

There are a number of other features in Plot. For example, you can plot to a file (*.plt) and take it with you to another computer. This is useful for large volume drawing duplication at stores that have no AutoCAD, only printers, such as FedEx or other local print shops. You can also turn on a plot stamp, so information such as date and time of the printing can appear in the border. Some plotting features are unique to 3D and solid modeling. We explore some of these topics as well in Chapter 19. Finally, there is a Preview window and a Number of copies to print selection, the functions of those being self-explanatory.

You can also select a printer that will generate PDF files (AutoCAD PDF or DWG to PDF) or a graphic format (Publish to Web PNG or Publish to web jpg). Rather than paper coming out of a printer, you will be asked where you want to save your PDF, JPG, or PNG file for viewing electronically. We will return to the PDF topic again in Chapter 16.

9.3 THE PLOT DIALOG BOX

Let us now take a close look at the Plot dialog box and identify the locations of all the features described earlier. Open the floor plan from the earlier chapters and bring up Plot via any of the following methods.

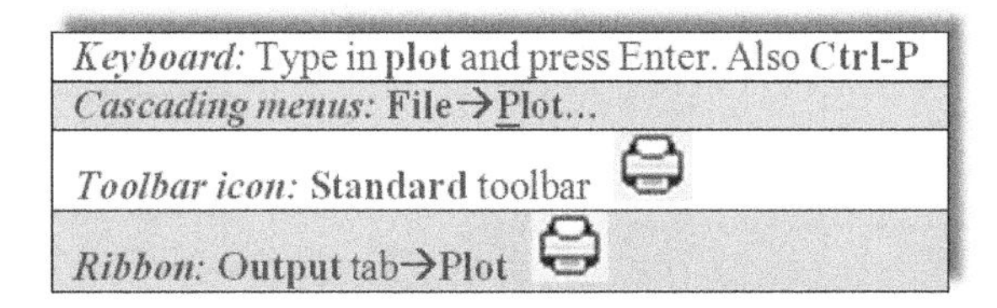

Keyboard: Type in **plot** and press Enter. Also **Ctrl-P**
Cascading menus: **File→Plot...**
Toolbar icon: **Standard** toolbar
Ribbon: **Output** tab→**Plot**

The Plot - Model dialog box appears, as seen in Fig. 9.1. In AutoCAD 2019, it should open to full size right away. If for some reason it is not full size, press the small arrow at the bottom right to expand it fully. The features we discuss are in Fig. 9.2.

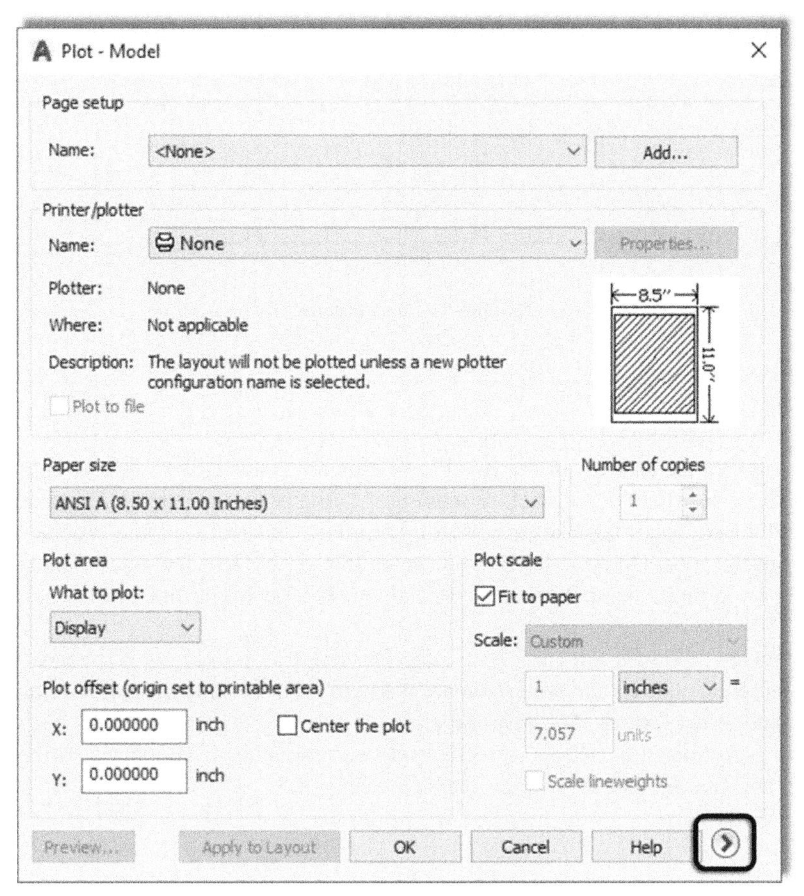

FIGURE 9.1 Plot dialog box, expand *arrow*.

Go through all the settings, referencing the explanations in the previous discussion, until you completely understand what each one does. Set the final settings as shown next.

- *What printer or plotter to use*: Set whatever printer is available at your location (it likely differs from what you see here). If you have a plotter, ignore it for now and use a printer instead, as we are using A-size paper for this exercise.
- *What paper size to use*: Select from the drop-down list. We want any A size available; Letter (8.5″ × 11″) or similar is fine.
- *What area to plot*: Select Extents.
- *At what scale to plot*: Check *Fit to paper* (for this exercise).
- *What pen settings to use*: Select acad.ctb, and click on Yes if prompted to use this for all layouts.
- *What orientation to use*: Select Landscape for this example.
- *What offset (if any) to use*: Check *Center the plot.*

Do *not* press OK yet. The Plot box should look like (or close to) Fig. 9.2. We need to first discuss an important feature of printing: the preview.

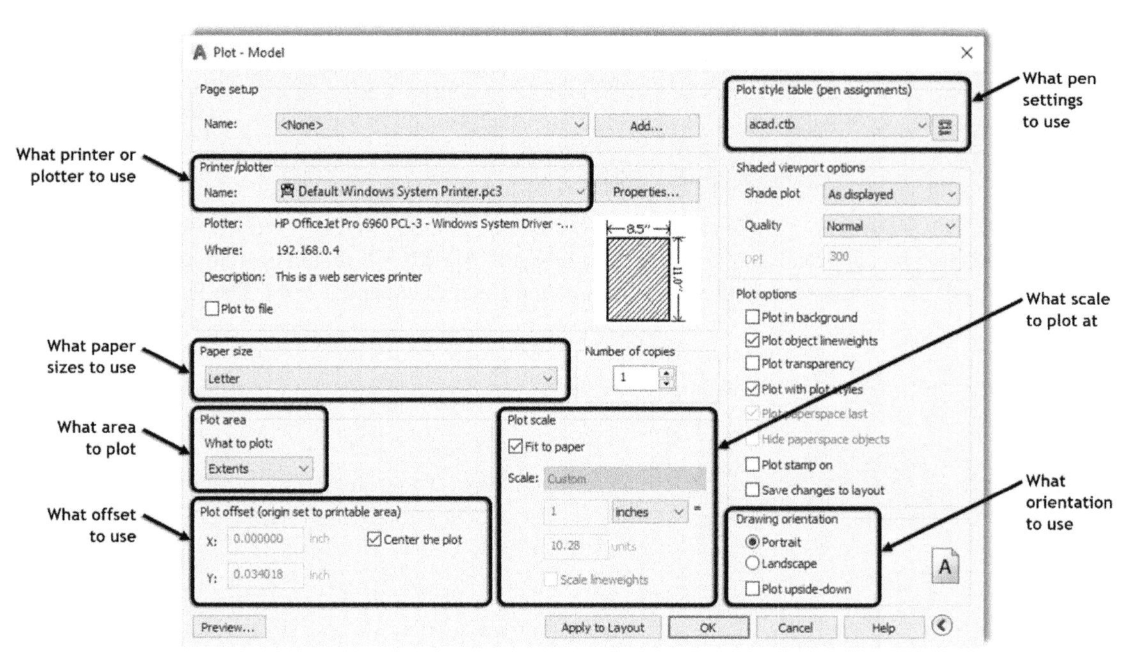

FIGURE 9.2 Plot features.

Preview

Much like the situation with creating an array or hatch, a Preview button is available. With printing and plotting drawings, it is crucially important to preview the drawing before sending it to the printer or plotter. Paper and ink cost money, and everyone regardless of skill level, makes mistakes in setting up printing. This really should become a habit, as errors can be costly, especially on large print jobs. On an industry-wide scale, this simple Preview button actually saves tons of paper and gallons of ink. Not only printing setup errors but even linework drawing errors can be caught during the preview process.

Go ahead and press Preview… (lower left corner), and the Preview screen in Fig. 9.3 appears. You can zoom and pan around using standard mouse techniques, and when you are ready to send the drawing to the printer (or return to Plot for adjustments), right-click and select Exit or Plot, respectively.

Go ahead and plot. If everything was set up correctly and your computer is connected to the printer, the floor plan should print pretty much as expected and seen in the preview of Fig. 9.3.

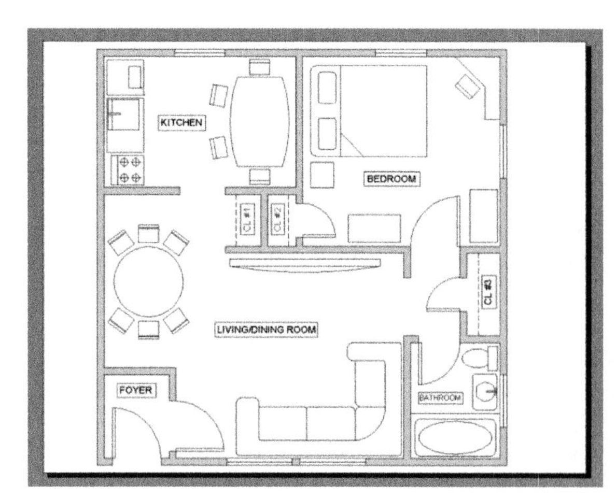

FIGURE 9.3 Plot preview.

If something does not look right anywhere in the process, review the complete set of steps, one by one, returning to the Preview screen to check up on the effects of the changes.

9.4 PAGE SETUP MANAGER

There is one last topic to learn in this chapter. The settings on the print job you just sent are not saved in memory. Test this by typing in plot again. Unless you happened to click the Apply to Layout button by chance, the settings have not been saved and you must reset them all again. Another technique to "save" your settings involves a feature of plotting called *Page Setup Manager*.

The idea here is as follows. You need to enter all the data for a print job in another, similar-looking print dialog box and save it under a descriptive name. Then, save the drawing with these settings, and when you do the actual plot, the name you saved it under appears under Page setup (upper left corner). No more entering settings one item after another; just select the Page setup (it may already be cued up), preview, and print.

This has two advantages. The first is obvious: It speeds things up a lot by keeping all the drawing's print settings ready for use. The second may not be as obvious. Because you can save as many setups as you want, you can have one drawing set up for multiple printers or multiple settings.

For example, your office may need the final output sent in color to a D-sized plotter at a certain scale, but in-house checks are done on 17″ × 11″ sheets, in black and white and no scale. You can easily set up two distinct setups and name them Client_Layout and Check_Layout. It is not unusual to see up to half a dozen setups on one drawing, each with its own purpose: presentation, archiving, optimized for color, check plots, and so forth.

A question usually arises when this topic is presented. Does this need to be done every time a new drawing is created from a blank file? The answer is yes, but it is something that needs to be done only once and saves you much time down the road. Obviously, using templates negates the need to do this at all, as all settings carry over.

The Page Setup Manager is found using the cascading drop-down menu: File → Page Setup Manager... or the Output tab on the Ribbon. It is shown in Fig. 9.4.

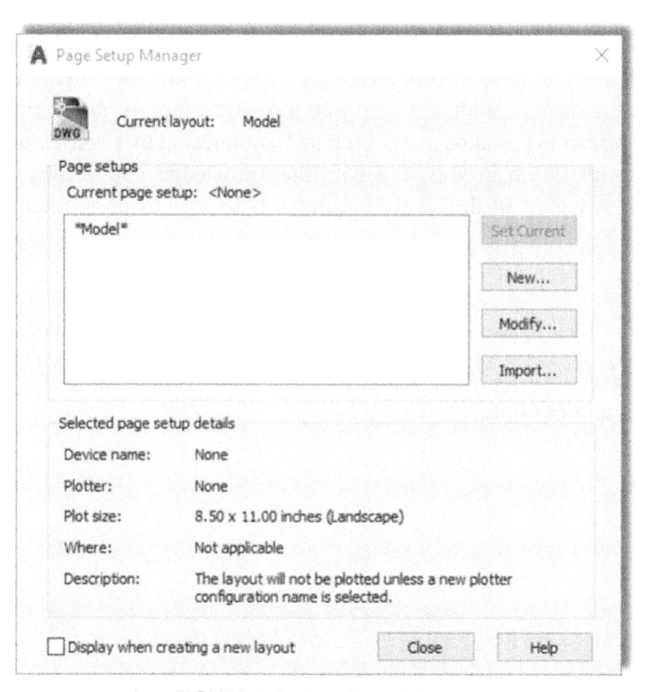

FIGURE 9.4 Page Setup Manager.

Press the New... button and the New Page Setup box appears, as shown in Fig. 9.5. Enter the name of the new setup and press OK.

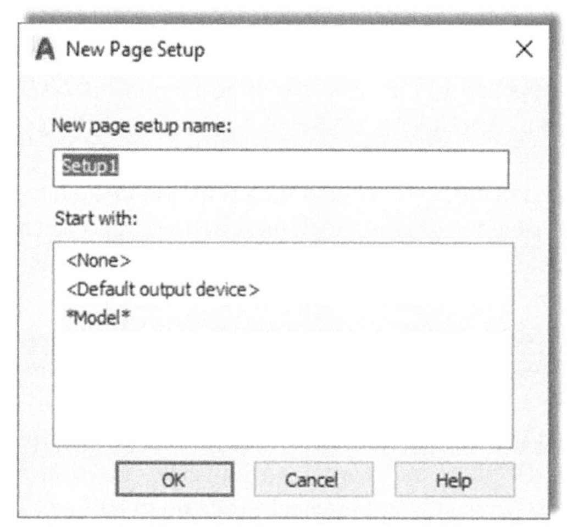

FIGURE 9.5 New Page Setup.

You are then taken to the now familiar-looking Page Setup dialog box. On completion of the settings, a preview (yes, still needed), and a press of the OK button, you are redirected to the Page Setup Manager, ready for another setup if needed. If not, press Close.

To see the benefit of this, go to Plot and select the name of the setup from the Page setup in the upper left corner; all your settings appear. Plot the drawing and save it. You never have to enter settings again (unless something changes, like the arrival of a new printer). So, finally, after learning this new material completely, you can now reduce printing to a pressing of the Preview and the OK buttons—our goal all along.

As stated before, there is still a lot we have not covered, and we revisit this topic again in Chapters 10 and 19. Printing is something you usually receive assistance with at a new job, training center, or consulting gig. However, the most you should expect to be told is what printer is commonly used (up to 10 connected to 1 computer is not unheard of) and what is the standard ctb file in use. The rest is up to you, and with the knowledge you gained in this chapter, you should be proficient in all printing matters just short of the level of AutoCAD management and administration.

SUMMARY

You should understand and know how to use the following concepts and commands before moving on to Chapter 10.

- What printer or plotter to use
- Differences between plotting and printing
- What paper size to use
  - A size
  - B size
  - C size
  - D size
- What area to plot
  - Extents
  - Window

  - Display
  - Limits
- At what scale to plot
  - 1:1
  - Fit to paper
  - Other scale
- What pen settings to use
- ctb files
- What orientation to use
- Portrait or Landscape
- What offset (if any) to use
  - Center the plot
  - X or Y offset
- Preview
- Page Setup Manager

REVIEW QUESTIONS

Answer the following based on what you learned in this chapter:

1. List the seven essential sets of information needed for printing or plotting.
2. List the five paper sizes and their corresponding measurements in inches.
3. What is the difference between printing and plotting?
4. List the four plot area selection options.
5. What scale setting ignores scale completely?
6. What is the extension of the pen settings file?
7. Explain the importance of Preview before printing and plotting.
8. Why is the Page Setup Manager useful?

EXERCISE

1. Open the mechanical device drawing from Chapter 8, Polar, Rectangular, and Path Arrays and practice printing with a variety of scales and other settings. (Difficulty level: Easy; Time to completion: <10 minutes.)
   - Select printer (specific to your school, home, or office).
   - Select paper size; stay with A size.
   - Select Extents for the area; observe the results in Preview.
   - Select Window for the area; select an area and observe the results in Preview.
   - Try a variety of scales such as 1:1, 1:2, and so on; observe the results in Preview.
   - Pick the monochrome.ctb file.
   - Try both Landscape and Portrait orientation.
   - Center the plot.

The following figures show the mechanical device in an Extents Preview (first part) and Windows Preview (second part).

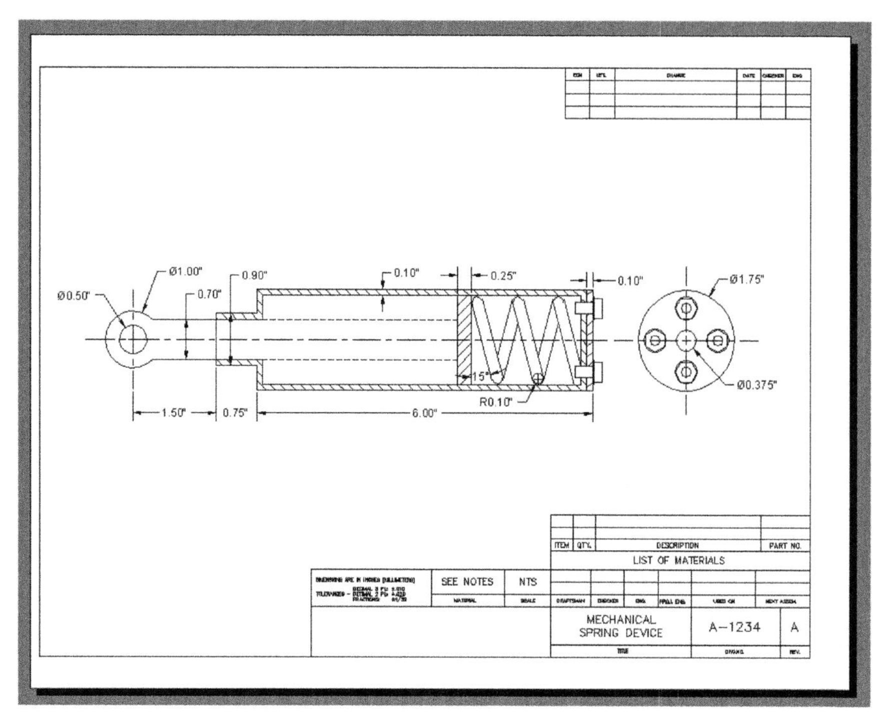

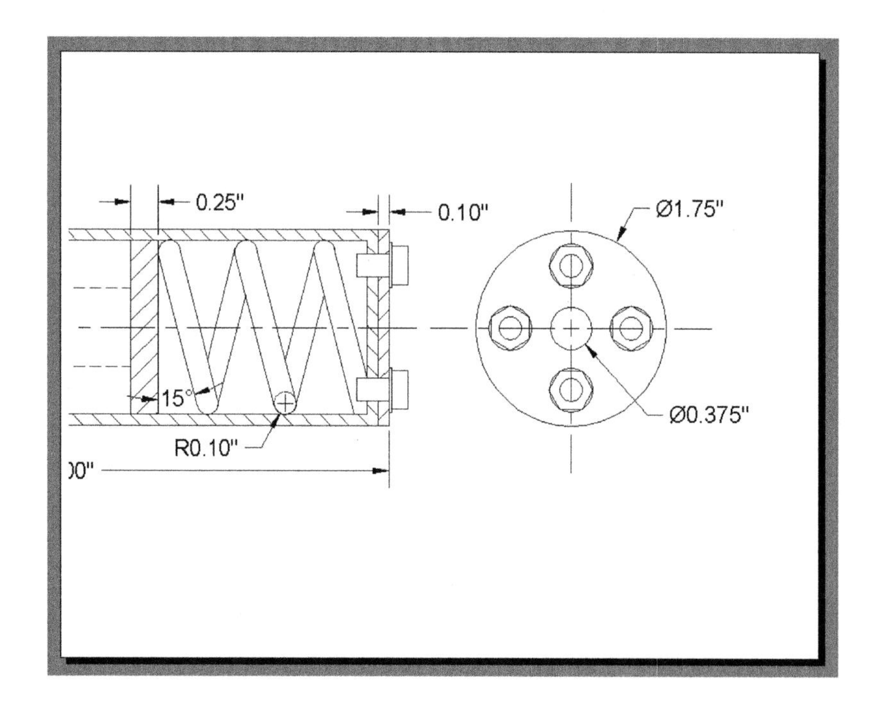

Chapter 10

Advanced Output—Paper Space

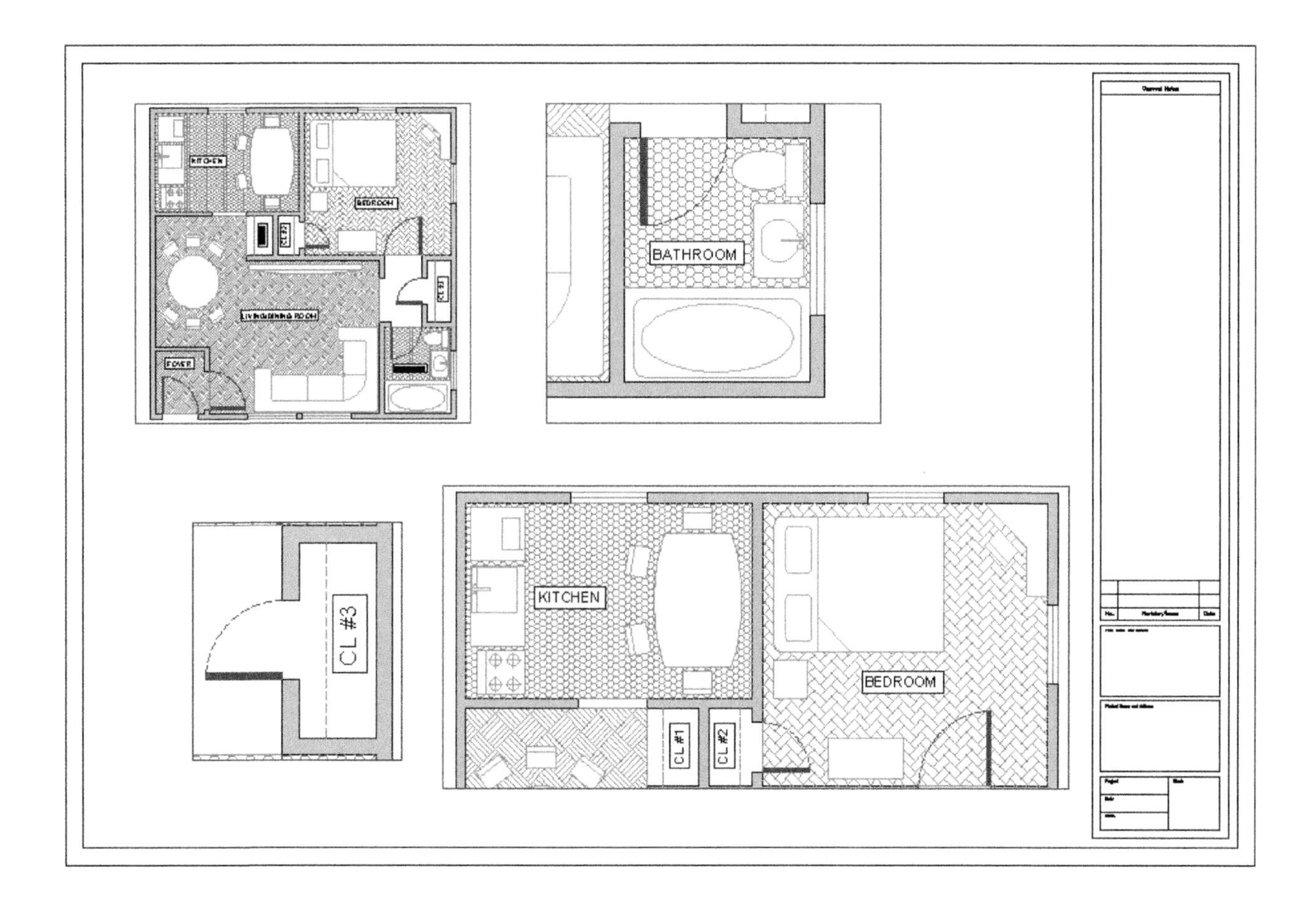

Learning Objectives

In this chapter, we introduce and thoroughly cover the concept of Paper Space. We specifically introduce

- The primary reasons for using Paper Space
- Entering Paper Space and setting up title blocks (Layouts)
- Viewports and their properties
- Viewport scale assignments
- Controlling what is visible in viewports
- Setting up text and dim styles based on the viewport scale
- The new Annotation feature that simplifies the setup of text and dims

By the end of this chapter, you will be well versed in applying this critical concept to your design work. Estimated time for completion of this chapter: 3 hours.

Up and Running with AutoCAD 2019. https://doi.org/10.1016/B978-0-12-816440-2.00010-0

10.1 INTRODUCTION TO PAPER SPACE

Paper Space is an extensive and critically important topic in AutoCAD. Learning it is a rite of passage for all AutoCAD designers, and proper setup and use of Paper Space is a mark of an expert user. The topic is not necessarily difficult but can be somewhat tedious, requiring care and attention from the designer who really has to "dot the Is and cross the Ts" to assure everything is done right. It is also a topic that needs to be understood well in theory, first and foremost. The actual "button pushing" process to set things up is much easier when you understand what Paper Space is and why it is used. The result of learning all this, however, is a professional drawing set, i.e., accurate and easy to read, understand, and print. This is a necessity and a goal well worth striving for. We explore Paper Space from all angles, one step at a time. Read and understand each section before moving on to the next, as the knowledge is cumulative.

What Is Paper Space?

When you create a design of a building with pencil and paper, this design needs to be drawn to a certain scale, as opposed to full-size, for the simple reason that it would not fit on a piece of paper if you did not do this. This is the reason why architects carry with them an architectural scale, a portion of which is shown in Fig. 10.1.

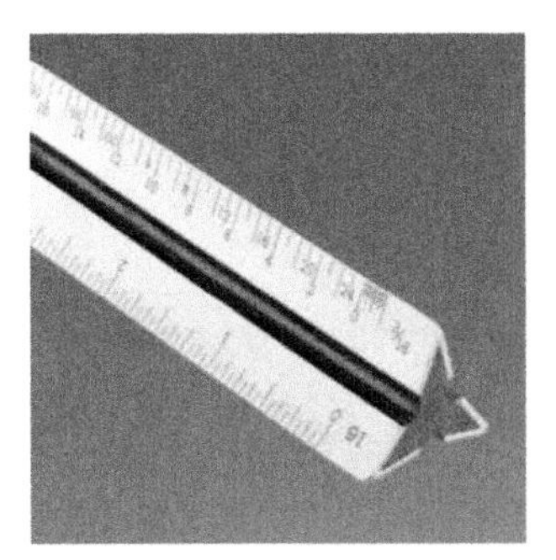

FIGURE 10.1 Architectural scale.

Then, they can simply pick the appropriate scale for the size of their intended design (1/8″ = 1′–0″ is common), and now, because 8 feet are represented by 1 inch, the building floor plan can fit nicely on a large sheet of paper.

In AutoCAD, however, you do not draw "to scale." Because you now have an infinite-size canvas on which to create your design, everything is drawn one to one (1:1) or "life size," meaning that, if you were to jump into your design through the computer monitor, you would find yourself in proportion to (and would be able to freely walk around inside) your building.

This is one of the Golden Rules of AutoCAD, and indeed in all of computer-aided design, and marks a significant departure in philosophy and method from the previous hand-drafting approach. In fact, attempting to draw "to scale" results in complications and potential for error when creating, and later changing, the design and is simply not done in practice.

Drawing one to one (1:1), however, creates a whole new set of problems. How do you output a design that may be dozens or hundreds of feet across, onto a paper that is generally only 3 or 4 feet wide? Furthermore, this design has to be framed by a title block or title page that itself has to be drawn one to one (1:1), so as to not violate the Golden Rule. The solution is elegant, simple in principle, and has stood the test of time.

The idea is as follows: Draw the design, full-size (1:1) in one area of AutoCAD, called *Model Space*, and then draw the title block, also to full size (1:1) in another unconnected and unrelated area, called *Paper Space*. Finally, connect the two together via a tool called a *viewport*, which is simply a "window" or a hole looking in from Paper Space onto Model Space. This viewport is then assigned an appropriate scale; and if everything is drawn (and printed) full-size, one to one (1:1) from Paper Space, then the design comes out to scale and the problem is resolved.

This in essence is Paper Space: Nothing more than a technique to separate the two worlds and reconcile them in a meaningful way to get around the tricky problem of a finite-size piece of paper holding a full-size, and usually much larger, design.

Be sure you understand the previous few paragraphs completely. Everything outlined from here on is easier to understand once you have a good idea of why we are doing what we are doing.

10.2 PAPER SPACE CONCEPTS

The devil, as they say, is in the details, and the study of Paper Space involves a number of separate concepts, each of which is an important part of the complete picture and needs to be looked at closely:

- *Layouts*: Entering Paper Space and setting up title blocks
- *Viewports*: Viewports and their properties
- *Scaling*: Viewport scale assignments
- *Layers*: Controlling what is visible in viewports
- *Text and dims*: Setting up text and dim styles based on viewport scale
- *Annotation*: A feature that simplifies the setup of text and dims
- *Summary*: Putting it all together

These topics are listed roughly in the order a designer would go through them as he or she sets up and does the project. As an example drawing, we continue using the architectural floor plan that you have been working on since the start of Level 1. It happens to fit nicely on a D-size (36″ × 24″) sheet of paper at ½″ scale to illustrate the concepts. If you want to use another design, perhaps one drawn in one of the end-of-chapter exercises or one your instructor assigned to you, by all means use that. If you stick with the apartment floor plan, then thaw all the layers. It may be messy, but we will fix that later. Let us begin with the first topic, layouts.

Layouts

The first step involves switching over to Paper Space and familiarizing yourself with the layout tabs. They are found at the bottom of your drawing area and resemble Microsoft Excel tabs. You will see Model, followed by Layout1, Layout2, and a plus for creating more tabs, as in Fig. 10.2.

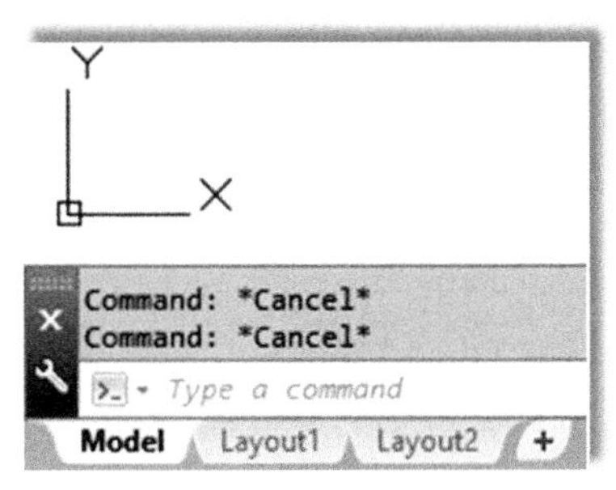

FIGURE 10.2 Layout tabs.

Go ahead and switch into Paper Space by clicking on Layout1. You notice a different environment, as shown in Fig. 10.3. For starters, the UCS icon is a triangle and the Layout1 tab is highlighted. Your background color is white, and finally you have a variety of background features (rectangles, shadows, viewports, and others). We modify this environment in a minute, but in the meantime, click on the Model tab and jump back into Model Space. You now know how to enter and exit Paper Space.

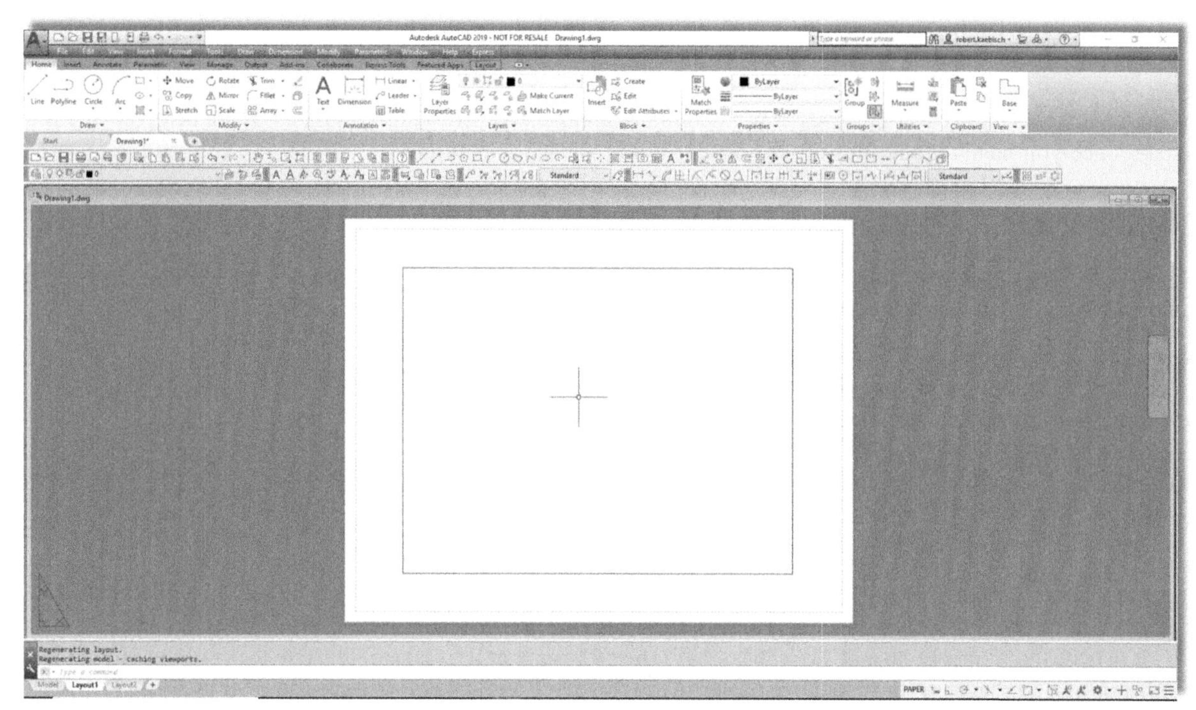

FIGURE 10.3 AutoCAD 2019 Paper Space.

Unless you really prefer looking at a "piece of paper" representation in Paper Space, there really is no need for it to look any different from Model Space. Nothing is inherently special about Paper Space. Remember, it is just a technique in AutoCAD to create two separate worlds. You can actually draw in Paper Space as if it is your regular drawing area with no adverse effects, but of course that would defeat the whole purpose of what we are doing.

Based on this, it is highly recommended that the Paper Space shadow, printable area, and other features are turned off. Indeed, most experienced users prefer their Paper Space to look as close to Model Space as possible. It is ultimately up to you if you want to do this, but all screen shots in the remainder of this chapter feature the plain Paper Space environment.

To do this, you need to right-click to Options..., and then select the Display tab and take a look at the Layout elements (lower left). Uncheck all but the very first choice, as seen in Fig. 10.4, and set the Sheet/Layout color to black in Colors... (though in the textbook the background remains white to conserve ink and for clarity). It is also a good idea to set your crosshairs to 100 for reasons to be explained later in this chapter. Ask your instructor for assistance if needed, and when done click OK.

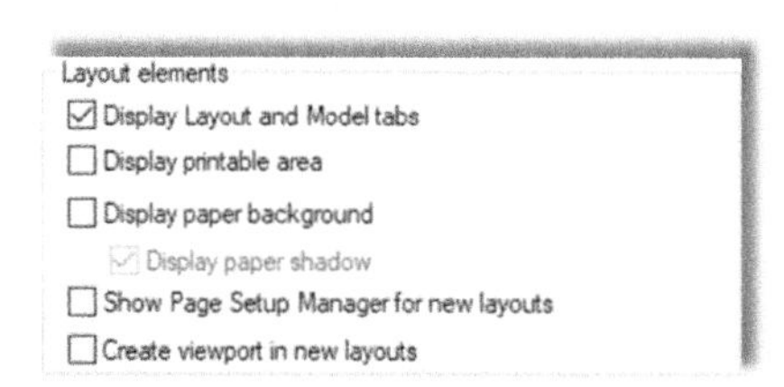

FIGURE 10.4 Recommended Paper Space preference settings.

After all these modifications, Paper Space should look like Fig. 10.5 (except of course for the screen color, which can be black if you prefer). Notice that now the only indication that you are in Paper Space is the triangle UCS icon and the highlighted Layout1 tab. You may or may not have seen a viewport rectangle present, but it was erased in this particular case. Of course, if you already had this environment (or a variation of it) set up, you could have skipped the previous few steps.

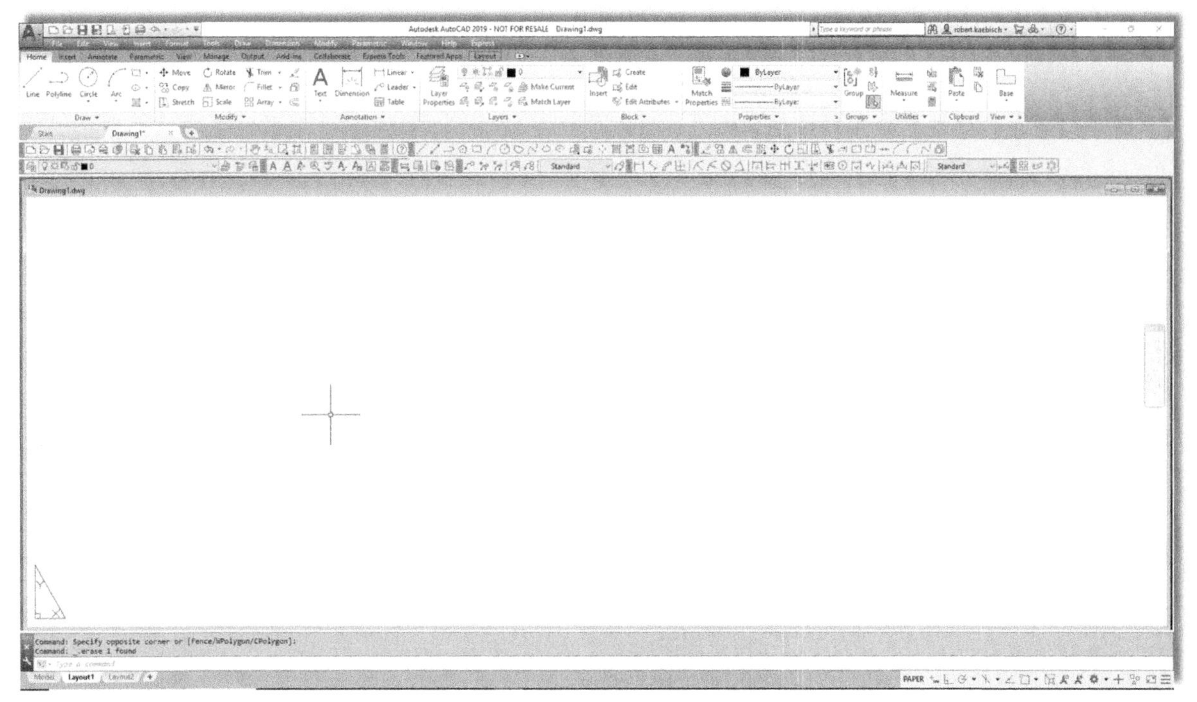

FIGURE 10.5 Paper Space, modified.

So, what are these layout tabs in real life? They are actually sheets of paper when the job is completed and ready for printing. The design (done in Model Space) is "presented" in Paper Space with each layout being one sheet of the total drawing set, i.e., the Architectural, Mechanical, Electrical, Details, and so on.

If you are thinking that we need to learn how to make title blocks at this point and perhaps create more layouts, then you are correct. We create a title block from a built-in AutoCAD template. This may or may not be the way you will do it at your job. If you are in an architecture office, there may be a predrawn title block that you can just retrieve and paste in, but for our learning purposes we create a new one.

Select the Layout1 tab (left-click) then *right-click* to bring up the menu shown in Fig. 10.6.

We work with this menu for the next few pages. Let us first of all try creating a title block from a template by clicking on the From template... choice. You get the dialog box seen in Fig. 10.7. Pick the Tutorial-iArch.dwt template and click on Open.

Another small dialog box, Insert Layout(s), appears, as shown in Fig. 10.8. It is simply informing you that the layout is a D-size one. Click on OK.

FIGURE 10.6 Layout tab 1, right-click menu.

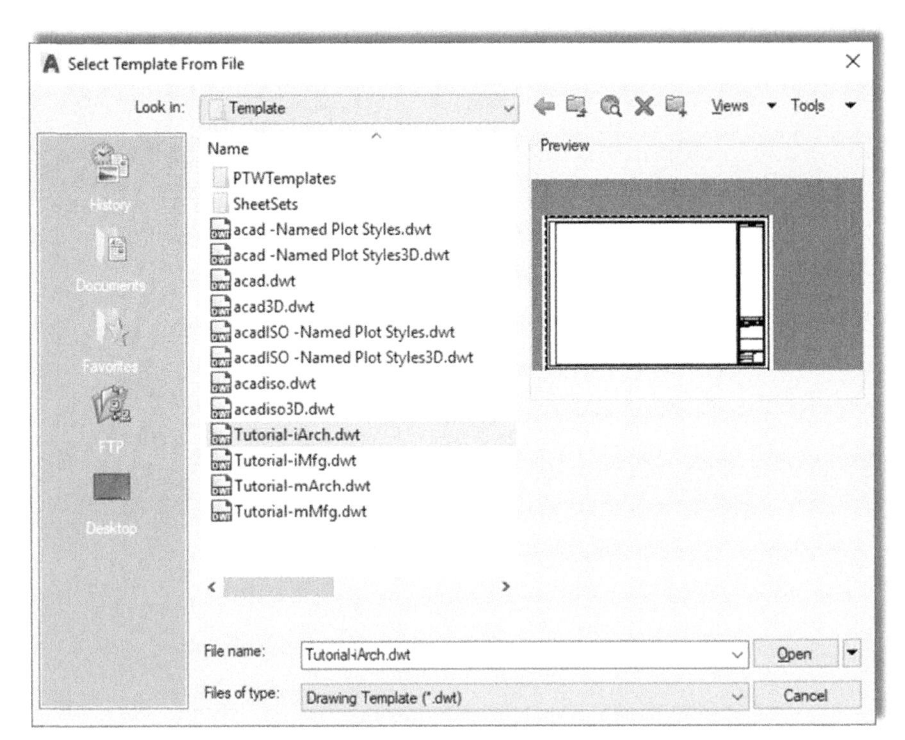

FIGURE 10.7 Select template from file.

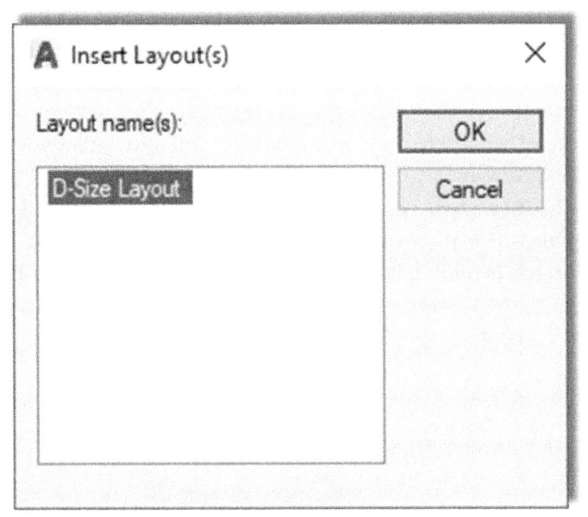

FIGURE 10.8 Insert Layout(s).

Notice what happened: You have a new tab, called *D-Size Layout*, positioned right after the other three layouts. Click on it and you see a title block, much improved in AutoCAD 2010 versus older releases (AutoCAD 2019 continues to use it). It is a reasonable representation, minus the fancy logos, of what a typical architectural office may use. Inside the title block is a blue rectangle through which you may or may not see your building; this is the viewport, and we cover that topic next. For now, go ahead and delete it, as we make our own soon. Also, explode the title block; we want to modify it.

What you have now is shown in Fig. 10.9 (except that your screen background likely is colored black).

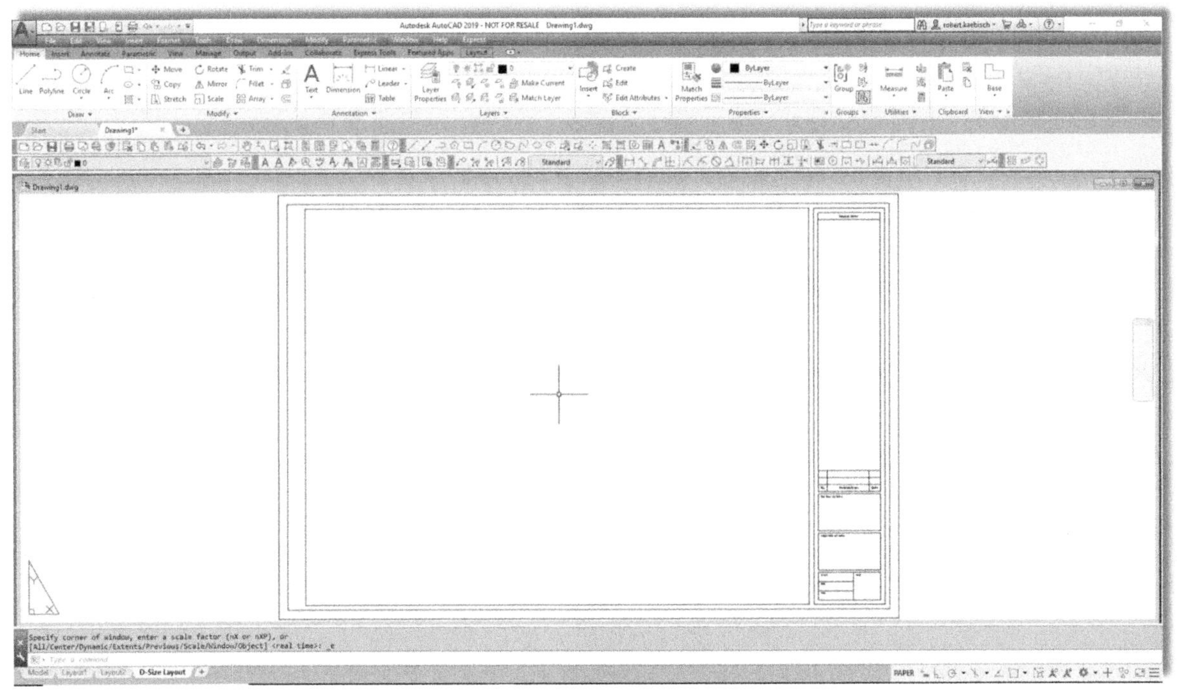

FIGURE 10.9 Basic Tutorial-iArch.dwt title block.

The next step is to modify the layout tabs by renaming them to something more descriptive and removing the unnecessary ones. Right-click on the new D-Size layout tab to get the menu to appear again. Select the Rename choice and the tab's name becomes editable; type in Architectural Layout. Now, get rid of the Layout1 and Layout2 tabs, as they are empty and not needed. Begin by clicking on Layout1, and then right-click to get the menu up again. Select Delete and press OK when the warning box pops up. Repeat the procedure for Layout2. After these steps, we have only two tabs remaining, Model and Architectural Layout, as seen in Fig. 10.10.

Take a moment to look over the title block in the Architectural Layout tab. As a reminder, delete the original viewport (the blue rectangle), as we will make our own from scratch. Then, explode the title block if you have not yet done so. Modify anything you may want to look different (e.g., delete the USER, REVDATE, and the FNAME text in the lower left corner), but this is optional.

Now, we need to create more of these title sheets, for the Mechanical and Electrical layouts. Go back to the menu (right-click on the Architectural Layout tab) and select the Move or Copy… choice. The dialog box in Fig. 10.11 appears. Note that you can also just click the plus sign found to the right of the tabs, create a generic layout, and then rename it. The disadvantage there is that the layout is blank; your title block is not carried over to this new tab.

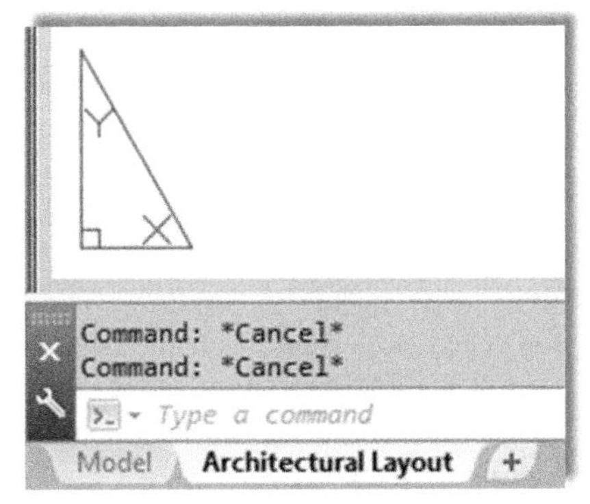

FIGURE 10.10 Model and Architectural Layout tabs.

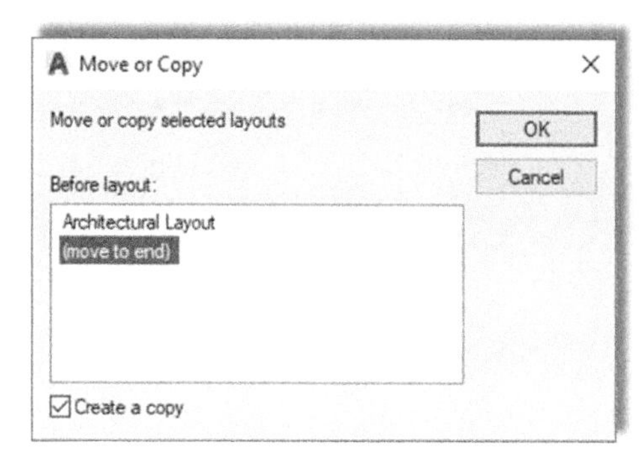

FIGURE 10.11 Move or Copy.

If you use the Move or Copy approach, be sure to do two things. First of all, check off the Create a copy box at the bottom left and click on the (move to end) choice, which simply places the new title sheet next in line after the existing Architectural Layout and may be important later when it comes to the printing order. Finally, click on OK and observe what happens with the tabs. A new one appears, saying Architectural Layout (2), and it is a perfect copy of the previous existing tab. Go ahead and rename it as Mechanical Layout. Repeat the whole procedure to create a third tab, called Electrical Layout, and finally a fourth tab, called Landscaping Layout. The final result is in Fig. 10.12.

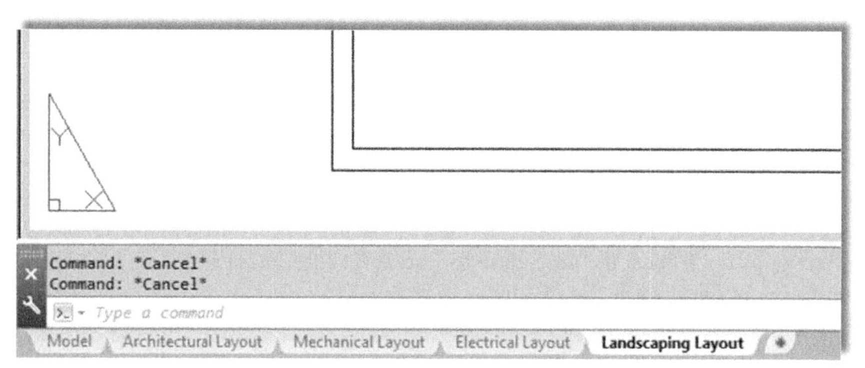

FIGURE 10.12 All tabs.

Review this entire section; it is an important first step to learning Paper Space and is a common procedure needed to set up a project that has multiple pages, which is pretty much the case with most projects. It should be clear as to what each sheet (tab) represents. Next, we work on the sheets to set up viewports, so the design can be viewed at the proper scale.

Viewports

Viewports, as mentioned in the introduction, are simply windows from the Paper Space world looking into the Model Space world. It is as if we ripped a hole in the fabric of Paper Space and observed what was behind there. We assign scales to these viewports later on and, in this manner, reconcile the Paper and Model Spaces. For now, we need to first learn how to create and modify the viewports and go over some of their properties. Your new layouts already have a premade viewport (the blue rectangle), but as mentioned earlier, you should have now erased it; we will learn to create a new one from scratch.

Just a reminder: To go through the following steps, make sure you have an accurate, full-size design in model space to look at, which you already should have, as suggested earlier in this chapter. For an example in this text, the apartment plan from the previous chapters continues to be used. It is slightly modified to include both furniture and floor hatching. You are welcome to duplicate the same thing. If all is well, then let us proceed to learning about the viewport or Mview window.

Click on the Architectural Layout tab, entering Paper Space, type in mview, and press Enter. This is the key command for creating new viewports. Alternatively, you can use the cascading menus View → Viewports → 1 Viewport.

Alternatively, a contextual tab of the Ribbon appears when you are in Paper Space called *Layout.* Go to Layout → Layout Viewports → Rectangular. Note here that *you can create more than one viewport at a time, but we do only one for now.*

- AutoCAD says: `Specify corner of viewport or [ON/OFF/Fit/Shadeplot/Lock/Object/Polygonal/Restore/LAyer/2/3/4]<Fit>`

Then simply click and drag the mouse across your screen, effectively drawing a large rectangle from one side of the sheet to the other (avoiding the actual title block in the lower right corner). AutoCAD says: Specify opposite corner: as you are doing this and Regenerating model as you complete the viewport. You should now be able to see the design inside the "rectangle" you created, as shown in Fig. 10.13.

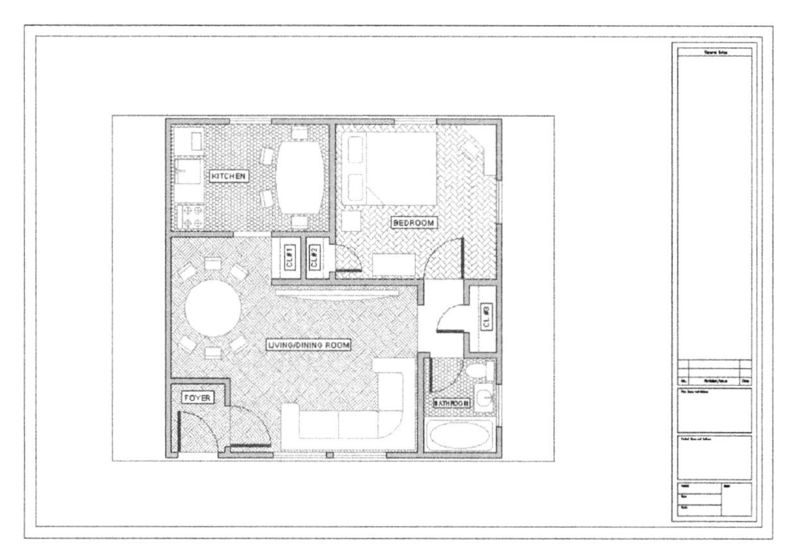

FIGURE 10.13 New viewport.

You can create more viewports in the other layouts if you like for practice, but we focus on this one, in the Architectural Layout.

Floating Model Space

Double-click inside the viewport. Its border becomes thicker and you notice your mouse is "inside" the viewport, as shown in Fig. 10.14. Note also how the UCS icon changes to the Model Space version. The advantage of having full width crosshairs (set at 100) is evident here, as the viewport "clips" the ends of the crosshairs and it becomes clear that you are in this mode. Pan and zoom while inside the viewport and you notice the design shifts around while the Paper Space title

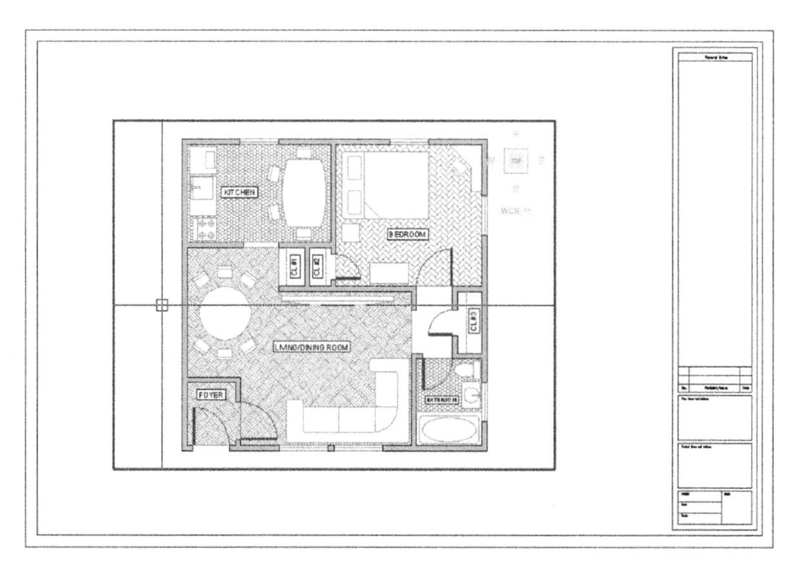

FIGURE 10.14 Floating Model Space.

sheet stays still. This Floating Model Space mode is sort of a "no-man's land" between the Model and Paper Spaces and is critical in scaling the design, as we shall soon see.

For now, double-click anywhere outside of the viewport to get back to Paper Space. Be careful not to double-click on the viewport itself. That takes you to the vpmax mode (maximum/minimum viewport size), which can be used for editing but is not needed here. If you do accidentally double-click on it, press the button shown in Fig. 10.15 to get back (bottom right of the screen).

FIGURE 10.15 Minimize viewport.

Practice jumping in and out of the viewport several times; this is very important for the rest of this chapter. Note also the big picture discussed so far. You are in Paper Space and created the viewport to look onto the design in Model Space. When you are inside the viewport, you are technically in Model Space but still in Paper Space as well. You need this for scaling purposes only. Design work on either the floor plan or the title sheet or block is done in their respective spaces.

Next, we need to learn how to modify the viewport(s). Note that they are not static displays but rather dynamic tools you can and should mold and shape them to your needs, as outlined in the next section.

Modifying the Viewport

First, make sure you are in Paper Space, and then click once on the viewport object (not inside of it but on the lines of the actual viewport rectangle). The viewport becomes dashed and grips appear. Using them, you can resize the viewport to any size desired by clicking on any grip, activating it (it turns red), and adjusting the viewport's size and shape. In general, you can:

- Scale the viewport(s) up and down to any size.
- Adjust the viewport(s) to any shape.
- Move the viewport(s) anywhere you like.
- Erase the viewport(s).
- Copy the viewport(s) as many times as needed.
- Make new ones as needed.

Take a few minutes to go through all these features. Make your viewport smaller, and copy it three times. Move all four around the screen, and erase one. Finally, create it again with mview. In the end, try to have what is shown in Fig. 10.16 and make sure you understand viewport manipulation, the mview command, and getting in and out of Floating Model Space. When you are inside the viewport, do not forget to also scale the drawing up or down and zoom to extents. We focus on a precise scale later.

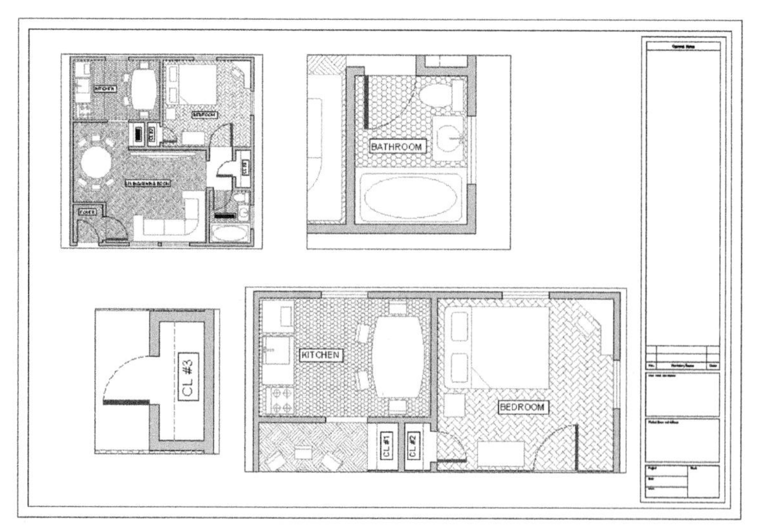

FIGURE 10.16 Four random viewports.

Polygonal Viewport

The shape of the viewport need not be rectangular or square; it can be any shape necessary by using the Polygonal option. To practice this, click over to the Mechanical Layout and create a new viewport using the mview command; however, before clicking to draw the actual viewport, select p for Polygonal in the lengthy menu.

- AutoCAD says: `Specify start point:` and after the first click `Specify next point or [Arc/Length/Undo]:`

Click your way around in some random fashion (do not overlap the clicks) and press c for Close to close the viewport. What you get after creating a starburst sort of pattern is shown in Fig. 10.17. You can see the design inside the oddly shaped viewport.

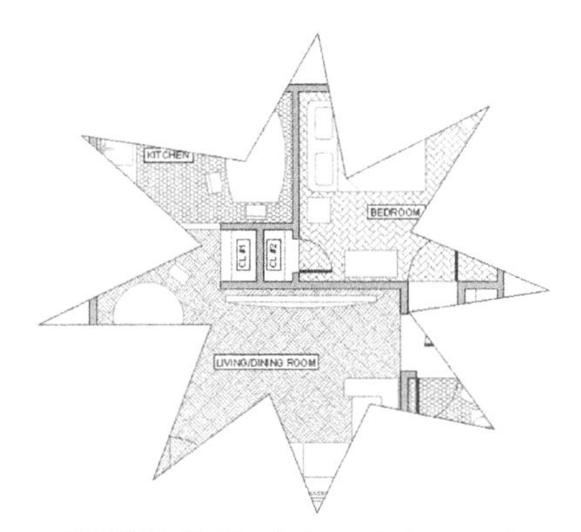

FIGURE 10.17 Polygonal viewport 1.

Of course, this is not really what this tool is for. Rather it is to make viewports that weave their way around possible revision blocks or notes on the drawing, thereby freeing you from being forced to use a rectangle and wasting space. An example is shown in Fig. 10.18, and this is often the case in real-life designs, making polygonal viewports very important for creating an efficient layout.

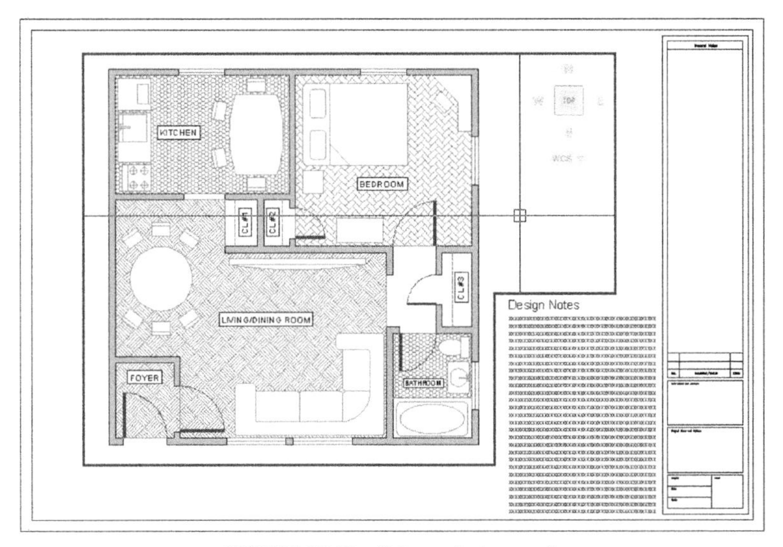

FIGURE 10.18 Polygonal viewport 2.

There are a few other important items to cover with viewports, such as locking them and making them invisible, but we introduce these features a bit later, as the need arises. What we need to look at next is the "core" of Paper Space: scaling the viewports and their contents.

Scaling

Here, we finally put everything together because really what we have done so far was just the setup. We have our design in Model Space and our title blocks on sheets in Paper Space, connected together by one or more viewports. Now, we need to give these viewports a scale, which was our goal all along.

To make things simple, let us just have one viewport in our Architectural Layout; so delete any others you may have, including polygonal ones, and leave one large viewport similar to Fig. 10.13.

So, what is scaling? Well, you may have already guessed it has something to do with zooming. After all, as the design is zoomed in and out (while in Floating Model Space), the scale continuously changes and always has some value. So scaling the design properly is just a matter of knowing how much to zoom it in the viewport. The big question of course is, How much? Another way to put it is, What scale factor do we apply?

There are two ways to assign scale to any viewport. The first one is a technique used for just about every AutoCAD release up until the last few. It is a bit more involved but really gets you to understand what is going on. Then, we cover the other, easier method.

Method 1. Double-click inside the viewport, type in zoom, and press Enter.

- AutoCAD says: `Specify corner of window, enter a scale factor (nX or nXP), or [All/Center/Dynamic/Extents/Previous/Scale/Window/Object]<real time>:`

Here is where you used to pick e for Extents and see the entire drawing fill up your screen. However, notice the `enter a scale factor (nX or nXP)` choice. This is what we are interested in right now, specifically nXP.

What nXP means is that you need to enter a ratio relative to Paper Space (which is the 1:1 "anchor," so to speak) and the design scales to that ratio. So, what is it? Well, for ¼″ = 1′–0″, the n is 1/48 (followed by the xP). So type in 1/48xp and press Enter. The design zooms to precisely ¼″ = 1′–0″ scale, as shown in Fig. 10.19.

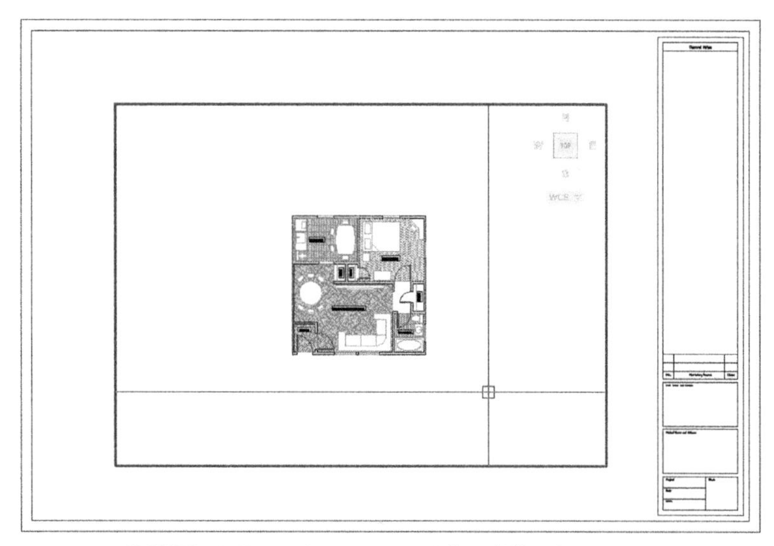

FIGURE 10.19 Design zoomed to ¼″ = 1′-0″ scale (1/48xp).

Now, this seems to be too small, doesn't it? No problem, just type in zoom and press Enter again. Then, type in 1/24xp and press Enter again. The design will zoom to ½″ = 1′–0″, as shown in Fig. 10.20, which seems to be a better size for this design and title sheet.

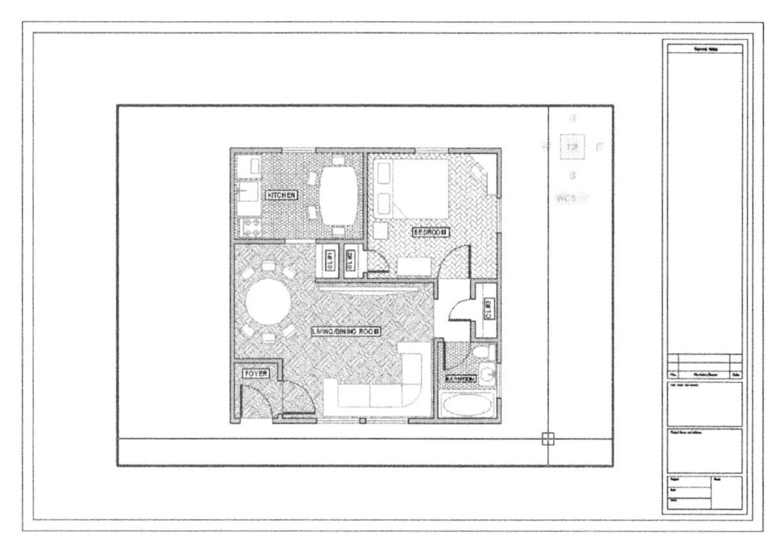

FIGURE 10.20 Design zoomed to ½″ = 1′–0″ scale (1/24xp).

So, where did 1/48xp and 1/24xp come from? They are just ratios developed from the fact that there are 12 inches in 1 foot and the scale is 1/4 or 1/2. Therefore, for 1/4, it is 1/48 (1/4 inch scale × 1/12 inches in a foot = 1/48 relative to the 1:1 Paper Space). For 1/2, it is 1/24 (½ inch scale × 1/12 inches in a foot = 1/24 relative to the 1:1 Paper Space) and so on. There are additional scales such as $\frac{1}{16}''$, $\frac{1}{8}''$, $\frac{3}{8}''$, $\frac{5}{8}''$, and others, all of which can be derived in a similar manner. Now that you understand basic viewport scaling, let us apply this method in a somewhat easier way.

Method 2. Double-click back to Paper Space and bring up the Viewports toolbar, shown in Fig. 10.21.

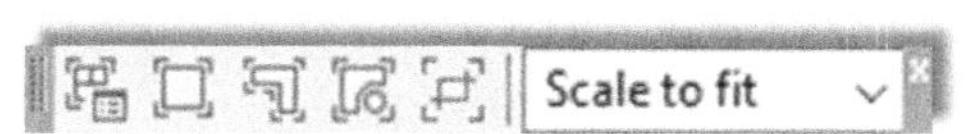

FIGURE 10.21 Viewports toolbar.

Now click on the viewport object once (not inside of it but on the lines of the actual viewport rectangle). When the viewport becomes dashed, the scale associated with that viewport is shown in the white field on the right side of the Viewport toolbar and in the drawing aid area. Click the down arrow and you see a wide variety of choices. Pick the scale you want, and the viewport takes on that scale, as shown in Fig. 10.22.

Layers

We have done a lot already in setting up the drawing in Paper Space. We now turn our attention to several important details that are either new or change significantly when dealing with Paper Space.

One of those is layers. While the fundamental layer concepts do not change, a new ability is added to the Layers dialog box. It turns out that, when viewports are present, you can freeze and thaw layers in those viewports independent of each other and the main drawing. This is very significant, as you want to be able to display different items as you go from viewport to viewport.

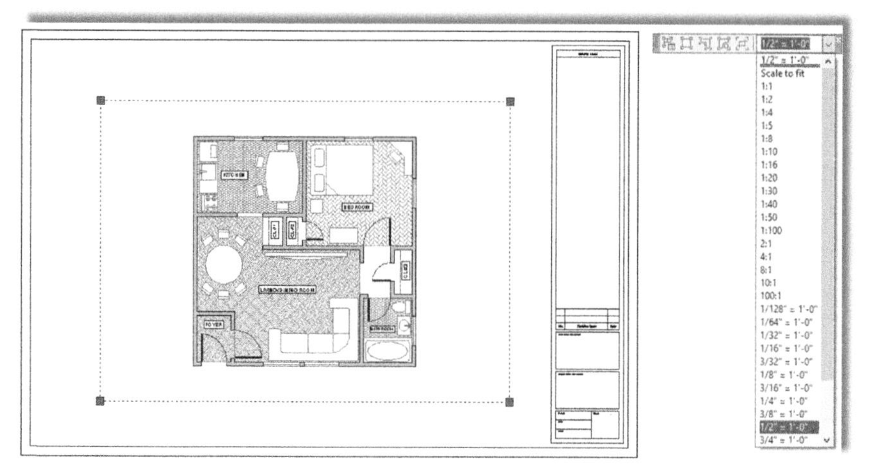

FIGURE 10.22 Viewport scaling via toolbar.

Perhaps the most important example of that is when you jump between the Architectural, Mechanical, and Electrical Layouts. Each tab is a new viewport, yet all views of the design originate from one actual copy of that design. In each layout, you need to see only what is relevant to that layout, and it would be a major problem if you could turn layers on and off only in the main Model Space drawing. Then, each tab would display the exact same thing.

Fortunately, this is not the case, and in each viewport, you can turn layers on and off as needed to get the design points across. Therefore, the Architectural Layout has only architectural layers and so on. It is the same idea with multiple viewports on one layout (such as the case with detail sheets). Each viewport is truly a world unto itself. To see an example of this, let us create four viewports in the Architectural Layout, as shown in Fig. 10.23. The scales assigned are random.

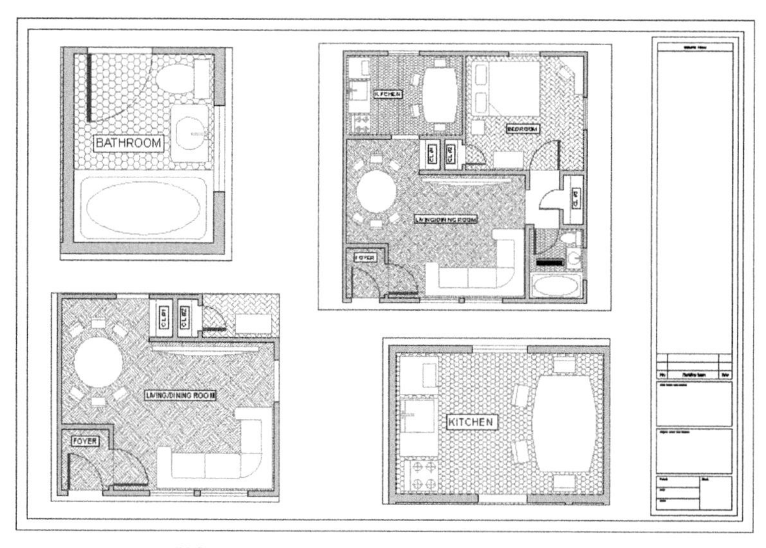

FIGURE 10.23 Four viewports for layers demo.

Double-click into the top-left viewport and bring up layers. The Layers dialog box (with some extra columns) appears, as shown in Fig. 10.24. All relevant columns have been expanded for maximum visibility of the contents. Look for the VP Freeze column (circled); this is the column in which we are interested right now.

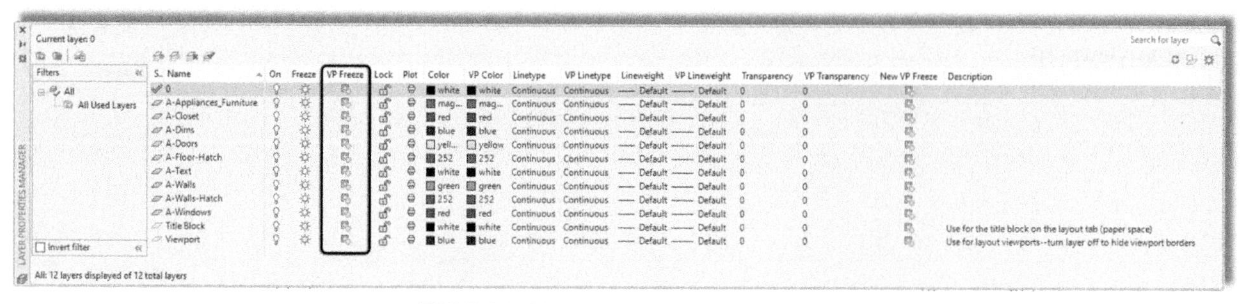

FIGURE 10.24 Layers with VP freeze column.

In that column, you can freeze all the unneeded layers for that layout without affecting either the main layout or the other viewports. Go ahead and freeze the A-Floor-Hatch layer. The result is shown in Fig. 10.25. Note how the neighboring viewports are not affected by what you did in the top-left one. Now click over to the top right one and freeze the A-Doors layer only. Click over to the bottom left viewport and freeze the A-Floor-Hatch and the A-Appliances_Furniture layers. Finally, in the bottom right viewport, freeze everything except the walls and windows. Note the distinctly different layer settings in each viewport; they are completely independent.

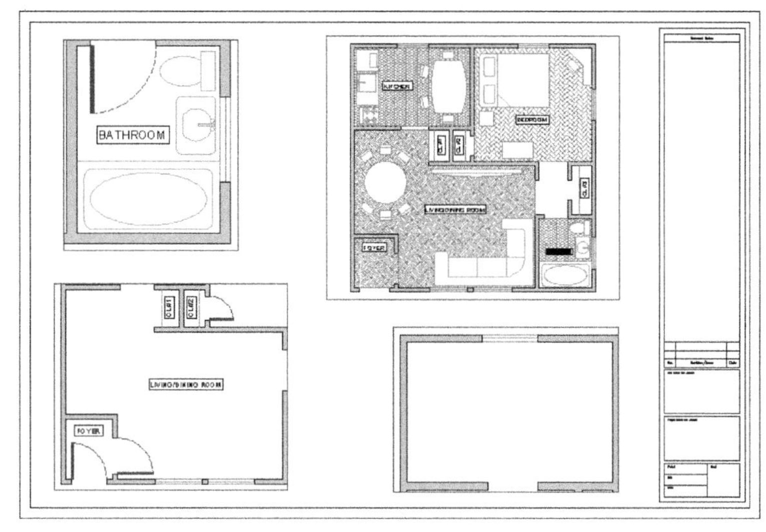

FIGURE 10.25 VP Frozen layers.

Lock VP

We now cover two other notable features of viewports. The first is the ability to lock the viewport, which simply means you cannot change the viewport's scale and all zooming attempts fail (the entire Paper Space drawing zooms instead). This can be useful to prevent accidental zooming while inside the viewport and should be set after the design is laid out and set up. To do this, bring up the Properties box. Then, click on the viewport once and select (under Misc.) the Display locked option, setting it to Yes, as shown in Fig. 10.26.

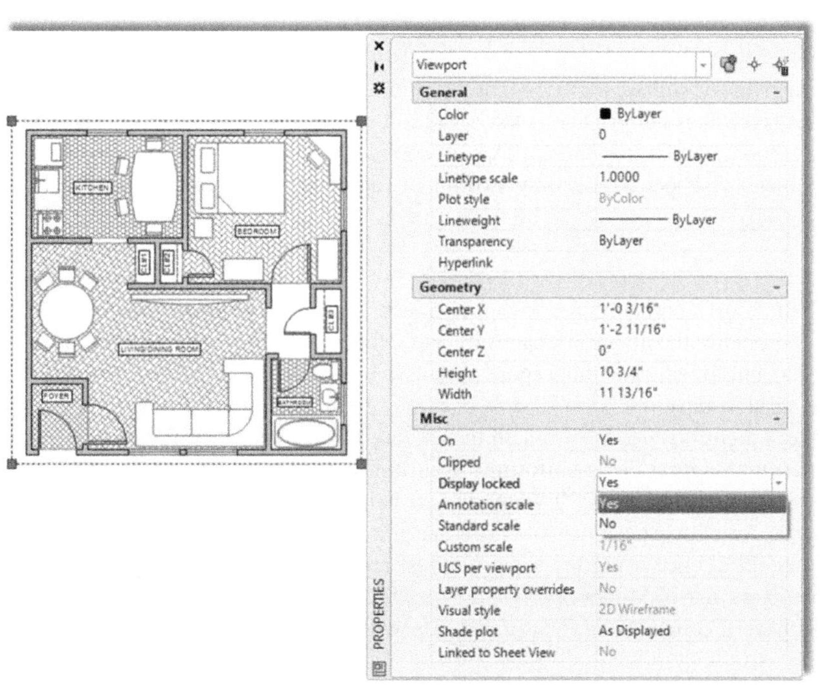

FIGURE 10.26 Locking the viewport.

Freeze VP

This next feature allows you to either "freeze" the viewport or simply prevent it from printing. The reason for this is that, generally, you do not want a big fat rectangle on your final plots, which will go to the client. This is especially true of the main layouts that feature just one large viewport. The visible viewport rectangle can be mistaken for part of the border and is unnecessary. Even with detail sheets, each of which may display many details, the trend is to just separate them adequately but not have any rectangles visible. If you "did not do it by hand in the old days," then there is no need to introduce it in AutoCAD.

Having said this, however, some designers prefer to see the viewports all the time and have them disappear only on printing. So you have several options in this regard.

First of all, we need to put the viewports on an appropriate layer if they are not there already. Designers generally create a_VP layer for them. Notice a trick here. The VP layer has an underscore in front of it. This forces the layer (or any layer that has that underscore) to rise to the top of the layer list, above the 0 layer. This makes it easy to find the title block and sheet or the VP layers, as long as you do not overdo it and stick too many up there. Go ahead and change the color of the layer to gray and freeze it. Then, finally, change all the viewports to that frozen_VP layer. If you did everything correctly, then the viewport boxes disappear, as shown in Fig. 10.27.

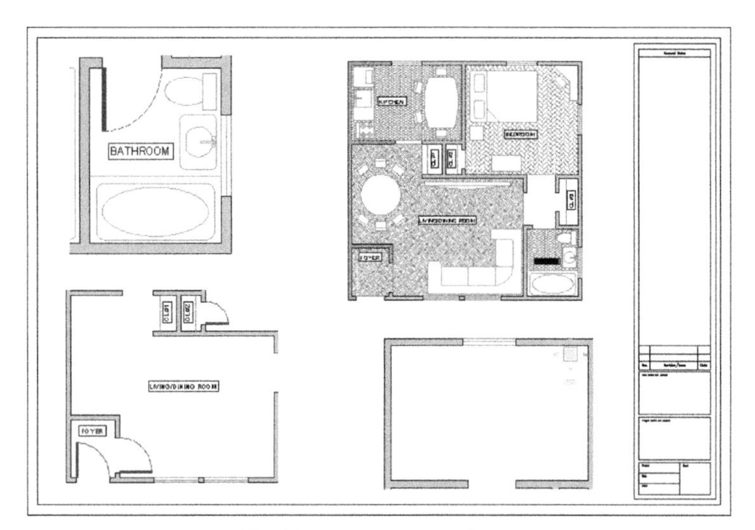

FIGURE 10.27 Viewports frozen.

There are other alternatives, of course, for those who want to see the viewports but not print them. You can put them on the Defpoints layer. This layer does not print (it is the definition points from dimensions). Another method is to just put a slash through the printer icon in the Layers dialog box corresponding to the viewport layer, which leads to the same result.

We are getting to the end of Paper Space; only two major topics are left: text and dims and the Annotation tools.

Text and Dims

This topic is probably the most confusing one in Paper Space and really requires the AutoCAD designer to not only understand everything thoroughly but also work carefully and deliberately. In practice, much of this is done only once and saved as part of a template, but for educational purposes, we assume you are being tasked with the text and dims setup from scratch and must know all the steps.

Our goal here is quite straightforward: We want all the text and dimensions to be the same size when printed, regardless of in what scale viewport they appear. This uniformity is a basic requirement of a professional drawing. Architects and engineers specify something to the effect of: "I want the text height to be ¼″ on all drawings, both in layouts and in all details," and it is up to the AutoCAD designer to ensure that this happens. With hand drafting, this was not an issue; the text and dims were drawn all at once using the same size test regardless of the drawing scale.

With AutoCAD, due to our use of Model Space and Paper Space, things are not so simple. The basic fact remains that all text and dimensions have a certain fixed measurable size when drawn in Model Space and appear as different sizes when viewed through the various scaled viewports, just as the design itself does. Some designers try to simplify everything by dimensioning the design on the actual sheet in Paper Space to achieve uniformity (AutoCAD lets you do this), but this is not advised. The dimensions and all text must remain with the design in Model Space. The reasons for this are many, not the least of which is that any movement of the design geometry in Model Space misaligns the corresponding annotation done in Paper Space.

So the text and dims need to stay where they are supposed to be, and we need to understand how to properly size text and dimensions based on the anticipated scale of the viewport in which they appear. This all really just boils down to a game of "match-up." The basic theory is as follows. You already know that viewport scale is a function of zooming. Therefore, when you have uniform text in Model Space and view the same set of text through the ¼″ = 1′−0″ versus the ½″ = 1′−0″ scale viewport in Paper Space, the text in the ½″ viewport appears two times as large, as the viewport is a larger scale. To counter this effect, you need to make sure that the text that appears in that ½″ viewport is drawn to be half as big, so it appears uniform with the ¼″ text (just as an example).

Here is another way to put it: imagine that we are going to print a floor plan at ¼″ = 1′−0″. The viewport scale for that plan would be set to ¼″ = 1′−0″ (or 1/48xp). Think of this as zooming a camera 48x away from the floor plan. If we want the labels for the rooms to print out at 1/8″ high, we would have to make the text 48X bigger, so when we zoom the viewport "camera" 48x away the text appears to be 1/8″ tall when printed; 48 x 1/8″ = 48/8″ or 6″. 6″-high text in model space yields 1/8″-high printed text in ¼″ = 1′−0″ viewports.

The trick is to be consistent and careful, so that the wrong size text does not appear with the wrong viewport scale. So, how is this done in practice? How do you know what size text and dims are appropriate for a particular scale? We have tables for this, such as the one shown in Fig. 10.28.

VIEWPORT SCALE	DRAWING TEXT HEIGHT			SCALE FACTOR	LT SCALE
	1/8	3/16	1/4		
Architectural					
1 1/2" = 1'-0"	1	1.5	2	8	3
1" = 1'-0"	1.5	2.25	3	12	4.5
3/4" = 1'-0"	2	3	4	16	6
1/2" = 1'-0"	3	4.5	6	24	9
3/8" = 1'-0"	4	6	8	32	12
1/4" = 1'-0"	6	9	12	48	18
1/8" = 1'-0"	12	18	24	96	36
1/16" = 1'-0"	24	36	48	192	72
1/32" = 1'-0"	48	72	96	384	144
Engineering					
1" = 10'	15	22.5	30	120	45
1" = 20'	30	45	60	240	90
1" = 30'	45	67.5	90	360	135
1" = 40'	60	90	120	480	180
1" = 50'	75	112.5	150	600	225
1" = 60'	90	135	180	720	270

FIGURE 10.28 Text and scale chart.

To use this chart, you have to know what scale viewports you are using. With the main layouts, such as the Architectural or Mechanical, done earlier, it is very easy. These layouts typically feature only one viewport of a known scale (⅛″ = 1′−0″ is commonly used for commercial architecture plans); therefore, if the architect desires all text output to be 3/16″ in height (also a common size), then by reading the chart from left to right until we meet the 3/16″ column, we arrive at a height of 18 inches, which is to be used for all text in that drawing. For a ¼″ = 1′−0″ scale detail view of the same plan, the text size is now 9 inches. We are zoomed in 2x closer, so the text is half as big to print at the same size.

With details, which may come in all the scales listed, the procedure is no different; you just have to be very careful in knowing ahead of time what scale viewport you are using, to use the appropriate text height. All this comes with experience, and after much practice, you will be able to quickly assess the necessary scale of a given detail to ensure that all aspects of it are visible and clear to whomever is reading the plans. This is the only factor in determining the appropriate text size.

The chart shows not only the text height but also the dimension sizes under the Scale Factor column. Here, you retrieve the appropriate size (24 in the ½″ = 1′–0″ example) and enter that value in the Use overall scale of: text field under the Fit tab of the DDIM dialog box. Also shown is the appropriate scale for linetypes (9 in the case of ½″ = 1′–0″). You need this to ensure you can see a hidden line, e.g., under this scale.

Most of the information contained in Fig. 10.28 needs to be entered into AutoCAD in the form of text styles and dim styles. This is the job of the CAD manager on initially setting up an architectural or engineering office's CAD standards. Yes, it is as tedious as it sounds; the chart is abbreviated and does not contain every scale used, so the work is substantial. Fortunately, it needs to be done only once, and a template is formed for all users to follow from there on out. Enforcing this usually turns out to be the hardest part of all, as the "system" is only as good as the people who follow it. You must be careful to use the proper style (Arial_.25, for example) and be sure to set that style as current while working on the relevant design or detail. The payoff comes later in the process, when you view the design in Paper Space. Uniformity and professionalism are the final goals. Go ahead and practice these concepts.

Step 1. Go to the Style dialog box and create three new text styles, all Arial fonts. They are set up for a final text height of ¼″ and are viewed through three viewports of the following scales:
½″ = 1′–0″ (therefore, size: 6″)
¼″ = 1′–0″ (therefore, size: 12″)
⅛″ = 1′–0″ (therefore, size: 24″)

Step 2. Go to the DDIM dialog box and create three new dim styles. All must use the appropriate text from the previous procedure (therefore, the ½″ = 1′–0″ dimension style uses the Arial_.5 style of 6″ height, etc.). Then, set the fit as shown on the chart and reproduced here:
½″ = 1′–0″ (therefore, Use overall scale of: 24″)
¼″ = 1′–0″ (therefore, Use overall scale of: 48″)
⅛″ = 1′–0″ (therefore, Use overall scale of: 96″)

Step 3. One by one, in Model Space, set each text and dimension style as current and create one set of text and one set of random dimensions next to each other, as shown in Fig. 10.29.

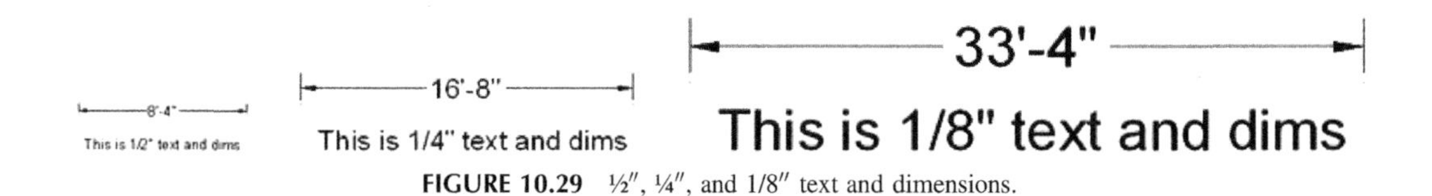

FIGURE 10.29 ½″, ¼″, and 1/8″ text and dimensions.

This may take you a few minutes to set up and execute. Be sure to work carefully, checking and rechecking all your settings. As expected, the three scales go up in size twice for each jump from ½″ to ¼″ to ⅛″. We now want to see them uniform in size in Paper Space.

Go into Paper Space and create three viewports next to each other. Use the tools learned previously to set the first one to the ½″ = 1′–0″, the second to ¼″ = 1′–0″, and the third to ⅛″ = 1′–0″. Pan around until the appropriate text and dims combination is in the respective viewport (you may have to go to Model Space and move them apart slightly). The result is shown in Fig. 10.30.

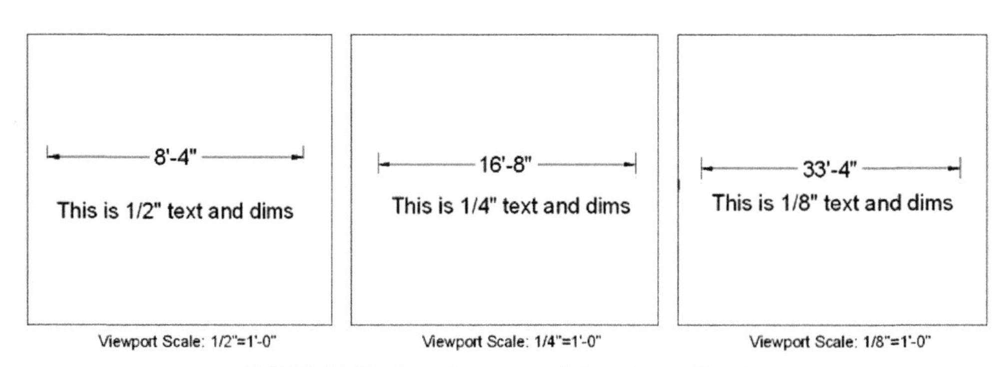

FIGURE 10.30 Paper Space text and dimensions uniformity.

Notice the complete uniformity of the text and dims in Paper Space, as opposed to how they looked in Model Space in Fig. 10.29. This is the essence of what we are doing here. If you can set up these three viewports and all the text and dims correctly, you can do any drawing. Be sure to review and learn everything thus far presented. Ask your instructor for assistance if needed.

Annotation

The final Paper Space feature we will discuss first appeared in AutoCAD 2008, called *Annotation*. Its purpose is to present an alternative approach to the methods of the previous section "Text and Dims." The new approach is different in that it allows you to set multiple scales for the same text, so the proper scales are visible in the proper viewports with no further input from the user. Several items can be annotated: styles, dimensions, blocks, and hatches. We explore annotative first with text, as outlined in the following procedure, and then present a more detailed example, this time with dimensions.

Open a brand-new file using ACAD.dwt, type in style, and press Enter. The familiar Style dialog box comes up. Create a new style, calling it *Arial_Anno*, font Arial. Check off the Annotative box under Size and leave the Paper Text Height as 0. The new text has a triangular symbol next to it, as seen in Fig. 10.31. Be sure to set it as Current.

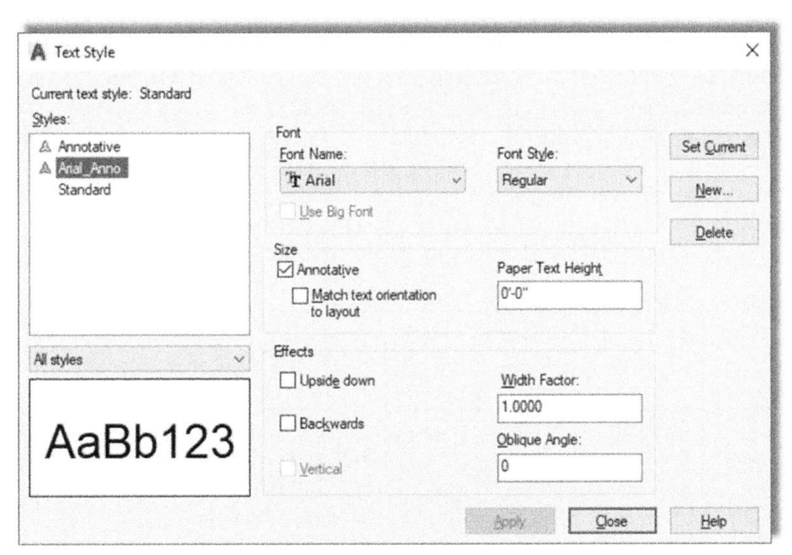

FIGURE 10.31 Annotative style.

Now switch to Model Space and create a line of text. The first time you do this, a dialog box, such as shown in Fig. 10.32, appears. Set the scale at which you anticipate this text to show up (such as inside a ¼″ = 1′—0″ viewport, as done here). Then press OK and finish typing the text.

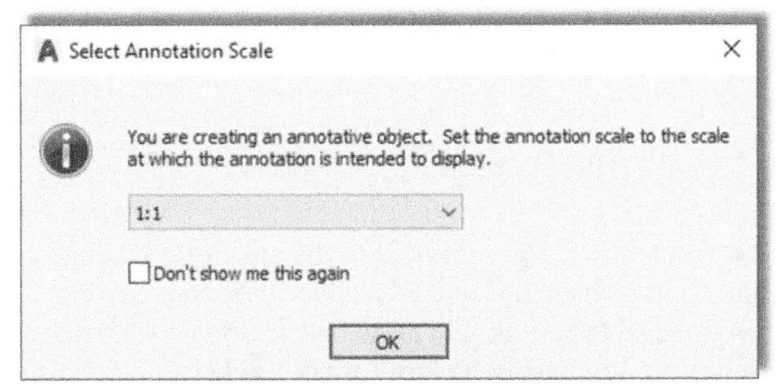

FIGURE 10.32 Select Annotation Scale.

Now, switch to Paper Space and create a viewport and set it to the ¼″ scale. Expand or contract it to fit nicely around the text, as shown in Fig. 10.33.

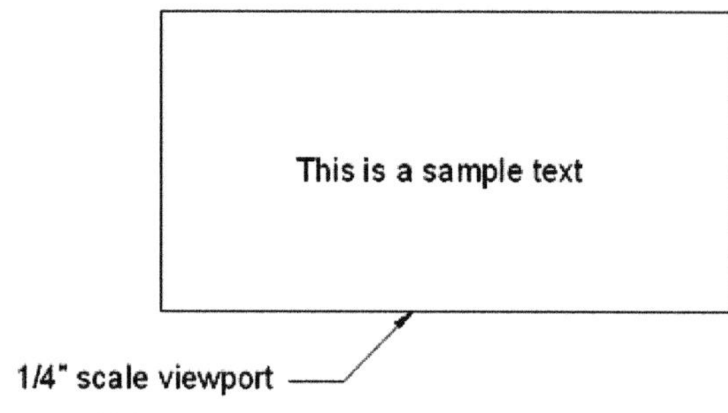

FIGURE 10.33 Annotation text in sample viewport.

Copy that viewport to the right and change its scale to ⅛″. The text briefly shrinks to a smaller size and then balloons back up to the same size you set (¼″), as shown in Fig. 10.34. Note that this only works with the AutoScale (annotative "lightning bolt," see Fig. 10.35) button turned on. If AutoScale is off, the text will disappear but only if the Annotation Visibility (annotative "light bulb," see Fig. 10.35) button is also off. If Annotation Visibility is on, the viewport will show smaller text because we zoomed farther away.

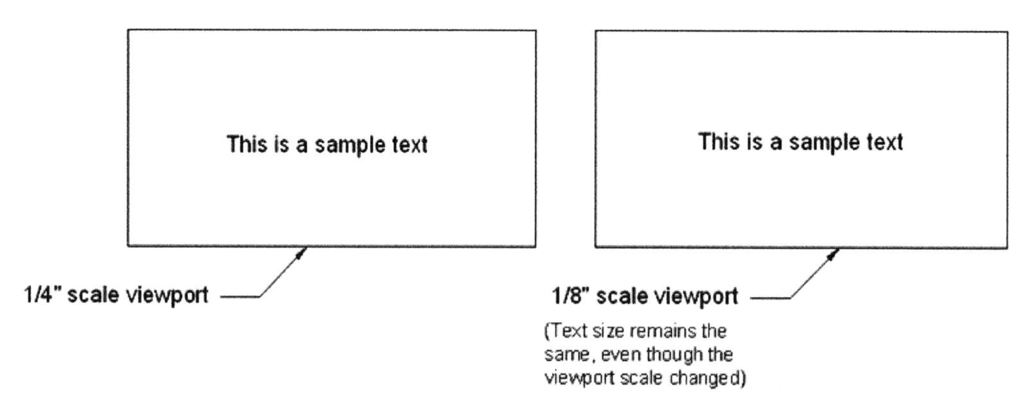

FIGURE 10.34 Annotation text in sample viewports.

This is the essence of annotation. The Annotation Scale is adjusted at the bottom right of the taskbar, as shown in Fig. 10.35, so when new text is being created in Model Space, you need to set its anticipated viewport scale before creating text or other annotative objects.

FIGURE 10.35 Annotation scale icons.

You can also set multiple annotation scales to the same text if you anticipate it being visible in two or more viewports set to different scales. In model space, click on the text and a "double" of it pops up (Fig. 10.36). This double is not a true copy but a temporary twin. With AutoScale turned on, go ahead and set the new annotation scale from the pop-up list in Fig. 10.35 and move the text to a new position using its grip (usually right on top of where the original was).

This is a sample text

This is a sample text

FIGURE 10.36 Annotation text copy.

Everything done here with annotative text can be done with dimensions, so let us go ahead and do one more example, this time with dimensions.

Step 1. Open Exercise 5 from Chapter 3. If you do not have it ready, go ahead and recreate it, leaving just the object (no dimensions or text) on your screen. It should look like what you see in Fig. 10.37.

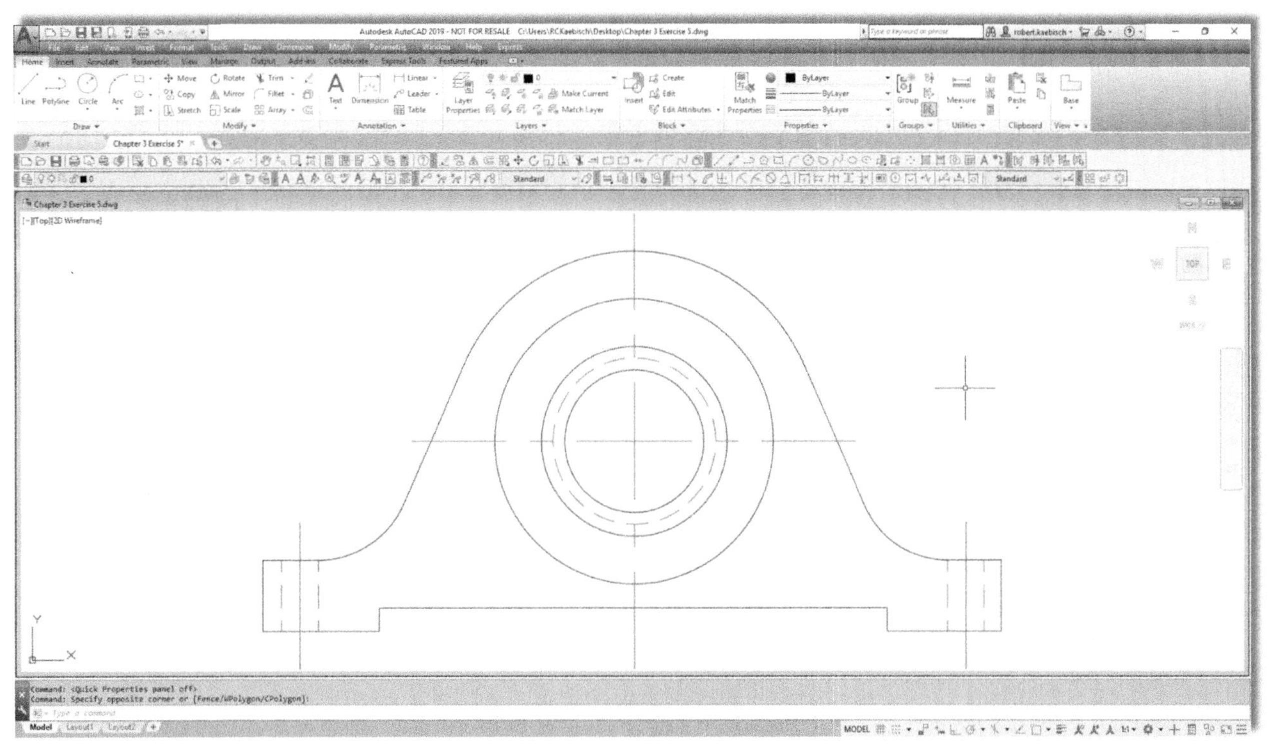

FIGURE 10.37 Exercise 5, Chapter 3.

Step 2. Set the *Annotative* dimension style current in the Home Tab > Annotation Panel drop-down as seen in Fig. 10.38.

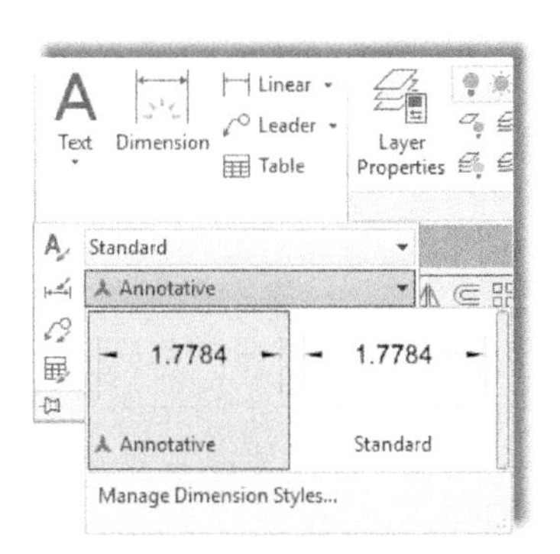

FIGURE 10.38 Set Annotative Dimension Style.

Step 3. Go to the Layout1 tab as your active view. Once in Layout1, erase the existing viewport that shows a view of your model space if it exists. You should now have a blank piece of paper, ready for new viewports as seen in Fig. 10.40. Note that in this screen shot the Paper Space has no border or sheet for clarity. Turn the Annotation Visibility and AutoScale off in the status bar (Fig. 10.39).

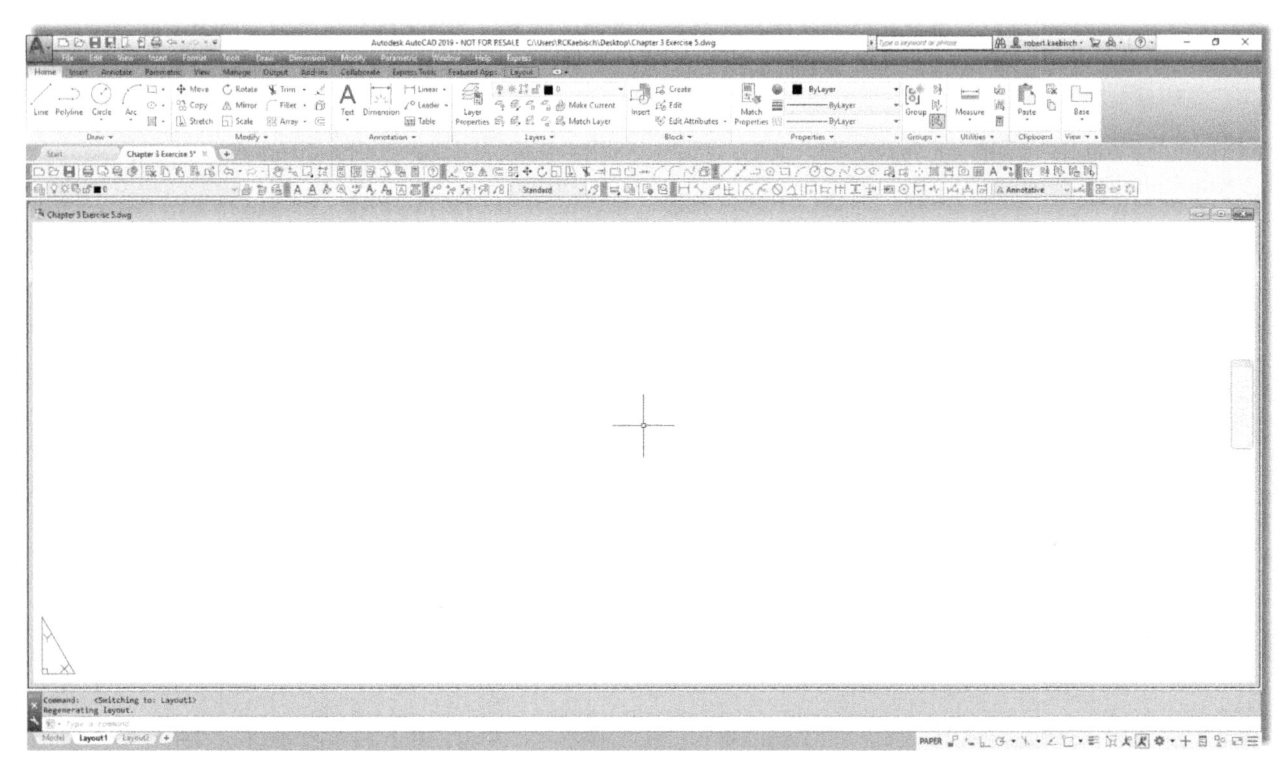

FIGURE 10.39 Blank Paper Space sheet.

Step 4. On the Layout tab of the Ribbon (a contextual tab that only appears when in a Layout/Paper Space), click the Viewports Rectangular button, and press "4" Enter for the create four viewports at once option, and then create an 11 × 8.5 rectangular viewport as seen in Fig. 10.40. The model will appear equally zoomed to extents in all four viewports.

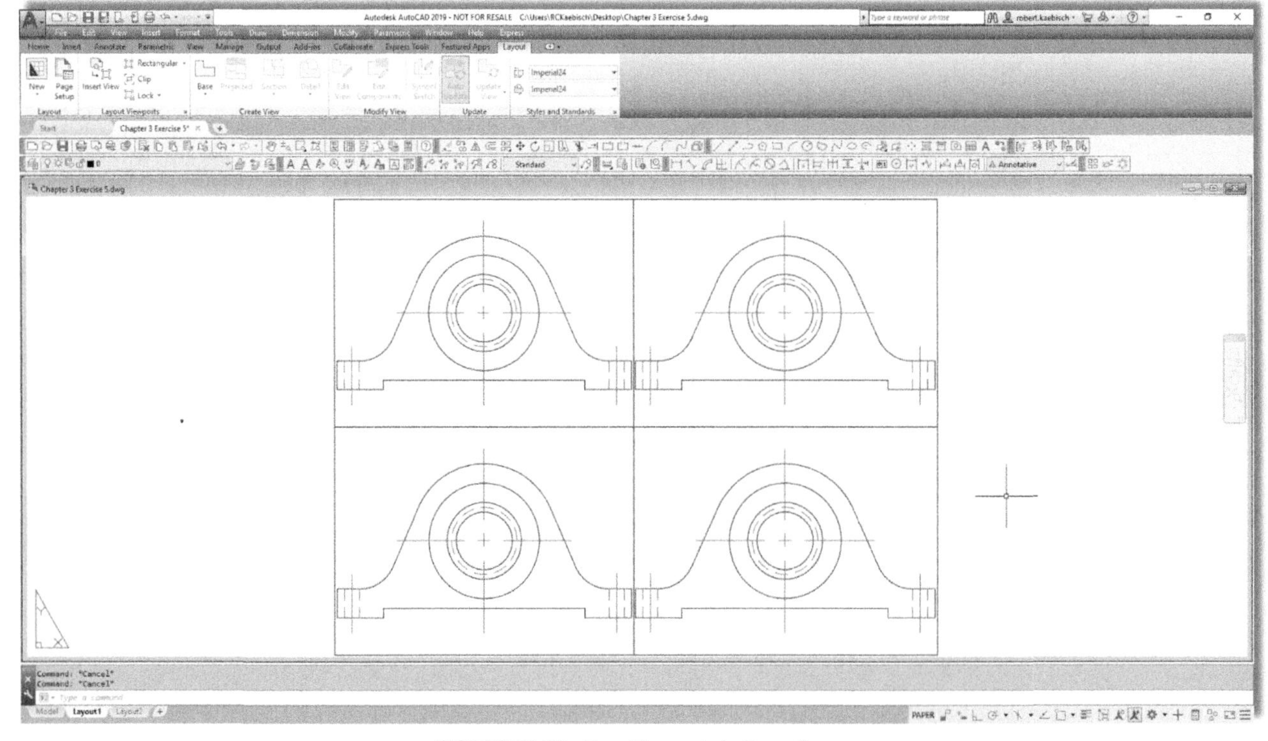

FIGURE 10.40 Four Viewports in Paper Space.

Step 5. Double-click within the top-left viewport to activate it. Then in the Options/Status bar (bottom right), find the viewport scale setting (it will show a decimal number) and click it. A lengthy selection menu will pop up. Set the scale to 1:2 (Fig. 10.41) and lock the viewport scale with the Viewport Lock button.

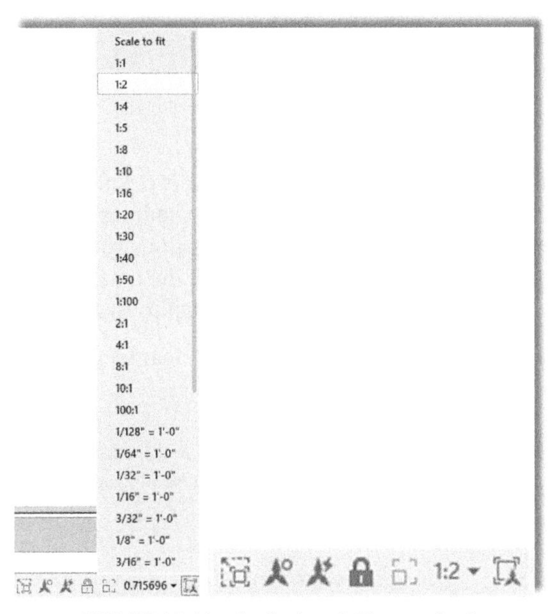

FIGURE 10.41 Set Scale and Viewport Lock.

Step 6. Click in the upper right viewport to activate it. Pan and zoom manually to focus on the lower left base of the part. Then, following the same procedure as Step 5, set the scale to 2:1 and lock the viewport scale.

Step 7. Click in the lower right viewport to activate it. Pan and zoom manually to focus on the lower right base of the part. Again, following the same steps, set the scale to 2:1, and lock the viewport scale.

Step 8. Finally, click in the lower left viewport to activate it. Pan and zoom manually to focus on the right quadrant of the center circles. Again, following the same procedure, set the scale to 4:1 and lock the viewport scale. Your result is shown in Fig. 10.42.

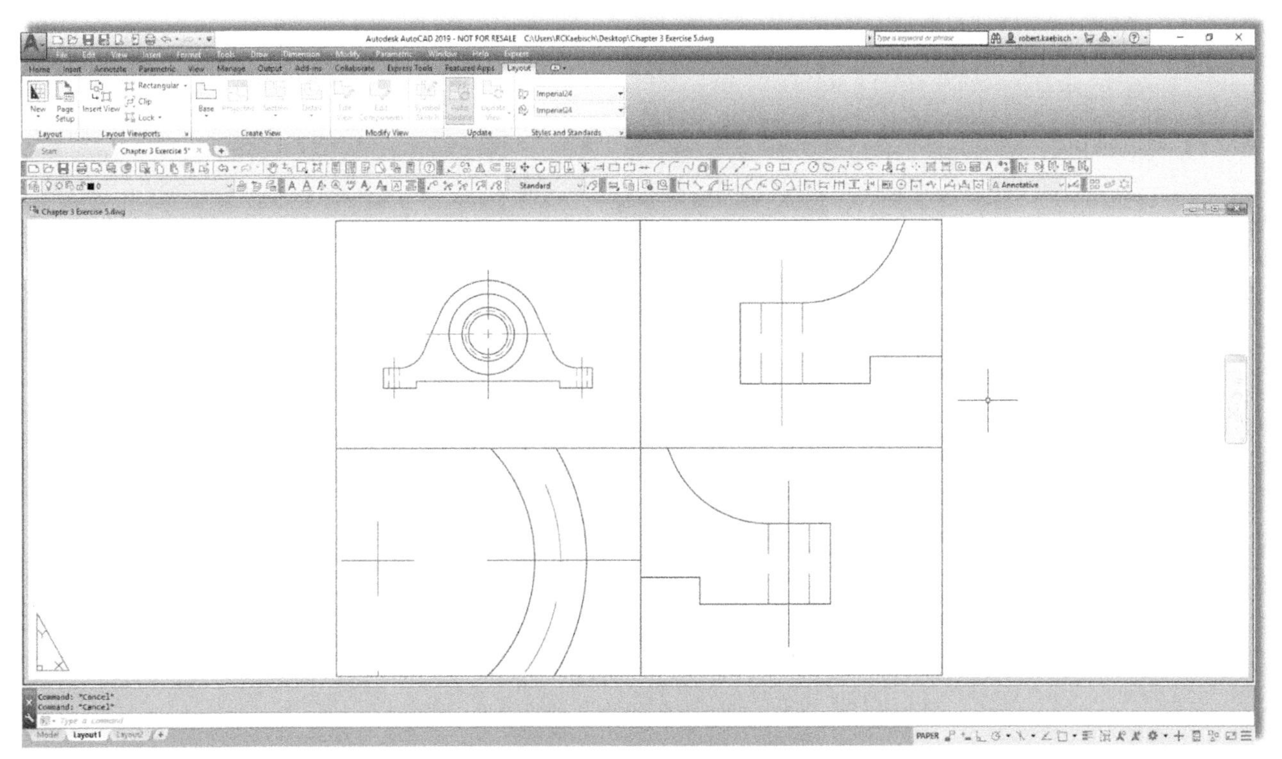

FIGURE 10.42 Final Viewport setup.

Step 9. Double check that the Dimension Style is set to Annotative and make sure both Annotation Visibility and AutoScale buttons are set to off (gray) as shown in Fig. 10.43 before we start adding dimensions through the Viewports.

FIGURE 10.43 Annotative settings.

Step 10. Add dimensions as follows by clicking into each viewport (result shown in Fig. 10.44):

- Using Linear Dimensions add an overall height and width, as shown, to the *upper left viewport.*
- Using Linear Dimension and Radius Dimension, add dimensions as shown to the *upper right viewport.*
- Using Linear Dimension and Radius Dimension, add dimensions as shown to the *lower right viewport.*
- Using Diameter Dimensions, add dimensions as shown to the *lower left viewport.*

When done, double-click outside of the boundary of the final viewport you are working on to get back onto Paper Space.

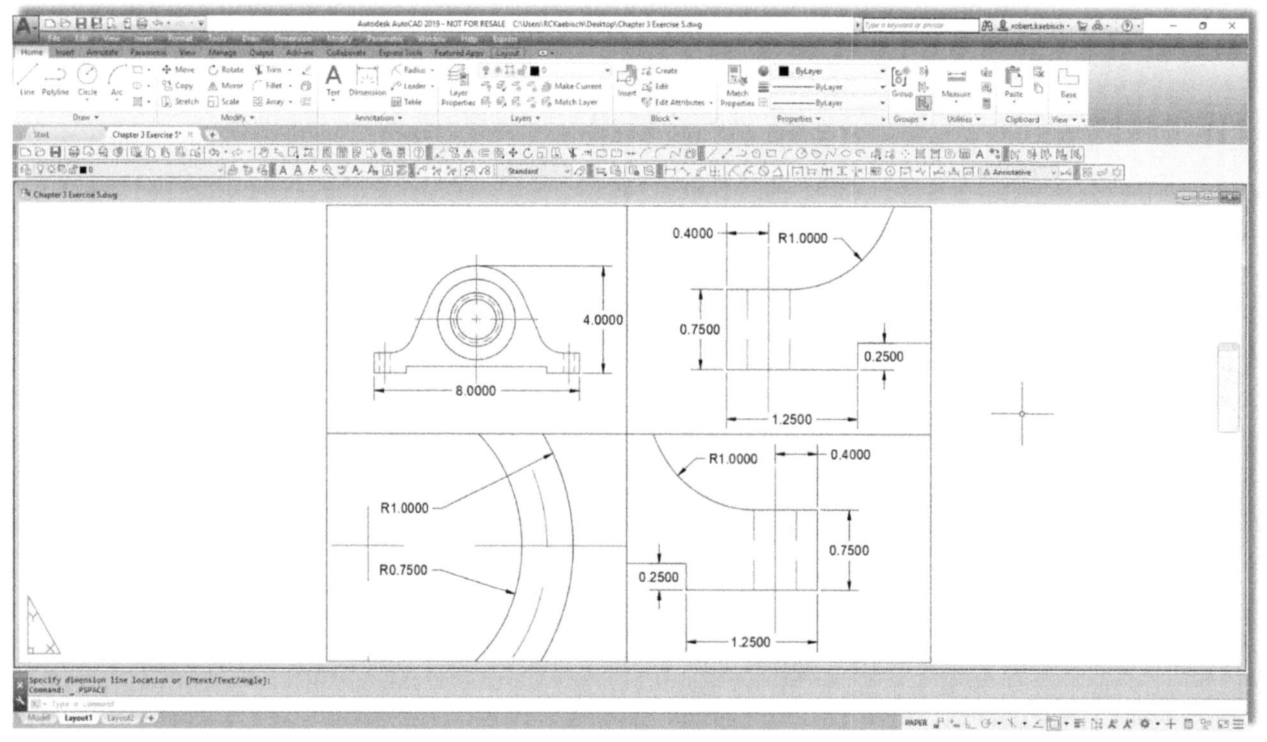

FIGURE 10.44 Final dimensioning.

So, let us do a quick analysis of what has just happened as it may not be obvious just yet. All your dimension features (text height arrow size, extension line gaps, etc.) are the same in each viewport even though the scale of each viewport is different in most of them. AutoCAD has essentially automated what would have been a tedious dimension setup process. This of course closely mirrors what you would have in hand drafting: no matter what scale of drawing you were producing, the annotation (text and dimensions) would be the same for all details.

To see how things "really are," click back into Model Space and turn on the Annotation Visibility button (blue is on as shown in Fig. 10.45). The Annotative Scale should remain at 1:1 for now.

FIGURE 10.45 Annotation Visibility on in Model Space.

You will see your full-size model with all the dimensions as "actual size" based on the viewport scale they were created in as seen in Fig. 10.46. Only the dimension features were scaled up or down according to the scale of the viewport (not the dimension distances or sizes). The closer the viewport was zoomed in to the object the smaller the dimension features became to account for that. The farther out the viewport was zoomed the bigger the dimension features became to account for that.

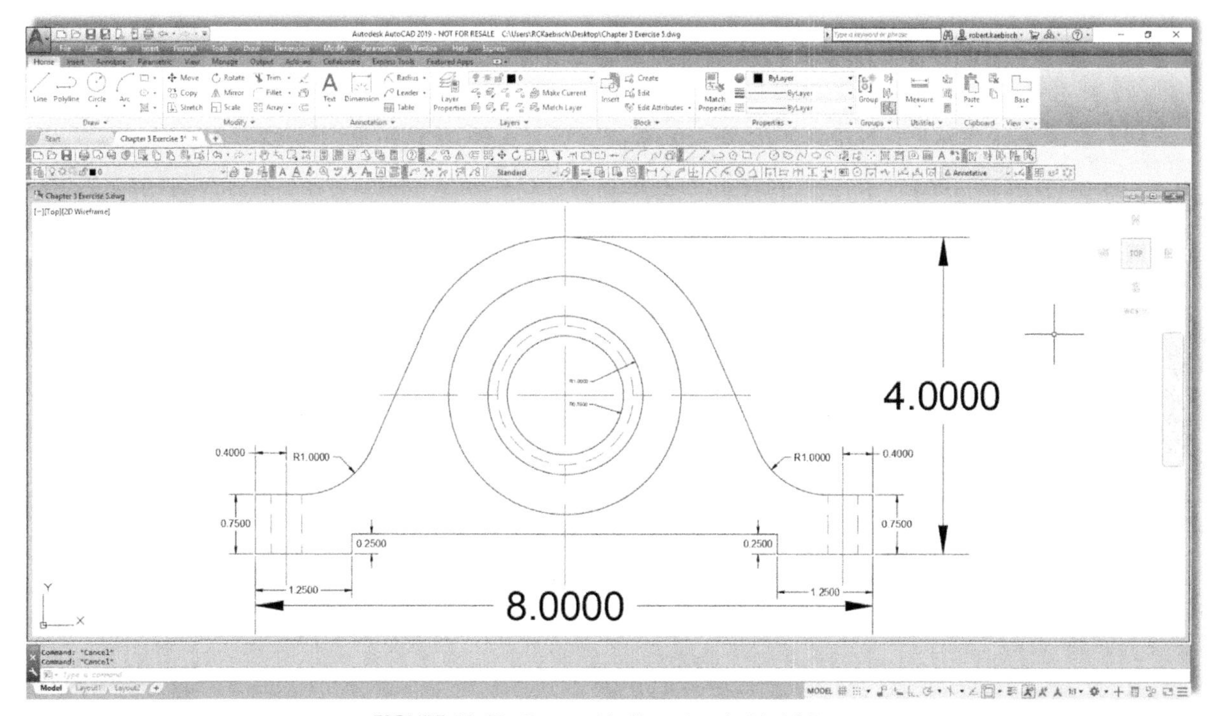

FIGURE 10.46 Image with dimensions in Model Space.

Now let us reverse that process and add Annotative Dimensions directly in Model Space. This would be a more typical procedure in industry because you would most likely be familiar with the scale of the viewport you were going to show your drawing in on the Layout/Paper Space. You would simply draw your objects full size as usual and then before adding any annotation, set your Annotative Scale for the anticipated viewport scale. Let us try that on the current drawing.

Step 11. Set the current Annotative Scale in Model Space to 1:2 and turn off Annotation Visibility (Fig. 10.47)

FIGURE 10.47 Annotation Visibility.

All dimensions should disappear except those created at the 1:2 scale. Now go ahead and add in additional dimensions as shown in Fig. 10.48. Note the new dimensions are shown bold and the existing dimensions are faded for clarity only.

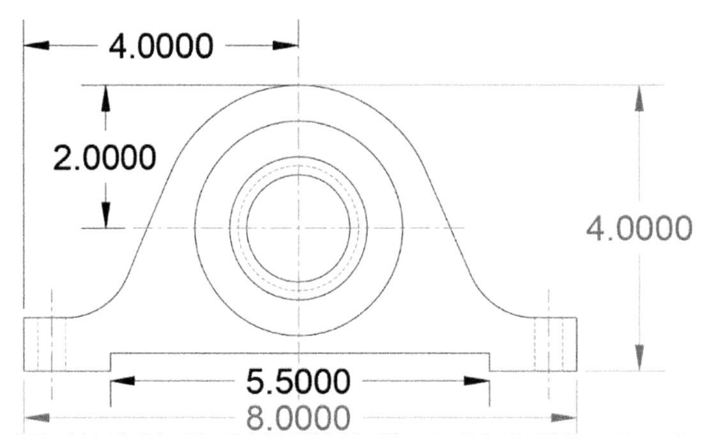

FIGURE 10.48 Additional dimensions added in at 1:2 scale.

Now change your Annotative Scale in Model Space to 2:1 and add some more dimensions as shown in Fig. 10.49. Note the new dimensions are shown bold and the existing dimensions are faded for clarity only.

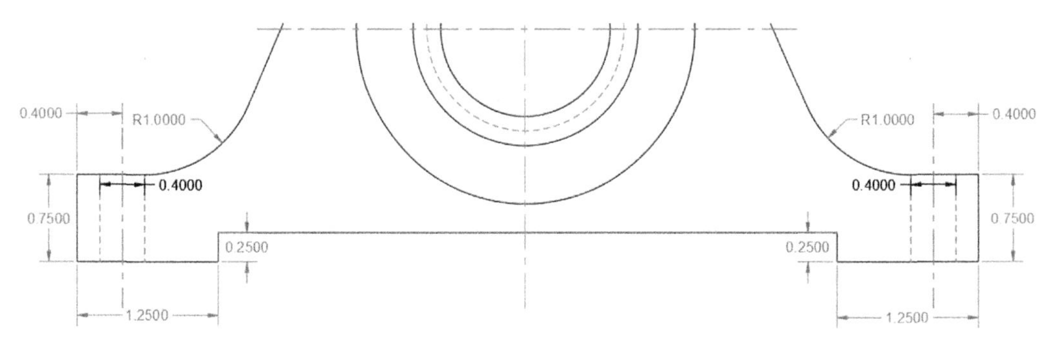

FIGURE 10.49 Additional dimensions added in at 2:1 scale.

Finally change the Annotative Scale in Model Space to 4:1 and add additional dimension as shown in Fig. 10.50. Note the new dimensions are shown in bold and the existing dimensions are faded for clarity only.

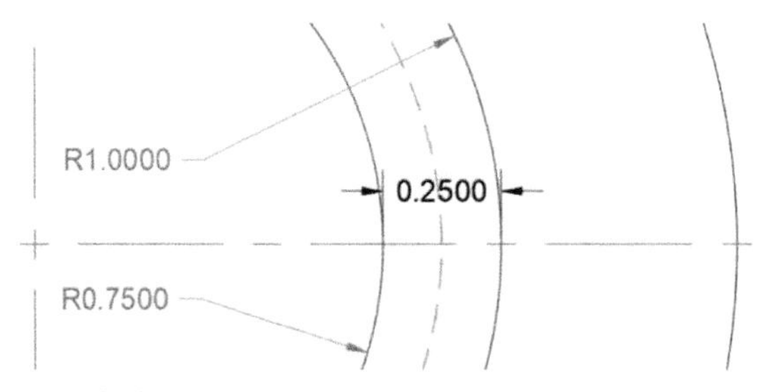

FIGURE 10.50 Final dimension added in at 4:1 scale.

Now go back to the Layout/Paper Space. You will notice that only the dimensions created at a scale that matches the viewport zooming are shown through that viewport (Fig. 10.51).

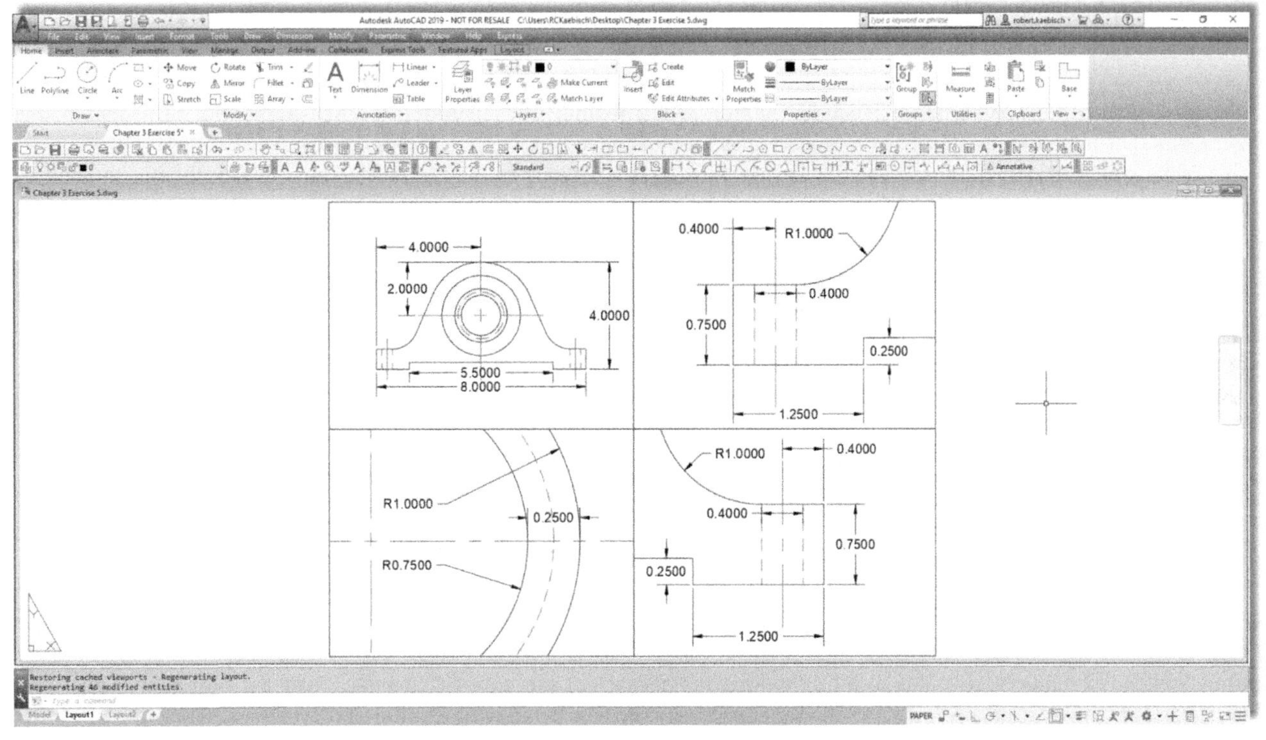

FIGURE 10.51 Final set of dimensions added in.

Now in Layout/Paper Space turn on the Annotation Visibility button as seen in Fig. 10.52.

FIGURE 10.52 Annotation Visibility turned on in Layout/Paper Space.

Now all dimensions at all annotative scales are shown in all viewports (Fig. 10.53). Not a desirable view, but it shows the drastic difference in the size of the dimension features required to show dimensions at different scales accurately as we saw in model space.

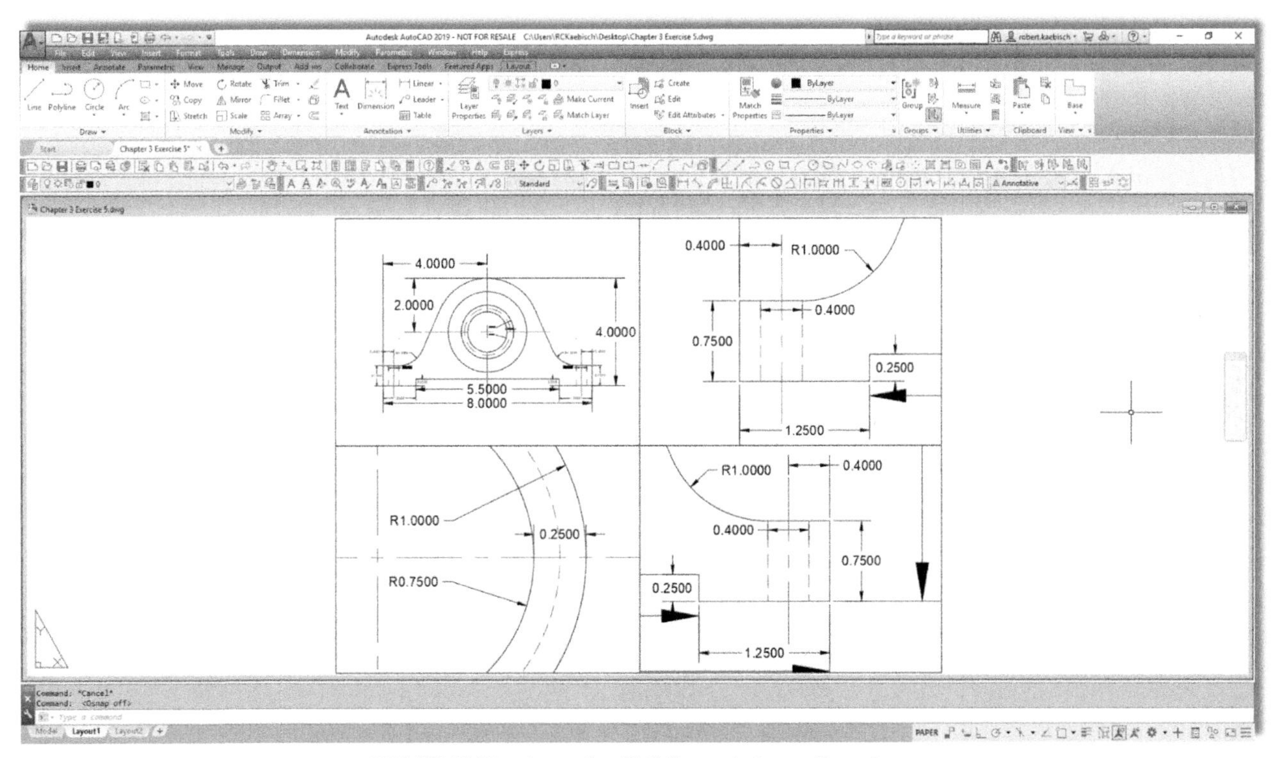

FIGURE 10.53 Annotation Visibility on in Layout/Paper Space.

Finally, turn off the Annotation Visibility button again to hide the "incorrect" dimensions and only display the dimensions that match the scale of the viewports they are displayed in.

So, what is the real benefit here? The first is that you could nearly get by with one annotative text style, one annotative dimension style and one annotative Mleader style for a given project. You also could nearly get by with only one layer for text, one layer for dimensions, and one layer for leaders because only the annotative objects that show through a viewport are the ones that match that viewport's scale.

When using annotative styles for text, dimensions and Mleaders in model space always draw your objects full size, 1:1 as usual. Then set the annotative scale to your desired printed scale/viewport scale before adding annotation. Changing the annotative scale in model space does not affect the drawing of your full size, 1:1 objects. If you are not sure of the annotation scale you will need for your final print, go ahead and make a viewport in your Layout/Paper Space sheet and set the scale desired. You can then go back to model space and set an annotation to match before adding your annotation or double-click in the viewport and start adding annotations.

If you choose to use non-annotative styles, you would have to create a separate style for each scale you would want to use. Then you would switch between styles and set a proper layer current before adding annotation at different scales. This would allow you to see the "properly" sized dimensions and turn off the layer for the "improperly" sized dimensions. In the end, annotative scaling of annotation created with annotative styles set properly can allow for fewer text styles, dimension styles, Mleader styles, and fewer layers to manage.

As discussed earlier, the AutoScale (Annotative Symbol with Lightning Bolt) button can be turned on (blue) to automatically add annotation scales to all annotative objects when the annotation scale changes. This means that you change the annotative scale any time and all annotative objects will be shown at additional scales including the ones that were created in initially. This can seem like a quick fix if you have not set your annotative scale correctly before adding text, dimensions, and Mleaders, but there are some dangers here. If trying to view an object, or parts of it at different scales through several viewports, you may start seeing these additional scaled annotations in viewports that you did not intend to see. If you need annotative objects to show in several viewports at different scales, it is a much safer choice to select those objects and use the quick Properties or Properties palette to add them.

SUMMARY

We go over a lot of information in this Paper Space chapter. Ideally, you have absorbed all of it and are able to use this valuable and important tool. Many use it improperly or not at all, and being an expert in Paper Space sets you apart from the crowd and adds a true level of professionalism to your designs. This summary serves as a capstone to this chapter and presents the big picture once again, now that the details have been mastered.

Do not think of Paper Space as a rigid or fixed system, where buttons are pushed and results pop out. While an element of this is present, the proper way of looking at it is as a flexible and loose collection of tools that allows you to shape and mold the presentation of your design in a way, i.e., efficient, clear, professional, and of course, pleasing to the eye. There is no *one* way to do things. Rather there is a constant tug of war between (sometimes conflicting) requirements.

For example, you have to consider

- *The size of the design*: This is the first and foremost concern. Your design takes priority over everything else, so focus on that in Model Space, creating your architectural layout or engineering device, as per client need.
- *The size of the available plotter and paper*: After the design is completed, output is considered. What do you have to work with? Most companies have a plotter capable of 36″ × 24″ plots. Some have the larger 48″ × 36″ plotters. Know your paper options. The larger plotters take E-sized paper, whereas the smaller ones only up to D size. You cannot print on what you do not have unless you outsource to a print shop.
- *The scale needed to clearly see all facets of the design*: This is the next important step. Take a look at how your design will look on the D- or E-size sheets. Can you see everything? Are any parts of the design not clearly visible? Generally, a 25′ × 25′ building (such as the one used as an example in this chapter) fits nicely on a D-size sheet at ½″ = 1′–0″ scale. If it is much bigger, then you can switch either to an E-size sheet (if possible) or to another scale. Another scale,

however, makes the small details of the design hard to see, so an option is to stay at the previous scale and "split" the drawing into halves or quarters and use match lines for alignment.

As you can see, many variables come into play in setting up Paper Space properly; and with experience, all this will be in your head at the start of a project and you will know instinctively what paper, scale, and viewports to go with, based on the size and expected overall shape of the design.

As of completion of this chapter, you have graduated from a beginner to an intermediate level of AutoCAD knowledge. If you have understood everything up to now and have not skipped any exercises or reading, then great job! If all you need AutoCAD for is basic drafting or just knowing enough to oversee other full-time AutoCAD designers, then you may stop here, you have what you need. If your work requires expert level knowledge, you need to go on to Level 2, Chapters 11–20. See you there!

You should understand and know how to use the following concepts and commands before moving on to Level 2:

- Paper Space
- Layouts
- Viewports
- Scaling a design inside of a viewport
- Working with layers in individual viewports
- Scaling text and dims in viewports
- The Annotation concept

REVIEW QUESTIONS

Answer the following based on what you learned in this chapter:

1. What is Paper Space?
2. What are layouts?
3. How do you create, delete, rename, and copy layouts?
4. What are viewports?
5. How do you create, scale, and shape viewports?
6. How do you scale a design inside a viewport?
7. How do you turn layers on and off in individual viewports?
8. How do you scale text and dims in viewports?
9. Describe the Annotation concept.

EXERCISES

1. Set up a total of five title blocks in Paper Space using the Tutorial-iMfg.dwt template file. Name the tabs as seen next and delete any extra tabs. (Difficulty level: Easy; Time to completion: <10 min)
    - Front View
    - Side View
    - Section A-A
    - Section B-B
    - Section C-C

2. Bring up the floor plan you did in Chapter 4, Text, Mtext, Editing, and Style, Exercise 7 (or redraw it). Then, go into Paper Space and create an Architectural Layout tab, followed by several viewports, as in the top figure that follows. Assign each of the viewports a scale as shown in the bottom figure. Freeze the dims layer in the two detail viewports. Note that the floor plan has been rotated 90 degrees clockwise for this exercise. (Difficulty level: Easy; Time to completion: 10–15 minutes if floor plan is completed, 55–75 if not.)

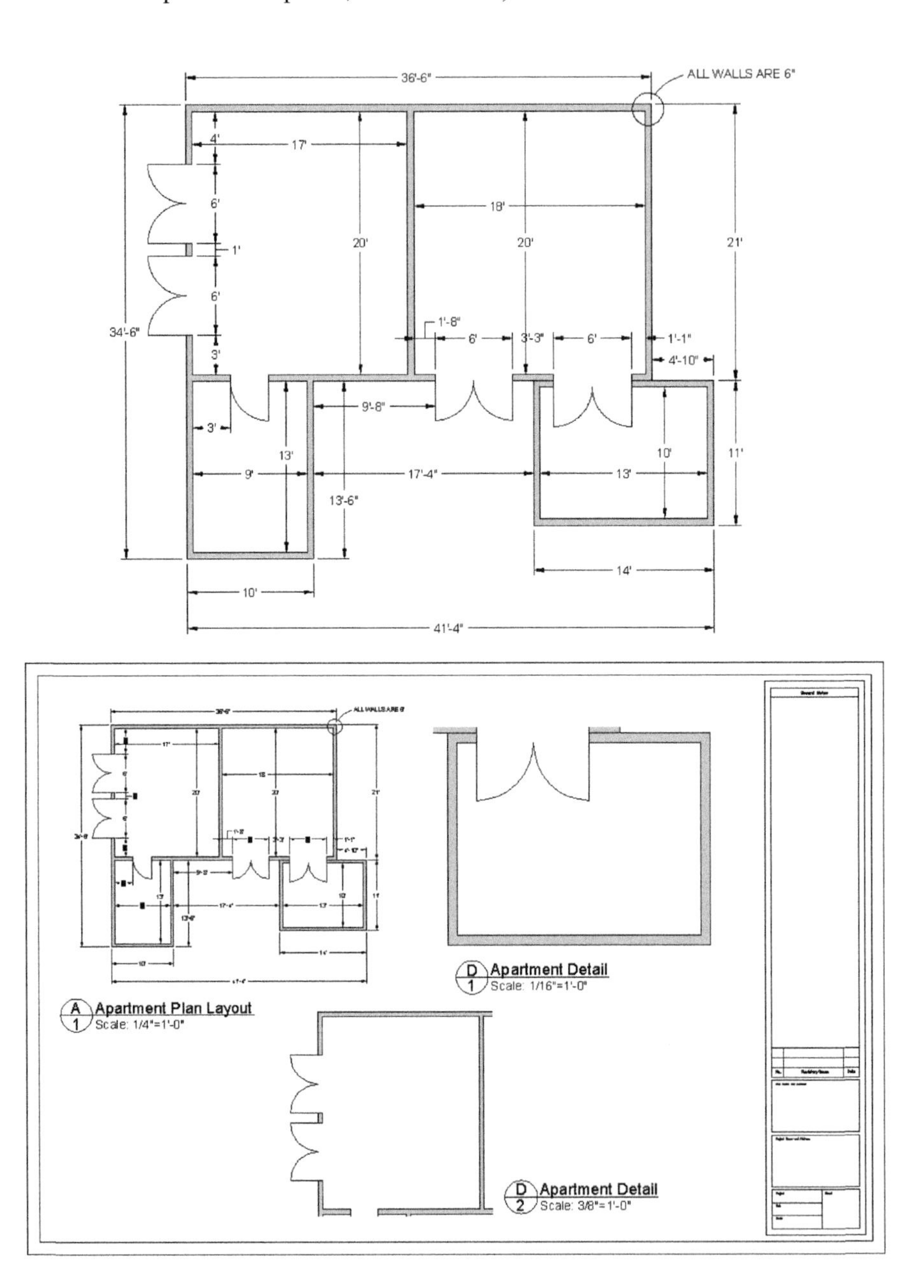

Level 1. Answers to Review Questions

Chapter 1 Review Questions

1. List the four Create Objects commands covered.
   Line, circle, arc, and rectangle.
2. List the 10 Edit/Modify Objects commands covered.
   Erase, move, copy, rotate, scale, trim, extend, offset, mirror, and fillet.
3. List the three View Objects commands covered.
   Zoom, pan, regen.
4. What is a workspace?
   A workspace is a customized screen layout.
5. What four main methods are used to *issue a command* in AutoCAD?
   Command line, toolbars, cascading menus, and the Ribbon.
6. What is the difference between a selection Window, Crossing, and Lasso?
   A Window is a solid rectangle produced by moving the mouse to the right of an imaginary line and selects only what is completely inside of it. A Crossing is a dashed rectangle produced by moving the mouse to the left of an imaginary line and selects anything it touches. A Lasso can behave like either a Window or Crossing, but is a free-form shape, not limited to a rectangle/square. It is invoked via a click-and-drag technique.
7. What is the name of the command that forces lines to be drawn perfectly straight, horizontally and vertically? Which F key activates it?
   Ortho, F8.
8. What six main OSNAP points are covered? Draw the snap point's symbol next to your answer.
   Endpoint, Midpoint, Intersection, Center, Quadrant, and Perpendicular. See the text for the symbols.
9. What command brings up the OSNAP dialog box?
   OSNAP.
10. Which F key turns the OSNAP feature on and off?
    F3.
11. How do you terminate a command-in-progress in AutoCAD?
    Esc key.
12. What sequence of commands quickly clears your screen?
    Erase → All, Enter, Enter.
13. What sequence of commands allows you to view your entire drawing?
    Zoom → E. Also double-click center mouse wheel (if equipped).
14. How do you undo a command?
    Type in u for undo or use the Undo arrow in the Standard toolbar.
15. How do you repeat the last command you just used?
    Press the space bar or Enter.

Chapter 2 Review Questions

1. How do you activate grips on AutoCAD's geometry?
   Click on any of the objects.
2. How do you make grips disappear?
   Press the Esc key.
3. What are grips used for?
   To modify the shape or position of an object.
4. What command brings up the Drawing Units dialog box?
   Units.
5. The snap command is always used with what other command?
   Grid.
6. Correctly draw a Cartesian coordinate system grid. Label and number each axis appropriately, with values ranging from −5 to +5.
   See the text, Fig. 2.4.
7. On the preceding Cartesian coordinate system grid, locate and label the following points: (3, 2) (4, 4) (−2, 5) (−5, 2) (−3, −1) (−2, −4) (1, −5) (2, −2).
   Student activity.
8. What are the two main types of geometric data entry?
   Dynamic Input and Manual Input.
9. Describe how Dynamic Input works when creating a line at an angle.
   With DYN activated, begin drawing a line, entering a distance value in the first highlighted box. Then, tab over to the second highlighted box, entering an angle value.
10. List the six Inquiry commands covered.
    Area, distance, list, ID, radius, and angle.
11. What command simplifies a rectangle into four separate lines?
    Explode.
12. What is the difference between an inscribed and a circumscribed circle in the polygon command?
    Inscribed means the polygon appears inside a given circle, with vertices tangent. Circumscribed means the polygon appears outside a given circle, with the sides tangent.
13. What is a chamfer as opposed to a fillet?
    A chamfer creates an angled edge, whereas a fillet creates a smooth circular edge or none at all.
14. What template should you pick if you want a new file to open?
    acad.dwt.
15. What is the idea behind limits?
    Limits define the outer boundaries of your work area.
16. Why should you save often?
    Because all software and computers are subject to crashes and malfunctions.
17. What is the graphic symbol for save?
    A computer floppy disk.
18. How do you get to the Help files in AutoCAD?
    By pressing F1 at any time.

Chapter 3 Review Questions

1. What is a layer?
   Layers is a concept that allows you to group drawn geometry in distinct and separate categories according to similar features or a common theme.
2. What is the main reason for using layers in AutoCAD?
   To maintain control over your drawn geometry.
3. What typed command brings up the Layers dialog box?
   Layer or just la.
4. List two ways to create a new layer.
   New Layer icon or, once a new layer exists, just press Enter while selecting any layer.
5. How do you delete a layer?
   The red X.

6. How do you make a layer current?
   Double-click on it or select it and press the green check mark symbol.
7. What three available Color tabs are available?
   Index Color, True Color, and Color Books.
8. What does freezing a layer do?
   It makes all objects on that layer invisible.
9. What does locking a layer do?
   It makes all objects on that layer uneditable.
10. What are some of the linetypes mentioned in the text?
    Hidden, Center, Dashed.
11. What are three ways to bring up the Properties toolbar?
    Double-click on any object, press Ctrl + 1, or type in `ch`.
12. What two other ways to change properties or layers are discussed?
    Match Properties and Layers toolbar.

Chapter 4 Review Questions

1. What are the two types of text in AutoCAD? State some properties of each, and describe when you would likely use each type.
   Text: single line, limited formatting. Mtext: paragraph form, advanced formatting.
2. What is the easiest method to edit text in AutoCAD? What was another typed method?
   Double-click on the text. `Ddedit`.
3. What feature allows you to add extensive symbols to your mtext?
   Character Map.
4. What are the three essential features to input when using Style?
   Font name, font, and size.
5. What is the width factor?
   How wide a font is; 1 is standard width.
6. What is an oblique angle?
   The "tilt" of the font.
7. How do you initiate spell check?
   Right-click on the underlined misspelled word, or type in spell.
8. What new OSNAP is introduced in this chapter?
   Nearest.
9. How do you get rid of the UCS icon?
   Type in `ucsicon`, *then* `off`.
10. What can you do if you cannot see the dashes in a Hidden linetype?
    Use ltscale.

Chapter 5 Review Questions

1. List the four steps in creating a basic hatch pattern.
   Pick hatch pattern, indicate where you want it to go, fine-tune the pattern via scale and angle, and preview pattern.
2. What two methods are used to select where to put the hatch pattern?
   Directly picking an object or picking a point inside that object.
3. What commonly occurring problem can prevent a hatch being created?
   A gap.
4. What is the name of the tool recently added to AutoCAD to address this problem?
   Gap tolerance.
5. How do you edit an existing hatch pattern?
   Double-click on it.
6. What is the effect of exploding hatch patterns? Is it recommended?
   The hatch breaks up into individual components. No.
7. What especially useful pattern to fill in walls is mentioned in the chapter?
   Solid Fill.

Chapter 6 Review Questions

1. List the 11 types of dimensions discussed.
   Linear, Aligned, Diameter, Radius, Angular, Continuous, Baseline, Leader, Arc length, Ordinate, and Jogged.
2. What command is best for editing dimension values?
   Double-click.
3. What is the difference between natural and forced dimensions?
   Natural dimensions show the actual value, whereas forced dimensions are added manually and may be false values.
4. How do you restore a natural dimension value when you forgot what it was?
   Type in <>.
5. List the four items that needed to be customized in our dimensions.
   Units, font, arrowheads, and fit.
6. What command brings up the Dimension Style Manager?
   Dimstyle or ddim.
7. What type of arrowhead do architects usually prefer?
   Architectural tick.

Chapter 7 Review Questions

1. What are blocks and why are they useful?
   A block is a collection of objects grouped together under a common name. They are useful to keep multiple objects together and for reuse via symbol libraries.
2. What is the difference between blocks and wblocks?
   A block is internal to a file, and a wblock is external.
3. What are the essential steps in making a block? A wblock?
   For a block, you need to specify a name, select objects, and pick a base point. For a wblock, you also need to specify a path.
4. How do you bring the block or wblock back into the drawing?
   Via the insert command.
5. Explain the purpose of purge.
   Purge rids the drawing of unseen elements present in the file, such as blocks, layers, or fonts.
6. Explain the basic idea behind dynamic blocks. What steps are involved? What is visibility?
   Dynamic blocks are intelligent blocks that can be modified without being redefined. To create a dynamic block, you add parameters and actions. Visibility allows you to combine many blocks together and control which is visible at any one time via visibility states and a drop-down menu contained in the main block.
7. What are the advantages and disadvantages of using groups?
   Advantages of a group: A group can have objects added and subtracted from it and can have individual objects (and their properties) edited. Disadvantages of a group: A group cannot be transferred between drawings (for "local" use only), and an update to one group does not update other groups with the same name.

Chapter 8 Review Questions

1. List the three types of arrays available to you.
   Polar, rectangular, and path arrays.
2. What steps are needed to create a polar array?
   You need to specify what object(s) to array, what the center point of the array is, how many objects to create, and whether you want the pattern to fill out a full circle or part thereof.
3. To array chairs around a table, where is the center point of the array?
   The center of the table.
4. What steps are needed to create a rectangular array?
   You need to specify what object(s) to array, how many rows and columns, and the distance between those rows and columns.
5. What steps are needed to create a path array?
   You need to specify what object(s) to array, the path, and how many objects to create.

Chapter 9 Review Questions

1. List the seven essential sets of information needed for printing or plotting.
   What printer or plotter to use, what paper size to use, what area to plot, at what scale to plot, what pen settings to use, what orientation to use, and what offset (if any) to use.
2. List the five paper sizes and their corresponding measurements in inches.
   A size: 11″ × 8.5″ sheet of paper (letter), B size: 17″ × 11″ sheet of paper (ledger), C size: 22″ × 17″ sheet of paper, D size: 36″ × 24″ sheet of paper, and E size: 48″ × 36″ sheet of paper.
3. What is the difference between printing and plotting?
   Printing usually refers to the output of drawings on A- or B-sized paper; and plotting usually refers to the output of drawings on C-, D-, or E-sized paper.
4. List the four plot area selection options.
   Extents, Window, Display, and Limits.
5. What scale setting ignores scale completely?
   Fit to paper.
6. What is the extension of the pen settings file?
   ctb.
7. Explain the importance of Preview before printing and plotting.
   Mistakes can be spotted before a print is sent to the printer or plotter, thereby saving paper.
8. Why is the Page Setup Manager useful?
   Because it allows you to set up and save print and plot settings for future use.

Chapter 10 Review Questions

1. What is Paper Space?
   Paper Space is an AutoCAD concept that allows you to draw your design in full size (1:1) units and still print it to scale using viewports.
2. What are layouts?
   Layouts are Paper Space pages that hold the title block and viewport.
3. How do you create, delete, rename, and copy layouts?
   All these tools are accessed by right-clicking on the Layout tabs.
4. What are viewports?
   A viewport is an AutoCAD concept that allows you to link together Model and Paper Space.
5. How do you create, scale, and shape viewports?
   You can create a viewport via the mview command. Scaling and shaping it are accomplished via the viewport grips.
6. How do you scale a design inside a viewport?
   You can use the icons on the Viewport toolbar or via keyboard entry using scale, then 1/xp.
7. How do you turn layers on and off in individual viewports?
   By clicking inside a viewport, calling up the Layer Manager, and freezing the layers under the VP Freeze column.
8. How do you scale text and dims in viewports?
   By using the appropriate charts and matching the text and dim sizes with the correct scale.
9. Describe the Annotation concept.
   Annotation gives text, dims, and other AutoCAD features the ability to scale to viewports automatically based on the assigned scaling of those viewports.

Level 2. Chapters 11–20

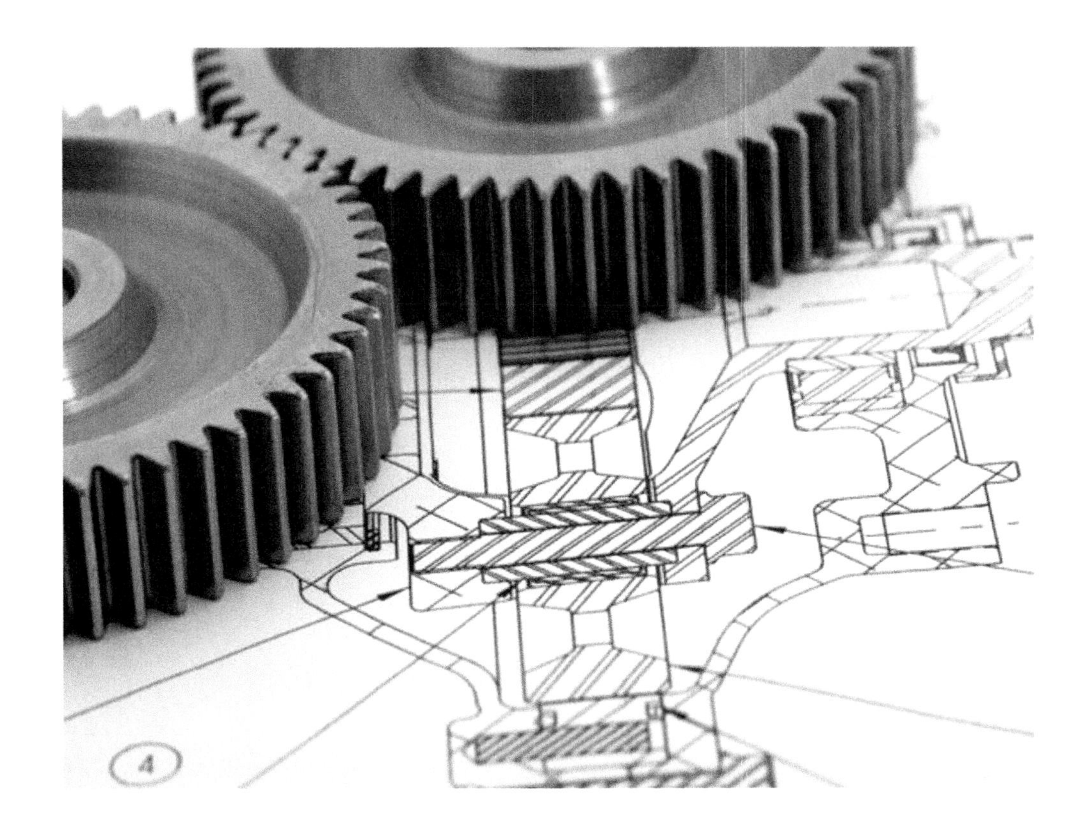

Now that you have completed Level 1, beginner to intermediate, you should have the basic skills necessary to work on a drawing of moderate complexity with a certain amount of speed and accuracy. Level 2 is the second part of your journey in becoming proficient in AutoCAD. The goal at the end of this level is for you to learn most of the remaining practical information about AutoCAD 2018 in two dimensions. This will allow you to not only work on very complex drawings but also set up and manage them. Then, with some additional work experience, you will have the skills of a CAD manager or administrator, a necessity in many engineering or architectural firms, even if your job is not just AutoCAD.

Starting with the next chapter, this textbook does not feature any of the basic introductions concerning AutoCAD and what it is. It is assumed you have already read through all the introductory information and are familiar with all the material in Chapters 1–10. There is minimal explanation or review of earlier material, so if you need to refresh anything from Level 1, do so before proceeding. With this level, we dive into the new material right away and do not look back, as there is much more to cover.

Level 2 consists of several parts. Chapters 11–13 are all about advanced features of basic concepts: lines, layers, and dimensions. We add significant depth to these Level 1 topics. Chapters 14–16 cover a broad set of topics that introduce various advanced features, such as interacting with other software, the Design Center, Express tools, the CUI, and many shorter topics. This clears the way for some remaining "big ideas" in AutoCAD: external references, attributes, and advanced output and pen settings, which make up Chapters 17–19. These are very important topics, especially external references, as they are critical to many architectural designs. We complete Level 2 with isometric design, which is a very useful tool in its own right and a great lead-in to 3D if you go on to learn that aspect of AutoCAD.

The pace of the course picks up somewhat as well, and more responsibility is placed on you to follow up and do your own research into additional AutoCAD functions. We also introduce a drawing project that spans the entire level from Chapters 11–20. It is a more in-depth floor plan that, while still greatly simplified, attempts to replicate a real-life architectural layout. There also are fewer generic drawings to do, in favor of more topic-specific exercises at the end of each chapter. One constant remains: You still need to practice to ensure success.

Chapter 11

Advanced Linework

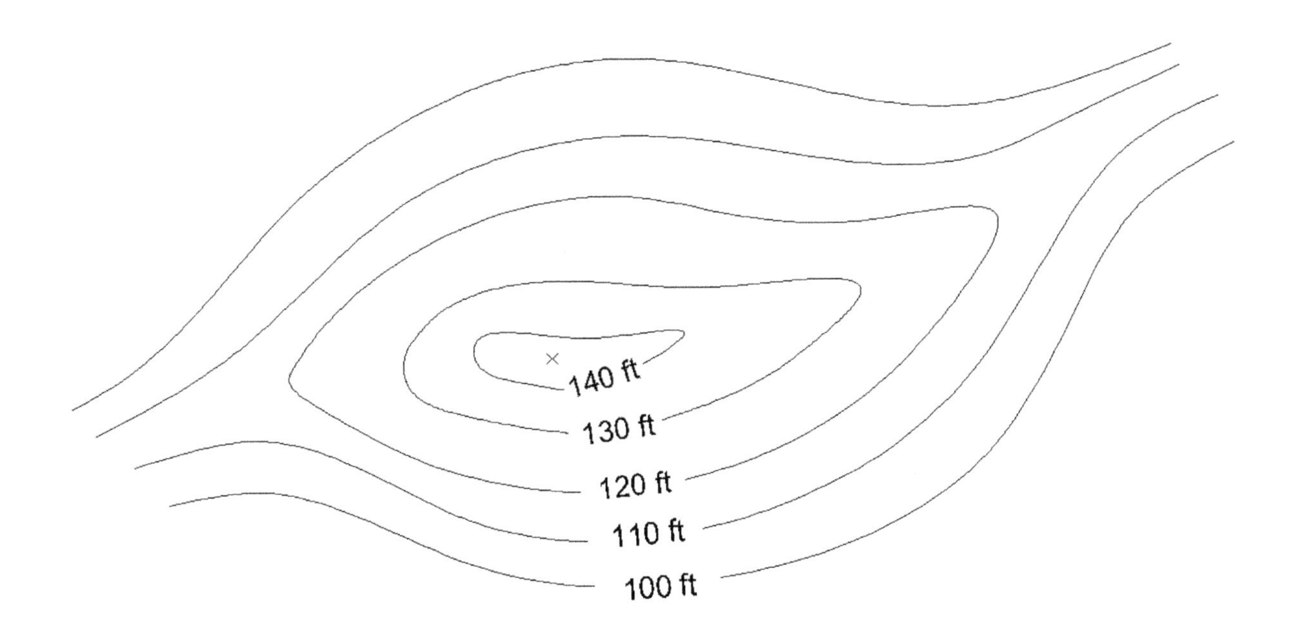

Learning Objectives

In this chapter, we revisit the line concept and expand on it by discussing the following:

- Polyline and pedit
- Bpoly
- Xline
- Ray
- Spline
- Mline, mlstyle, and mledit
- Sketch

By the end of this chapter, you will learn how to use and apply these new types of lines and be able to incorporate them into future design work.

Estimated time for completion of this chapter: 2–3 hours.

11.1 INTRODUCTION TO ADVANCED LINEWORK

We begin Level 2 in the same way we started Level 1, with the simple line command. At the time, this was enough to start with, and no mention was made of the many other types of lines AutoCAD has. Classroom experience had shown that introducing all the other types of lines at that stage was a sure way to scare off brand-new students, and the other lines were not needed anyway. Now that you are more advanced, we can come back and look at lines in depth.

Up and Running with AutoCAD 2019. https://doi.org/10.1016/B978-0-12-816440-2.00011-2

As it turns out, while the line command is useful for most drawing situations, it is inadequate or inconvenient for a small but important number of other applications. Here, we look at other variants of the line including the pline (polyline), the xline (construction line), its close relative the ray, a special case of the polyline called a *spline*, an mline (multiple line), and the most unusual one of all, the sketch command.

11.2 PLINE (POLYLINE)

A pline is a regular line with some unique properties:

Property 1. A pline's segments are joined together.
Property 2. A pline can have thickness.

These two properties add tremendous importance and versatility to the old line command, which creates only individual segments that maintain a constant zero thickness. Fig. 11.1 provides an illustration of the first property. The top "W" is drawn with four line segments. As you can see, they are all separate, with only the left segment clicked on, and the grips reflect that. The "W" below it is drawn using the pline command and is all one unit, as seen by clicking it once. Notice the slightly compressed center grips.

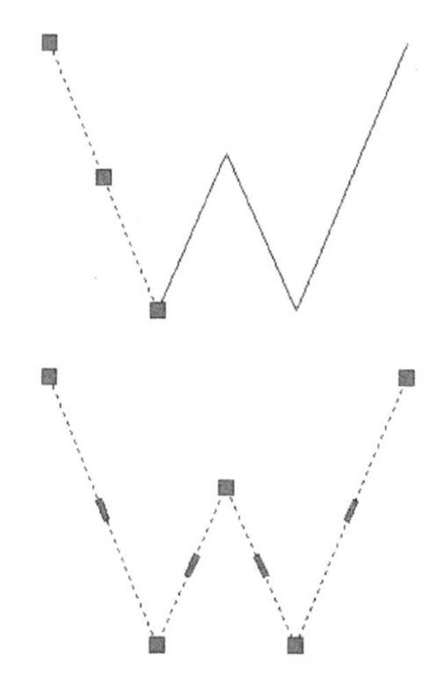

FIGURE 11.1 Line versus pline command (Property 1).

One example of applying this feature may include drawing lengthy runs of cabling or piping. Then, if the run needs to be deleted, only one click is necessary to remove all the segments. Let us try to draw a pline.

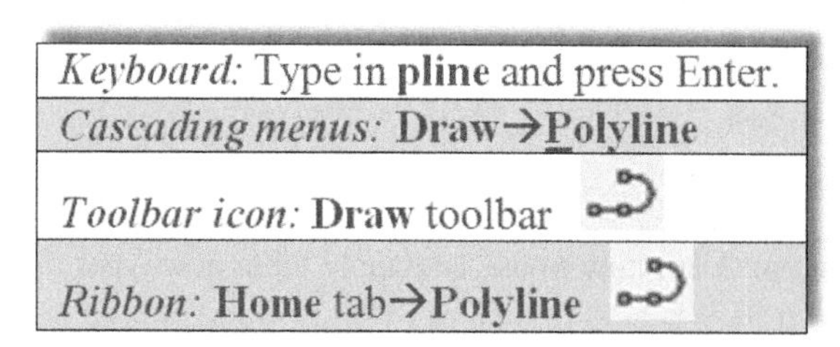

Keyboard: Type in **pline** and press Enter.
Cascading menus: **Draw→Polyline**
Toolbar icon: **Draw** toolbar
Ribbon: **Home** tab→**Polyline**

Step 1. Start the pline command via any of the preceding methods.
- AutoCAD says: `Specify start point:`

Step 2. Left-click anywhere on the screen.
- AutoCAD says: `Current line-width is 0.0000 Specify next point or [Arc/Halfwidth/Length/Undo/Width]:`

Step 3. Ignoring the lengthy menus for now, click around until you get the same W shape seen earlier.
- AutoCAD says: `Specify next point or [Arc/Close/Halfwidth/Length/Undo/Width]:`

When done, press Enter or Esc. Then, click on your new pline to see that it is all one unit not separate lines.

Pedit

What if you wanted to apply thickness to the pline (Property 2) or create a pline out of a bunch of lines? For this, you need pline's companion command, called *pedit* or *polyline edit*. Let us try this out on the pline you created and give it thickness. You need to add another toolbar to your collection, called *Modify II* (Fig. 11.2).

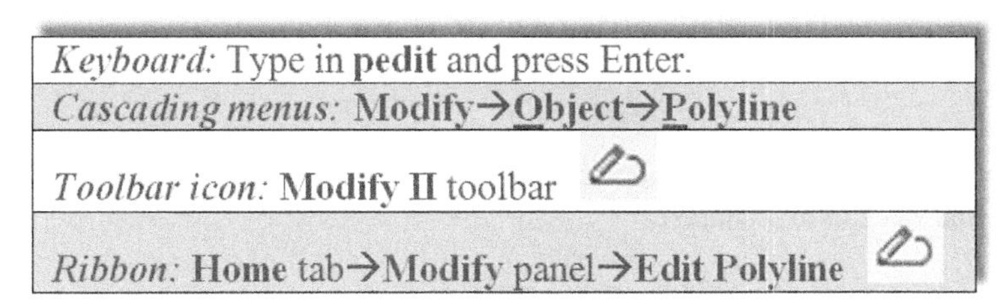

Keyboard: Type in **pedit** and press Enter.
Cascading menus: **Modify→Object→Polyline**
Toolbar icon: **Modify II** toolbar
Ribbon: **Home** tab→**Modify** panel→**Edit Polyline**

FIGURE 11.2 Modify II toolbar.

Step 1. Start the pedit command via any of the preceding methods.

- AutoCAD says: `Select polyline or [Multiple]:`

Step 2. Click on any of the W polyline segments.

- AutoCAD says: `Enter an option [Close/Join/Width/Edit vertex/Fit/Spline/Decurve/Ltype gen/Reverse/Undo]:`

Step 3. Press `w` for `Width` and press Enter.

- AutoCAD says: `Specify new width for all segments:`

Type in a small value, say, 0.25, and press Enter. The pline adopts the new width for all segments, as shown in Fig. 11.3. You can then press Esc to completely exit out of the pedit command. Now, try making plines out of lines. Draw the W once again out of ordinary lines, then follow the procedure shown next:

FIGURE 11.3 Pline with thickness (Property 2).

Step 1. Start the pedit command via any of the previous methods.

- AutoCAD says: `Select polyline or [Multiple]:`

Step 2. Click on any of the W line segments.

- AutoCAD says: `Do you want to turn it into one? <Y>`

Step 3. Press Enter to accept "yes."

- AutoCAD says: `Enter an option [Close/Join/Width/Edit vertex/Fit/Spline/Decurve/Ltype gen/Reverse/Undo]:`

Step 4. Select `j` for `Join` and select all the lines of the W, pressing Enter when done. The collection of separate lines is now a polyline and you may press w for Width as before if you wish. Press Esc to completely exit out of the pedit command.

Another notable option in the pedit command is spline, and if you press s while in the pedit command, your W will look like Fig. 11.4.

FIGURE 11.4 Pline with the Spline option.

To undo this you can press d for Decurve, yet another pedit option. Spline is an important case of a polyline. We look at spline as a separate command later in this chapter.

Exploding a Pline

Plines can be easily exploded using the explode command. The effect is twofold:

- The pline loses its thickness.
- The pline is broken into individual lines.

Putting it back together requires the pedit command, so explode and pedit are opposites of each other. Be careful though and do not explode plines unless absolutely necessary, as there may be many individual lines to account for once exploded.

Additional Pline Options

A few more options are worth mentioning with the pline command. First of all, you can start out drawing a pline with a preset width (or half width). Here is the command sequence for that, presented in a shortened "actual" format, as seen from the command line by the user:

Command: pline (via any of the previously shown methods)

```
Specify start point:
Current line-width is 0.0000
Specify next point or [Arc/Halfwidth/Length/Undo/Width]:w
Specify starting width<0.0000>:.25
Specify ending width<0.2500>:.25
Specify next point or [Arc/Halfwidth/Length/Undo/Width]:
Specify next point or [Arc/Close/Halfwidth/Length/Undo/Width]:
```

As you can see, the W option was chosen and a width of 0.25 was entered for both start and end. If those lengths are not similar, you get something akin to Fig. 11.5. In that example, a zero start width was followed by a 0.75 end width to create the triangular first segment of the pline. You can change these widths "on the fly" to create some interesting shapes.

FIGURE 11.5 Pline with varying width.

Another option is L for Length. This is similar to Direct Distance Entry from Level 1. Finally, there is the A for the Arc option. This creates arcs on the fly and is very useful for drawing organic-looking shapes for everything from landscape design to furniture to cabling and wiring, as shown in Fig. 11.6.

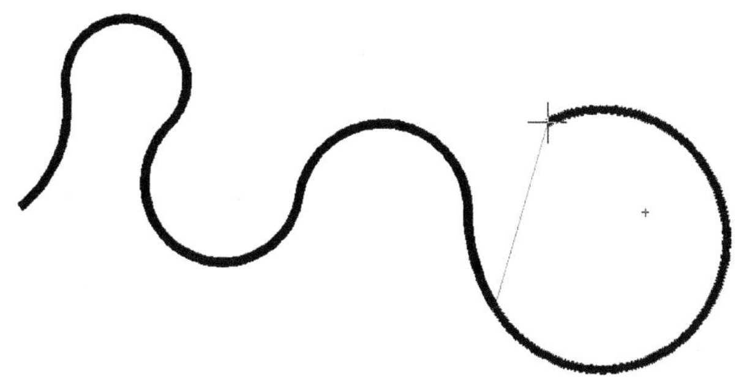

FIGURE 11.6 Pline of 0.2 width with Arc option.

Finally, several releases ago, AutoCAD introduced some additional tools to work with plines (including unexploded rectangles, which are plines by definition). These tools come in the form of a new menu, which appears when you hover the mouse over a grip belonging to a pline or a rectangle (Fig. 11.7).

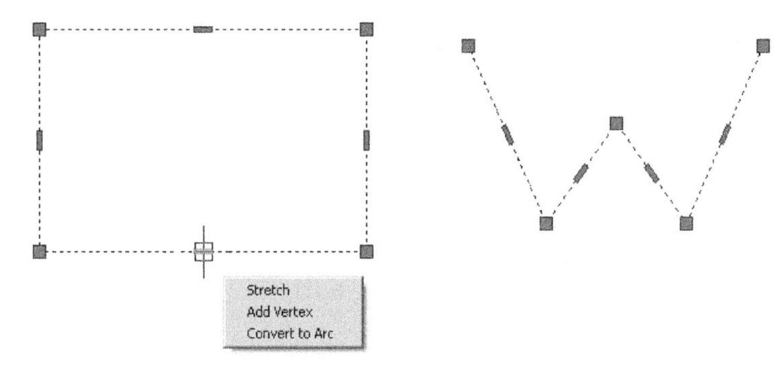

FIGURE 11.7 Grip menu.

As you can see, you can stretch, add a vertex, or even convert the line to an arc, allowing some new design possibilities and even saving some time.

Stretch

This option, also covered as a separate command in Chapter 15, Advanced Design and File Management Tools, allows for stretching that particular line and anything attached to it in one of several ways, with one possibility seen in Fig. 11.8.

Add Vertex

This option allows you to "break" the line with the addition of another control point (grip), as seen in Fig. 11.9. You can then create two angled lines out of one straight line.

Convert to Arc

This somewhat similar option also adds a vertex but one that allows an arc to be formed, as seen in Fig. 11.10.

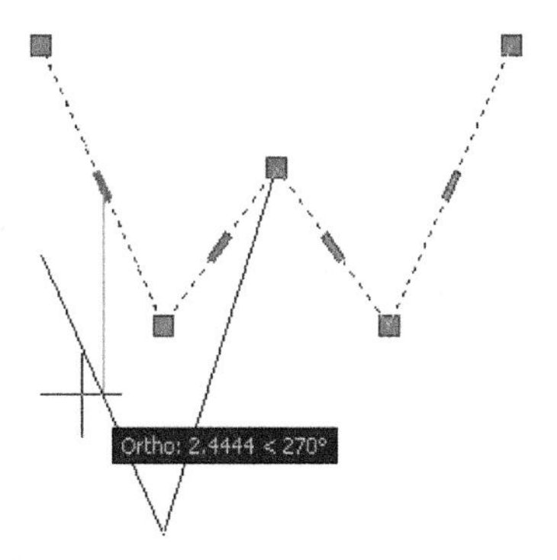

FIGURE 11.8 Stretch option (with Ortho on).

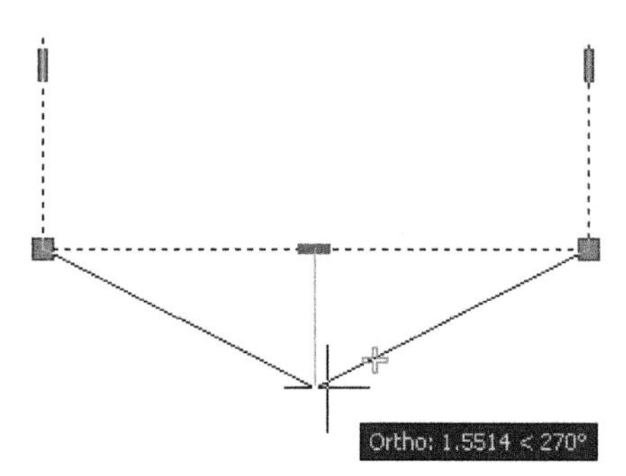

FIGURE 11.9 Add Vertex option (with Ortho on).

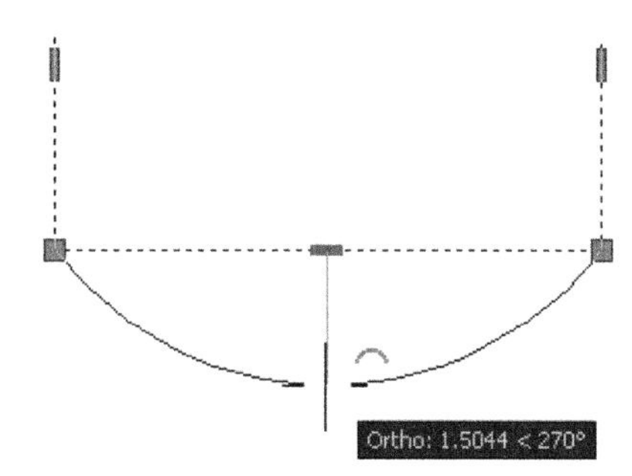

FIGURE 11.10 Convert to arc option (with Ortho on).

Boundary

The boundary command is part of the pline family and is a bit of a "hidden gem" from a long time ago, when it was called *bpoly*. It is no longer listed under that name in the Help files and has been permanently renamed *boundary*, but bpoly still works if typed in on the command line (Autodesk never seems to permanently get rid of any old commands).

The premise behind boundary is very simple. If you have a collection of regular lines forming a closed area, then you can use this command to just click inside that area and form a pline on top of all the regular lines. You need not select anything because the bpoly command works similar to the hatch command's *Add: Pick points option*, in the sense that the command "flows" outward, like a bucket of spilled paint, until it hits the edges of the area. Every element of the edge that it touches contributes to the new pline.

Keep in mind, the command does not touch any of the original linework, it merely forms the new pline on top of it. You can then use this new pline border for area calculations, hatching, or any other purpose. If needed, you can delete the pline on completion of any subsequent tasks. To try this out, clear your screen and draw a simple set of ordinary lines to form a closed shape, as seen in Fig. 11.11. These lines can represent the outline of a room in an architectural plan or the property lines on a civil engineering site plan.

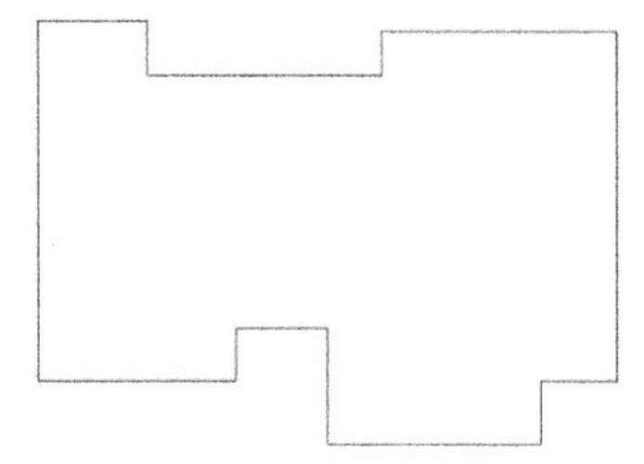

FIGURE 11.11 Set of lines for bpoly.

Start the command via any of the following:

Keyboard: Type in **boundary** or **bpoly** and press Enter.	
Cascading menus: **Draw→Boundary...**	
Toolbar icon: none	
Ribbon: **Home** tab→**Draw→Boundary...**	

The dialog box shown in Fig. 11.12 appears.

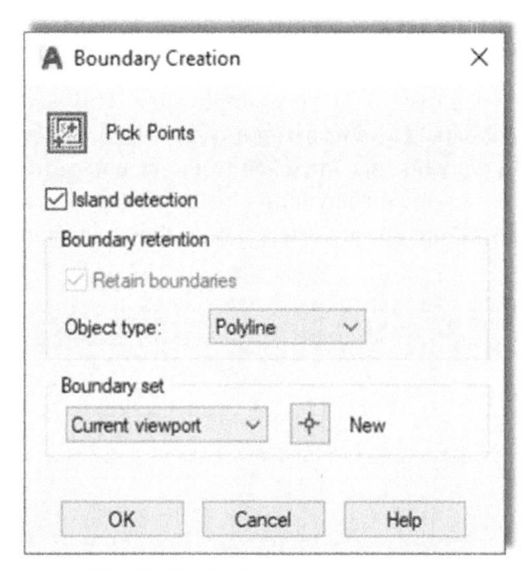

FIGURE 11.12 Bpoly boundary creation.

Note that the command features a very "hatchlike" *Island detection* option, which you can uncheck if you wish to ignore any elements inside the boundary but in most cases, the only thing you need to do is just click on the Pick Points button.

- AutoCAD says: `Pick internal point:`
  Go ahead and click anywhere on the inside of the shape. AutoCAD runs a short analysis, as seen here:
  ```
  Selecting everything...
  Selecting everything visible...
  Analyzing the selected data...
  Analyzing internal islands...
  ```
  Press Enter.
- AutoCAD says: `BOUNDARY created 1 polyline`
  At this point you are done. Click on the newly formed boundary and note how it is a separate entity. In Fig. 11.13, this new pline boundary has been moved slightly down and to the right, and one of the lines underneath has also been clicked to show that the original structure remains as a collection of ordinary lines.

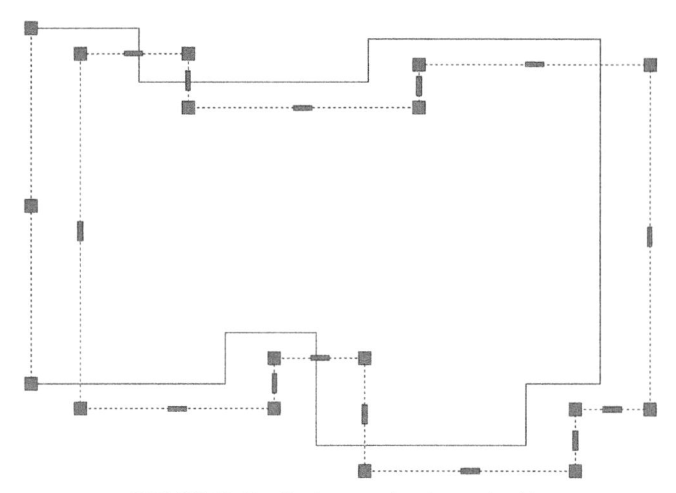

FIGURE 11.13 Bpoly created and moved aside.

11.3 XLINE (CONSTRUCTION LINE)

From the relatively involved pline, we swing over to the very simple xline. This is a continuous line that has no beginning or end. The zoom command has no effect on it. The construction part of the name refers back to the days of hand drafting, when you would draw, in light pencil, a guideline that stretched from one end of the paper to the other and was used as a reference starting point for a string of elevations (or anything else) to be drawn one after another.

To use it, make sure Ortho and OSNAP are not activated and do the following:

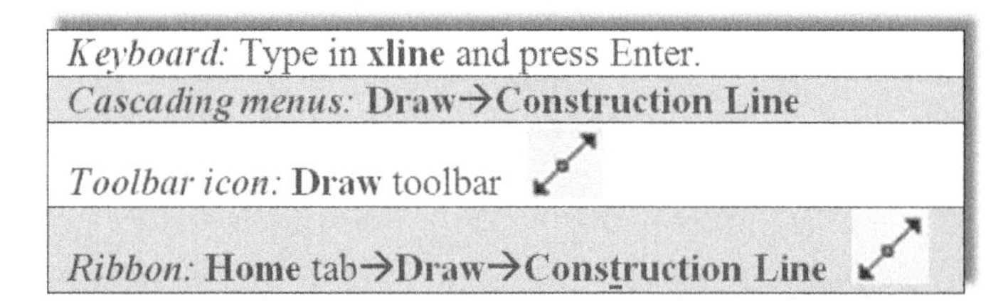

Keyboard: Type in **xline** and press Enter.
Cascading menus: **Draw→Construction Line**
Toolbar icon: **Draw** toolbar
Ribbon: **Home** tab→**Draw→Construction Line**

Step 1. Start the xline command via any of the previous methods.

- AutoCAD says: `Specify a point or [Hor/Ver/Ang/Bisect/Offset]:`

Step 2. Click anywhere and an xline appears.

- AutoCAD says: `Specify through point:`

Step 3. Click randomly anywhere along the screen and multiple xlines appear. Fig. 11.14 shows one result of clicking in a circular manner.

Fig. 11.14 is perhaps not the most useful of creations, but landscape designers may find some use for the result of placing a circle at the intersection of those lines and trimming. Then, via some creative scaling, rotating, and copying, you get the reasonable representation of trees or shrubs in Fig. 11.15.

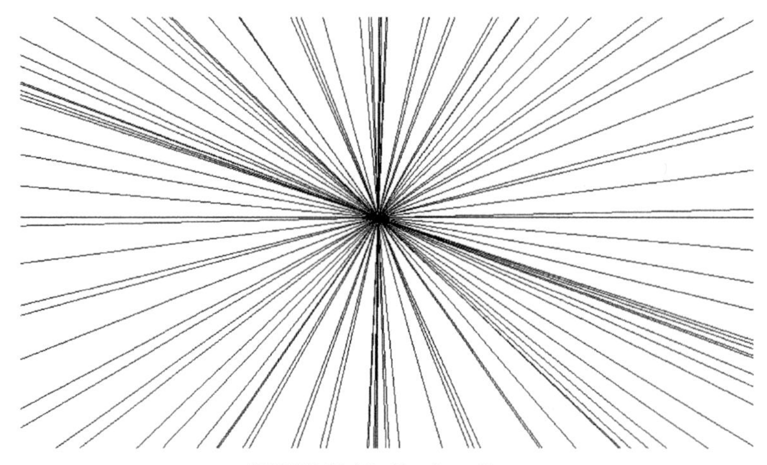

FIGURE 11.14 Random xlines.

Xline
Circle

Trim

Copy
Scale
Rotate
Move

Copy
Scale

FIGURE 11.15 An application of xline.

As seen in the menu that follows the xlines, you can also create perfectly horizontal, vertical, or angular xlines. The same can be accomplished by having Ortho on, although the options allow you to create multiple xlines more easily. Clear your screen, start xline again, and when you see the following menu, try out the various options (results are seen in Fig. 11.16): `Specify a point or [Hor/Ver/Ang/Bisect/Offset]:`

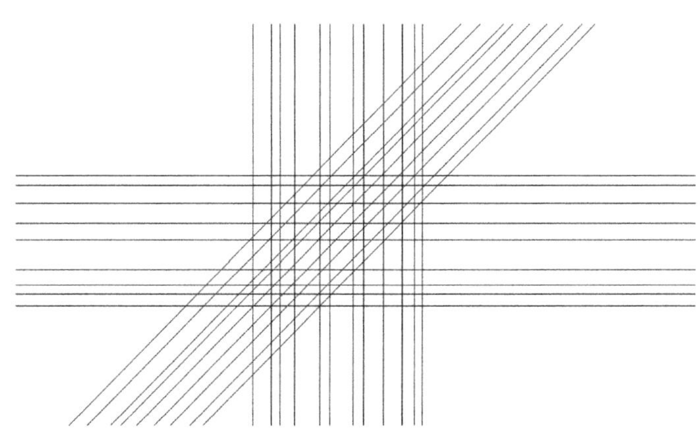

FIGURE 11.16 Hor, Ver, and Ang (set to 45 degrees) options in xline.

11.4 RAY

A ray is simply an xline but infinite in only one direction. It is easy to use and quite similar to xline, except there are no options for horizontal, vertical, or angular; to get a perfectly straight ray, you just turn on Ortho.

Keyboard: Type in **ray** and press Enter.
Cascading menus: **Draw→Ray**
Toolbar icon: none
Ribbon: **Home** tab→**Draw→Ray**

Step 1. Start the ray command via any of the preceding methods.

- AutoCAD says: `Specify start point:`

Step 2. Click anywhere and a ray is created (Fig. 11.17).

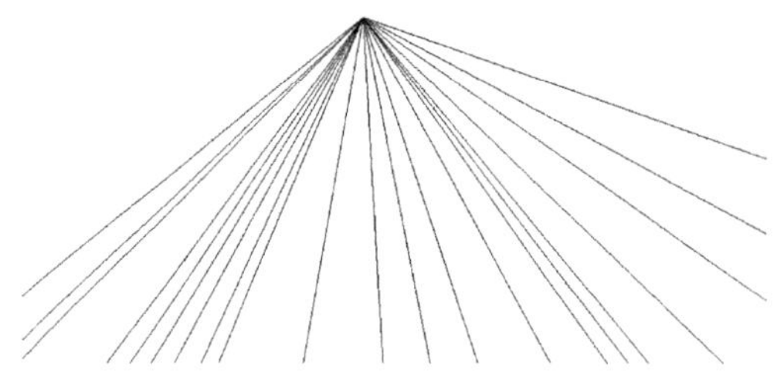

FIGURE 11.17 Ray command.

11.5 SPLINE

Spline is a unique and valuable command. Its origins are in the mathematical field of numerical analysis, and it is called a *piecewise polynomial parametric curve* (or a NURBS curve, nonuniform rational B-spline). While knowing the details is not necessary for using this command, what is important is that a spline is ideal for approximating complex shapes through curve fitting. What you do is create a path made out of point clicks, and AutoCAD interpolates a "best-fit" curve through those points.

The result is a smooth, organic shape useful for a great number of applications, from contour gradient lines in civil engineering to landscape design to electrical wiring and many other shapes and objects. The term *spline*, by the way, comes from the flexible spline devices used by shipbuilders and draftspersons to draw smooth shapes in pre-computer days. To draw a spline, do the following:

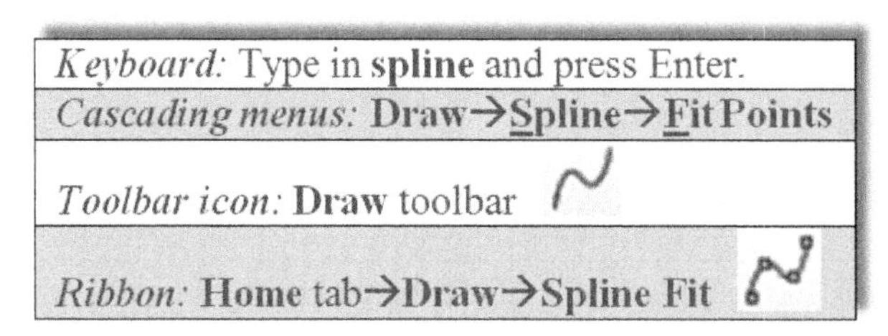

Keyboard: Type in **spline** and press Enter.
Cascading menus: **Draw→Spline→Fit Points**
Toolbar icon: **Draw** toolbar
Ribbon: **Home** tab→**Draw→Spline Fit**

Step 1. Start up the spline command via any of the preceding methods.

- AutoCAD says: Current settings: `Method = Fit Knots = Chord`
  `Specify first point or [Method/Knots/Object]:`

Step 2. Click anywhere and a spline appears.

- AutoCAD says: `Enter next point or [start Tangency/toLerance]:`

Step 3. Click on another point somewhere on the screen, in an up and down wave pattern, as seen in Fig. 11.18.

- AutoCAD says: `Enter next point or [end Tangency/toLerance/Undo/Close]:`
  As early as this step, you can select `c` for `Close` to close the spline tool. In this case, however, continue with the spline curve to add more points.

Step 4. Click on a few more points in a similar wave pattern. AutoCAD keeps saying the same thing as in Step 3. Continue clicking several more times (refer again to Fig. 11.18). When done, simply press Enter.

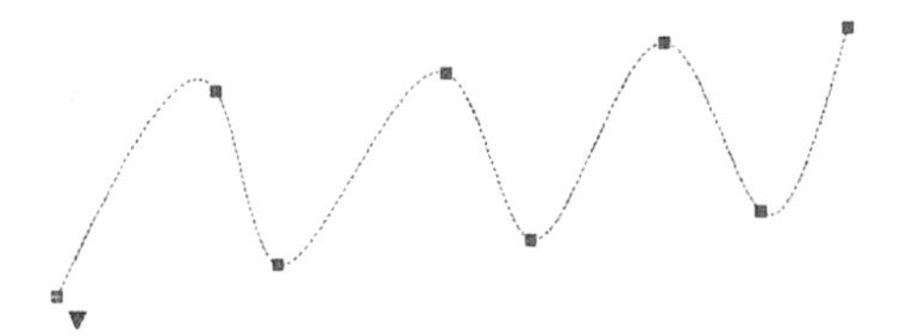

FIGURE 11.18 Spline command, grips enabled.

This is your basic "out-of-the-box" spline. Notice how, as you click on points, AutoCAD fits a smooth curve that best approximates the path. Each grip in Fig. 11.18 represents a click of the mouse, a control point. You can manipulate the positions of these grip points in the standard manner, with the spline following suit. This makes the spline a very flexible tool, adjustable to the exact demands of the designer. Try activating the grips and moving them around.

The command was redesigned somewhat back in AutoCAD 2011, and many of its advanced features are hidden away in submenus, which is a good thing, as most users need only the basic spline and not extensive options for tangencies and other details.

That is not to say they are not there if needed. Click on the down arrow on the bottom left of the spline (grips enabled) and select Show Control Vertices, which is another way to design the spline, useful for 3D work. Go over the various other options on your own with the Help files; there are too many to go into here, but learning those helps you understand this very intriguing command.

Through the examination of spline, you should now have a good idea of how to create and control it. The points of the spline do not have to be random. You can set up a framework for the shape and use OSNAPs to precisely place where you want the spline to go, as shown in Fig. 11.19. What you see here are four lines, representing a framework. Then, using OSNAP points, a spline is added in click by click, from an endpoint to a midpoint to another midpoint, then to the final endpoint, at the bottom left.

A spline can then be offset using the standard offset command, although be careful in areas where the spline takes a sharp turn. If the offset distance is too great and the spline cannot make a smooth turn after the offset, it breaks down into separate lines. The offset spline also has many more grip and control points than the original spline. A 3″ spline offset, with all grips enabled, can be seen in Fig. 11.19.

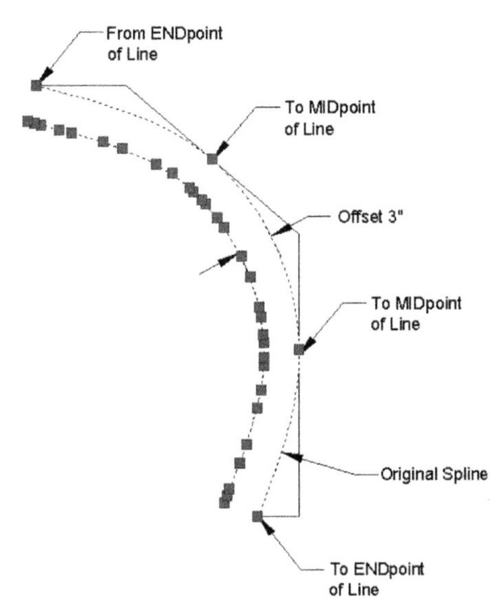

FIGURE 11.19 Spline with framework, offset, and grips.

Examples of spline use are numerous. Already mentioned were contour lines (this chapter's cover image), electrical wiring, piping, network cables, landscape design (golf courses, curved driveways, and walkways), interior design (Jacuzzis, swimming pools, furniture), and much more can be constructed with this tool.

11.6 MLINE (MULTILINE)

A multiline is a very useful tool that does pretty much what you would expect from the name: It is used to draw multiple lines all at once. This, in theory, should save you time when drawing everything from architectural walls to multilane highways. As we see soon, the basic mline is easy to draw but not too useful. The trick is to set it up for useful applications using additional tools, as outlined in the next sections. Try out the basic mline to get a feel for it, as shown next.

Keyboard: Type in **mline** and press Enter.
Cascading menus: **Draw→Multiline**
Toolbar icon: none
Ribbon: none

Step 1. Start up the mline command via any of the preceding methods (no toolbar or Ribbon is available).
- AutoCAD says: `Current settings: Justification = Top, Scale = 1.00, Style = STANDARD Specify start point or [Justification/Scale/STyle]:`

Step 2. A generic mline appears. Click again.
- AutoCAD says: `Specify next point or [Undo]:`

Step 3. Click somewhere else again.
- AutoCAD says: `Specify next point or [Close/Undo]:`

You can continue in this manner until finished. The default distance between the lines is 1.0, and of course, if the mline is too small or too large, you can zoom in or out. Fig. 11.20 shows what you should see after properly executing the previous steps.

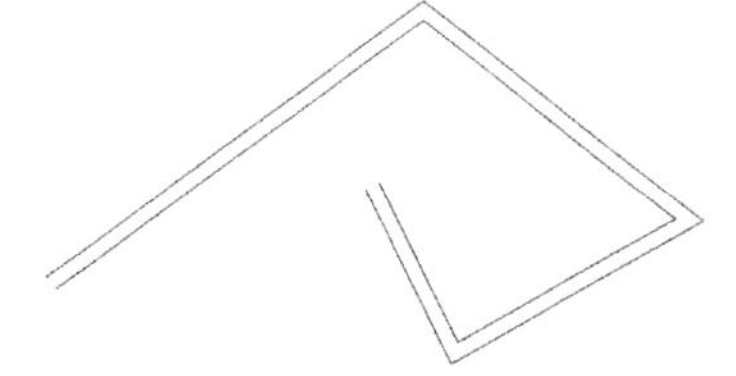

FIGURE 11.20 Basic mline.

Modifying the Mline

While simple, the preceding was not too useful. Specifically, you may want to change one or more of the following properties:

- Number of lines making up the mline.
- Distances between those lines.
- Linetypes making up the lines.
- Colors of the lines.
- Fills, end caps, and other details.

All these and more can be input and modified, thereby making the mline more useful and sophisticated. The key command for setting up mlines is mlstyle.

Mlstyle (Multiline Style)

The multiline style command opens a dialog box that allows you to set all the preceding parameters. Type in `mlstyle` and press Enter or select Format → Multiline Style... from the cascading menu (there is no toolbar or Ribbon alternative). The dialog box shown in Fig. 11.21 appears.

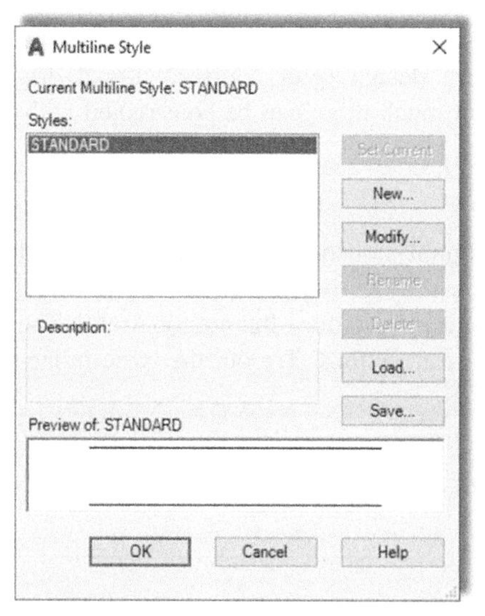

FIGURE 11.21 Multiline Style dialog box.

Let us create a new style. Press New... and you see a smaller Create New Multiline Style box. Type in a name for your mline (no spaces allowed in the name; use underscores if needed) and press Continue. You then see the main dialog box shown in Fig. 11.22. The name created for this example was a generic MY_NEW_MLINE.

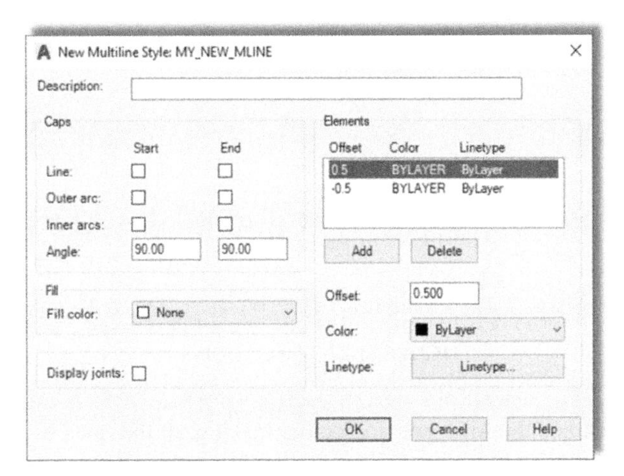

FIGURE 11.22 New Multiline Style dialog box.

We can now go through the essential changes to the mline. Our goal is to create a typical wall used in construction (see Fig. 11.23), often shown in cross section in architectural elevations. For simplicity, it is 6″ wide and features a 1″ drywall panel on either side, with insulation in between, and room for a wooden stud (not shown here).

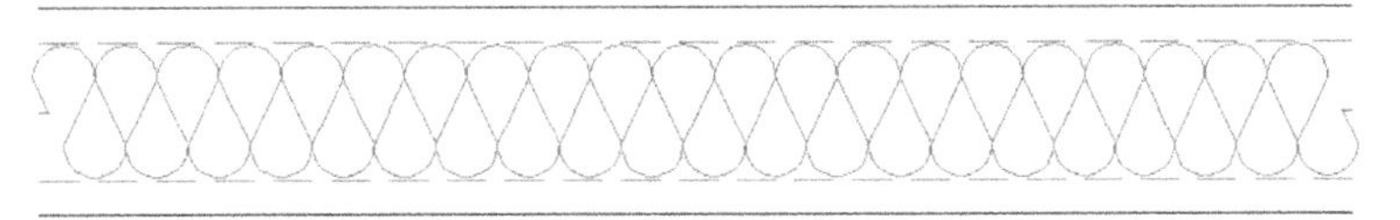

FIGURE 11.23 Architectural wall with insulation.

This mline, once set up, can be used to draw building walls faster and easier than offsetting and changing properties one line at a time. The process to do this follows.

Step 1. You first need to determine the *number of lines* making up the mline and the distances between those lines. Both of these can be set inside the Elements area (right-hand side) of the New Multiline Style dialog box using the Add button and the Offset field. The idea is to add more line elements (via the Add button) and enter an offset distance for each of these line elements (via the Offset field). This offset distance spaces the lines at the desired interval to create the wall.

Before you do all this, however, you need to decide what the origin of the offset is. This is your starting point for the offsets. The two choices are positioning zero all the way on the leftmost line element (then, all values are positive) or in the middle element (then, half the values are negative). With the first choice, the outer walls are at 0 and 6, and with the second choice, they are at −3 and +3, as shown graphically in Fig. 11.24.

For this example, choose Method 1, and after adding additional lines, for a total of six, set the walls to 0, 1, 3, 5, and 6 units (there are no feet or inches in these settings; you use decimal units, representing inches, throughout).

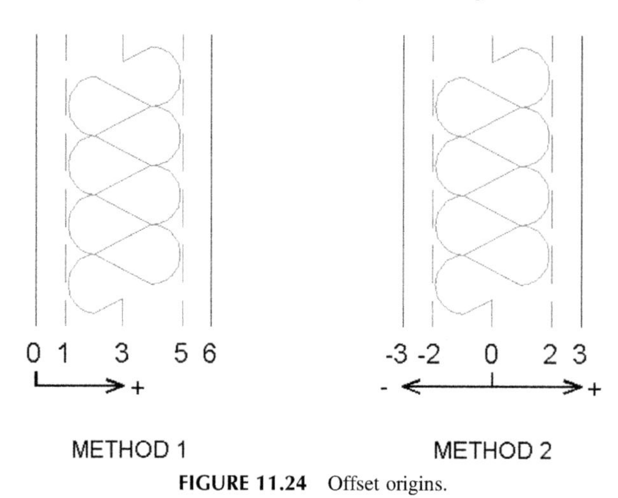

FIGURE 11.24 Offset origins.

Step 2. Next you need to determine the linetypes and colors making up the lines of the overall mline.

Both of these can be set in the Elements area using the Color: drop-down menu and the Linetype… button, as shown in Fig. 11.25. Simply highlight the line element you wish to change and select a new color or linetype. Remember that you may not have the linetypes loaded yet, so you may need to do this (refer back to Chapter 3 for a reminder).

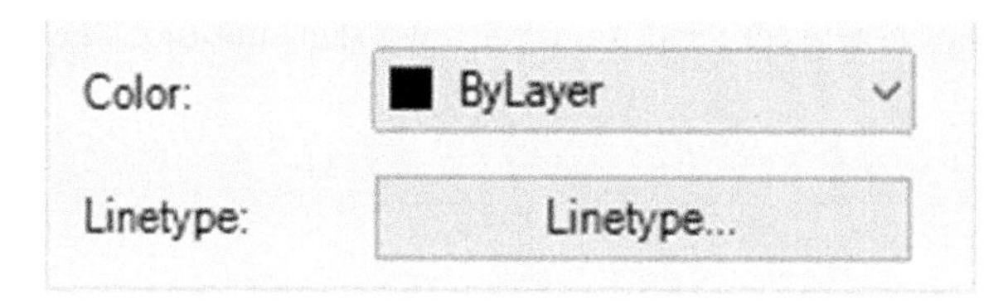

FIGURE 11.25 Color and Linetype settings.

In Figs. 11.23 and 11.24, the outer wall color was selected to be blue and the inner walls red with a hidden linetype. The centerline is a special linetype, often used for insulation, called BATTING. Feel free to choose your own colors and linetypes; those shown are just for illustration, but they are a good starting point. Fig. 11.26 shows the Elements area as you go through Steps 1 and 2.

Elements

Offset	Color	Linetype
6	blue	ByLayer
5	red	HIDDEN
3	8	BATTING

Add Delete

Offset: 6.000

Color: Blue

Linetype: Linetype...

FIGURE 11.26 Elements settings.

Step 3. As the final step, you can add in fills and start/end caps.

End caps, as the name implies, are various ways to finish the start and end of your mlines. You can add regular lines or two types of arcs. Often mlines are joined to each other, so these may not be necessary. You can also add color fills to the mlines, although this is rarely done in practice. All of these can be done via the check-off boxes and drop-down menu on the left-hand side of the New Multiline Style dialog box, as seen in Fig. 11.27. You can also display the joints of a mline via a checked box, but that is also rarely called for. This example has no caps or fills, but feel free to experiment on your own.

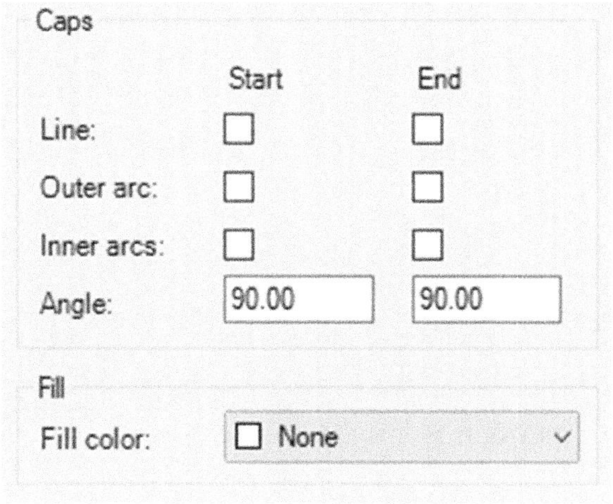

Caps
Start End
Line:
Outer arc:
Inner arcs:
Angle: 90.00 90.00
Fill
Fill color: None

FIGURE 11.27 Caps and Fill.

After your new mline has been created, go ahead and click on OK; you are taken back to the dialog box of Fig. 11.21, with your new mline previewing in the window at the bottom. The other choice should be Standard, the default mline you created by just typing in mline and saw in Fig. 11.20. Set your new mline as the current one (Set Current button) and click on OK. Now, go ahead and type in mline and draw part of a rectangle, as seen in Fig. 11.28.

FIGURE 11.28 The new mline.

The linetypes may not look right in this example. This is due to a low linetype scale being set. If you recall from Chapter 3, Layers, Colors, Linetypes, and Properties, the way to fix this is to type in ltscale and enter a new, more appropriate value. The value used is 4.8. The result is shown in Fig. 11.29.

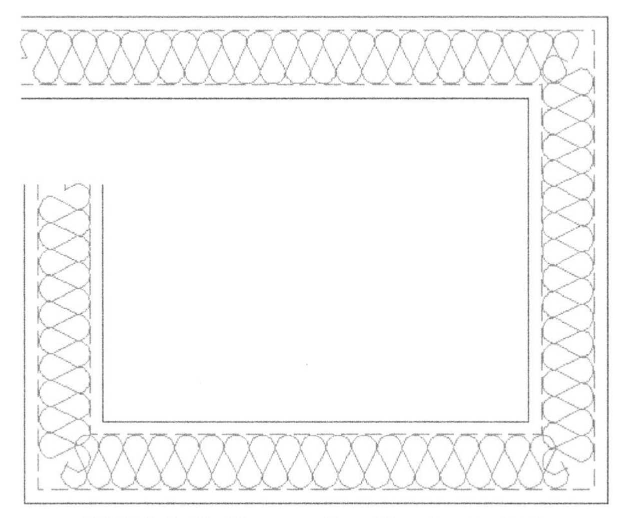

FIGURE 11.29 The new mline with an ltscale of 4.8.

Mledit (Multiline Edit)

Multiline Edit is a dialog box that deals with the practical applications of mlines, such as how to trim intersecting mlines and how to deal with joints, cuts, and vertexes. To try out some of the features, draw two intersecting mlines, as shown in Fig. 11.30.

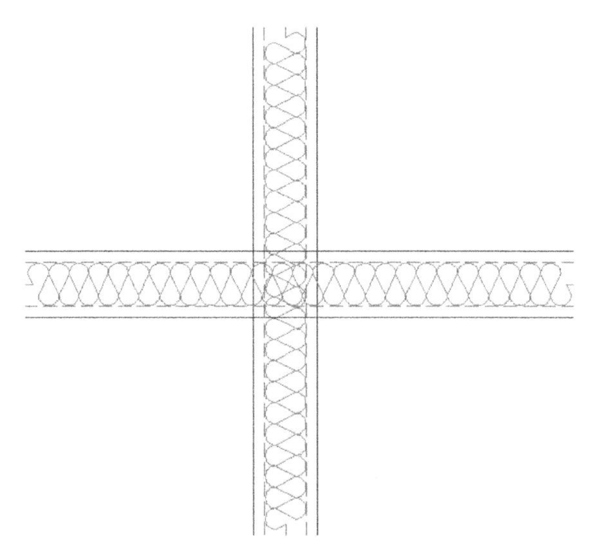

FIGURE 11.30 The new mline with an ltscale of 4.8.

Then, type in mledit, press Enter (or select Modify → Object → Multiline… from the cascading menus), and the dialog box shown in Fig. 11.31 appears.

Select Closed Cross in the upper left. The dialog box disappears, and you are asked to select the first mline, click on it, then the second mline, and click on that also. After the second click, Fig. 11.32 appears. As you can see, the mline's intersection has been modified.

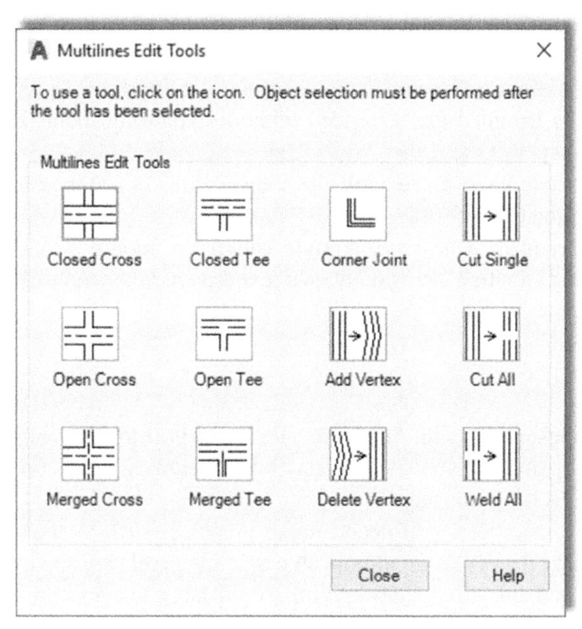

FIGURE 11.31 Multiline Edit Tools.

In the same manner, try out all the features of mledit at your own pace; they are all relatively straightforward and self-explanatory. A cut cuts the line; a vertex adds a "kink" in the mline; Close, Open, and Merge deal with intersections; and so on.

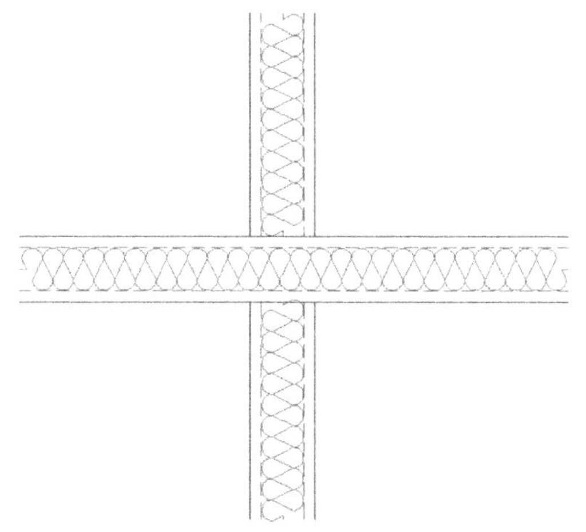

FIGURE 11.32 Multilines edited.

Other Mline Properties

- *Grips*: All mlines have grip points at either the top or bottom of the overall line (this is set using Justification when first creating the mline). Feel free to use these grips to stretch the line around as needed.
- *Exploding the Mline*: Mlines can be exploded, which separates them into individual component lines. This may not necessarily be a bad thing. An architectural floor plan may be quite complex, and you may not be successful in getting a mline to do exactly what you want all the time. If the mline is exploded, then you can use the fillet, trim, and extend commands to coax it into the shape you want. Much time has already been saved by using a mline in the first place, so exploding it may be the right solution to finish off the floor plan linework.
- *Trim and Extend*: Mlines can be trimmed and extended without exploding them (although filleting is not possible). However, an option that pops up is this: `Enter the mline junction option [Closed/Open/Merged]<Open>:` The Closed, Open, and Merged options have to do with the intersections, so you need to pick one option to continue the trim or extend. Go through them all as practice.
- *Modification of Mlines*: Once created, a new mline style cannot be modified once you start using it (but can be beforehand). If you wish to change a mline after you have a few drawn, you need to create a new style that incorporates your desired changes.

11.7 SKETCH

This is perhaps one of the most unusual tools in AutoCAD. It allows you to draw objects in freehand. This may seem counterproductive and unnecessary. After all, you spent so much time learning precision and accuracy; but sketch actually finds a wide array of applications, as we soon see.

To truly sketch in freehand on a computer you need bitmap graphics (such as those found in Photoshop). Then, the maximum resolution is determined by the pixels. In almost all cases, the pixels are so small that it truly seems like you are drawing smooth, continuous freehand lines and shapes. You would have to zoom in pretty close to see what is called *pixilation* or the breakdown of smooth lines into the little squares (pixels) that make up the drawn geometry.

This is all fine and would be useful for AutoCAD if not for one problem: AutoCAD is a "vector program," meaning it does not use pixels. Instead, all the geometry is represented by mathematical formulas, which completely describe the shape, size, and location of everything seen on the screen. This approach is necessary for many reasons, not the least of which is to ensure the integrity of the geometry when zoomed in close. So then, how would you sketch in freehand? The solution is simple: Instead of pixels, use very short lines, which look continuous if they are small enough and you are not too close to them. This is exactly how sketch works. The command uses small lines, with their size determined by something called a *record increment*, which we discuss in a moment.

The sketch command has another twist to it. A few releases ago AutoCAD allowed you to turn sketched linework into a spline or a polyline. This can occasionally be useful if you are drawing simple freehand shapes and want maximum smoothness, although the pline option is not recommended, as it adds a tremendous amount of control points to the linework. We will discuss mostly the "classic" sketch command here. First, we need to set the record increment to a very small value (use 0.01). Let us try this out.

Step 1. Type in `sketch` (there is no toolbar icon, cascading menu, or Ribbon for the classic sketch command) and press Enter.
- AutoCAD says: `Type = Lines Increment = 0.1000 Tolerance = 0.5000 Specify sketch or [Type/Increment/toLerance]:`

Step 2. First, make sure `Type = Lines`, as that is what we want to practice first. If not, press `t` for `Type` and select `l` for `Lines`. Then, press `i` for `Increment` and press Enter.
- AutoCAD says: `Specify sketch increment <0.1000>:`

Step 3. Type in 0.01 and press Enter.
- AutoCAD says: `Specify sketch or [Type/Increment/toLerance]:`

Step 4. Nothing further needs to be set so press Enter.
- AutoCAD says: `Specify sketch:`

Step 5. Move the mouse around to begin drawing.
- AutoCAD says (and continues to say): `Specify sketch:`

You will see what is essentially a smooth line, green, and completely freehand generated. To "lift the pen," you need to left-click once; then click again to "put it down." You can also click and hold the left mouse button down when you want to draw, which is more realistic for some. To "set" the drawing, press Enter; the lines turn white (or some other color depending on the current layer). Be careful not to press Esc until you set the lines, as everything disappears. Fig. 11.33 shows what you can get after a few minutes of doodling.

As mentioned before, you can also turn the sketched lines into a spline by selecting the Spline option. You will still sketch freehand at first, but on pressing Enter, the lines smooth out into the familiar spline shapes. The process is the same with the Pline option. You should spend a few minutes experimenting with both. You can also access the Spline option directly via the Ribbon's Surface tab (look for the spiral on the Curves panel), but for that you need to change the workspace to 3D Modeling.

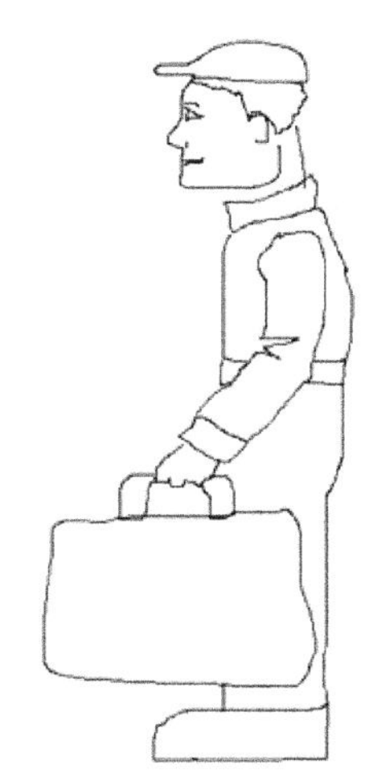

FIGURE 11.33 Basic sketch.

Applications of Sketch

I hope whatever you tried to draw was better than the odd-looking little man with a briefcase. It is not that easy to draw something well with a mouse; occasional erasing and fixing up is needed to get a good sketch. Sometimes, a graphic design tablet and a pen-like stylus may be used. Zoom in close on whatever you created and you notice the figure is made up of the small lines we talked about. In the close-up of the hand and briefcase handle in Fig. 11.34, you can clearly see the individual line segments. Be careful making a dense drawing. All these lines increase the size of the file very quickly.

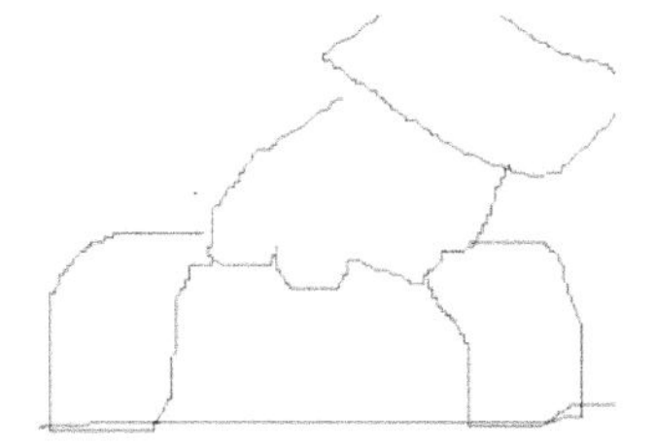

FIGURE 11.34 Line increments visible in the sketch.

Sketch has many applications, which are further expanded via the spline variation of the command. Some examples include

- Architectural elevations featuring sketched people for comparative sizing.
- Sketched people for traffic sign design and civil engineering studies.
- Trees, shrubs, and other landscape design objects.
- Realistic drawings of small cables and wiring (via spline).
- Anytime an organic, nonrigid shape is needed.

There are numerous websites that can provide free blocks, which are usually uploaded and shared by other users. Just a few examples of these sites include cad-blocks.net, cadforum.cz, cadblocksfree.com, cad-block.com, and many others. Many of these have blocks that feature the sketch command in creation of vegetation for landscaping or people for traffic and civil engineering as well as interior design and architecture layouts. Fig. 11.35 shows some examples pulled from these sources.

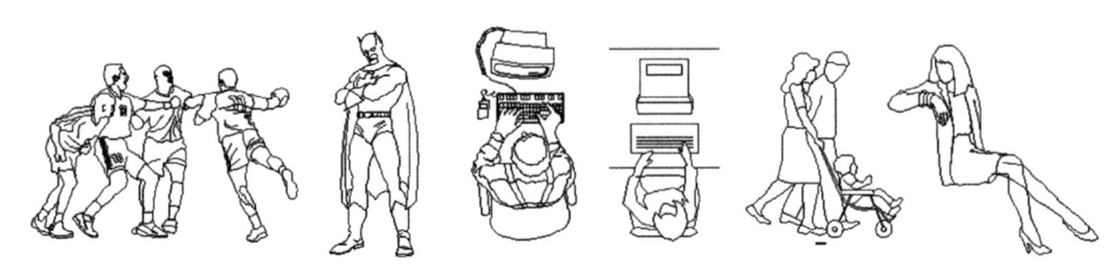

FIGURE 11.35 Creative use of the sketch command.

GEOmetric Center OSNAP

This brand-new OSNAP point, introduced a few years back with AutoCAD 2016, was requested by past AutoCAD users and really should have been included a long time ago. It allows you to snap from (or to) the geometric center of objects such as triangles, multisided polygons, and various nonsymmetrical shapes. In technical terms, you are snapping from (or to) the mathematical "centroid" of that shape. This can be quite useful; in the past, you needed to draw guidelines and find that center yourself. The only rule is that this shape has to be a closed polyline, hence we delayed discussing this OSNAP point until the pline, pedit, and spline commands were introduced.

To try out this OSNAP point, first draw a small collection of shapes, as shown in Fig. 11.36. In all cases, make sure they are closed polylines via the pedit command. You need to do this with the first and third shapes. The second one (octagon) is already closed, and the fourth shape is a spline, also closed.

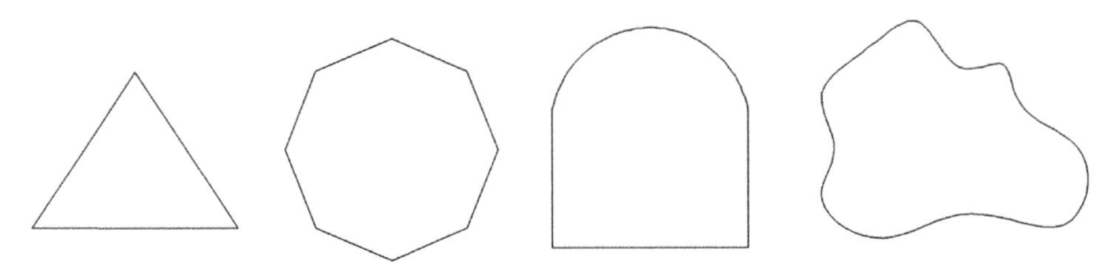

FIGURE 11.36 Shapes for the Geometric Center OSNAP.

Now open up the Drafting Settings dialog box and set the Geometric Center OSNAP. Its symbol is a circle, and it is located just below the Center OSNAP. For learning purposes, turn off all the other ones. Now, begin a line command, hovering the mouse in the center of each shape. AutoCAD recognizes these as closed shapes and places a special symbol at the centroid (as seen in Fig. 11.37). Click to begin a line; that is all there is to it. Fig. 11.38 shows all the shapes with a line beginning from the geometric center of each.

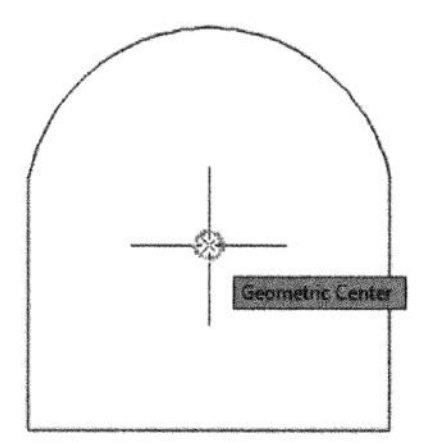

FIGURE 11.37 Geometric center recognized.

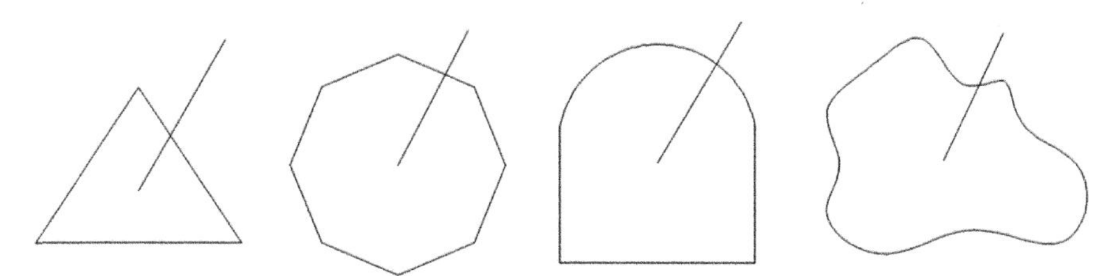

FIGURE 11.38 Lines starting from geometric centers of each shape.

11.8 LEVEL 2 DRAWING PROJECT (1 OF 10): ARCHITECTURAL FLOOR PLAN

Level 2 features a drawing project that is designed chapter by chapter, beginning with this one and completing in Chapter 20. It is an architectural layout, similar to what you did in Level 1, although a bit more involved. The project enables you to put to use all of the cumulative knowledge you acquired up until now and includes features you learn from Level 2 chapters.

The project features many of the basic elements of a "real-world" design and definitely is a step closer to the real thing. You will see a moderately complex floor plan and electrical, HVAC/mechanical, and landscaping plans. You then set up Paper Space and printing with lineweights and other details. A real house design also is likely to feature numerous details, riser diagrams, and sections of doors, windows, and walls. We include some of these elements here and there to increase realism.

Follow all the steps closely, but feel free to embellish the design and get creative. This is excellent practice for the real thing, and if you get through the project smoothly, you will feel right at home with a real design.

Step 1. Open a new file and save it as House_Plan.dwg.

Step 2. Set up units as Architectural and create the following initial layers. The drawing is monochrome in this textbook for clarity, but you need to follow CAD standards as required in the industry (color suggestions shown). For now, you need

- A-Deck (Blue).
- A-Deck-Hatch (Gray_9).
- A-Dims (Cyan).
- A-Walls-Exterior (Green).

Step 3. Draw the floor plan shown in Fig. 11.39. The dimensions are not just for your reference anymore; you also need to draw them in. They serve as checks on your design. Use the proper layers and consistent colors. When done and you are sure everything is correct, freeze the dimensions. It should take you about 30 minutes or less to set up and complete this design.

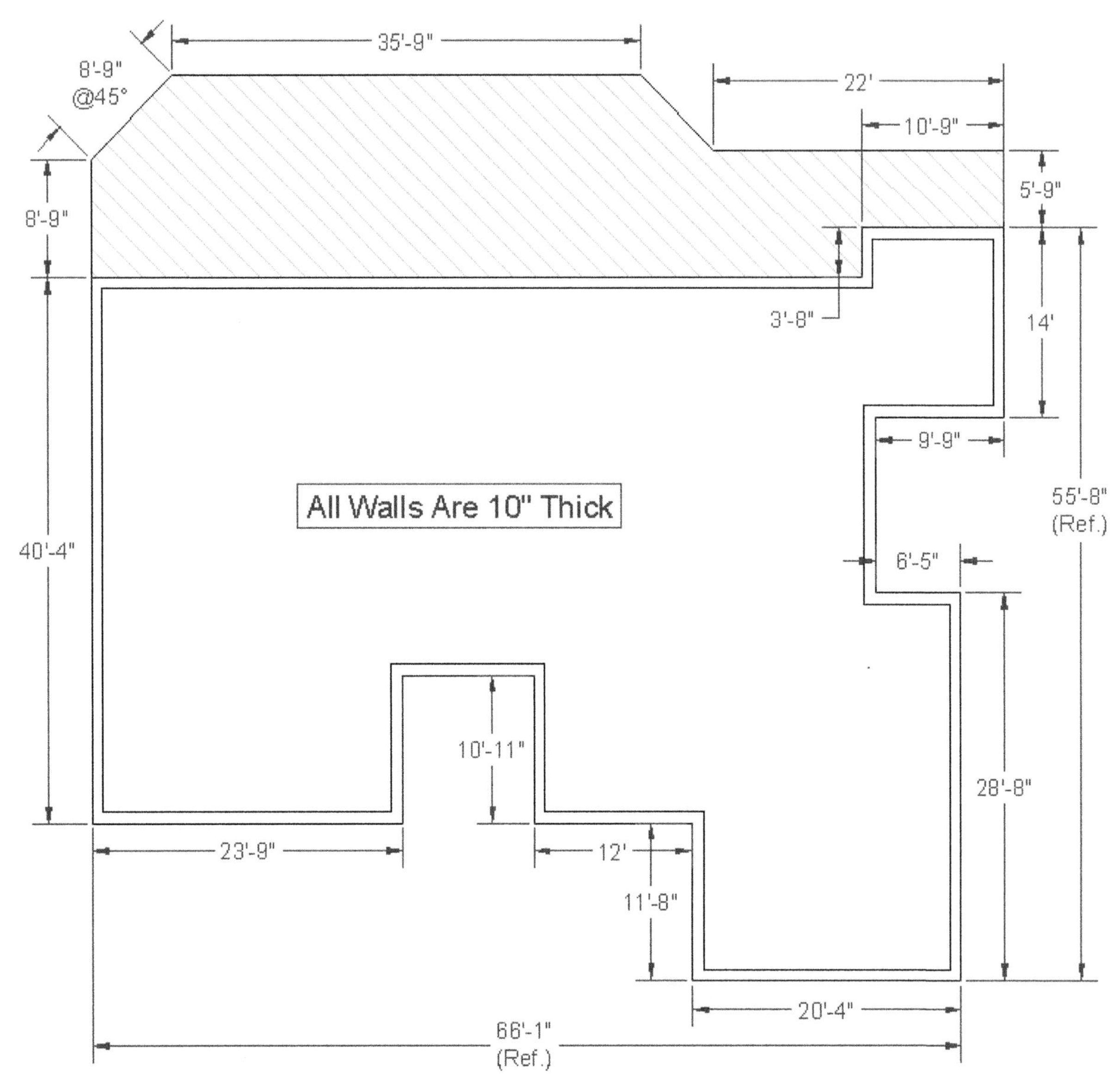

FIGURE 11.39 Level 2 project—overall layout.

SUMMARY

You should understand and know how to use the following concepts and commands before moving on to Chapter 12:

- Pline (polyline)
  - Creating a pline
  - Pedit
  - Setting constant width (thickness)
  - Setting variable width (thickness)
  - Changing lines into plines

  - Adding spline curves and decurving
  - Other pline options (Length, Arc, Offset)
  - Exploding a pline
  - Bpoly command
  - Xline (construction line)
  - Creating a random xline
  - Creating a horizontal, vertical, or angular xline
- Ray
  - Creating a random ray
  - Creating a horizontal or vertical ray (with Ortho)
- Spline
  - Creating a random spline
  - Offsetting the spline
  - Grips and modifying the spline
- Mline (multiline)
  - Creating a generic mline
  - Mlstyle
  - Mledit
  - Other mline properties
- Sketch
  - How sketch works
  - Setting a record increment
  - Applications of sketch
- Geometric center

REVIEW QUESTIONS

Answer the following based on what you learned in this Chapter 11:

1. What are the two major properties a pline has and a line does not?
2. What is the key command to add thickness to a pline?
3. What is the key command to create a pline out of regular lines?
4. What option gives the pline curvature? What option undoes the curvature?
5. What two major things happen when you explode a pline?
6. What other options for pline were mentioned?
7. What is the bpoly command?
8. Define an xline and specify a unique property it has.
9. How is a ray the same and how is it different from an xline?
10. Describe a spline and list its advantages.
11. List some uses for a spline in a drawing situation.
12. What is a mline? List some uses.
13. What are some of the items likely to be changed using mlstyle?
14. What are some of the uses of mledit?
15. Describe the sketch command. What is a record increment? What size is best?
16. Describe the basic function of a Geometric Center OSNAP point.

EXERCISES

1. In a new file, set your units to Architectural and do the following pline exercise. Step A: Draw a pline with 10 segments, each 2″ long and at right angles to each other as shown. Step B: Give the pline a thickness of 0.2″. Step C: Spline the pline. Step D: Explode the spline. (Difficulty level: Easy; Time to completion: <5 minutes.)

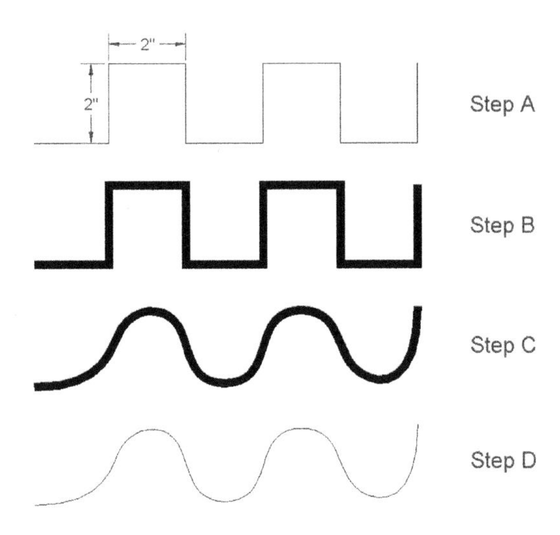

2. In a new file, set your units to Architectural and do the following pline exercise. Begin at the bottom and proceed to the right, following the given pline length and width information. (Difficulty level: Easy; Time to completion: <5 minutes.)

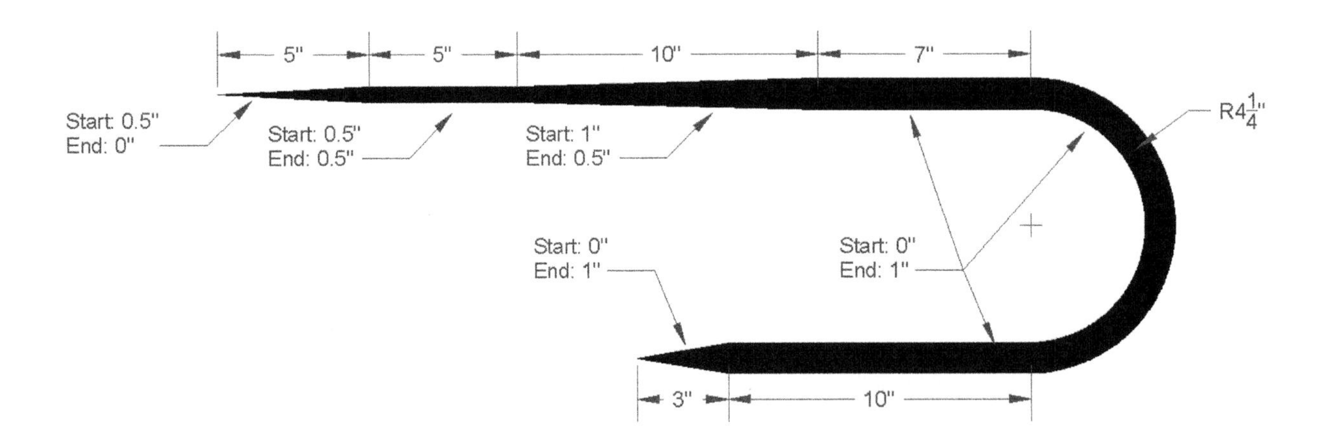

3. Using xline, circle, trim, copy, and rotate, draw a tree symbol as seen in Fig. 11.15, reproduced here. Repeat the same procedure using the ray command. (Difficulty level: Easy; Time to completion: <5 minutes.)

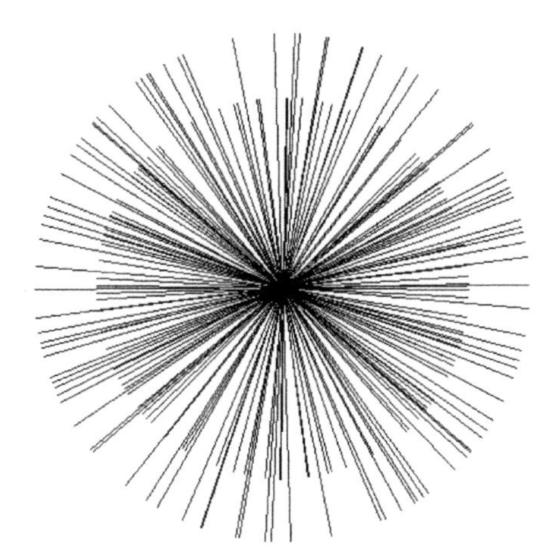

4. Using the spline and text commands, draw contour lines as seen in this chapter's cover image and reproduced here. To create the "gaps" in the splines where the height values are positioned, draw two long vertical lines and trim between them. We cover another method to do this in Chapter 15, Advanced Design and File Management Tools. (Difficulty level: Easy; Time to completion: 5–7 minutes.)

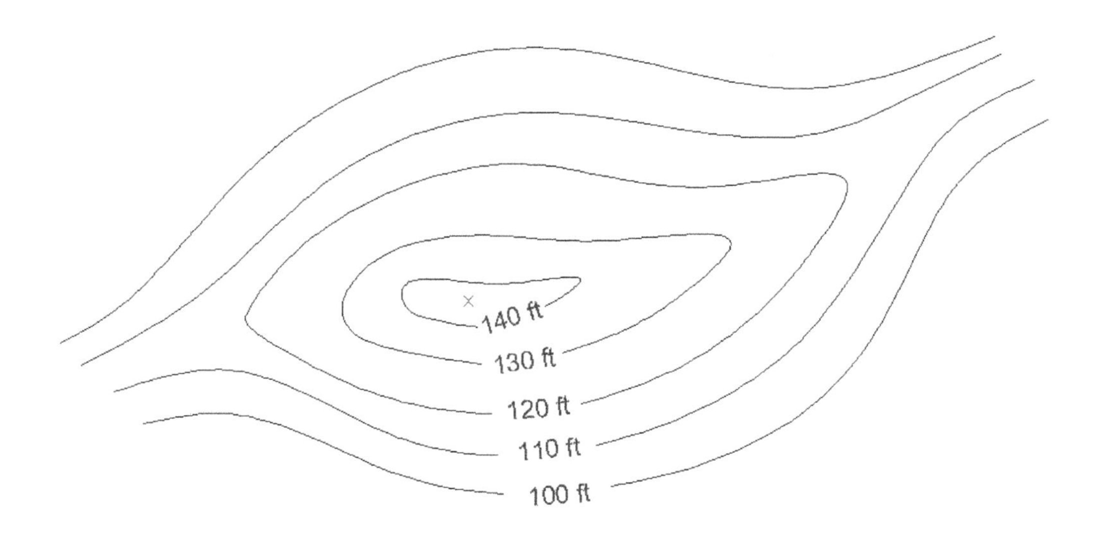

5. Use the mline command to create the following wall cross section. Be sure to replicate the linetypes shown, including the central BATTING linetype. (Difficulty level: Easy; Time to completion: 5–7 minutes.)

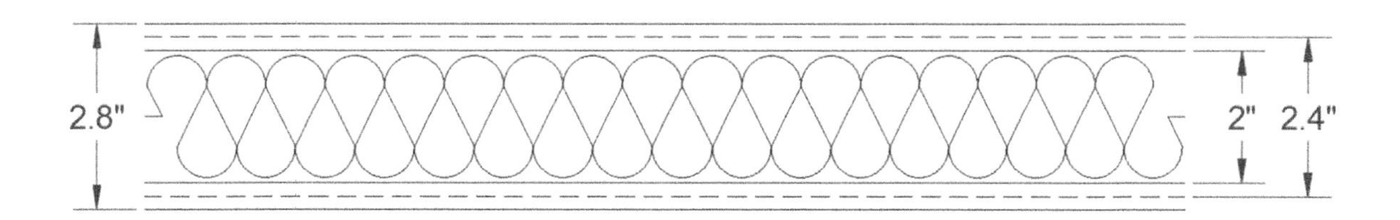

6. Using the mline, line, and hatch commands, draw a representation of a divided highway, as shown. You have to convert feet and inches into just inches, such as 4′–10″ = 58″, to set up the mline offsets. (Difficulty level: Intermediate; Time to completion: 15–20 minutes.)

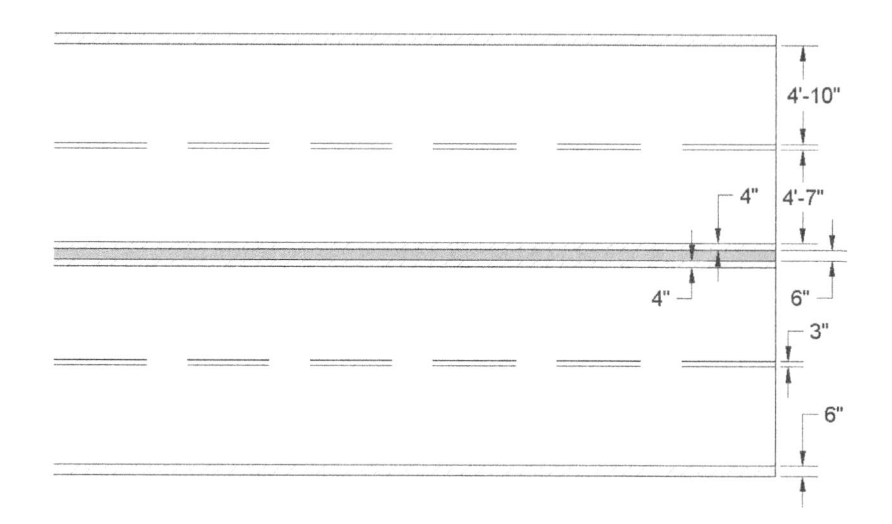

7. Using the pline or spline command, create the following cross-sectional contour of an airplane wing leading edge. Then, add in the additional hardware detail, text, and shading, as shown. (Difficulty level: Intermediate; Time to completion: 30–45 minutes.)

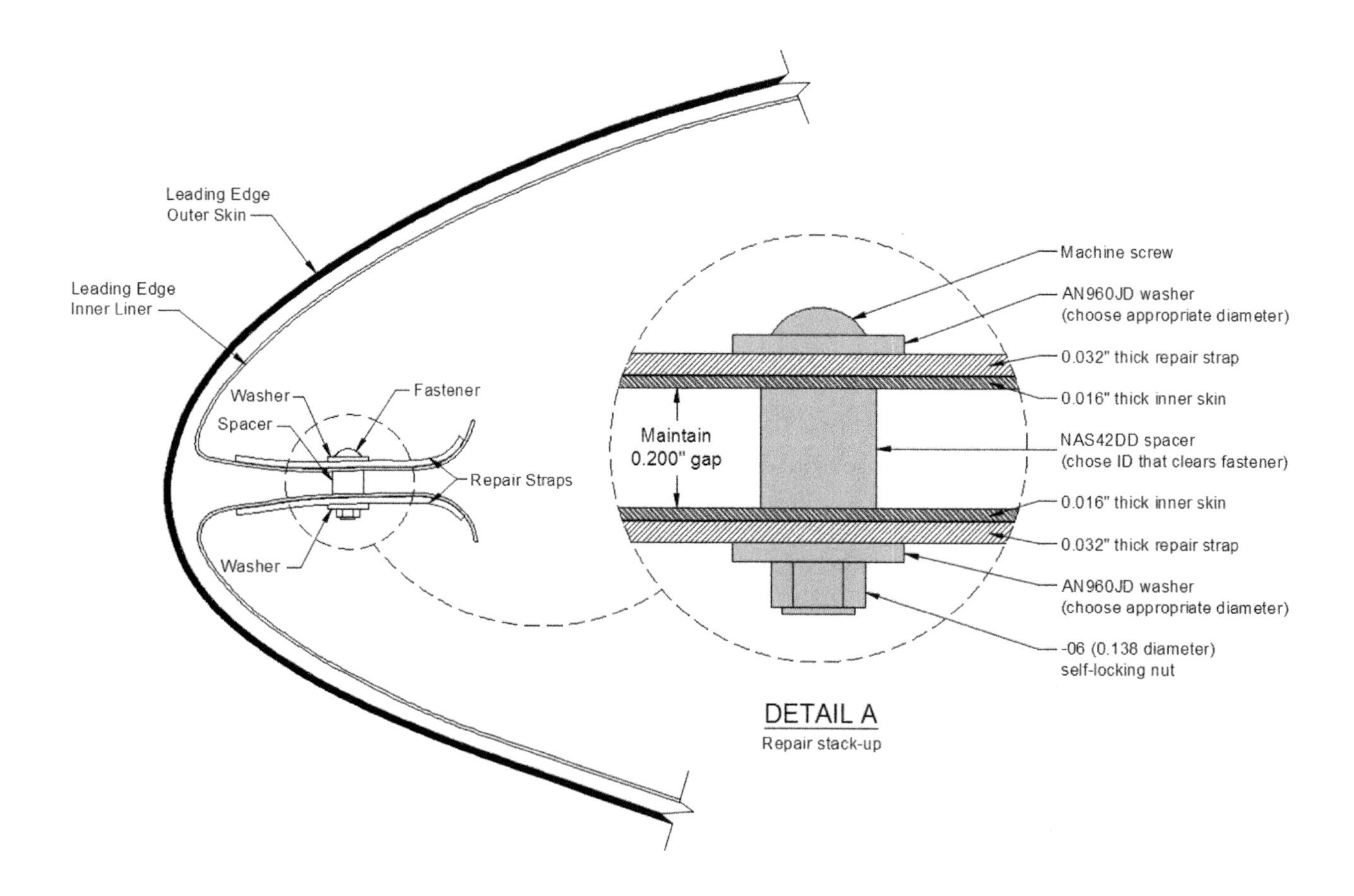

Spotlight On: Civil Engineering

Civil engineering specializes in design and construction of bridges, roads, buildings, dams, canals, and other infrastructure. Civil engineering is the oldest of the engineering specialties, and some of its principles were already well established during the building of the pyramids in ancient Egypt and aqueducts in Rome. Civil engineering has a number of subspecialties, including environmental engineering, geotechnical engineering, structural engineering, transportation engineering, municipal or urban engineering, construction, and water resources among them.

FIGURE 1 *Image source: www.megacivilengineering.blogspot.com.*

Education for civil engineers starts out similar to that for the other engineering disciplines. In the United States, all engineers have to attend a 4-year ABET-accredited school for their entry-level degree, a Bachelor of Science. While there, all students go through a somewhat similar program in their first 2 years, regardless of future specialization. Classes taken include extensive math, physics, and some chemistry courses, followed by statics, dynamics, mechanics, thermodynamics, fluid dynamics, and material science. In their final 2 years, engineers specialize by taking courses relevant to their chosen field. For civil engineering, this includes courses in asphalt, concrete, soil, geotechnical, water resources, traffic engineering, seismic, bridge, and advanced structural design.

FIGURE 2 *Image source: www.ciprianicharlesdesigns.wordpress.com.*

On graduation, civil engineers can immediately enter the workforce or go on to graduate school. Many civil engineers choose to pursue a Professional Engineer (P.E.) license. The process first involves passing a Fundamentals of Engineering (F.E.) exam, followed by several years of work experience under a registered P.E., and finally sitting for the P.E. exam itself. The P.E. license allows the civil engineer to approve and sign off on project plans, and many engineers offer their consulting services to architects on large building projects, where the expertise of a civil engineer is a major requirement.

Civil engineers can generally expect starting salaries (with a bachelor's degree) in the $60,000 per year range, which is roughly in the middle-to-lower range among engineering specialties. The median pay is about $82,000. This, of course, depends highly on market demand and location. The highest starting pay of all, $88,000 and up, is for petroleum engineering graduates. This specialty is a close cousin of civil engineering that specializes in locating and extracting petroleum. A master's degree is highly desirable and required for many management spots. The job outlook for civil engineers is excellent, as the profession is very much needed as the world's infrastructure ages and population growth creates a need for new development.

How do civil engineers use AutoCAD and what can you expect? Industry wide, AutoCAD enjoys a significant amount of use among civil engineers, as many civil projects are 2D in nature. This is especially true with site plan work and some building design. An example of a civil site plan is shown in Fig. 3, with many of them quite complex and detailed in nature. They use Xrefs for the building footprint(s), plines or splines for the gradients, and other advanced tools. The figure features the architectural layout (footprint), parking spots, landscaping, and site boundaries among other items. Depending on the plan, utility and power lines may be shown, as well as outdoor lighting, gradient lines, and various legal boundaries, such as easements and offsets.

As you move more into classic civil engineering projects (bridges, roads, etc.), 2D AutoCAD gives way to dedicated 3D software applications specially geared to this type of work, such as MicroStation, GEOPAK, and InRoads from Bentley. When AutoCAD is used, it is sometimes paired with additional software such as Civil 3D, Land Desktop, or various GIS add-ons.

Layering in civil engineering work is highly specialized and depends on the project. Site plan work may include some AIA layers, but mostly they are descriptive and unique to the company; there is no universal standard. Road, Easement Line, Gas Line, and Property Line are all valid layer names.

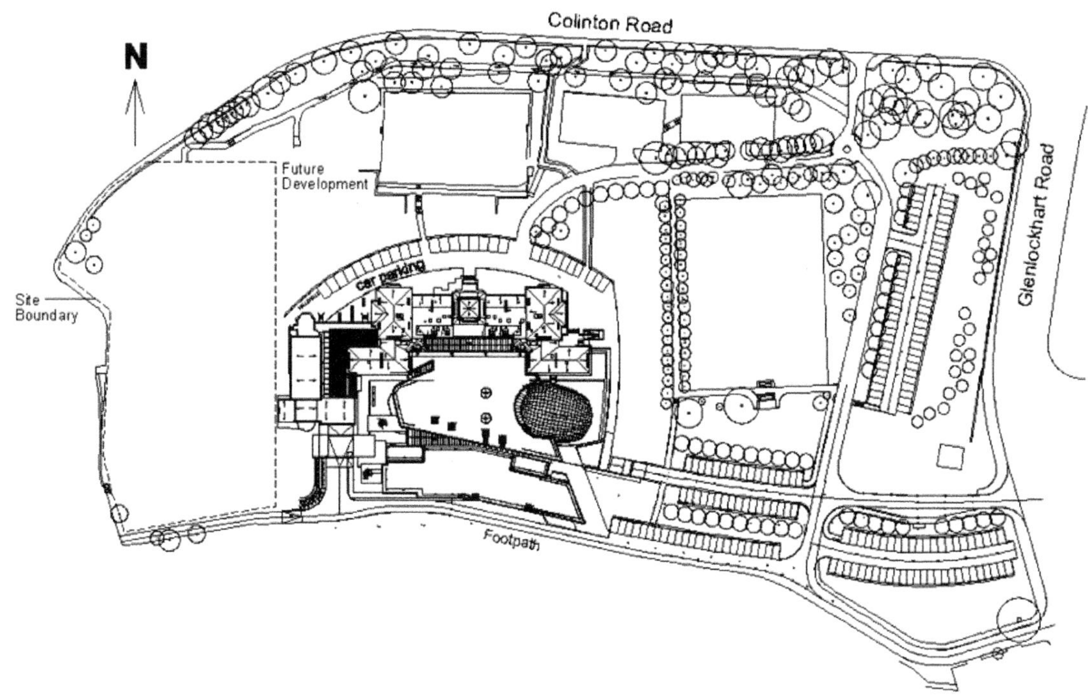

FIGURE 3 An example of a 2D civil engineering site plan. *Image source: Elliot Gindis.*

Chapter 12

Advanced Layers

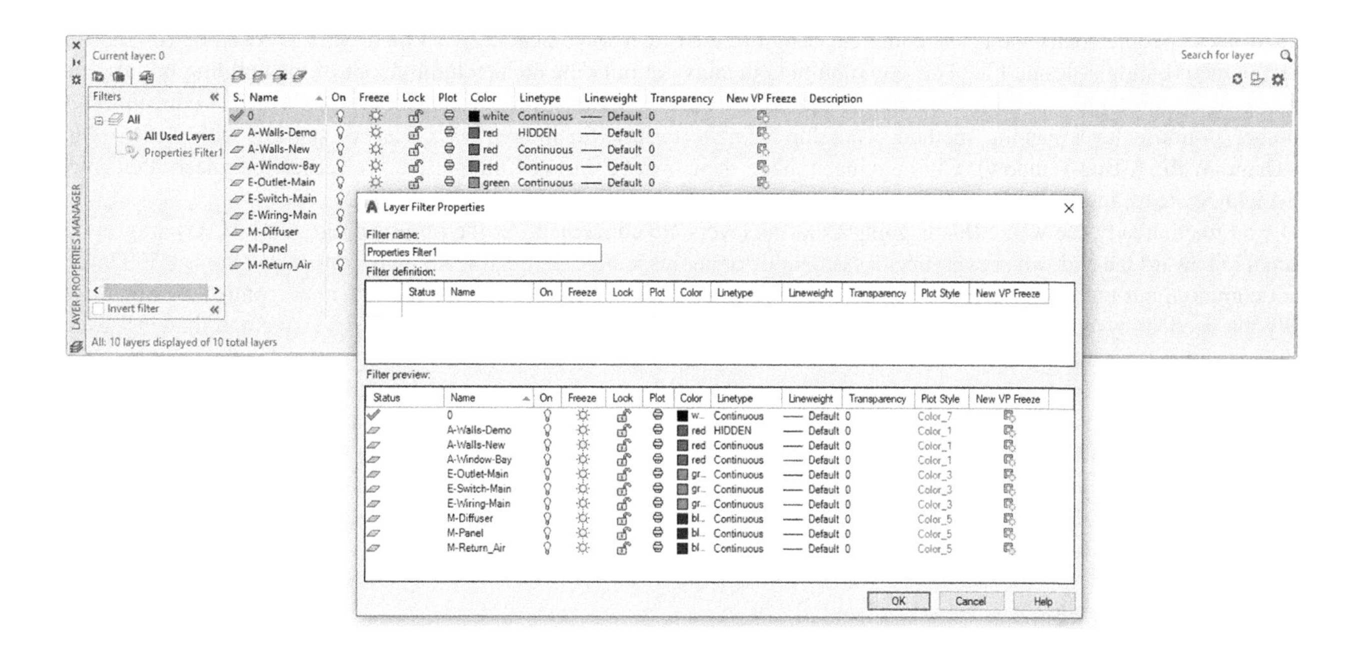

Learning Objectives

In this chapter, we introduce advanced layers and discuss the following:

- The need for automation
- Script files
- Layer State Manager
- Layer filters
- Layer Freeze, Isolate, and Walk.

By the end of this chapter, you will learn how to use and apply these new types of layer tools and be able to incorporate them into future design work where a large number of layers are present in the file.

Estimated time for completion of this chapter: 1–2 hours.

12.1 INTRODUCTION TO ADVANCED LAYERS

After the basic commands in Level 1, we introduced the layer concept. This was just a basic primer, showing you how to create them and set a variety of items, such as color and linetype. In addition, several methods of control were covered, such as freezing and locking. Here, we look at additional powerful features that allow for unprecedented control over your layers. They are script files, the Layer State Manager, and layer filtering.

Up and Running with AutoCAD 2019. https://doi.org/10.1016/B978-0-12-816440-2.00012-4

So, why do you need advanced layer control tools? Well, imagine the following scenario. You work at a large architecture or engineering consulting firm that specializes in complete commercial building and site design. Among your staff are a number of

- Architects and interior designers to lay out the overall exterior and interior of the building.
- Civil engineers to assist in structural design and site plans.
- Mechanical engineers to design the HVAC and mechanical systems.
- Fire protection and security engineers to design the electrical, sprinkler, and security systems.
- Landscape designers to integrate landscape design into the site plan.
- Additional design professionals as needed for specialized subsystems.

All these people contribute to the complete design in their own way. Generally, all their work is based on (or ends up in) one main building design file. This common file contains not only the architectural layout of the building but also the data from all these other disciplines. Each discipline has its own set of layers, usually following the AIA convention or another in-house layer naming method. Therefore, architects have their own set of layers starting with an A- (e.g., A-Demo-Wall, A-Bay-Window). Civil engineers have their own (C-Struct-Column, etc.), electrical engineers their own (E-Light-Switch), and so on.

You may already see where this is going as far as layers are concerned. As the project progresses, the layers grow in number. Toward the end, when everyone has added his or her respective design, the file may contain as many as 500 layers (not common, but not entirely unusual either!). There needs to be a way to impose some order and control over them, so only the needed layers are visible to each discipline. This can certainly be done by manually freezing and thawing layers but is very time-consuming. If you guessed that some sort of simple automation is available to you for this purpose, then you are correct.

This, in essence, is advanced layer management, a process where the layers, once set, are saved under a descriptive name, such as Arch_Layout or Elect_Layout, and recalled when necessary by simply clicking on the name. AutoCAD then freezes and thaws the layers automatically, exactly the way you initially told it to. This is what script files, the Layer State Manager, and layer filtering do. These tools allow sorting, classifying, and filtering of layers according to a variety of properties and are indispensable when working on a large project.

12.2 SCRIPT FILES

A script, in most general terms, is simply a text file with one command on each line. Its purpose is to automate a sequence of commands when executed. It is perhaps the simplest type of programming you can do. All you need is a text editor (such as Notepad). You next list the commands you want the script to go through, one per line. Then you save it under a name and a .scr extension (e.g., MyScript.scr), close the file, and run it from AutoCAD's command line by typing in script and pressing Enter (alternatively via the Ribbon's Manage tab → Run Script). The Script dialog box appears, the script is double-clicked on (executed), it does its → thing; and if you got the command sequence correct, a short task is automatically performed.

Scripts are generally not used as much now for purposes of layer control as they once were, but we briefly cover them because knowing how they perform gives you insight into the other remaining tools. It is also a good skill to have overall. You can automate a variety of AutoCAD routines. It may be worth your time to read up on further script file details in the Help files, as we do not go too much in depth here.

We present a very simple example of a script file. It freezes all layers after first thawing them and setting 0 as the current layer. As scripts do not work with dialog boxes, everything has to be command-line driven, such as this version of the layer command [generally any commands that are preceded by a hyphen (-) go to the command line, bypassing any dialog boxes]. After the -layer is a set of answers to prompts: t means thaw, * means all, s means set, and f means freeze. This exactly answers everything AutoCAD asks for if you were to do this by hand at the command line. The last two keystrokes (→ →) are Enter and Enter.

```
-layer
t
*
s
0
f
*
→→
```

This reads as Layer, Thaw, All, Set, 0, Freeze, All, Enter, Enter. Try this out with an AutoCAD file that has a bunch of layers. In a real application, you of course have to manually type in the layers deemed not relevant to the particular state (which therefore need to be frozen) in place of the last * and before the two Enters. The steps prior to that (t, *, s, 0) are necessary in case a layer that needs to be frozen is set to current (you cannot freeze the current layer), so the 0 layer is set as current.

This was a lengthy process and is the main reason why script files gradually fell out of favor, replaced by the Layer State Manager for layer control, as shown next. Keep in mind, however, that the Layer State Manager (LSM) works in pretty much the same way, except you select layers and do everything graphically through the Layers dialog box. Now that you have read through the underlying theory, we can rapidly go through the functions of the LSM.

12.3 LAYER STATE MANAGER

As described previously, the LSM is essentially a graphical version of script files. As you should now have a basic understanding of what you need to do, let us just set up a collection of layers and go through the mechanics of how to set up several layer states. While our example is simple—we set up only three layer states, an Arch_Layout, Mech_Layout, and Elect_Layout—the theory can be easily extended to a large number of layers and states. Open up a new AutoCAD file and set up the following layers:

A-Walls-Demo
A-Walls-New
A-Window-Bay
E-Outlet-Main
E-Switch-Main
E-Wiring-Main
M-Diffuser
M-Panel
M-Return_Air

Assign color to these layers as shown in Fig. 12.1. What we want to do now is to create a layer state called Arch_-Layout. This means only the A- layers are needed, with the electrical and mechanical layers frozen. So go ahead and freeze the six layers beginning with E- and M-, as shown in Fig. 12.1. For those of you who notice every detail, the Transparency category has been moved out of the way to the end of the layer palette.

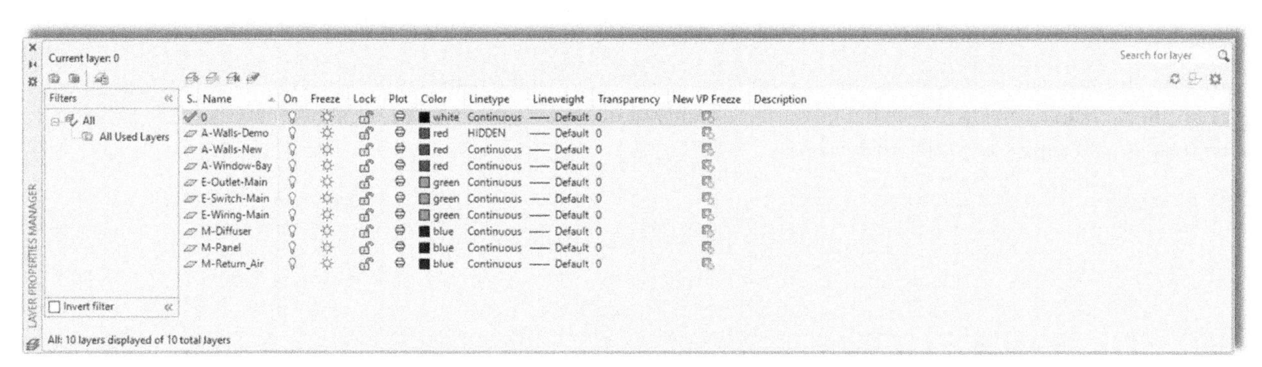

FIGURE 12.1 Layer setup for Layer State Manager exercise.

Now that the layers have been frozen, we need to save this "state" under the appropriate name. Click on the Layer State Manager button in the upper left of the Layers dialog box, as seen in Fig. 12.2. The LSM then appears (Fig. 12.3). It can be expanded (if it is not already) by pressing the arrow in the lower right corner.

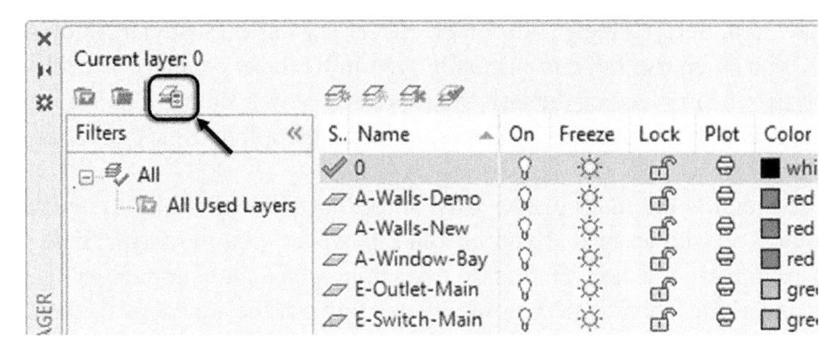

FIGURE 12.2 Layer State Manager icon.

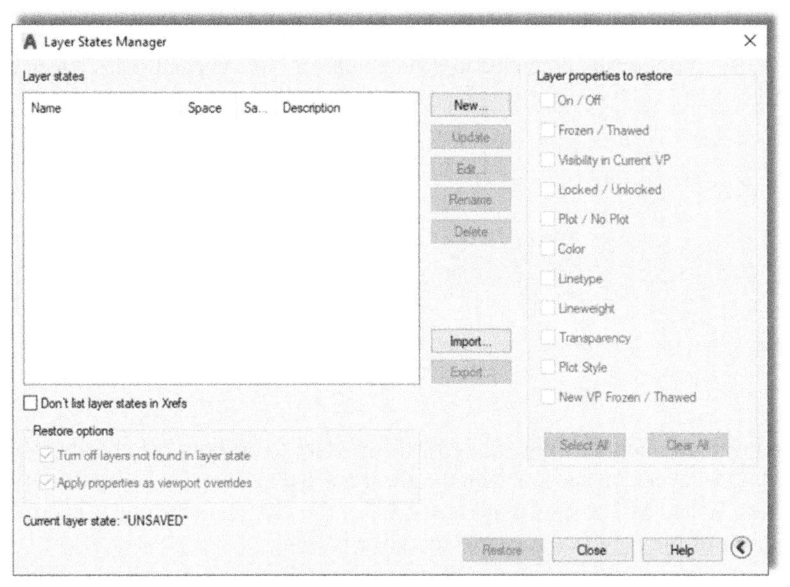

FIGURE 12.3 Layer State Manager.

We now create a new layer state by clicking on New… and filling in the name and description (if desired), as shown in Fig. 12.4, then clicking on OK. Once the layer state is entered, it appears in the LSM, as shown in Fig. 12.5, and you can press Close to exit out of the LSM dialog box.

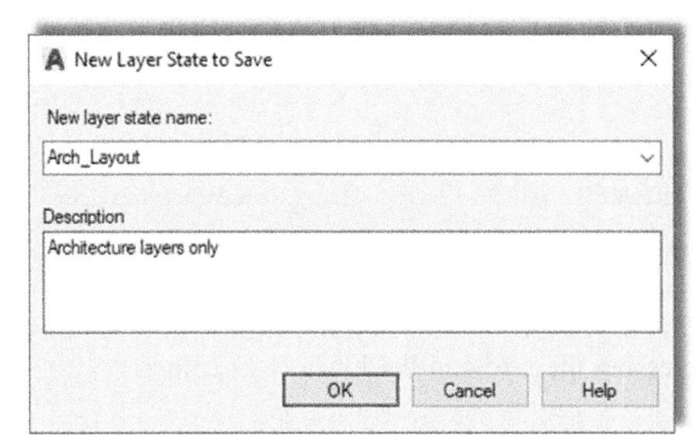

FIGURE 12.4 New Layer State to Save.

In the same manner, go back to the Layers dialog box and create the other two states: Elect_Layout (only E- layers visible) by turning off the unneeded layers and activating the LSM and Mech_Layout (M- layers visible). The result is shown in Fig. 12.6.

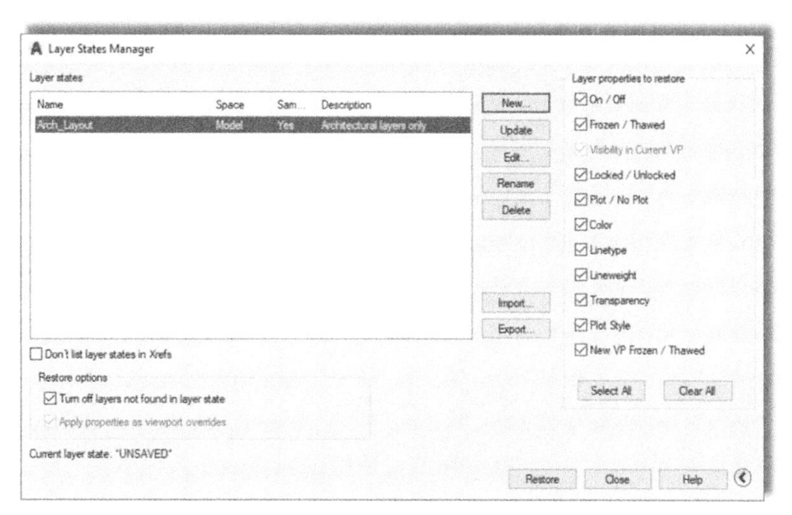

FIGURE 12.5 Layer State Manager with new layer state.

To activate these states when needed, you simply go to the Layers dialog box, click on the LSM icon, and when it appears, double-click on the state name or highlight it with one click and press Restore. Instantly, the correct layers are frozen or thawed (you can see this happen in real time in the Layers dialog box as you click on one state or another).

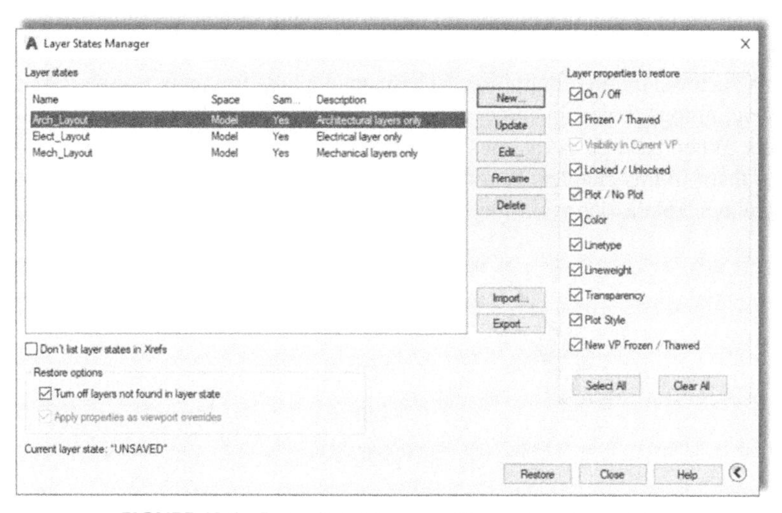

FIGURE 12.6 Layer State Manager with remaining layer states.

The purposes of the buttons found to the right of the main LSM window are as follows:

- *New*: Displays the New Layer State to Save dialog box, which we already used.
- *Save*: Saves the selected named layer state.
- *Edit*: Displays the Edit Layer State dialog box, where you can modify the layer state.
- *Rename*: Allows editing of the layer state name.
- *Delete*: Removes the selected layer state.
- *Import*: Layer states (*.las extension) can be imported from another file by means of this option, although this is not done often in practice.
- *Export*: Layer states (*.las extension) can be exported to another file by means of this option, although this is also rarely done in practice.

The additional options available in the LSM dialog box can be made visible or hidden if you click on the small arrow in the extreme bottom right, as previously mentioned (Fig. 12.7).

Layer properties to restore are simply a listing of every property a layer typically has, and you can check off different properties if you want them restored. If all are checked, as is more typical in practice, all the layer's properties are affected.

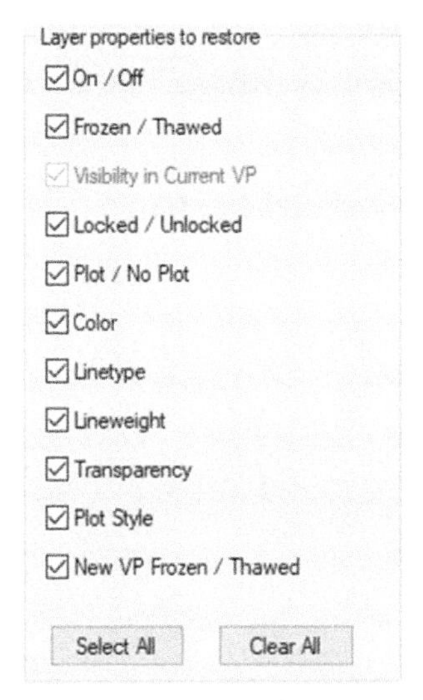

FIGURE 12.7 Additional Layer State Manager options.

12.4 LAYER FILTERING

Layer filtering is a tool that allows you to filter out (or selectively choose) layers that fall under a certain description. You can create filters of your own choosing and assign them names. For example, if for some reason you want only the red layers to show or ones assigned a hidden linetype, this can be done. Layer names can also be filtered. If you want only those layers that start with an A- to show, this can be arranged, although in this case you are essentially doing what was described previously with the LSM.

The key to layer filtering is to set up filter definitions. These can include full names or partial names (using the wildcard * character) and property definitions (color, linetype, etc.). The entire left side of the Layers dialog box, since AutoCAD 2005, is dedicated to filters. When used properly, these are very effective tools. We highlight some of the more important features essential to using them in this chapter. You are encouraged to explore them further on your own as well.

Two types of filters are available as icons on the top left of the Layers dialog box, as shown in Fig. 12.8.

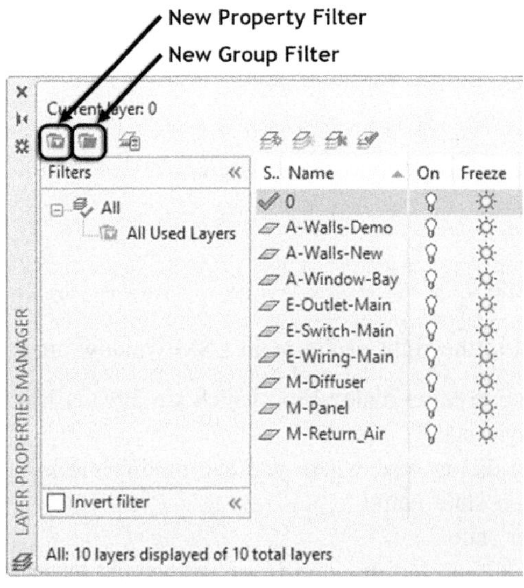

FIGURE 12.8 Layer filters.

The New Property Filter is what you use to set up the definitions. If you press that icon, the box in Fig. 12.9 appears, with the previously entered layers visible.

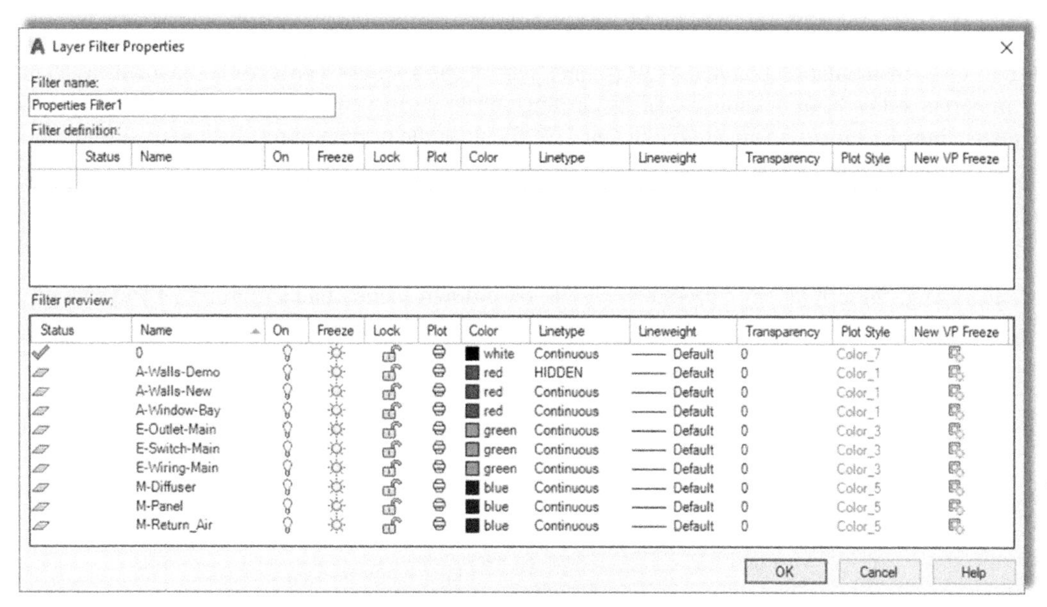

FIGURE 12.9 Layer Filter Properties.

The Layer Group Filter is a general filter that includes all the layers put into the property filter when you define it, regardless of their names or properties. Selected layers can then be added from the layer list by dragging them into the filter. This type of filtering is generally not used but can be employed to make a new filter based on the layers of another filter. To begin, click on the icon, create a name, and click and drag layer names from the right side of the Layers box into the group filter.

Remember an important point with filters: By themselves, they do little except sort the layers. It is up to you to then do something useful with this, such as freeze or lock them as needed. Therefore, think of filtering as simply a way to get the layers in some sort of order for further action on them.

Here, you first enter the name of your filter in the text field in the upper left. Then, it is just a matter of careful selection of the properties or names (or both) of the layers you want to see. You may use the wild character * in conjunction with typed parts of the layer name to indicate to the filter to include anything before or after the specific letters. Any property can also be used as a filter; color and linetype are common ones.

Go through the filter at your own pace and experiment with the settings. Shown in Fig. 12.10, as a simple example, are the layers that are green (grey in print version) and the letter W somewhere in the name.

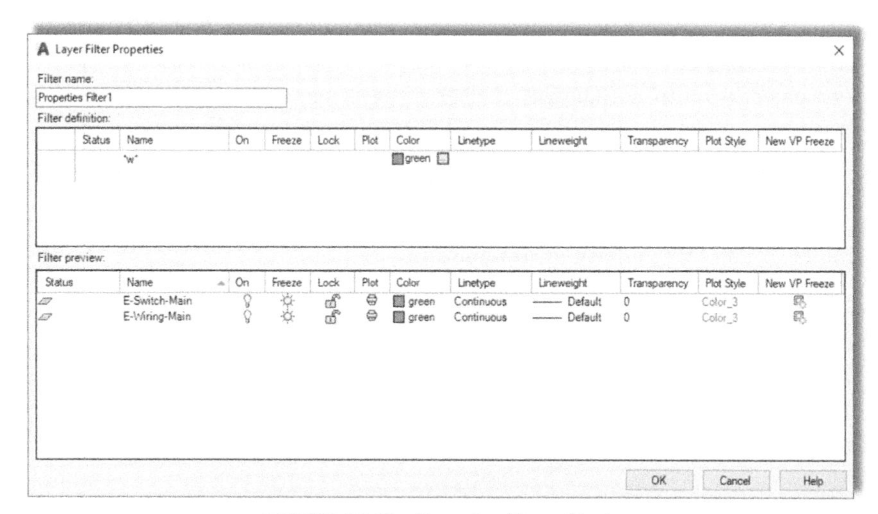

FIGURE 12.10 Example of layer filtering.

12.5 LAYER FREEZE, ISOLATE, AND WALK

These very useful layer manipulation commands were once part of the Express Tools (to be covered in detail in Chapter 14), but their importance has been acknowledged by an upgrade to permanent status within regular AutoCAD menus. Although there are a bunch of them (many are redundant), we focus on the primary three that are of significant importance:

- Layer Freeze (command line: `layfrz`)
- Layer Isolate (command line: `layiso`)
- Layer Walk (command line: `laywalk`)

In addition to typing, these tools are accessible via the drop-down menus under Format → Layer tools, as seen in Fig. 12.11.

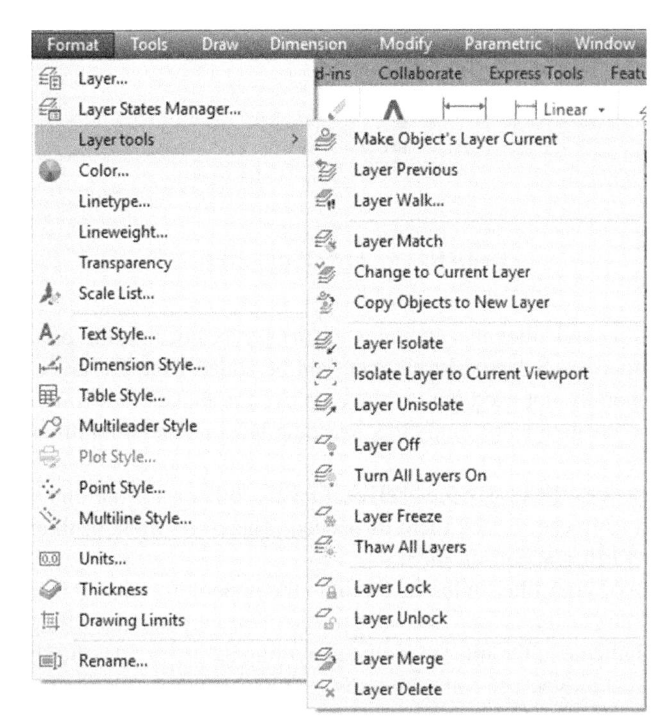

FIGURE 12.11 Layer tools.

Layer Freeze

In a complex drawing, layers may number in the dozens or even hundreds. Any tool that can simplify dealing with them is a welcome addition. Layer Freeze allows you to freeze a layer merely by selecting it. This bypasses the tedious process of finding out what layer an object is on (if not known), then freezing it manually via the layer dialog box or the toolbar. Simply type in `layfrz` and press Enter or select Format → Layer tools → Layer Freeze.

- AutoCAD says: `Select an object on the layer to be frozen or [Settings/Undo]:`

Pick an object that represents the layer you want frozen, and it disappears. This can be repeated as often as needed and is a good way to clear a lot of junk from a busy, dense drawing.

Layer Isolate

Layer Isolate allows you to isolate a layer merely by selecting it. The command turns off (not freezes in this case) all the other layers, leaving the selected one by itself. This bypasses the tedious process of doing it manually. Simply type in `layiso` and press Enter or select Format → Layer tools → Layer Isolate.

- AutoCAD says: `Current setting: Hide layers, Viewports = Vpfreeze`

  `Select objects on the layer(s) to be isolated or [Settings]:`

Pick an object that represents the layer you want isolated, press Enter, and all the others disappear. It is another good way to clear a dense drawing.

Layer Walk

This is perhaps the most interesting of all the layer controls and a good way to shuffle through an entire drawing layer by layer. Layer Walk presents a list of layers and isolates one of them. Then, you use the arrow keys on your keyboard to select your way up or down the list. AutoCAD turns layers on or off in succession, allowing you to inspect each layer, one at a time (walk through them). To try this, open up any file that has a number of layers. Then, type in `laywalk` and press Enter or select Format → Layer tools → Layer Walk....

A dialog box appears, listing all the layers visible from layer 0 until the last one, as shown on a sample file (Level 2 project) in Fig. 12.12. Click on layer 0; it highlights that layer and all the rest disappear. Use the arrow keys to cycle down through each layer and notice how each one appears and disappears as its turn comes up in the listing.

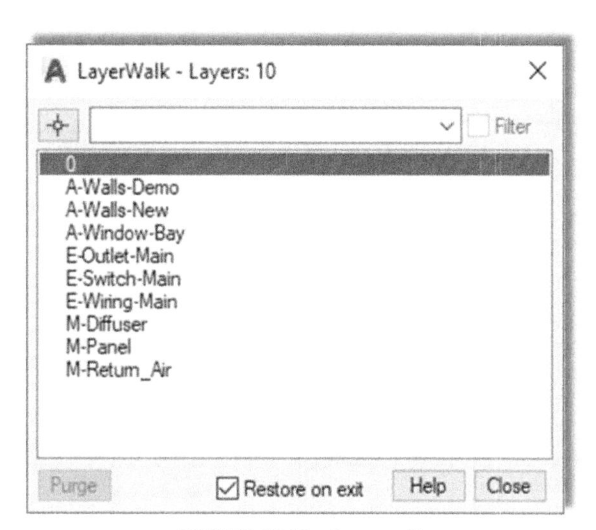

FIGURE 12.12 Layer walk.

This is a great tool to review all the layering on an unfamiliar drawing as well as one of your own to ensure accuracy. Sometimes AutoCAD designers let their layering accuracy slip as the project gains complexity, and by the end, layers are sloppy, with objects residing where they are not supposed to be.

After inspecting all the layers, you can stop on any one of them for editing; if the Restore on exit selection box is checked, then all the layers come back on when you press Close.

Most of the other layer commands listed in Fig. 12.11 are redundant, as you can already perform these functions in other familiar ways (via the LA dialog box or toolbar), so we do not discuss them further, but review them at your own pace.

12.6 LEVEL 2 DRAWING PROJECT (2 OF 10): ARCHITECTURAL FLOOR PLAN

For Part 2 of the drawing project, you continue to add to the walls of the floor plan, with a lot of new work on the interior, as well as adding doors, windows dimensions, and text. We can now also hatch the exterior and interior walls.

Step 1. As always, before starting to draft something new, create the appropriate layers, as follows:

- A-Doors (Yellow).
- A-Text (Cyan).
- A-Walls-Hatch (Gray_9).
- A-Walls-Interior (Green).
- A-Windows (Red).

Step 2. Add in the exterior windows and all doors. Be sure to follow the layering guidelines carefully. Make all windows and doors blocks for convenience. The exterior dimensions are shown in Fig. 12.13. Note that the deck and deck hatching is frozen, as it is not relevant to the new material you are adding.

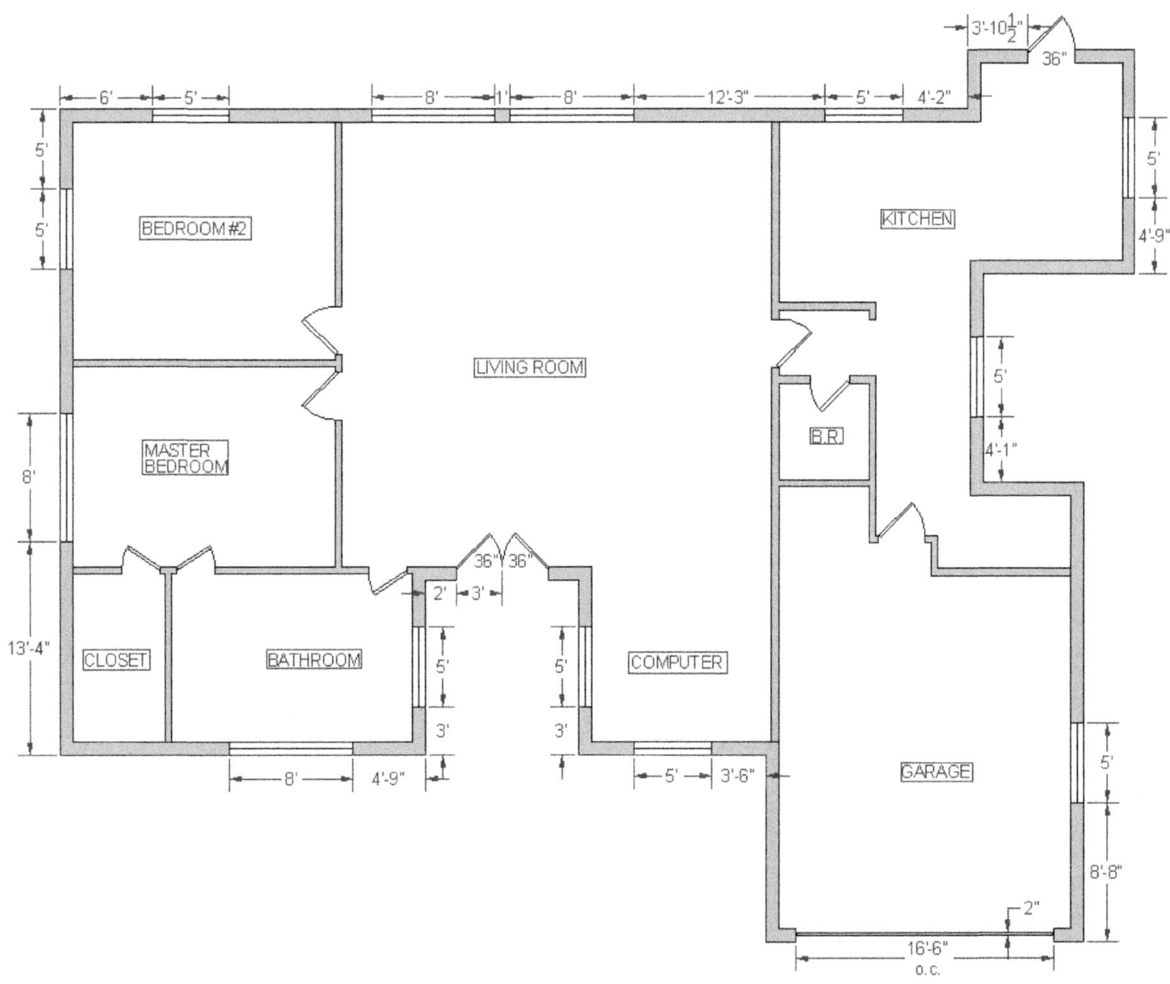

FIGURE 12.13 Exterior windows and doors.

Step 3. Add in the interior walls and doors as shown in Fig. 12.14. All interior walls are 5″ thick. Also add in the actual dimensions themselves. As mentioned before, they are very useful as a check of your drafting accuracy.

Step 4. Finally, add in the hatch (using the light gray color 9). This should be the very last step because you need to erase the hatch if you find a mistake in the layout.

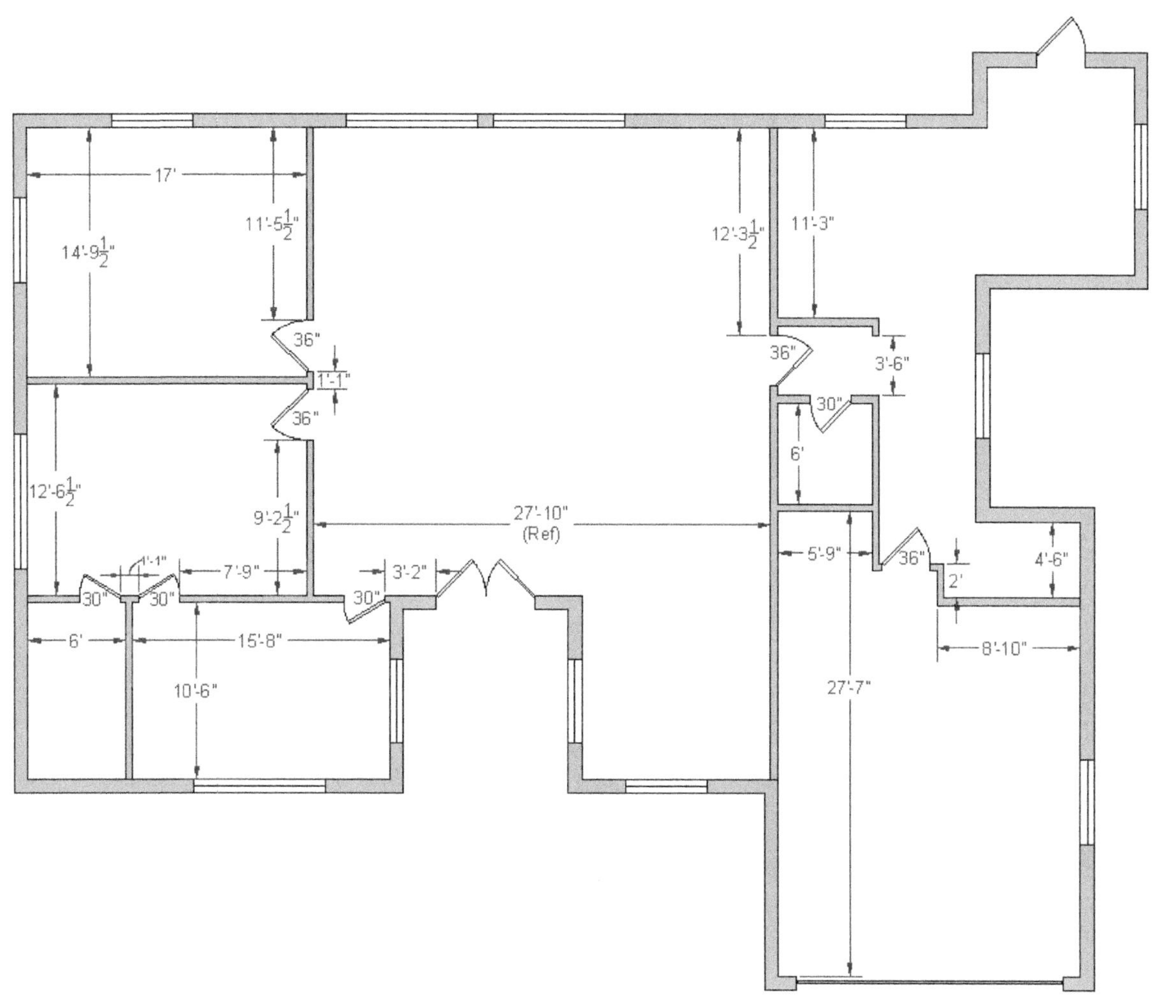

FIGURE 12.14 Interior walls and doors.

SUMMARY

You should understand and know how to use the following concepts and commands before moving on to Chapter 13:

- Script files
  - Basic concept
  - Writing a simple script file

- Layer State Manager (LSM)
  - New
  - Save
  - Edit
  - Rename
  - Delete
  - Import
  - Export
  - Restore
  - Layer properties
- Layer filters
  - New Property filter
  - Layer Group filter
- Layfrz
- Layiso
- Laywalk

REVIEW QUESTIONS

Answer the following based on what you learned in this Chapter 12:

1. Describe what is meant by *layer management*. Why is it needed?
2. Describe the basic premise of a script.
3. What can a script file do for layer management?
4. Describe what the Layer State Manager does.
5. What are the fundamental steps in using the Layer State Manager?
6. What is layer filtering? What are the two types?
7. Describe layfrz, layiso, and laywalk.

EXERCISES

1. Create six layers (and colors) as shown next. Then, write a script file to thaw all the layers; set layer 0 as current. Finally, freeze all and thaw just the even numbered layers. (Difficulty level: Easy; Time to completion: 10 minutes.)
   - A-Walls_1 (Green)
   - A-Walls_2 (Red)
   - A-Walls_3 (Green)
   - A-Walls_4 (Red)
   - A-Walls_5 (Green)
   - A-Walls_6 (Red)
2. Using the same six layers, set up layer states Odd_Layers and Even_Layers and run them. (Difficulty level: Easy; Time to completion: <5 minutes.)
3. Using the same six layers, create filters to sort and capture only the green layers. (Difficulty level: Easy; Time to completion: <5 minutes.)

4. To maintain your drafting skills, draw the following two sets of mechanical valves. Not all dimensions are shown, so improvise when needed. The upper and lower valve assemblies have many features in common. Use the detail images to assist in drafting, but there is no need to draw them separately. (Difficulty level: Intermediate; Time to completion: 40–50 minutes.)

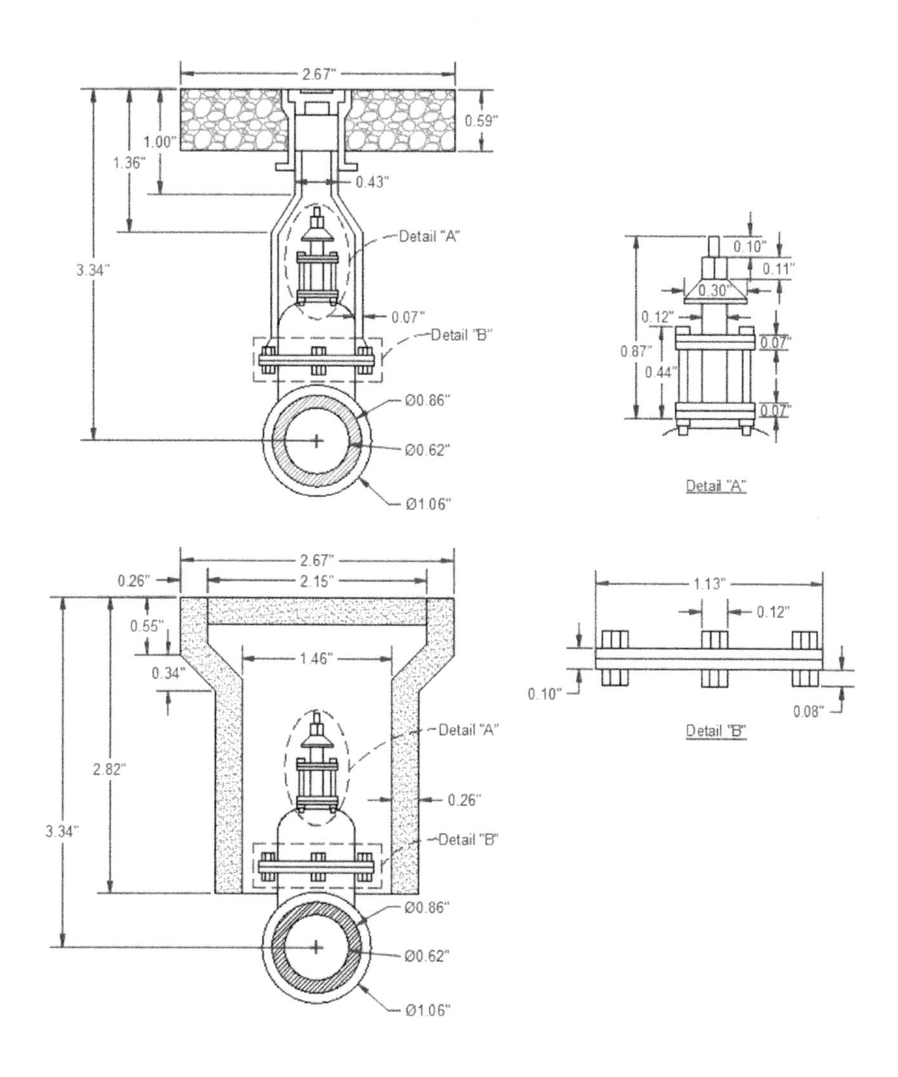

Chapter 13

Advanced Dimensions

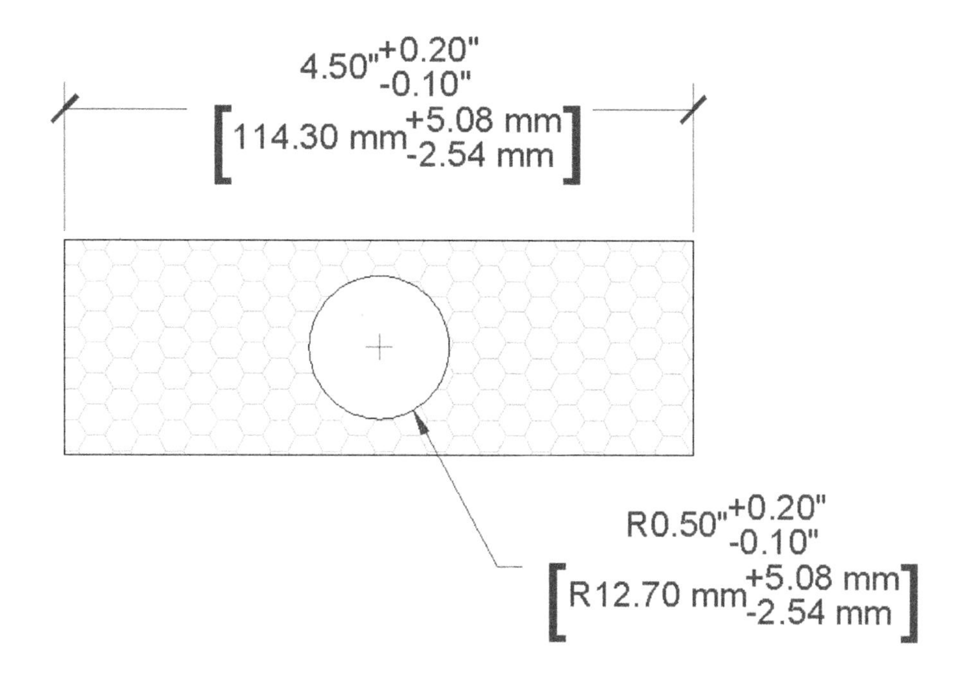

Learning Objectives

In this chapter, we introduce advanced dimensions and discuss the following:

- Lines tab
- Symbols and Arrows tab
- Text tab
- Fit tab
- Primary Units tab
- Alternate Units tab
- Tolerances tab
- Geometric constraints
- Dimensional constraints

By the end of this chapter, you will learn additional, advanced features of the Dimension Style Manager and be capable of setting up complex dimensions. We also discuss two advanced dimensioning features called *geometric* and *dimensional constraints*.

Estimated time for completion of this chapter: 2–3 hours.

Up and Running with AutoCAD 2019. https://doi.org/10.1016/B978-0-12-816440-2.00013-6

13.1 INTRODUCTION TO ADVANCED DIMENSIONS

This chapter is meant to complete your knowledge of AutoCAD's dimensioning features. Specifically, we look at the New Dimension Style (NDS) dialog box and Parametrics. In Level 1, we focused primarily on defining the types of available dimensions and how to properly dimension geometry. Little was mentioned of this dialog box except for the four essential features deemed most important: Arrowheads, Units, Fit, and Text Style (not needed as much anymore, as the better-looking Arial is the default font). In this chapter, we go through the entire NDS dialog box and discuss other options and features, some more than others, according to their usefulness.

The reason you need this knowledge is because AutoCAD allows for a tremendous amount of power, flexibility, and variation with its dimensioning. As a beginner user, you may not come to appreciate what is available because much of it may already be "set up" for you. As a CAD manager and advanced user (for whom Level 2 is geared), you may be charged with doing this "setting up" and need to know what to do. We conclude this chapter with a feature introduced back in AutoCAD 2010 called *Parametric Dimensions and Constraints*. These are routinely found in advanced 3D software but are relatively new for AutoCAD.

13.2 DIMENSION STYLE MANAGER

To start off, let us take a closer look at the Dimension Style Manager, which was previously described in Chapter 6. So as a reminder, here is how we got it up on our screen.

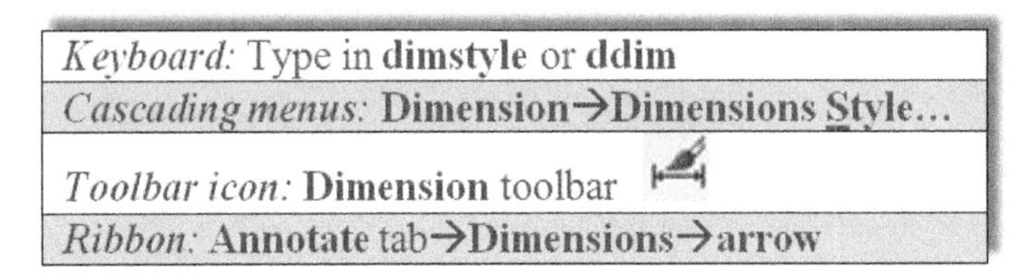

Keyboard: Type in **dimstyle** or **ddim**
Cascading menus: **Dimension→Dimensions Style...**
Toolbar icon: **Dimension** toolbar
Ribbon: **Annotate** tab→**Dimensions**→**arrow**

Create a new style, by pressing the New... button, and name it as Sample Style (Fig. 13.1). Click on Continue when done. You then see the Create NDS dialog box. We now take a much closer look at the seven tabs: titled Lines, Symbols and Arrows, Text, Fit, Primary Units, Alternate Units, and Tolerances.

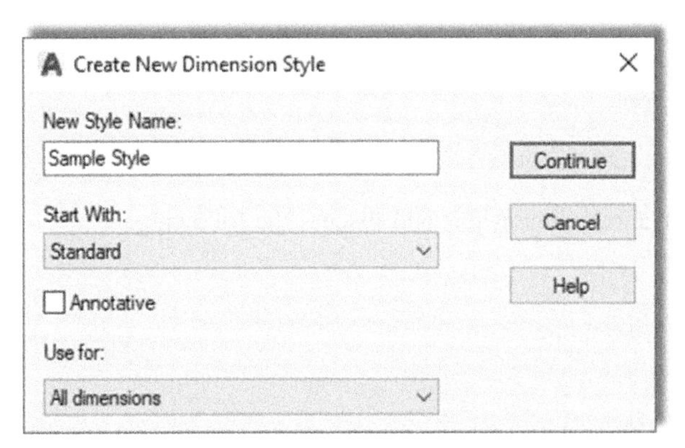

FIGURE 13.1 Create New Dimension Style.

Keep in mind as we go through these that our goal is not to cover every possible button and menu in extreme detail. Many of the settings are rarely used, and to describe them all completely would take dozens of pages and bewilder even the most dedicated AutoCAD student.

A much better approach is to give a basic overview of each tab and focus only on the features that are of utmost importance and widely used. You will see those features identified and described in more detail, with emphasis on how they fit in with the design process and how they benefit you as an AutoCAD user. If you wish to learn more, use this information as a starting point and consult the Help files.

Let us take a look at the first tab, all the way on the left of the NDS dialog box, called *Lines* (Fig. 13.2). If it is not already visible, click on it to bring it up. Throughout the process, carefully observe the Preview window at the upper right to see the results of changing or adjusting whatever we discuss.

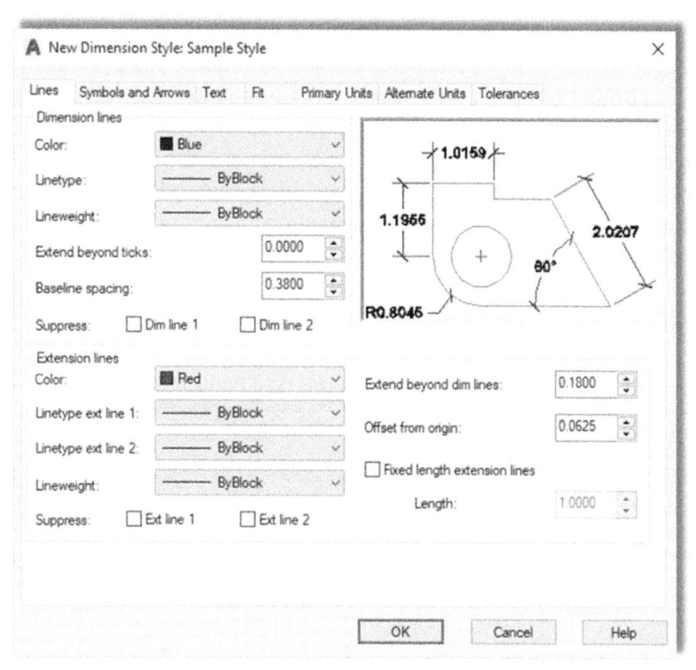

FIGURE 13.2 Lines tab.

Lines Tab

This first tab is one that most users leave "as is" and generally do not modify. Here, we can make mostly cosmetic changes to two items specifically: Dimension lines (upper left) and Extension lines (bottom). It is, of course, instructive to know what they are. Dimension lines extend from the numerical value and contain arrowheads or another symbol. The color Blue is selected for them in Fig. 13.2. Extension lines extend from the part you are dimensioning and "frame" the rest of the dimension. The color Red is selected for them in Fig. 13.2.

The Lines tab is all about working with these two features. For both the dimension and extension lines, you can change their color, linetype, and lineweight. You simply click on the down arrow and change the property. You can also suppress either one or both of the extension or dimension lines, which make them disappear from view. This has some value if, e.g., a dimension is sitting between two walls and the extension lines are not visible anyway. You can then suppress them. This, however, would be a "local" thing, applied to just that one dimension via the Properties dialog box. There is little use in suppressing either extension or dimension lines for the entire drawing. Next, a few values are presented:

- Extend beyond ticks: (grayed out).
- Baseline spacing.
- Extend beyond dim lines.
- Offset from origin.

All these values refer to the various spaces between dimension elements or length of elements. They can all be fine-tuned one at a time. However, it is very important to understand that you do not want to do that here in this tab. If you wish to change the size of a dimension, you need to do this by increasing the overall scale, not one element or space at a time. We discuss this in great detail later, when we cover the Fit tab. If you try to adjust the values one at a time, you end up with a lopsided dimension, so leave them all as default.

Finally, we have a check box for fixing the length of an extension line (and assigning a permanent value). This has the effect of making the extension line a specified length regardless of how far the dimension line is placed away from the objects being dimensioned. It is rarely used, as it takes away some flexibility in dimension placement. Just for practice, go ahead and change the colors of the dimension lines to blue and the extension lines to red, as seen in Fig. 13.2.

Symbols and Arrows Tab

The most important feature found under this tab is the ability to change the arrowheads of your dimensions (upper left of tab). You already did this briefly in Chapter 6: changed your arrowheads to tick marks. Go ahead and do this again, making a note of the other available choices. You can make the second arrowhead different from the first, of course, but that is rarely done. The leader is generally left as an arrow because its main purpose is to point at something. Just below the leader is a field to change arrow sizing. This also falls under the "do-not-do-it-here" category, as discussed in the previous tab. Arrowheads need to be sized according to the overall scale, not individually. We discuss this under the Fit tab.

Moving below the Arrowheads, we have the Center marks. This has to do with measurements of a circle and what you want to see in the center of one when a diameter or radius dimension is added. Mark or Line is typically selected; let us go with Line. Note how the Preview window reflects both this and the arrowhead changes.

Below Center marks and to the right, we have some relatively minor options. The Arc length symbol concerns the placement of the symbol, and the Radius jog dimension has to do with the angle of the jog "wiggle." Leave them, as well as the Linear jog dimension, as default all around. Do the same thing for the Dimension break field. It sizes up or down according to overall scale settings, so leave it as is for now. Your Dimension Style dialog box should look like Fig. 13.3.

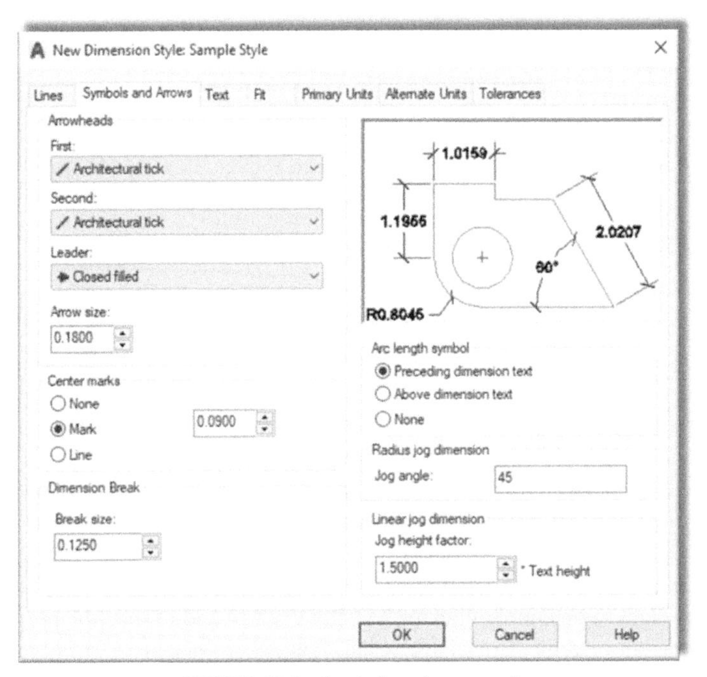

FIGURE 13.3 Symbols and arrows tab.

Text Tab

This tab is all about what dimension text looks like (appearance) and where it goes (placement and alignment). Take a look at the Text appearance category at the upper left. You already practiced changing the text under Text style. The goal was to make the dimension text the same as the rest of the text in the drawing: same font and usually the same size. In Chapter 6, when this was done, the text you selected on the floor plan was Arial 6″.

In this case, create another font from scratch by pressing the button to the right of the Text style: field (more options button with the three dots). Create a nonannotative Text Style called *HelveticaLight 0.5*, setting the Font Name to HelvLight and Size, Height to 0.5000. This technique is first outlined in Chapter 4, so review that chapter if you have any problems doing this.

Moving down the Text appearance category, we find

- *Text color*: Usually not set separately and remains "By Layer." Here, for practice, we set it to the color Magenta, as seen in Fig. 13.4.

- *Fill color*: Creates a color highlight behind the text. Try it out for practice, but in this example, no fill is set permanently. To add a frame around the text, there is a check box for this further on down, which we check off. Often a fill and a box are used together.
- *Text height (grayed out)*: This is usually not adjusted here. Use the previously mentioned Style dialog box.
- *Fraction height scale (grayed out)*: This applies to the size of fractions. The field can be set only when architectural units are being used. We look at this setting in detail soon.

These settings take care of pretty much every text appearance issue. Next, we look at how the text is placed in the drawing.

The Text placement category (bottom left) allows you to position the dimension value in every conceivable way near the rest of the dimension. You can place the values above, below, or centered on the dimension line. You can also move the values around the extension line, lining them up. Experiment with all the possible settings, observing the results in the preview window to the upper right. The default setting for both Vertical and Horizontal is Centered, as seen in Fig. 13.4. View Direction: flips the values around, and Offset from dim line: is best left alone and is adjusted as part of the overall scale settings.

With Text alignment (bottom right), the common default is Horizontal because people generally do not want to tilt their heads to the side every time they check a dimension on a printout. Do try the other two options (aligned and ISO) to see what they do, though. Your Dimension Style dialog box should look like Fig. 13.4.

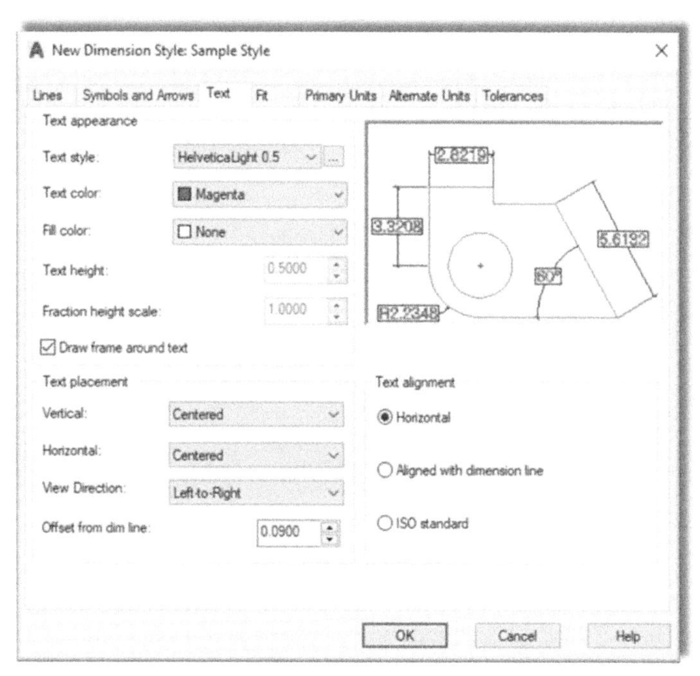

FIGURE 13.4 Text tab.

Fit Tab

The contents of this tab can be a bit confusing at first; after all, did not we just focus on text placement under the Text tab? Yet, here is another set of categories to fine-tune text placement. You can think of the Fit tab as a continuation of the Text tab with additional Fit options, Text placement, and Fine tuning categories.

Sometimes when you dimension small areas, the extension lines are so close together that the dimension value just does not fit in between them. The Fit options category, at the upper left, simply asks: "If the dimension does not fit, what do you want to move?" The best answer here is to select the first option, *Either text or arrows (best fit)*, and let AutoCAD decide.

You can manually force other choices as seen in the listing (arrows, text, etc.), of course, but then this choice is applied to all cases uniformly, which may not always be correct. You can even suppress arrows if they do not fit between the extension lines, but that is rarely a good idea and makes for some odd-looking dimensions.

The Text placement category (bottom left) is really just a minor setting and should be left in the default first choice. At the very bottom right, we have the Fine-tuning category. These two boxes are not all that important, as you rarely want to place text completely manually (adds extra workload) and asking AutoCAD to always draw a dim line between extension lines makes no sense, as the numerical value is then "crossed out."

The most important category to discuss here is the *Scale for dimension features* area just above the Fine-tuning category. You already saw the all-important Use overall scale of: field in Chapter 6 (you entered the value 15). Then, in Chapter 10 there was a full discussion of what values go in that field and why. Recall that this value boosts (or lowers) the overall sizing of the dimension, and there is no need to adjust all the individual components that we mentioned in the previous tabs. For now, just enter the value 5.0000.

Your Dimension Style dialog box should look like Fig. 13.5.

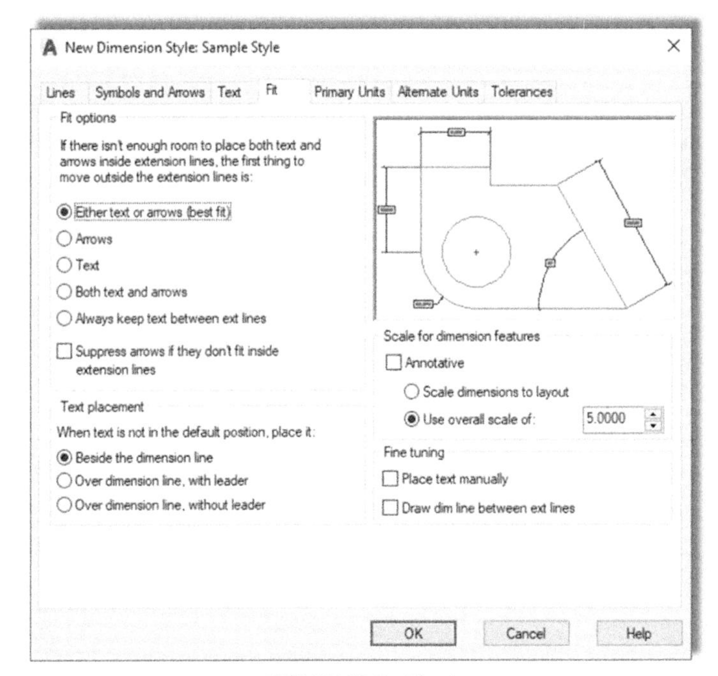

FIGURE 13.5 Fit tab.

Primary Units Tab

This tab features quite a lot of useful settings. In the Linear dimensions category (upper left), you already practiced changing the Unit format: from Decimal to Architectural in Chapter 6. Generally, the dimension units are the same as the drawing units, so if you draw in architectural feet and inches, you would want to dimension in similar units. For this exercise, we leave the units in Decimal mode. In both cases, the Precision setting just below the units can be set to whatever value is appropriate. The range is 0–0.00000000 for Decimal and 0′–0″ to 0′–1/256″ for Architectural. Make a note also of the other units that can be selected, although they are rarely used.

Just below Precision you find the Fraction format: Decimal separator: and Round off: fields. The first two should be relatively clear in their function. Fraction format gives you a few choices (Horizontal, Diagonal, and Not Stacked) in how you want the fractions to look if using architectural units. This field is grayed out if using decimal dimensions.

The Decimal separator gives you a few choices (Period, Comma, and Space) in how you want the decimal values to look (European engineers do not use a comma when writing 1,000, for example). This field, of course, is grayed out if using architectural dimensions. You do not want to use Round off: at all. This field, of course, is grayed out if using architectural dimensions. Some other features further down the column are outlined next.

- *Prefix/Suffix*: Very useful features allowing you to add text, symbols, or values before and after the dimension values. We add the inch symbol in the suffix as an example.
- *Measurement scale*: Scales dimension values; not necessary in common usage.

- *Zero suppression*: Another very useful feature allowing you to suppress zero values either before or after the main significant digits when using decimal units. For example, a 0.625 value becomes just 0.625 and a 1.250 becomes 1.25. This is used in conjunction with the Precision setting to "clean up" dimension values, removing extra zeros. With the Architectural units setting, the idea is the same but now 0 feet and 0 inches are suppressed.
- *Angular dimensions*: This is the final category (bottom right). Here, you can select other units for angles such as Gradians, Radians, and Degrees/Minutes/Seconds (useful for surveyors). For most uses, however, Decimal Degrees are sufficient and the Precision: field below that allows for setting 0–0.00000000. Angular dimensions can also have zero suppression, leading and trailing.

Run through all the options available and, when done, set your Dimension Style dialog box to what is shown in Fig. 13.6.

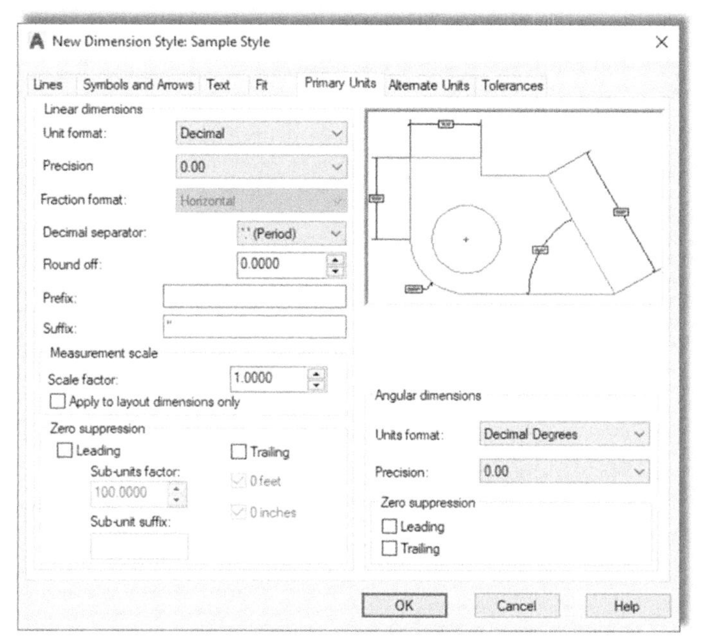

FIGURE 13.6 Primary Units tab.

Alternate Units Tab

This is a tab we did not explore at all in Level 1. Alternate units are a secondary set of dimensions that reside inside parentheses attached to the main set of units. Their purpose is to present the main units in an alternate form, such as metric versus English (just one example).

The first thing you must do, as you click over to the Alternate Units tab, is to turn on the alternate units if they are all grayed out. Check off Display alternates units at the upper left, and the entire tab activates and the fields become editable. Most of our discussion centers on the Alternate units category (upper left). The Zero suppression and Placement categories should be familiar by now.

Under this category, we have

- Unit format:
- Precision:
- Multiplier for alt units:
- Round distances to:
- Prefix:
- Suffix:

Most of these should also be familiar to you, such as Unit format and Precision. These values generally match the main set of units. Skipping Multiplier, we have Round (never a good idea to use) and finally the previously explained Prefix: and Suffix: fields.

Understanding alternate units then boils down to fully grasping the Multiplier for alt units: concept. It is a powerful and simple idea. Instead of presenting a collection of "canned" multipliers, AutoCAD allows you to set your own. So, if the main unit is 1 mile, to express this in feet (as an alternate unit), type in 5280 as a multiplier. You are in no way limited as to what sort of value you can enter (inches to miles or millimeters to yards is completely acceptable, if not necessarily practical), but of course you have to know exactly what the multiplier is or else, as they say, "garbage in, garbage out." Therefore, an incorrect multiplier gets you false alternate unit readings.

Explore all the features of Alternate units and try entering several multipliers. For example, the number of yards equal to 1 mile (1760). Note also the default multiplier: 25.4. That is how many millimeters there are in an inch; this is useful for the most common application of alternate units, which is presenting metric units next to English ones.

In our case, let us go with the amount of centimeters in an inch (2.54) and enter that as the alternate unit. Also add cm in the suffix. Your Dimension Style dialog box should look like Fig. 13.7.

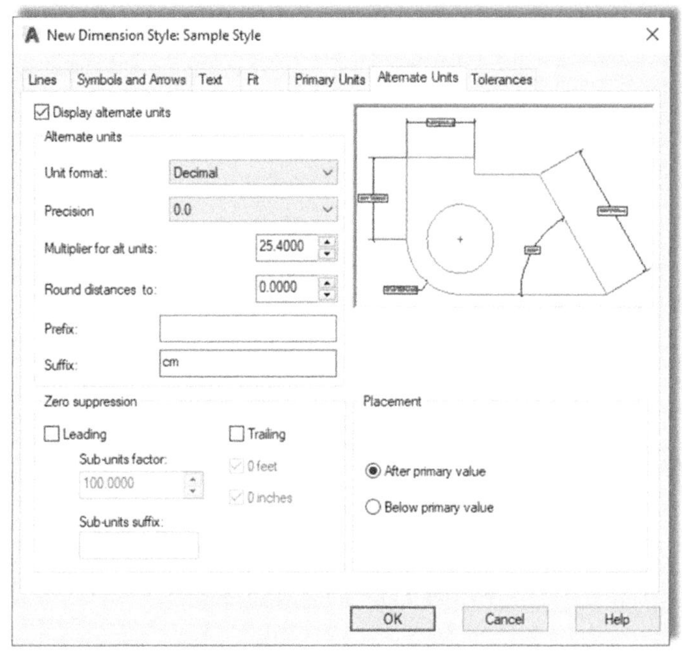

FIGURE 13.7 Alternate Units tab.

Tolerances Tab

This final tab also has not been examined yet. Tolerances are generally defined as deviations from an ideal stated value, as used in manufacturing specifications of engineering devices. Tolerances are relevant mostly to engineering and manufacturing and less so to architecture, but knowledge of this tab is useful for all AutoCAD students.

The underlying need for specifying tolerances hinges on the fact that, in manufacturing, closer tolerances may be costlier to achieve, whereas larger tolerances, although cheaper, may adversely affect performance of a part. Tolerances are a way of expressing just how much a part may be off the mark and still be acceptable, based on engineering need and cost analysis. Aerospace, automotive, and biomedical engineering tolerances are closer than most consumer products, for example.

Tolerances are a familiar topic to mechanical engineers and machinists. To them, no further explanation is needed, and AutoCAD's tolerance choices are familiar: Symmetric, Deviation, Limits, and Basic. Other options under this tab, such as Zero suppression and Placement, should look familiar to all. A short description of each type of tolerance is presented next.

- *Symmetric tolerances*: These indicate the overall deviation, which is symmetric for upper and lower values; therefore, you set only one (the upper value).
- *Deviation tolerances*: These indicate nonsymmetrical deviations in either direction with a stacked ±.

- *Limits tolerances*: These are similar in theory and indicate the acceptable limit in one or both sets of values, but no ± is shown.
- *Basic tolerances*: These are the same as none, the first choice.

Tolerances are an interesting topic to explore further, but because a large percentage of this textbook's readers will not need them in their current or future work (as related to architecture), we do not go any further in depth. If you wish to learn more, there is a tremendous amount of information online. For now, feel free to experiment and set your tolerance to Deviation, as seen in Fig. 13.8.

So what did we end up with after all these settings? Well, in the interest of trying out as much as possible, we did not really create a very realistic looking dimension. Let us take a look at it. Click on OK in the Dimension Style box; it will disappear and take you back to the Dimension Style Manager. Then, press Set Current in the next box, followed by Close. Now draw a 10″ × 10″ rectangle and dimension it using the basic Horizontal dimension. It should look similar or close to Fig. 13.9.

Let us try another sample dimension on a hypothetical section of roadway. Draw something similar to Fig. 13.10, with the two vertical lines 3 units apart. Run through all the tabs again, using the following data:

- Architectural ticks.
- Red-colored extension lines.
- Blue-colored dimension lines.
- Arial font, 0.25″, color: Red.
- Fit: 1.000.
- Primary units: Decimal. Precision = 0.
- Suffix added on main units → mile(s) and alternate units → feet.
- Alternate units, use 5280 as the multiplier.
- No tolerances.

You should get a result similar to Fig. 13.10. Note how simple it is to use alternate units if you know the multiplier.

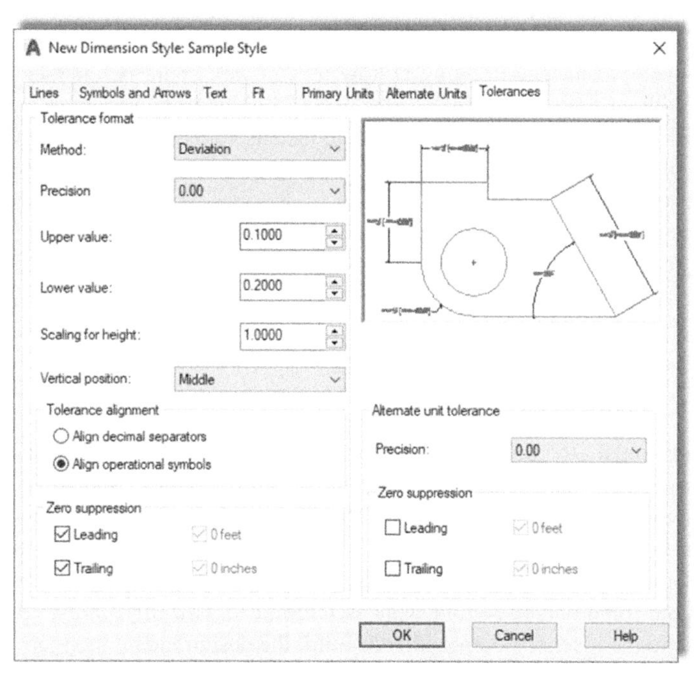

FIGURE 13.8 Tolerances tab.

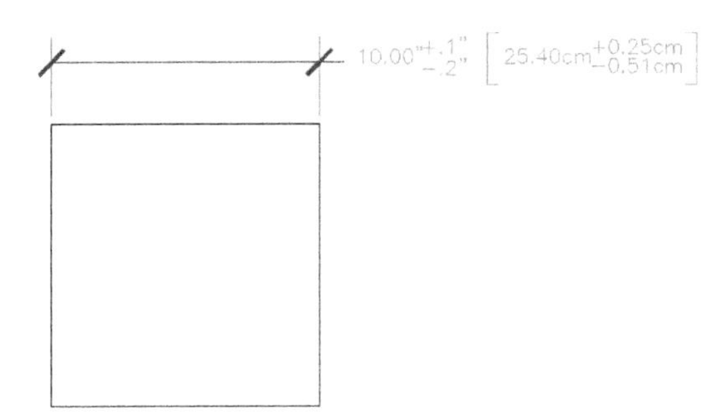

FIGURE 13.9 The new dimension style.

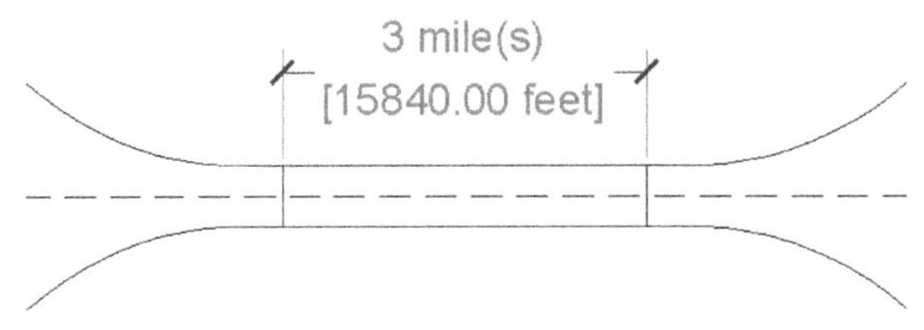

FIGURE 13.10 Sample dimension.

13.3 INTRODUCTION TO CONSTRAINTS

AutoCAD was always a top-notch 2D drafting software application but one with a bit of an inferiority complex. Not in regard to any other 2D competitors but rather toward 3D solid modeling and design software. These sophisticated products were developed in their own world, apart from AutoCAD, via development firms that never competed against or had anything to do with Autodesk or architecture. These programs included CATIA, NX, Pro-Engineer (now called *Creo*), and SolidWorks among others. Their developers invented quite a few revolutionary concepts and methods that transformed the way engineering design is done. Autodesk eventually jumped on board with Inventor and later Revit, both of which are excellent 3D products (one for engineers and the other for architects).

That left the AutoCAD development team looking at how its software could be enhanced by incorporating some of these sophisticated 3D tools without fundamentally altering AutoCAD's identity or formula for success. If you go on to study AutoCAD 3D, you will notice how many "borrowed" concepts found their way into the software as AutoCAD reached ever further into the high-end 3D world.

Some of these concepts also found their way into 2D, such as the ones we are about to discuss, namely, Parametrics, which is really two topics, geometric constraints and dimensional constraints; the latter is also referred to as *dimension-driven design*. If you are an engineer or an engineering school student who used the previously mentioned 3D solid modeling software, you quickly recognize these concepts. AutoCAD literally took a page out of their playbook. Let us discuss each concept separately.

13.4 GEOMETRIC CONSTRAINTS

The fundamental idea behind geometric constraints, regardless of what software we deal with, is to restrict (constrain) the possible movement of drawn geometry and force that geometry into certain positions. Ortho is a horizontal and vertical constraint on lines. So, if you want two parallel or perpendicular lines, a constraint can be placed on them that forces them to always stay parallel or perpendicular as well as to be that way in the first place. Expanding to more general terms, geometric constraints set "relationships" between geometries, such as parallel, perpendicular, tangent, and coincident.

This overall concept is quite important in setting what is referred to as *design intent*. If a drilled hole in a bracket absolutely has to be concentric to the bracket's corner fillet to fulfill the design intent, then that is the main driving factor in its design (*concentric* means the circles and arcs share the same center point). Other aspects of the design, such as perhaps notches and other geometry on the bracket, however, may be less critical. The constraints then "hold" the critical geometry, while you design the rest around it.

This is a very simple example; with a bracket, you can probably keep track of the critical relationships without help. Geometric constraints really earn their keep with more complex designs, where many constraints are needed to preserve the design intent. Without them, it would be easy to forget one or two important relationships and discover later down the road that the part you are working on does not function properly because a tangency somewhere was not preserved.

Almost since the beginning of high-end 3D software, geometric constraints were an important part of the design approach, but they were not included in AutoCAD until recently (release 2010); and AutoCAD designers needed to keep track, and check over, their geometric relationships manually. With these new tools, AutoCAD's engineering and architecture design capabilities have been greatly enhanced.

Types of Geometric Constraints

Listed next are most of the constraints available to you. We do not go through every single one but cover only the more commonly used ones, leaving the rest for you to explore. The best way to set constraints is via the Geometric Constraints toolbar (see Fig. 13.11) or the Ribbon's Parametric tab.

FIGURE 13.11 Geometric constraints toolbar.

The critical constraints we discuss are

- *Perpendicular*: Constrains two lines or polylines to perpendicular 90 degrees angles to each other.
- *Parallel*: Constrains two lines to the same angle.
- *Horizontal*: Constrains two lines to lie parallel to the X axis.
- *Vertical*: Constrains two lines to lie parallel to the Y axis.
- *Concentric*: Constrains circles, arcs, or ellipses to maintaining the same center point.
- *Tangent*: Constrains a tangency between curves and lines.
- *Coincident*: Constrains two points together (to be discussed along with dimensional constraints).

Adding Geometric Constraints

Adding geometric constraints involves simply picking the constraint you wish to add and selecting the first object (which serves as the reference point) followed by the second object. The constraint then appears next to the objects, reminding you it is set. The constraint tool also forces the objects into that constraint, so if you draw two lines that are not quite parallel, once you set the constraint, they become parallel right away.

Having read this, you may be tempted to completely do away with Ortho and even OSNAP to create accurate geometry, allowing constraints to "fix" drawn shapes. This is not generally recommended, of course, as sloppy drafting catches up to you sooner or later (though we do try this in the next example). While geometric constraints are very useful new tools, use them when appropriate, without forgetting or throwing away basic accuracy habits.

Let us try this out with the first constraint on the previous list: Perpendicular.

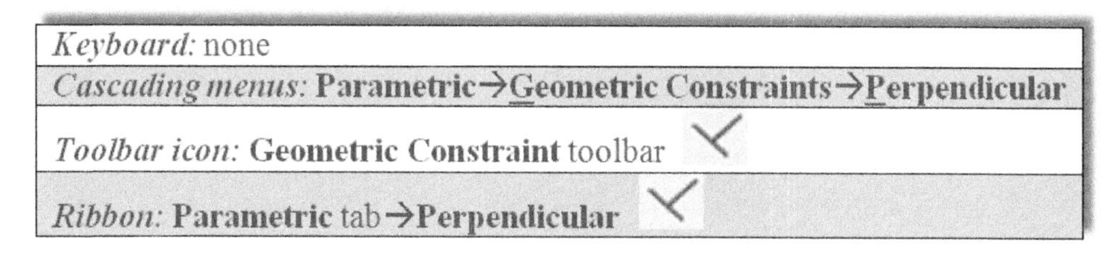

Keyboard: none
Cascading menus: **Parametric→Geometric Constraints→Perpendicular**
Toolbar icon: **Geometric Constraint** toolbar
Ribbon: **Parametric** tab→**Perpendicular**

Step 1. Draw two lines that are roughly perpendicular, as seen in Fig. 13.12 (left).

Step 2. Select the perpendicular geometric constraint via any of the preceding methods.

- AutoCAD says: `_GcPerpendicular`
  `Select first object:`

Step 3. Select the first line (the vertical one).

- AutoCAD says: `Select second object:`

Step 4. Select the second line.

Immediately after the selection of the second line, you see it snap to perpendicular and two sets of geometrical constraints are added next to each line, as seen in Fig. 13.12 (right). Note that, if you were to select the bottom line first, then the vertical one would snap to it, not the other way around. Selection order matters.

Let us try one more, the concentric one. We leave the rest as an in-class exercise.

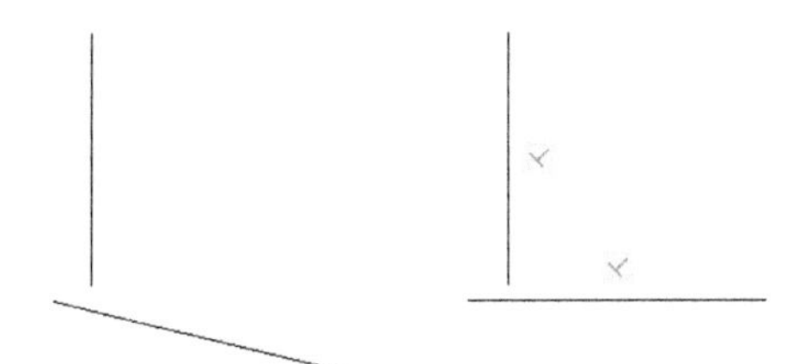

FIGURE 13.12 Perpendicular geometric constraints.

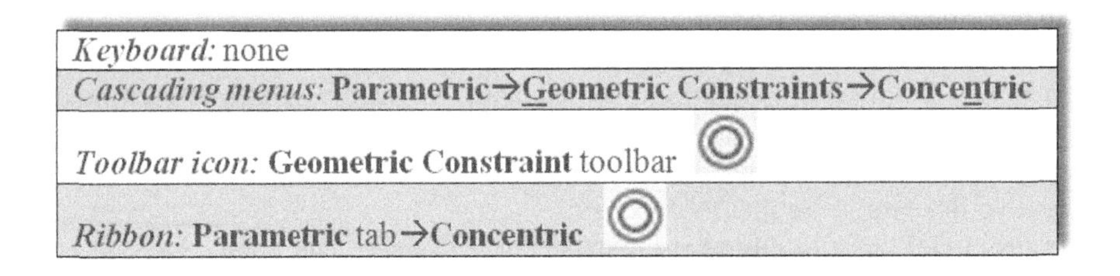

Keyboard: none
Cascading menus: **Parametric→Geometric Constraints→Concentric**
Toolbar icon: **Geometric Constraint** toolbar
Ribbon: **Parametric** tab→**Concentric**

Step 1. Draw a circle, then an arc above it, as seen in Fig. 13.13 (left).

Step 2. Select the concentric geometric constraint via any of the preceding methods.

- AutoCAD says: `_GcConcentric`
  `Select first object:`

Step 3. Select the circle.

- AutoCAD says: `Select second object:`

Step 4. Select the arc.

Immediately after the selection of the arc, you see one or both snap to a concentric setting (their centers are the same), as seen in Fig. 13.13 (right). Note that, if you were to select the arc first, the circle would adjust to it, not the other way around. Once again, selection order matters.

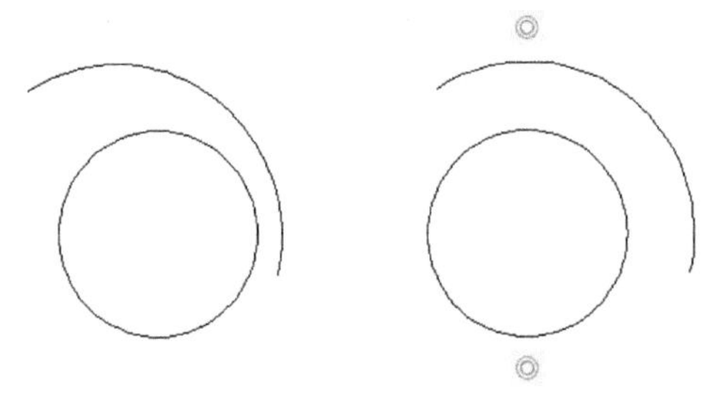

FIGURE 13.13 Geocentric geometric constraints.

Try out the remaining geometric constraints on your own as an in-class exercise.

Hiding, Showing, and Deleting Geometric Constraints

There are two ways to access these additional options to work with constrains. You can either right-click on one of the constraints to reveal a small menu, as seen in Fig. 13.14 or find the same options on the Ribbon, as seen in Fig. 13.15, along with the rest of the constraints.

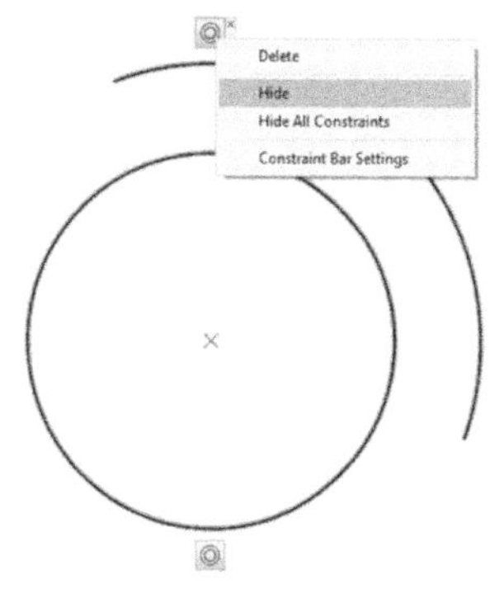

FIGURE 13.14 Geometric constraints options, right-click menu.

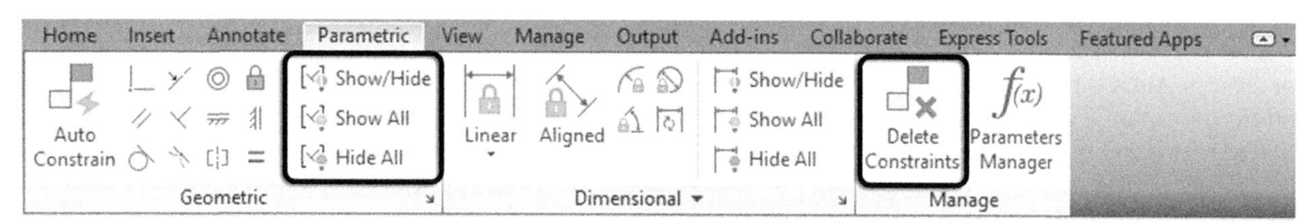

FIGURE 13.15 Geometric constraints options, Ribbon.

The Constraint Settings dialog box, which can be accessed via the small drop-down arrow at the bottom right of the Geometric tab, is actually just for global settings, for which geometric constraints to show. Leave all the choices selected and the transparency at 50%, as seen in Fig. 13.16. Finally, Fig. 13.17 shows all seven of the constraints mentioned, as you should see them on your screen after completing the in-class exercise.

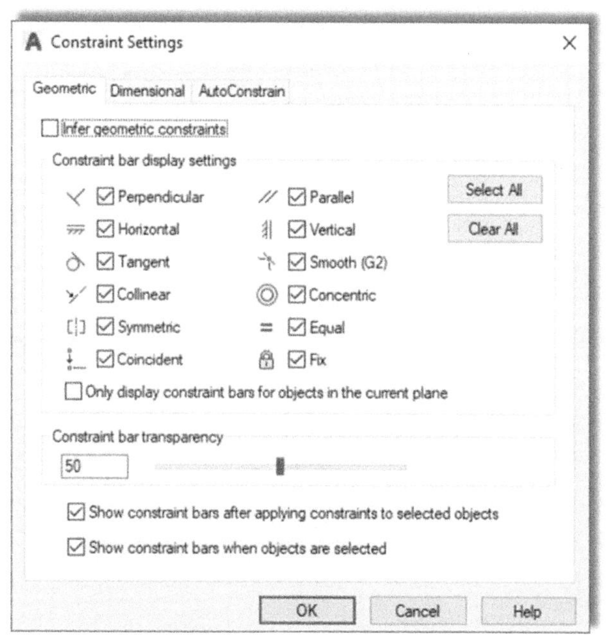

FIGURE 13.16 Constraint settings.

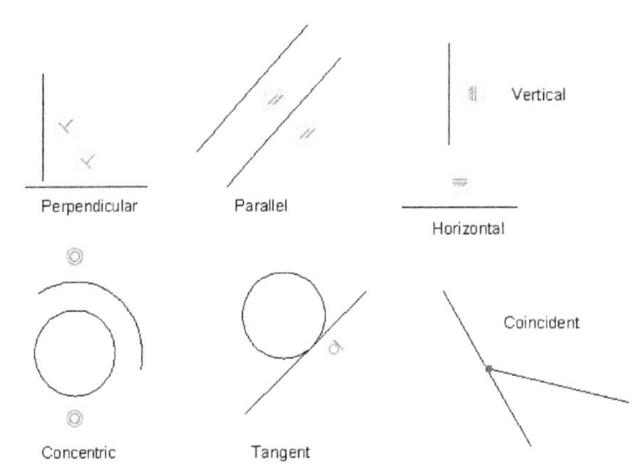

FIGURE 13.17 Geometric constraints.

13.5 DIMENSIONAL CONSTRAINTS

As we already discussed the basic idea of constraints, this introductory section is shorter. The idea here is to move from constraining pieces of geometry (and their positions relative to each other) to constraining the actual dimensions of the drawn design. So, if two circles, representing drilled holes, absolutely need to be 3″ apart, then constraining that dimension ensures it does not change regardless of what other design work goes on around them.

Dimensional constraints are also a major part of 3D parametric design and can be easily transferred to 2D work, as was done here in AutoCAD. Furthermore, because the geometry must obey the constrained dimension, you can change the geometry by simply changing the dimension value, which is quite a departure from the normal order of business with AutoCAD. This ability to change geometry by updating dimensions is called *dimension-driven design*. With high-end 3D solid modeling software, this is really the only way to change the size of an object (usually called a *feature*). AutoCAD lacks this ability in 3D but, as just mentioned, can do this with 2D shapes.

At this point, you may start to see why all the preceding is referred to as *parametric design*. We are designing by changing or assigning parameters to our geometry. The geometry itself is almost secondary to the relationships and data that drive its existence and creation. This type of philosophy and approach is one of the several steps critical to assuring that what we have is a valid engineering design, not just a pretty picture. The data behind the design must make sense.

Working With Dimensional Constraints

Learning the basics of dimensional constraints are straightforward. You can either add them in right away to your design or convert existing regular dimensions to constraints. Dimensional constraints have their own toolbar (Fig. 13.18), or they can be accessed through the Ribbon (Fig. 13.19).

FIGURE 13.18 Dimensional constraints toolbar.

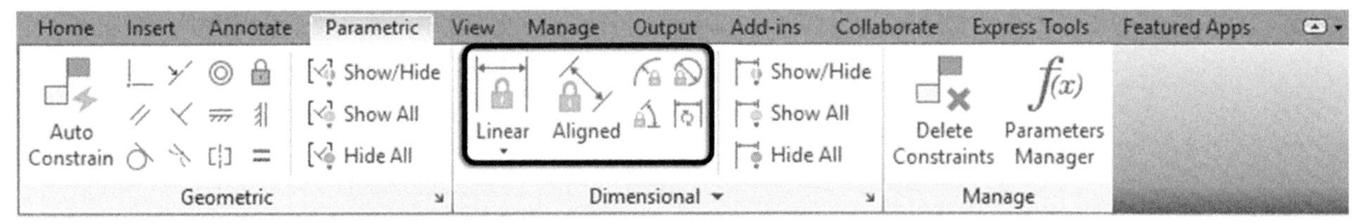

FIGURE 13.19 Dimensional constraints using the Ribbon.

The available dimensional constraints mirror the regular dimensions closely:

- Horizontal
- Vertical
- Aligned
- Radius
- Diameter
- Angle

To practice them, draw the bracket seen in Fig. 13.20 but do not dimension it; use the shown values only to create the geometry.

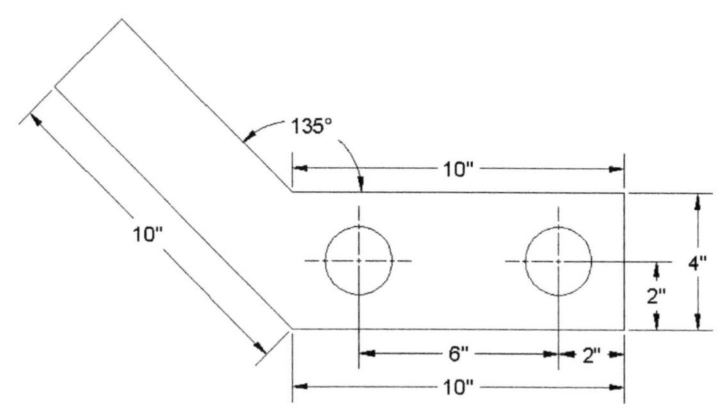

FIGURE 13.20 Bracket.

Keyboard: none
Cascading menus: **Parametric→Dimensional Constraints→Horizontal**
Toolbar icon: **Dimensional Constraints** toolbar
Ribbon: **Parametric** tab→**Linear**

Step 1. Start up the dimensional constraint via any of the preceding methods.
- AutoCAD says: Specify first constraint point or [Object] <Object>:

Step 2. Pick one end of the bottom part of the bracket (a red circle with an X appears).
- AutoCAD says: Specify second constraint point:

Step 3. Pick the other end of the bottom part of the bracket (a red circle with an X appears again).
- AutoCAD says: Specify dimension line location:

Step 4. Locate the dimensional constraint some distance away from the part, like a regular dimension. AutoCAD displays the value (d = 10.0000) with a graphic of a lock, as seen in Fig. 13.21. Press Enter.

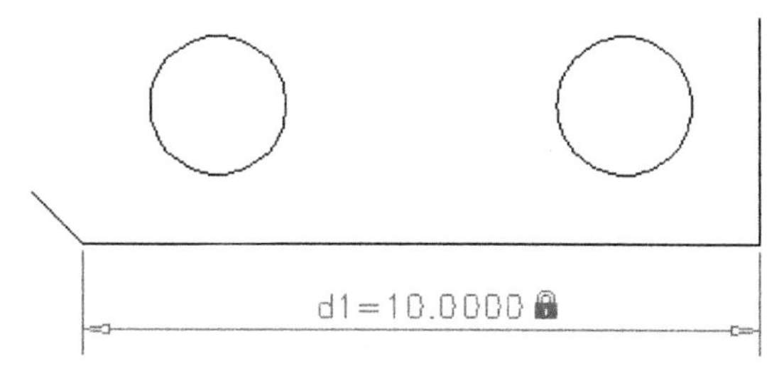

FIGURE 13.21 Dimensional constraint.

In a similar manner, you can add additional dimensional constraints to the rest of the bracket or convert existing dimensions (if you had some) via the Convert button in the Ribbon's Parametric tab. Note the central idea behind this tool. The dimensional values are "locked in," as represented by the lock graphic and do not budge regardless of how other surrounding geometry may change.

13.6 DIMENSION-DRIVEN DESIGN

Let us put the preceding information together into a concise approach to design, as well as add in some comments about how to work smoothly with dimensional constraints. First of all, you can change the value of the dimensional constraint by just double-clicking on it and typing in something else (Fig. 13.22).

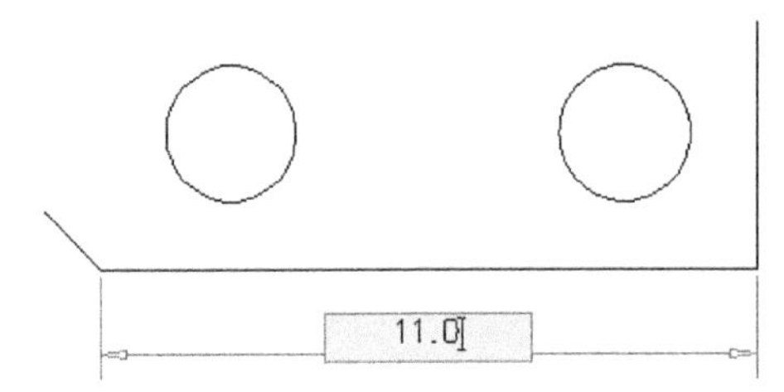

FIGURE 13.22 Altering linear dimensional constraint.

The value forces the line to move to the right, leaving a gap, which is not quite what you intended. Try the same with a circle and diameter constraint (Fig. 13.23).

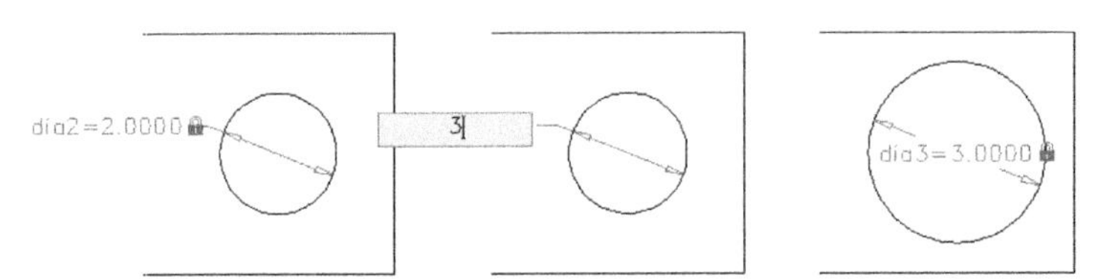

FIGURE 13.23 Altering diameter dimensional constraint.

This leads to much better results; the circle has changed its size. There is a very important point to understand here in order to take full advantage of these tools. With linework and angles, dimensional constraints need to work together with geometric constraints. Used just by themselves, dimensional constraints change only the dimension value and drag the lines along for a ride; but if you add a coincident geometric constraint, the lines are "glued" together so you can make entire sections of the design properly constrained. A coincident constraint asks you to select one of the lines, then the second, and create a point where they intersect. Try this out on our bracket's bottom pieces and try changing the bottom line's size again, leading to quite a different effect.

A lot more can be written about constraints in general; it is a broad and important topic. You are encouraged to explore further and practice setting up dimensional constraints on critical parts of a design. It will take some getting used to, and applying dimensional constraints intelligently can be awkward at first. The key is to apply these constraints only to the critical geometry and not constrain every single line and circle on the screen. Fortunately, AutoCAD prevents you from doing this and issues a warning that you are over constraining or there are conflicts, similar to high-end solid modeling applications.

13.7 LEVEL 2 DRAWING PROJECT (3 OF 10): ARCHITECTURAL FLOOR PLAN

For Part 3 of the drawing project, you complete the architectural/interior design by adding furniture and appliances to the house.

Step 1. You again need new layers:

- A-Appliances (Magenta).
- A-Furniture (Red).

Step 2. Dimensions for the furniture and appliances are not specifically given but draw in reasonable-size designs, as seen in the layout. Feel free to add more than is shown. The full plan is in Fig. 13.24. Your drawing features the colors listed in Step 1.

Step 3. Add in text describing the furniture and appliances. Put it on the text layer. It should take you about an hour to an hour and a half to add in the new features. Use copy often!

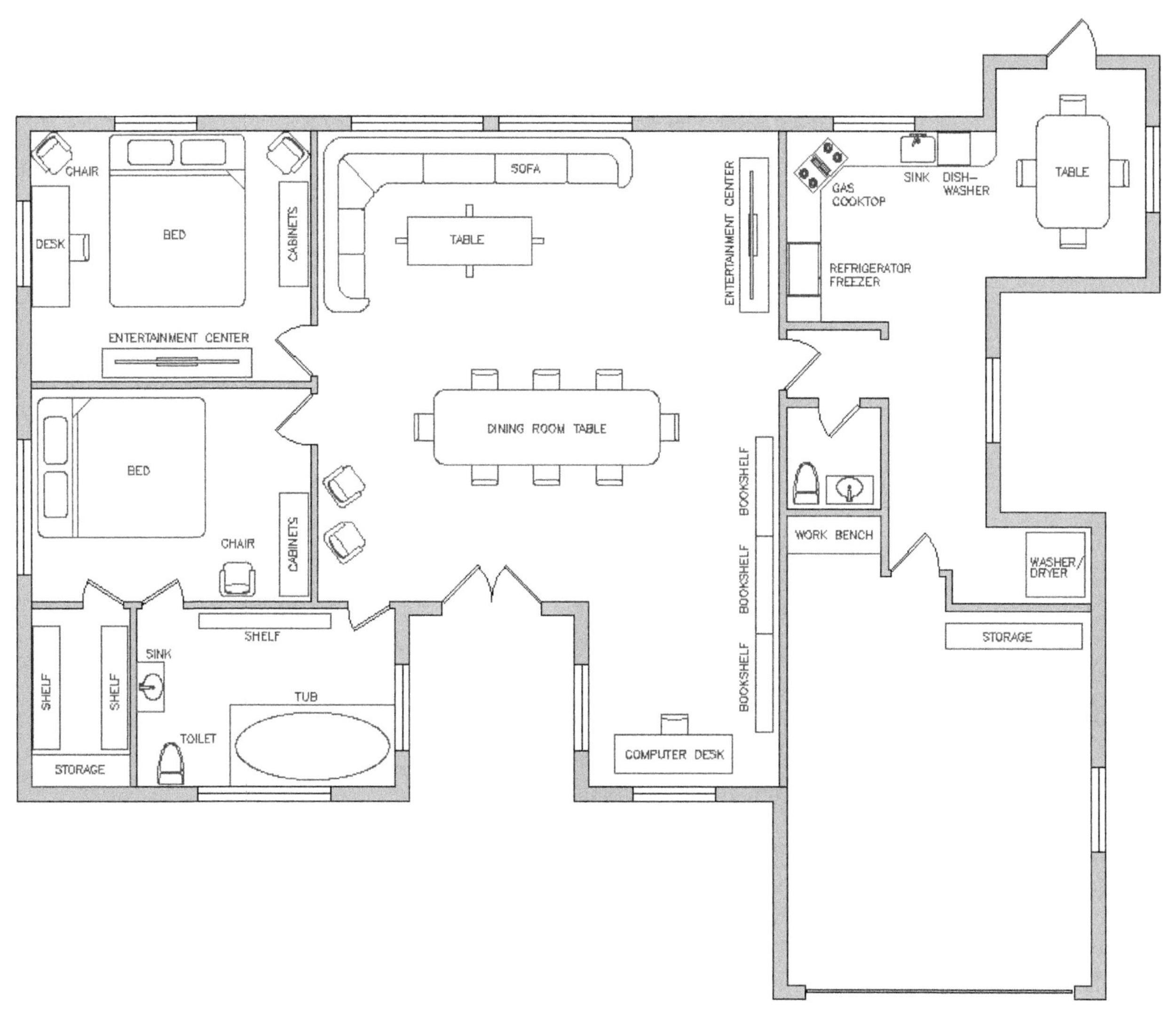

FIGURE 13.24 Furniture and appliances in floor plan.

SUMMARY

You should understand and know how to use the following concepts and commands before moving on to Chapter 14:

- Creating a new dim style (ddim)
  - Lines tab
  - No permanent settings, but know functions of all
  - Symbols and Arrows tab
  - Architectural tick setting; know functions of all others

- Text tab
- Text font and size settings; know functions of all others
- Fit tab
- Fit setting; know functions of all others
- Primary Units tab
- Units setting
- Precision setting
- Prefix setting
- Suffix setting
- Zero suppression settings; know functions of all others
- Alternate Units tab
- Multiplier settings; know functions of all others
- Tolerances tab
- Tolerance settings (if needed)

- Geometric constraints
  - Perpendicular
  - Parallel
  - Horizontal
  - Vertical
  - Concentric
  - Tangent
  - Coincident
- Dimensional constraints
  - Horizontal
  - Vertical
  - Aligned
  - Radius
  - Diameter
  - Angle

REVIEW QUESTIONS

Answer the following based on what you learned in Chapter 13:

1. Describe the important features found under the Lines tab.
2. Describe the important features found under the Symbols and Arrows tab.
3. Describe the important features found under the Text tab.
4. Describe the important features found under the Fit tab.
5. Describe the important features found under the Primary Units tab.
6. Describe the important features found under the Alternate Units tab.
7. Describe the important features found under the Tolerances tab.
8. Describe the purpose of geometric constraints.
9. What seven geometric constraints are discussed?
10. Describe the purpose of dimensional constraints.
11. What six-dimensional constraints are discussed?

EXERCISES

1. Draw the following shape and label it exactly as shown. Be sure to note all the modifications to the dimensions, including arrowheads and the presence of alternate units and tolerances. You may have to get creative with the leader. (Difficulty level: Intermediate; Time to completion: 10−15 minutes.)

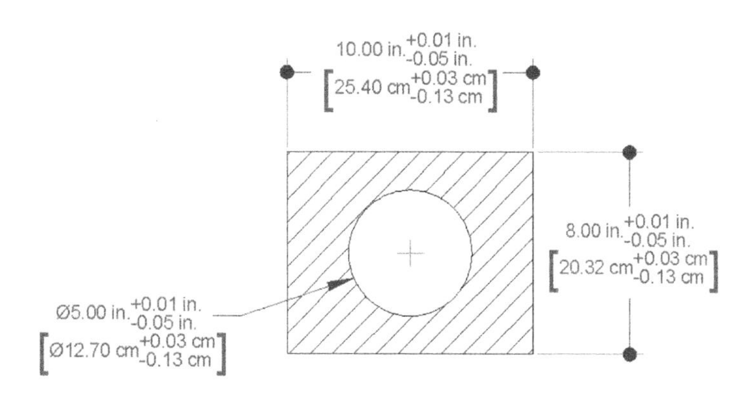

2. Draw a rough parallelogram shape, with one side 20″, as shown on the left. Parallel constrain all four sides, filleting if necessary. Then, carefully, coincident constrain all four corners. Finally, dimension constrain one side, using a value of 30, as seen on the right. If everything is done right, the rectangle should increase in size uniformly. (Difficulty level: Easy; Time to completion: 5 minutes.)

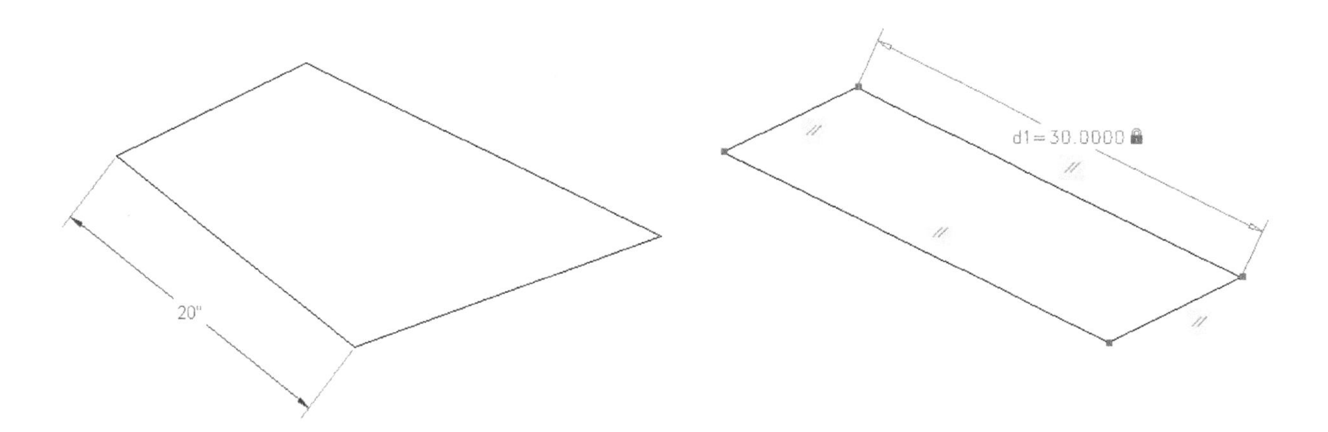

3. Draw the following plate and dimensions, including the tolerances as shown. The tolerance format method used here is Deviation. Carefully place all dimensions in the positions shown for maximum clarity as this is a dimension-dense drawing. (Difficulty level: Easy; Time to completion: 10–15 minutes.)

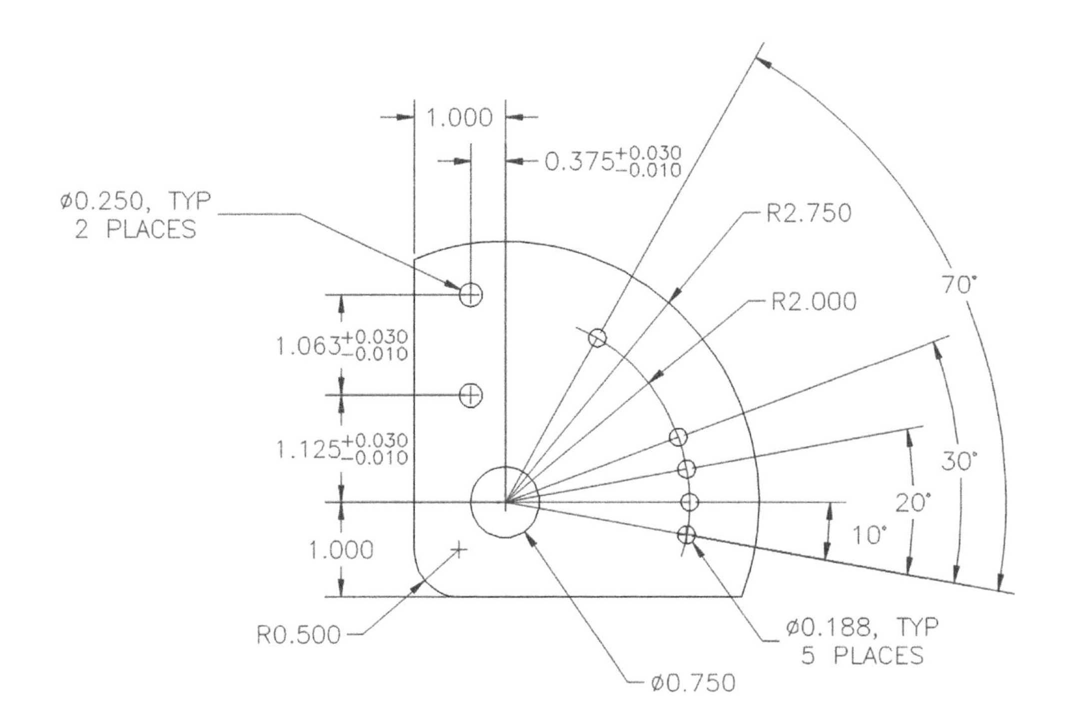

Spotlight On: Aerospace Engineering

Aerospace engineering is a specialized branch of mechanical engineering that evolved relatively recently in our history, as humankind took to flight, first in the atmosphere then into space. Formally defined, it is the science behind the design and construction of aircraft (aeronautical engineering) and spacecraft (astronautical engineering). These engineers conceptualize, build, and test everything from gliders to missiles to jet fighters and space rockets. A few also work in naval design, as some of the principles guiding the flow of air (a fluid) around an aircraft readily transfer to the flow of water (also a fluid) around a ship or submarine.

FIGURE 1 Grumman X-29 Experimental. *Image source: Northrop Grumman.*

Aerospace engineers need a mastery of a wide variety of engineering sciences, such as fluid dynamics, structures, control systems, aeroelasticity, thermodynamics, materials, and electrical engineering. It is perhaps one of the more diverse engineering disciplines, and specialization is necessary. Three broad areas of specialization include aeronautics, propulsion, and space vehicles.

FIGURE 2 Airbus A380 airliner. *Image source: Airbus.*

Education for aerospace engineers starts out similar to that for the other engineering disciplines. In the United States, all engineers typically have to attend a 4-year ABET-accredited school for their entry-level degree, a Bachelor of Science. While there, all students go through a somewhat similar program in their first 2 years, regardless of future specialization. Classes taken include extensive math, physics, and some chemistry courses, followed by statics, dynamics, mechanics, thermodynamics, fluid dynamics, and material science. In their final 2 years, engineers specialize by taking courses relevant to their chosen field. For aerospace engineering, this includes aerodynamics, structures, controls, propulsion, and orbital mechanics, among others.

On graduation, aerospace engineers can immediately enter the workforce or go on to graduate school. Although not required and not common in the aerospace industry, some engineers choose to pursue a Professional Engineer (P.E.) license. This would typically be in related mechanical engineering, as there is no aerospace P.E. exam. The process first involves passing a Fundamentals of Engineering (F.E.) exam, followed by several years of work experience, and finally the P.E. exam.

Aerospace engineers can generally expect somewhat higher starting salaries than civil, mechanical, and industrial engineering graduates but on par with electrical and lower than petroleum engineering and some computer science degrees. A master's degree is usually the path to higher starting pay, which with a bachelor's, averaged around $70,000 nationwide. Your specific starting salary, of course, depends highly on locality, current market demand, and even GPA on graduation. The median pay for all aerospace engineers was about $110,000, and the national average pay for all aerospace engineers is about $91,000.

Aerospace engineers often work for government or defense contractors, and as such need to get and maintain a security clearance (Secret, Top Secret, TS/SCI, or TS/SAP), a somewhat unique requirement that may not be necessary in other fields. Keep in mind that the security checks will go as far back as your 18th birthday, so this may not be the field to get into for those who "experimented" extensively with illegal substances or activities while in college.

So how do aerospace engineers use AutoCAD and what can you expect? Industry wide, AutoCAD use is not widespread due to the nature of the profession, which relies on more sophisticated software to model parts directly in 3D. The industry standard is CATIA, though NX and Pro/Engineer (Creo) are also used. Aerospace engineering also relies heavily on software for initial testing or design validation. Examples include finite element analysis and software such as NASTRAN or ANSYS to model loads, deflections, and heat propagation. MATLAB can be used to model vibrations and stresses, whereas computational fluid dynamics and software such as Fluent are used to model the flow of air around the design at various speeds and temperatures (also applicable to the flow of water). See Appendix B if you would like to learn more about these techniques.

AutoCAD, however, can still be found for small part design, aerospace electrical schematics, and overall system layouts. Smaller companies that have little to no modeling, testing, and manufacturing needs also may use AutoCAD due to the associated cost savings. Many home kit airplane designers or builders turned to AutoCAD to produce drawings or just to document ideas. The general trend is to move away from AutoCAD as design requirements become more sophisticated and testing and manufacturing are involved.

AutoCAD layering is far simpler in aerospace engineering than in architecture. There is no AIA standard to follow, only an internal company standard, which may not be as detailed. As always, the layer names should be clear as to what they contain. Fig. 3 shows an aerospace application of AutoCAD for a set of composite wings for a small unmanned aerial vehicle design.

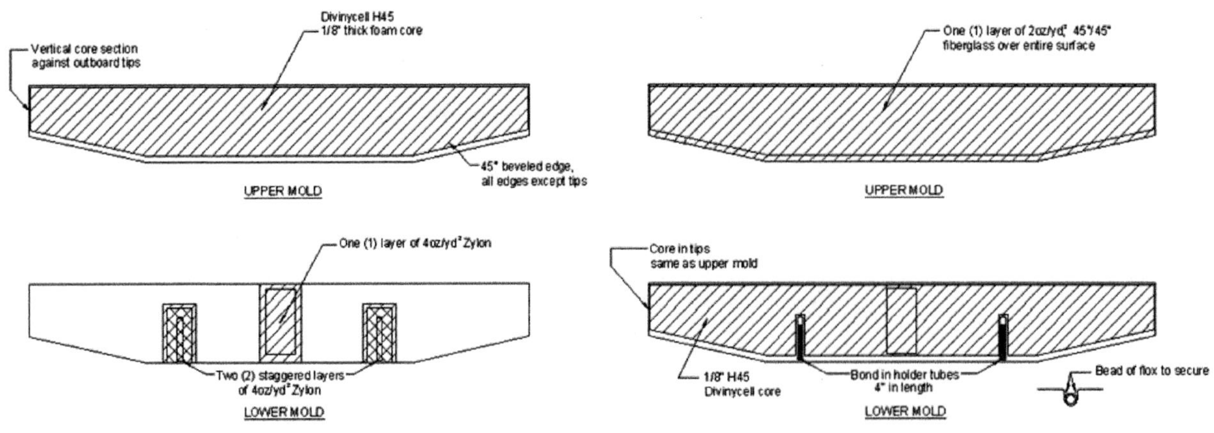

FIGURE 3 Wings for a small unmanned aerial vehicle drawn with AutoCAD. *Images source: Elliot Gindis.*

Fig. 4 shows a typical design of a small general aviation aircraft wing, with overall dimensions and specs.

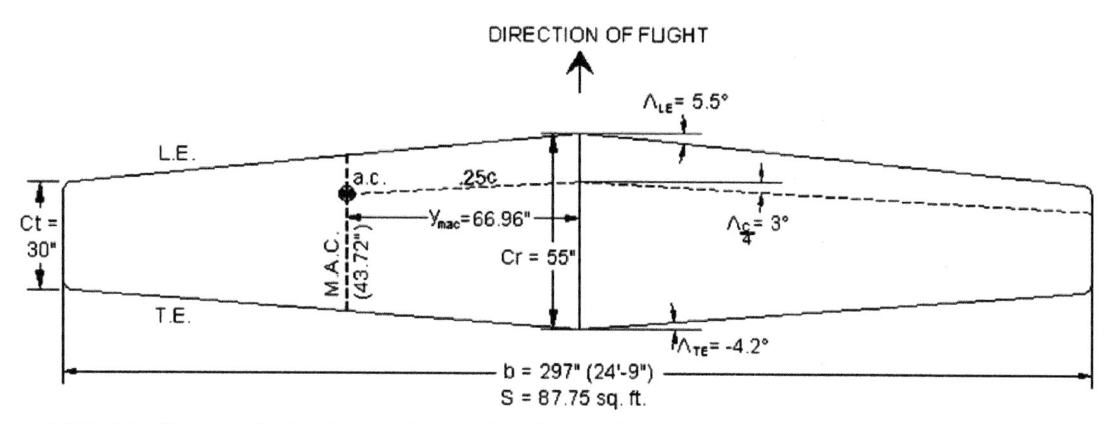

FIGURE 4 Wing specifications for a small, general aviation aircraft drawn with AutoCAD. *Image source: Elliot Gindis.*

Finally, Fig. 5 shows a side profile of a small private plane (a Lancair Legacy) drawn in AutoCAD for a student project. This side profile study was used for sizing while learning design; the vertical and horizontal numbers are inch position from the tip of the propeller cone and the ground. Note the use of a thicker polyline in the outline to accentuate the profile.

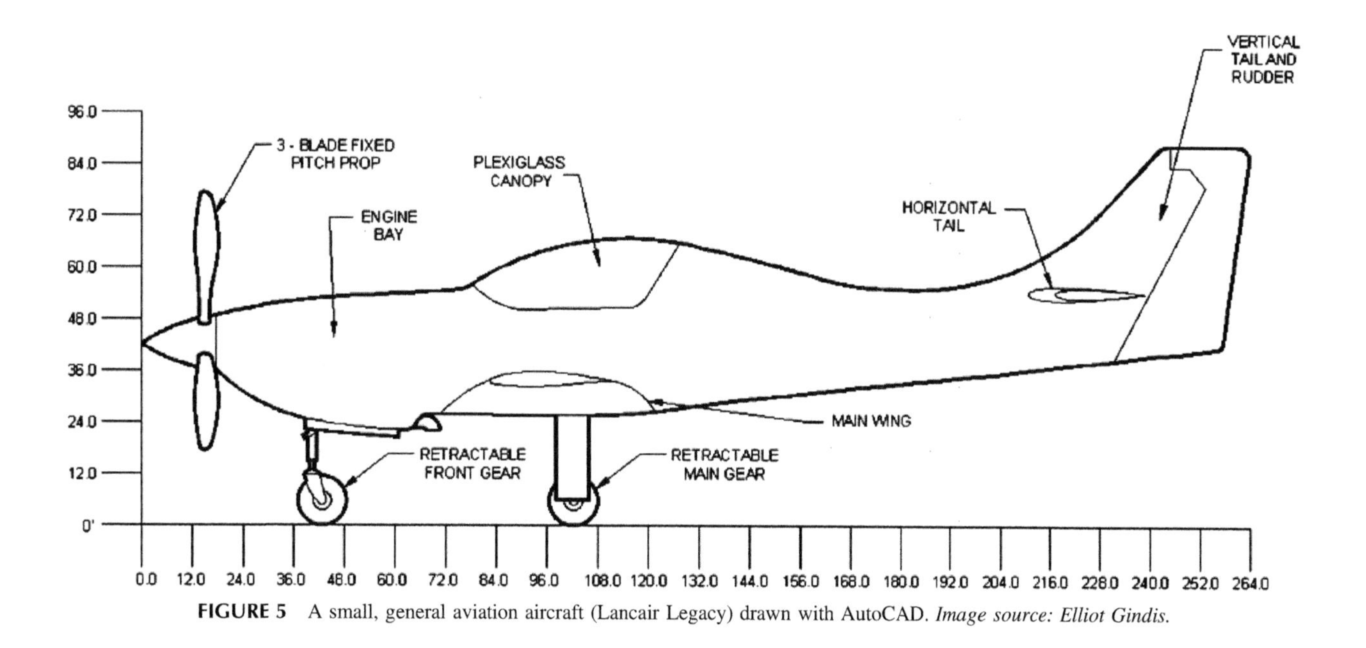

FIGURE 5 A small, general aviation aircraft (Lancair Legacy) drawn with AutoCAD. *Image source: Elliot Gindis.*

Chapter 14

Options, Shortcuts, Customize User Interface, Design Center, and Express Tools

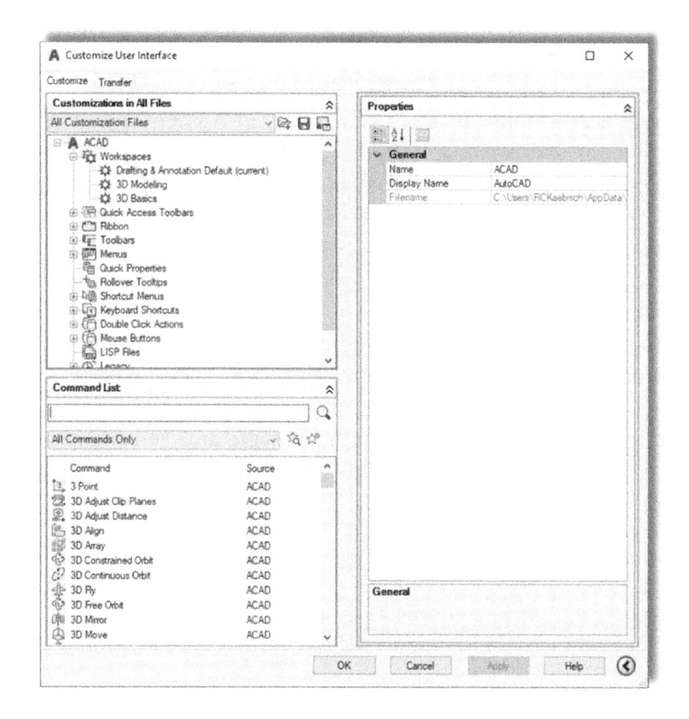

Learning Objectives

In this chapter, we introduce a variety of advanced tools and discuss the following:

- Options dialog box
  - Files tab
  - Display tab
  - Open and Save tab
  - Plot and Publish tab
  - System tab
  - User Preferences tab
  - Drafting tab
  - 3D Modeling tab
  - Selection tab
  - Profiles tab

Up and Running with AutoCAD 2019. https://doi.org/10.1016/B978-0-12-816440-2.00014-8

- Shortcuts and the acad.pgp file
- Customize User Interface (CUI)
- Design Center (DC)
- Express Tools

By the end of this chapter, you will learn how to significantly customize your environment and speed up your work via importing predrawn blocks and using shortcuts to accelerate command input.

Estimated time for completion of this chapter: 2–3 hours.

14.1 OPTIONS

Our topics for this chapter are the Options dialog box, followed by the concept of Shortcuts (the acad.pgp file), and the CUI. We then look at the DC and the Express Tools. These topics introduce you to basic customization, greatly enhance your efficiency, and will hopefully dispel any remaining frustration with making AutoCAD look and behave how you want it to.

The Options dialog box is your access point for basic and advanced AutoCAD customization menus, and through it, you can change many essential settings and tweak AutoCAD to your personal taste and style. Almost every user finds a few items worthy of changing, and knowing this dialog box is absolutely essential. Oftentimes, while covering this topic, students remark how they have been annoyed by some feature (or lack of it) and now know how to turn it off or on. This is exactly why we introduce this dialog box or cover it in more detail if you have already seen it.

Note that our goal is not to go over every single item in Options; it is far too big for that, and it is unnecessary anyway. Instead, we outline the most important features and explain why they are necessary. There are instances where you will be advised against changing the default values of one item or another. If you want to know more, press F1 (Help files) while in each of the tabs for a detailed outline of every single feature. Note, however, that you should first have a clear understanding of the selected items this chapter focuses on before going into more detail.

To access the Options dialog box, right-click anywhere in the drawing area and a menu appears, as shown in Fig. 14.1. Click on the last choice: Options... Alternatively, you can also just type in `options` and press Enter.

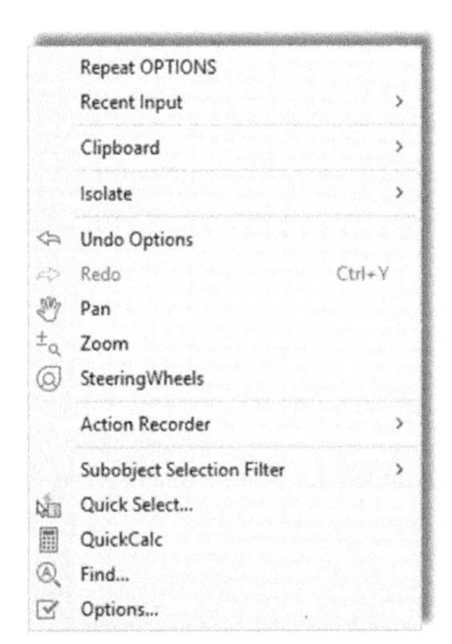

FIGURE 14.1 Accessing Options....

A sizable dialog box called *Options* appears. Across the top you will see the following tabs: Files, Display, Open and Save, Plot and Publish, System, User Preferences, Drafting, 3D Modeling, Selection, Profiles, and Online. You will likely see Display as the active default tab. With the exception of the 3D Modeling tab (which we leave for the online 3D book), we go over each one and introduce all the significant features of each. In some cases, it is a thorough discussion, and in others, just glance at. Go ahead and click on the very first tab (Files) and let us begin.

Files Tab

This first tab is the easiest to go over for the simple reason that you rarely have to change anything here. What are all these listings, then? When AutoCAD, or any software for that matter, installs in your computer, it spreads out all over a certain area of the disk drive (where you indicated it should go on installation) and drops off necessary files as it sees fit.

What you are looking at in Fig. 14.2 is a collection of paths to some of those files as they appear after an installation. They are deemed to be appropriate for a user to access. You can view each path and modify it by clicking on the plus signs. There are quite a few of them, indicating locations of a variety of important AutoCAD functions: Help files, automatic saves, and the like. Some of these you may recognize, some you may not.

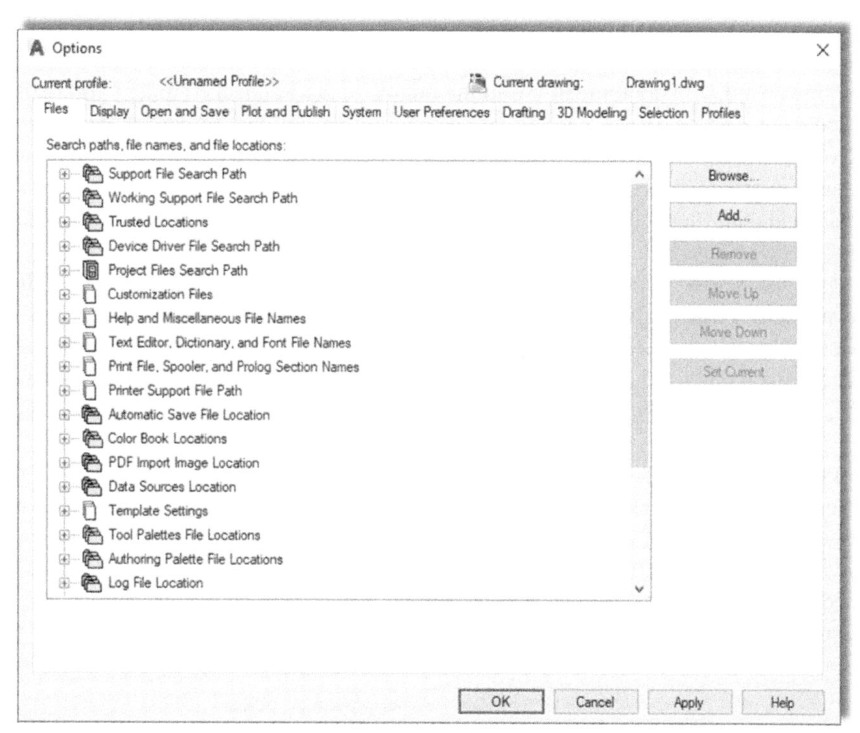

FIGURE 14.2 Files tab.

Be careful though; default settings are a good thing in this case. If you change the end location of some file, often you (and sometimes AutoCAD itself) may not find it, and difficulties result, such as when you try to move the device driver files. It is strongly advised to leave most of these as they are, as there really is no compelling reason to change the preset locations. There are really just a few file paths that *may* need changing, as listed next:

- *Automatic Save File Location*: The default path is C:\users\owner\appdata\local\temp (or something similar). You may want to change that to an easier-to-find locale if you plan on using Autosave.
- *Log File Location*: The default path for these saved text window data logs reside at C:\users\owner\local\autodesk directory (or somewhere near there). If you plan to log certain activity on your project, you may want to relocate these.
- *Action Recorder Settings→Action Recording File Location*: This path should be set to your file's project folder if you plan on using the Action Recorder (a Chapter 15 topic).
- *Plot and Publish Log File Location*: These data logs are for plots, and they also reside in the C:\users\owner\local\autodesk neighborhood. Reset them to some easier-to-find locale if you plan on keeping logs of print jobs.

Other than these four items, the average user will have little to change among these paths, and I can count on just one hand the amount of times I needed to do this in all my years of drafting. If you do wish to learn, via a detailed description, of what each path means, press F1 while you have the Files tab open.

Display Tab

From here on forth, we group and discuss everything by category, similar to how AutoCAD groups everything within these tabs—by a named topic, with frame drawn around it. We list them from top to bottom and left to right. The Display tab (Fig. 14.3) has Window Elements, Layout elements, Display resolution, Display performance, Crosshair size, and Fade control. Among them are quite a few items of interest.

Window Elements

The first thing you will notice is the Light/Dark theme selector. By default AutoCAD 2019, just like its predecessor, has the Dark theme as a default. It was changed to Light way back in Chapter 1 for better visibility on a printed page and to save ink. You can certainly set it to Dark if you wish to achieve that somewhat menacing, cutting-edge look. Note that this has nothing to do with the screen's background color (which we discuss shortly), but merely the overall theme of the surrounding Ribbon, menus, toolbars, etc.

Working your way downward, you do not really need the first three boxes checked unless you really like the scroll bars (a holdover from AutoCAD's early days) or need to see larger toolbar buttons; however, those *are* very useful for the visually impaired. Below that is the Ribbon icon resizing, just a minor adjustment, which you can also leave unchecked. The next five boxes feature various ToolTip-related settings, and these are actually good to leave checked to explain the meaning of new toolbars and other items. Included in there is the delay time, with 1 or 2 seconds adequate. Leave Display File Tabs checked as well.

The Colors... button just below ToolTips refers to the color of your screen background. The default dark gray background color has an RGB value of 33,40,48. Generally users change this color to black, and it is preferred by most (but not all) users. Black is a good choice for the simple reason that the dark background throws less light at your eyes from the screen and lessens fatigue (some of this was discussed as far back as Chapter 1). The only catch is that printed paper output is always white, and some people prefer seeing the design as it would appear in real life, on white paper, hence the white background.

Obviously, it is easy to get used to the black background and pretty much disregard the "paper" aspects of the output, and this is what most users do. In this and most textbooks, however, the screen background color is always white to save ink and blend images smoothly with the white paper.

Other colors are available, of course (actually all 16 million colors are available as a background), but as we joke in class, if you see someone using a bright neon green background, stay away—this is probably not a sane individual! To change the color scheme for practice, press the Colors... button, and the dialog box shown in Fig. 14.4 appears.

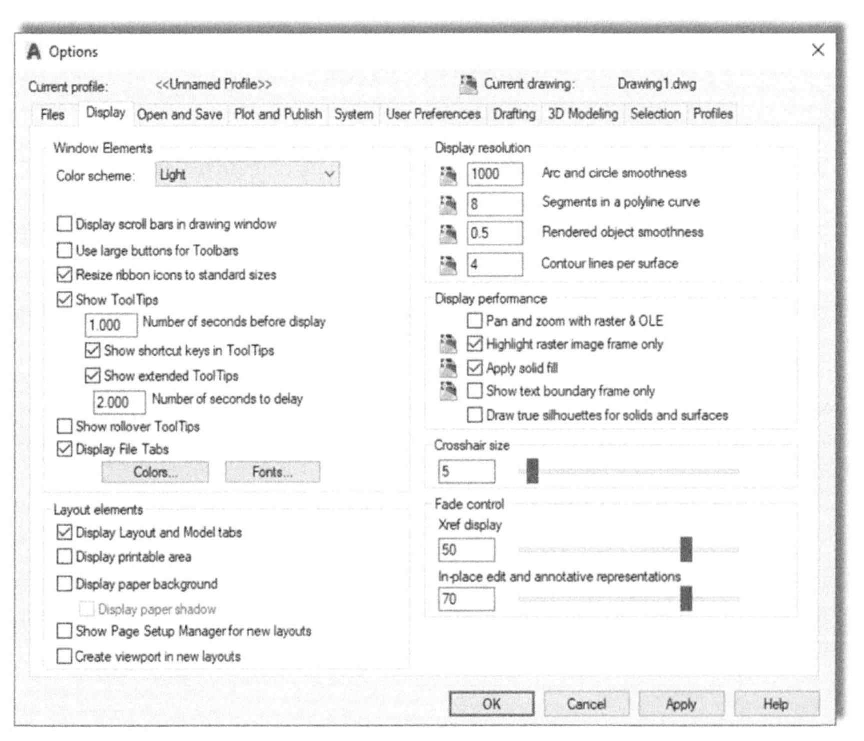

FIGURE 14.3 Display tab.

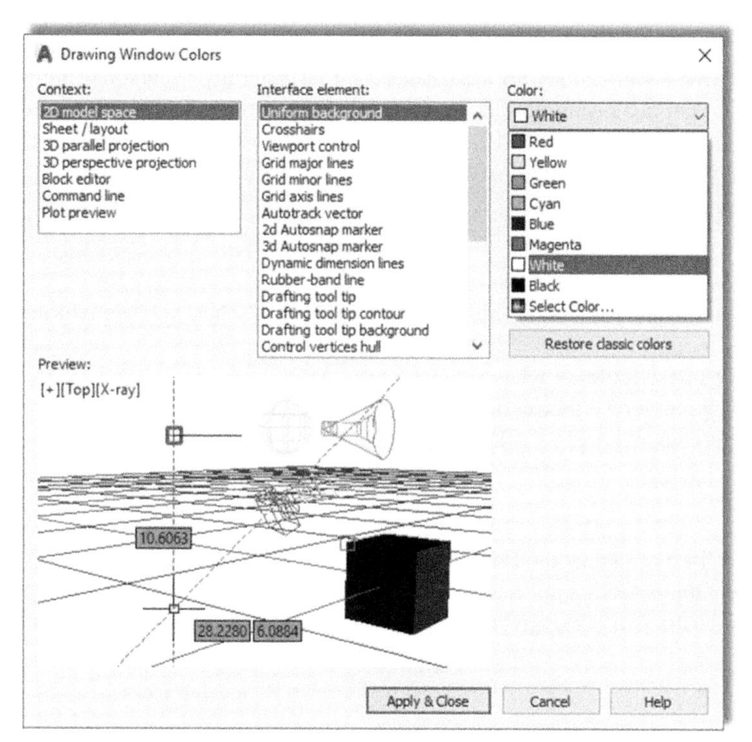

FIGURE 14.4 Drawing Window Colors with White selected.

Here, you can select from the general Context: menu on the left (only the first choice, 2D model space, for now) and select some element in the Interface element: menu on the right (e.g., the Uniform background). Then, pick the color you want this element to have. Fig. 14.4 shows white being chosen. Note the sizable number of choices in both the left and middle columns. AutoCAD allows for the color of just about anything to be changed; browse through all the choices to get familiar. If you change so much that you forget what you did, you can always press the Restore classic colors button on the right of the dialog box. When done, press Apply and Close.

The Fonts... button, to the right of Colors... in the Display tab, refers to the font present on the command line text, not the font in the actual drawing, so unless you have an objection to the Courier New font (the default), do not change anything.

Layout Elements

In this category at the bottom left, it is recommended to uncheck all the boxes except the top one: Display Layout and Model tabs. Generally, all those options (area, background, etc.) are distractions and unnecessary in Paper Space, as was discussed in more detail in Chapter 10. As with many such suggestions, it does come down to personal taste. Check out Paper Space with and without these buttons checked to determine how you prefer it.

Display Resolution

Moving to the upper right of the Display tab, we have the Display resolution category. Here, it is suggested to leave all but one field as default. That field is the Arc and circle smoothness value, which is 1000 by default. That was set low for historical reasons that have to do with accelerating regenerations on slower, old computers by having arcs and circles be made up of small lines. You can set that to the 20,000 maximum, which is fine for today's fast machines.

Display Performance

This category is also a bit of a throwback from the old days of slow computers. These five checked boxes combine to accelerate regeneration of various elements. It is best to leave everything as default, with box 2 and 3 checked and the others unchecked.

Crosshair Size

Moving down we have the Crosshair size setting. Crosshair size is generally a personal preference. Many leave it at a "flyspeck" setting, which is 5, but increasing the size to a full 100 is recommended, as this makes life easier when you are in a Paper Space viewport, as detailed in Chapter 10, or checking if a drawn line is straight by comparing it to the straight-edge crosshairs. In the end, it is up to you, and experience has shown that newer users prefer the flyspeck, whereas the veterans stick to the full size.

Fade Control

The last category at the bottom right has to do with how dark the xref is when you are editing it in place, a Chapter 17, External References (Xrefs) topic. For now, the default settings are fine.

If you want a more detailed description of what each selection means, press F1 to read the Help files while you have the Display tab open. Alternatively, hover the mouse over each element for a brief description. If you are done, do a final check to see that your settings reflect what is in Fig. 14.3 and let us move on to the next tab.

Open and Save Tab

The Open and Save tab (Fig. 14.5) contains the File Save, File Safety Precautions, File Open, Application menu, External References (Xrefs), and the ObjectARX Applications categories. Several important features are worth discussing here, most of them contained inside the two categories on the left-hand side: File Save and File Safety Precautions.

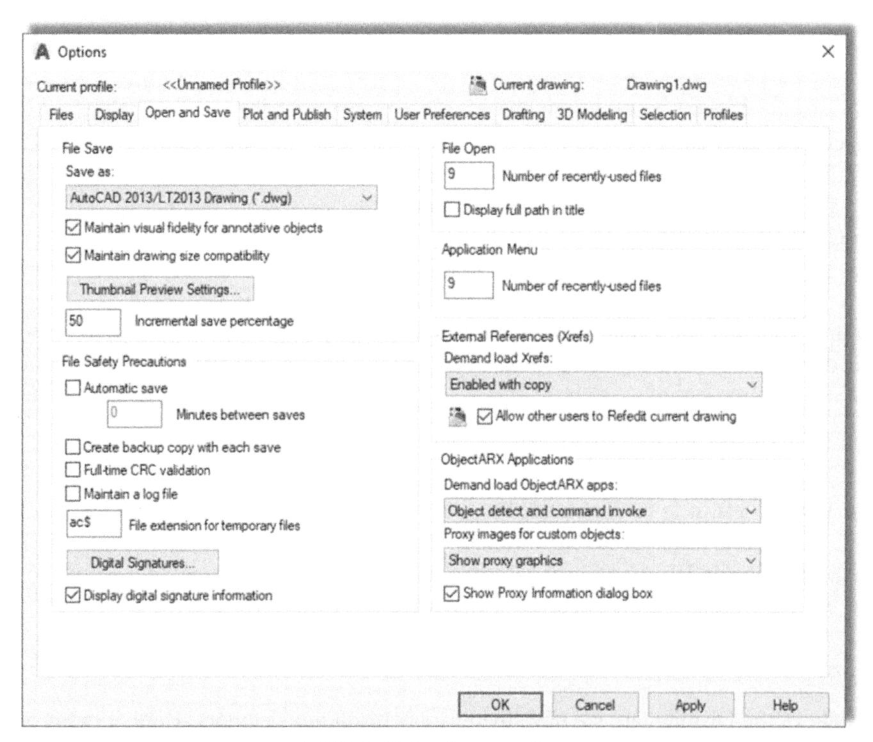

FIGURE 14.5 Open and Save.

File Save

In this category, let us take a look at the top option, labeled *Save as:* This drop-down menu is very important, as it allows you to permanently save your files down to an older version of AutoCAD. The significance of this cannot be overstated because you simply cannot open a file created on a more recent version using an older release. AutoCAD is an often updated software, with new releases every year (since 2004), and you can rest assured that not everyone is running the same version. I have seen AutoCAD 14 (from the mid-1990s) running on a company PC as late as 2006. While that may have been an isolated case, AutoCAD 2013, 2014, 2015, 2016, and 2017 are still in widespread use, and nothing is more frustrating to someone than getting a copy of your files for collaboration and not being able to work with them.

Autodesk of course realizes this and allows you to save files down on a case-by-case basis during the save procedure. AutoCAD even automatically saves to the last DWG file format change, in this case 2018. But to be safe, if it is not already set this way, select the AutoCAD 2013/LT 2013 Drawing (*.dwg) choice, as seen in Fig. 14.5; that way you "undercut" about five releases and assure that the file can be accessed by just about everyone. If you need to, you can save individual files down even further for collaborators still using versions older than 2013 (all the way back to R14, LT98 LT97 if needed). We can skip over the rest of the options under the File Save category (leave everything as default) and move down to the File Safety Precautions category.

File Safety Precautions

Let us work our way top to bottom in this category. The Automatic save check box is the first thing we encounter, and it is about as close to controversial as something in AutoCAD gets. The suggestion is this: Do not use it. Others may say, why not, it is there. The argument against it is simple: As a computer user, you need to save your work often, no matter what application you use. To rely on automatic saves is to invite trouble when you use a program that does not have an automatic save (or it is not turned on). As an AutoCAD user, you should be saving your work every 5–10 minutes, something students often do not do in class.

There are also some practical difficulties with Automatic save. For one thing, AutoCAD needs to exit a command to save a drawing. AutoSave does this for you, so even if you are in the middle of something, you are kicked out for AutoSave to run (except while in block editing mode). Other issues are where AutoSave files go and what extension they have. You can set where they go using one of the paths under the File tab, but then all AutoSave files go there, regardless of what actual drawing is open, and overwrite existing files. The extension of AutoSave files is .sv$.

In summary, while AutoSave can be set up, it is not advisable to rely on it; often, you see veteran programmers and computer users ignore this feature in all computer use and save instinctively every few minutes. If you do wish to use AutoSave, simply turn it on by checking the box and setting a time threshold (Minutes between saves) just below that.

The Create backup file with each save check box can also be mildly controversial, as it seems like a good idea initially, yet does not stand up under scrutiny. Backup files (if the option is checked) are created every time you save a file. A backup file (.bak) is essentially a regular drawing file (.dwg) that is cloaked with an extension Windows does not recognize, but you do (see Appendix C, File Extensions for more info on all extensions). If the .dwg file is lost, then the .bak file can be renamed and, voila, the drawing is there again. So, what is wrong with this? Well, in theory nothing, but in real-life use there are a few issues.

The main problem is that backup files actually double the size of your project folders; it is like having a copy of every drawing file next to the original. They are just not necessary. The combination of cheap storage media and a heightened sense of risk in the last decade and a half have prompted just about every company to do daily and weekly backups. Also, AutoCAD's drawings rarely "go bad" to the point of being unrecoverable by the Audit and Recover commands. Finally, backup files reside right next to the drawing files themselves, and if you lose the entire folder, they all go. Rarely do you lose just the drawing; even then, going to the previous night's backups are enough to get them back. So, while some would argue that no harm is done by keeping the .bak files, one could also say, why add the extra junk to your company's server? Ultimately, the choice is up to you, although sometimes company policy may dictate the use of backup files, and this is beyond your control.

Moving further down, most users do not need Full-time CRC validation. CRC stands for Cyclic Redundancy Check and is an error checking mechanism for when your drawings seem to crash after you bring something in. This does not happen often, so you generally do not need this feature on.

Maintain a log file makes AutoCAD write the contents of the text window to a log file, also an unnecessary feature for most users. File extensions for temp files can also be left as the default ac$.

Digital Signatures... is a relatively new feature in AutoCAD that addresses the need to secure or authenticate the drawings. This is not often used, as most AutoCAD drawings are not secret in nature, but the idea is intriguing and worth a look. When you click on the button, the box in Fig. 14.6 appears.

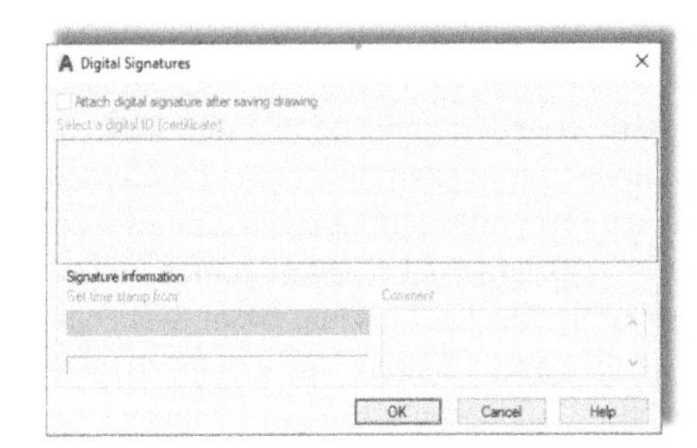

FIGURE 14.6 Security Options.

The first tab allows you to add a password to the drawing. Then, the drawing can be opened only by someone who knows the password; although be careful—if the password is forgotten, the drawing is lost. The second tab has to do with a digital signature. This is a lengthy topic in itself and is sometimes used when collaborating with others. In short, a user installs a certificate from a digital security vendor (VeriSign in this case), which allows for a "digital signature," which in turn prevents changes to the drawing or keeps track of them. This is a very simplified version of the concept, of course, but if your company already uses this, then no further explanation is necessary; if not, you will likely not care about this feature anyway.

File Open

Moving to the top right of the Open and Save tab we have File Open category, which simply shows a number of recently open files, up to nine. Leave as the default max, with Display full path in file unchecked as that may take up a lot of room across.

Applications Menu

This is the same idea as File Open but refers to the application menu's recent documents. Leave it as default.

External References

These fine points of external reference loading and usage are best left as default.

ObjectARX Applications

ObjectARX is a relatively complex topic that requires some programming background to fully understand. ARX stands for AutoCAD Runtime Extension and is an application programming interface (API) for customizing and extending AutoCAD. An API specifies how software components should interact with each other. ObjectARX Software Development Kits are available to third-party programmers and software engineers to develop plug-in applications for each release of AutoCAD. See Appendix F, AutoLISP Basics and Advanced Customization Tools for more information on this. As far as this category is concerned, leave everything as default and allow AutoCAD to show and load the associated apps and proxy graphics, which are visual objects that are not "proxied" and have no graphical equivalent.

If you want a more detailed description of what each selection means, press F1 to read the Help files while you have the Open and Save tab open. Alternatively, hover the mouse over each element for a brief description. If you are done, do a final check to see that your settings reflect what is in Fig. 14.5 and let us move on to the next tab.

Plot and Publish Tab

The Plot and Publish tab (Fig. 14.7) contains the Default plot settings for new drawings, Plot to file, Background processing options, Plot and publish log file, Auto publish, General plot options, Specify plot offset relative to categories, and also two separate buttons at the bottom right: Plot Stamp Settings… and Plot Style Table Settings. Except for a few items, not a tremendous amount is important under this tab for the average user. We run through everything with some general comments and descriptions.

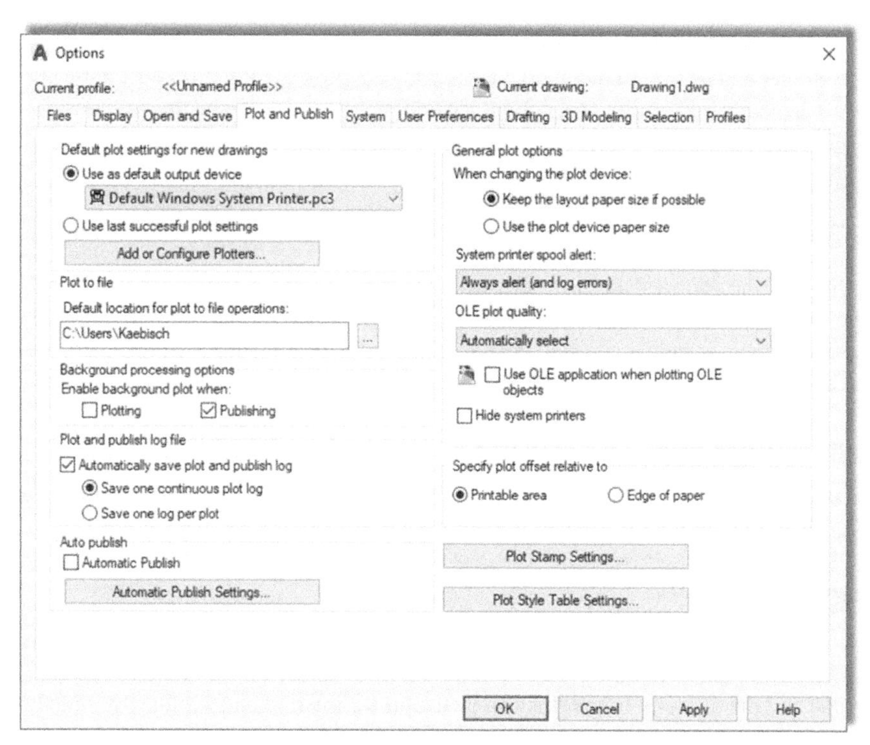

FIGURE 14.7 Plot and Publish.

Default Plot Settings for New Drawings

This category selects the default plotter or adds a new one. However, this should not be done from here but rather from the Windows control panel. When new plotters or printers are added, AutoCAD notices them, as well as what the chosen default is. In the end, however, this is of little significance. Recall from Chapter 9 that the printer settings for each drawing are set individually. Therefore, leave everything here as default.

Plot to File

In this category, a drawing can be plotted to a .plt (plot) file and saved in this location. While this is generally an outdated legacy command from the old days of .plt files being sent (or physically carried on a floppy disk) to plotters, it has found some use today when taking drawings to a printing company. The company may not have AutoCAD, or anyone qualified to operate it, so instead they accept the .plt files, which are nothing more than ready-to-print code that can be interpreted by any plotter.

Background Processing Options

This category has to do with being able to plot in the background while working on a drawing. With the Plotting option unchecked, you must wait until the plotting is complete before continuing. Publishing may take longer, so you want to allow this by checking it.

Plot and Publish Log File

This category is an option that creates (or turns off) the log feature, which is nothing more than a record of the Job name, Date and time started and completed, Full file path, Selected layout name, Page setup name, Device name, Paper size name, and a few others. It is recommended to disable (uncheck) this feature unless there is a specific reason to create a record of all this information.

Auto Publish

This category is an option that allows you to automatically create design web format (DWF) files. DWF is a topic all unto itself. The basic idea is to generate vector files that can be viewed by others who have the Autodesk Viewer or, if not, then through a browser. You need this box checked only if you planning on generating these files to give to someone or to put your drawings up on the web.

General Plot Options

Moving to the upper right of the Plot and Publish tab, we have the General plot options category. This has to do with system printers and the output quality of object linked embedded (OLE) items, a topic to be discussed in Chapter 16. The OLE plot quality drop-down choices can be adjusted depending on what is embedded. In this case, leave everything in this category as default, as these are the most reasonable choice for the average user.

Specify Plot Offset Relative To

Leave this category as default, as you generally want to have the printable area as the start point of any plot offsets, not the edge of the paper, to ensure extra room.

Plot Stamp Settings...

This button, at the bottom right, is an item of interest. A plot stamp is simply a text string that appears on all output when activated in the Plot dialog box. This text string has information on a variety of parameters, but the ones of most interest include the drawing name and the date and time. This can be useful for internal check plots (obviously not for final output) to show where the drawing is on the company server or to indicate which paper output is the latest based on the date and time. Bring up the Plot Stamp (Fig. 14.8) by clicking the button and use the Advanced button in the lower left to further refine the plot stamp if need be; these settings mostly have to do with the size and position of the stamp.

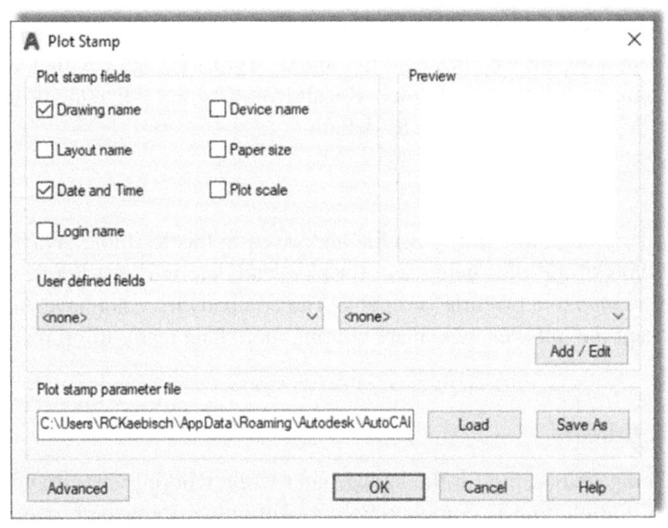

FIGURE 14.8 Plot Stamp.

Plot Style Table Settings

Leave everything under this button as default. These settings are addressed elsewhere, such as the plot dialog box and the plot style manager. See Chapter 19 for more on this.

If you want a more detailed description of what each selection means, press F1 to read the Help files while you have the Plot and Publish tab open. Alternatively, hover the mouse over each element for a brief description. If you are done, do a final check to see that your settings reflect what is in Fig. 14.7 and let us move on to the next tab.

System Tab

The System tab (Fig. 14.9) contains the Hardware Acceleration, Current Pointing Device, Touch Experience, Layout Regen Options, General Options, Help, InfoCenter, Security, and the dbConnect Options categories. Much like the previous tab, not a tremendous amount is important under this tab for the average user except for a few items. We run through everything with some general comments and descriptions.

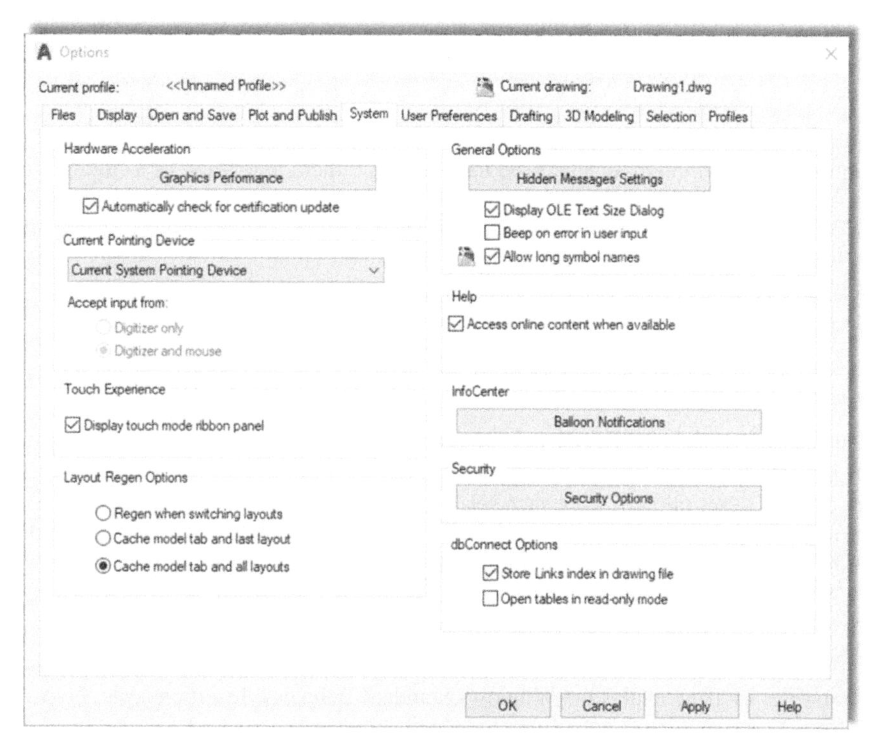

FIGURE 14.9 System tab.

Hardware Acceleration

This category, and the associated Graphics Performance button, is tailored toward 3D performance tweaks, and the category was even called 3D Performance in a previous version of AutoCAD. It details various graphics settings and system descriptions. These settings are already optimized automatically and should be just fine for the average user. Feel free to run through the Tuner Log, which can be seen after a few clicks—lot of interesting information is in there for the tech savvy.

Current Pointing Device

This category allows for setting up digitizer operations, a topic not relevant to most users, who have no digitizer.

Touch Experience

The Touch Experience has to do with a touch pad and is also irrelevant to most users.

Layout Regen Options

This category should be left as default; it refers to regeneration of layouts when you switch to them. With today's fast machines an extra regen here and there does not affect productivity.

General Options

Leave everything in this category as default. Here resides perhaps one of the most annoying features of AutoCAD: The "Beep on error in user input" option. This may push a student just starting out and making lots of errors over the edge. You definitely do not want that box checked!

Help

Unless your computer is not connected to the internet, keep this box checked to access online content.

InfoCenter

The InfoCenter and its Balloon Notifications button refer to the various information balloons that appear with helpful updates or tidbits of information. Although an advanced user may find them annoying, as a student you should keep them around. Briefly explore the button to see what options it allows you in regard to these notifications.

Security

The category refers to dealing with *.exe (executable) files and *.lsp (AutoLISP) files. Leave everything as default.

dbConnect Options

This category should also be left as default. The *db* refers to database connectivity, which is also not relevant to most users but may be important in some high-end custom applications, where AutoCAD is linked to a database.

If you want a more detailed description of what each selection means, press F1 to read the Help files while you have the System tab open. Alternatively, hover the mouse over each element for a brief description. If you are done, do a final check to see that your settings reflect what is in Fig. 14.9 and let us move on to the next tab.

User Preferences Tab

The User Preferences tab (Fig. 14.10) contains the Windows Standard Behavior, Insertion scale, Hyperlink, Fields, Priority for Coordinate Data Entry, Associative Dimensioning, Undo/Redo categories and the Block Editor Settings..., Lineweight Settings..., and Default Scale List... buttons. Some significant items lie under the first category. We describe it in detail and run through everything else with some general comments and descriptions.

Windows Standard Behavior

This category concerns speed, which is a topic near and dear to the hearts of most AutoCAD users. As time equals money, whatever technique accelerates drafting speed is always welcome. One of the more important ones (shortcuts) is discussed later in this chapter. Another important "trick" is to set your mouse buttons to the most likely and needed settings for fast drafting. As such, the Right-click Customization... button is very important. Keeping the boxes above that button checked as a default, click on it, and the dialog box in Fig. 14.11 appears.

Your mouse settings should be as shown in Fig. 14.11, with the last one, Command Mode, set to ENTER. This subtle but powerful setting speeds up your basic drafting by allowing you to instantly finish commands by using the right mouse button instead of having to press Enter on the keyboard. Also set Edit Mode to Repeat Last Command and Default Mode to Shortcut Menu. The time sensitivity settings at the very top refer to the mouse button "reaction times." The default 250 ms value happens to be a good time for most people, so nothing else needs to be adjusted here. Press Apply and Close.

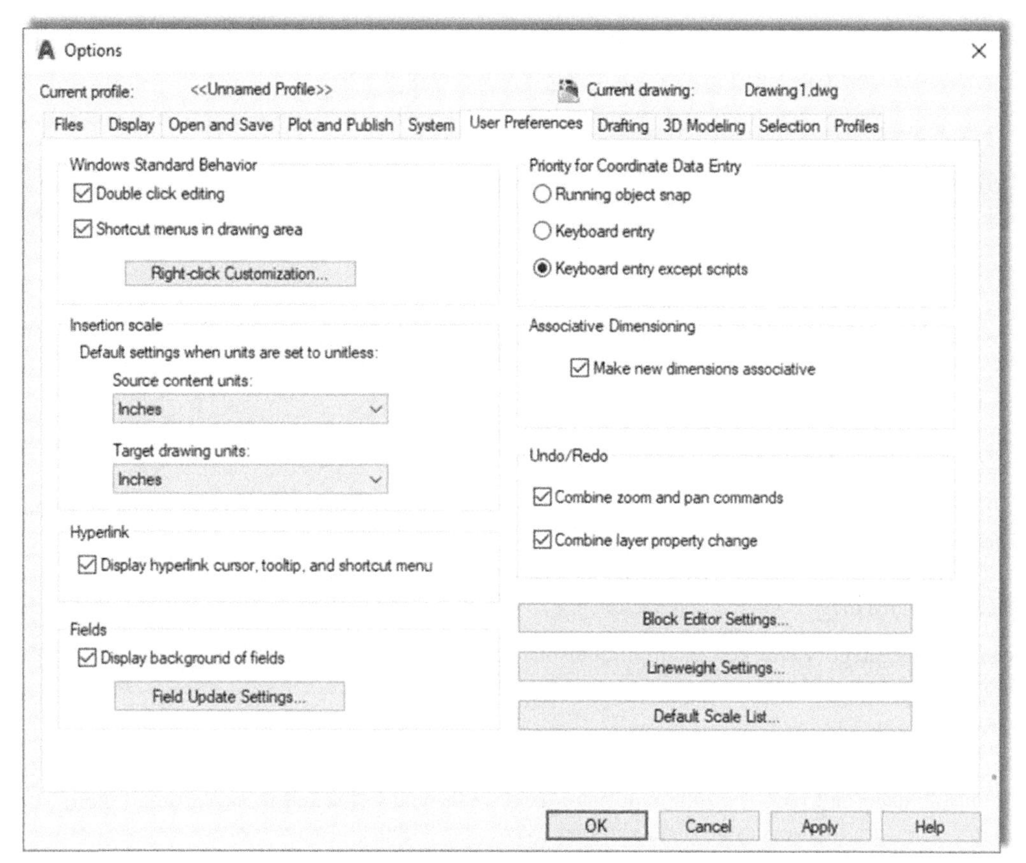

FIGURE 14.10 User Preferences tab.

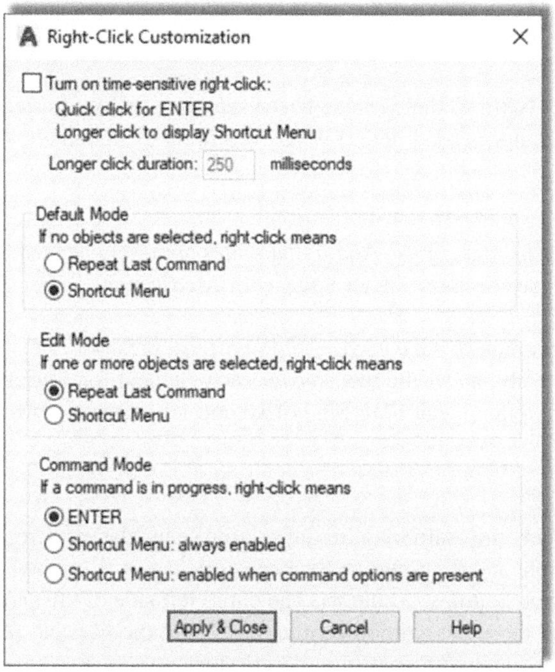

Right-Click Customization
Turn on time-sensitive right-click:
Quick click for ENTER
Longer click to display Shortcut Menu
Longer click duration: 250 milliseconds
Default Mode
If no objects are selected, right-click means
Repeat Last Command
Shortcut Menu
Edit Mode
If one or more objects are selected, right-click means
Repeat Last Command
Shortcut Menu
Command Mode
If a command is in progress, right-click means
ENTER
Shortcut Menu: always enabled
Shortcut Menu: enabled when command options are present
Apply & Close
Cancel
Help

FIGURE 14.11 Right-Click Customization.

Insertion Scale

This category controls the scale for inserting blocks; leave the default values.

Hyperlink

This category controls visual settings of hyperlinks; leave this box checked as the default is fine. We cover hyperlinks in the next chapter.

Fields

This category sets field preferences, which you should leave as default. A field is text in AutoCAD drawings that automatically updates itself (after a regen command) whenever there is a change in the values to which the field is connected. I have not found them to be widely used.

Priority for Coordinate Data Entry

Moving to the top right of the User Preference tab, this category controls how AutoCAD responds to coordinate data input also not necessary to change for most users.

Associative Dimensioning

Leave the box under this category checked, as all dimensions should be associative (meaning, if the geometry changes, so do the dimensions associated with it).

Undo/Redo

This category controls undo and redo for zoom and pan; leave both boxes checked to save some time if you have to go back a lot of steps. They are then combined and skipped over.

Block Editor Settings...

This button changes the environment of the block editor. You worked with that editor when you first learned dynamic blocks. Clicking on it brings up the Block Editor Settings dialog box, where you can change just about every color, size, and font parameter. For most users, it is best to leave everything as a default.

Lineweight Settings...

This button displays the Lineweight Settings dialog box. These settings can be adjusted when setting up the .ctb files, as described in Chapter 19; so they are not needed here. Leave everything as default.

Default Scale List...

This button displays the Edit Scale List dialog box. Additional scales are rarely needed but can be added here, if you have some unusual requirements. This may be more of an issue with Metric units, as engineers may use more scales than architects.

If you want a more detailed description of what each selection means, press F1 to read the Help files while you have the User Preferences tab open. Alternatively, hover the mouse over each element for a brief description. If you are done, do a final check to see that your settings reflect what is in Fig. 14.10 and let us move on to the next tab.

Drafting Tab

The Drafting tab (Fig. 14.12) contains the AutoSnap Settings, AutoSnap Marker Size, Object Snap Options, AutoTrack Settings, Alignment Point Acquisition, and the Aperture Size categories, as well as the Drafting Tooltip Settings..., Lights Glyph Settings..., and Cameras Glyph Settings... buttons. The entire left side of this tab is devoted to snap point markers, those colored geometric objects that represent endpoints, midpoints, and the like. No really critical items are here, so we run through everything with some general comments and descriptions.

AutoSnap Settings

Leave everything in this category as default, including Colors, unless you prefer something different.

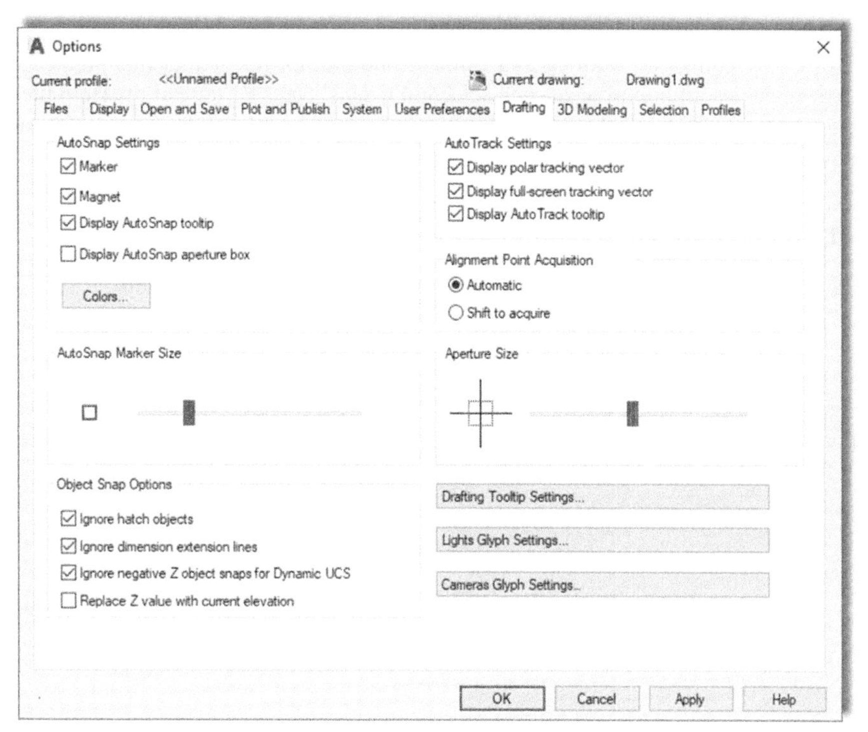

FIGURE 14.12 Drafting tab.

AutoSnap Marker Size

Leave this category as default, unless you have trouble seeing the symbols or find them too big. The sizing of the marker is easily adjusted via the slider.

Object Snap Options

Also leave everything in this category as default. You especially want to make sure that the Ignore hatch objects (the first box) is checked off, as that would make the object snap go crazy trying to snap to every piece of a complex hatch pattern.

AutoTrack Settings

Moving to the upper right of the Drafting tab, we have the AutoTrack Settings category. It has to do with AutoCAD's OTRACK tracking tool, which is briefly discussed in Chapter 15, along with other miscellaneous topics. Leave these settings as default for proper operation of that tool.

Alignment Point Acquisition

Leave the setting in this category as default.

Aperture Size

Aperture Size is perhaps the one item that may occasionally be adjusted to suit the user's taste but generally keep it the same size or slightly larger than the AutoSnap marker. If the aperture is too big, it may be distracting or may actually be confused with small rectangles on the screen.

Drafting Tooltip Settings...

This button features minor nuanced adjustments to drafting tool tips (such as distance or angle displays) and their visibility to the users (the transparency). Leave all them as default.

Lights Glyph Settings... and Cameras Glyph Settings...

These buttons are minor color settings for lighting and cameras and are not relevant at this point; leave them as default.

If you want a more detailed description of what each selection means, press F1 to read the Help files while you have the User Drafting tab open. Alternatively, hover the mouse over each element for a brief description. If you are done, do a final check to see that your settings reflect what is in Fig. 14.12 and let us move on to the next tab.

3D Modeling Tab

Settings and adjustments under this tab are not discussed in this 2D-only level of the book.

Selection Tab

The Selections tab (Fig. 14.13) contains the Pickbox size, Selection modes, Ribbon options, Grip size, Grips, and Preview categories. Some important items are under the Visual Effects Settings... button under the Preview category, so we cover that in detail, and run through everything else with some general comments and descriptions.

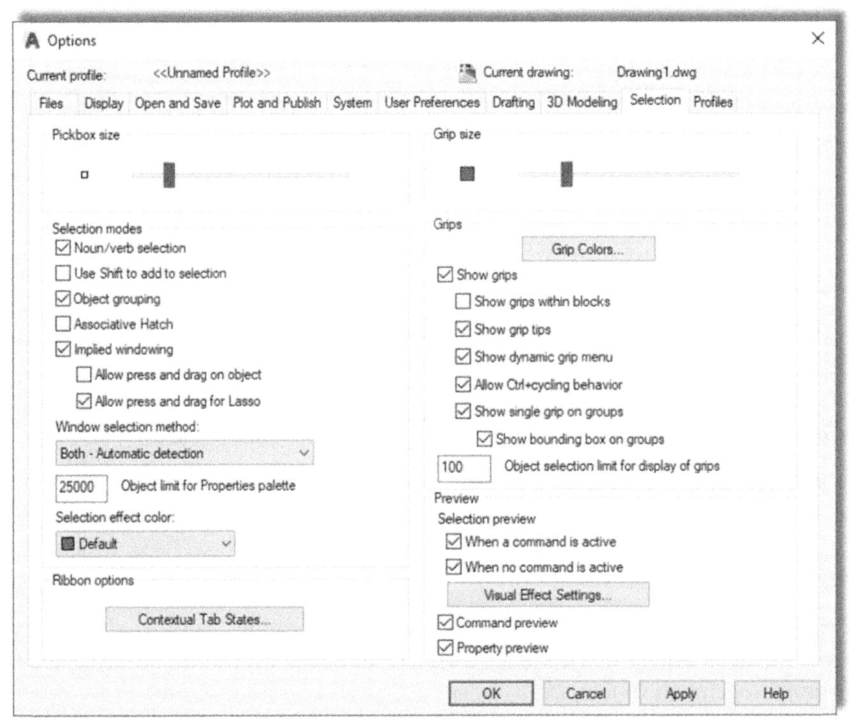

FIGURE 14.13 Selection tab.

Pickbox Size

This category is simply what the cursor becomes when you are in the process of picking something. Set it too small and you have trouble selecting the geometry. Set it too big and you select more than you wanted. Generally, leave it at the default size shown in Fig. 14.13.

Selection Modes

This category is just more adjustments to object selection. Leave them all as default, as seen in Fig. 14.13, including the Window selection method choice. The Object limit for Properties palette can be left as the ridiculously high default of 25,000. Rarely will you have that many objects selected or even present in your drawing.

Ribbon Options

This category provides some options regarding contextual tab states. Leave them as default.

Grip Size

Moving to the upper right of the Selection tab, the Grip size category with the slider adjustment should be self-explanatory. Leave the grips as default size unless you prefer bigger or smaller ones.

Grips

This category has yet more grip settings for colors and other adjustments. There is really no reason to change grip color. Blue means cold (unselected), red means hot (selected), and hover can be any color (default Color 11 is just fine). Leave the other settings as default; you want the grips enabled but not inside blocks. Leave the maximum amount of grips that can be shown at 100 so as to not overwhelm the drawing if you select more.

Preview

This category is a relatively new feature in AutoCAD that causes objects to "light up" when your mouse hovers over them. In AutoCAD 2015, this effect was made more subtle, eliminated dashing the objects, and only added a gray highlight to them. If you find this feature distracting, you can turn it off by unchecking the first two buttons. To leave it on but modify the settings, click the Visual Effects Settings... button. When you press it, you see what is shown in Fig. 14.14.

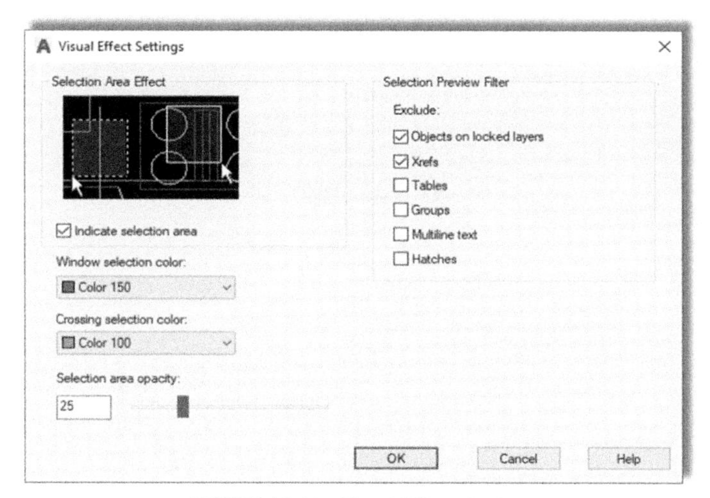

FIGURE 14.14 Visual Effects Settings.

On the left side of this dialog box are various controls for the internal colors of the familiar Window/Crossing and the new Lasso tool. You can change those colors and their intensities by varying the Selection area opacity slider. You can also eliminate them, as the colors are technically unnecessary; what is important is whether the selection rectangle is dashed or solid. If you prefer minimum fuss with your selection methods, you can remove the colors by setting the opacity at zero or uncheck the Indicate selection area button.

If you want a more detailed description of what each selection means, press F1 to read the Help files while you have the User Selection tab open. Alternatively, hover the mouse over each element for a brief description. If you are done, do a final check to see that your settings reflect what is in Fig. 14.13 and let us move on to the next tab.

Profiles Tab

This tab (Fig. 14.15) is simply there to save your personal settings, so if someone else is using AutoCAD on your computer and prefers other settings (maybe that neon green background color) that person can set his or her own and save them as well. Simply press the Add to List… button and enter a name for the profile, as seen in Fig. 14.16; finish by pressing Apply and Close.

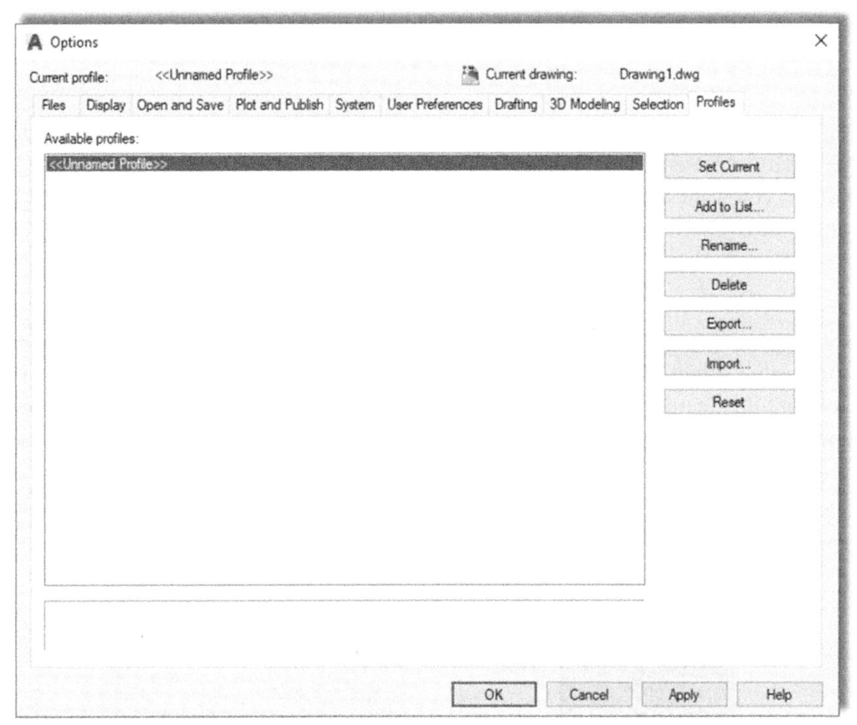

FIGURE 14.15 Profiles tab.

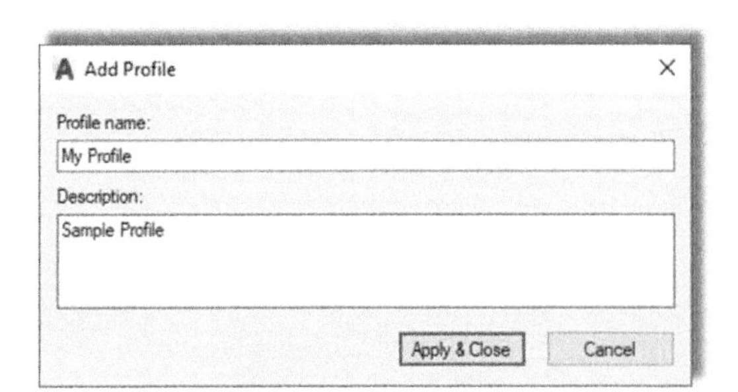

FIGURE 14.16 Add profile.

The profile is added to the list (as many as you want can be added) and can be restored by selecting it and pressing Set Current. Profiles can be exported (as .arg files), imported, renamed, and deleted as needed. Profiles were more relevant in the days when PCs were more expensive and each one in a company may have had a few users, but some people still use them to avoid office arguments about exactly how AutoCAD "should look" if one were to "borrow" a PC for the day, then change AutoCAD to suit his or her taste, and forget to reset it to how the original owner or user likes it.

14.2 SHORTCUTS

Recall that it was mentioned in the previous pages how important speed is to an AutoCAD designer. One of the tools to help you work faster was already mentioned, the mouse Right-Click Settings under the User Preferences tab. Shortcuts, however, are another important way to accelerate your input. They are useful only if you type. If you are a diehard Ribbon or toolbar user and abhor the keyboard, then you have no need for them. Many students, however, end up typing quite a lot, even if they also integrate some toolbar and Ribbon use.

The reasons are simple: Typing is a fast, very direct method to interact with AutoCAD and remains quite popular despite other methods. However, to take advantage of this, you need shortcuts, as typing out the full command defeats the speed advantage.

You may have already been using shortcuts all along. Instead of typing in `arc`, you may have just pressed the letter `a`; instead of `move`, the letter `m`. This is exactly what is meant by shortcuts: It is the process of shortening the typed commands so they can be entered faster. What we want to do here is put all this on solid footing and outline exactly how to set your own shortcuts to make best use of them.

PGP File

Here is a basic primer on shortcuts. They have been a part of AutoCAD since the early days, when typing commands was the best (and, in the very beginning, the only) way to interact with the software. Abbreviating the commands was the logical next step to speed up typing, and that capability was integrated into AutoCAD as well.

The abbreviations are stored in a file called acad.pgp, which resides in the Support folder of your AutoCAD installation. AutoCAD comes with a wide range of preinstalled shortcuts, but the file is easily found and meant to be modified to suit the user's taste and habits. The key here is that you need to have your shortcuts memorized (or it makes no sense to use them in the first place), and of course, you need to pick the most efficient abbreviations of the commands. To that end, AutoCAD does not care what you use—it can be names of your relatives if you so choose—but obviously you want to use something that makes sense to you and shortens the typing effort. Ultimately, the first letter, first and second letters, or first and last letters are used for all but a few of the commands.

Here are some suggestions for the Modify commands:

Erase: e
Move: m
Copy: c
Rotate: ro
Scale: s
Trim: t
Extend: ed
Fillet: f
Offset: o
Mirror: mi

The idea here is to abbreviate the most often used commands in the shortest way possible. If, for example, there is a conflict between two commands that start with the same letter (e.g., circle and copy), then you determine which command gets used more often and give that command fewer letters. Copy is generally used more often, so it would be `c` and circle would be `ci`.

Sometimes, doubling up of letters is done, such as with array (aa), because a is taken by arc, and so on. Listed in Fig. 14.17 is a sample acad.pgp file. It is presented here only for illustration purposes and may or may not match what you may want in yours, although it can be a good starting point to creating your own.

```
A,     *ARC
AA,    *ARRAY
B,     *BHATCH
BR,    *BREAK
BL,    *BLOCK
C,     *COPY
CF,    *CHAMFER
CH,    *CHANGE
CI,    *CIRCLE
D,     *DTEXT
DD,    *DDEDIT
MT,    *MTEXT
DI,    *DIST
DD,    *DDEDIT
DDC,   *PROPERTIES
DDO,   *DSETTINGS
DDS,   *DDSTYLE
DDU,   *DDUNITS
DV,    *DVIEW
E,     *ERASE
ED,    *EXTEND
EX,    *EXPLODE
EL,    *ELLIPSE
F,     *FILLET
I,     *INSERT
ISO,   *LAYISO
L,     *LINE
LA,    *LAYER
LF,    *LAYFRZ
LW,    *LAYWALK
LI,    *LIST
ISO,   *LAYISO
M,     *MOVE
MM,    *MATCHPROP
MI,    *MIRROR
MI3,   *MIRROR3D
O,     *OFFSET
P,     *PLOT
PE,    *PEDIT
PL,    *PLINE
PP     *PURGE
Q      *QSAVE
R      *REGEN
R3     *ROTATE3D
RO     *ROTATE
RE     *RECTANG
RR     *RTZOOM
RT     *RTPAN
S      *SCALE
ST     *STRETCH
T      *TRIM
U      *UNDO
UU     *UCSICON
W      *WBLOCK
X      *EXTRUDE
Z      *ZOOM
```

FIGURE 14.17 Sample acad.pgp file.

Altering the PGP File

So, how do you find and alter the pgp file to suit your tastes? The long way is to find it via this path on your "C" drive: C:\Users\owner\AppData\Roaming\Autodesk\AutoCAD 2019\R21.0\enu\Support, but you really do not need to do this, and the shorter way is to access the file automatically through the drop-down menus as follows: Tools → Customize → Edit Program Parameter (acad.pgp). The pgp file then appears on your screen (opened as a Notepad file, which is just a simple ASCII text editor, if you never used it before). Scroll down the file until you see the rather lengthy built-in listing of shortcuts. At this point, you may do one of three things:

1. Leave the entire file as it is and use the shortcuts without modification.
2. Erase or change the shortcuts for individual commands that you either do not use or would like to modify.
3. Erase the entire list and start fresh, entering the shortcuts as you would like them to be, one by one. If this is the case, be sure to follow the format given, where you enter the shortcut, then a comma, then a few spaces, and finally the command preceded by a *, as seen in Fig. 14.17. You need not use all capital letters—that is just a preference.

When done, save the acad.pgp file and close Notepad. At this point, for the changes to take effect, you can either restart AutoCAD or (if you are in the middle of a drawing and do not want to do that) type in reinit and press Enter. The dialog box in Fig. 14.18 appears.

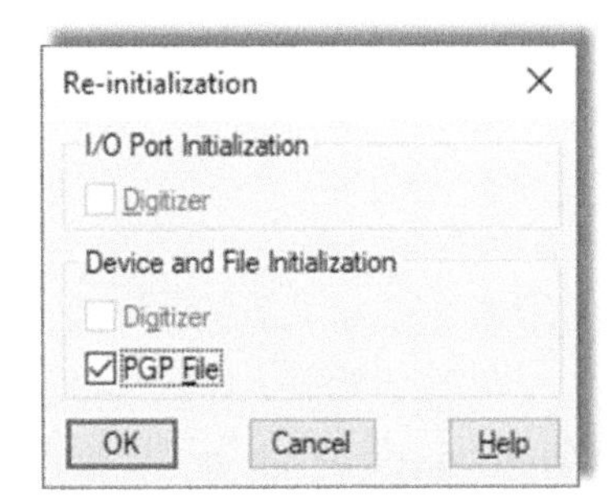

FIGURE 14.18 Reinitialization of the acad.pgp file.

Check off the PGP File box and press OK. The pgp file is ready to go, and you can use it right away. To add or subtract commands, just repeat the steps at the beginning of this section, save the pgp file, and reinitialize.

It is strongly recommended that you at least try to modify and use the pgp file. It is almost always faster to type the shortcuts than to click on icons. You can eventually learn to type quickly and without even glancing at the keyboard (similar to a good secretary or court reporter), assuming you do not type fast, as many students do. Another advantage to typing commands is that you can free up space on your screen by removing duplicated commands in toolbars and the Ribbon.

14.3 CUSTOMIZE USER INTERFACE

The CUI dialog box (referred to from here on as just the CUI) is the next thing we examine in this chapter. The CUI has to do with changing the look and functionality of menus, toolbars, and other items and, as such, can be of importance to a small set of users who may need to do this for aesthetic or functional reasons.

The CUI is an advanced and somewhat mysterious concept and one not taken advantage of by most users, as either they do not know much about it or do not wish to change their toolbars or menus. A few though find this customization ability of great interest and spend time customizing the look of AutoCAD to the point of almost having AutoCAD look like another program entirely.

We take the middle ground here and present some CUI basics, demonstrating the creation of a new custom toolbar and leaving the student with some CUI ideas to pursue further if he or she is interested in doing so.

Let us introduce the basic CUI as it opens in default mode. Type in `cui`, press Enter (or alternatively use the Ribbon's Manage tab → User Interface), and after a few seconds, the dialog box in Fig. 14.19 opens up. Press the expansion button (small arrow, bottom right) to make it full size if it does not do this automatically.

In general, the CUI is your window into customizing virtually anything in AutoCAD that you can push, select, or type in. The full list of candidates is in the upper left of the CUI box: Workspaces, Toolbars, Menus, Dashboard Panels, and all the other choices. You may need to minimize the Workspace menu to see all the rest of them. Just to get your feet wet, let us create a custom toolbar that has only the buttons you want and none of the ones you never use. This is good idea for 3D, where toolbar use is more of a necessity than in 2D.

Step 1. In the upper left-hand corner, find the Toolbars expandable menu and right-click, selecting New Toolbar. Give your new toolbar a name in the name field (`My_New_Toolbar`, in this example). Once it is done, you can click on it and a Toolbar Preview Properties appears on the right of the CUI window, along with an expanded Properties palette.

Step 2. On the bottom left begin selecting the commands you want in your new toolbar (you can filter the commands to find them more easily) and drag them into the new toolbar name on the upper left. As soon as you begin doing this, a toolbar preview and Button Image category appear on the right-hand side, as seen in Fig. 14.20, when Rectangle, Revcloud, Arc, and Spline were added in.

Interestingly, you can even modify the image itself by pressing Edit... in the Button Image category at the upper right, which takes you into the dialog box shown in Fig. 14.21. Here, you can use pixel-by-pixel editing tools to edit the image. You have some additional tools, such as a grid and a Preview window, to see what you have created. Finally, you can save your work and return to the CUI. You can draw some rather creative-looking buttons if you so desire (a few practical jokes of smiley-face icons from years gone by come to mind here). Admittedly, this is not done very often, as most users have better things to do with AutoCAD or at least one would hope.

Finally, after adding a few more commands (Revcloud, Circle, and Spline) and pressing Apply and OK, the new toolbar is created, as shown in Fig. 14.22; this is certainly a good trick if you like toolbars and want to see only what you use on yours.

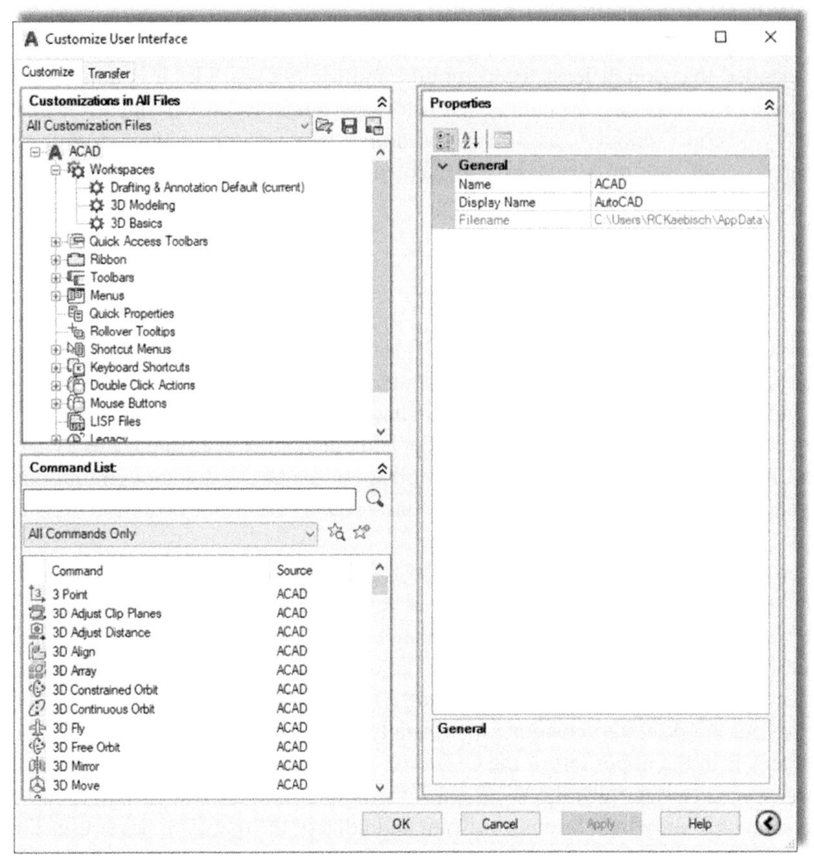

FIGURE 14.19 The Customize User Interface (CUI) dialog box.

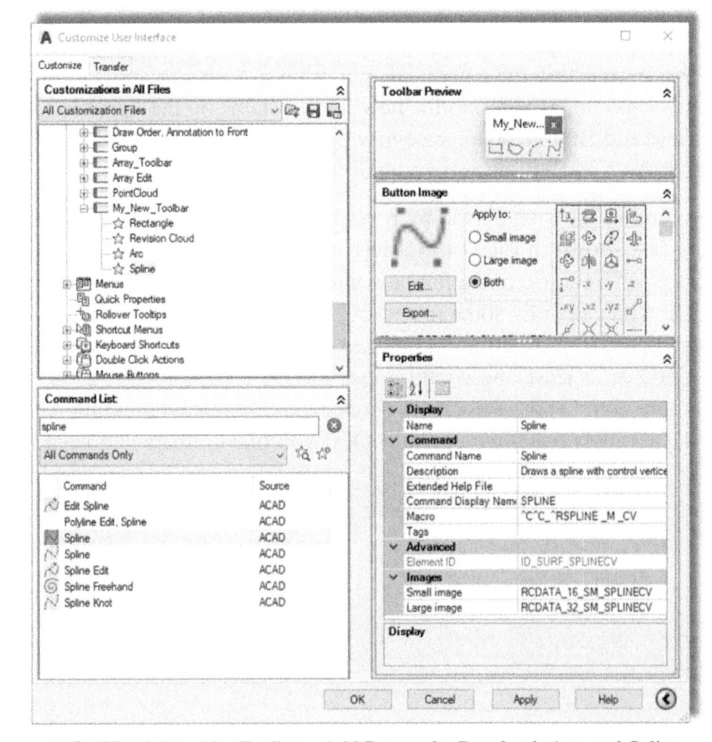

FIGURE 14.20 My_Toolbar—Add Rectangle, Revcloud, Arc, and Spline.

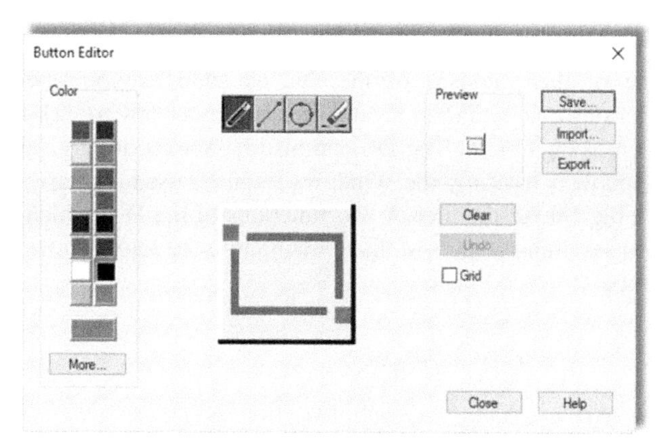

FIGURE 14.21 Button Editor.

FIGURE 14.22 The new My_New_Toolbar.

This is just the tip of the iceberg, of course. You can modify all the rest of the choices. Shortcut Keys under the Keyboard Shortcuts are quite useful. This refers to shortcuts such as Ctrl+C for Copy Clip (also known as *accelerator keys*). You can set up your keyboard to accelerate many commands. Just some of the preset ones are shown in Fig. 14.23 at the upper right of the CUI dialog box.

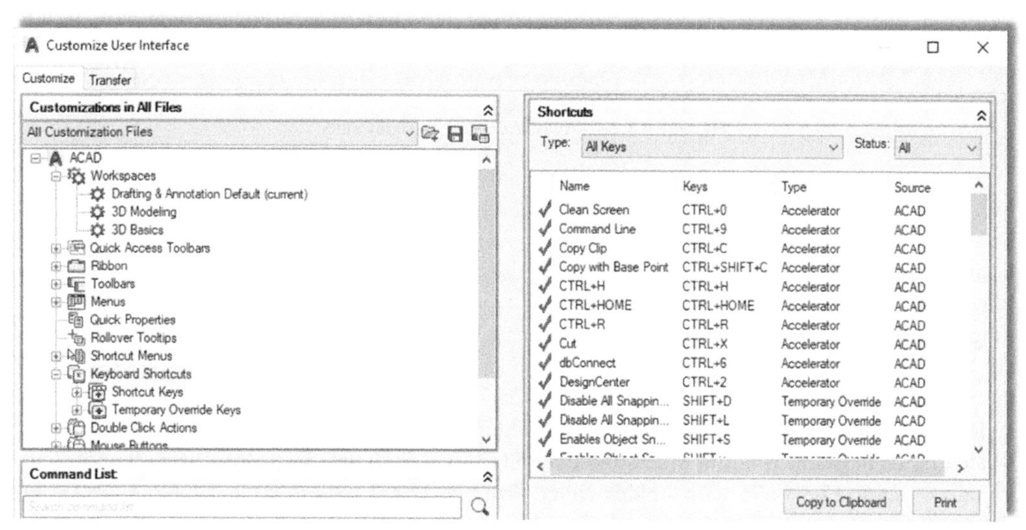

FIGURE 14.23 Accelerator keys.

It may be worth your time to look through the Help files and investigate the CUI further; it really is a bottomless pit of customization, although be careful not to change something just for the sake of change but only if you feel it may help your productivity.

14.4 DESIGN CENTER

You can probably guess at one of the purposes of the DC as soon as you open it via the cascading menus Tools → Palettes → DesignCenter (also via typing in dc and pressing Enter or Ctrl+2). Be sure to adjust all the window panes for good visibility, as shown in Fig. 14.24. The DC can also be docked to any side of the screen.

What you see in the far left window is basically the Windows Explorer used regularly to view files (such as when you click on My Computer). This is a big clue to the first of two functions of the DC, which are to view the contents of one drawing on the left and, as we soon see, transfer parts of that drawing into the active drawing on the right. Let us give this a try using the process outlined next.

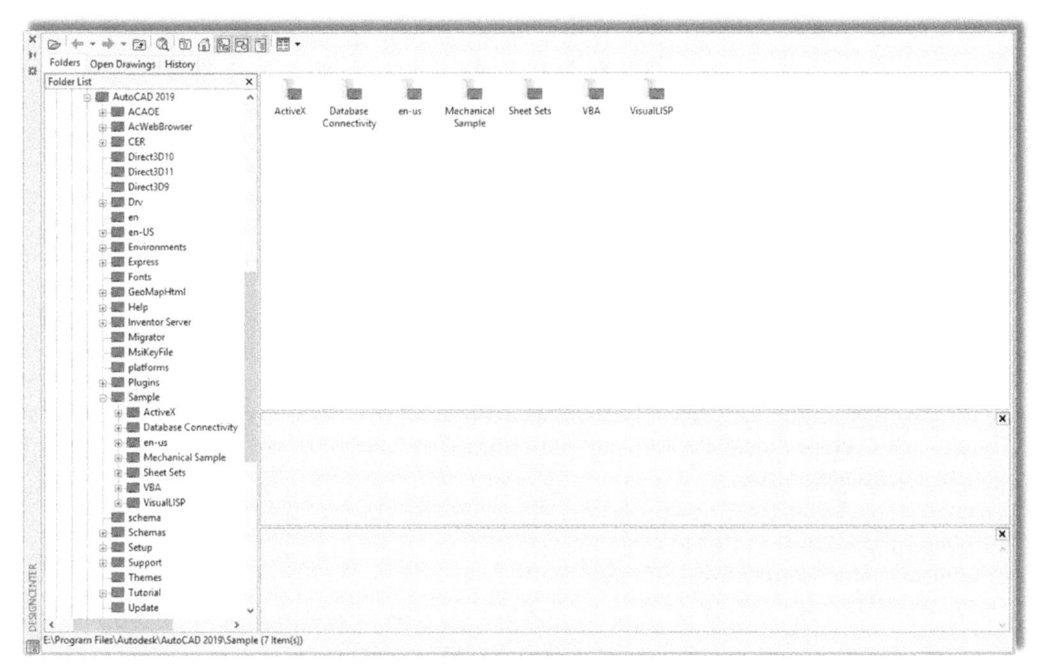

FIGURE 14.24 AutoCAD 2019 Design Center.

By default, the DC will likely open to the//AutoCAD 2019/sample file folder on the left. Go ahead open an additional few folders called *en-us*, followed by *Design Center*. There you see a sample drawing called HVAC—Heating Ventilation Air Conditioning.dwg. You can browse for and choose an actual work drawing you created, of course (that is the whole point, after all), but for our purposes right now, one of these sample drawings is just fine.

Click on the plus sign next to the HVAC file name and expand the folders, revealing a collection of useful data, such as Blocks, Dimstyles, Layers, and Layouts. Now click on a category, such as Blocks. A collection of blocks present in this drawing is revealed in the adjacent window. Click and drag any block from the choices given on the left into the drawing space on the right, as seen in Fig. 14.25 with a block called *Fan and Motor with Belt Guard*. Note that, in Fig. 14.25, the DC is fully docked to the left side of the screen and the window panels are adjusted to give equal spacing to the DC and the drawing area. The Toolbars are turned off to maximize the area.

In the same manner, you can transfer layers and anything else lifted from one drawing to another. This is very useful when you need only a few items, as opposed to everything, and is a great visual alternative to the usual Copy/Paste (Ctrl+C and Ctrl+P).

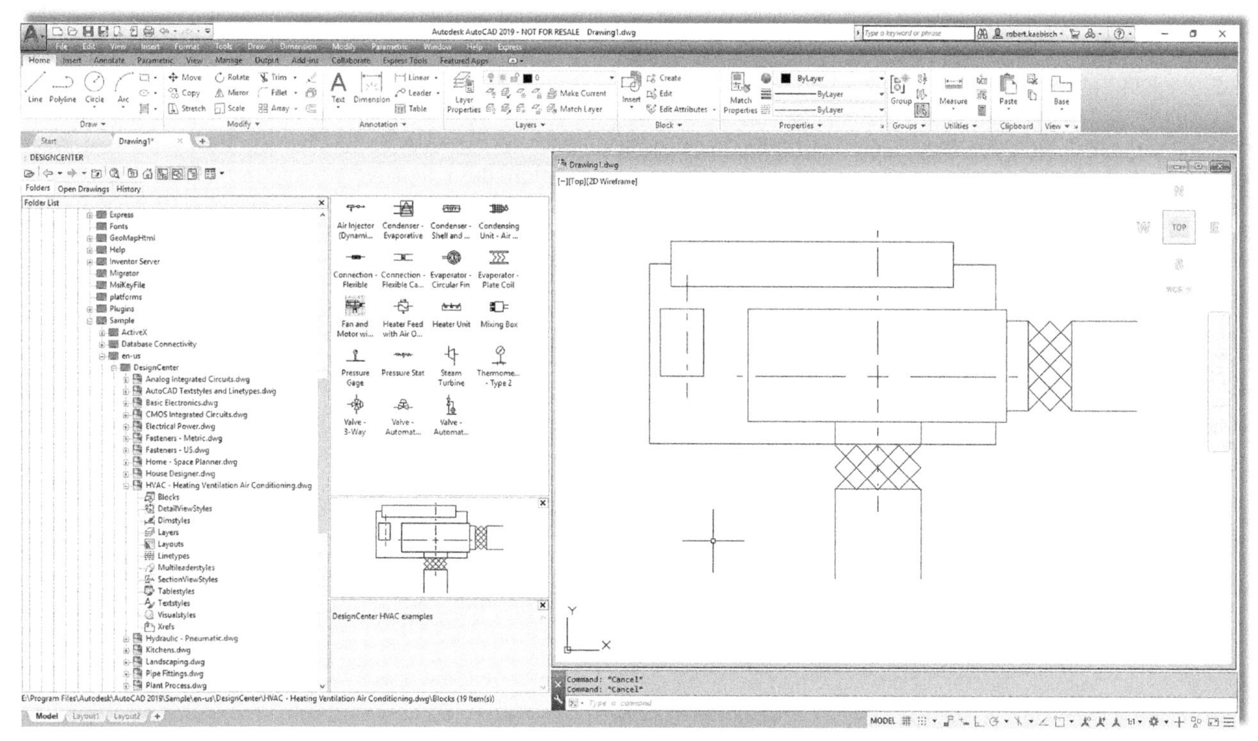

FIGURE 14.25 Design Center transfer of blocks (toolbars are turned off).

The next two tabs of the DC are not all that significant. Open Drawings simply looks at the content of the current active drawing, and History is exactly that the history of all the open drawings thus far in the drawing session.

For the 2019 release Autodesk has discontinued *Autodesk Seek*. That service and content is now hosted and controlled by BIMobject. Assuming you have internet access on your computer, you can access their database of manufacturer produced design files, blocks, and 3D models in various file formats. The web address is https://bimobject.com. You can see the home screen in Fig. 14.26.

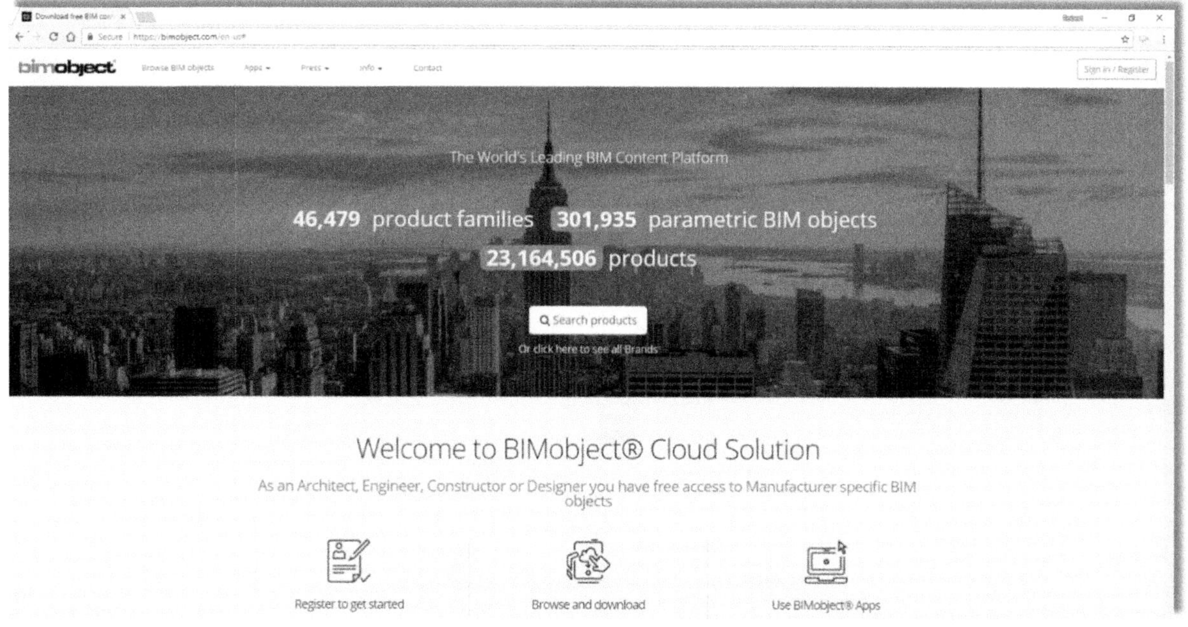

FIGURE 14.26 BIMobject web page in Google Chrome.

When you see the web page load, click on the Browse BIM Objects link at the upper left, and you are taken to a splash page that scrolls endlessly and contains numerous Building Information Modeling (BIM) items. To the left are subcategories and you can choose by Brand, Category, File Type, Region, and Types. Click around to see what is available. For example, if you go under Types, Kitchen Appliances and then select the "60 CM GAS HOB" you see what is in Fig. 14.27. If you then click the Download button, you get a number of file choices to save the file/block as a Revit, ArchiCAD, AutoCAD, Sketchup, and other file types (Fig. 14.28). Explore this extensive website further; it is truly a great resource with 301,935 BIM objects advertised as available as of this writing. You must create an account to access the downloadable content.

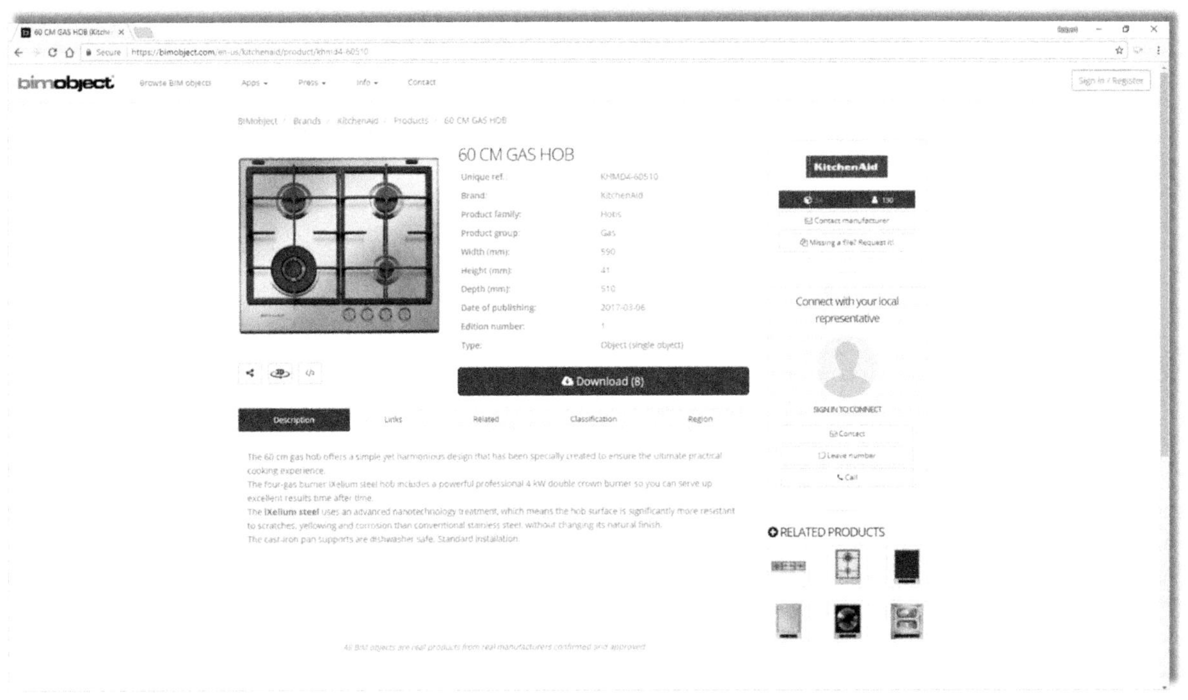

FIGURE 14.27 Example product page.

FIGURE 14.28 Example download file(s) page.

14.5 EXPRESS TOOLS

Express Tools are a set of commands grouped together under the Express drop-down menus. These commands add extra functionality to AutoCAD, and whereas some of them are redundant or not that useful, others became so indispensable that they were later incorporated into regular AutoCAD menus and command sets. It really is a mixed bag, and our goal here is to highlight only the most useful ones, leaving the rest for you to explore if you so choose.

Express Tools, originally called *Bonus Tools*, got their start around Release 14 in the mid-1990s as a collection of useful routines and small programs. Where they originally came from is a matter of debate, but Autodesk likely acquired them from other third-party developers and possibly regular programming-savvy users who felt that AutoCAD just did not have that perfect command and developed their own as a "plug-in."

However, it happened, these tools found their way to Autodesk, and the company began to offer them as an "extra" to AutoCAD (not technically part of the core software). Sometimes, these tools were offered for sale; sometimes, they were free to VIP subscribers; and eventually they became free to all customers, although up until recently they often still needed to be installed in a separate and deliberate step while installing AutoCAD itself.

The first step in learning Express Tools is to check that you have them (you should with AutoCAD 2019), and we proceed from here on with the assumption that all is well and you are ready to go over them. Fig. 14.29 shows what the topmost listing looks like when dropped down from the top cascading menu. This is the easiest way to access these tools, although there are a few toolbars as well.

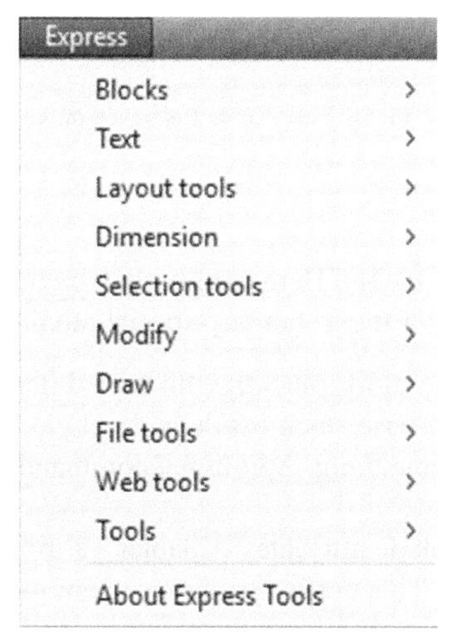

FIGURE 14.29 AutoCAD 2019 Express Tools cascading menu.

Next is a rundown of the most useful Express Tools. If you would like info on those commands that are not specifically covered,refer to the Express Tools Help files for detailed description of each one. The Help files for the Express Tools are the same as the regular AutoCAD Help files (first seen in Fig. 2.26). They are accessed via the usual F1 key and then just do a search for "Express Tools Reference." The Express Tools Help menu is shown in Fig. 14.30. Sometimes, you may find the functionality of an express tool already a part of AutoCAD in some other way, shape, or form. Occasionally, typed commands call up these tools as well. These are discussed as they come up.

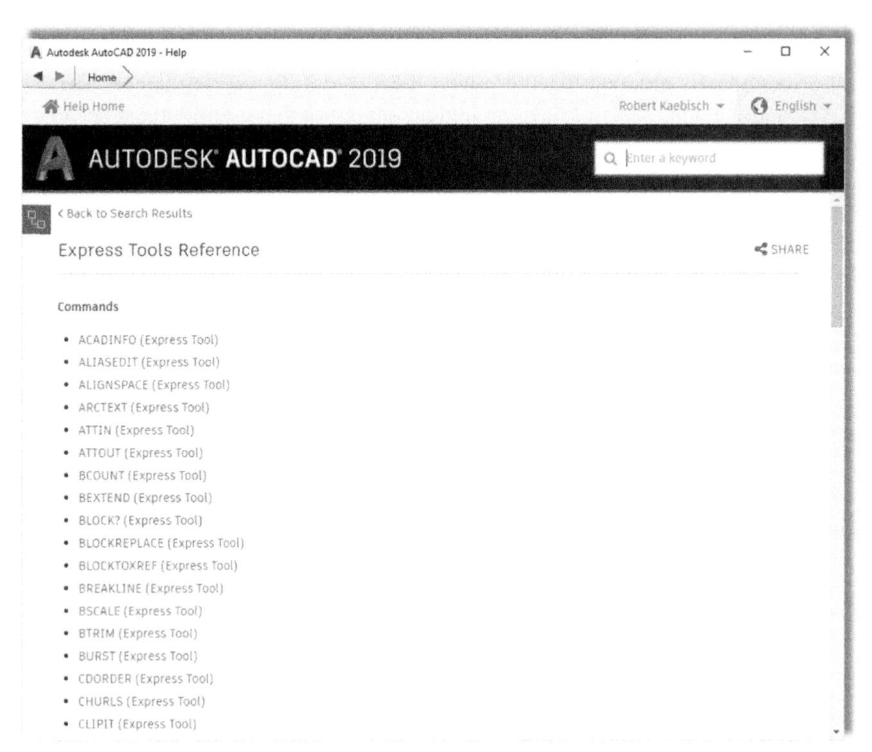

FIGURE 14.30 Express Tools Reference in the Help files.

Blocks

The first category from the top is Blocks (Fig. 14.31). Here, you find 10 menu choices, all having to do with blocks, nested objects, shapes, and attributes. Do explore all of them, but pay special attention to the ones described in more detail next.

- *Copy Nested Objects*: This command is useful for plucking out nested items, such as those embedded in xrefs (Chapter 17 topic) or blocks. These objects are otherwise stuck inside and not easily accessible for copying. The command line equivalent for this is ncopy and does the same thing. Additional commands below this one allow you to trim and extend to these nested objects as well.
- *Explode Attributes to Text*: When you explode attributes (Chapter 19 topic), they do not turn into regular text; instead they take the shape of attribute definitions. This useful tool forces them to revert to ordinary text so you need not retype anything.

Text

The second category from the top is Text (Fig. 14.32). Here, you find 12 menu choices, all having to do with text and mtext. Do explore all of them, but pay special attention to the ones described in more detail next.

- *Convert Text to Mtext*: This command pretty much does what it advertises: It converts lines of text to a paragraph of mtext. It is faster and more efficient than copying the text one line at a time or redoing mtext completely. Simply activate the command, pick the lines of text you want to convert, and press Enter.
- *ArcAligned Text*: This command is used to tackle a problem that has never been addressed by AutoCAD prior to the Express Tools existence: How to create text that stretches around an arc. The procedure is to draw an arc and select this tool (or just type in arctext), then select the arc. The dialog box seen in Fig. 14.33 appears. Enter the actual text and adjust parameters such as fonts, size, and anything else you want (they all have explanatory tool tips if you float your mouse over them); finally, press OK when done. Your text should look similar to that shown in Fig. 14.34. If the text is too squeezed or spread out, adjust the sizing. Note that the text can be over or under the arc. Also remember that you need to type in arctext to go back and edit it. Regular ddedit or double-clicking does not work here.
- *Enclose Text with Object*: This command encloses your text with rectangles, slots, or circles. Run through the menu options, setting offsets, shapes, and sizes.

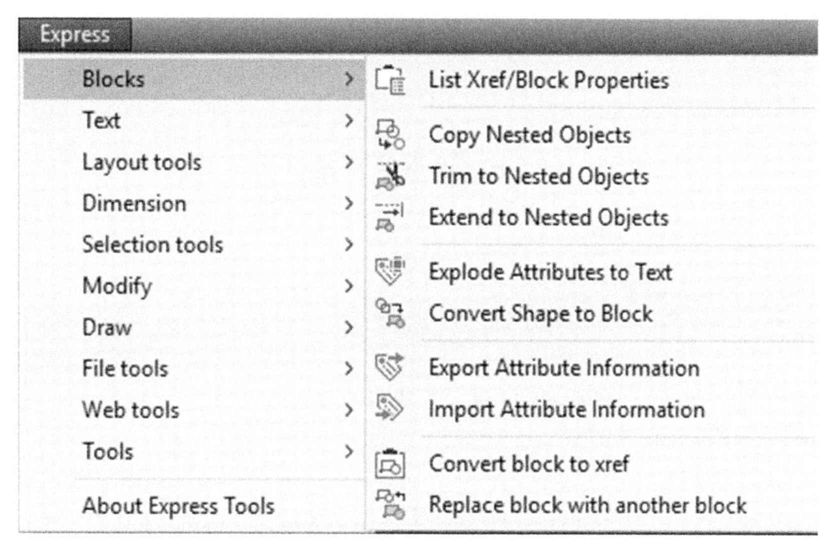

FIGURE 14.31 Express Tools (Blocks).

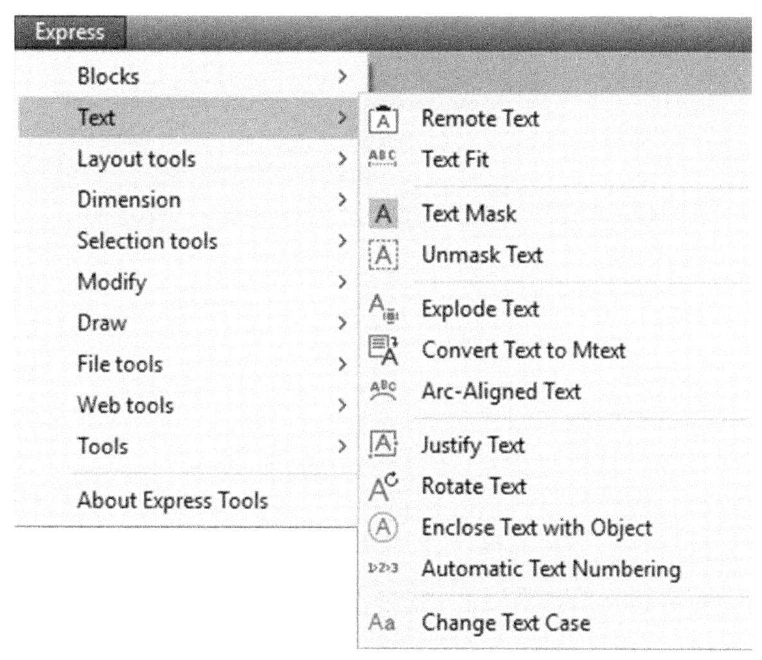

FIGURE 14.32 Express Tools (Text).

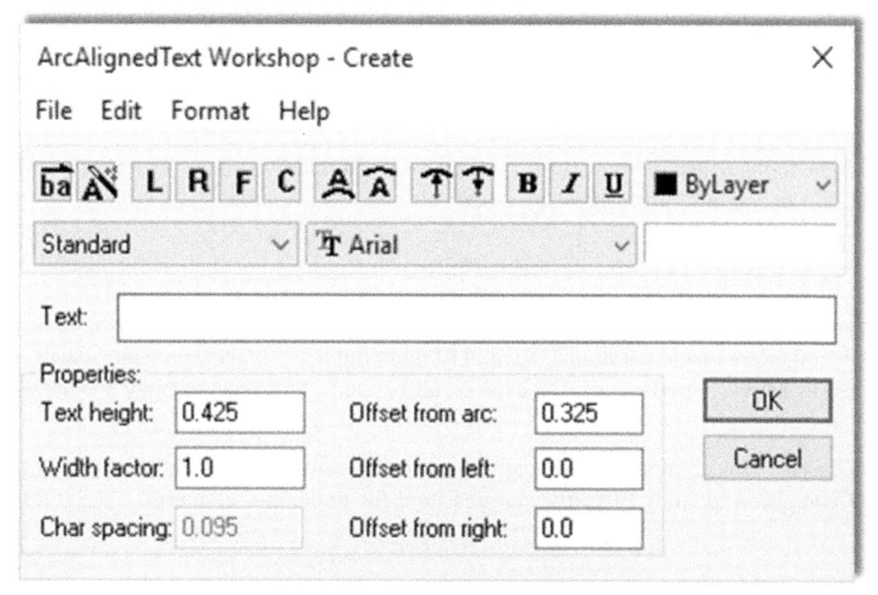

FIGURE 14.33 ArcAligned Text.

FIGURE 14.34 ArcAligned text sample.

Layout Tools

The third category from the top is Layout tools (Fig. 14.35). Here, you find four menu choices, all having to do with Paper Space viewports. These commands are rarely used or are duplicates of existing commands but do take a few minutes to briefly explore them. We do not go into further detail.

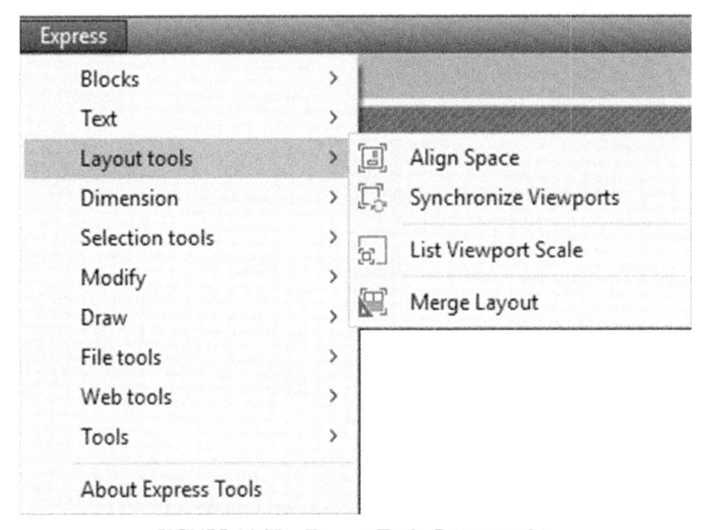

FIGURE 14.35 Express Tools (Layout tools).

Dimension

The fourth category from the top is Dimension (Fig. 14.36). Here, you find six relatively obscure dimstyle and leader tools. Take a few minutes to explore them, but we do not go into further detail.

Selection Tools

The fifth category from the top is Selection tools (Fig. 14.37). Here, you find two relatively obscure commands that have to do with selection sets. We do not go into further detail here.

Modify

The sixth category from the top is Modify (Fig. 14.38). Here, you find nine menu choices having to do with various move, copy, rotate, stretch, and other command variations. Several of them duplicate other commands (such as Multiple copy) or are of limited use (such as Move/Copy/Rotate). Do explore all of them, however, but pay special attention to the ones described in more detail next.

- *Delete Duplicate Objects*: This command is basically the overkill command, which we cover in the next chapter.
- *Flatten Objects*: If you go on to study 3D, this is a great tool for flattening (converting) a 3D image into a 2D one.

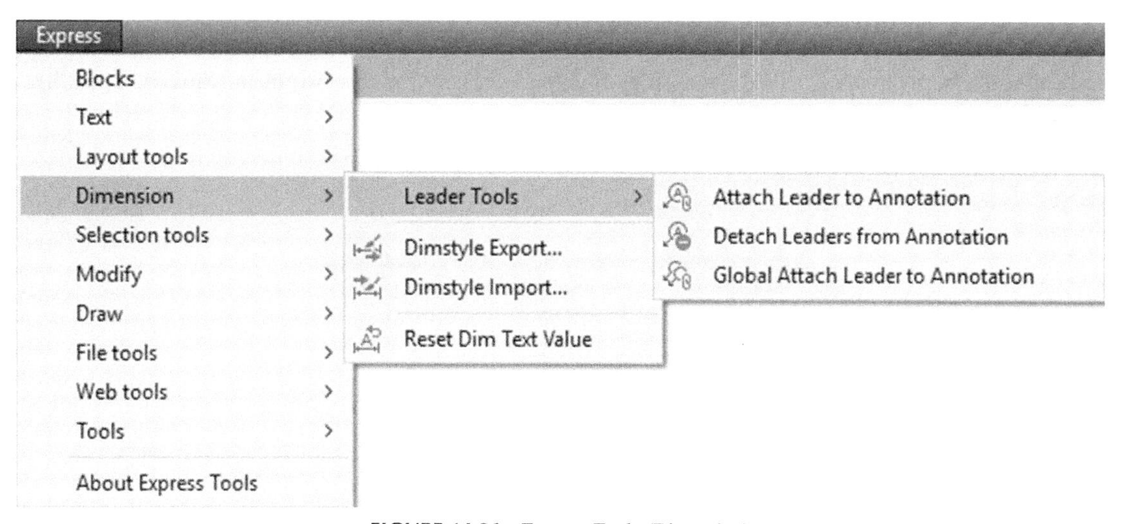

FIGURE 14.36 Express Tools (Dimension).

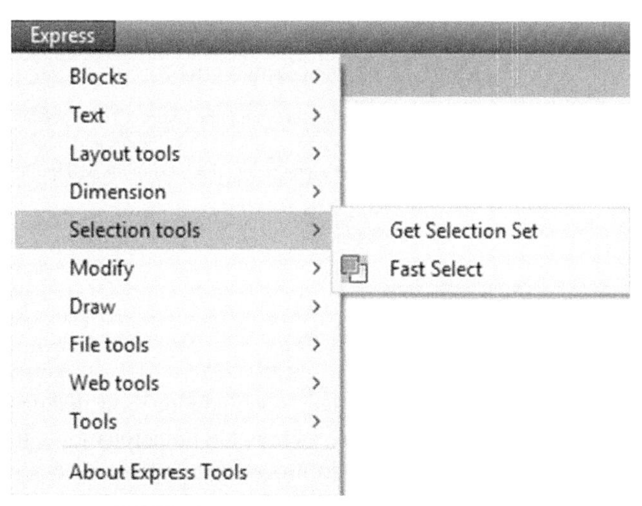

FIGURE 14.37 Express Tools (Selection tools).

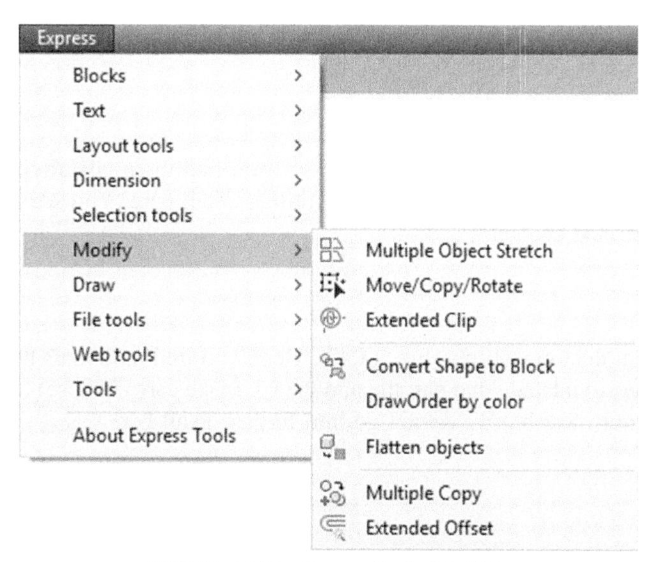

FIGURE 14.38 Express Tools (Modify).

Draw

The seventh category from the top is called Draw (Fig. 14.39). Here, you find just two menu choices, Break-line Symbol and Super Hatch.... We go over both next.

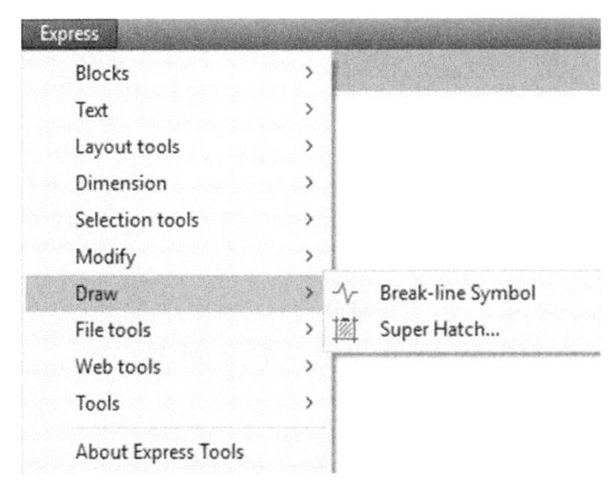

FIGURE 14.39 Express Tools (Draw).

- *Break-line Symbol*: Students have found this command one of the most useful among the Express Tools because it seems everyone would like to avoid drawing this symbol from scratch. Here, the process of making one is as easy as clicking two points and indicating a midpoint for the actual break line. Try it by selecting the command and specifying a first and second point along a line; then, using the MIDpoint OSNAP, place the break symbol in the middle. The Size option allows you to alter the break size.
- *Super Hatch...*: This command creates hatch patterns out of ordinary images by basically copying them many times over to fill an area. It was covered in detail toward the end of Chapter 5.

File Tools

The eighth category from the top is File tools (Fig. 14.40). Here, you find nine tools that deal mostly with file management. Much of what is here can be done manually in other ways, such as the numerous tools for opening, closing, and saving multiple drawings. We go over one tool in particular and do not go into depth with the rest.

- *Convert PLT to DWG*: This converts plot files to useable drawing files. The plt files are printer-ready data files that allow a printer to print an AutoCAD drawing without AutoCAD being loaded. They are rarely used anymore, unless you take a file to a professional printer. If you find any stray plt files lying around from the old days, you can use this express tool to convert them back to see what you have.

Web Tools

The ninth category from the top is Web tools (Fig. 14.41). Here, you find three URL/hyperlink-related tools. AutoCAD hyperlinks have not yet been discussed, so we do not go into further detail here. You see a bit more on these tools when hyperlinks are introduced in the next chapter.

Tools

The 10th and final category from the top is Tools (Fig. 14.42). Here, you find nine tools that serve a variety of functions, from editing aliases and variables (outdated—the pgp file and the CUI take care of those) to making shapes and linetypes. Look over the commands on your own, but we do not go into further detail here.

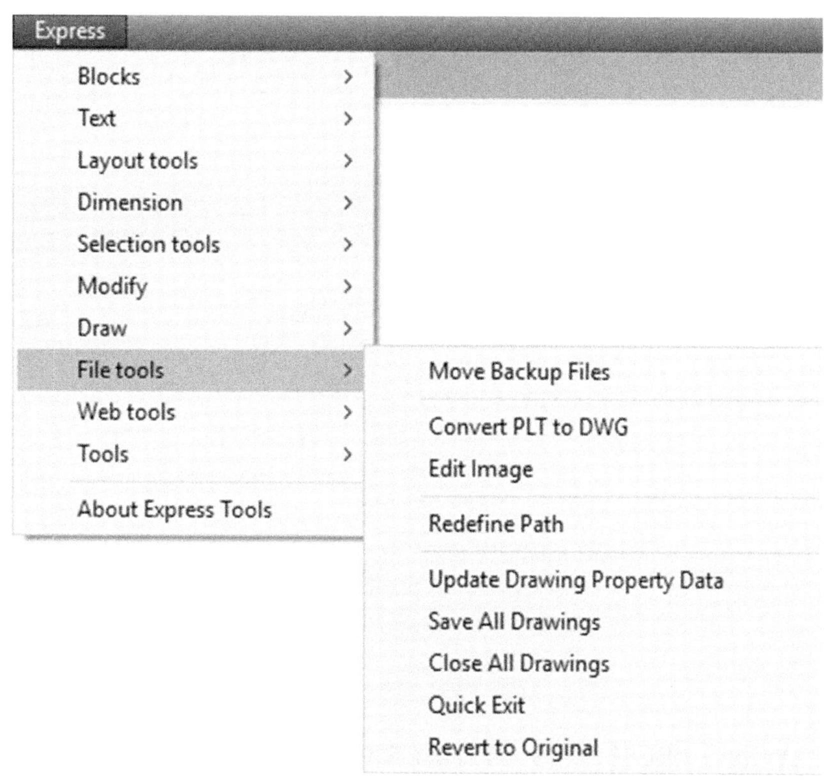

FIGURE 14.40 Express Tools (File tools).

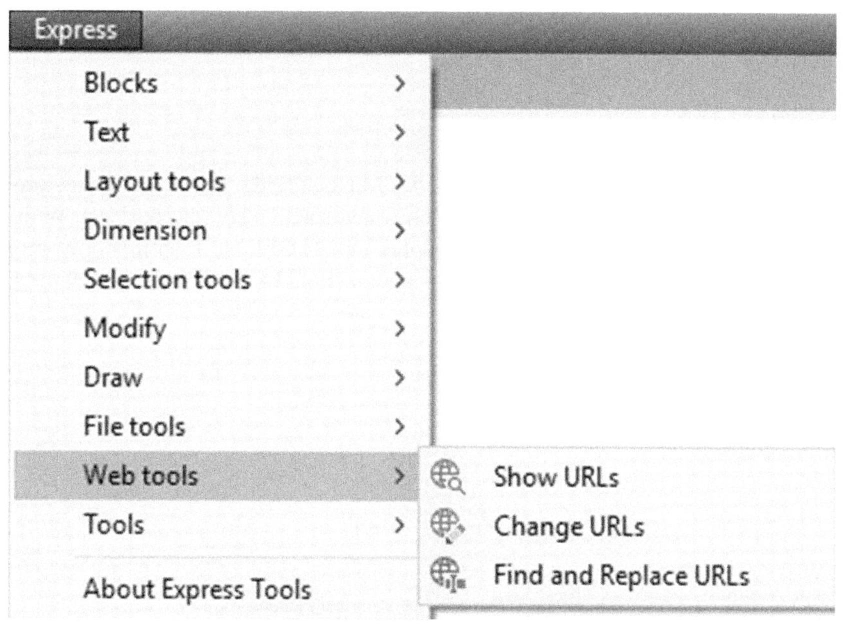

FIGURE 14.41 Express Tools (Web tools).

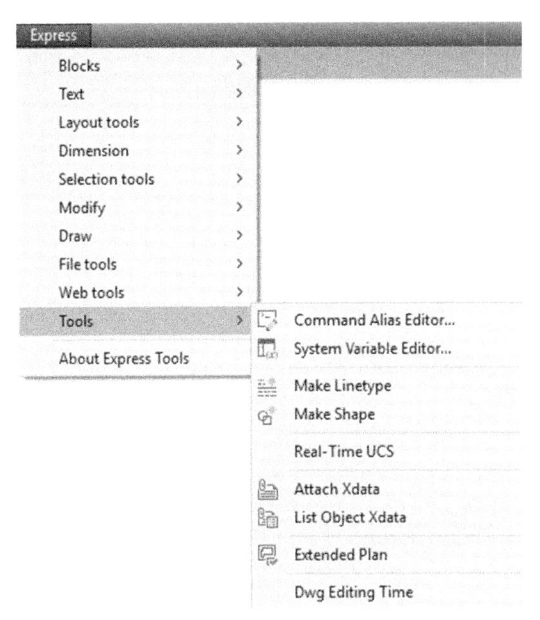

FIGURE 14.42 Express Tools (Tools).

14.6 LEVEL 2 DRAWING PROJECT (4 OF 10): ARCHITECTURAL FLOOR PLAN

For Part 4 of the drawing project, you add in electrical wiring, switches, lights, panels, and outlets. This design is similar in principle, if not in complexity, to a real electrical design.

Step 1. You need the following layers (with suggested colors) for the electrical pieces. The pieces are monochrome in Figs. 14.43 and 14.44.

- E-Light (Green).
- E-Outlet (Red).
- E-Panel (White).
- E-Switch (Magenta).
- E-Wiring (Blue).

Step 2. Create the electrical symbols as shown in Fig. 14.43. Make sure the layers and colors are appropriate. Finally, make a block out of each symbol, so you can just copy and rotate them into position. With switches and outlets, there are numerous types (dimmer, three-way, duplex outlet, GFI, etc.), so this only scratches the surface.

Step 3. Lay out the electrical symbols as seen in the floor plan in Fig. 14.44. Some of the wiring runs are splines and some are arcs.

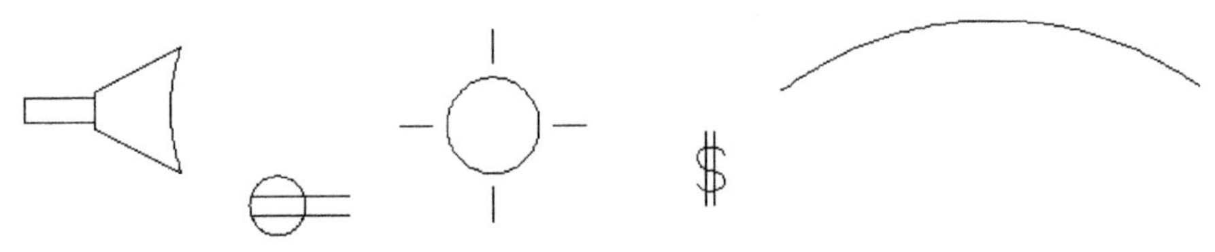

FIGURE 14.43 Various electrical symbols.

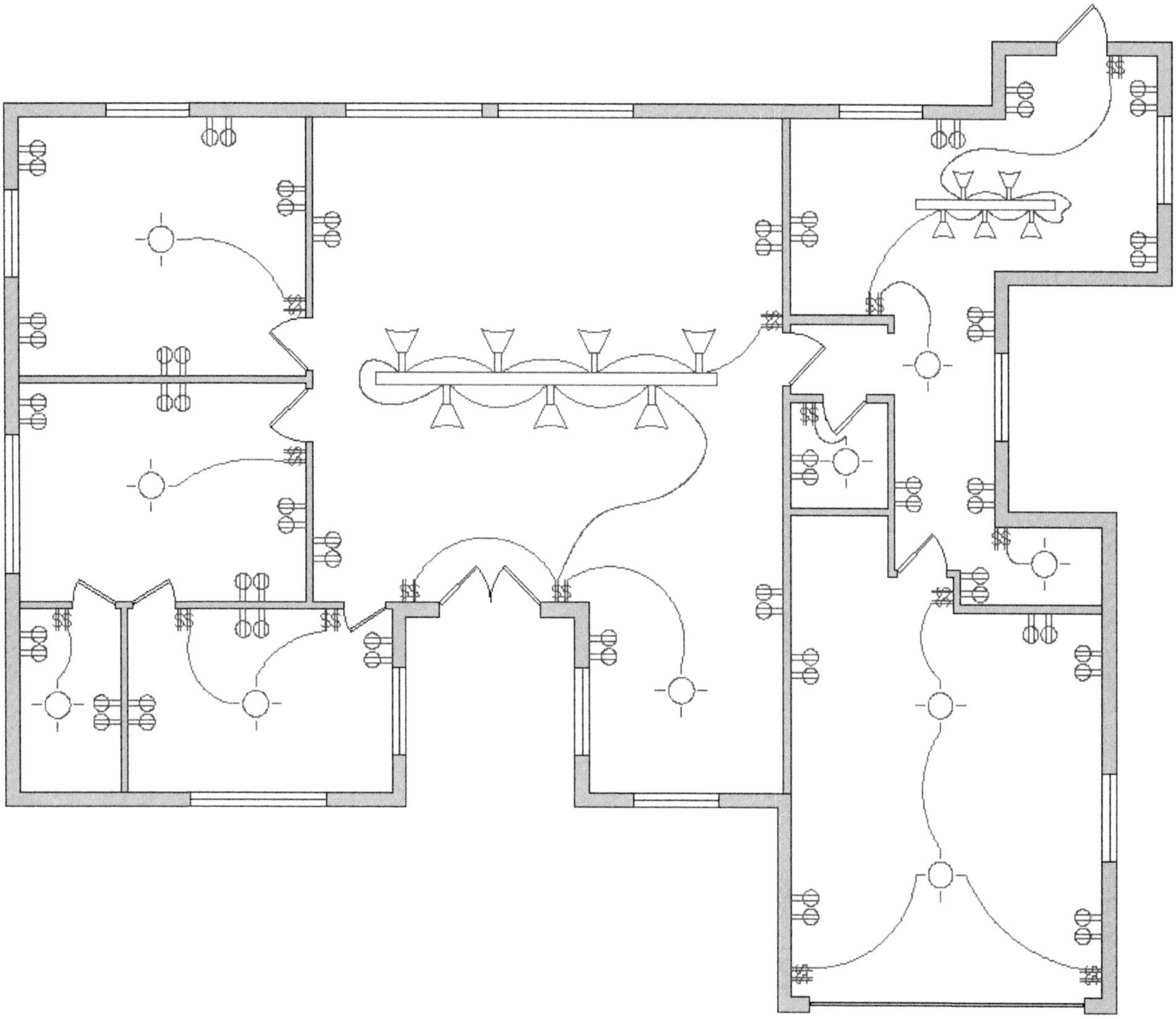

FIGURE 14.44 Electrical layout.

SUMMARY

You should understand and know how to use the following concepts and commands before moving on to Chapter 15:

- Files tab
- Display tab
- Open and Save tab
- Plot and Publish tab
- System tab
- User Preferences tab
- Drafting tab
- 3D Modeling tab

- Selection tab
- Profiles tab
- Shortcuts
- pgp file
- CUI
- DC
- Express Tools

REVIEW QUESTIONS

Answer the following based on what you learned in Chapter 14:

1. What is important under the Files tab?
2. What is important under the Display tab?
3. What is important under the Open and Save tab?
4. What is important under the Plot and Publish tab?
5. What is important under the System tab?
6. What is important under the User Preferences tab?
7. What is important under the Drafting tab?
8. What is important under the Selection tab?
9. What is important under the Profiles tab?
10. What is the pgp file? How do you access it?
11. Describe the overall purpose of the CUI.
12. Describe the purpose of the DC.
13. Describe the important Express Tools that are covered.

EXERCISES

1. Based on what you learned about the Options dialog box, set up your own profile. Be sure to include or at least consider all the variables and settings discussed. Save the profile as Profile_Your_Name. (Difficulty level: Easy; Time to completion: <10 minutes.)
2. Based on what you learned about the pgp file, create your own based on whatever abbreviations you prefer. Be sure to utilize the commands shown in Fig. 14.17 as well as add any of your own that you may want included. (Difficulty level: Easy; Time to completion: 20 minutes.)
3. Based on what you learned about CUI, create your own toolbar that features the following command icons:
   - Line
   - Circle
   - Rectangle
   - Move
   - Copy
   - Rotate
   - Offset

   (Difficulty level: Easy; Time to completion: 15 minutes.)
4. Review the Express Tools. Practice creating another arc-aligned text and break line. Open up any multilayer file and review the Layer tools: layfrz, layiso, and laywalk. (Difficulty level: Easy; Time to completion: 20 minutes.)

5. To keep your Level 1 drafting skills sharp, draw the following 3-view layout of a nut design. Make use of array as much as possible. (Difficulty level: Easy; Time to completion: 20–30 minutes.)

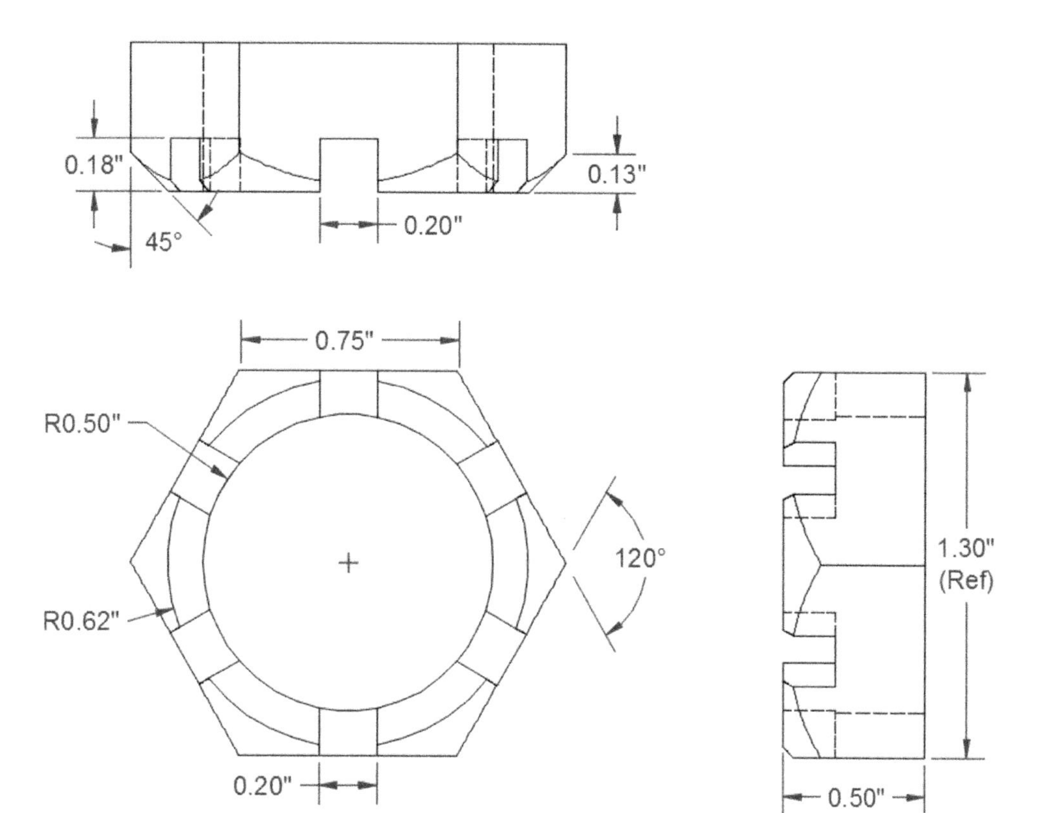

6. To keep your Level 1 drafting skills sharp, draw the following 2-view mechanical spindle layout. (Difficulty level: Easy; Time to completion: 20–30 minutes.)

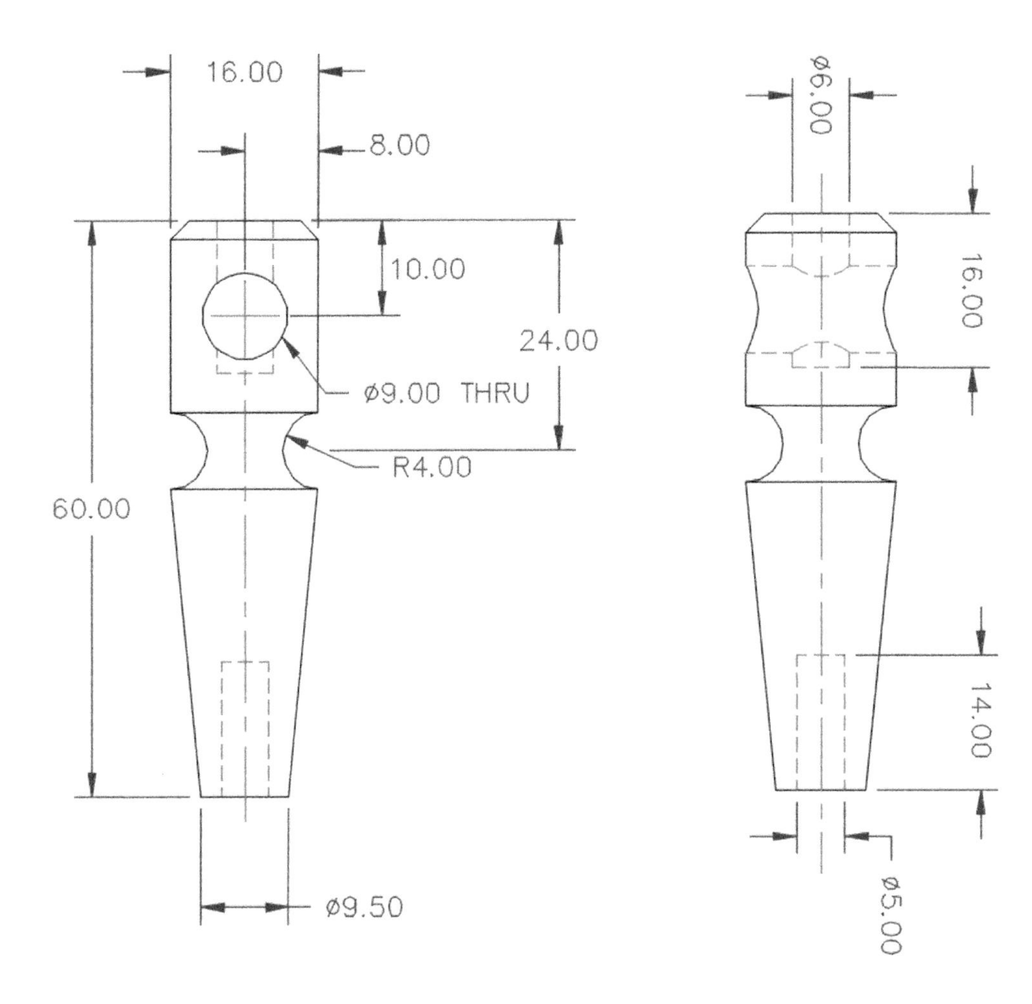

Chapter 15

Advanced Design and File Management Tools

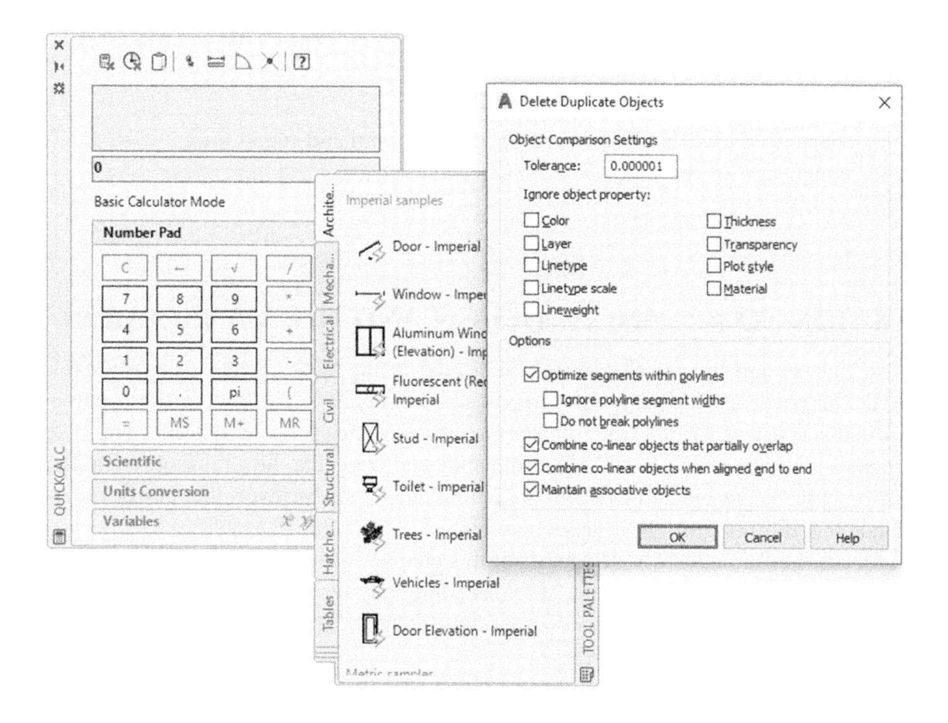

Learning Objectives

In this chapter, we continue to introduce a variety of advanced tools and discuss the following:

- Align
- Action Recording
- Audit and recover
- Blend
- Break and join
- CAD Standards
- Calculator
- Defpoints
- Divide and point style
- Donut
- Draw order
- eTransmit
- Filter
- Hyperlink
- Lengthen
- Object snap tracking (OTRACK)

Up and Running with AutoCAD 2019. https://doi.org/10.1016/B978-0-12-816440-2.00015-X

- Overkill
- Point and node
- Publish
- Raster
- Revcloud
- Sheet sets
- Selection methods
- Stretch
- System variables (SVs)
- Tables
- Tool palette
- UCS and crosshair rotation
- Window tiling
- Wipeout

By the end of this chapter, you will add significantly to your command repertoire.
Estimated time for completion of this chapter: 3 hours.

15.1 INTRODUCTION TO ADVANCED DESIGN AND FILE MANAGEMENT TOOLS

This is easily the most diverse and unique chapter you will read in the book. Here, we really explore the "nooks and crannies" of AutoCAD and cover a wide range of very diverse commands. The reason these particular commands are grouped here is because this text was written from the start to simplify and isolate core topics and essential ideas, without cluttering the main themes with additional AutoCAD commands and concepts. Several more "big ideas" are left, such as xrefs, attributes, and isometric. Those are addressed in upcoming chapters. However, that leaves out a great deal of useful concepts and commands that did not quite fit in anywhere else.

These bite-sized chunks of information form the basis of this chapter. Many of these topics are just random commands, standing alone, not part of any other major concept, whereas some may be part of a larger family but were not included earlier to avoid getting distracted from a much more important theme. All are useful to one extent or another and are necessary for a well-rounded knowledge of AutoCAD.

Obviously, this is not an exhaustive list; some items have been left out due to space limitations or very limited usefulness, but for the most part, the important ones are here. The list is in alphabetical order. Go through it at your own pace, noting how each concept or command can help you in your design work. Some are whimsical, some are rather odd, and surely you will find one or two that you cannot do without!

15.2 ALIGN

This command does exactly what it advertises, aligns one object to another. In Fig. 15.1, we have a rectangle drawn and rotated to some random angle and below it a straight line. While one can always measure the angle of the rectangle relative to the line and use the rotate command, there is a much easier and accurate way to align the rectangle to the line (or vice versa).

Step 1. Type in `align` and press Enter. You can also use the cascading menus Modify → 3D Operations → Align, or the Ribbon via the Modify tab.

- AutoCAD says: `Select objects:`

Step 2. Select the object to be moved into alignment (the rectangle). Press Enter.

- AutoCAD says: `Specify first source point:`

Step 3. Pick the lower left corner of the rectangle using the ENDpoint OSNAP (Point 1 in Fig. 15.2).

- AutoCAD says: `Specify first destination point:`

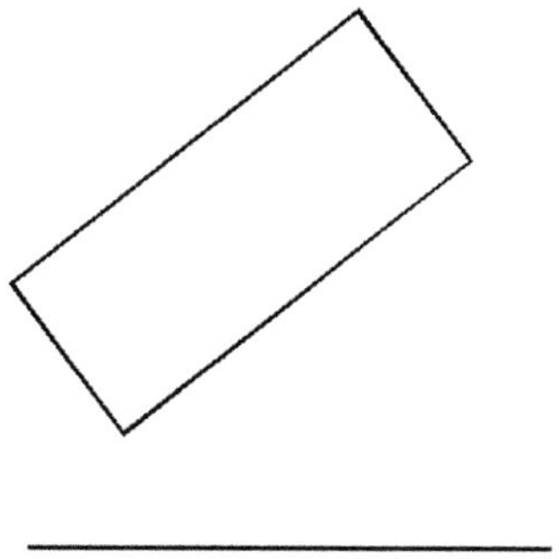

FIGURE 15.1 Rectangle to be aligned with line.

Step 4. Pick the left point of the line using the ENDpoint OSNAP (Point 2 in Fig. 15.2).
- AutoCAD says: `Specify second source point:`

Step 5. Pick the lower right corner of the rectangle using the endpoint OSNAP (Point 3 in Fig. 15.2).
- AutoCAD says: `Specify second destination point:`

Step 6. Pick the right point of the line using the ENDpoint OSNAP (Point 4 in Fig. 15.2).
- AutoCAD says: `Specify third source point or <continue>:`

Step 7. You have no need for a third set of points, so press Enter.
- AutoCAD says: `Scale objects based on alignment points? [Yes/No] <N>:`

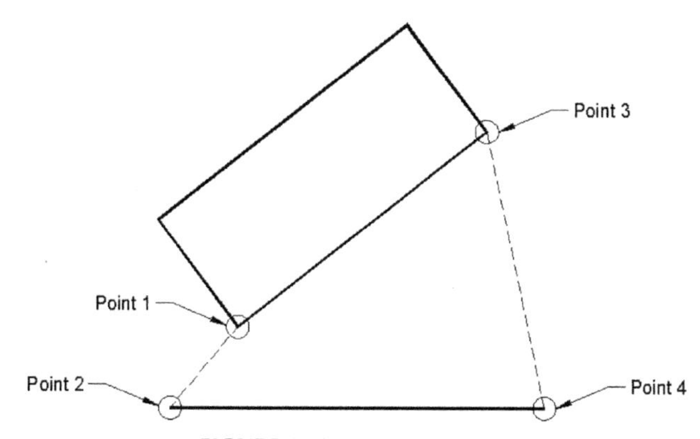

FIGURE 15.2 Alignment points.

You can do something interesting here. Not only can you align the rectangle to the line, you can also scale it up to the line if needed by selecting `Yes`. We do not try this now but go back after learning the basic align command and try this variation of it. For now, just press Enter to accept the default `No` option. The rectangle is aligned to the line, as seen in Fig. 15.3.

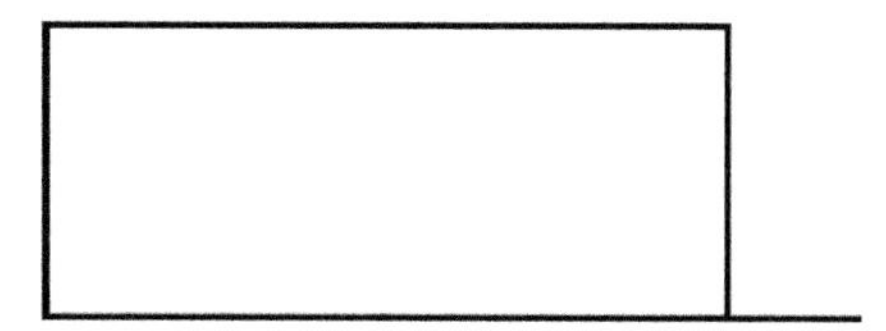

FIGURE 15.3 Alignment, final result.

15.3 ACTION RECORDING

Action recording is a relatively new tool in AutoCAD that is used to build Action Macros and save them as .actm files. *Macros* is short for *macroinstructions*, a computer term that can be defined as a short set of instructions to perform a task. The idea here is to actually record your on-screen actions and save them for future use. Needless to say, these actions should be some sort of useful steps to build a tedious or repetitive object and not just any simple steps. You can then use this macro to make your life easier later on.

To give this a try, open up a brand-new file with the Ribbon present. You can use the command without the Ribbon, by simply typing in `actrecord` to start the process and `stop` to end it, but it is easier and more visual with the Ribbon present. Either way, start the command via any of the following methods:

Keyboard: Type in **actrecord** and press Enter
Cascading menus: **Tools→Action Recorder→Record**
Toolbar icon: none
Ribbon: **Manage** tab→**Record**

You then see what is shown in Fig. 15.4. Notice the Ribbon tab's record button became a stop button. The mouse has also acquired a red "Recording" dot. If for any reason the Ribbon did not open to the right tab, just manually select the Manage tab.

Begin drawing anything you like, such as a piece of furniture, as seen in Fig. 15.5. Notice how each action of your mouse, keyboard, menu, Ribbon, and so on is recorded in the log on the upper left. You can even add comments to the macro that will pop up on playback, as well as more advanced scripts and macros via AutoLISP or a more advanced language such as Visual Basic.

When done, press the Stop button and the Action Macro dialog box appears. Give the macro a descriptive name (no spaces allowed) and press OK, as seen in Fig. 15.6. Note that, if the Folder Path is darkened out and not editable, it can be easily changed via Options.... Go into the first tab, Files, then find Action Recorder Settings. There you can change the destination folder for your action macros.

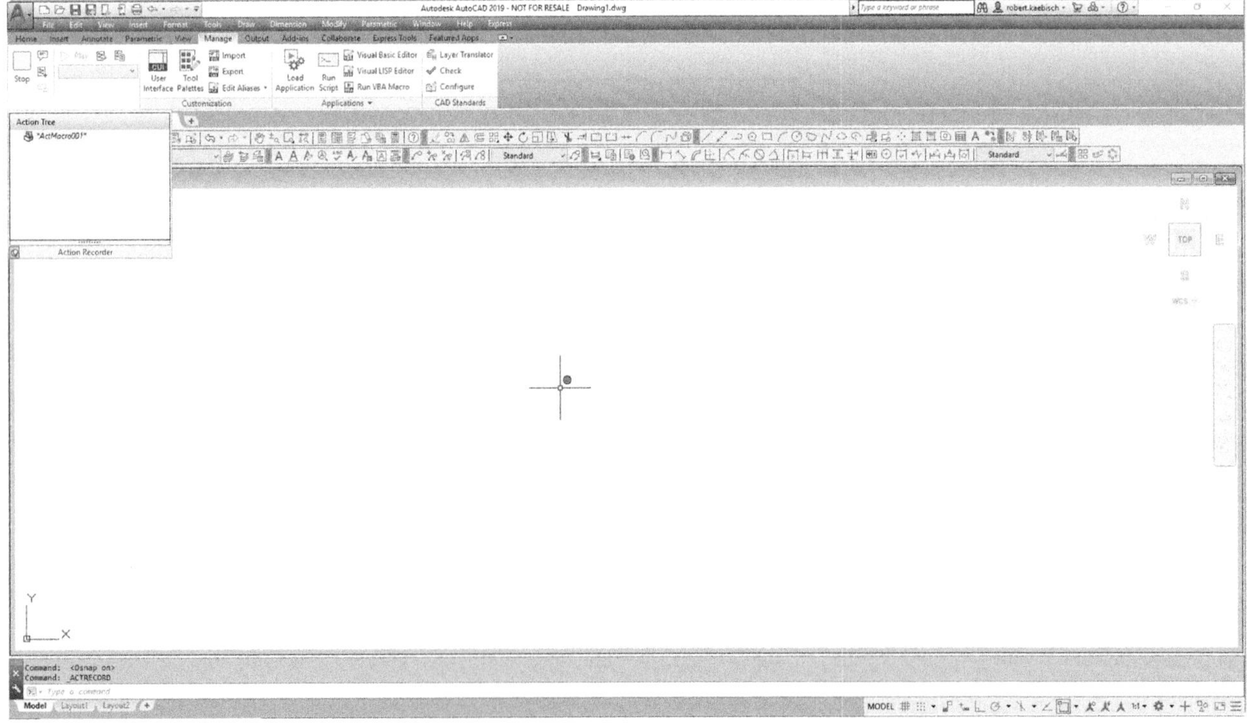

FIGURE 15.4 Action recorder.

The macro has now been completed and stored as an .actm file in a default location (which can be changed via Options… → Files). You can put it to use by erasing the chair and pressing the Play button. The macro executes and the chair is drawn. You then see the message shown in Fig. 15.7.

This is a very simple example to illustrate the basics. You can get quite complex with macros. For example, you can insert user input or a message. Macros recognize only a few dialog boxes (like layers), so if you need to use one, then make it a command line dialog box via the - (hyphen) prior to typing it in. So, hatch would be `-hatch`, and so on.

Experiment with this on your own; there are many possibilities.

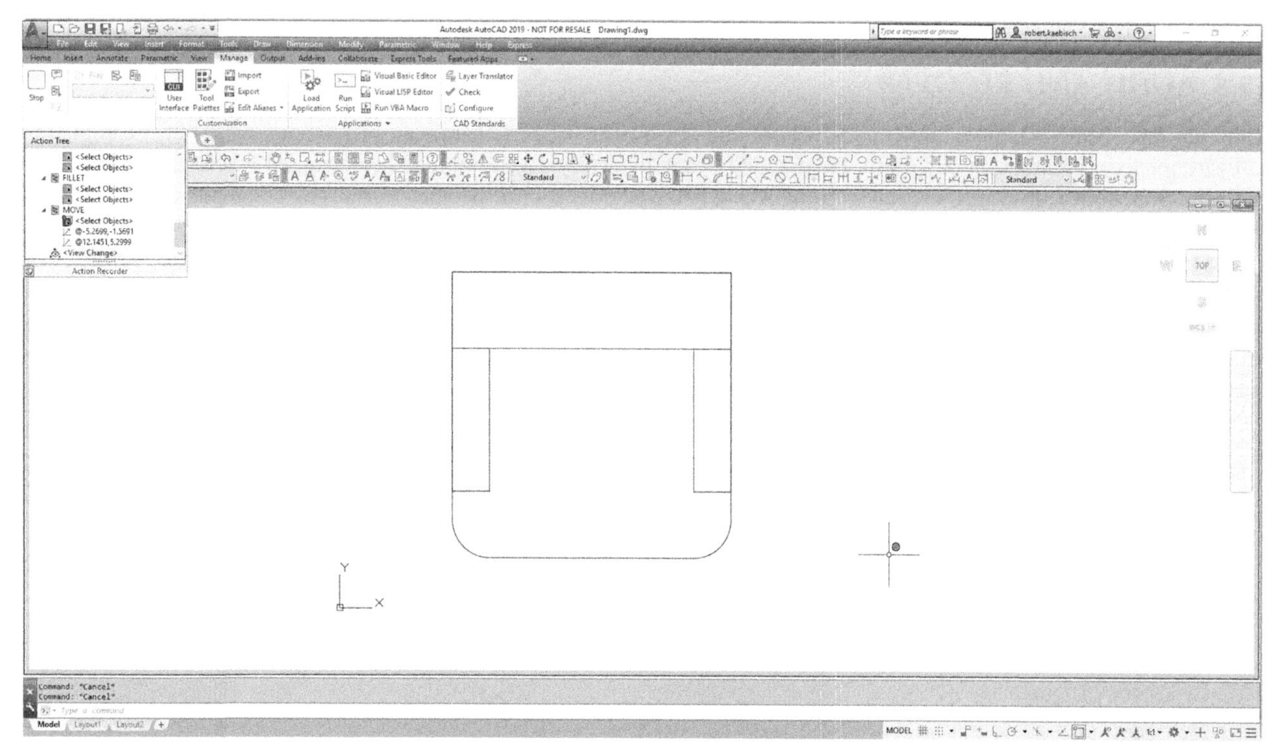

FIGURE 15.5 Recording log.

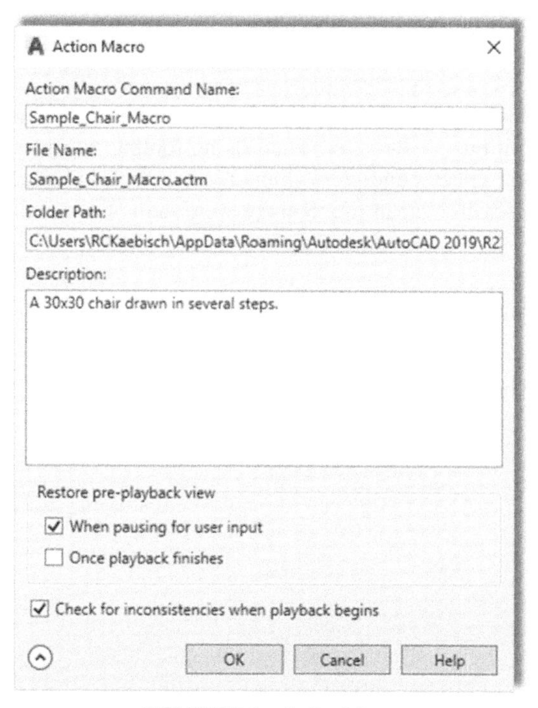

FIGURE 15.6 Action Macro.

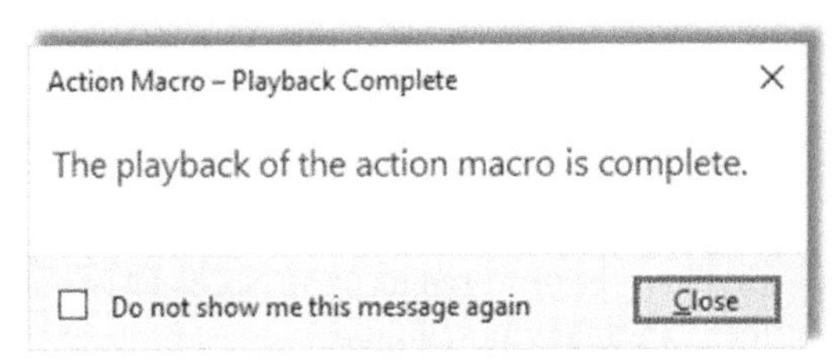

FIGURE 15.7 Action Macro - Playback Complete.

15.4 AUDIT AND RECOVER

Although rare, AutoCAD does sometimes lock up and crash (usually a PC hardware or memory issue but it can also be software related). In the event of a crash, the drawing is automatically saved but may be corrupted when you try to open it up again. Sometimes, drawings just cause AutoCAD to behave oddly for no specific reason (i.e., commands do not work as expected or just fail). In either case, the drawing needs to be audited or recovered.

Auditing is the lower-level check. AutoCAD evaluates the integrity of the drawing and corrects some errors. You simply type in `audit` and press Enter; alternatively, use the cascading menus File → Drawing Utilities → Audit. There is no toolbar or Ribbon equivalent. AutoCAD asks about fixing any detected errors (say `Yes`), cycles through drawing entities, and in a few seconds generates a log similar to what you see next (assuming nothing wrong was found or fixed, as seen next):

```
Auditing Header
Auditing Tables
Auditing Entities Pass 1
Pass 1 100 objects audited
Auditing Entities Pass 2
Pass 2 100 objects audited
Auditing Blocks
1 Blocks audited
Total errors found 0 fixed 0
Erased 0 objects
```

Of course, if something is found, fixes are reported. Sometimes, however, the audit is not enough and you must recover the drawing. This is not as drastic as it sounds. The recover command is powerful and almost always brings the drawing back as a matter of routine. I can count on one hand the number of drawings that were lost forever due to corruption in 18 years of CAD design. And, even in those cases, going to the previous night's server backup files restored the drawing, in spite of a few hours of lost work.

To use this command, open a blank drawing file, type in `recover`, and press Enter. Alternatively, you can use the cascading menus File → Drawing Utilities → Recover. There is no toolbar or Ribbon equivalent. You are prompted to browse and find the drawing. Once you select it and press Open, recover runs and generates its own log, as seen next. This particular example was also generated with a clean file. If there were items to fix, recover would have shown the actions taken.

```
Drawing recovery.
Drawing recovery log.
Scanning completed.
Validating objects in the handle table.
Valid objects 755 Invalid objects 0
Validating objects completed.
Salvaged database from drawing.
Auditing Header
Auditing Tables
Auditing Entities Pass 1
Pass 1 900 objects audited
```

```
Auditing Entities Pass 2
Pass 2 900 objects audited
Auditing Blocks
19 Blocks audited
Total errors found 0 fixed 0
Erased 0 objects
```

Following the log, you see the message box shown in Fig. 15.8.

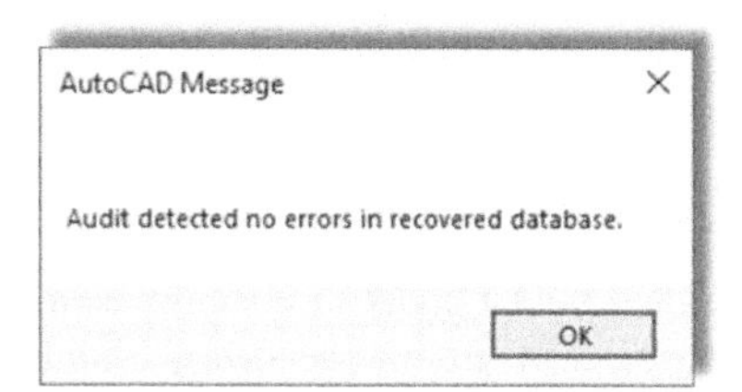

FIGURE 15.8 Recover message.

If something were indeed wrong (and then fixed), the drawing would have just opened as normal. Immediately save it.

15.5 BLEND

The blend command is a relatively new one; introduced in AutoCAD 2012. It allows you to join, or blend together, linework that is otherwise separate. This linework typically is straight lines or curves, such as arcs or splines. To a lesser extent, some shapes (such as rectangles) can also be blended, although that does not really yield useful results. In all cases, what the blend command does is create a spline between these shapes, curves, or lines. The spline shape can then be modified if needed. To try it out, create two arcs as seen in the top part of Fig. 15.9, and then follow the steps shown next.

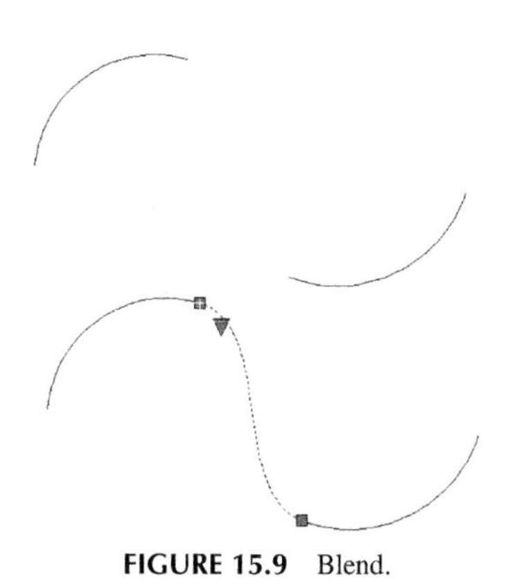

FIGURE 15.9 Blend.

Keyboard: Type in **blend** and press Enter
Cascading menus: Modify→**Blend Curves**
Toolbar icon: Modify toolbar
Ribbon: Home tab→Modify

Step 1. Start up the blend command via any of the preceding methods.
- AutoCAD says:
`Continuity = Tangent`
`Select first object or [CONtinuity]:`

Step 2. Pick the first arc, close to where the blend will start.
- AutoCAD says: `Select second object:`

Step 3. Pick the second arc, close to where the blend will end. You see a live preview of the new spline just as you are about to click. A new spline then is created to connect the arcs.

The bottom part of Fig. 15.9 shows the results. The new blend has been selected, and the grip points and a down arrow are visible. The down arrow reveals a menu if clicked on. This menu lets you choose between showing control vertices or fit points, which presents you with different methods for altering the blend. Experiment with both to see the effects.

15.6 BREAK AND JOIN

The break command takes a geometric object and puts a break in it (removes a section, in other words) or, if applied at an intersection with another object, breaks that object at the intersection without actually removing any material. The join command is the exact opposite. It adds material to close the break, but works only in certain circumstances. Let us examine each command closer.

Break, Method 1

To try out the break command, open a new file and draw a line of any size or direction on your screen.

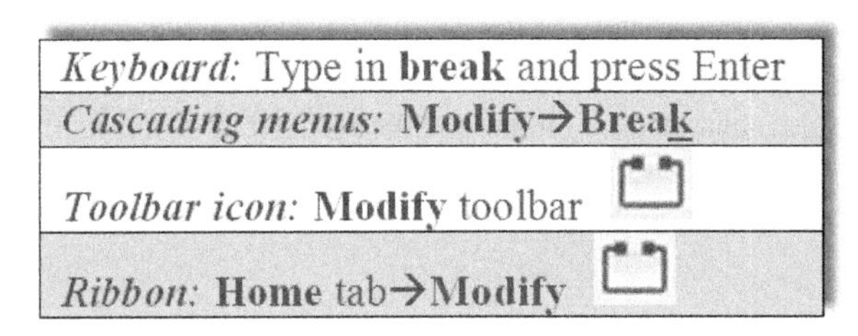

Keyboard: Type in **break** and press Enter
Cascading menus: **Modify→Break**
Toolbar icon: **Modify** toolbar
Ribbon: **Home** tab→**Modify**

Step 1. Start up the break command via any of the preceding methods. Be sure to turn off OSNAP.
- AutoCAD says: `Select object:`

Step 2. Go ahead and click to select the line at some random spot. It turns dashed.
- AutoCAD says: `Specify second break point or [First point]:`

Step 3. Pick another nearby point on the line, and a section between these two points is removed (Fig. 15.10). Note that you can preview the result of the break command as you execute it, similar to many other modify commands.

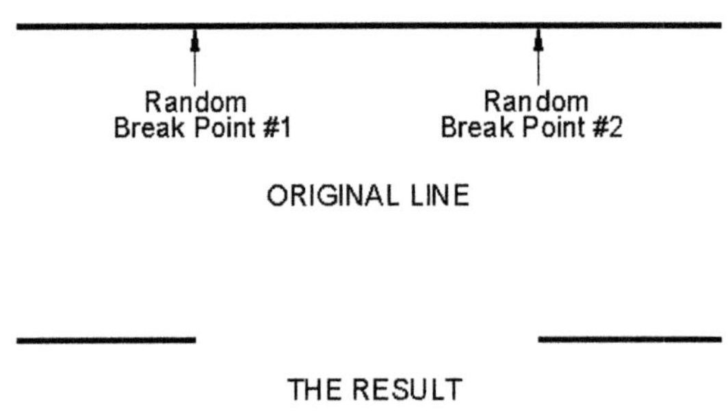

FIGURE 15.10 Break command, Method 1.

In this method, the first point of the break is somewhat random. When you selected the line, the first point was automatically chosen. However, as shown in the next method, you can select the line to be broken, then, using the first point option, specify exactly where to start the break. We use this to our advantage to create four lines out of two, as seen next.

Break, Method 2

Draw a horizontal line of any size, followed by a perpendicular vertical one that is of equal size. They should cross perfectly in the middle. You have a plus sign. One good trick to do this is to first draw a circle and just connect the four quadrant points with two lines, as seen in Fig. 15.11.

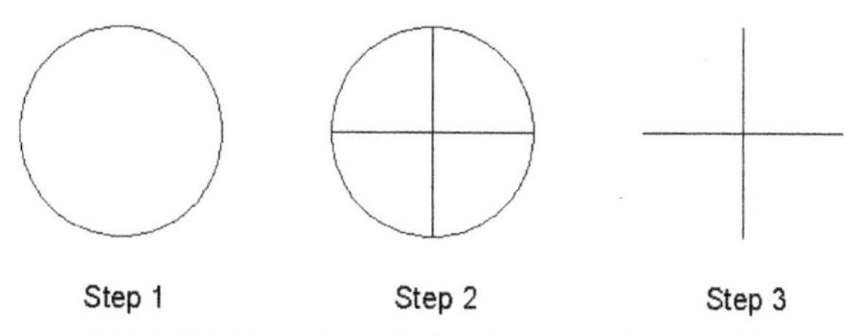

FIGURE 15.11 Setup for break command, Method 2.

Step 1. Start the break command via any of the previous methods.
- AutoCAD says: `Select object:`

Step 2. Go ahead and click to select the horizontal line at some random spot. It turns dashed.
- AutoCAD says: `Specify second break point or [First point]:`

Step 3. Type in `f` for `First point.`
- AutoCAD says: `Specify first break point:`

Step 4. Using OSNAPs, select the midpoint of the horizontal line.
- AutoCAD says: `Specify second break point:`

Step 5. Once again, using OSNAPs, select the same midpoint of the line.

Now, repeat these steps with the vertical line. The result is four lines where there were only two just moments earlier, as seen in Fig. 15.12, when the grips are revealed. All four are broken at their intersection.

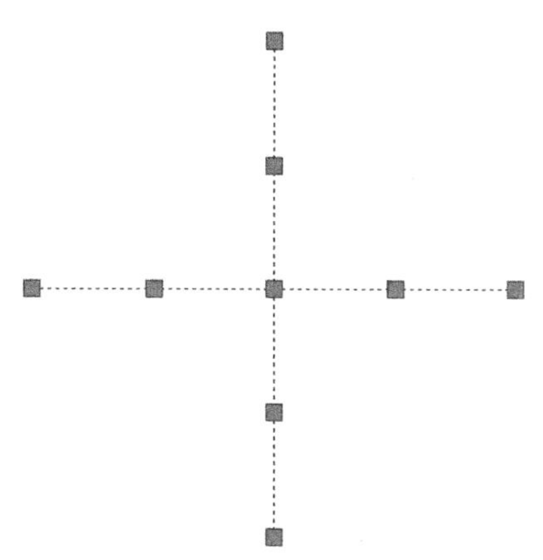

FIGURE 15.12 Break command, Method 2.

Join

The join command is included in this discussion because it is the exact opposite of break. It closes the gap between coplanar lines. Note the important disclaimer, however. The join command does not unite lines that are not on the same plane. This means that, if they are not pointing in the exact same direction and are not actually one after the other, then it does not work. To try out this command, draw two coplanar lines (or use break to put a gap in), as seen in Fig. 15.13. Note that join works on an unlimited number of coplanar lines; we are using only two in this example for simplicity.

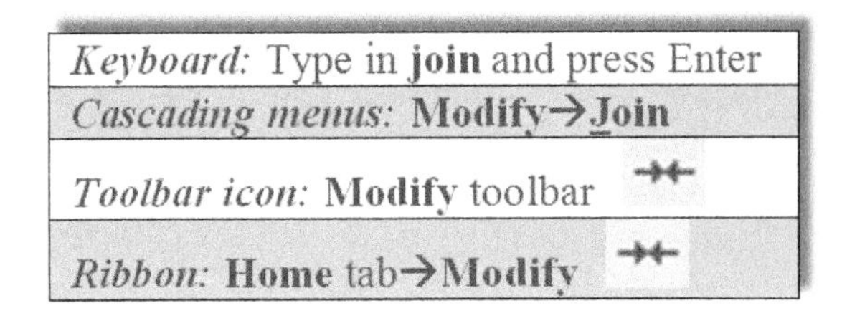

Keyboard: Type in **join** and press Enter	
Cascading menus: **Modify→Join**	
Toolbar icon: **Modify** toolbar	
Ribbon: **Home** tab→**Modify**	

FIGURE 15.13 Setup for join command.

Step 1. Start the join command via any of the preceding methods.
- AutoCAD says: `Select source object:`

Step 2. Select the first line.
- AutoCAD says: `Select lines to join to source:`

Step 3. Select the second line.
- AutoCAD says: `Select lines to join to source:`

Step 4. You are done with the two segments, so press Enter.
- AutoCAD says: `1 line joined to source`

A solid line takes the place of the two segments.

15.7 CAD STANDARDS

This rather obscure procedure may be valuable to a few organizations, and I have on occasion recommended it to prime contractors who often find their design files bouncing around from subcontractor to subcontractor. After all the respective additions to the drawings are finished, they are sent back to the main contractor. There, it is found that the standards have deteriorated, as everyone did his or her own thing in regard to layers and text or dim styles. A way to identify and clean up the mess would be useful, hence, the existence of this tool.

The CAD Standards procedures can be a bit confusing, and the process is not effortless nor does it guarantee perfect results. Depending on the damage, you may still need to do manual cleanup of layer properties, errant fonts, and the like. The best defense is to hold all others who may work on your drawings to a high standard from the start and outline what is acceptable and what is not. Obviously this can be challenging in just one office, never mind implementing it for designers at other, remote locations.

Here is the basic idea. Come up with one file that is in great shape. This can be a regular job that is singled out for its perfect or near perfect use of company standards and is of the highest quality and integrity. Then, do a Save As for this job under a .dws (drawing standards) extension. This is your standards file. All others are compared to it as they come in from subcontractors or just other designers.

To implement the standards, open up a job that just came in and is of questionable CAD quality. Type in `standards` and press Enter; alternatively use the cascading menu Tools → CAD Standards → Check.... There is no toolbar or Ribbon equivalent. The Configure Standards dialog box appears, as seen in Fig. 15.14.

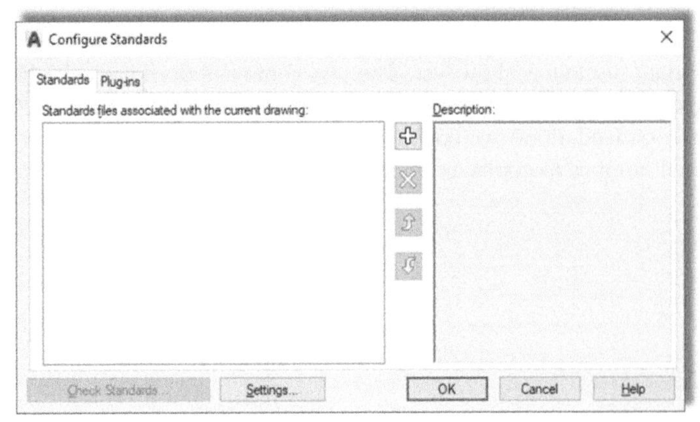

FIGURE 15.14 Configure Standards.

Next use the plus (+) sign to open up the saved .dws file. The file appears in the window on the left with some descriptions on the right. The Check Standards button (lower left) also lights up, at which point you press it to execute the check. The Check Standards dialog box appears, as seen in Fig. 15.15.

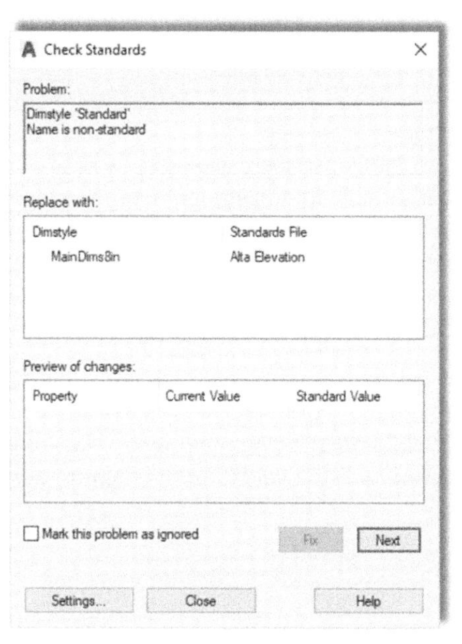

FIGURE 15.15 Check Standards.

Next, you need to run through the fixes, item by item. Continue pressing the Fix button as AutoCAD runs through the fixes and matches up text, layers, and other items. When done, the Check Complete notice shown in Fig. 15.16 displays. In this particular example, three problems were found but none fixed automatically.

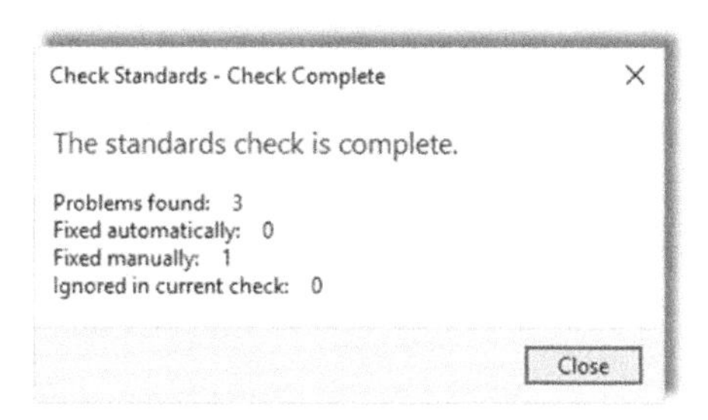

FIGURE 15.16 Check completed.

15.8 CALCULATOR

The calculator is a surprisingly powerful and versatile tool within AutoCAD, and one that is often overlooked by users. The function's power and versatility should come as no surprise, as AutoCAD itself is highly mathematical software, a calculator in itself, performing complex operations "under the hood" as you draw, copy, and rotate objects.

The calculator function can perform all regular and scientific calculations you would expect from a handheld device; in addition, it can assist you in the construction of geometry and even allows you to use AutoLISP, a programming language. A full description of the calculator command can take up dozens of pages, so we explore only the basics here. However, we take a look at one example of construction assistance using the transparent line command cal and one example of the actual graphic calculator tool (QuickCalc) to do unit conversions. You are encouraged to read the Help files for a wealth of additional information.

First, a brief overview of the basic cal command. This was the original version of the calculator. It is a command line tool enabled by typing in `cal` and pressing Enter. AutoCAD then says: `>>Expression:` at which point you can enter some math function using the familiar characters such as +, −, *, and /. Pressing Enter again yields an answer.

Not too long ago the limit of the calculator was +32,767 and −32,768 as the largest displayable final integer, but that restriction was lifted somewhere around the AutoCAD 2005 release, and you can now work with integer numbers between +2,147,483,647 and −2,147,483,648 (that is 2 to the 15th and 31st powers, respectively, if you really wanted to know where those seemingly random numbers came from).

Let us use the previously described cal command to assist in design, because crunching numbers alone can be done by your handheld or Windows calculator. Suppose you need to offset a line one-half of some number, but you do not know what that value is (say, one-half of 17.625). Instead of reaching for that calculator, let AutoCAD do both: calculate one-half of 17.625 as well as offset the line, all in one command. To do this,

1. Draw a line of any reasonable size.
2. Start the offset command via any method.
3. You do not know the offset distance, so let us use the calculator.
4. Type in `'cal` (the apostrophe makes cal, or any command, temporarily transparent, meaning you can enter it without exiting out of another, in-progress command, which in this case is offset). Press Enter.
5. The `>>Expression:` prompt asks for the expression. Enter `17.625/2` and press Enter.
6. Your answer (`8.8125`) appears, and you may resume the offset command.
7. Continuing as usual, select the line and pick a direction to complete the offset and Esc to exit.

This was a very simple case of how the calculator assists in design tasks. It can do a lot more in that regard, and you are encouraged to examine the Help files for additional examples.

The calculator function is not just a typed command line but also now comes in a graphic version. If you use this graphic version, you are using what is referred to as a *QuickCalc*. It is almost a whole new command all to itself and has quite a bit more functionality.

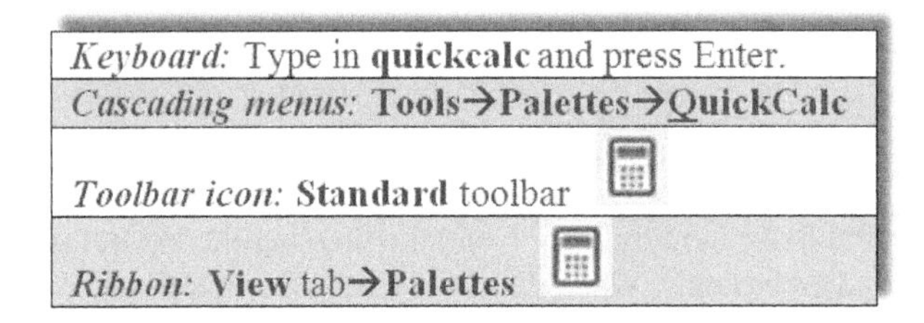

Keyboard: Type in **quickcalc** and press Enter.
Cascading menus: **Tools→Palettes→QuickCalc**
Toolbar icon: **Standard** toolbar
Ribbon: **View** tab→**Palettes**

Start up the QuickCalc command via any of the preceding methods and you see Fig. 15.17. The Number Pad and Scientific keypads are self-explanatory. If you ever used a handheld calculator, you should be right at home with this one.

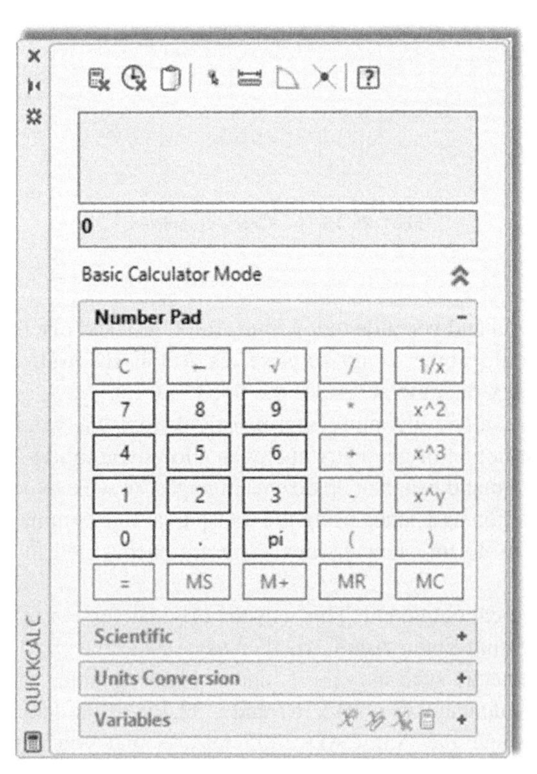

FIGURE 15.17 QuickCalc.

Variables, at the bottom, are not covered, but Units Conversion deserves a special mention. It is a very useful tool for converting between a wide variety of Length, Area, Volume, and Angular values. Simply select from the fields on the right (drop-down arrows appear) and enter the values to convert just below that. A converted value appears according to what you requested.

15.9 DEFPOINTS

Defpoints is not a command, but rather a concept that appears often and bears a short mention. Defpoints stands for definition points, and they appear in the drawing (and mysteriously in the Layers dialog box, where most users first notice them) as soon as you add any dimension to your drawing. What are they exactly? They are the tiny points where a dimension attaches itself to the object being dimensioned. The Defpoints layer does not go away easily and usually prompts questions from students as to its nature and intent. The layer fortunately does not plot, and you can pretty much disregard its existence if you choose. One benefit of it not plotting is that you can put any objects that you do not want to see on paper on it (such as viewports, as is discussed in Chapter 10). In fact, that is exactly what many users do, though a dedicated VP layer is recommended instead. In conclusion, do not worry too much about Defpoints; just understand what they are and continue on.

15.10 DIVIDE AND POINT STYLE

Divide does exactly what it advertises; it divides an object into equally spaced intervals. Note, however, that the object is not actually broken into those segments (that is for a break command to do) but rather just divided by equally spaced points. Then, you can use these points as guides for further work. Let us give this a try.

Step 1. Draw a horizontal line of any size.
Step 2. Type in `divide` and press Enter. Alternatively, you can use the cascading menu Draw → Point → Divide. There is no toolbar or Ribbon equivalent.
- AutoCAD says: `Select object to divide:`

Step 3. Select the line.
- AutoCAD says: `Enter the number of segments or [Block]:`

Step 4. Enter some value, `10`, for example, and press Enter.

The line looks the same, with no apparent divisions. It seems as though the command has not worked. It actually worked just fine, and the line is divided into the 10 segments using 9 points, as expected. The problem is that you cannot see the points; they are obscured by the line, and we need to change the point style to something more visible. Using the cascading menus, select Format → Point Style.... You then see the dialog box shown in Fig. 15.18.

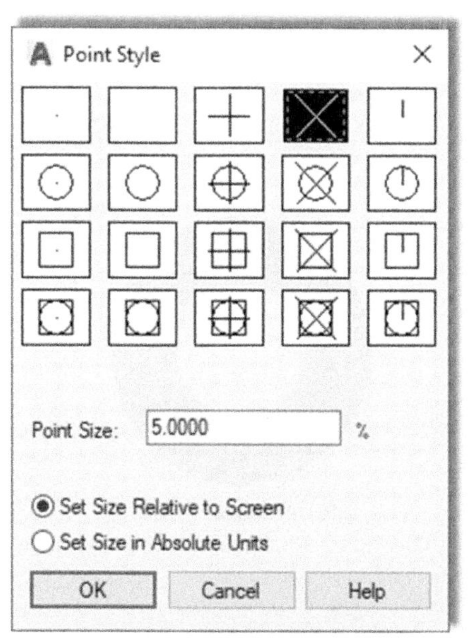

FIGURE 15.18 Point Style.

Select a point style that is not too complex; the X format in the top row (second from right) is a popular one. Press OK, and the divided line appears as you were expecting it, divided into the 10 visible segments (Fig. 15.19).

FIGURE 15.19 Divided line.

So, how exactly can you use this to your benefit? Well, the Node OSNAP, which is discussed later in this chapter, allows you to connect to points. If you activate it, you can attach new geometry, perhaps new lines, to those Xs. This is quite useful in basic construction and layout. Note an interesting quirk. If you zoom in or out, the points change size, but revert back to "standard" size upon typing in regen.

15.11 DONUT

Although this command sees only occasional use, it is still worth a mention. It creates exactly what it advertises: a donut of varying size. Broken down into its components, a donut is two circles of different size, one inside the other, and a solid hatch fill. You can technically create this shape with existing techniques, but the donut command just makes it easier and much faster. You have control over the sizes of both circles and, as a result, the size of the exterior and interior of the donut. The inner fill between them is always a solid hatch—and if you specify an inner circle diameter of zero, you will create a solid filled circle with just one command. Let us give it a try.

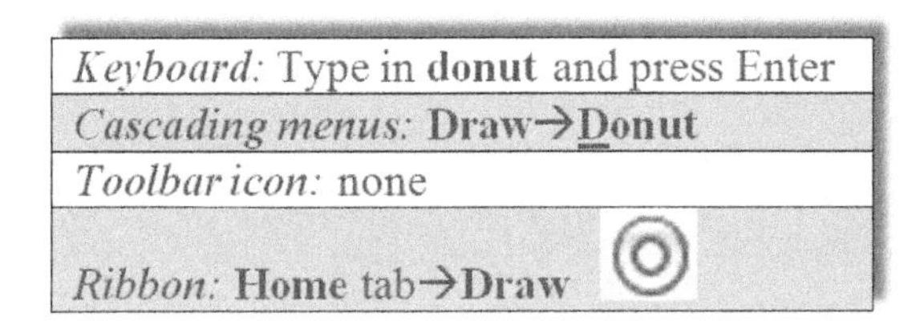

Keyboard: Type in **donut** and press Enter
Cascading menus: **Draw→Donut**
Toolbar icon: none
Ribbon: **Home** tab→**Draw**

Step 1. Start the donut command via any of the preceding methods.
- AutoCAD says: `Specify inside diameter of donut <0.5000>:`

Step 2. You can accept this by pressing Enter or type in another value (we accept the 0.5 in this example).
- AutoCAD says: `Specify outside diameter of donut <1.0000>:`

Step 3. Once again, you can accept this value or type in your own (we accept the 1.0 in this example).
- AutoCAD says: `Specify center of donut or <exit>:`

Step 4. Click anywhere on the screen and the donut appears, as seen in Fig. 15.20.

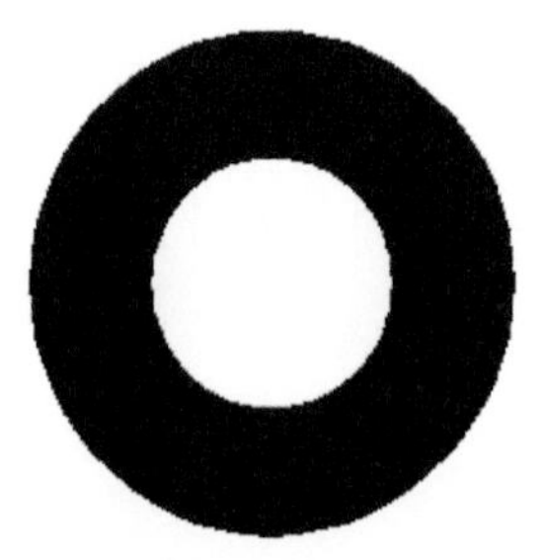

FIGURE 15.20 Donut.

15.12 DRAW ORDER

Draw Order refers to the stacking order of objects positioned one on top of another. This feature is commonly used in most graphic design programs and has an occasional use in AutoCAD as well. The stacking order is usually not of major concern in day-to-day basic drafting. If items overlap, the one drawn most recently is displayed on top; and in cases of two intersecting lines, the intersection point is 1 pixel in size, hardly worth a second thought.

Sometimes, however, such as when two or more hatch patterns are stacked, the order is critical. We use this feature later in this chapter to great effect when we discuss Wipeout. For now, just know where the Draw Order command is. It is found in the cascading menu under Tools → Draw Order, with options for Objects, Text, Hatches, and Dimensions.

Bring to Front and Send to Back send objects strictly to and from the top and bottom of the order, whereas Bring Above Objects and Bring Below Objects are more selective, moving items one step at a time. With these four tools, you can arrange any amount of objects in any order.

15.13 ETRANSMIT

The purpose of this command is to assist with emailing drawings. The problem is that, when drawings are sent out, items can be inadvertently omitted. These omissions may include something major, such as associated xrefs, or something secondary, such as fonts and styles. As a project grows more complex, it is easy to forget to send xrefs and even easier to forget about the unique fonts, styles, and .ctb files that have been generated over time. The recipient of the email may have only some of these, and what they download and reconstruct on their end may not be quite what you may have intended to be seen.

The eTransmit command is a go-getter of sorts. It sniffs out every associated file and everything else that may be relevant to recreate the file(s) in a new location as originally intended. Then, it packages everything together in a folder and sends it out. While veteran users have remarked that an experienced and aware designer needs to always have a clear idea of what to send (eTransmit is a relatively recent addition to AutoCAD after all), it is still a valuable tool for novices and experts alike.

So, what is (and what is not) included in the eTransmit transmittal? Here is a partial list of what is included. You should know of most of these files. If not refer to Appendix C.

***.dwg**: Root drawing file and any attached external references.
***.dwf**: Design web format files that are attached externally to the root drawing or xrefs.
***.xls**: Microsoft Excel spreadsheet files that are linked to data extraction tables.
***.ctb**: Color-dependent plot style files used to control the appearance of the objects in the drawings of the transmittal set when plotting.
***.dwt**: Drawing template file associated with a sheet set.
***.dst**: Sheet set file.
***.dws**: Drawing standards file associated to a drawing for standards checking.

Here is a partial list of what is not included. Do not worry if you do not recognize all these formats. You will know most of them if you read the Appendices in addition to completing Level 2.

***.arx,*.dbx,*.lsp,*.vlx,*.dvb,*.dll**: Custom application files (ObjectARX, ObjectDBX, AutoLISP, Visual LISP, VBA, and .NET).
***.pat**: Hatch pattern files.
***.lin**: Linetype definition files.
***.mln**: Multiline definition files.

To use eTransmit, save your work but keep the file open. Type in `etransmit` and press Enter. Alternatively, you can use the cascading menu File → eTransmit…. There is no toolbar or Ribbon alternative. You are asked to save the file if it has not been saved yet. Then, the dialog box in Fig. 15.21 appears.

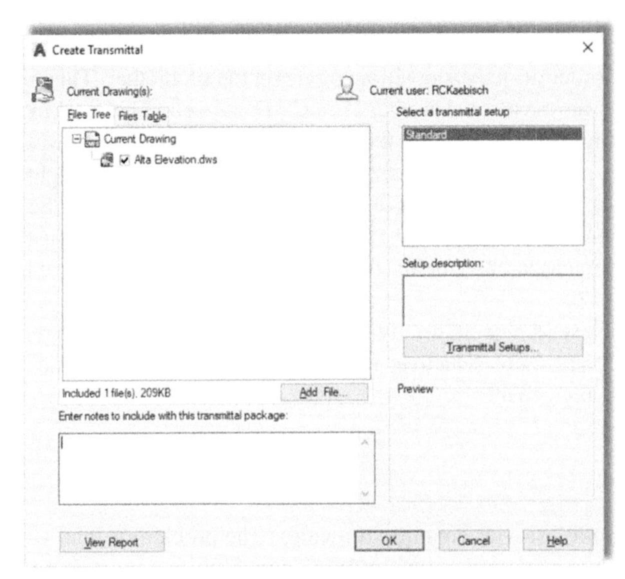

FIGURE 15.21 Create transmittal.

The procedure is highly automated, but you do have to select a few parameters. The current drawing that is open is the one that is packaged together. Some of the included items are shown in the file tree. Expanding the + signs shows specific files (which you can deselect if desired). You also can add more drawings to the transmittal using the Add File... button. If you are fine with just the current drawing (and what is included in it), then press the Transmittal Setups... button. You then see the dialog box shown in Fig. 15.22.

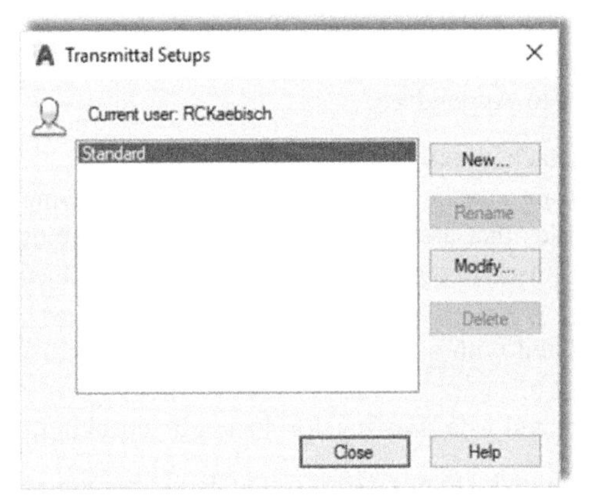

FIGURE 15.22 Transmittal Setups.

Let us create a new setup as an exercise. Press the New... button, and the dialog box in Fig. 15.23 appears. Give the transmittal a name and press Continue. The dialog box in Fig. 15.24 then appears.

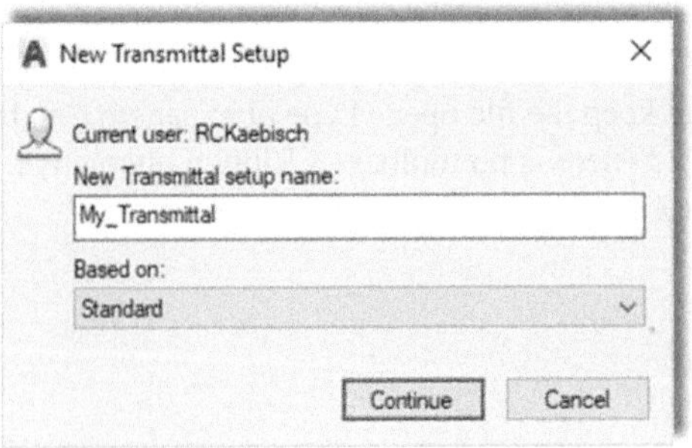

FIGURE 15.23 New Transmittal Setups.

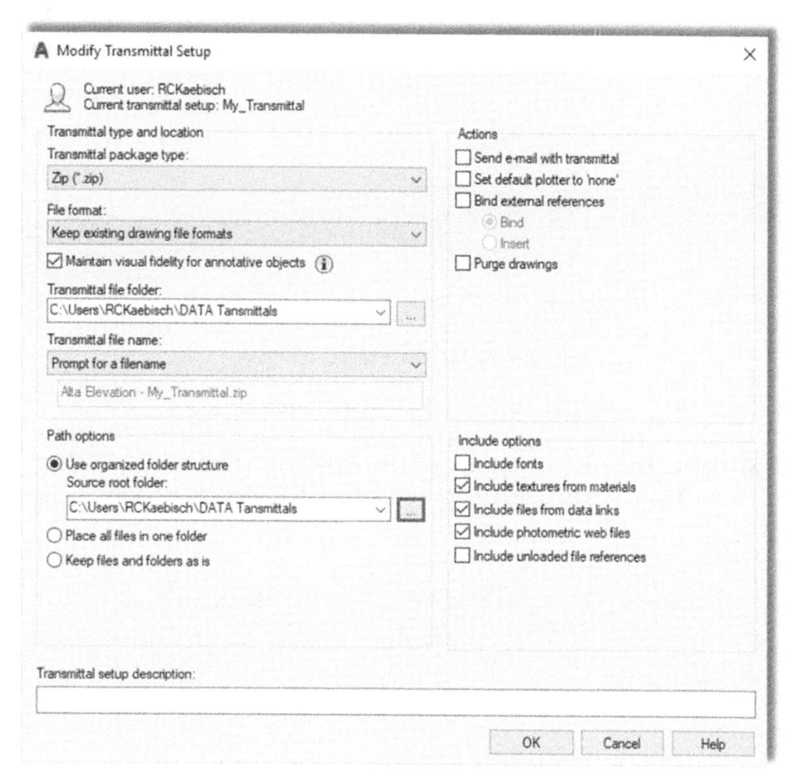

FIGURE 15.24 Modify Transmittal Setup.

All you really have to do here is visually check all the settings. Looking at the fields from the left, top to bottom, we see the Zip option for actually sending the files (recommended), followed by a variety of information related to formats, names, and locations. Unless you have a specific reason to change something, leave everything as default. On the right-hand side, top to bottom, leave everything under the Actions category as default (unchecked) as well as everything under the Include options category (middle three boxes checked).

When done press OK, then Close in the next dialog box (Fig. 15.22), and finally OK in the main setup box (Fig. 15.21). You are asked where to put the Zip file and for any last-minute name changes. Pressing Save drops the Zip file into its place, and it is ready for emailing.

15.14 FILTER

Filters were briefly discussed in the context of advanced layers (see chapter: Advanced Layers). While the ideas introduced here are roughly similar, the application is slightly different, so let us review them from the beginning.

Filters are nothing more than a method to isolate and select objects using their features or properties. So, e.g., you can isolate and select all the circles by virtue of them being circles, and find green objects because they all have the color green in common. You can also add together properties and create requests such as "Find me circles that are red only" and so on. This is a good tool for quickly finding stuff in a complex drawing. Once you find these objects, you can change their properties, delete them, or do whatever else is needed.

To experiment with filters, let us first create a simple drawing that is just a collection of shapes. Draw the set of circles, rectangles, and arcs seen in Fig. 15.25. Then, change the shapes' colors to blue or red.

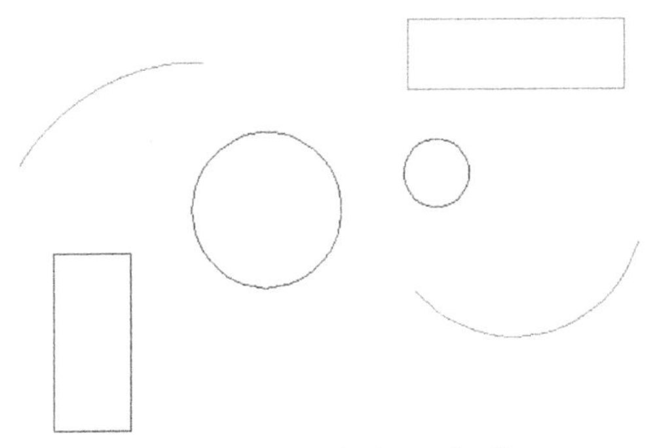

FIGURE 15.25 Basic shapes for filters.

Type in `filter` and press Enter. There is no cascading menu, toolbar, or Ribbon equivalent. Object Selection Filters appears, as shown in Fig. 15.26. The procedure here is to first select the filter, then add it to the list, and finally apply it.

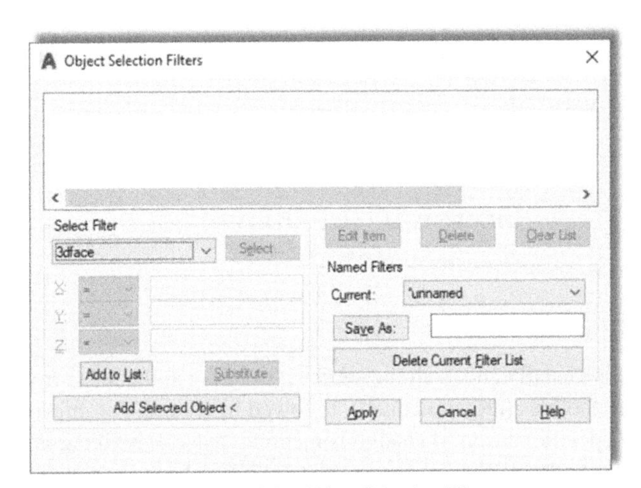

FIGURE 15.26 Object Selection Filters.

Select Arc from the Select Filter drop-down menu on the left. Then, press the Add to List: button just below that. For good measure, let us also add the color red. Select the color from the same Select Filter menu. Notice, at that point, the Select button lights up. Go ahead and select the color red. Then, press Add to List: Your filter should look like Fig. 15.27.

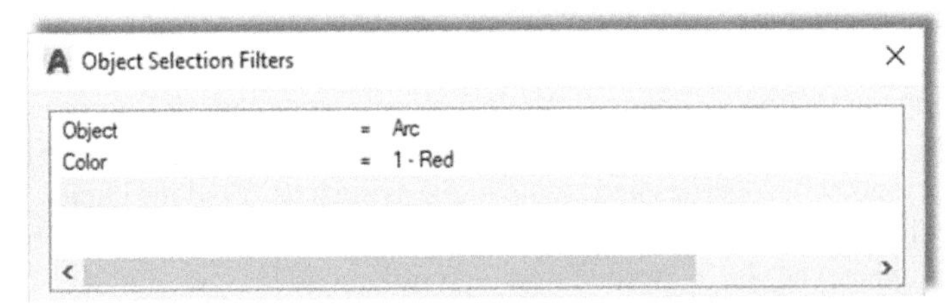

FIGURE 15.27 Objects and colors selected.

Now, press the Apply button. AutoCAD asks you to select objects. Select everything by typing in all. Every red arc in the drawing becomes dashed after the first Enter. Press Enter again and the grips come on, as seen in Fig. 15.28. Now type in `erase` and press Enter. Because they are selected, all the red arcs disappear. Filters can be saved for future use via the Save As: button and the listing can be modified or deleted via the Edit Item, Delete, and Clear List buttons.

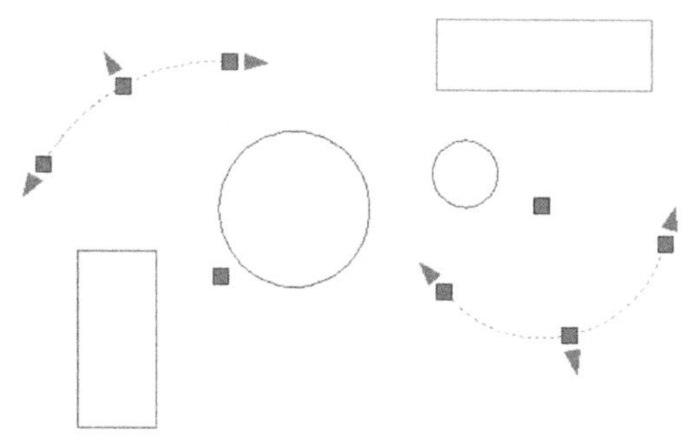

FIGURE 15.28 Segmented arcs selected by filter.

15.15 HYPERLINK

The hyperlink command attaches a hyperlink to either an object or text. During this process, you are asked to specify what this hyperlink opens when activated. It can be a website (usually) or a document (sometimes). In either case, it is a great command to quickly access additional useful information. Someone, e.g., can see a drawing of a product and follow the link to a website of the manufacturer or call up an Excel file that lists all the specs of a design. To try out the hyperlink features, draw a symbol for a chair (make a block) and a text website address, as seen in Fig. 15.29.

FIGURE 15.29 Chair and web address.

We would like to link the chair (an *object*) to a furniture website (IKEA, for example) and link the Google text (a text *string*) to the Google website.

Type in `hyperlink` and press Enter. Alternatively, you can use the cascading menu Insert → Hyperlink…. There is no toolbar or Ribbon equivalent. AutoCAD asks you to select objects. Go ahead and pick the chair block. The dialog box in Fig. 15.30 then appears.

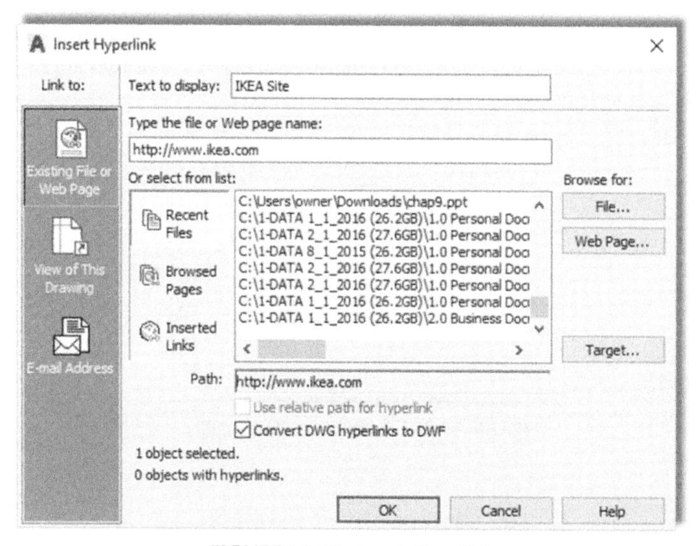

FIGURE 15.30 Insert Hyperlink.

Enter the name of the desired website in the appropriate field. That is where it says: Type the file or Web page name, not Text to display. Not much else needs to be done, so press OK. Repeat the process for the Google text.

Hover your mouse over the chair and then the Google text and notice there is something new. A hyperlink symbol appears, with text prompting you to hold down the Ctrl key and click to follow the link (Fig. 15.31). If your computer is currently connected to the internet, go ahead and do that in both cases (chair and Google website) to bring up the respective websites.

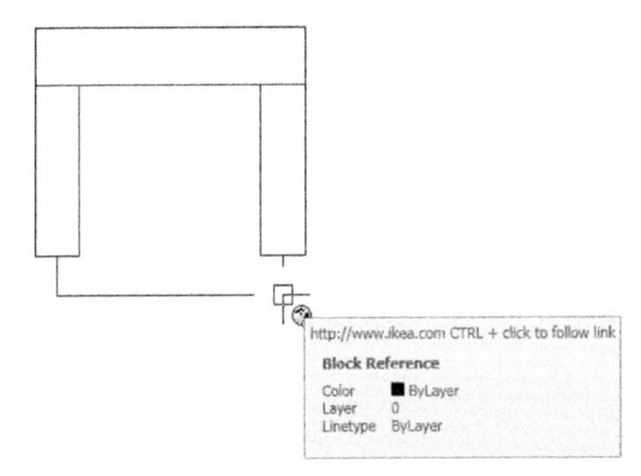

FIGURE 15.31 Hyperlink.

There is an interesting variation to this, as mentioned earlier. You need not attach a website to the link. You can also attach a regular document, so it is called up when the hyperlink is clicked on, as before. Pretty much anything can be attached if it can be found and the computer can run the application: Word, Excel, and PowerPoint documents; pictures; videos—you name it.

To do this, follow the same steps as before, but this time select the File... button under the Browse for: header on the right side of the dialog box. Find the file of interest, double-click on it, and OK the Hyperlink dialog box. That is it; the document is now attached to your object or text and supersedes any other links you may have.

As a final note, recall that Express Tools has a category called Web Tools. We skipped discussing them in Chapter 14 because the hyperlink command was not yet introduced. Take a few minutes now to look over what is available. These express commands include tools to Show, Change, and Find and Replace URLs. These are useful when you have multiple URLs in your drawing and need to manage them. The three tools are very straightforward to use, and we do not spend any further time on them.

15.16 LENGTHEN

The lengthen command is quite a bit like the extend command in its effect but goes about doing it differently and is worth adding to your AutoCAD vocabulary. The command works by lengthening nonclosed objects, such as lines or arcs (circles and rectangles do not qualify). You can lengthen the objects by change in size `[DElta]`, by percentage added or subtracted `[Percent]`, by total final size `[Total]`, or dynamically "on the fly" `[Dynamic]`. All this applies to angles (of arcs) as well. To try it out, draw an arc and a line (Fig. 15.32).

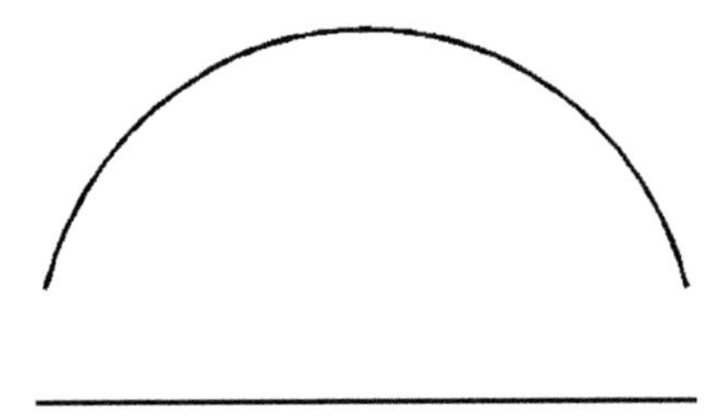

FIGURE 15.32 Arc and line for lengthen.

Step 1. Type in `lengthen` and press Enter. Alternatively, use the cascading menu Modify → Lengthen. There is no toolbar or Ribbon equivalent.

- AutoCAD says: `Select an object or [DElta/Percent/Total/DYnamic]:`

Step 2. Pick the `Percent` choice by typing in `p` and pressing Enter.

- AutoCAD says: `Enter percentage length <100.0000>:`

Step 3. Enter a value such as 150 and press Enter.

- AutoCAD says: `Select an object to change or [Undo]:`

Step 4. Pick the line and it lengthens 150%. Press Esc to exit the command.

The process is essentially the same with the arc and other lengthening options. Repeat the command several times and run through each option.

15.17 OBJECT SNAP TRACKING

OTRACK is a drawing aid that allows you to find various OSNAP points of an object and connect to them without actually drawing anything that touches that object, a sort of remote OSNAP. This function can be activated when the OTRACK button is pressed (turns blue) at the bottom of the AutoCAD screen and is deactivated when the button is pressed again (turns gray). Alternatively, you can also press F11 on the keyboard. The best way to see the purpose and function of this command is to try it. Draw a series of rectangles or squares spaced some distance apart, as shown in Fig. 15.33. You need not number them; that is only for identification purposes.

What we want to do is connect a series of lines from (and to) the midpoints of each of these shapes via right angles (not directly). An example of this may be a cable diagram or an electrical schematic. One way would be to start a line from the midpoint of the right side of shape 1, stop somewhere randomly, and repeat the process from the top of shape 2. Then, you can use trim, extend, or fillet to connect the lines. This, however, is cumbersome and slow.

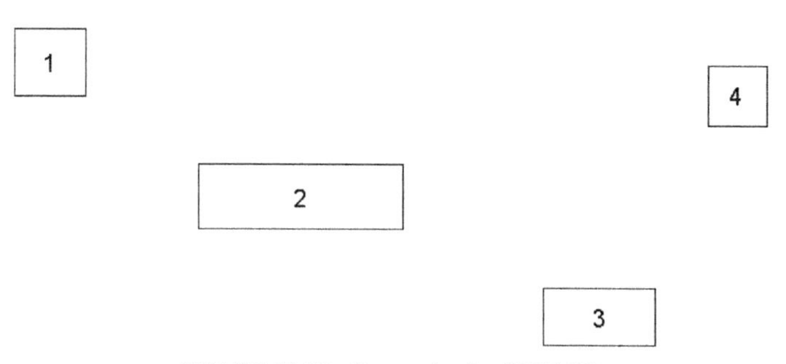

FIGURE 15.33 Rectangles for OTRACK.

A better way is to begin the line from the right side midpoint of shape 1 as before (turn on ORTHO also) but then activate OTRACK and, without clicking, position the mouse over the midpoint of the top of shape 2. A straight dashed vertical line appears. Follow that dashed line back until the mouse meets the straight horizontal dashed line. Then and only then, click once and continue to draw the line to shape 2. This is shown in Fig. 15.34. Notice what OTRACK does. It senses where that midpoint of shape 2 is and gives you a position to confirm and click before proceeding. Complete similar lines from the bottom of shape 2 to the top of shape 3 and then 4.

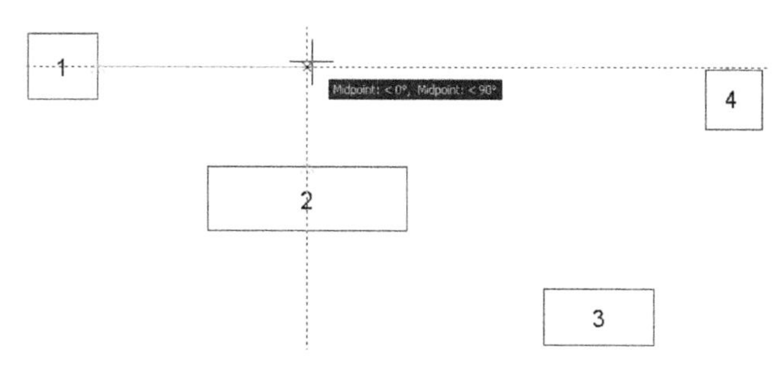

FIGURE 15.34 OTRACK in progress.

15.18 OVERKILL

This interesting command was once under the Express Tools but now is part of the main AutoCAD commands, found under Modify → Delete Duplicate Objects (you can still type its more dramatic original name: `overkill`). What this command does is quite useful. It deletes duplicate objects that overlap each other. In the 2D flat world of AutoCAD, it can be quite challenging and time-consuming to find these overlaps, especially if similar shapes and colors are involved. With overkill, you simply select the offending objects and AutoCAD figures out which to keep and which to delete, effectively merging them all into one.

To try it out, draw five lines, one on top of another. Although it may not be obvious to a novice, an experienced user notices these lines are a bit darker and "fuzzy," indicating the presence of an overlap; it is subtle but there. Now type in `overkill` and press Enter. AutoCAD asks you to select objects. Select all five lines using a Window or Crossing (not one at a time) and press Enter when done. The dialog box in Fig. 15.35 appears. Note that, if you have seen this command in older versions of AutoCAD, it has been redesigned somewhat.

You can leave all the defaults as they are, pressing OK to have all the lines merge into one line. This is probably the simplest application of the overkill command. However, what if the lines are on different layers?

In that case, you have to check off Ignore Layers and the command merges the layers into one. Numeric fuzz works by comparing the distances between nearly overlapping objects and acts on them if the distance falls inside the fuzz value. Overkill has other useful features as well, so go through the command, referring to the Express Tools Help files for more information.

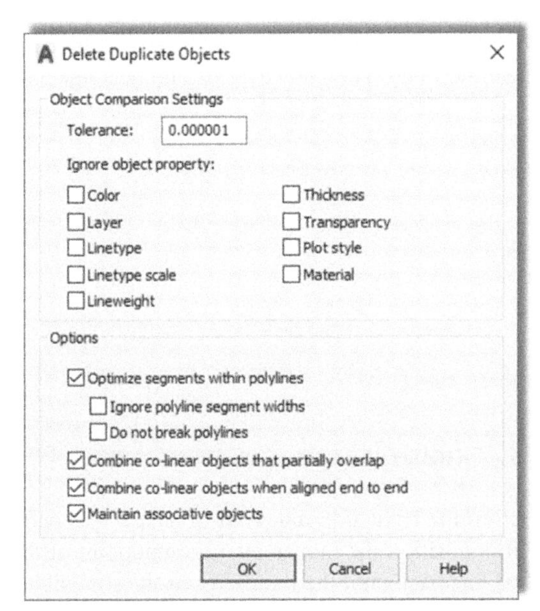

FIGURE 15.35 Delete Duplicate Objects (Overkill).

15.19 POINT AND NODE

You have already seen and used points when we discussed the divide command earlier in this chapter. In that example, points were created as a means to divide an object. To see them clearly, you needed to modify the form of the point, from a dot to something more visible, using Format → Point Style....

The point command can also be used to create points from scratch, as described next.

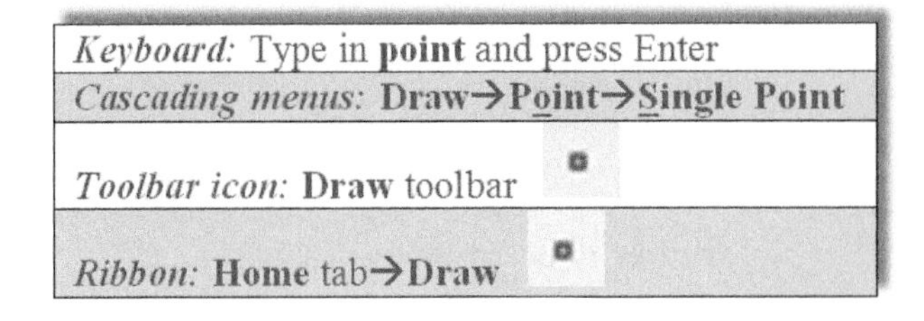

Keyboard: Type in **point** and press Enter
Cascading menus: **Draw→Point→Single Point**
Toolbar icon: **Draw** toolbar
Ribbon: **Home** tab→**Draw**

Step 1. Start the point command via any of the preceding methods.

- AutoCAD says: `Current point modes: PDMODE = 3 PDSIZE = 0.0000Specify a point:`

Step 2. Click anywhere and a point appears.

As you may imagine, points sitting there just by themselves are not too useful, but change them to an X (for visibility) and they can be used to create "anchors" to which you can attach geometry. For this, you need an OSNAP point that can do that. Called *Node*, it is one of the 14 OSNAPs. The symbol is a circle with an X superimposed on it. Go ahead and activate Node and practice drawing some lines that originate from the points you just created.

15.20 PUBLISH

The Publish command is a convenient way to print multiple sheets (layouts) all at once. This command depends on knowing the basics of Paper Space, so you may want to review Chapter 10 if needed. Assuming, however, that you have done so, the idea here is quite simple. When a file with multiple layouts is opened, the Publish command sees them and allows you to print them all at once, as opposed to clicking on each tab one at a time and using the plot command. For this example, a file called Alta Elevation is opened. It features a design spread out over two layouts: a Plan View and an Elevation View. The Publish command is then started via any of the following methods.

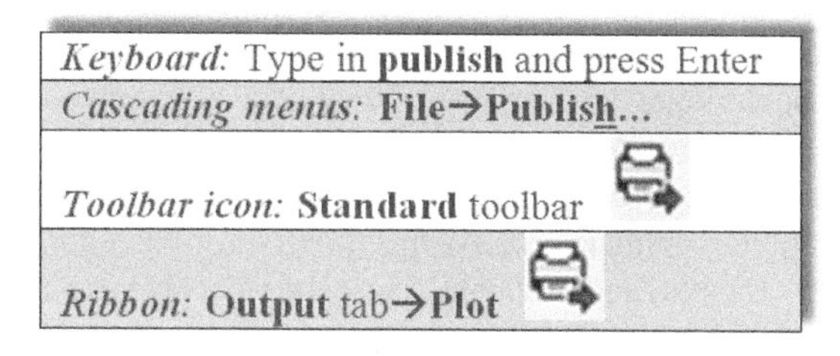

Keyboard: Type in **publish** and press Enter
Cascading menus: **File→Publish...**
Toolbar icon: **Standard** toolbar
Ribbon: **Output** tab→**Plot**

The dialog box in Fig. 15.36 then appears. Note that it can be expanded, if it is not, by pressing the Show details button on the lower left.

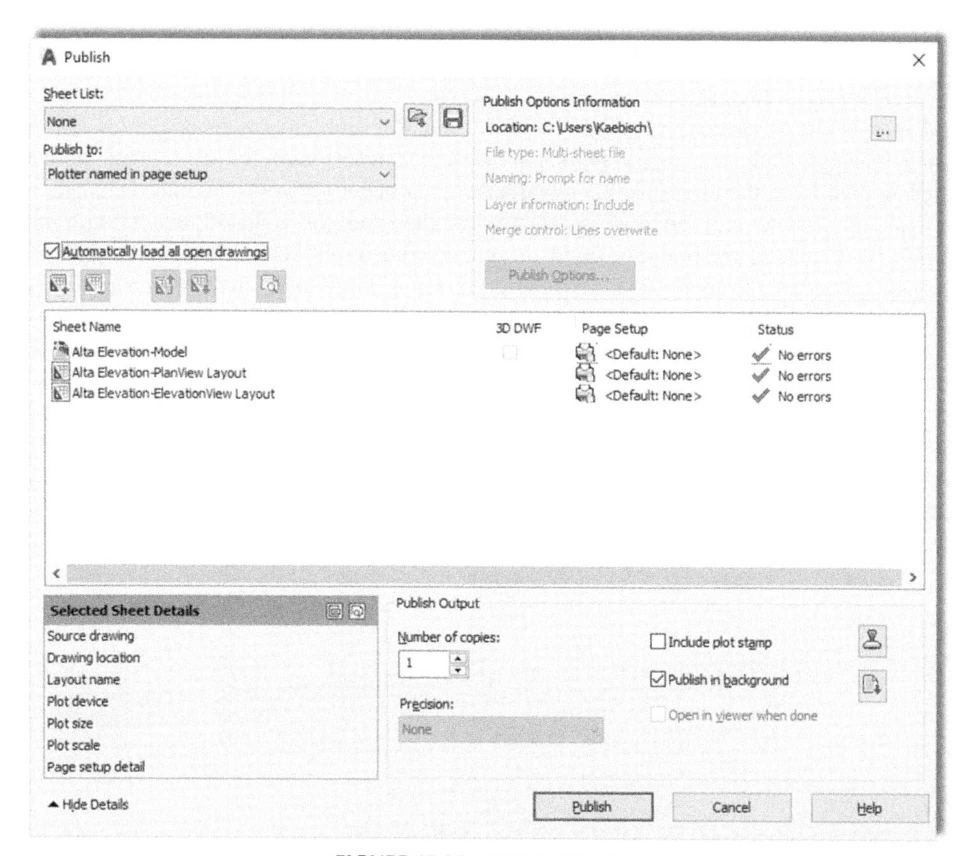

FIGURE 15.36 Publish dialog box.

The dialog box features some options to fine-tune publishing, such as adding additional sheets, plot stamp, deciding how many copies, and a few others. All you have to do at this point is just press the Publish button on the bottom, and AutoCAD prints out one copy of every single layout the file contains. It is quick and easy and saves time. Go ahead and explore this very useful command on your own for additional features.

15.21 RASTER

Raster is a term you may hear in AutoCAD circles, and it is not so much a command as a concept in graphic design. A raster graphic is nothing more than a bitmap, which in turn is a grid of pixels viewable on any display medium, such as paper or a computer monitor. This method of presenting graphics stands in marked contrast to vector graphics, which are based on underlying mathematical formulas. Vectors can be scaled up or down with no loss of clarity, whereas bitmaps cannot. So raster images are basically pictures. These can easily be inserted into AutoCAD, as covered in a previous chapter.

Raster graphics hold a unique place in AutoCAD history. They were needed to implement a massive worldwide conversion from paper-based, hand-drawn designs to CAD files. When it became obvious that computer-aided design was the future, millions upon millions of hand-drawn documents, spanning decades of design work, were scanned into the computer, became backgrounds, and were patiently redrawn. Government and private industries around the industrial world collectively spent large sums of money and many work hours to transform these archives into viable CAD files for future use. Most of this work was done during a 10-year span (1988–98), although some of it was started prior and some work is still ongoing. I recall putting in some significant screen time doing these conversions early in my career, and for a new AutoCAD user, this was excellent, if boring, practice: scan, draw, print, check, and repeat.

Raster scanning software can be quite sophisticated. At the time, we used regular large sheet scanners that simply pulled up the graphic on the screen. We then tugged at it until it was the right size (by comparing a drawn line to a scanned known dimension). Later, we acquired more sophisticated software that actually recognized shapes and attempted recreating these pieces of geometry with somewhat mixed results. Any smudges, imperfections, or odd shapes could confuse the system. Noise level (sensitivity) settings were used to control how much was picked up, and it was usually better to set it lower, so you could add what was missing rather than constantly erase accidentally picked up junk. As an AutoCAD user today, you may never need these tools, but it is good to know what they are just in case.

15.22 REVCLOUD

This is a command that architects everywhere were screaming for. Revision clouds, or *revclouds* for short, are used to indicate changes on a drawing. While you can use regular ellipses, rectangles, or circles just as well, these shapes blend too easily with actual design work, so the unmistakable cloud shape is a better choice to highlight changes on a print.

Revclouds are made up of small arcs, and that is exactly how they were once drawn, one arc at a time, making it an extremely tedious process. Finally, some program-savvy designers wrote a routine to automate the process. Autodesk then acquired the routine and upgraded it to semiofficial Express Tools status. Even there, it was much too useful to be optional, and it eventually became its own native AutoCAD tool, and the modern revcloud was born.

Some significant new enhancements to the command were introduced several releases ago, and it is easier than ever to use. In the past, you either freehanded the cloud or turned a predrawn geometric object (usually a rectangle or circle) into a revcloud. You can still do this, but now you can create instant rectangular and multipoint clouds as well. In all cases, you can set the size of the arcs that make up the revcloud and tell them which way to face, in or out.

Take some time to explore all the possible options—what follows are just some basic instructions to get you going; fully describing this command would take a few pages.

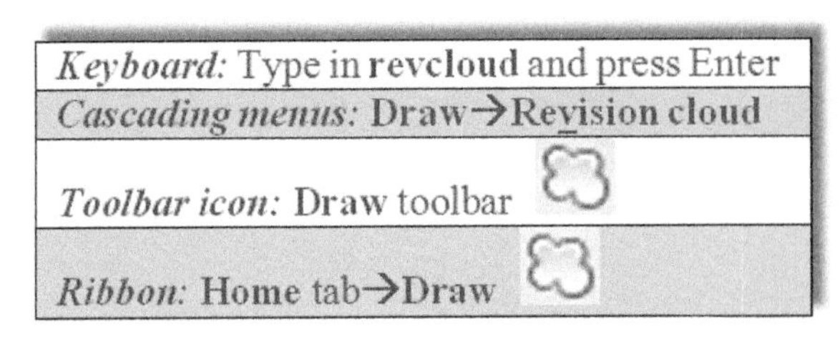
Keyboard: Type in revcloud and press Enter
Cascading menus: Draw→Revision cloud
Toolbar icon: Draw toolbar
Ribbon: Home tab→Draw

Step 1. Start up the revcloud command via any of the preceding methods.

- AutoCAD says:

```
Minimum arc length: 1.0000Maximum arc length: 1.0000
Style: Normal Type: Rectangular
Specify first corner point or [Arc length/Object/Rectangular/ Polygonal/Freehand/Style/
Modify] <Object>:
```

Step 2. Click anywhere once.

- AutoCAD says: `Specify opposite corner:`

Step 3. Draw a revcloud exactly the same way you would draw a rectangle. When you drag your mouse across the screen, a rectangular revcloud appears. When you click a second time the revcloud is complete

Your result should look similar to Fig. 15.37 (you may have to zoom in or out to see the revcloud clearly).

FIGURE 15.37 Rectangular revcloud.

As mentioned before, there is a lot more to this command. When you first started revcloud you noticed the menu consisting of `Arc length`, `Object`, `Rectangular`, `Polygonal`, `Freehand`, `Style`, and `Modify`. Let us briefly mention each, and then go ahead try all of them one by one. Many of these options were recently added in AutoCAD 2016, so you may not see them if you end up using an older release at work.

- *Arc length*: This option sets the size of the arcs that make up the revcloud. You can enter a value or click through several distances.
- *Object*: This option makes a revcloud out of a geometric shape, such as a rectangle or a circle.
- *Rectangular*: This is now the default method for creating revclouds and was the method used in the multistep example previously.
- *Polygonal*: This option is the same basic idea as the Rectangular option, but you now are free to click around to create as many sides to the revcloud as you want. Useful for creating a complex cloud with many twists and turns, as befitting real-life design changes. This process is reminiscent of creating a Lasso Window or Crossing.
- *Freehand*: This option is the "classic" way of making revclouds, by freehand sketching them around the parts of your drawing that changed. This approach requires a steady hand on the mouse, in regards to not just direction but speed also. The size of the arcs change if you accelerate the mouse across the screen.
- *Style*: This option allows you to change the style of the revcloud from the regular one seen in Fig. 15.37 to one that looks like it was drawn with a calligraphy pen. This is done via a special routine that uses pline arcs of decreasing thickness and is a throwback to the really "old days," when pen and pencil hand drafters took great pains in making their revclouds look fancy, using (you guessed it) calligraphy pens.
- *Modify*: This option was a badly needed one; it allows you to easily modify already-drawn revclouds. You select the revcloud you wish to modify, draw the revised section, and following the given prompts, erase what is not needed.

15.23 SHEET SETS

Like many new features in AutoCAD, these are not revolutions but rather evolutions of previous ideas. This tool, introduced in AutoCAD 2005, was meant to combine some aspects of Paper Space layouts and the Publish command with a few new twists, to create an "all under one roof" method of setting up the presentation and printing of a project. The idea is to set up sheet sets (1 Layout = 1 Sheet) that define exactly what is needed to plot the entire job correctly. These sheets can come from different drawings, even different jobs, and are grouped together by their respective categories. The idea is that they are now easier to handle and organize.

The industry has shown some reluctance to use a tool that somewhat duplicates existing functionality (even if new features are added), but a few organizations, not surprisingly ones that produce jobs that span dozens of sheets, took to this sheet set idea. We provide a brief overview, and you can decide if you want to pursue this further.

To create a sheet set, you need to set it up. Go to File → New Sheet Set... and you see what is shown in Fig. 15.38. Select the first choice, an example sheet set, and press Next >. What is shown in Fig. 15.39 appears.

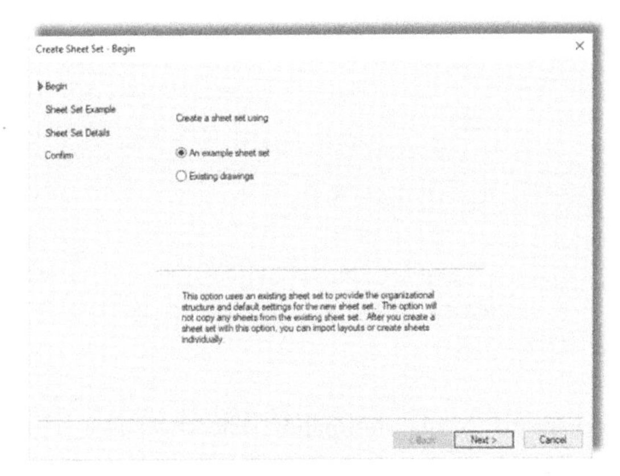

FIGURE 15.38 Create Sheet Set - Begin.

Select the appropriate sheet set, Architectural Imperial, e.g., and press Next >. Fig. 15.40 appears. Here, you can give the set a name and location. Press Next >, and what is shown in Fig. 15.41 appears. The various sheet sets, sorted by category, are visible. Press Finish, and the Sheet Set Manager appears, as seen in Fig. 15.42.

This was all just to set up the basic sheets (placeholders). They are still blank and need drawings associated with them. Right-click on the Architectural folder and select Import Layouts as Sheet.... What is shown in Fig. 15.43 appears.

You now have to browse for the right drawing associated with the Architectural layout. Continue down the categories until all are filled. This is just the basic idea of sheets sets. It is an extensive tool that can span many more pages of descriptions, and you should explore some of it on your own.

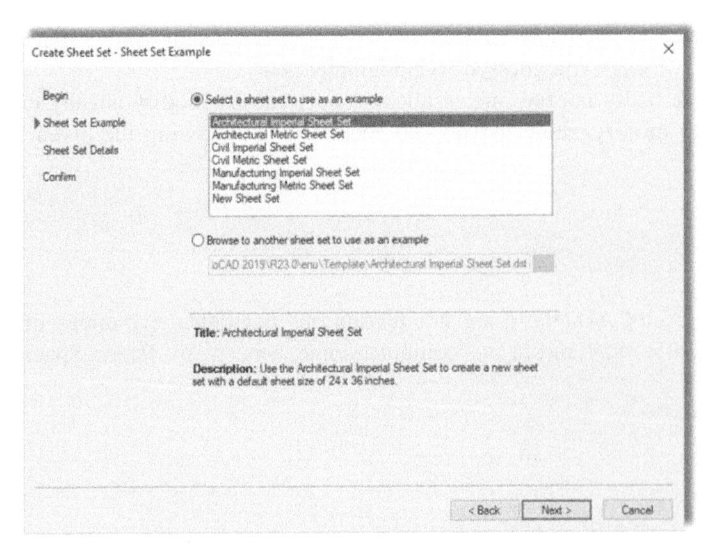

FIGURE 15.39 Create Sheet Set - Sheet Set Example.

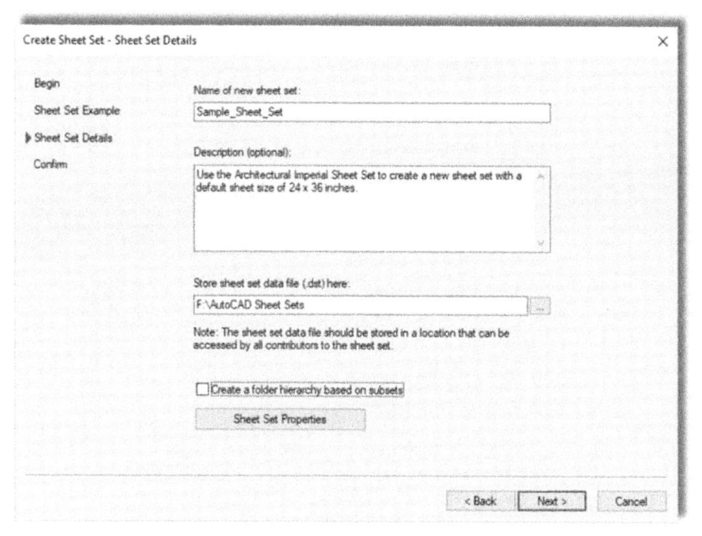

FIGURE 15.40 Create Sheet Set - Sheet Set Details.

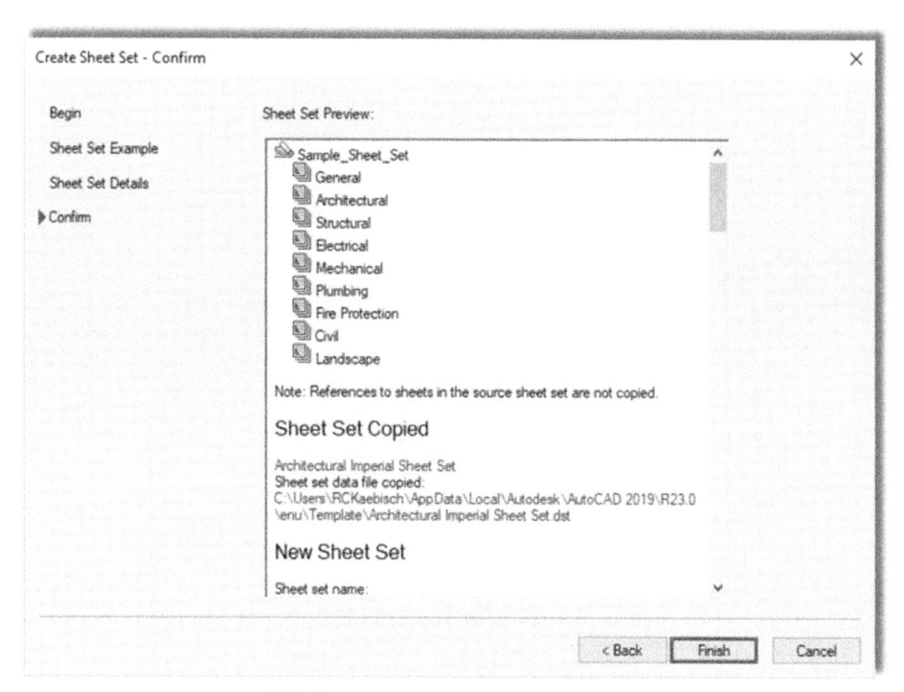

FIGURE 15.41 Create Sheet Set - Confirm.

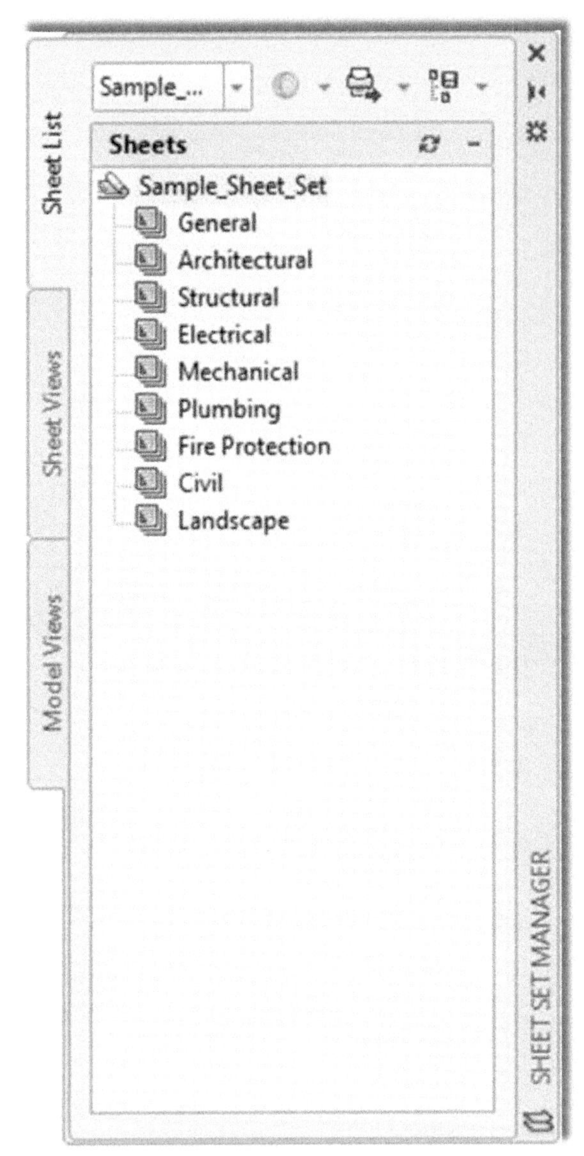

FIGURE 15.42 Sheet Set Manager.

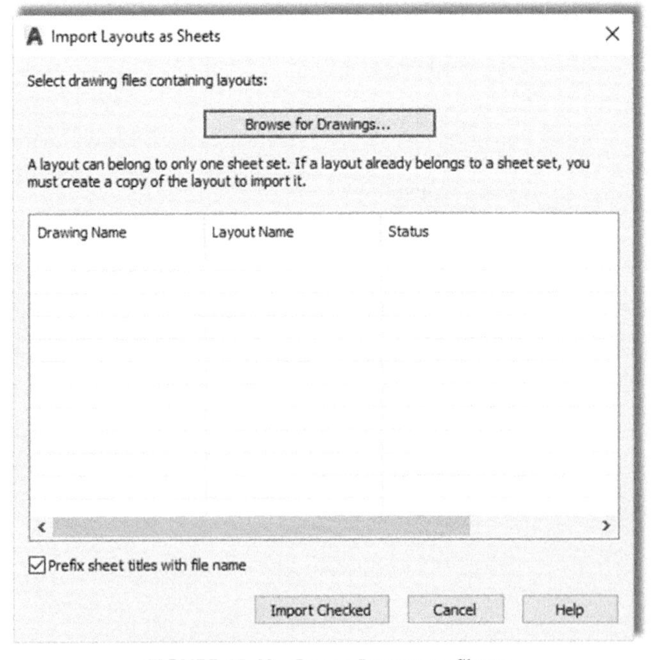

FIGURE 15.43 Import Layouts as Sheets.

15.24 SELECTION METHODS

What we outline here are a few other methods of selecting items that you may have not yet seen: the Window Polygon (WP), Crossing Polygon (CP), and Fence (F). All these are applicable when selecting objects for a number of operations such as erase, move, copy, and trim. These options would be secondary to the relatively new Lasso tool, and are similar in many ways, but they are still good to know. To try out the WP,

Step 1. Draw a set of circles, as seen in Fig. 15.44.
Step 2. Begin the erase command via any method you prefer.
Step 3. When it is time to select objects, type in `wp` for `Window polygon` and press Enter.
- AutoCAD says: `First polygon point:`

Step 4. Click anywhere.
- AutoCAD says: `Specify endpoint of line or [Undo]:`

Step 5. Continue to "stitch" your way around the circles you wish to select, and right-click when done.

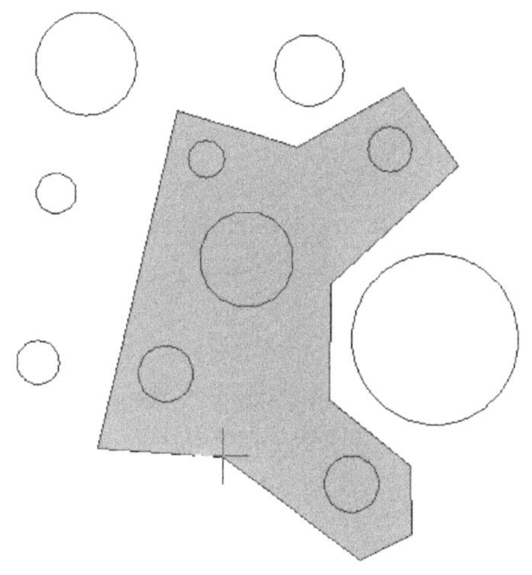

FIGURE 15.44 Window polygon.

Notice the difference here between the standard Window selection tool (strictly rectangular in nature) and the flexible "any shape you want it" WP (Fig. 15.44). Its action, of course, is similar; everything it encloses is selected.
To try out the CP,

Step 1. Draw another set of circles, as seen in Fig. 15.45, or just undo your previous steps.
Step 2. Begin the erase command via any method you prefer.
Step 3. When it is time to select objects, type in `cp` for `Crossing polygon` and press Enter.
- AutoCAD says: `First polygon point:`

Step 4. Click anywhere.
- AutoCAD says: `Specify endpoint of line or [Undo]:`

Step 5. Continue to "stitch" your way through the circles you wish to select, and right-click when done.

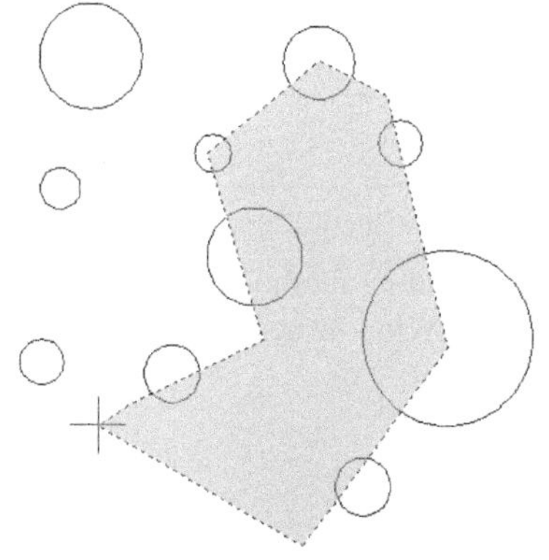

FIGURE 15.45 Crossing polygon.

As you can see, the CP (Fig. 15.45) works the same exact way as the WP, except anything it touches gets selected, much like how the regular Crossing operates.

The Fence option works by having a drawn line by the cutting tool during a trim command and selects multiple lines to be trimmed. Draw a bunch of vertical lines, and one horizontal line, as seen in Fig. 15.46. Then execute the trim command, but when it is time to select the lines to be trimmed, type in f for Fence, and click twice to draw a line across the lines to be trimmed, finally pressing Enter to make them disappear.

In case you discovered it on your own, yes, you can accomplish the exact same thing using a regular Crossing to eliminate the lines once a cutting edge is selected. However, the Fence command comes in handy when going around corners, as the flexible line-based approach can bend around any shape, whereas the rigid rectangular-shaped Crossing cannot. Try out both methods, as well as using Fence around corners.

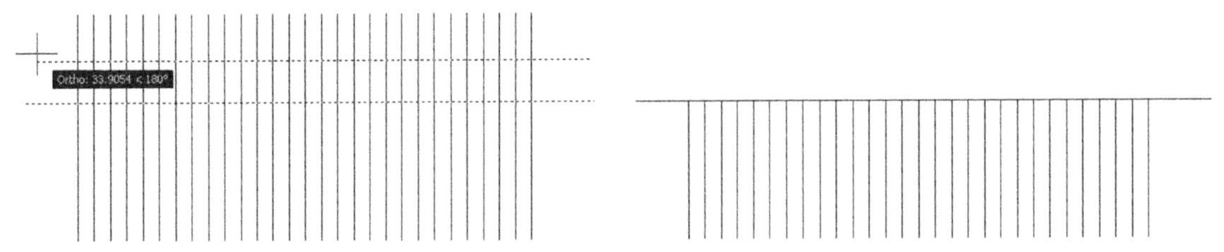

FIGURE 15.46 Fence command and the results.

15.25 STRETCH

This is an easy command to learn and use, and it can work wonders for quick shape modifications. The stretch command uses the Crossing tool to literally stretch shapes. Create two rectangles, as seen in Fig. 15.47.

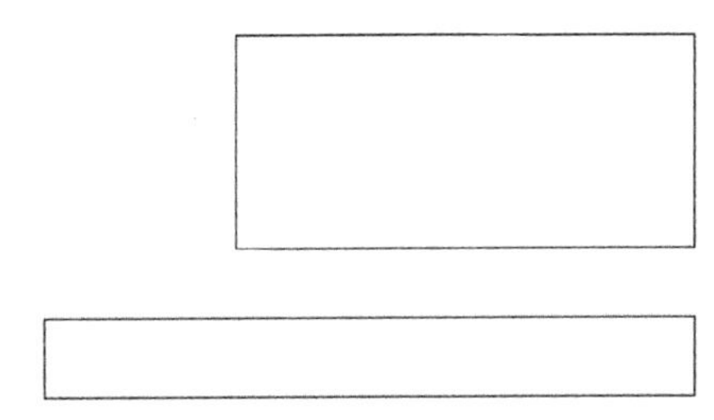

FIGURE 15.47 Rectangles for stretch command.

We want to stretch the right sides of both rectangles some distance.

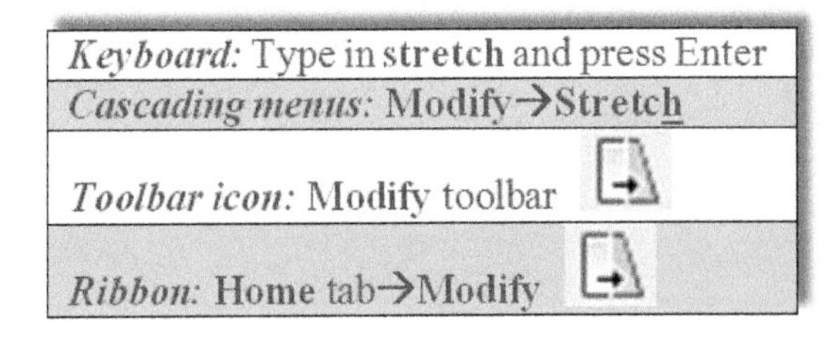

Keyboard: Type in **stretch** and press Enter
Cascading menus: Modify→**Stretch**
Toolbar icon: Modify toolbar
Ribbon: Home tab→Modify

Start the stretch command via any of the preceding methods.

Using a *Crossing* (this does not work with the Window selector), select the first quarter or so of the rectangles, as seen in Fig. 15.48.

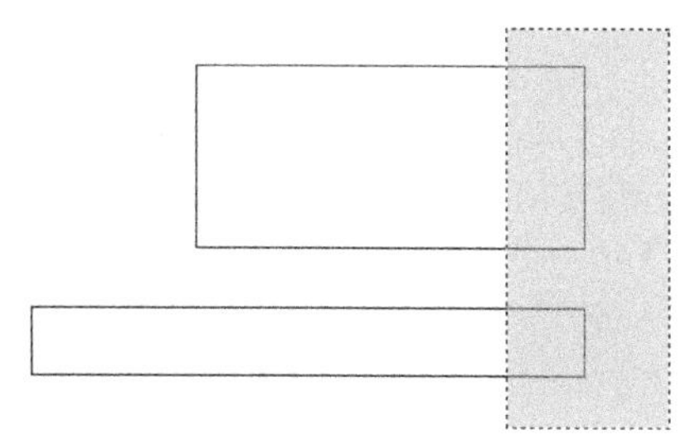

FIGURE 15.48 Starting the stretch command.

They become dashed. Press Enter and click anywhere to begin stretching them, as seen in Fig. 15.49. You need to have Ortho on and all OSNAPs off.

FIGURE 15.49 Stretch command in progress.

Finally, click elsewhere and the rectangles have a new length. You did not have to explode and redraw them or use grips. You can use the `Displacement` option to key in specific X, Y, or Z distances or stretch to another OSNAP point. Note that stretch does not work on partially selected circles and it moves, not stretches, fully selected circles and other objects.

15.26 SYSTEM VARIABLES

There is not much to say about SVs except to make you aware of what they are, so if they come up in reading other texts or the AutoCAD Help files, you know with what you are dealing. SVs are Yes/No, On/Off, or 0/1 types of commands used to indicate a *preference* or a *setting*. AutoCAD 2019 has over 850 of them for every conceivable variable (something that changes) in the software, and listing them all is futile, although if you wish to see them all, type in `setvar`, press Enter, then type `?`, and press Enter twice.

Many SVs are set graphically via a dialog box, and users may not even be aware or know they are changing an SV. On perusing the Help files and AutoCAD literature, especially on an advanced level, you find many references to them. Knowing SVs is important for those wishing to learn advanced customization (such as AutoLISP). Here is a useful example that is typed in on the command line. Suppose you are mirroring text and the new set is backward, as seen in Fig. 15.50.

FIGURE 15.50 Mirrtext.

That is not quite right, as you want text to be readable even after mirroring, so some sort of SV needs to be set. The responsible SV is called *mirrtext*, and it is of the 0/1 variety, with the two possible states represented by a 0 and 1. Somehow, it got set to 1, and needs to be 0. Type in `mirrtext`, press Enter, type in the new value of `0`, and press Enter again. Try the mirror command again—it should work fine now.

As mentioned before, this is just one example; there are hundreds of others. Keep an eye out for them. To help you do that, AutoCAD now features a very handy tool in this latest release called the *SV Monitor*. Type in `sysvarmonitor` and press Enter. You can also double-click the relevant icon all the way on the far right of the Drawing Aids strip (it looks like a sine wave with an orange triangle). The new SV Monitor dialog box appears and lists a number of SVs and their currently set values, along with their preferred ones. You can then manually change them to the preferred values as (and if) needed, or if you really messed up something, press Reset all.

15.27 TABLES

If you have not yet used tables, here is a quick primer on them. They are pretty much what you might expect, a bunch of rows and columns, with a header on top, ready for you to enter information. You can specify the number of rows and columns and add or take them away at a later time as well. Tables are sort of like low-tech Excel spreadsheets and can be useful in certain situations (like when you need data displayed in a hurry). We cover some basics here and let you explore more on your own.

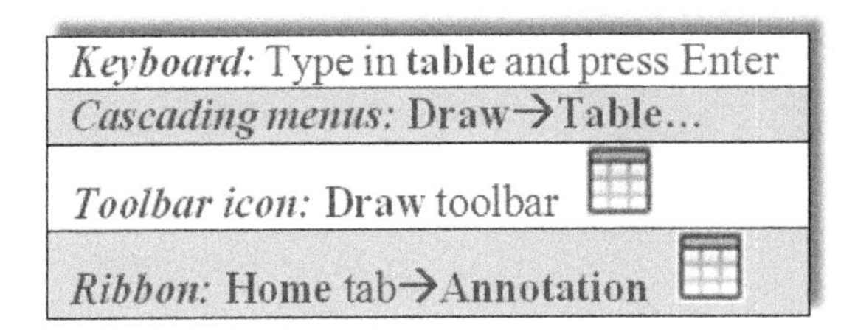

Start the table command via any of the preceding methods. You see Fig. 15.51.

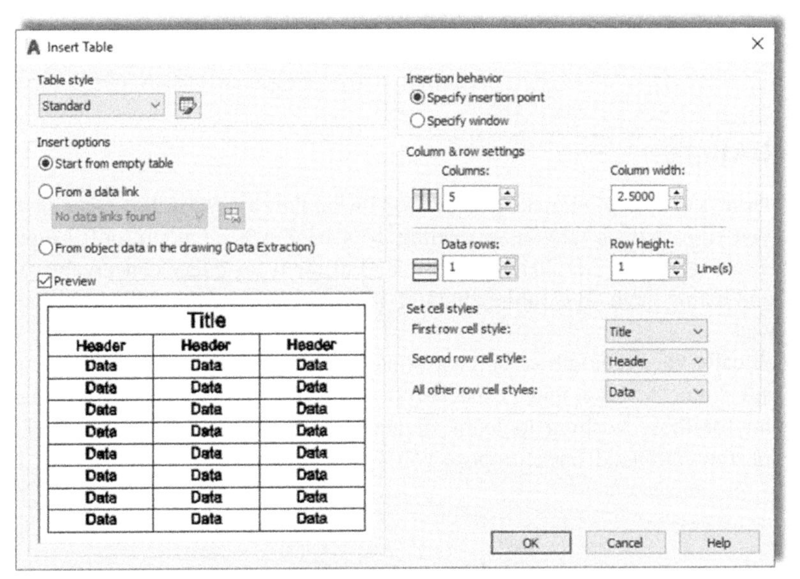

FIGURE 15.51 Insert Table.

Under the Column & row settings category, set your columns and rows as well as their respective widths and heights (we do a 10 × 10 here). When done, press OK, and you are asked to click where you want the table to go. When you position it, you see what is shown in Fig. 15.52.

Here, you are asked to set up the header. Type in "Sample Data Table" and press OK. You see the completed table in Fig. 15.53. You can now fill in the table with data by clicking in each cell and typing in values, as seen in Fig. 15.54.

There is more to this command, such as sizing and modification of the table, as well as some ability to link up with Excel spreadsheets, so go ahead and explore it further. You can also set up styles of tables by first using the tablestyle command, which gives you the dialog box for "designing" a table (Fig. 15.55).

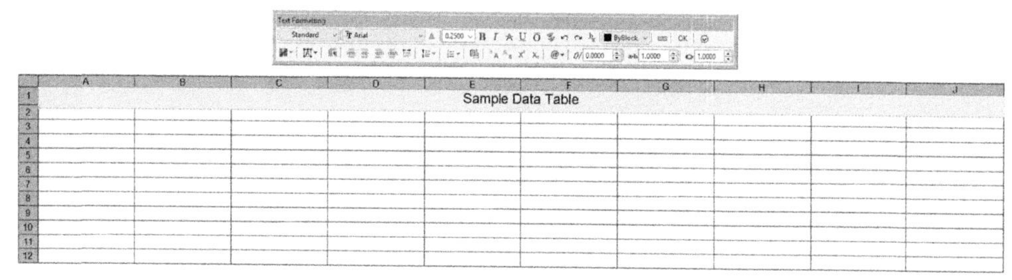

FIGURE 15.52 Insert Table.

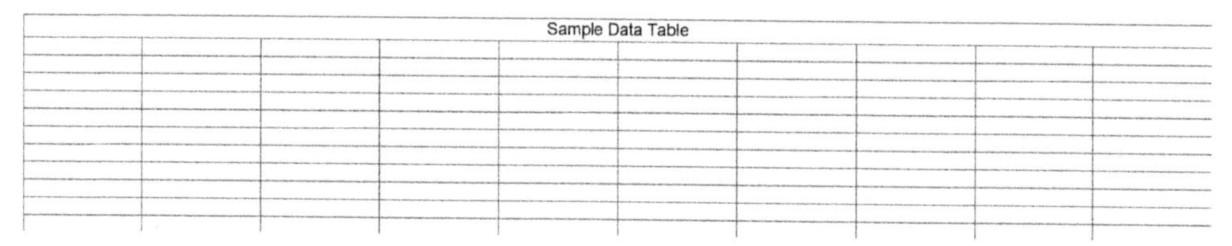

FIGURE 15.53 Fill in header.

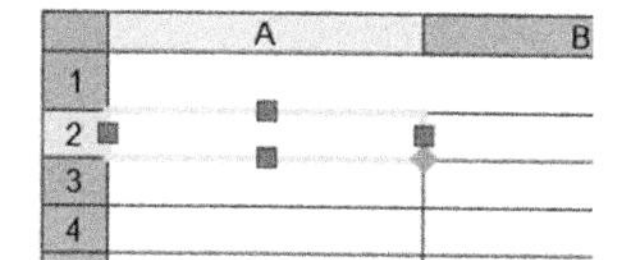

FIGURE 15.54 Fill in individual cells.

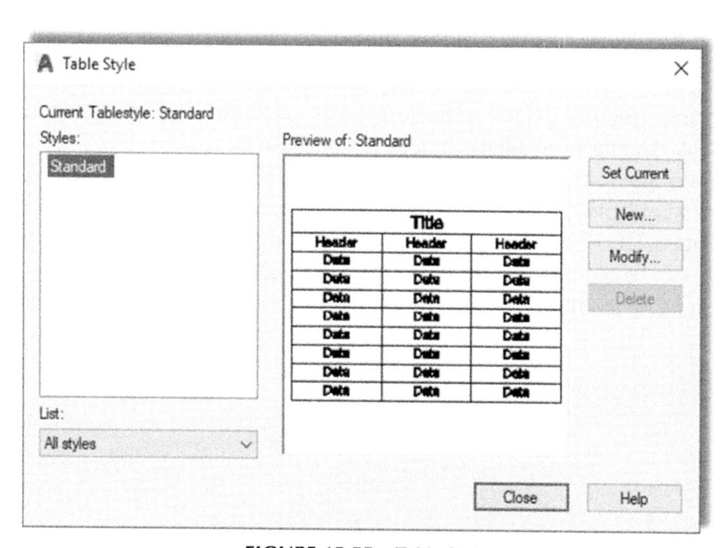

FIGURE 15.55 Table Style.

Press New... and give your style a name. You then get the New Table Style dialog box (Fig. 15.56). Here you can work with the general properties, text, and borders of the new table, similar to an Excel spreadsheet, and design a custom style that fits your needs.

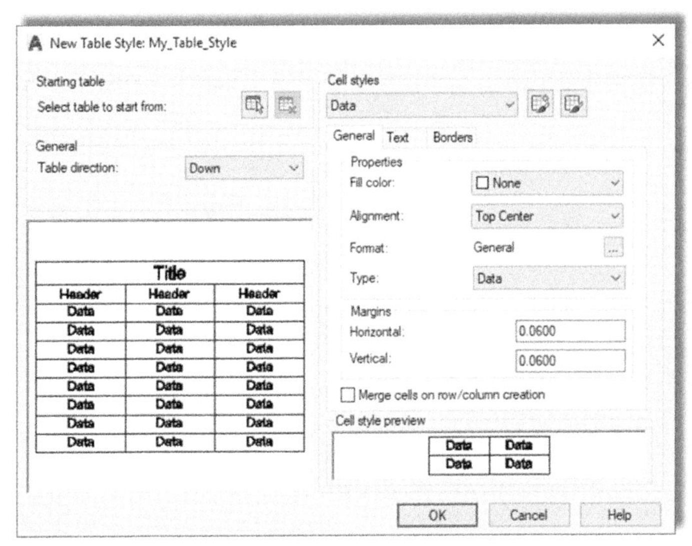

FIGURE 15.56 New Table Style.

15.28 TOOL PALETTE

This useful palette, shown in Fig. 15.57, can be accessed by typing in `toolpalettes`, via the cascading menu Tools → Palettes → Tool Palettes, or by pressing Ctrl +3. It may also be on your screen already, but hidden away off to the side. To see it, you have to hover the mouse over the palette's spine for it to reveal itself. This "autohide" feature can be disabled via a right-click (on the spine) menu, seen in Fig. 15.58.

The tool palette represents a collection of commonly used commands and predefined blocks. What you actually see when the palette comes up may be different from Fig. 15.57, as it depends completely on what menu is pulled up or what tab is clicked. In all cases, to use this palette, just click on a command (or click and drag), and it is executed as though you typed or selected a toolbar icon. To access more menus, simply right-click on the gray "spine." As mentioned before, the menu in Fig. 15.58 appears.

Even more tabs are available in this palette. Go to the bottom left of the tabs (where it looks like several tabs are bunched together) and left-click. Another extensive menu appears, as shown in Fig. 15.59. Take the time to go through each of the menu choices. The amount of available blocks is quite extensive.

Although we do not go into this further, you may also customize what tabs you have available by creating your own and adding custom commands and blocks to it. This is similar to creating new toolbars using the CUI, as covered in Chapter 14, Options, Shortcuts, CUI, Design Center, and Express Tools. You can also modify existing commands in the tabs; finally, the entire palette can be hidden, docked, and even made transparent, so it is always there but you can see what you are doing underneath it, a neat, if distracting, trick. Explore all the options; most are easy to figure out and self-explanatory.

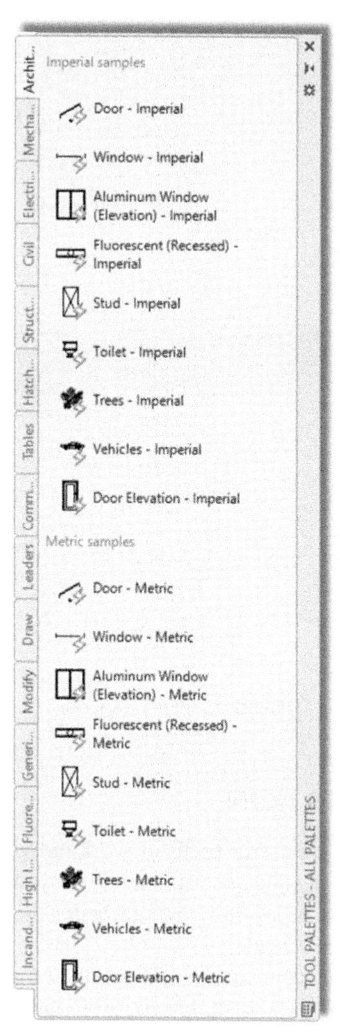

FIGURE 15.57 Tool Palettes, All Palettes - Architecture tab.

FIGURE 15.58 Spine menu.

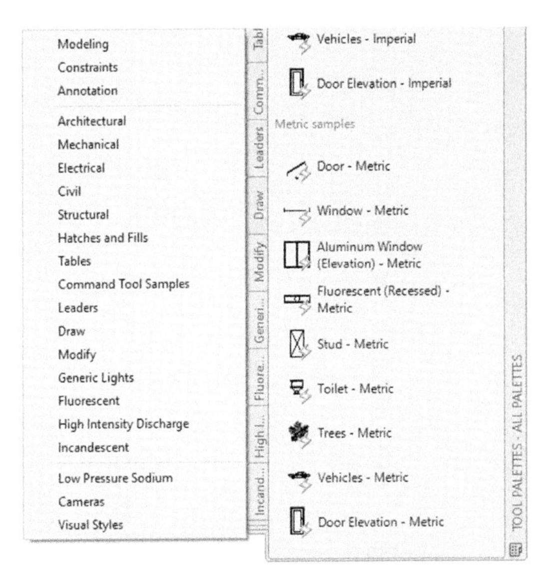

FIGURE 15.59 Tabs menu.

15.29 UCS AND CROSSHAIR ROTATION

Rotating the crosshairs or the UCS icon is a very useful trick for working with tilted shapes. Not everything you work on will be perfectly horizontal and vertical. Some designs, or parts thereof, will be rotated at some angle; and it would be nice to align the crosshairs to them. This allows for easier drawing and better orientation for the designer. Commands such as Ortho can now be implemented at the new tilted angle.

Rotating the UCS icon is technically not the same as rotating just the crosshairs, but we do not split hairs (so to speak) over the conceptual differences, and we treat both methods as equivalent for now. There are two ways to rotate the Crosshairs/UCS icon: the legacy method prior to AutoCAD 2012 (using snap or the UCS command) and the new method via the UCS icon grips menu, which we cover first.

Method 1

In AutoCAD 2012, the UCS icon gained some new functionality in the form of grips. It seems like an odd place to add grips—the icon is not a construction object after all—but it works well. To demonstrate, draw a rectangle tilted at a 45 degrees angle and click on the UCS icon. You see three grips appear. The outer ones can be clicked again to manually rotate the icon to any desired angle. The inner one can be used to move the icon around. Each of the grips reveals a menu as well. We ignore the inner grip for now and focus on the outer ones. Select the Rotate Around Z Axis choice (Fig. 15.60), and AutoCAD prompts you to enter a rotation angle (also 45 degrees in this case).

Once you enter the value, the crosshairs rotate and turn a colorful green and red. See the rotation of the crosshairs in Fig. 15.61. If you have Ortho on, you can now draw lines that are aligned to your rectangle.

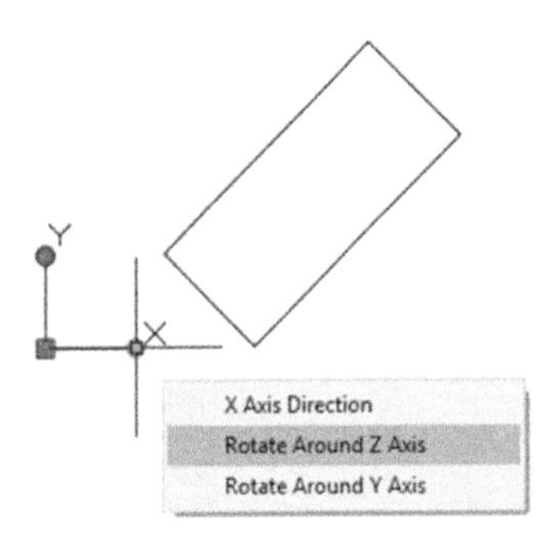

FIGURE 15.60 UCS grip menu.

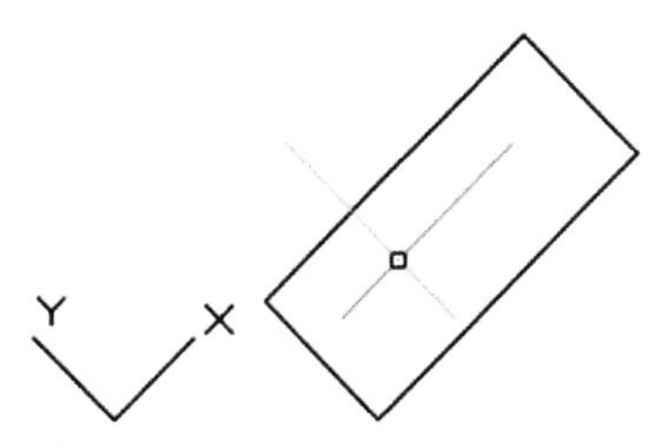

FIGURE 15.61 Rotated crosshairs.

Note that you can achieve the same result by the following steps:

Step 1. Type `ucs` and press Enter.

- AutoCAD says: `Specify origin of UCS or [Face/NAmed/OBject/Previous/View/World/X/Y/Z/ZAxis] <World>:`

Step 2. Type in `z` and press Enter. You always want to rotate around the Z axis in 2D.

- AutoCAD says: `Specify rotation angle about Z axis <90>`

Step 3. Type in 45 degrees for this example. The UCS icon is rotated along with the crosshairs.

To return to the usual orientation of the UCS/crosshairs, you can type `ucs`, press Enter, and type in w for World, or select World from the center grip menu, as seen in Fig. 15.62.

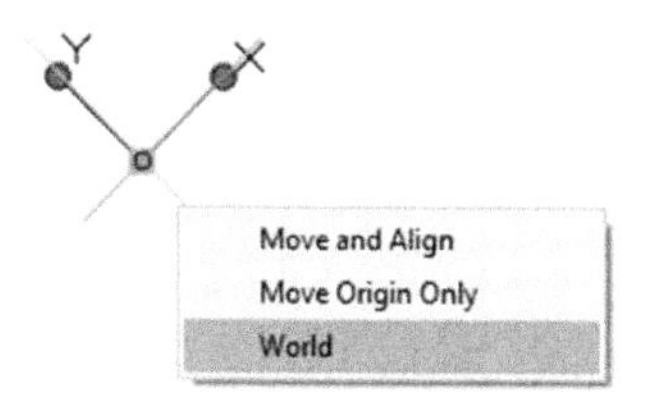

FIGURE 15.62 UCS World.

Method 2

This legacy method does something different: It rotates the crosshairs but not the UCS icon. It is a minor variation on the same idea but one of which you should be aware. In 2D AutoCAD, this makes no practical difference, as stated earlier. In 3D, there is a more significant consequence, and the UCS method is used.

Step 1. Type in `snap` and press Enter.
Step 2. Type in `ro` and press Enter.
Step 3. For the base point, type in `0,0`.
Step 4. For angle of rotation type in per for perpendicular.
Step 5. Select an angled line.
Step 6. Turn off the SNAP command.

To get back to normal, repeat all six steps, except type in `0` at step 4. Finally, to close out this section, an honorable mention needs to go to yet another way to rotate the *view* to match the current UCS. Type in `plan` and press Enter twice.

15.30 WINDOW TILING

By now you should be quite familiar with the multiple drawing tabs that can be found on top of the AutoCAD drawing area if you have several files open. At a minimum you will see the Start Tab with the Learn and Create screens, and then whatever files you may have open. All this was first discussed in Chapter 1, and these relatively new techniques represent a welcome jump forward in opening and managing files as well as easily accessing other design tools.

However, with this approach, each file remains under its own tab and therefore on its own separate screen. What if you wanted to see the files side by side? A method, called *Window Tiling*, addresses this. This is just a simple tool for allowing the viewing of two (or more) drawings side by side in a convenient manner. While each file still is managed under its own tab, they can now be located together in your viewing area. You can stack these vertically or horizontally, with AutoCAD spacing the drawings out evenly.

Open up two drawings in one session of AutoCAD and press Restore Down (middle button) at the upper right for one of the drawings, so the two drawings are tiled one on top of the other. You see something similar to Fig. 15.63, with the Start menu in the background. First of all click and minimize that Start menu and then select Window → Tile Vertically from the drop-down menu, as seen in Fig. 15.64. After the button is clicked on, you see what is shown in Fig. 15.65. The active drawing is highlighted.

Both drawings are lined up neatly on the screen. You can now click in and out of each and use copy and paste commands to transfer items if necessary. The same thing, of course, can be done for a horizontal tiling.

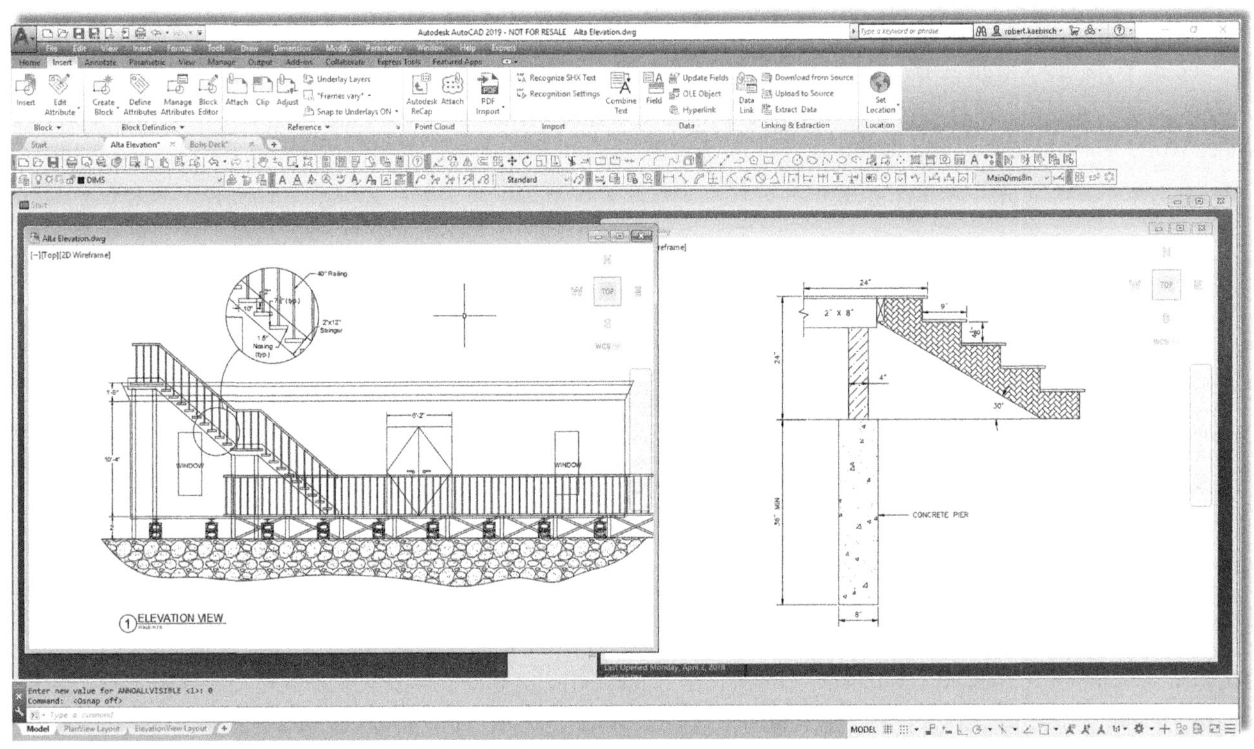

FIGURE 15.63 Window tiling, start.

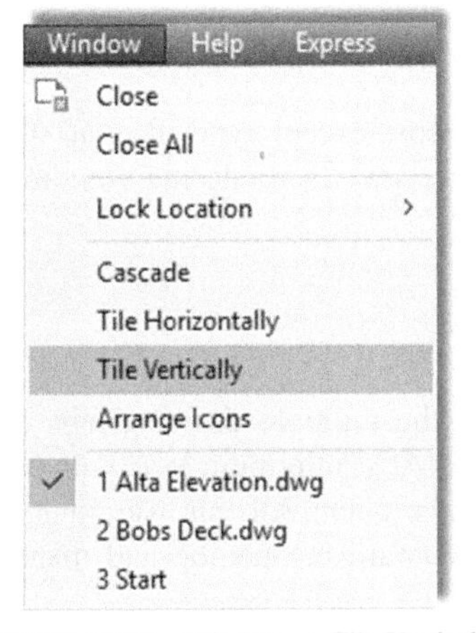

FIGURE 15.64 Window → Tile Vertically.

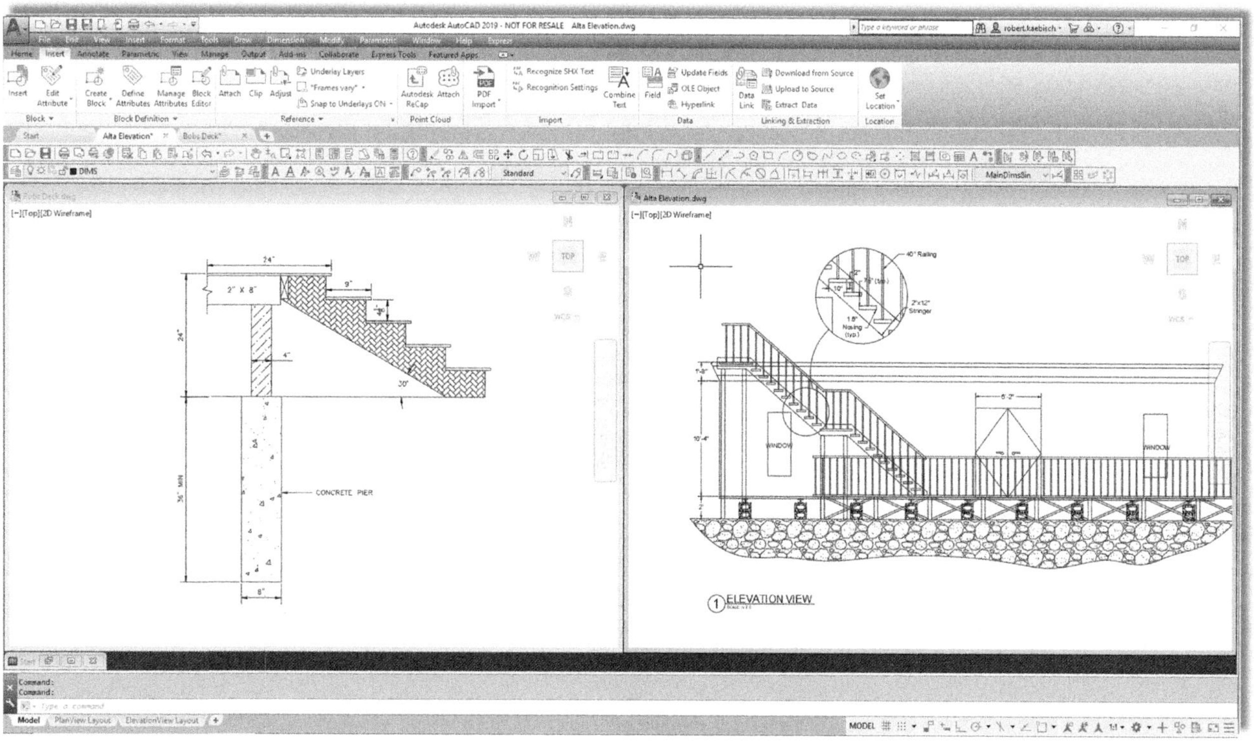

FIGURE 15.65 Window tiling, end result.

15.31 WIPEOUT

For reasons that may not be immediately apparent but are explained in detail, this command is one of the most important in 2D, to produce drawings that not only possess extraordinary clarity but are also easy to work with and modify. When used properly, wipeout is truly an "insiders'" secret, not known by beginner students, and one of the reasons why drawings done by AutoCAD experts have that intangible "something" that sets them apart from those done by less-experienced designers.

First, let us define what a wipeout is. It is simply a tool that creates a mask (or screen) to block out the viewer's ability to see other drawn geometry covered by the wipeout. To use it, draw a shape of some sort. Then type in `wipeout`, press Enter, and click your way around all or part of the shape. When you press Enter or right-click, the shape disappears behind the wipeout, as seen in Fig. 15.66.

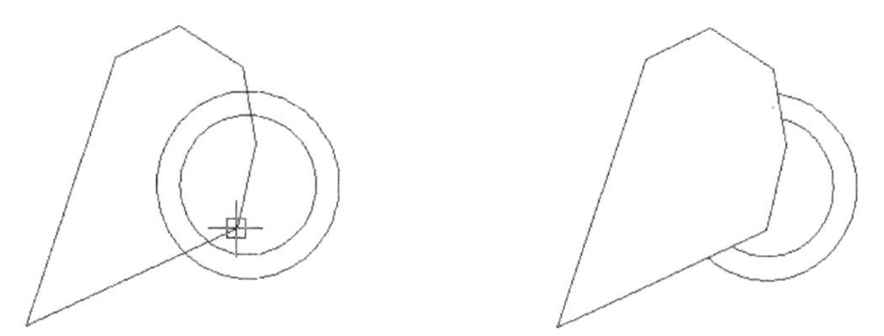

FIGURE 15.66 Creating a basic wipeout, and the result.

Notice the wipeout itself is still visible (something to be addressed later), but the double circle shape has been partially blocked, as expected. Practice using this command a few times just to become familiar with it.

Remember that a wipeout itself is an object. It can be moved, erased, copied, rotated, scaled, and just about anything else you can do to a regular geometric object. You can also click on the wipeout (click on the lines, not the blank center) to activate its grips. Then, the grips can be used to reshape it into any shape whatsoever, as seen in Fig. 15.67.

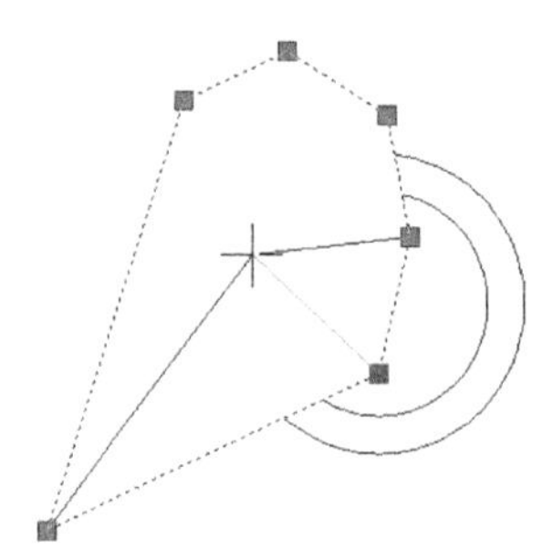

FIGURE 15.67 Wipeout grips.

So, what is so important about wipeout? Well, this command allows us to do something very special in 2D work. We can give our designs *depth*.

When you look around you in the real world, objects are generally not transparent. So, if you are looking at a bus in front of a building, part of the building is obscured by the bus. If another car were to park between you and the bus, it would obscure part of the bus, and so forth. In 3D, this is no problem at all. All objects are solid, and while they can be viewed in transparent wireframe form, a click of the button gives you rendered models that create an instant illusion of depth. This is important, as it matches our everyday experience, and we quickly identify what we are looking at. In the image in Fig. 15.68, the chair and the computer monitor both block a part of the computer desk. It is easy to tell what is in front of what in the rendered image on the left, but not as easy with the wireframe image on the right.

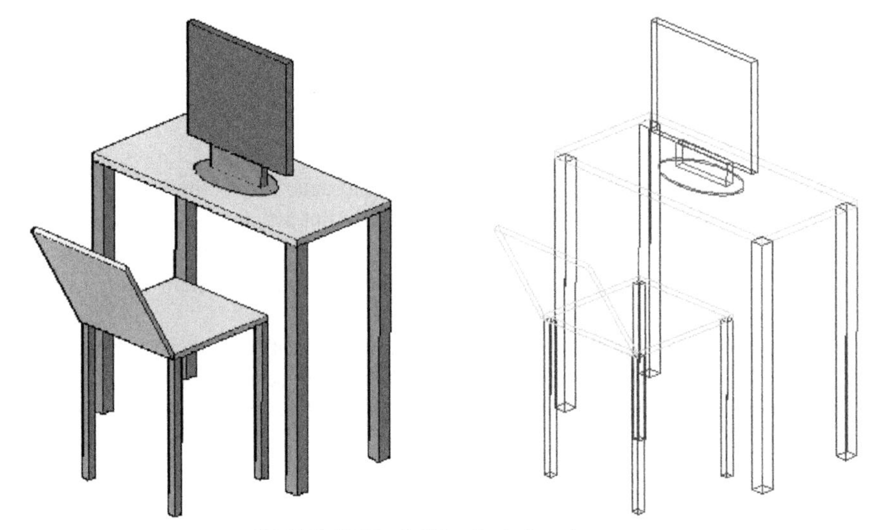

FIGURE 15.68 Solid and wirefame images.

Let us try the same thing with the bus and car, as seen in Fig. 15.69. Which is in front of which? You really cannot tell without shading the image, as in Fig. 15.70.

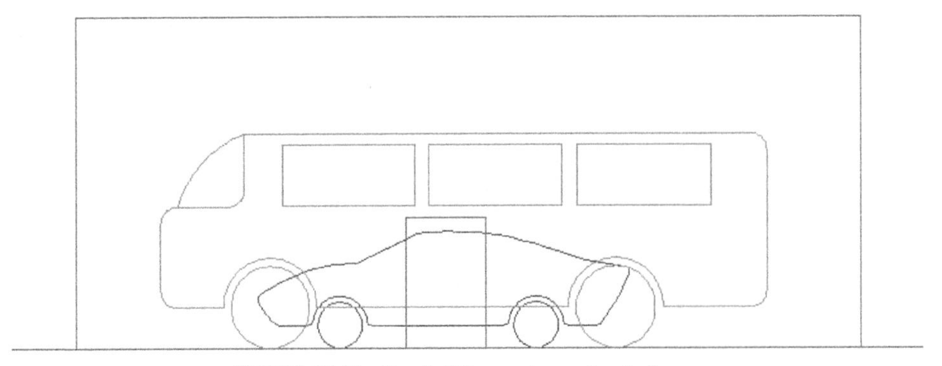

FIGURE 15.69 Bus, building, and car - in wireframe.

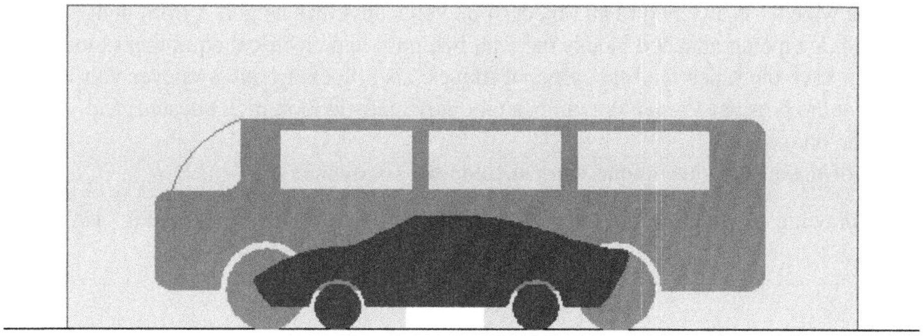

FIGURE 15.70 Bus, building, and car - solid.

In 2D, though, we have a problem: We cannot shade in the conventional sense. Everything is drawn using lines and nothing is solid, even though it may be in real life. In a plan view of an architectural floor plan, this is not an issue, as we are looking straight down, but what about in an elevation view, such as the previous images? A section through a room in an elevation view may feature furniture obscuring walls and walls obscuring other items behind them. Or, in electrical engineering, we have equipment on a rack obscuring the mounting holes. To provide realism and a sense of depth, we need to somehow obscure our view of items not seen. Of course, you can trim linework behind the object, but what if the object moves? You now have a gaping hole. We need a way to temporarily obscure and hide geometry that is tied directly to the object doing the hiding, so if it moves, the linework reappears. Wipeout is the secret to that trick.

In Fig. 15.71, notice how the equipment bolted to the rack obscures part of the rack. Now, if the equipment is moved, such as in Fig. 15.72 (notice the numbers to the left change from 20U–22U to 8U–10U), the wipeout moves with it and it acts like a solid object. The move is successfully done with no trimming necessary.

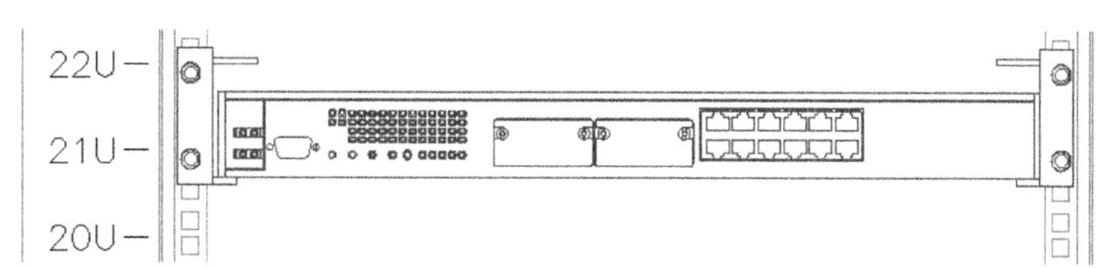

FIGURE 15.71 Electrical gear on a rack with a wipeout.

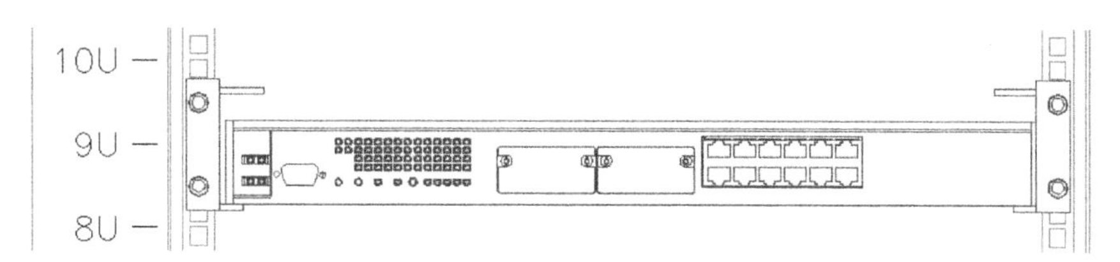

FIGURE 15.72 Electrical gear relocated.

What if you did not have a wipeout? Such is the case in Fig. 15.73. Notice that the rack is visible underneath the equipment, giving no indication of depth. You would have to trim lines to achieve this.

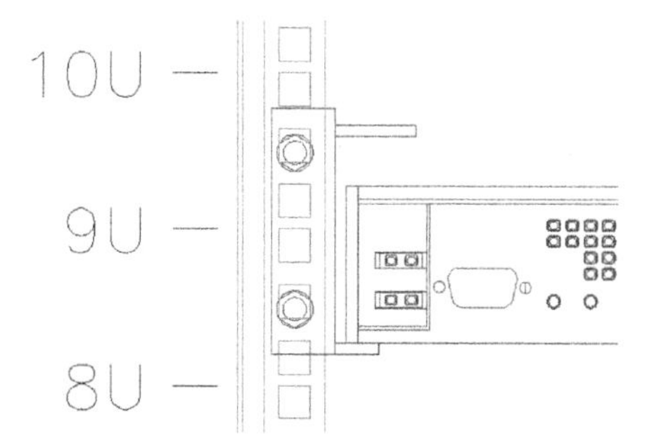

FIGURE 15.73 Same electrical gear with no wipeout.

In summary, when a wipeout is attached to an object, it prevents anything behind it from being seen, exactly as in real life. So, a bookshelf with a wipeout attached blocks the wall behind it and electrical equipment blocks the mounting holes to which it is attached. Move the object and the wipeout moves, also blocking out whatever you happen to be blocking. This is the technique that the pros use to give drawings a uniquely realistic look in 2D design, and you should learn this as well. Let us go through some details.

The essential steps of making this technique work include the following:

Step 1. Create the drawing of what you are designing, such as the equipment rack in Fig. 15.74.

FIGURE 15.74 Wipeout, Step 1.

Step 2. Using wipeout, click to trace the outline of the drawing point by point (just the four corners in this case).
Step 3. Using Tools → Draw Order → Send to Back, put the wipeout behind the design.
Step 4. Make a block out of the design and wipeout.

The object is now "solidified" and can be placed in front of other objects. If you create more objects in this manner, you can simply use the Draw Order tool to shuffle them around top to bottom as needed (as long as all of them are blocks). In this manner, you can create a rather impressive sense of depth and clarity to an otherwise busy drawing.

Take a close look at Fig. 15.75. It is a larger section of the same cabinet shown in the previous figures. It could not have been done this way without extensive use of the wipeout command. Notice how it is easy to tell which equipment is in front and which is behind the rack. There are, in fact, several layers of objects: the equipment bolted to the front of the rack, then the rack itself, and then the equipment in the back of the rack.

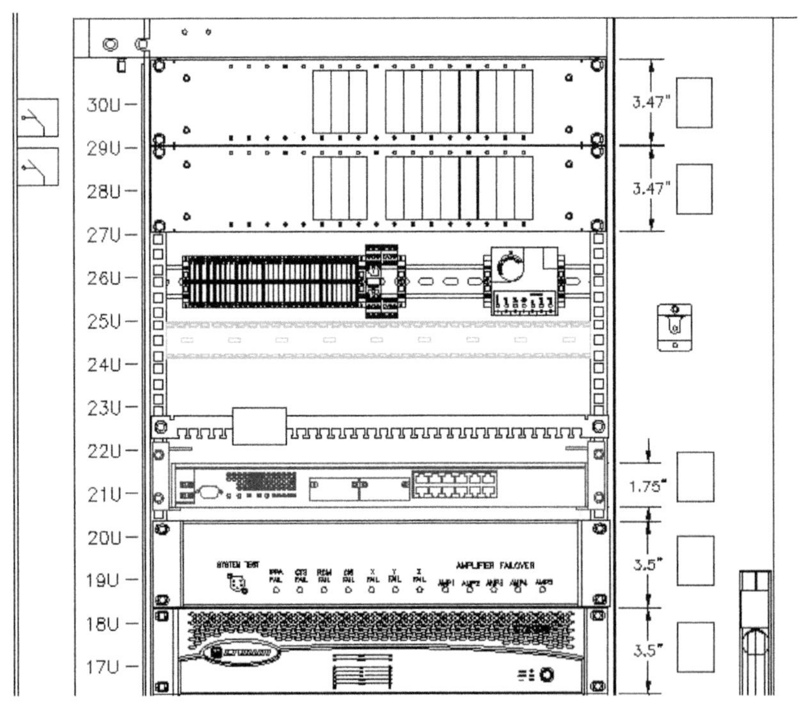

FIGURE 15.75 Full equipment rack with multiple wipeouts.

The final item to mention with wipeouts is that they themselves can be hidden (while their ability to hide objects remains in effect). To do this, simply type in wipeout, press Enter, and select f for Frames in the submenu. AutoCAD then says: `Enter mode [ON/OFF]<ON>:`. If you type in `off`, AutoCAD turns off all wipeouts on the screen, while preserving their effects on other objects.

15.32 LEVEL 2 DRAWING PROJECT (5 OF 10): ARCHITECTURAL FLOOR PLAN

For Part 5 of the drawing project, you add carpeting and flooring to your floor plan. This is somewhat arbitrary and may or may not be the type of carpeting, tile, or flooring an architect would select on another design. You may use a different pattern, of course, if you wish. As always, be sure to use the proper layering.

Step 1. At a minimum use the following layer convention:

- A-Carpeting (Gray_9).
- A-Flooring (Color_32).

Step 2. Freeze all layers that are not relevant to the carpeting, leaving only walls, wall hatch, appliances, and windows. Draw a set of temporary lines to block off doorways so the hatch does not bleed from one room into another. Finally, create the carpeting/flooring plan shown in Fig. 15.76.

FIGURE 15.76 Carpeting and flooring layout.

SUMMARY

You should understand and know how to use the following concepts and commands before moving on to Chapter 16:

- Align
- Action Recording
- Audit and recover
- Blend
- Break and join
- CAD Standards
- Calculator
- Defpoints
- Divide and point style
- Donut
- Draw Order
- eTransmit
- Filter
- Hyperlink
- Lengthen
- Object snap tracking (OTRACK)
- Overkill
- Point and node
- Publish
- Raster
- Revcloud
- Sheet sets
- Selection methods
- Stretch
- System variables
- Tables
- Tool palette
- UCS and crosshair rotation
- Window tiling
- Wipeout

REVIEW QUESTIONS

Answer the following based on what you learned in this chapter:

1. What are the purposes of the audit and recover commands? What is the difference?
2. For join to work, lines have to be what?
3. The Defpoints layer appears when you create what?
4. Does the divide command actually cut an object into pieces?
5. What is the idea behind eTransmit?
6. To what can you attach a hyperlink? What can you call up using a hyperlink? How do you activate a hyperlink?
7. What is the idea behind overkill?
8. Can you make a revcloud out of a rectangle? A circle?
9. What three new selection methods are discussed?
10. Do you use a Crossing or a Window with the stretch command?
11. Name a system variable mentioned as an example.
12. Explain the importance of the wipeout command. What does it add to the drawing?

EXERCISES

1. Draw the following two rectangles according to the dimensions given. Pedit the width to 0.25″ and shade each separately with a solid hatch, color 9. Then, use align to place the 10″ box on top of the 4″ box. (Difficulty level: Easy; Time to completion: 5 minutes.)

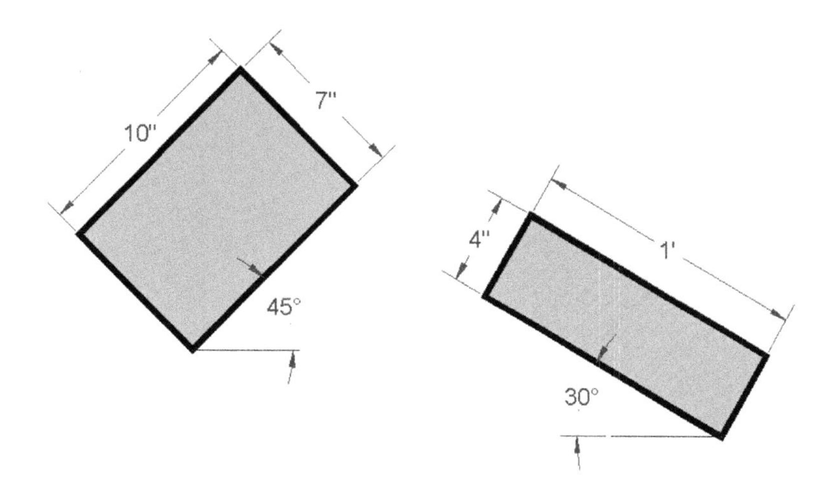

2. Draw the following line and divide it into seven sections using six points of the style shown. Then, pedit the line to a thickness of 0.25″ and break it using the break command (First point option) at each of the points. Finally, offset each alternate section 0.5″ up and 0.5″ down, as shown. (Difficulty level: Easy; Time to completion: 5–7 minutes.)

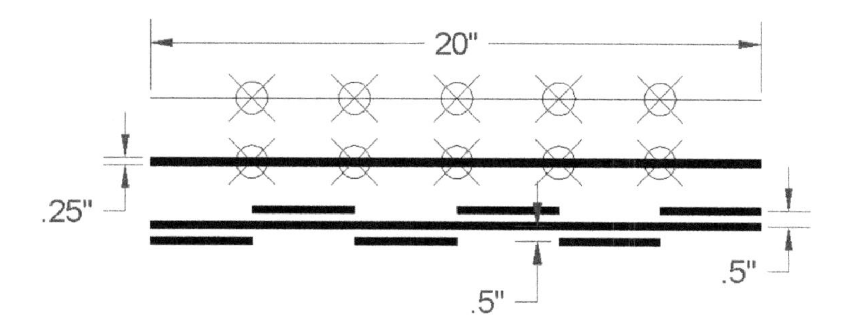

3. Create the following donuts according to the sizing shown. Include the linework for the truck itself. (Difficulty level: Easy; Time to completion: 5–10 minutes.)

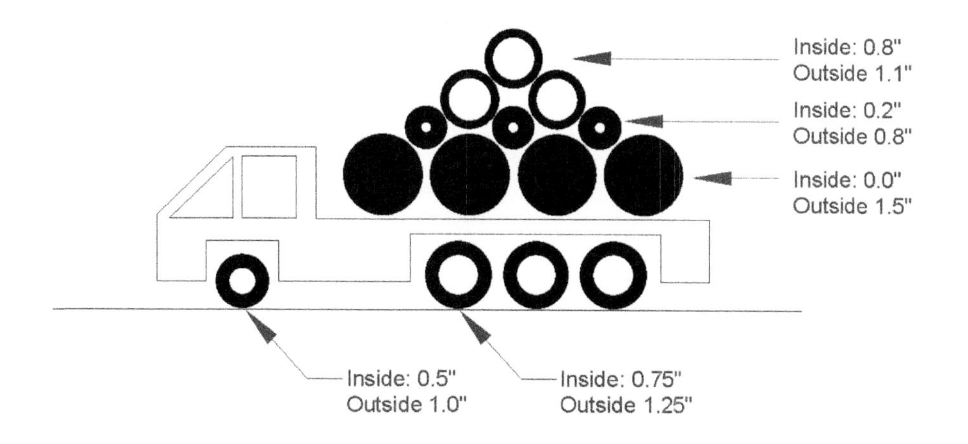

4. Draw the following rectangle, circle, and triangle according to the dimensions shown. Then, use revcloud and the Object option to turn each into a revcloud. Finally, pedit the new shapes to a thickness of 0.25″. (Difficulty level: Easy; Time to completion: <5 minutes.)

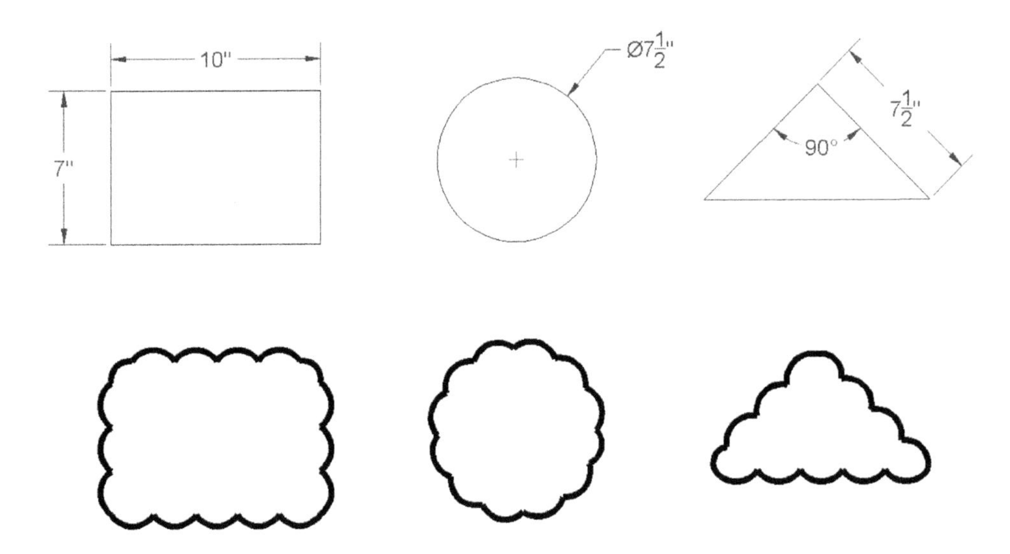

5. Use the OTRACK tool to connect the following shapes as shown. The exact sizing of the shapes is not important, but do try to locate them relative to each other as shown in the exercise. (Difficulty level: Easy; Time to completion: 5 minutes.)

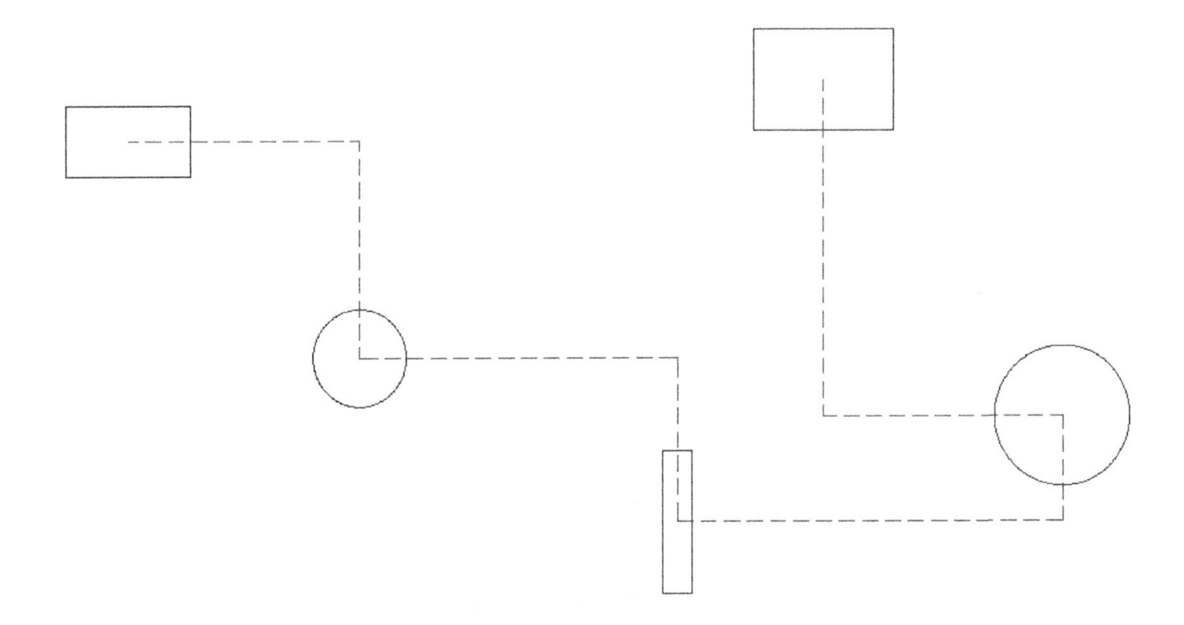

6. Draw a 4 × 4 rectangular array made up of 5″ circles. Then, scale every other one by 50% and shade the smaller circles with a solid hatch, as shown in the image on the left. Next, erase the shaded circles all at once using the erase command in combination with a Window Polygon. Your selection should look somewhat similar to the image on the right. Repeat again with a Crossing Polygon, then a Fence. (Difficulty level: Easy; Time to completion: 5–10 minutes.)

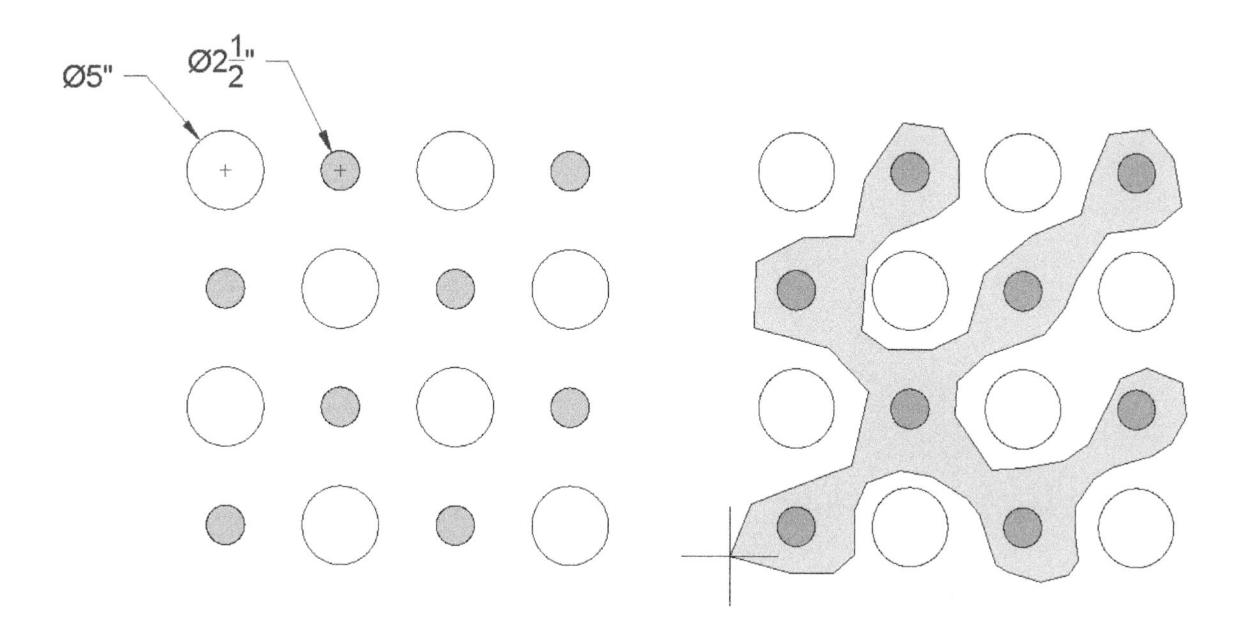

7. Recreate Exercise 1 of Chapter 4, (a table) using the table command as opposed to the basic lines and text approach of the original exercise. The table is reproduced here for your convenience. (Difficulty level: Intermediate; Time to completion: 10–15 minutes.)

		LIST OF MAIN PARTS		
ITEM	QTY	DESCRIPTION	SIZE DWG	DWG. #
1	1	SKID ASSEMBLY	A1	83006
2	1	STRAINER ASSEMBLY	A1	83008
3	1	GENERATOR ASSEMBLY	A1	59470-001
4	1	GENERATOR ASSEMBLY	D1	59470-002
5	2	POWER SUPPLY (OIL IMMERSED)	B1	58717-001
6	1	POWER SUPPLY (OIL IMMERSED)	B1	58717-002
7	1	LOCAL CONTROL PANEL ASSEMBLY	D1	83017-001
8	2	LOCAL CONTROL PANEL ASSEMBLY	A1	88017-002
9	1	INSTRUMENT AIR ASSEMBLY	A1	83022
10	1	AIR PUMP	A1	86584

Spotlight On: Chemical Engineering

Chemical engineering is a unique branch of engineering that combines chemistry, physics, and sometimes biology to process raw materials (often chemicals) into useful products. Chemical engineers are broadly divided into two categories. The category most people associate with the profession is "process engineering." Here, the focus is on the design and operations of industrial plants. The second category, less known, but just as important, is "product engineering." Here, the engineers are tasked with development of chemical substances for products such as cleaners, pharmaceuticals, and other items, including food and beverages.

Often students ask how exactly chemical engineering is different from chemistry. The difference is that chemists generally work with basic compositions of materials, on the atomic and molecular levels, using this knowledge to synthesize products. Chemical engineers (and some *applied* chemists), however, are more interested in developing higher-level "end user" products and industrial processes. These processes utilize scientific principles but do not develop them. In short: chemists come up with our fundamental knowledge; chemical engineers apply it in ways useful to society.

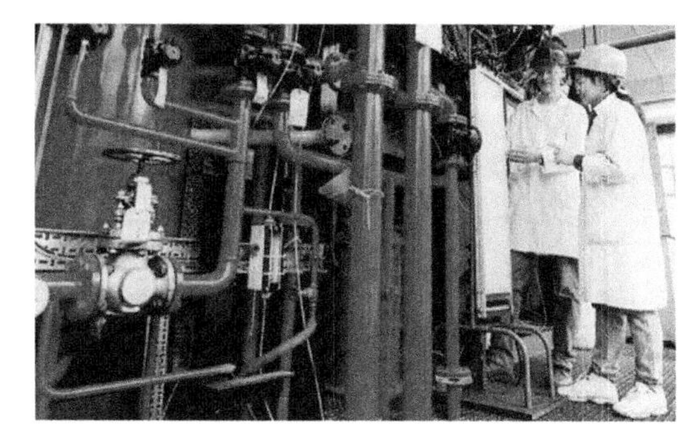

FIGURE 1 *Image source: www.omanengineering.blogspot.com.*

Education for chemical engineers starts out similar to education for the other engineering disciplines. In the United States, all engineers typically have to attend a 4-year ABET-accredited school for their entry-level degree, a Bachelor of Science. While there, all students go through a somewhat similar program in their first 2 years, regardless of future specialization. Classes for chemical engineers include extensive math, physics, chemistry, and thermodynamics. In their final 2 years, chemical engineers specialize by taking courses in biochemistry, kinetics, transport processes, polymer science, and materials, among others.

FIGURE 2 *Image source: www.chem-eng.blogspot.com.*

On graduation, chemical engineers can immediately enter the workforce or go on to graduate school. Although not required, some engineers choose to pursue a Professional Engineer (P.E.) license. The process first involves passing a Fundamentals of Engineering (F.E.) exam, followed by several years of work experience under a registered P.E., finally sitting for the Chemical Engineering P.E. exam itself.

Chemical engineers can generally expect starting salaries in the $65,000/year range with a bachelor's degree and $77,000 with a master's degree, which is one of the highest among engineering specialties, just behind petroleum engineering, which in itself is a field that attracts some chemical engineers. The pay with 10 years of experience is over $95,000. This, of course, depends highly on market demand and location. A master's degree is highly desirable and required for many management spots.

So, how do chemical engineers use AutoCAD and what can you expect? One major application of AutoCAD in this profession is P&ID. This is a topic that was mentioned in several Level 1 chapters, with a few corresponding exercises. P&ID stands for "piping and instrumentation diagrams," and these are commonly used in chemical and process engineering with respect to plant design. A basic P&ID diagram is shown in Fig. 3.

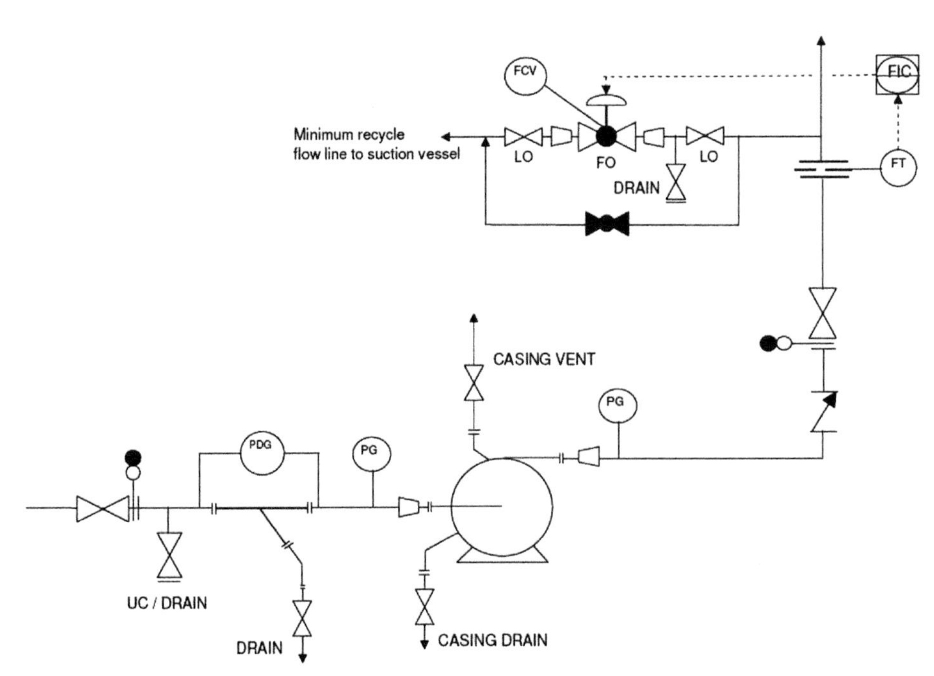

FIGURE 3 Basic P&ID diagram. *Image source: Elliot Gindis.*

Much like electrical schematics, most P&IDs are 2D in nature (though 3D models are used on occasion) and inherently simple from an AutoCAD perspective, though probably far from simple in the engineering sense. The key here is accuracy and precision! The schematics have to be easy to understand and read, and that may be the biggest challenge and responsibility of drafting—to present a complex design in an easy-to-read manner. The layering system is relatively straightforward, with layer names reflecting the functionality of the pieces. Expect to see modular design, symbol libraries, and data-embedded blocks, a.k.a. *attributes*, in this field.

Autodesk has an add-on package to automate and simplify working on P&IDs called, unsurprisingly, *AutoCAD P&ID*. It has tools to assist with error checking, managing data, and symbol libraries. Autodesk also makes Plant 3D, for design and visualization of plant design. There are of course other companies in the market, such as PROCAD, Bentley and its OpenPlant software, Siemens with Cosmos P&ID, and Intergraph's SmartPlant. If your career takes you into this field, you will likely encounter these design and testing packages.

Chapter 16

Importing and Exporting Data

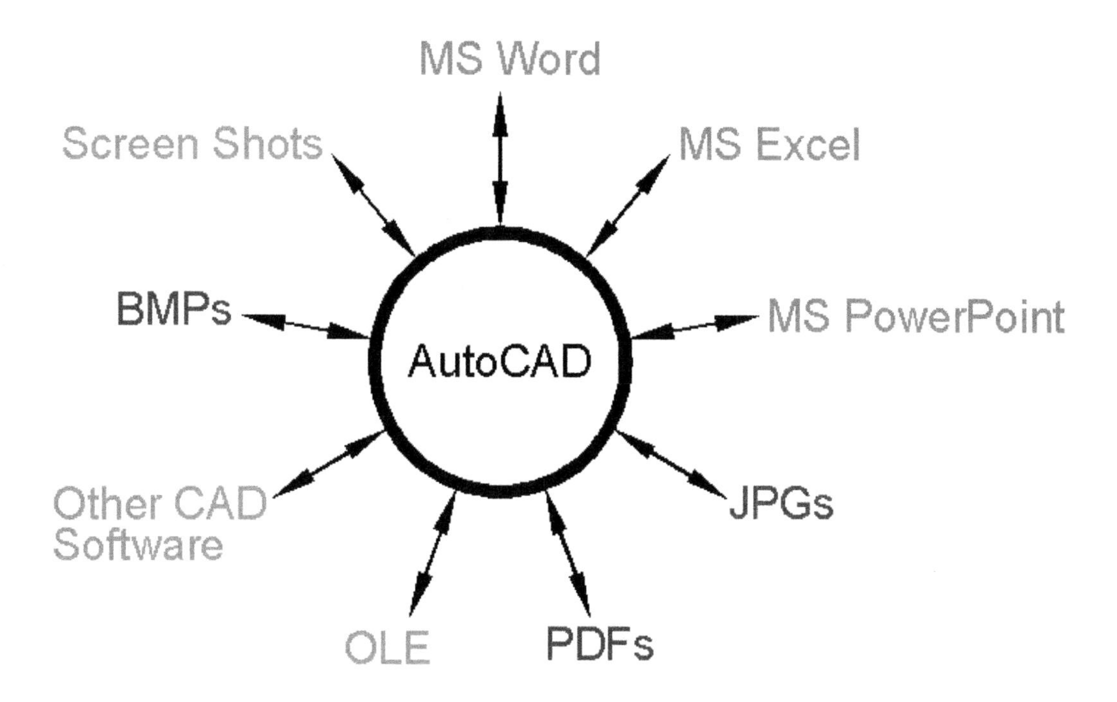

Learning Objectives

In this chapter, we learn how AutoCAD interacts with other software you are likely to use in the course of design work. We cover the following topics:

- AutoCAD and MS Word
- AutoCAD and MS Excel
- AutoCAD and MS PowerPoint
- AutoCAD and JPGs
- AutoCAD, PDFs, and screen shots
- AutoCAD and other CAD software
- Exporting and Save As
- Importing and OLE (object linked embedded)

By the end of this chapter, you will smoothly import and export data among a variety of common office and design applications.

Estimated time for completion of this chapter: 2 hours.

Up and Running with AutoCAD 2019. https://doi.org/10.1016/B978-0-12-816440-2.00016-1

16.1 INTRODUCTION TO IMPORTING AND EXPORTING DATA

AutoCAD and the designers who use it usually do not work in a "software vacuum." They typically interact with not only AutoCAD but also a number of other applications, some of them generic and used by many other professions, others unique to their own field. As such, there is often a need to either import data into or export data out of AutoCAD. Examples are numerous, such as an architect who wishes to transfer notes typed in MS Word into a project details page in AutoCAD or an engineer who needs to drop in an Excel spreadsheet. Just as often, AutoCAD drawings may need to be inserted into MS Word reports, PowerPoint presentations, and other applications. You may then need to create PDFs of your drawing or insert a PDF or an image into one.

Swapping files among the various software applications has been tricky in the past. Software companies are under no obligation to make such tasks flow smoothly, as it is not their business what else you may have on your PC (which may include programs from a direct competitor). Of course, such thinking scores no points with the customer, so developers do try to minimize compatibility issues whenever possible. Today, such tasks flow much smoother but still require knowledge of procedures that can be unique to each application.

The sheer quantity of software on the market is staggering, so we limit the discussion to

- Software or file types that are used or encountered by virtually everyone who uses a computer (Word, Excel, PowerPoint, JPG, and PDF).
- Software that has special relevance to AutoCAD users (such as other CAD programs).

We focus on two distinct tasks, how to *import* files *from* these applications and how to *export* AutoCAD files *into* them, analyzing each software pairing bidirectionally, and conclude with a brief overview of the export and insert features and OLE.

16.2 IMPORTING AND EXPORTING TO AND FROM MS OFFICE APPLICATIONS

We begin with MS Word, Excel, and PowerPoint, as those are probably the ones most important and relevant to the majority of users. Throughout, unless noted otherwise, we make extensive use of Copy and Paste. These Copy/Paste to clipboard tools are some of the most useful in computing, allowing you to shift around just about anything from one spot to another. The keyboard shortcuts are Ctrl+C for copy and Ctrl+V for paste or you can just right-click and select Copy and Paste.

Be aware that the "copy" is not AutoCAD's standard copy command, so do not use that. Yes, some students mix these up, so let us be clear from the start. Open up Word, Excel, and PowerPoint and keep them handy as we go through each one.

Word Into AutoCAD

An example of this may be when you would like to bring in extensive typed construction notes onto one of the pages of the AutoCAD layout.

Go ahead and click and drag to highlight the text in Word first and right-click or Ctrl+C to copy to the clipboard. Go to the AutoCAD screen but do not right-click and paste in the text. If you do this, it appears as an embedded object, with a white background (seen in the top line of Fig. 16.1) and cannot be edited. Instead, create an mtext field of an appropriate size and paste the copied text into it. The result is easily editable text that can be further formatted, as seen in the bottom line of Fig. 16.1.

The quick brown fox jumped over the lazy dog.
The quick brown fox jumped over the lazy dog.

FIGURE 16.1 Insertion of MS Word text.

AutoCAD Into Word

An example of this may be when you are preparing a report and would like to include some of the graphics spread out among the text to enhance clarity or support documentation.

The easiest way is to use the PrtScn (print screen) key, Microsoft's Snipping Tool, or other screen capture software. Make sure your screen background is set to white before doing any screen capture. You can also use the PublishToWeb JPG or PublishToWeb PNG printers that are built into AutoCAD.

Once inserted, some cropping may be necessary to get rid of excess white space. Select the image and then right-click to bring up a shortcut menu with Wrap, Size, and Format options. A typical view of this procedure is shown in Fig. 16.2 in Microsoft Word 2016.

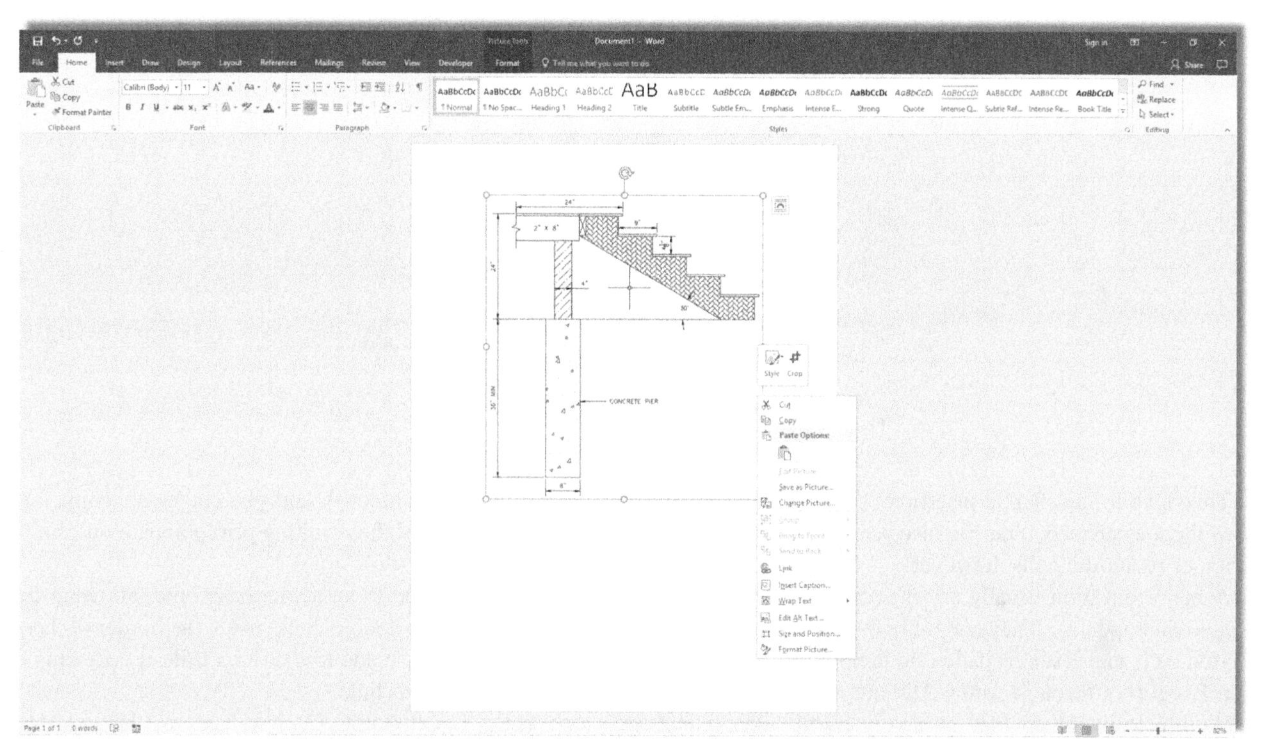

FIGURE 16.2 Insertion of an AutoCAD design into an MS Word document.

Excel Into AutoCAD

An example of this may be when you would like to bring a spreadsheet into your electrical engineering drawing that details the type of connections and wiring used in a design.

Once again, you use Copy/Paste. In Excel, create a small group of cells, number them (adding borders if desired), then highlight all, right-click Copy, go into AutoCAD, and right-click Paste. As soon as you do this, pick an insertion and the spreadsheet appears as shown in Fig. 16.3.

1	2
3	4
5	6
7	8
9	10

FIGURE 16.3 Copy/Paste Excel into AutoCAD.

If you double-click the border of the inserted Excel sheet, the properties palette will appear as shown in Fig. 16.4.

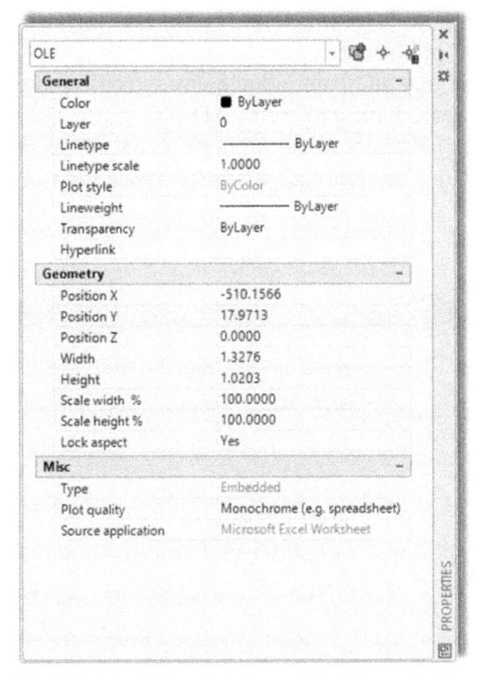

FIGURE 16.4 Object linked embedded Properties Panel.

This is your basic Excel insertion. The cells are regular objects (very much like blocks), and you can erase, copy, and move them as needed. You can also scale them by clicking once and moving around the resulting grip points. You cannot mirror or rotate the cells, however.

A major question usually arises at this point. Are the inserted Excel cells linked to an actual Excel file, allowing for changes and updates? They are not in the true sense of the word. What you can do is double-click inside the border of these inserted cells and Excel is called up temporarily for editing purposes. Any changes made to that Excel file if accessing it from Excel (not through AutoCAD) are not reflected in AutoCAD. There is no real link.

Having truly linked files is a very useful idea. There are some fancy Excel/AutoCAD interfaces in the industry, where extensive Excel data are used to generate corresponding AutoCAD drawings, but this trick requires additional programming and extensive customization and is beyond the scope of this book.

AutoCAD Into Excel

This action basically pastes an image into Excel with the same Copy/Paste procedure as for Word. This, however, is rarely done, as Excel is primarily for spreadsheet data; and unless you need a supporting drawing, there is just no overwhelming reason to do this.

PowerPoint Into AutoCAD

This action is not very common, as PowerPoint is almost always the destination for text and graphics, not an intermediate step to be inserted into something else. However, if it is needed, you can insert slides into AutoCAD via regular Copy/Paste.

AutoCAD Into PowerPoint

This makes a lot more sense, as you may want to insert a drawing into a PowerPoint presentation. Fortunately, PowerPoint behaves very much like Word in this particular case, and one screen capture followed by some cropping is all it takes to insert an AutoCAD design, and no new techniques need to be learned.

16.3 SCREEN SHOTS

Screen shots or screen captures refer to capturing a snapshot of everything on your screen as seen by you, the user, and sent to memory, to be inserted somewhere. This is done by simply pressing the Print Screen button, which on most computers is called *PrtScn*, *SysRq*, or sometimes just *F12*. You can capture what is on your screen this way and insert it into Word or PowerPoint and then crop out what you do not need to see. While many computer users may know of the Print Screen function to actually print something, they may overlook this useful trick for inserting data into other applications.

16.4 JPGS

This file compression format (used most often with photographs) can be inserted directly into AutoCAD and is a great idea if you would like a picture of what you drafted next to the drawing. Examples include an aerial photo of a site plan for architectural and civil applications and a photo of an engineering design after manufacture for as-built or record drawings. Inserting JPGs is a straightforward process.

Step 1. In Windows file manager, click on the nameof a file you want to insert, right-click for the pop-up menu, and select Copy.

Step 2. Then in AutoCAD right-click for a pop-up menu, select Clipboard > Paste it into AutoCAD.

- AutoCAD asks for an insertion point: `Specify insertion point <0,0>:`

Step 3. Click anywhere you like.

- AutoCAD then asks for the image size: `Base image size: Width: 0.00, Height: 0.00 Inches`
  `Specify scale factor or [Unit] <1>:`

Step 4. Scale the JPG by moving the mouse to any size you wish (or type in a value or simply press Enter).

- AutoCAD then asks for the rotation angle: `Specify rotation angle <0>:`

Step 5. If you want the image rotated, enter a degree value; otherwise, press Enter.

The result of a JPG insertion is shown in Fig. 16.5 (all toolbars are removed for clarity). The edge has then been selected (one click), revealing the Ribbon's Image Editor. The image can be erased, copied, moved, rotated, mirrored, and just about anything else you can do to an element. You can also do some very basic image adjusting via the Ribbon or by double-clicking on the picture. The dialog box shown in Fig. 16.6 appears, allowing you to adjust Brightness, Contrast, and Fade, as well as reset everything. Any changes apply only to that JPG, not copies of it (if other copies exist).

FIGURE 16.5 Embedded JPG.

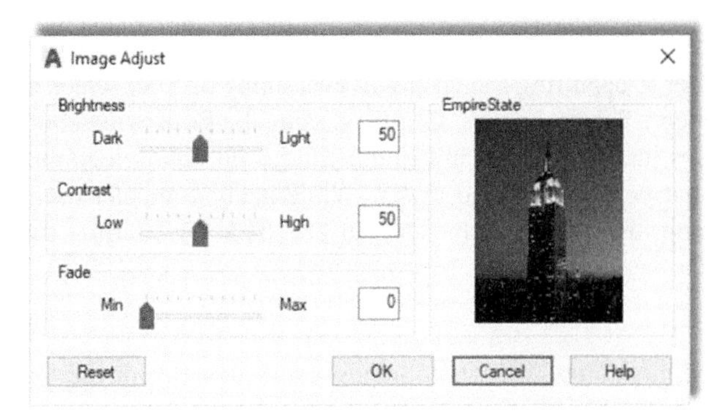

FIGURE 16.6 Image Adjust.

16.5 PDFS

The Portable Document Format (PDF) is, of course, Adobe's popular software for document exchange. AutoCAD always had a good relationship with PDF, allowing users to easily convert drawings by simply printing to PDF—in other words, selecting a PDF "printer" built into AutoCAD, installed with Windows, other PDF printers or, of course, Adobe Acrobat. The Print function would then execute, and AutoCAD asks you where you want the file saved. Once the destination was selected, the PDF file was created. This process was much the same with Word and other software, and continues, with even more enhancements introduced in first AutoCAD 2018.

Another way to generate a PDF, especially if you need to customize the output (or you do not have Acrobat installed), is via the Ribbon's Output tab → Export approach. If you select the PDF option, a "Save As PDF" dialog box appears and allows some tinkering with options and output to finesse the results, as seen in Fig. 16.7.

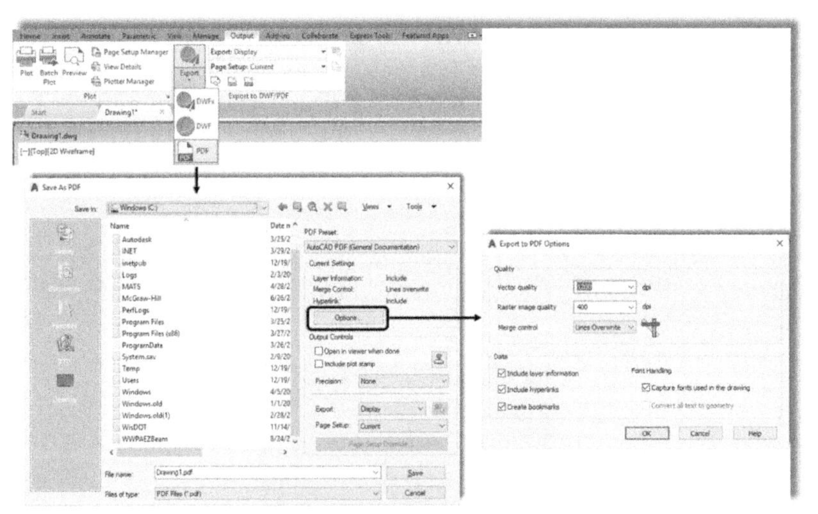

FIGURE 16.7 Save As PDF with Options.

AutoCAD 2010 gained the ability to bring in a PDF file as an underlay. This meant that it came into AutoCAD as a "background," which you could see but not edit (at least not back then). This is not unlike raster images, mentioned in Chapter 15, and was quite useful if all you have of a design is a PDF. After insertion you drew over it or used it as a supporting image, such as a key plan. You can still do all this with the current AutoCAD release.

To try it out and bring in a PDF, use the cascading menu Insert → PDF Underlay…. You then are prompted to look for the file you want to insert (that dialog box is not shown), and once you select the PDF, you will see the dialog box in Fig. 16.8.

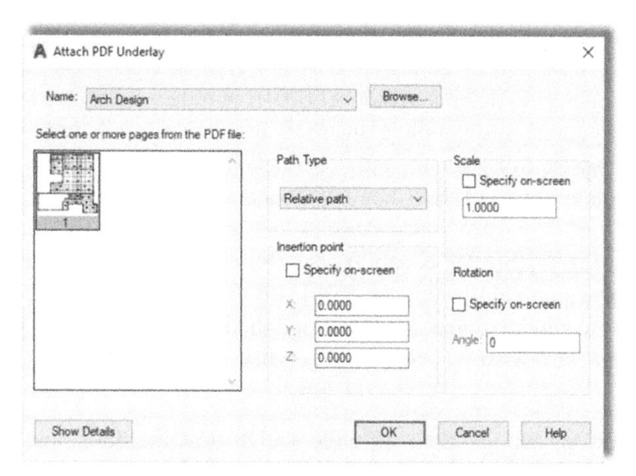

FIGURE 16.8 Attach PDF Underlay.

Once you press OK, the PDF inserts into the AutoCAD file and the box prompts you for an insertion point and scale factor. Once the image is embedded, you can click on it and the Ribbon changes to show you some of the image editing options, as seen in Fig. 16.9, with Change to Monochrome and some Fade and Contrast adjustments among the more useful tools. The PDF is now part of your drawing and you can draw over it and delete it later if needed. It remains an embedded feature and is not editable itself, so the biggest use for this is to serve as a background to redraw the image.

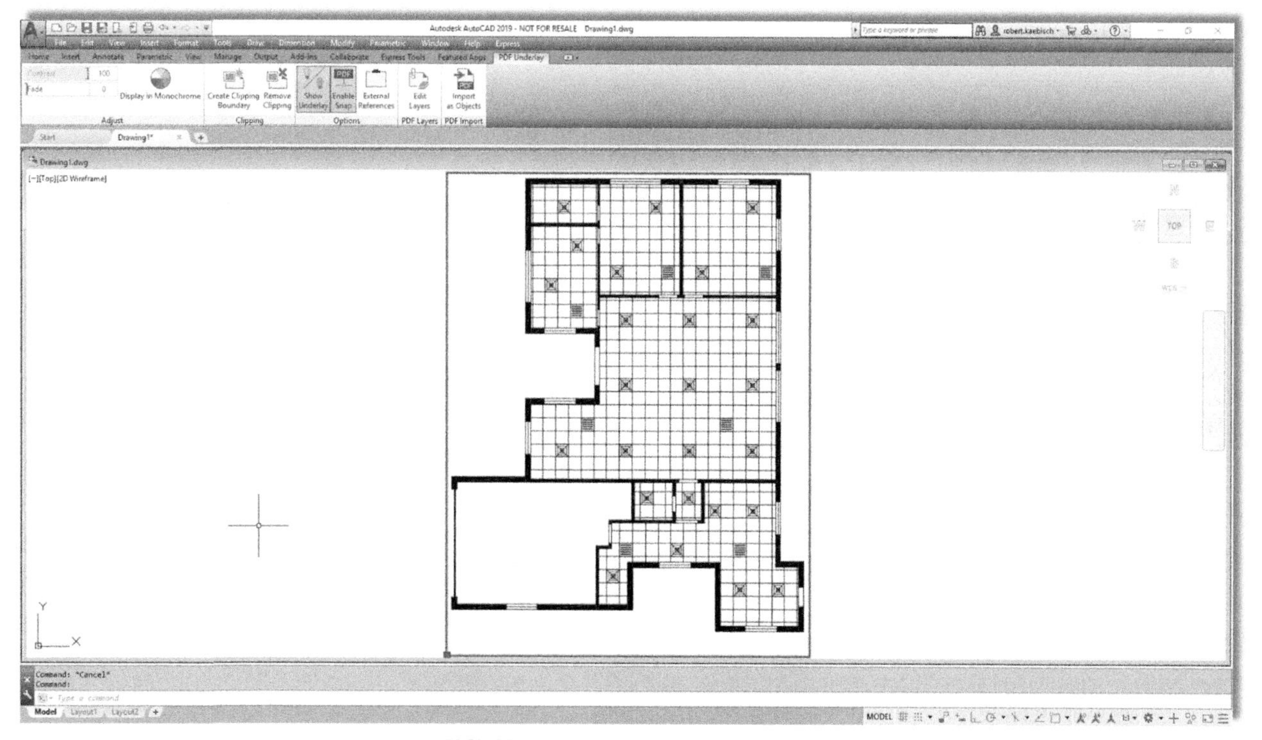

FIGURE 16.9 PDF Underlay Ribbon menu.

A question that arises often is how to go backward, in other words convert a PDF document back into AutoCAD. There is considerable need for this in the industry. Some drawings exist only in PDF format with the original dwg files lost or unavailable, and you may need to convert a PDF into a useable drawing form. Prior to AutoCAD 2017, there were basically two ways to do this: via online websites or via desktop applications. In all cases, the quality of your conversion depends on the method used to create the PDF in the first place. If the PDF was a scan of a hand drawing, then the conversion software may not be able to accurately tell what is what. If the PDF is a print of a CAD file, things are much easier, and you can get close to a 100% correct conversion.

If this is something you need done, and you have AutoCAD 2016 or older, research a bit online for both methods. Web-based (and usually free) converters require you to upload the PDF and they return a dwg. This method is usually limited as the website is trying to get you to buy the full software and is interested in having you only try it out online. If you purchase a PC software package (most are pretty low cost, less than $100), then you can convert directly on your computer, without going online.

Autodesk must have seen a glaring gap in PDF capability and introduced some new tools starting with AutoCAD 2017. You can now convert PDFs directly into dwg files via the PDFIMPORT command. To be fair, it is not a full converter but rather more of a handy new import feature that goes a step beyond the basic underlay and recognizes features such as lines, circles, and text. Keep in mind, however, what was mentioned earlier—the better the source file, the better the conversion. Let us give this a try.

Step 1. Open up a blank AutoCAD file and also set aside a PDF of some floor plan in a folder.

Step 2. You can start up the command via the Ribbon → Insert → Import → PDF Import command or the Application menu (the big red "A" at the top left corner of the screen). Drop that down and select "Import" then "PDF." Alternatively, you can just type in `pdfimport` and press Enter.

- AutoCAD will say: `Select PDF underlay or [File] <File>:`

Step 3. Go ahead and select `F` for `File`.

Step 4. Browse around and select the PDF file you wish to import. Once you click the file name, the Import PDF dialog box (Fig. 16.10 appears).

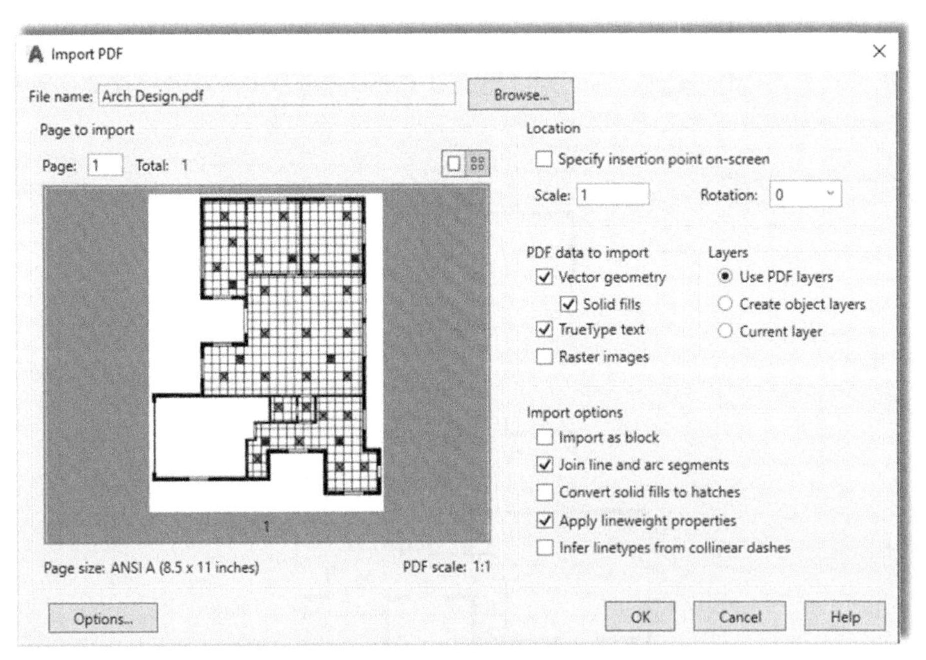

FIGURE 16.10 Import PDF.

Step 5. Review the information on the right of the dialog box and click OK. If all is well, and AutoCAD likes the PDF file, it will be converted into vector elements (lines, circles, arcs, text, hatching, etc.).

- AutoCAD will say: `Importing page 1 of PDF file:` (your file path and name...)

Take a close look at the new CAD file and try to erase or otherwise manipulate some of the features; it is indeed a true AutoCAD drawing now. The layering generally is not to be brought in; and if you bring up the layer dialog box, you will see generic PDF_Geometry, PDF_Solid Fills, and similar type of layers instead of the custom ones the original drawing may have had.

While this new feature is certainly a welcome addition to AutoCAD, keep in mind that it is not a 100% solution. Text does not always import correctly and may just come in as splines. Hatches are also made up of pieces as opposed to solid fills. These issues may or may not be present with more advanced software available from third-party vendors—it is not possible to evaluate them all. Autodesk may also enhance this import feature in future releases. However, the bottom line is that importing, converting, and then "fixing up" a few items in a PDF is still preferable to redrawing the entire file!

16.6 OTHER CAD SOFTWARE

Initially, as AutoCAD gained market dominance, it did not really attempt, nor had any incentive, to "play nice" with other CAD software on the market. Being the 2D industry leader allowed AutoCAD to get away with not accepting any other files in their native format. It was the company's way or the highway, but that slowly changed. One major concession in regard to importing and exporting was with MicroStation, AutoCAD's nemesis and only serious competitor. MicroStation always opened and generated AutoCAD files easily but not the other way around.

Finally, in recent releases, AutoCAD listed the .dgn file format as something you can import and export. You were then able to bring in MicroStation files, more or less intact, as well as create ones from AutoCAD drawings. In AutoCAD 2009, the settings box was expanded and that was carried over into the current AutoCAD version. You can now compare layers and other features side by side and make adjustments so the target file is similar to the original.

To access this, use the cascading menu File → Import… and search for the MicroStation dgn file you are interested in importing. Another way around all of this is just to have the MicroStation user generate the AutoCAD file. You can also export .dwg files directly to .dgn as well by using the cascading menu File → Export… and selecting .dgn as the target file. You then see the Settings dialog box in Fig. 16.11. Of course, MicroStation accepts the AutoCAD file with no problem, so this may not be necessary. If you would like to learn a bit more about MicroStation in general, see Appendix B.

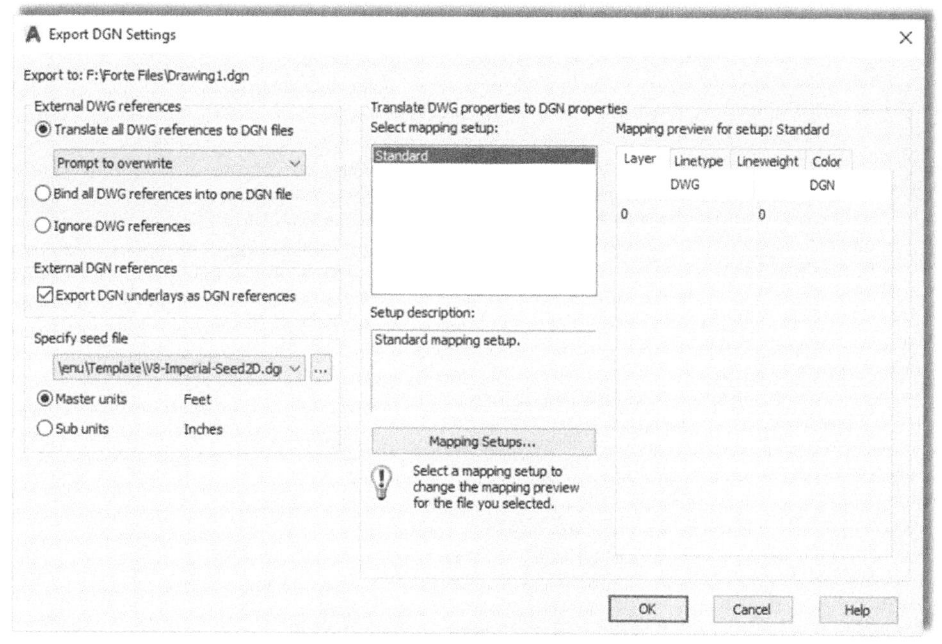

FIGURE 16.11 Export DGN Settings.

Let us expand the discussion to other CAD software. What if someone is using another CAD program and needs to exchange files with an AutoCAD user? The .dxf file is the solution, as it is a format agreed on by most CAD vendors as the go-between that can be understood by all applications for easy file sharing. AutoCAD easily opens a .dxf or generates its own .dxf files for any drawing. This is discussed momentarily.

A few observations on sharing files: Among architects, the issue is not much of a problem, as AutoCAD is the main application in this profession, and competitors such as ArchiCAD easily open AutoCAD's files and send a .dxf right back if needed. In engineering, however, AutoCAD has a lesser presence. While some electrical, mechanical, and civil engineers use AutoCAD, many others, especially industrial, rail, aerospace, automotive, and naval engineers, use 3D modeling and analysis software instead, such as CATIA, NX, Pro/Engineer, or SolidWorks.

Until recently, these applications did not interact well with AutoCAD. Their respective "kernels" (the software's core architecture) are significantly different in design and intent from AutoCAD's ACIS 3D kernel. Their native and export file types (such as IGES and STEP) may also be unfamiliar to AutoCAD users. Some of the previously mentioned software can generate .dxf files, but those files are "flattened" and what you get is a 2D snapshot of what used to be a 3D model. AutoCAD was also able to import images from CATIA and SolidWorks using the OLE command (to be discussed soon), but once again, they were just flattened images.

Things improved somewhat with AutoCAD 2012. The 3D models can now be imported into AutoCAD with little loss in fidelity. Much more is said about this in the free downloadable 3D section of this book.

All this can be somewhat confusing. Because so many different CAD packages are on the market, AutoCAD designers need to be aware and up to date on what others in their industry are using and think on their feet when it comes time to exchange files, as there usually is a way to access most file types. Let us now take a look at the specifics of exporting, importing, and the OLE command to explore the available options.

16.7 EXPORTING AND THE SAVE AS FEATURE

As previously with the MicroStation example, exporting data can be done by selecting File → Export... from the cascading menu. What you see is the dialog box in Fig. 16.12.

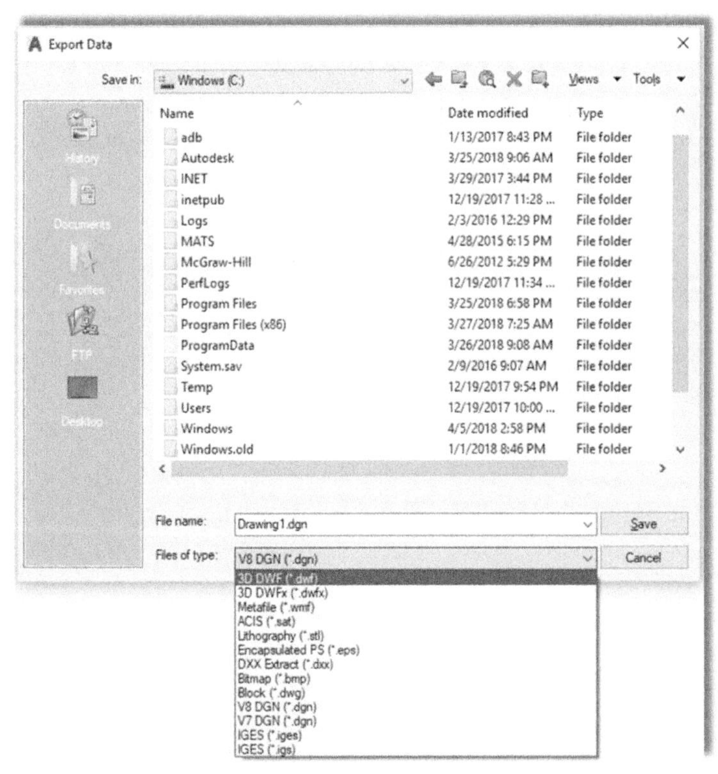

FIGURE 16.12 Export Data dialog box.

Your drawing can be exported to the following formats:

- *3D DWF* or *DWFx*: This is to convert the drawing to web format for use on viewers.
- *FBX*: FBX stands for Filmbox and is a proprietary file format developed by Kaydara and now owned by Autodesk. It is used to provide interoperability between digital content creation applications.
- *Metafile*: This is to convert the drawing to a Windows Meta File, a Microsoft graphics format for use with both vectors and bitmaps.
- *ACIS*: ACIS is the 3D solid modeling kernel of AutoCAD as well as a format to which a drawing can be exported. Few other software currently use the ACIS kernel (SolidEdge comes to mind), as Parasolid is the industry standard, so this is a rarely used export format.
- *Lithography*: Stereo lithography files (*.stl) are the industry standard for rapid prototyping and can be exported from most 3D CAD applications, including AutoCAD. Basically, it is a file that uses a mesh of triangles to form the shell of your solid object, where each triangle shares common sides and vertices.
- *Encapsulated PS*: The eps is a standard format for importing and exporting PostScript language files in all environments, allowing a drawing to be embedded as an illustration.
- *DXX Extract*: DXX stands for Drawing Interchange Attribute. This is a .dxf file with only block and attributes information and is not relevant to most users.
- *Bitmap*: This converts the drawing from vector form to bitmap form, with the resulting pixilation of the linework; it is not recommended.
- *Block*: This is another way to create the familiar block covered in Level 1.
- *V8 DGN*: MicroStation Version 8, as described previously.
- *V7 DGN*: MicroStation Version 7, as described previously.
- *IGES*: IGES stands for Initial Graphics Exchange Specification and is a vendor-neutral format for 3D CAD solid modeling files. The files can end in either .iges or .igs.

The Save As dialog box can be accessed anytime you select File → Save or File → Save As… from the drop-down cascading menus. You can also type in saveas and press Enter. In either case, you get the dialog box shown in Fig. 16.13.

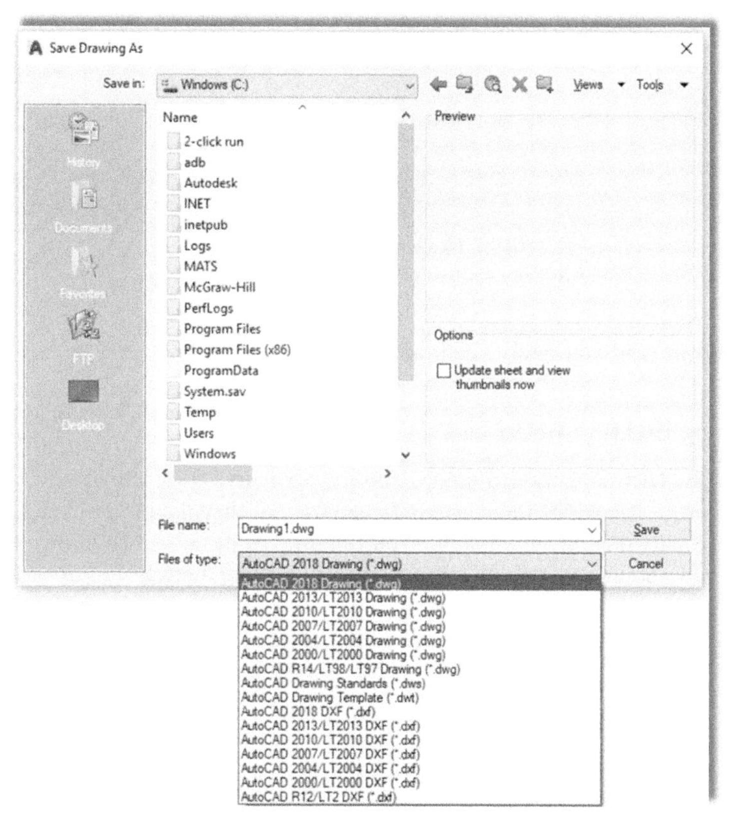

FIGURE 16.13 Save Drawing As dialog box.

The extensions presented here are described in detail in Appendix C, File Extensions, but in summary, you can save your drawing as an older version of AutoCAD (an important step that is automated via a setting in Options, as covered in chapter: Options, Shortcuts, CUI, Design Center, and Express Tools) or as a .dxf file, as described earlier. Note that the .dxf files can also be created going back in time, all the way to Release 12 (that was a mostly DOS-based AutoCAD). The other extensions, .dws and .dwt, are used less often and are also mentioned in Appendix C.

16.8 INSERTING AND OBJECT LINKED EMBEDDED

The final discussion of this chapter features the Insert menu option in general and the OLE concept or command in specific. The Insert cascading menu is shown in Fig. 16.14. You may already be familiar with most of what is featured in this drop-down menu, or it is covered soon as a separate topic:

- Block insertion is the same as typing in `insert` (from Level 1).
- Hyperlinks and raster images are covered in Chapter 15.
- External references are all of Chapter 17.
- Layouts are a Paper Space topic from Chapter 10.

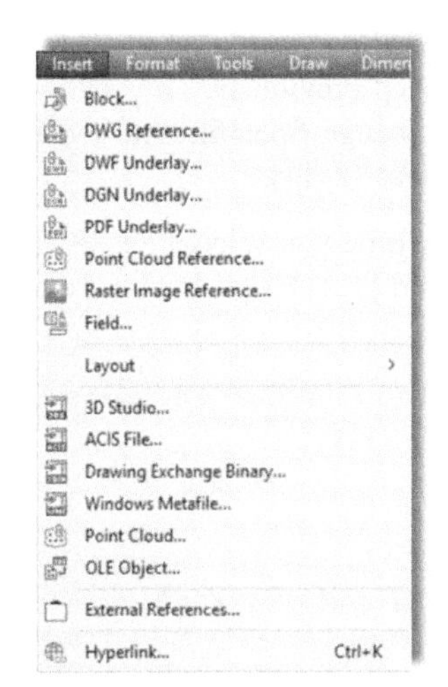

FIGURE 16.14 Insert cascading menu.

Other items we do not cover (and you rarely need) are

- Fields.
- 3D Studio, ACIS, Drawing Exchange Binary, and the like.

Let us then focus on OLE. OLE is a way to embed (insert) nonnative files into AutoCAD on a provisional basis, meaning these files are not part of AutoCAD (they cannot be anyway, as they are of non-AutoCAD formats), but they are merely dropped in and are visually present. In principle, you can drop in just about anything, with just about any results. Some files sit nicely, others do not appear, and still others may even cause a crash. Let us take a look at the dialog box.

Select OLE Object... from the Insert menu, and the dialog box in Fig. 16.15 appears.

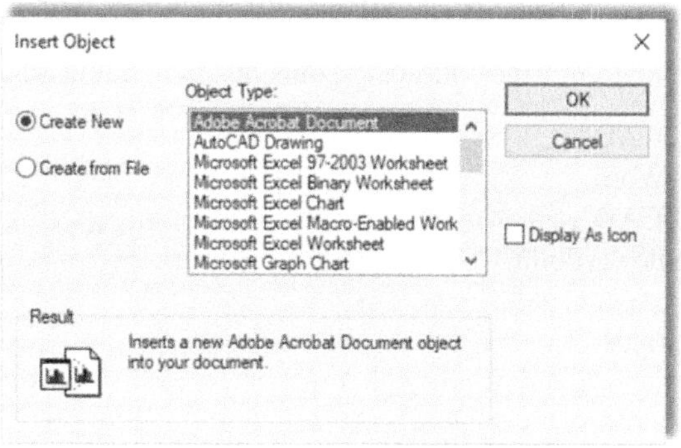

FIGURE 16.15 Insert object.

These, in theory, are the various software file items that can be inserted into AutoCAD. Looking through it, top to bottom, you can see that a wide variety of Adobe, Microsoft, and even Dassault (they own CATIA and SolidWorks) files can be inserted. You can also, in principle, insert video and audio clips (one can envision music files playing on the opening of an AutoCAD drawing) and even Flash animations. Simply pick what you want to insert and select Create from File. Some of the more useful OLE insertions are related to the Microsoft products listed, but much of that topic was already covered in this chapter. To insert a file, select Create from File and browse for the file in which you are interested.

16.9 LEVEL 2 DRAWING PROJECT (6 OF 10): ARCHITECTURAL FLOOR PLAN

For Part 6 of the drawing project, you complete the internal design of the house by adding a reflected ceiling plan (RCP) and a heating, ventilating, air-conditioning (HVAC) system that is made up of diffusers and a few return air ducts. Note that a residential home is not likely to have an RCP, which is so named because it looks exactly as if you held a mirror in your hand and looked down at it to see the ceiling. RCPs are mostly part of commercial space design and include what you generally see in a typical office: white ceiling tiles, lights, and HVAC diffusers. Here, for educational purposes, we add this ceiling into the residential design.

Step 1. Create some appropriate layers, such as
- M-RCP (Gray_8).
- M-Diffuser (Yellow).
- M-Ret_Air (Cyan).

Step 2. Create some HVAC symbols, as shown in Fig. 16.16. Make blocks out of them for easier handling. Be sure to be on that object's layer before making a block.

Step 3. Create the RCP plan by drawing arbitrary vertical and horizontal "starter" lines in the middle of each room and offsetting 2 feet in either direction until you reach the walls. Then, trim as necessary.

Step 4. Add in the diffuser and return air duct symbols as shown in the plan (Fig. 16.17).

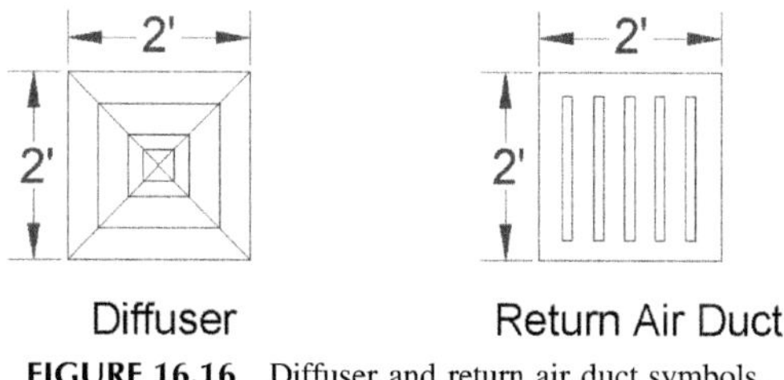

FIGURE 16.16 Diffuser and return air duct symbols.

SUMMARY

You should understand and know how to use the following concepts and commands before moving on to Chapter 17:

- Data transfer from Word to AutoCAD
- Data transfer from AutoCAD to Word
- Data transfer from Excel to AutoCAD
- Data transfer from AutoCAD to PowerPoint
- Screen shots, JPGs, and PDFs
- Interaction with other CAD software
- Exporting and Save As features
- Inserting and OLE

REVIEW QUESTIONS

Answer the following based on what you learned in Chapter16:

1. What is the correct way to bring Word text into AutoCAD?
2. What is the procedure for bringing AutoCAD drawings into Word?
3. What is the correct way to bring Excel data into AutoCAD?
4. What is the procedure for bringing AutoCAD drawings into PowerPoint?
5. What is the procedure for importing JPGs and generating PDFs?
6. Describe the basic ideas of interacting with other CAD software.
7. What are Exporting and Save As?
8. What is OLE?

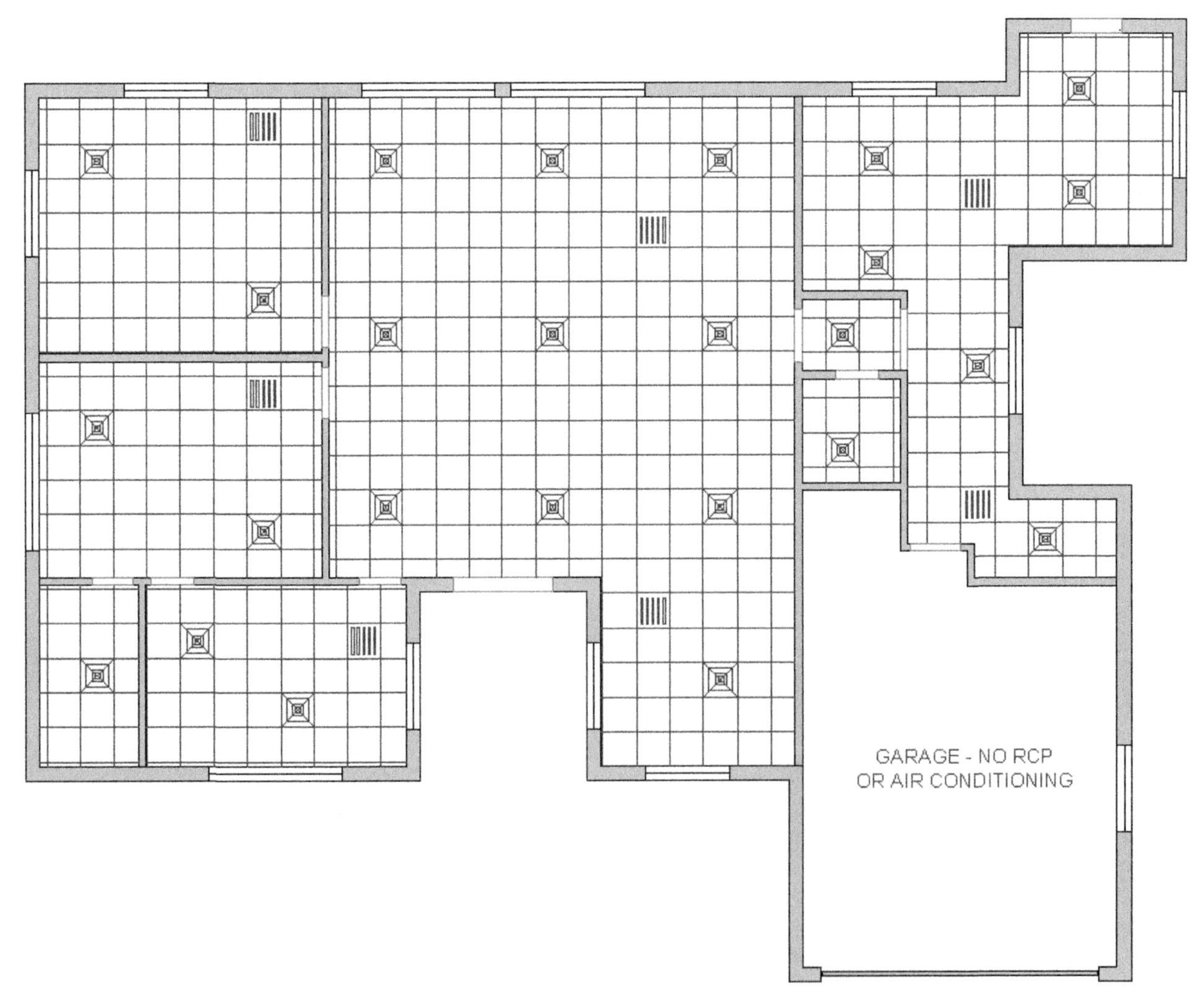

FIGURE 16.17 Reflected ceiling plan and heating, ventilating, air-conditioning layout.

EXERCISE

1. At your own pace, review everything presented in this chapter and do the following:
   - Type any short paragraph in MS Word and import it into AutoCAD correctly, using Copy and Paste.
   - Type in several columns of data in MS Excel and then import them into AutoCAD. Adjust them by scaling up and down.
   - Import any image from AutoCAD into MS Word using:
     - The Print Screen function.
     - Copy and Paste.

- Export any drawing to a .dxf format.
- Bring in any JPG (picture) using Copy and Paste. Adjust it in AutoCAD.

(Difficulty level: Easy; Time to completion: 15 minutes.)

Chapter 17

External References

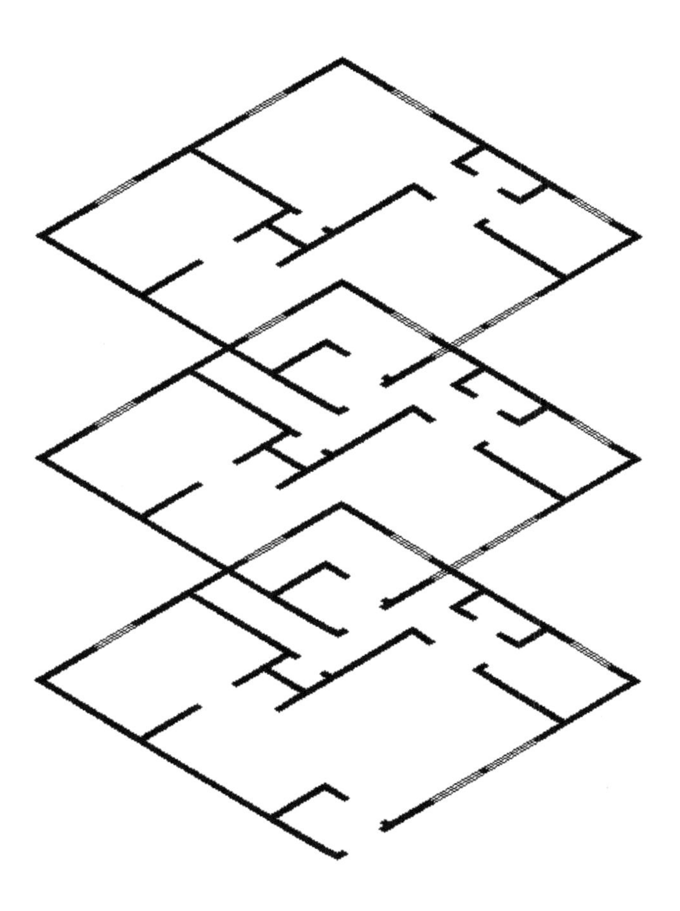

Learning Objectives

In this chapter we introduce and thoroughly cover the concept of external references (xrefs). We specifically introduce the following topics:

- The primary reasons for using xrefs
- Loading xrefs
- Unloading xrefs
- Binding xrefs
- Updating and editing xrefs
- Layers in xrefs
- Multiple xrefs

By the end of this chapter, you will be well versed in applying this critical concept to your design work. Estimated time for completion of this chapter: 1–2 hours.

Up and Running with AutoCAD 2019. https://doi.org/10.1016/B978-0-12-816440-2.00017-3

17.1 INTRODUCTION TO EXTERNAL REFERENCES

External references, *xrefs* for short, are a critically important topic for architectural AutoCAD users and very useful knowledge for all others. Along with Paper Space (see Chapter 10), xrefs form the core of advanced AutoCAD knowledge. Like many advanced topics, you need to understand fundamentally what it is and when to use it. The "button pushing," the mechanics of how to make it work is the easy part. In this introduction, we will discuss what an xref is and why there is a need for this rather clever concept, which has existed since AutoCAD's early days and is a necessary tool with any drafting software.

What Is an External Reference?

As defined, an xref is a file that is electronically attached (referenced) to another new file. It then appears in that new file as a fully visible background, against which you can position new design work. This xref is not really part of the file, but it is merely "electronically paper clipped" to it and cannot be modified in a traditional way; it serves as only a background. The word *xref* is also an action verb, as in "I need to xref that file in."

Why Do We Need an External Reference? What Is the Benefit?

The preceding paragraph was just a strict definition and may not have yet illuminated the real reason for xrefs. To see why we have this concept in AutoCAD, let us propose the following scenario.

You are designing a tall office building. It is boxy in shape (similar to the Aon Center, Chicago), and as such, the exterior, stair towers, and elevator shafts do not change from floor to floor. The interior, however, does. The first floor is a lobby, the second is a restaurant, the third is storage, and floors 4 through 100 are office space. In short, for a designer, once the exterior is done, the rest is all interior-space work. Therefore, going from floor to floor, the exterior design file can be xrefed in and just "hangs out," providing a background for placement of interior walls and other items. This dramatically reduces the size of each file, as a copy of the exterior is merely referenced and never part of the interior files. Got 100 floors to do? No problem; one xref file of the exterior is all that is needed if those floors do not change. To summarize: *Xref is used to reduce file size by attaching a core drawing to multiple files.*

If this were all there was, xref would be a neat trick and would have surely fallen out of favor as computer speed, power, and disk storage space increased over the years. However, the preceding was just a warm-up. The main reason for the xref concept is design changes. If you need to change the shape of the exterior walls, you need to do it only once, and the change propagates through all files using that xref. Think about the implications for a moment. In the old hand-drafting days, each floor had to be drawn separately. Imagine now a last-minute change to the exterior design. Hundreds of sheets would have been updated by hand. Score one for AutoCAD. To summarize: *Xref is used to automate design change updates to core drawings.*

Need another benefit? How about security? The xref file can be placed in a folder that is accessible only to the structural engineer and lead architect, not the interior designer or electrical engineer. Granted, this "need to know" basis for access may not be necessary, but it is good to have a way of keeping someone from shifting structural walls to accommodate the furniture. To summarize: *Xrefs are used to enhance critical file security.*

Ideally, you are now sold on the benefits of an xref. So, how is this all done in practice? The xref is typically not just only the exterior walls but also columns and the interior core (elevators and stairs), as those also tend to not change from floor to floor. The designer creates these files, naming them something descriptive, such as Exterior_Walls_Ref or Columns_Core. Next, the xrefs are attached together and then to the first floor file or to the first floor file individually.

The first floor file is the active file, of course, and is named Floor_One_Arch. At first, it is blank, showing just the xref. Then, the interior design is completed. The designer opens another blank file, attaches the same xref or set of xrefs and names this file Floor_Two_Arch and so on. Xrefs can be attached, detached, refreshed, and bound to the main drawing file. We cover all this shortly. We also cover some xref-related topics, such as layering and methods to edit xrefs in place if absolutely needed.

17.2 USING EXTERNAL REFERENCES

To demonstrate the features of the xref command, we first need to have a building exterior file with which to work. You can select one of your own or one assigned by your instructor, if taking a class. Alternatively (and for good practice), draw the floor plan in Exercise 1 at the end. We use that floor plan for the rest of the xref discussion. In Fig. 17.1, the plan is rotated horizontally and the left corner is moved to point 0,0. All the dimensions are frozen and the file is saved as ExteriorWalls.dwg.

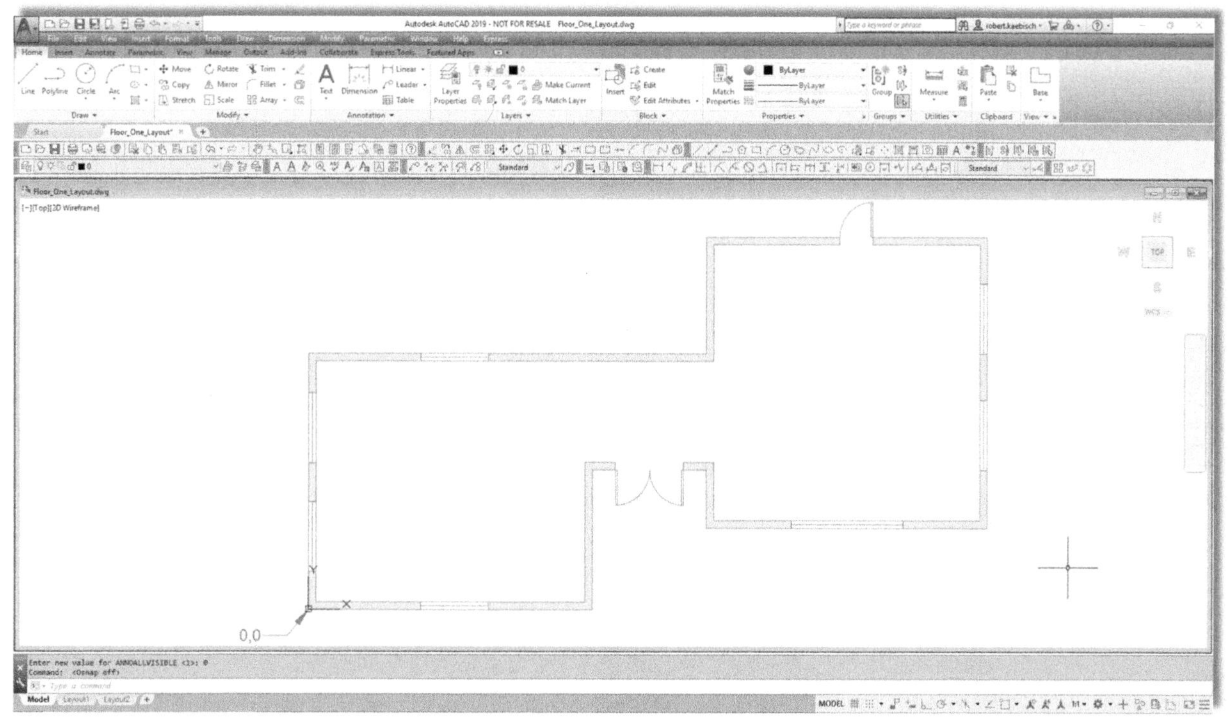

FIGURE 17.1 The external reference file.

As of now, nothing is special about this file. We still need to xref it to the working file. Open up a blank file and save it as Floor_One_Layout.dwg. Then, type in `xref` and press Enter. Alternatively, you can use the cascading menu Insert → External References…. The palette in Fig. 17.2 appears. At the upper left is a white rectangle with a paper clip (hence, the earlier paper clip reference). That is the Attach drawing button. You can click on the down arrow to reveal further attachment options (Image, dwf, dgn), but we do not need those. Simply click on the Paper icon and the Select Reference File browsing window opens. Browse to find the xref file (ExteriorWalls.dwg) and click on Open. The External Reference insertion box appears, as shown in Fig. 17.3.

Examine what is featured here, most of which should be familiar to you (such as Insertion Point, Scale, and Rotation). Uncheck Insertion Point; the fields turn from gray to white. We want the xref to insert at 0,0,0. Finally, press OK.

The xref appears with its lower left corner at 0,0,0. Zoom out to see the entire drawing if not apparently visible. Notice the new addition to the xref palette, as seen in Fig. 17.4. The ExteriorWalls file is visible and indicated as Loaded with the appropriate date, time, and path.

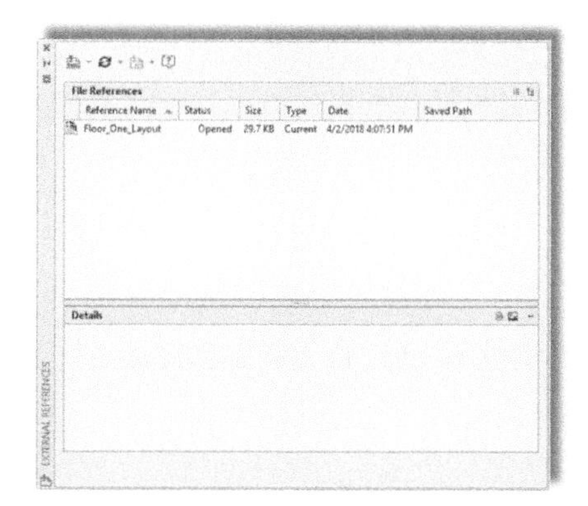

FIGURE 17.2 The external reference palette.

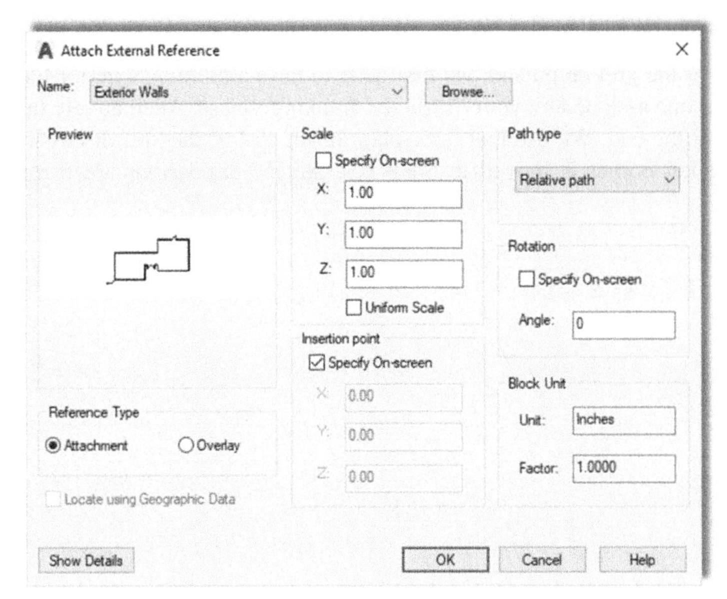

FIGURE 17.3 Attach external reference.

You can now proceed to add new design work to the file, as the xref procedure is complete. Let us skip ahead and explore some of the other available options, now that the xref is loaded. By right-clicking on the ExteriorWalls paper icon, you get access to the important menu in Fig. 17.5.

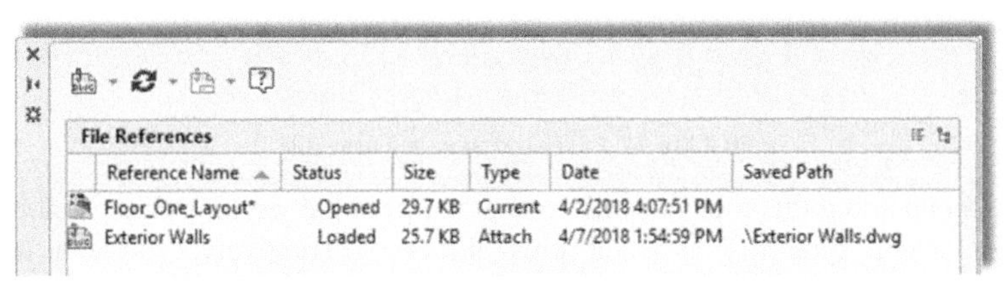

FIGURE 17.4 External reference loaded.

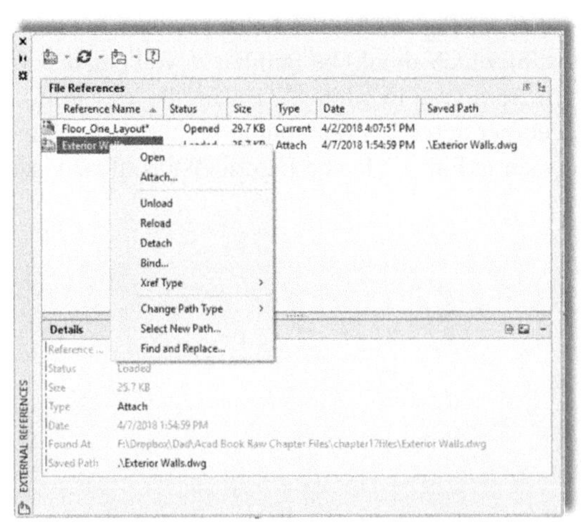

FIGURE 17.5 External reference menu.

The Xref menu includes the following:

- *Open*: This option allows you to open the original xref if you wish to make changes.
- *Attach*: This option allows you to attach another xref to your working file, using the same procedure as outlined earlier in this chapter.
- *Unload*: This option removes the xref from your working file but preserves the last known path, so if you wish to bring back the xref, you need not browse for it again but just click on Reload.
- *Reload*: The option just referred to for bringing back an unloaded xref.
- *Detach*: This option is more permanent and removes the xref from the working file, requiring you to browse and find it if reattachment is called for again.
- *Bind*: This option allows you to attach a permanent copy of the xref to your working file and severs all ties with the original. The xref then becomes a block; and if it is exploded, it completely fuses together with the working file. This action is potentially dangerous and is rarely used, as you are now stuck with the xref permanently and changes to the original external file will not be shown here. Two possible uses include permanent drawing archiving and emailing the drawing set (so as to not forget the xrefs), though eTransmit can take care of that.

17.3 LAYERS IN EXTERNAL REFERENCES

A very common problem may appear when using xrefs. The xref file itself has layers, often many of them. There may also be several xrefs in one active drawing. And, of course, the drawing itself may have numerous layers of its own. It is almost guaranteed that, on a job of even medium complexity, layer names can and will be duplicated. This is not allowed in AutoCAD, and some method is needed to separate layers in the xref from mixing and being confused with layers in other xrefs or the main design drawing. The solution is quite simple. In the same manner that people have first and last names to tell them apart (especially helpful if the first names are the same), so do layers when xrefs are involved.

All layers that are part of a certain xref have that xref's file name preceding the layer name with a small vertical bar (and no spaces) between them. All layers that are native to the working file have no such prefix, of course.

This simple method ensures no direct duplication by having the xref file name act as the "first name" and the layer name as the "last name." An example is shown in Fig. 17.6. New layers have been added in the host file. Notice the small vertical bar in the xref layer name and the inclusion of the xref file name.

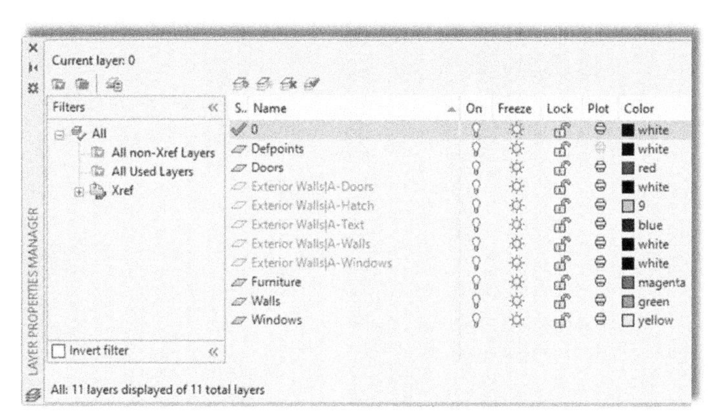

FIGURE 17.6 External reference layering.

When an xref is bound to the drawing, the layer's vertical bar changes to a 0 set of symbols, indicating they are now bound and part of the working drawing, as seen in Fig. 17.7. You should be familiar with this if you have to examine a bound set.

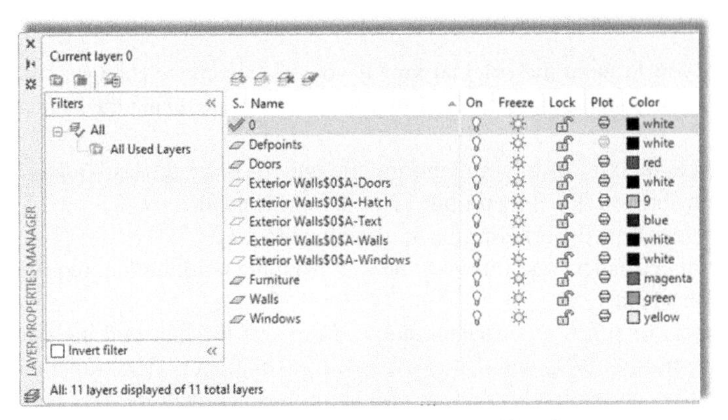

FIGURE 17.7 External reference bound layering.

17.4 EDITING AND RELOADING EXTERNAL REFERENCES

At one time xrefs could not be changed in place. This means you had to open the original xref file to edit it and could not do this from the main working file. This made sense, as you generally want to limit access to the original xref, so it cannot be casually edited.

All things (and software) change, and a number of releases ago, AutoCAD started allowing xrefs to be edited in place using a technique similar to editing blocks (somewhat diminishing the security benefit of xrefs, as mentioned earlier). Let us give it a try. Double-click on any of the lines making up the xref while in the main file. The Reference Edit box appears, as seen in Fig. 17.8, with the Ribbon also changing to the (nonfunctional at this point) xref tab.

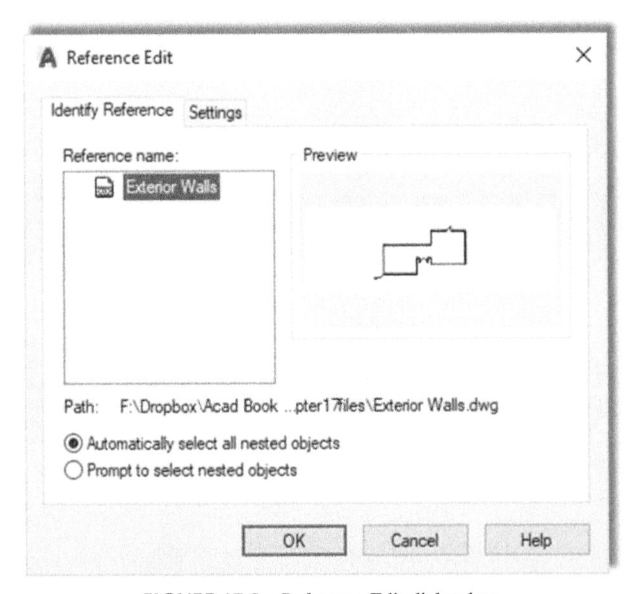

FIGURE 17.8 Reference Edit dialog box.

The xref you clicked on is listed under Reference name. Press OK and you are taken into Reference Edit mode. You may notice a few new features or items of interest:

- The xref has remained brightly lit up while everything that is on the main drawing file has faded down. This is a feature of Ref Edit. You can now clearly tell apart the xref geometry from the rest. The intensity of the fade can be adjusted under right-click, Options, Display tab, Fade control, Xref display.
- The xref can now be edited in place; go ahead and change a few items around.
- You can bring it up the *Refedit* toolbar via Tools → Toolbars → AutoCAD or use the Edit Reference Ribbon panel that appears (both shown in Fig. 17.9).

FIGURE 17.9 Refedit toolbar and Ribbon Edit Reference panel.

When you finish editing the xref, press the Save Changes button in either location. A warning icon appears; press OK, and you are done. The xref has been edited in place.

Just as a reminder, this procedure is only for editing the xref while you have the main working file open, and what you just did can also be done in a far simpler manner by just opening the xref as its own file (although with Ref Edit you have the advantage of seeing the internal design while editing). In fact, when changes are extensive, the designer may just open the xref itself and work on it all day, without resorting to Ref Edit. Every time the designer saves his or her work, the changes become permanent and accessible to any other designers who may be working on a job that uses that xref.

A good question may come up at this point. How do other designers know that the xref is changed so they can adjust their interior work accordingly? Many releases ago, the only way to know was for the person working on the xref file to simply tell others via a visit to their cubicle or office, phone, or email. The other users would then press Reload All References, as seen in Fig. 17.10.

Of course, you can still do this, and designers often refresh every few minutes on their own if they know someone is actively working on the xref. However, a new tool appeared recently. It is simply a "blurb" announcement that appears in the lower right corner of AutoCAD's screen, as shown in Fig. 17.11. Press the hyperlink and the xref reloads and refreshes automatically.

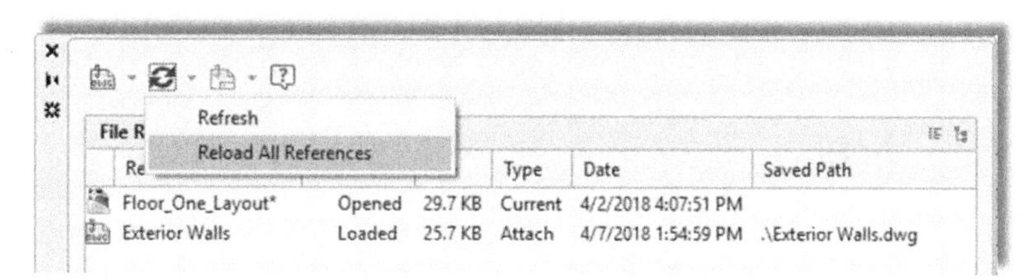

FIGURE 17.10 Reload external reference.

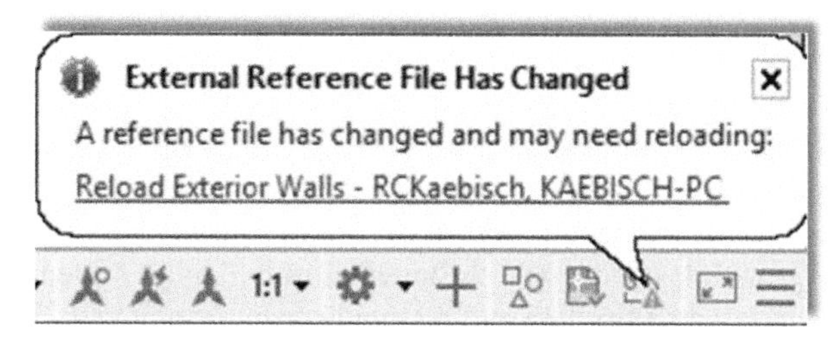

FIGURE 17.11 Reload external reference announcement.

Another brief topic in this chapter concerns how to chain together xrefs if there is more than one of them. Some methods are suggested next.

17.5 MULTIPLE EXTERNAL REFERENCES

As mentioned earlier in this chapter, many complex drawings use more than one xref. Typically, you may find the exterior walls, columns, and the core as three separate xrefs. The ceiling grid (RCP) can also be xrefed in for extensive ceiling work, as can the demolition plan and many other combinations.

There are essentially two options. Let us say you work with a total of five xrefs. You can attach them one to another then to a master working file (daisy chain) or one at a time, as shown in Figs. 17.12 and 17.13.

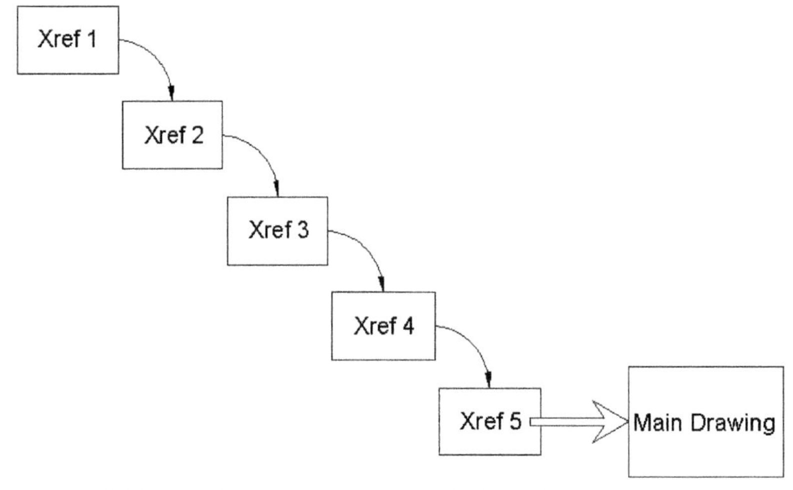

FIGURE 17.12 Multiple xref attach, Method 1. xref, external reference

Which method you use is up to you (there may be company policy or a convention in use at your job). Each method has minor pros and cons, and you should expect both to be used in the industry.

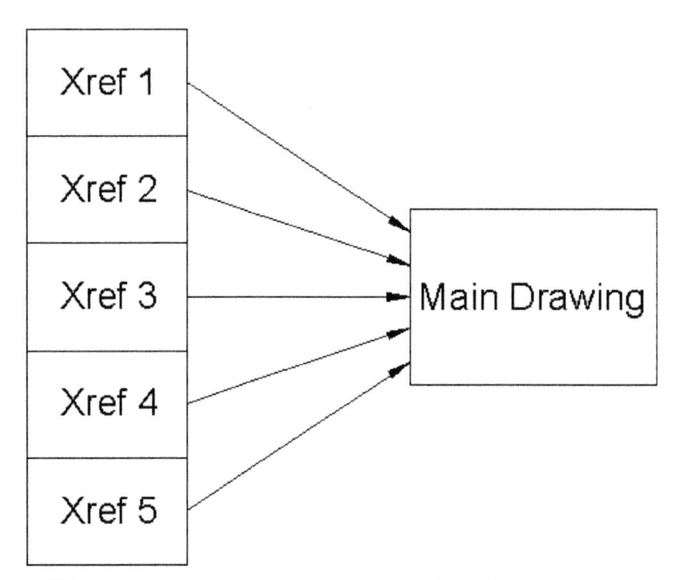

FIGURE 17.13 Multiple xref attach, Method 2. xref, external reference

One final word on when not to xref, as this feature gets abused occasionally. Xrefing of furniture, fixtures, and text is not what xref is for. It is intended for multilayered, complex drawings of buildings, systems, or infrastructure and NOT for individual items normally inserted as blocks (furniture and plumbing fixtures) and text. The logic behind xref in the first place is to attach objects that are not likely to change and can serve as the backdrop, so use the tool wisely.

17.6 RIBBON AND EXTERNAL REFERENCES

Some xref functions can be accessed via the Ribbon, and we discuss them separately in this section. You can find them under the Insert tab, Reference panel, as seen in Fig. 17.14.

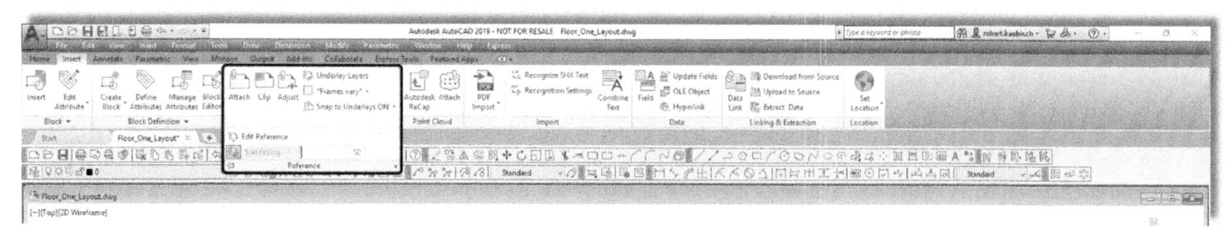

FIGURE 17.14 External reference via Ribbon, 1.

One important control is the xref fading slider you see toward the bottom of the tab. This controls the darkness of the xref and was mentioned a few paragraphs ago as also being under the right-click, Options..., Display tab. Some users prefer the xref to be dark, similar to the surrounding design, whereas some prefer it very light, so they clearly know what is an xref and what is the design.

Finally, the Ribbon switches to a dedicated xref tab if you click on the xref, as seen in Fig. 17.15, although there is much duplication of the previous tab's commands.

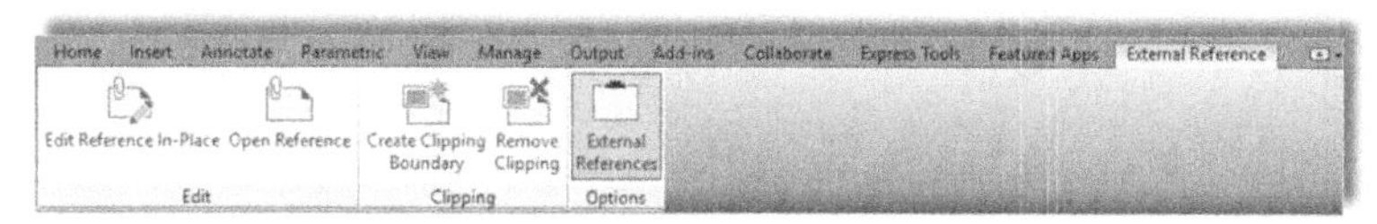

FIGURE 17.15 External reference via Ribbon, 2.

17.7 LEVEL 2 DRAWING PROJECT (7 OF 10): ARCHITECTURAL FLOOR PLAN

For Part 7 of the drawing project, we shift to the outside of the building and create some landscaping in the form of trees, shrubs, and drive/walkways. You need some Chapter 11 tools for the greenery.

Step 1. Create some appropriate layers as shown next. Freeze all layers that are not relevant to the task at hand.

- L-Driveway (Gray_8).
- L-Trees_Shrubs (Color_108).
- L-Walkway (Gray_8).

Step 2. Create the landscaping plan as seen in Fig. 17.16. You need to create plines for the borders of the driveway and the walkways. Then, after adding the hatch patterns (Sand and Brick), erase some or all the plines. For the trees and shrubs, you need to call on the xline and spline.

SUMMARY

You should understand and know how to use the following concepts and commands before moving on to Chapter18:

- Xref command
- Inserting
- Detaching
- Unloading
- Reloading
- Binding
- Layers in xref
- Nesting xrefs

REVIEW QUESTIONS

Answer the following based on what you learned in Chapter 17:

1. What are several benefits of xref?
2. How is an xref different from a block?
3. How do you attach an xref?
4. How do you detach, reload, and bind an xref?
5. What do layers in an xrefed drawing look like?
6. What do layers look like once you bind the xref?
7. What are some nesting options for xref?

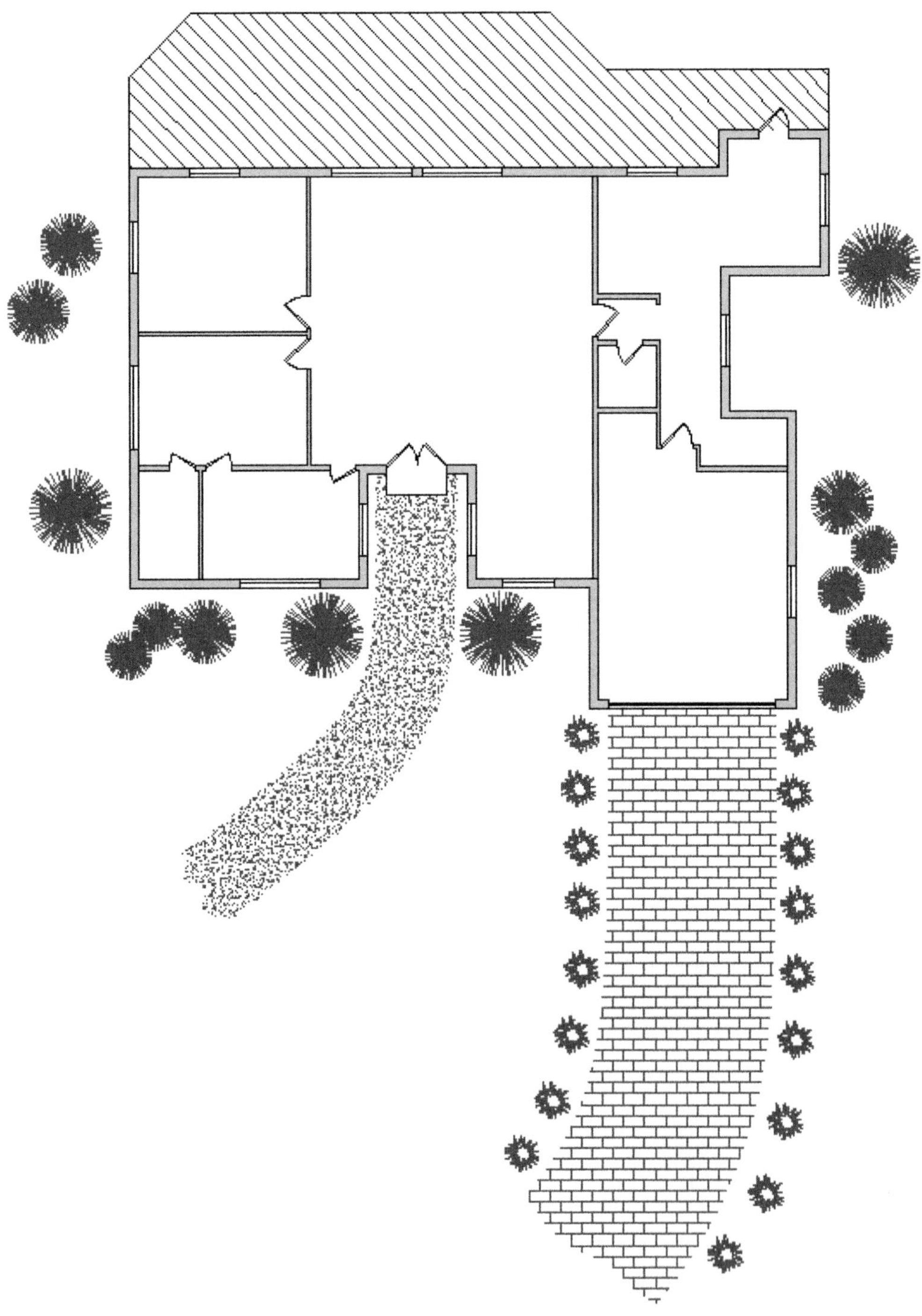

FIGURE 17.16 Landscaping plan.

EXERCISE

1. Draw the following floor plan to practice the xref command. (Difficulty level: Easy/Moderate; Time to completion: 45–60 minutes.)

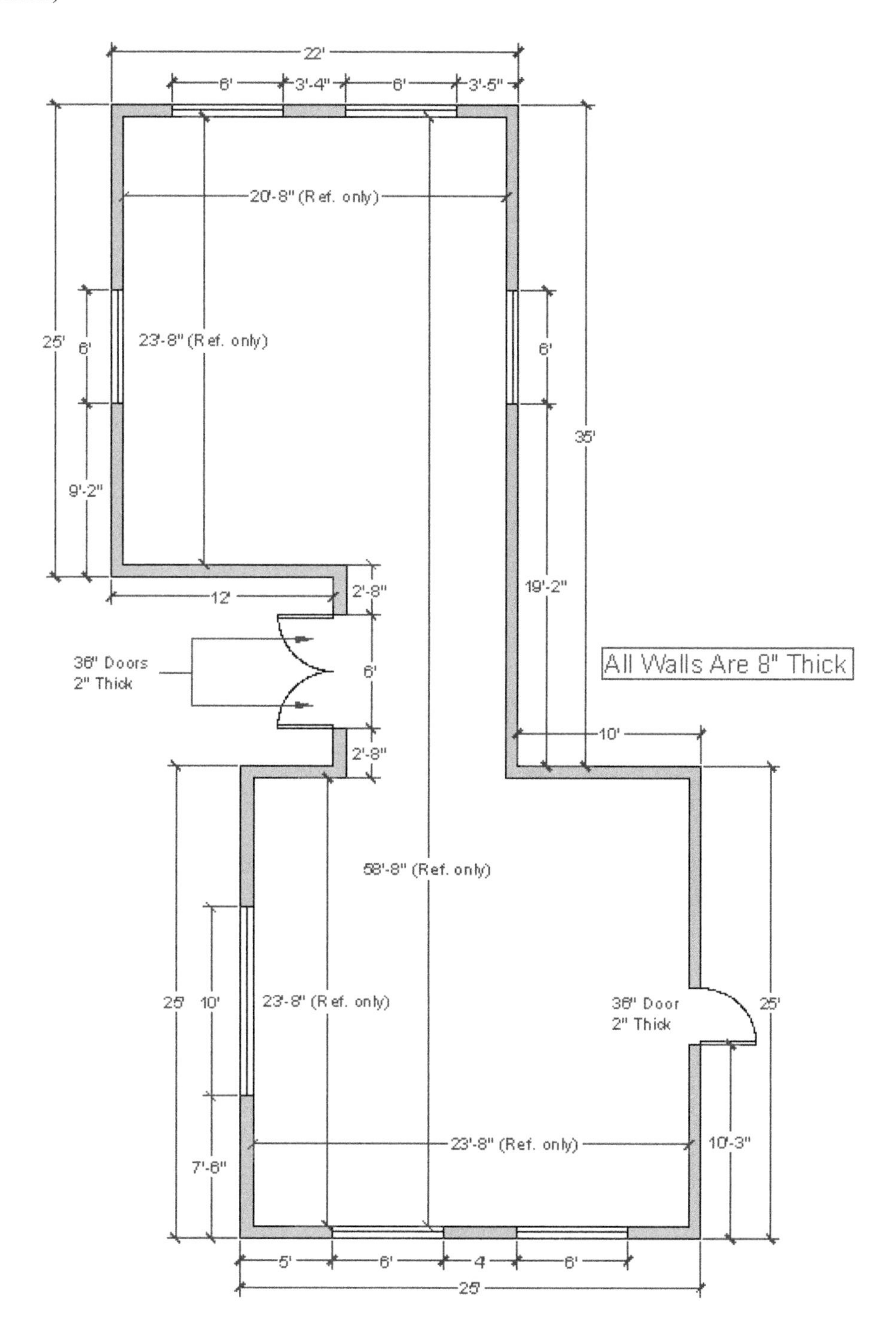

Spotlight On: Biomedical Engineering

Biomedical engineering is the merging of engineering and medicine (or biology). It is the newest of the engineering disciplines and has just recently become a field in its own right. Engineers who entered this profession were usually mechanical, electrical, or chemical engineers. They collaborated side by side with medical researchers and doctors or attended medical school themselves, becoming a physician/engineer. Today, a number of schools offer specialized biomedical engineering degrees that combine all these fields.

FIGURE 1 *Image source: www.osoti.net.*

Biomedical engineers often work in research, and depending on their field of interest, they may specialize in bionics, genetic engineering, tissue engineering (artificial organs), or pharmaceutical engineering. Probably, the one product most often associated with biomedical engineering is artificial body parts, such as the artificial heart or hip and knee replacements. Although old TV shows such as the *Six Million Dollar Man* create an image of an exotic half man, half machine with superhuman strength and speed, the vast majority of such applications are quite down to earth and commonplace. You may even know a grandparent or other relative with an artificial implant. Biomedical engineers help to design and manufacture these parts.

Another major field for these types of engineers is medical devices and tools. From heart valves to stents to surgical equipment to scanners of all types, biomedical engineers create devices that interact with the human body. It is a very interesting and diverse field, and an active area of research and innovation.

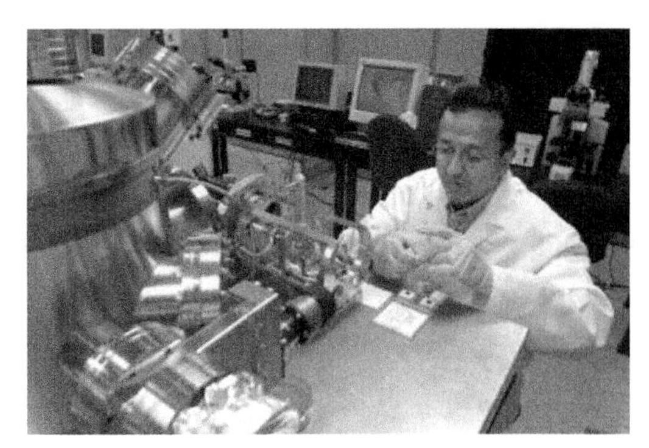

FIGURE 2 *Image source: Associated Press.*

Biomedical engineering is not an easy career to enter and be educated in. You are merging together two already challenging fields—engineering and medicine—into one profession. There are two approaches here. In the classic approach, you carry a dual major in engineering (usually mechanical or electrical) and biology or chemistry. You then go on to graduate school for an advanced degree in engineering or to medical school to become a physician, sometimes both. A more recently available (and somewhat easier) approach is to major in biomedical engineering from the start. More and more universities now offer this degree, and you will find yourself taking courses such as anatomy, biomechanics, materials and imaging, chemistry, and biomedical instrumentation. It is truly a hybrid of engineering and medicine. Additional courses include robotics, circuit analysis, and industrial systems, with additional medical classes such as molecular biology.

On graduation, biomedical engineers almost always need to go to graduate or medical school. In fact, some programs discourage students from majoring in this field unless they plan, at the very least, on getting a master's degree, as they are just not taught enough on the undergraduate level for such a complex field. As of 2015, there was no PE licensing for a biomedical engineer. In the past many biomedical engineers were only engineers and picked up the biological or medical knowledge on the job. As the field sees more growth and specialization, however, the need to have dually trained professionals—part engineer, part medical doctor, especially for research—is greater than ever.

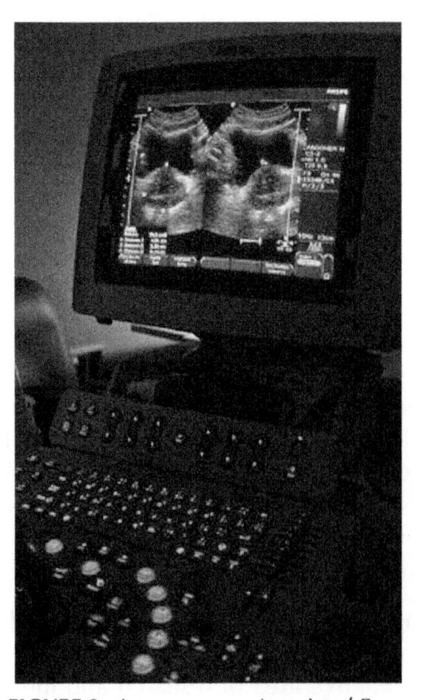

FIGURE 3 *Image source: Associated Press.*

Biomedical engineers (with just a 4-year biomedical degree) can generally expect starting salaries in the $50,000/year range, which is toward the low end among engineering specialties. The median pay nationwide was $64,000 in 2017. However, as stated earlier, many go on to graduate school or even further, as required for research; and the median salary for 2014 was a very respectable $94,000 for those with advanced degrees.

So, how do biomedical engineers use AutoCAD and what can you expect? Probably the top area in AutoCAD is to be found biomedical instrumentation. The "engineering" of biomedical engineering is often of the electrical kind, and you will find a great deal of schematics and wiring for scanners and imaging devices designed in AutoCAD. Things get a bit more complex as you move into implant design. Here, the "engineering" tends to lean more toward the mechanical side, and the implants themselves feature smooth organic curvature and complex material requirements. Often the appropriate software is a 3D parametric solid modeling package, not AutoCAD. One well-known company in that exact specialty, Stryker Howmedica Osteonics of New Jersey, uses Pro/Engineer (Creo) for its design work.

Chapter 18

Attributes

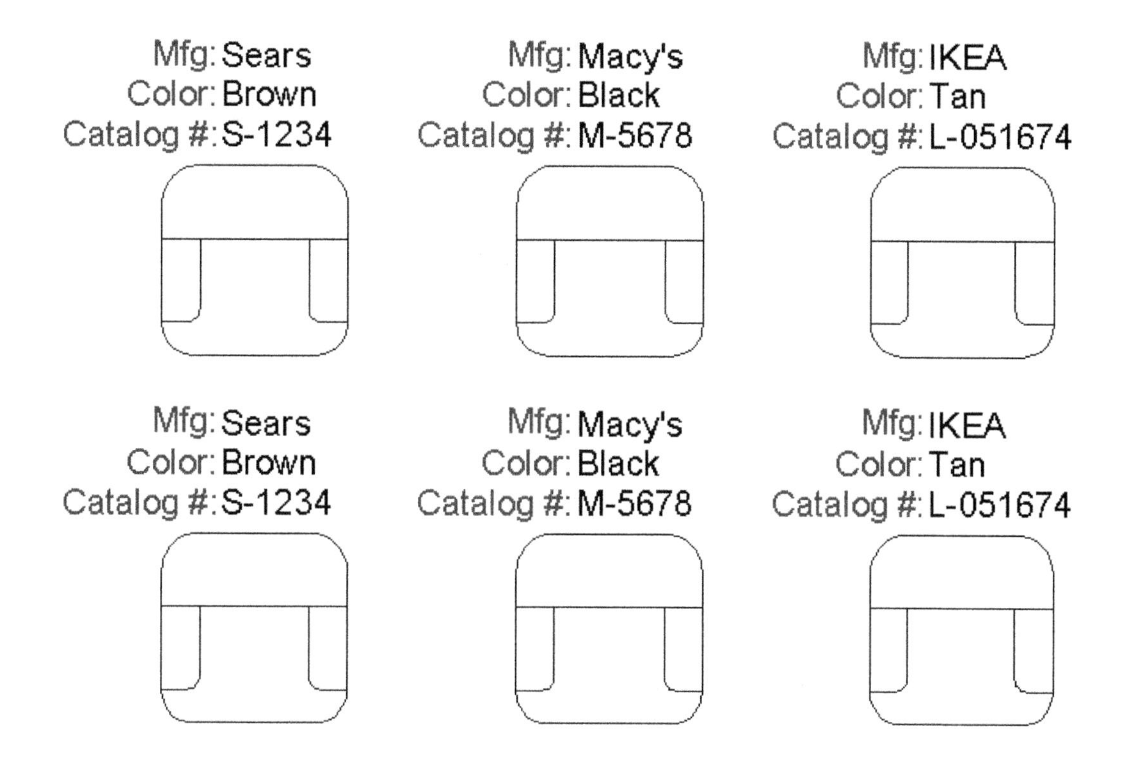

Learning Objectives

In this chapter, we introduce and thoroughly cover the concept of attributes. We specifically discuss

- The purpose of attributes
- Defining attributes
- Editing attributes and properties
- Extracting data from attributes
- Invisible attributes

By the end of this chapter, you will be well versed in applying this critical concept to your design work.
Estimated time for completion of this chapter: 1–2 hours.

18.1 INTRODUCTION TO ATTRIBUTES

Attributes is an important advanced topic. An attribute is defined in AutoCAD as information inside a block. The key words here are *information* and *block*. You can have a block without information—you have done this many times before; it is called a *block* or *wblock* but not an attribute. You can also have information in the form of text sitting in your drawing.

Up and Running with AutoCAD 2019. https://doi.org/10.1016/B978-0-12-816440-2.00018-5

That is just text or mtext but also not an attribute. However, put the two concepts together and you have what we are now about to explore: Essentially an "intelligent" block, one that has useful, editable, embedded information.

What is the reason for creating attributes in the first place? Why would you want information inside of blocks, and what can you do with this information? The information itself can be anything of value to the designer or client. For an office chair in a corporate floor plan, one may want to know its make, model, color, and catalog number. Then, if there are a variety of chair types for this office, the designer can keep track of all of them. Another example is a list of connections going into an electrical panel. It would be nice to know what and how many. All this information can be entered into the attribute and displayed on the drawing if you wish.

However, this is only half the story, and displaying information is not the main reason for creating attributes. As a matter of fact, this information is often completely hidden, so as not to clutter up the drawing. So, what is the main reason? The key is that this information can be extracted out of the attributes and sent to a spreadsheet (among other destinations). Now you have a truly useful system in place, where extensive information is hidden unobtrusively in the drawing yet can be accessed if needed at the touch of a few buttons and put into an Excel file or other very useful form for cost analysis or just plain old keeping track and organizational purposes.

You may have already heard of these tools in use or envisioned a need for them in architecture with such items as doors and windows schedules. The software keeps track of each door and window in the drawing and links the information to a database that is updated as you add or remove the objects from your design. While basic AutoCAD does not do this automatically, other add-on software does. Underneath it all is the concept of embedded attributes, similar to those we discuss here.

You should have a good understanding of the preceding few paragraphs. As always, the button pushing is far easier once the concept is crystal clear. We now go ahead and do the following:

- Create a simple design.
- Create a basic attribute definition.
- Make a block out of the simple design and the attribute definition.
- Extract information out of our new attribute.

The information is extracted via a multistep process to two locations, an Excel spreadsheet and a table. In both cases, the information is in an organized tabular form and, while not linked bidirectionally, is an exact reflection of drawing conditions. Let us give it a try.

18.2 CREATING THE DESIGN

We need nothing complicated drawn to learn the basics of attributes. Go ahead and draw a simple chair made up of several rectangles and rounded corners, as shown in Fig. 18.1. Make the chair 30″ × 30″ and approximate the rest of the dimensions relative to those.

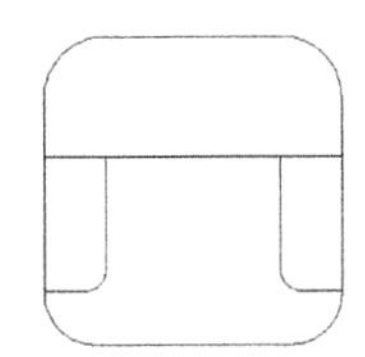

FIGURE 18.1 Basic chair design.

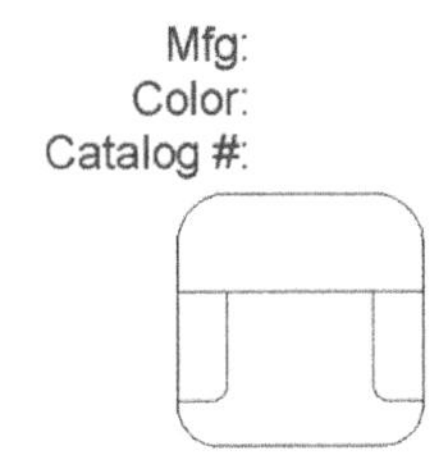

FIGURE 18.2 Basic chair design text.

We would like to embed the following information in it—manufacturer, color, and catalog number—and put these three items just above the chair. Let us first add some text, a header of sorts, indicating what the three categories are, as seen in Fig. 18.2. Use regular text of any appropriate size, font, and color.

This is not yet the attribute definition, merely the name of the categories, so we know what we are looking at. You do not have to do this, but it helps clarify things initially, as the attributes are just plain Xs. You now have a simple design and the start of some useful information. We are now ready to make an attribute definition.

18.3 CREATING THE ATTRIBUTE DEFINITIONS

Here, we define the new attributes that contain the actual information. The idea for now is to simply assign them all a value of X and change the X to meaningful data later in the process, so we are creating generic definitions, which allow us to create multiple sets of data from one template.

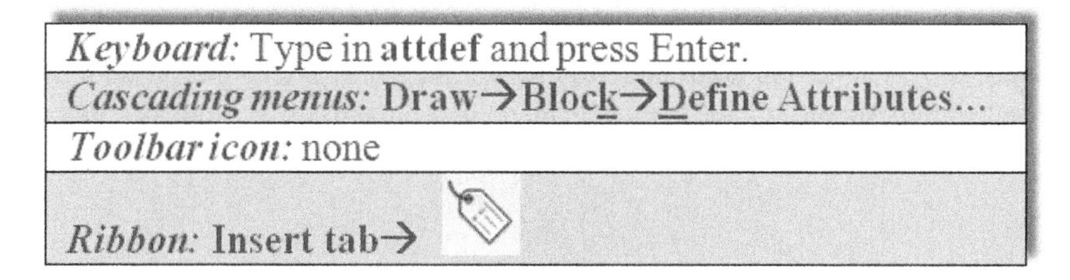

Keyboard: Type in **attdef** and press Enter.
Cascading menus: **Draw→Block→Define Attributes...**
Toolbar icon: none
Ribbon: **Insert tab→**

Start the attribute definition via any of the preceding methods. The dialog box in Fig. 18.3 appears.

FIGURE 18.3 Attribute Definition.

We are interested in only the upper right of the dialog box for now. Under the Attribute category, let us focus on Tag: Prompt: and Default:

- For Tag, type in Mfg for Manufacturer (this is the actual category of the attribute definition).
- For Prompt, type in Please enter manufacturer (this is just a reminder, or question to answer, for the user in a dialog box).
- For Default, type in X (this is a generic value that can be changed by the user later), as seen in Fig. 18.4.

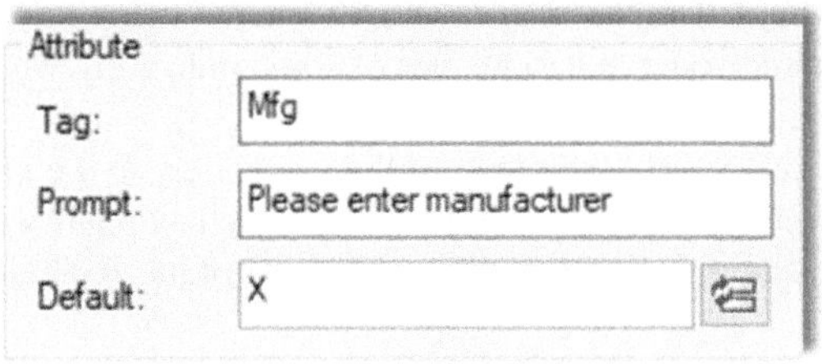

Attribute
Tag: Mfg
Prompt: Please enter manufacturer
Default: X

FIGURE 18.4 Attribute fields.

The rest of the information in the dialog box can be left as is for now, although we discuss a feature under Mode (upper left) later in this chapter. One thing you can adjust, however, is the text height, matching the attribute size to the size of the text over the chair. The text is 4″ in height, so you can enter that value into the field.

Press OK and your attribute definition appears attached to your crosshairs. Go ahead and place it next to the Mfg: text. Note that it is in capital letters by default. Go ahead and go through the same procedure for the remaining two categories, changing the Tag value and slightly modifying the Prompt. The default remains as X. Note also that you cannot have blank spaces in the tag, hence the CATALOG_# underscore. The end results are shown in Fig. 18.5.

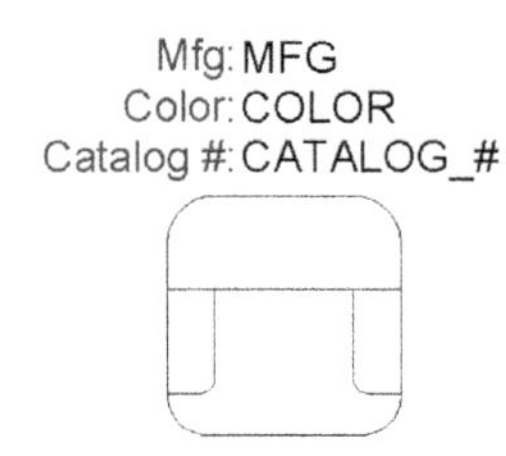

FIGURE 18.5 Attribute fields completed.

We are almost done creating the attribute definitions. Note the differences between the text and the attribute. Double-click on the original text, and you get the standard text editing field. Double-click on the attribute definition and you get a different, Edit Attribute Definition, field, as shown in Fig. 18.6.

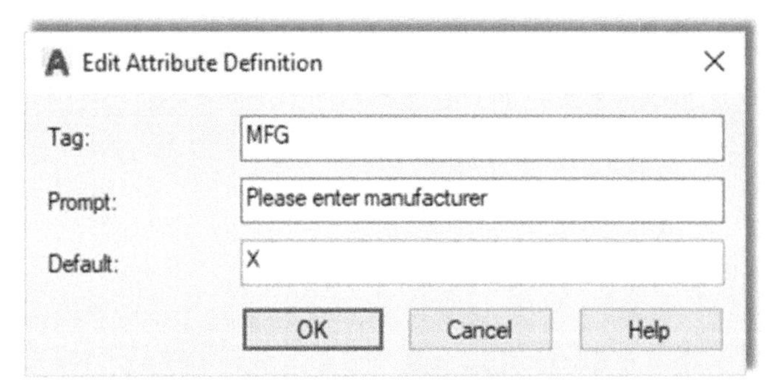

FIGURE 18.6 Edit Attribute Definition.

At this point, it may not be obvious why we have two identical sets of categories next to each other or why we entered the X. After the next step, what we are doing should become much clearer.

18.4 CREATING THE ATTRIBUTE BLOCK

This next step should be familiar: Create a regular block named Sample_Chair via any method you prefer. Give it a name, select the objects (both the chair and all the text), and pick an insertion point. Finally, click on OK, and the dialog box shown in Fig. 18.7 appears. This dialog box was upgraded last year to allow for more attributes to be listed without having to flip to another page—the previous version if AutoCAD had just 8 lines total; this one now has 15. In the interest of time and clarity we stick to just the three attributes shown here, but keep in mind you can enter many more if needed.

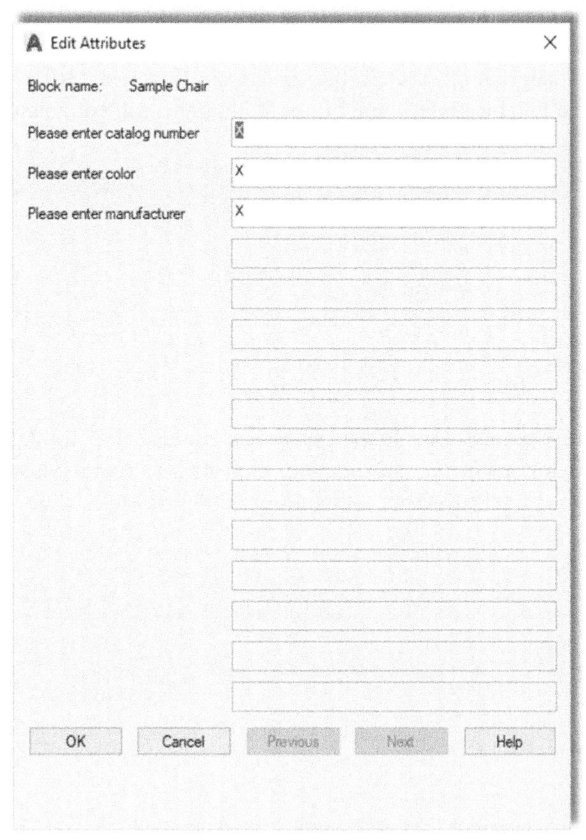

FIGURE 18.7 Edit Attributes.

The Edit Attributes dialog box allows you to enter specific values for those placeholders (marked as X), but we do not do that at this point. Simply press OK and your new attribute is officially born (Fig. 18.8). Notice all the previous categories are now just X and ready for customization. This is the process for creating a generic attribute. In the example of the office chairs, you create several copies (the distinct types of chairs) and customize the information for each. Then, copy and distribute the types as needed on the floor plan, as we do next.

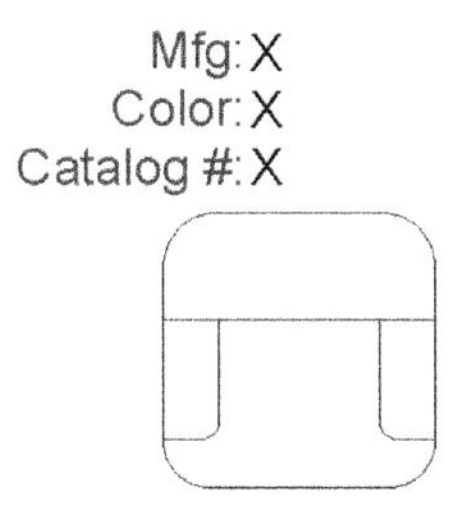

FIGURE 18.8 Edit Attribute.

18.5 ATTRIBUTE PROPERTIES AND EDITING

Now that you have a block with attributes, you need to learn how to edit its fields to add or delete information from an attribute as well as change its color, font, and just about any other property. Some of these techniques overlap each other and duplicate effort, so we need to clearly state what is used for what.

First, if you want to change only the content of an attribute (in other words, what it says), you need to get back to the dialog box shown in Fig. 18.7. It is basic and simple but does not change fonts and colors. In addition, the only way to get to it is by typing. The command, attedit, takes care of the content. Indeed, this is the way you should "populate," or fill in, the fields once we start copying the chair. Try it right now and edit the chair to say Sears for the manufacturer, Brown for color, and S-1234 for catalog number.

Now, what if you wanted to change an attribute's properties, as discussed earlier? Well, the method here changes slightly, as you need to deal with the Enhanced Attribute Editor. Although a full command matrix is shown next, the easiest way to get to it is just to double-click on the attribute.

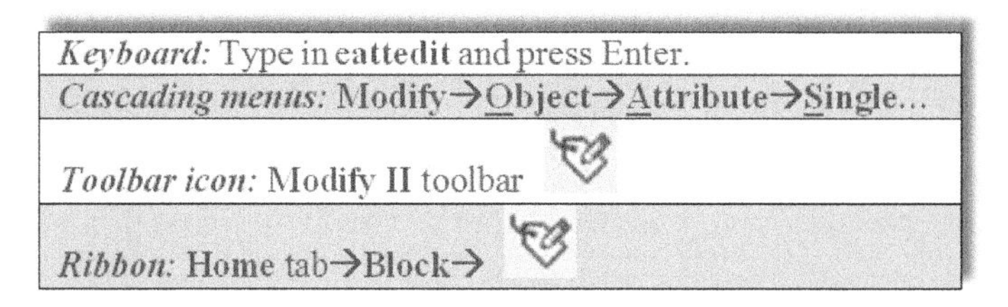

Keyboard: Type in **eattedit** and press Enter.
Cascading menus: Modify→Object→Attribute→Single...
Toolbar icon: Modify II toolbar
Ribbon: Home tab→Block→

What you get after either double-clicking or using the previous methods is Fig. 18.9. Here, you can do quite a lot. This Editor not only allows you to fill in the fields but also a variety of properties (colors, layers, fonts, etc.). Click through the three tabs, Attribute, Text Options, and Properties, to explore all the options. Here are some additional pointers on attributes.

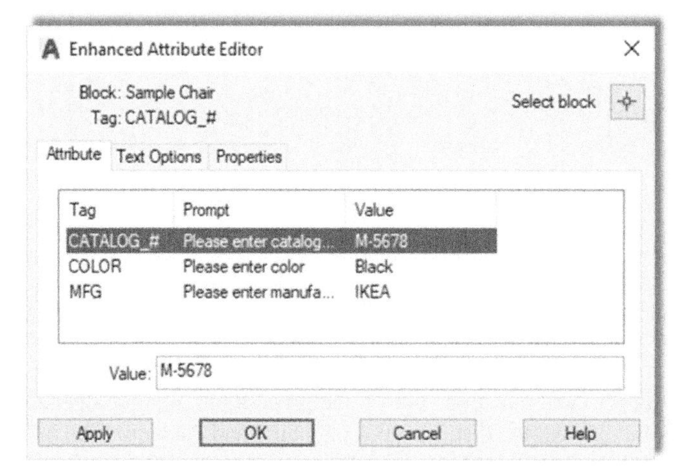

FIGURE 18.9 Enhanced Attribute Editor.

Exploding Attributes

Attributes can be easily exploded, but this action destroys the blocks and turns them into what you see in Fig. 18.8, where you have just generic placeholders unattached to a physical design. You can change a few fields and reassemble the block, which is a perfectly valid technique, especially if the attribute is giving you a tough time. Be careful though; exploding complex attributes can suddenly fill your screen with an enormous amount of data, so keep one finger on that Undo button.

Inserting Attributes

As you learned in this chapter, attributes are blocks; therefore, you can insert them as needed, as an alternative to copying them. When you insert a block that contains an attribute, you are prompted to fill in the X placeholders at the command line. Indeed, this is how title blocks are often filled out when a generic template block is inserted into a specific job file.

Ideally, everything makes sense now that you see the big picture of what was done. Review this chapter up to now if not. Our next step is to learn how to extract the useful information. For this, you need to copy the block to simulate having a number of chairs in a real drawing. Then, go ahead and change some of the attribute values. The result is shown in Fig. 18.10.

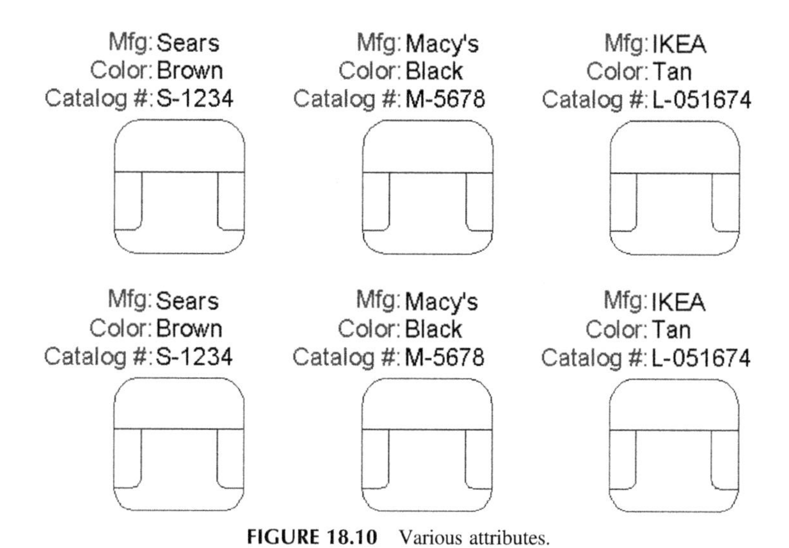

FIGURE 18.10 Various attributes.

18.6 ATTRIBUTE EXTRACTION

Extracting the information is a multistep process, necessarily so because of the number of options offered along the way. Our target extraction is both an Excel spreadsheet and a table embedded in AutoCAD. Remember, of course, that this is a basic generic extraction and more complex spreadsheets and tables can be created to hold such data, which is left to you to explore.

Step (Page) 1. Begin

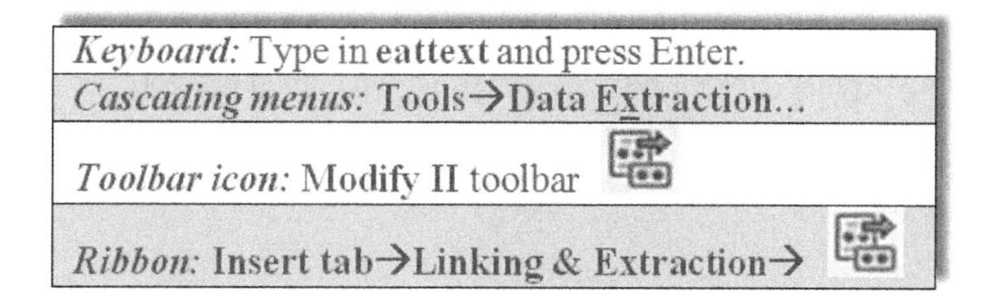

Keyboard: Type in **eattext** and press Enter.
Cascading menus: Tools→Data Extraction...
Toolbar icon: Modify II toolbar
Ribbon: Insert tab→Linking & Extraction→

Start up the data extraction process via any of the methods shown in the command matrix. The Data Extraction - Begin (Page 1 of 8) dialog box shown in Fig. 18.11 appears. You are going to create a new data extraction, so leave the default top choice but uncheck the Use previous... box. Then, press Next > and you are taken to the Save Data Extraction As dialog box. Save your file as Sample.dxe anywhere you will find it easily on your drive. You are then taken to the next page.

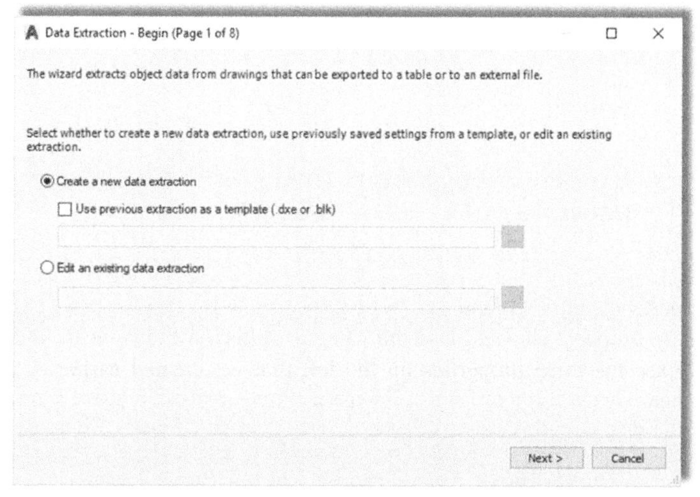

FIGURE 18.11 Data Extraction - Begin (Page 1 of 8).

Step (Page) 2. Define Data Source (Fig. 18.12)
Here, we are interested in the data source, or location, of the file to be looked at. Leave the default choice with the Include current drawing box checked off. Press Next >.

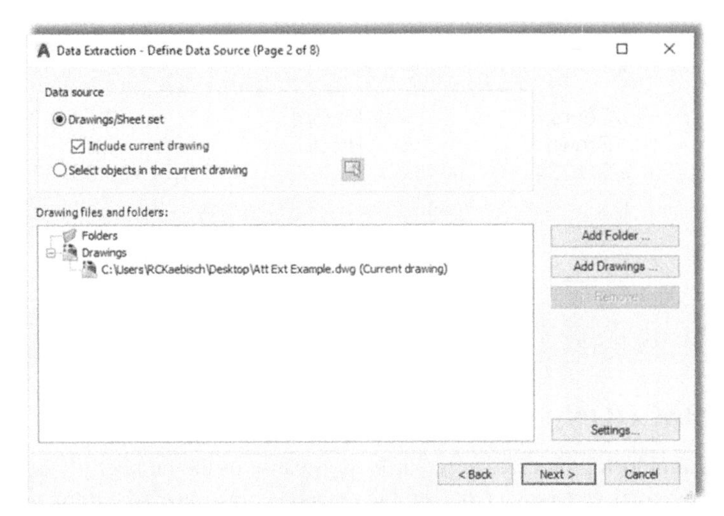

FIGURE 18.12 Data Extraction - Define Data Source (Page 2 of 8).

Step (Page) 3. Select Objects (Fig. 18.13)
Here, we are interested in selecting from what we want to extract the data. In our case, there is only one choice, the sample chair, as it is the only available block (as seen under the type column). You can also click on Display blocks only, under the Display options, to isolate the chair. In either case, select it, then press Next >.

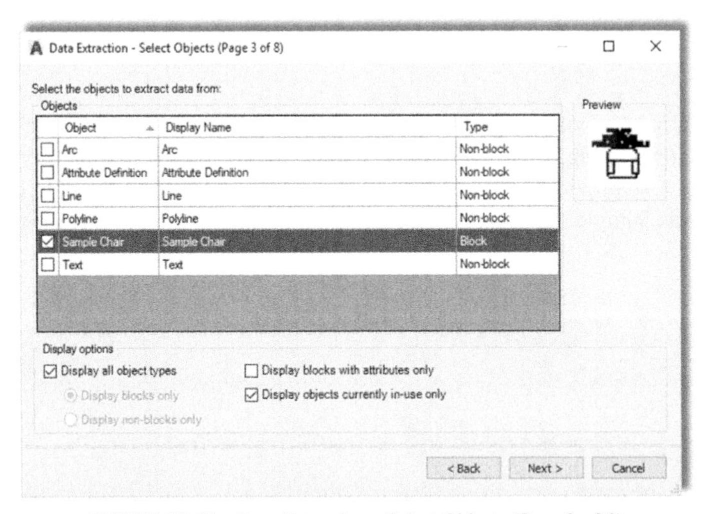

FIGURE 18.13 Data Extraction - Select Objects (Page 3 of 8).

Step (Page) 4. Select Properties (Fig. 18.14)
Here, you initially see many property choices. Use the category filter on the right to uncheck all the choices except Attribute. You then easily see the three properties on the left that we created earlier. Check off all three and press Next >.

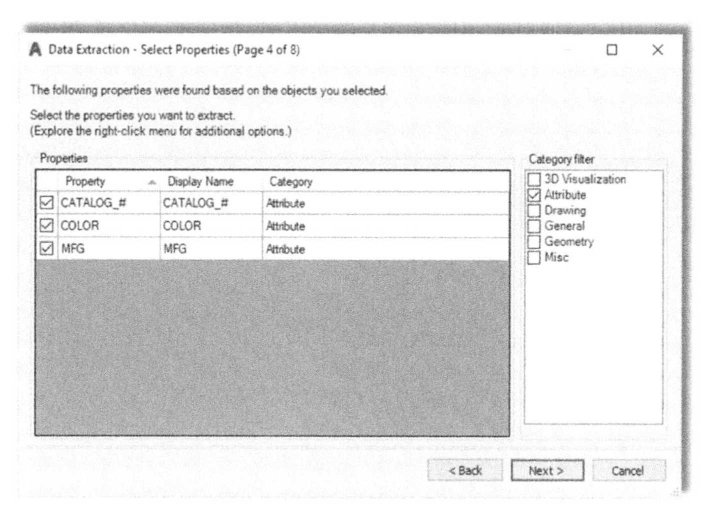

FIGURE 18.14 Data Extraction - Select Properties (Page 4 of 8).

Step (Page) 5. Refine Data (Fig. 18.15)
This step is for you to refine the data, if needed. It contains some interesting features, such as Link External Data… (to tie in this output to another Excel file) as well as some tools for adjusting and tweaking the columns. Much of this and more can be done in Excel itself after extraction is completed, but it can be useful to adjust the look and content of the table extractions, as tables are less flexible. If you want to change nothing, press Next > for the final few steps.

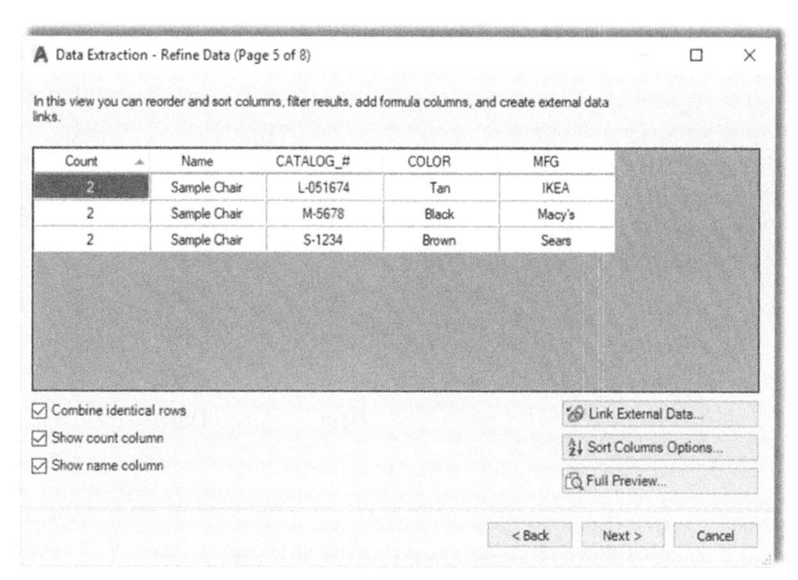

FIGURE 18.15 Data Extraction - Refine Data (Page 5 of 8).

Step (Page) 6. Choose Output (Fig. 18.16)
Here, we are interested in our destination output. Although the Excel spreadsheet gets all the attention, you can also extract to a table that will be inserted into the drawing itself. Tables are first introduced in Chapter 15, Advanced Design and File Management Tools, so review that section, if necessary. Check off both boxes if you have not done so yet, and finally select the location of the Excel file by using the three dots browse button. When ready, press Next >.

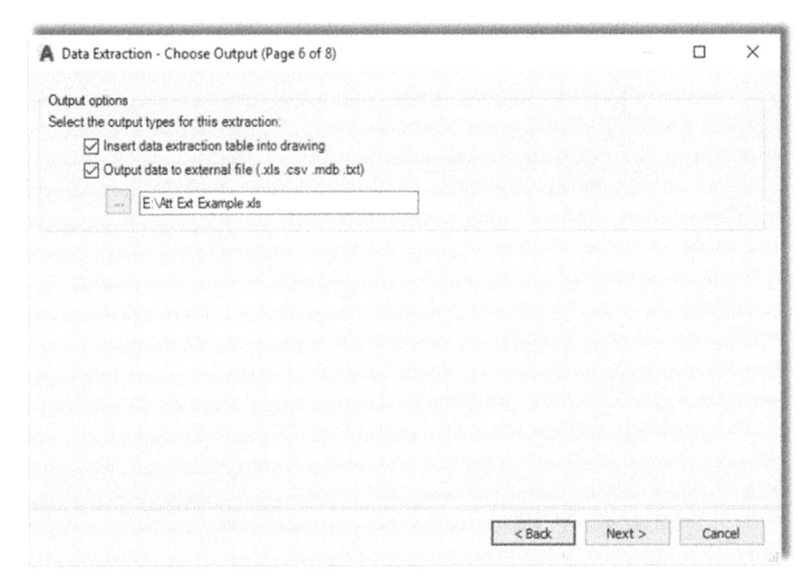

FIGURE 18.16 Data Extraction - Choose Output (Page 6 of 8).

Step (Page) 7. Table Style (Fig. 18.17)
We do not do anything to the table, leaving it as is, but you should review the options available and press Next > when done.

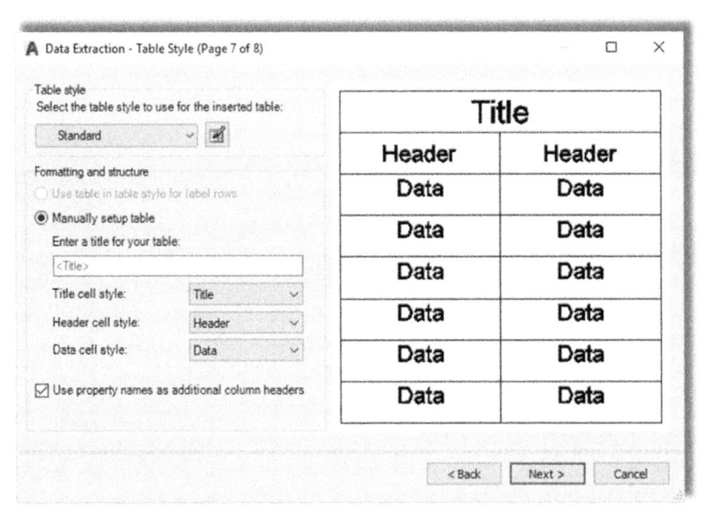

FIGURE 18.17 Data Extraction - Table Style (Page 7 of 8).

Step (Page) 8. Finish (Fig. 18.18)
On the final page, there is little to do except read what AutoCAD is telling you and press Finish. You are asked for an insertion point and notice the table attached to the mouse. You may have to zoom in to see it better. Finally, when you can see it, click to place the table. The table is inserted into your drawing, as shown in Fig. 18.19.

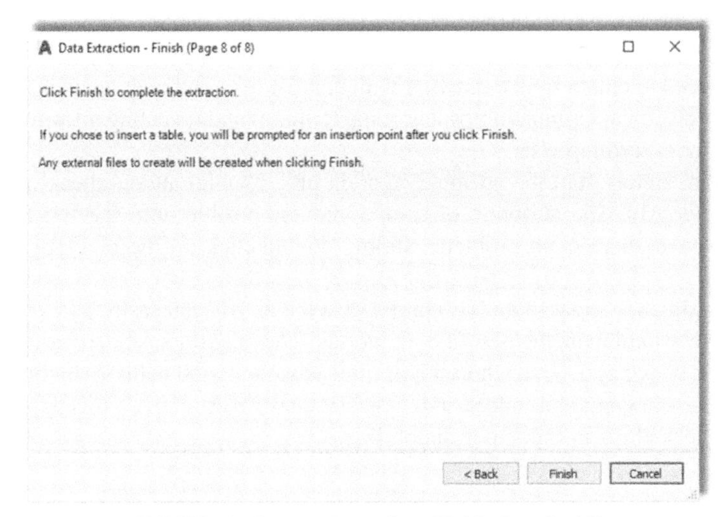

FIGURE 18.18 Data Extraction - Finish (Page 8 of 8).

It is not fancy, but it gets the information across. The main output, however, is the Excel file. Look for it wherever you may have dropped it off and double-click on it to start Excel. Then, after some cleanup (cell borders, column width, fonts, colors, etc.), select the cells and Copy/Paste them into AutoCAD, as you did in Chapter 16. It looks like Fig. 18.20.

This is attribute data extraction. You should recognize the value in what you did and the potential it holds for easy data access and accounting.

Count	Name	CATALOG_#	COLOR	MFG
2	Sample Chair	L-051674	Tan	IKEA
2	Sample Chair	M-5678	Black	Macy's
2	Sample Chair	S-1234	Brown	Sears

FIGURE 18.19 Table insertion (AutoCAD).

Count	Name	CATALOG #	COLOR	MFG
2	Sample Chair	L-051674	Tan	IKEA
2	Sample Chair	M-5678	Black	Macy's
2	Sample Chair	S-1234	Brown	Sears

FIGURE 18.20 Table insertion (Excel).

There are other uses for attributes. One common use is formatting a full title sheet and block with generic attributes and saving it as a company standard. In this case, the text is the fixed permanent categories, such as Date, Sheet, Drawn by, and other fields; and the attribute definitions are the variables, such as the actual date, sheet number, and the initials of whoever created the drawing. Then, everything is blocked out and saved for future use, with the Xs representing to-be-filled-in variables. As the need comes up for a new title block, it is inserted as a block into the design's Paper Space with the variables filled out on the spot. We conclude the attributes chapter by briefly going over an important option in the attdef command dialog box.

18.7 INVISIBLE ATTRIBUTES

This option appears as the first check-off box in the upper left of the attdef dialog box and is the most important of all the ones in that column. It is used to make attributes invisible on creation.

Now, why would you do this? As mentioned at the beginning of this chapter, this is a relatively common way of creating attributes for the simple reason that you do not need to see the text and the attribute information. If you know what you created, then there is no need to clutter up the drawing with that list over and over again. This is especially true if there is a great deal of information (electrical termination panels come to mind) and revealing all would create a dense blanket of text on the drawing, obscuring everything else.

To use the invisible option, simply start the attribute creation process with attdef, check off the Invisible option, and then proceed as outlined previously. The attribute, of course, remains visible until the block is created and disappears afterward. To see what you have, there is no problem: Simply double-click on the block or type in attedit, as outlined previously.

As you can see, being invisible for an attribute is no problem at all, as you can easily access it for editing if needed, and it is unobtrusively out of the way until then, the best of both worlds. The only downside, of course, is that you do not know that the attribute is there if you were not the one who set it up. It is always a good habit to double-click on blocks just to see what pops up on an unfamiliar drawing; and indeed, this is a standard investigative procedure in these cases. Fig. 18.21 shows the Invisible option checked off in the attdef dialog box.

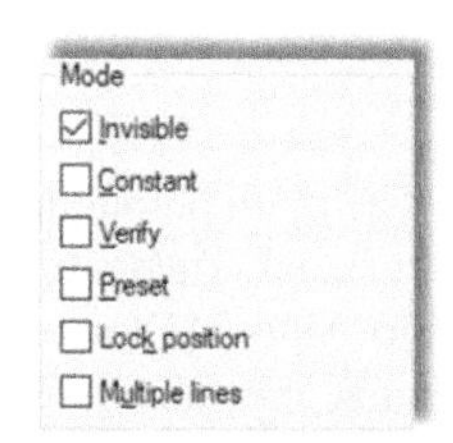

FIGURE 18.21 Invisible option.

18.8 LEVEL 2 DRAWING PROJECT (8 OF 10): ARCHITECTURAL FLOOR PLAN

For Part 8 of the drawing project, we finish up the design by creating an exterior elevation. Elevations are the "head on" views of a building. They are projected from overhead or plan views and are meant to give the client and builder a visual of the architect's design. These elevations feature walls, doors, windows, any other exterior design features (such as columns or fireplace chimneys), and of course the roof design.

Step 1. You need no new layers for this. Open up the floor plan, freeze all layers except the walls and windows, and draw guidelines straight down from each bend or feature in the wall (such as a door or window), as seen in Fig. 18.22.

Step 2. Raise the elevation and add the doors and windows according to the provided dimensions, as seen in Fig. 18.23. For simplicity, some details, such as gutters, are missing.

Step 3. Complete the elevation by adding exterior wall hatch and some other finishing touches. Fig. 18.24 shows the final result. Repeat the same procedure for at least one other side of the house or complete all four.

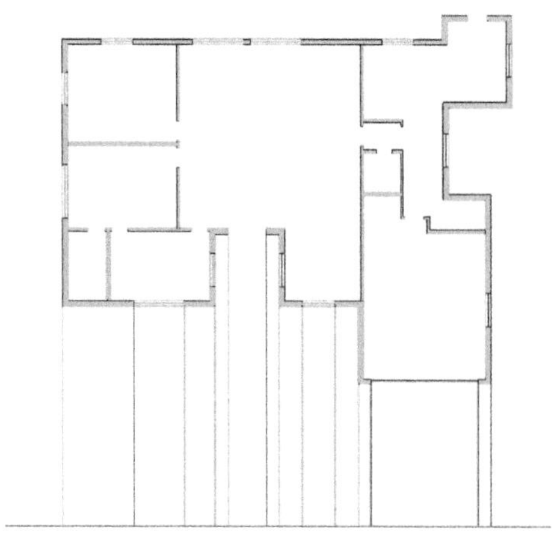

FIGURE 18.22 Elevation, Step 1.

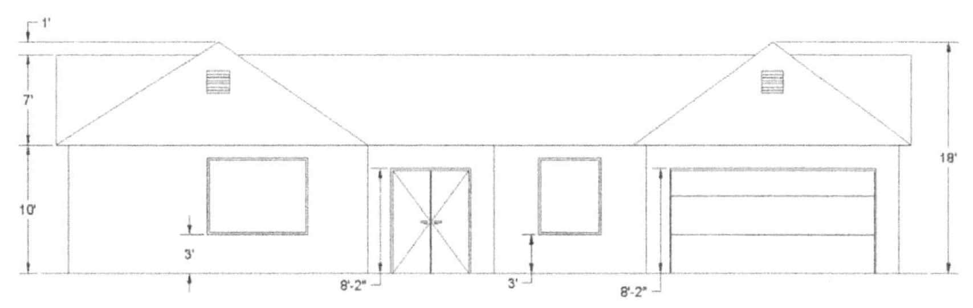

FIGURE 18.23 Elevation, Step 2.

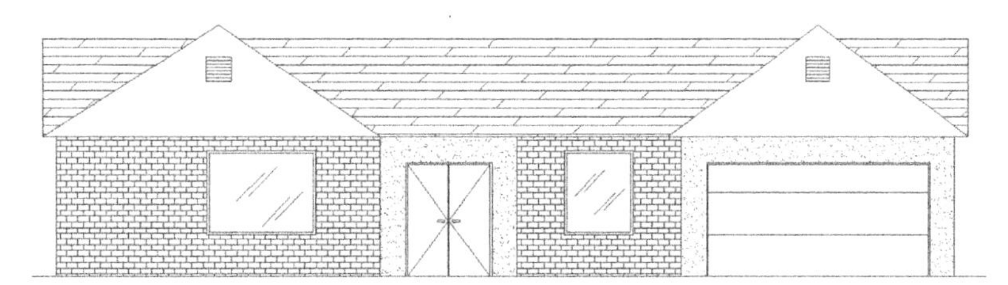

FIGURE 18.24 Elevation, Step 3.

SUMMARY

You should understand and know how to use the following concepts before moving on to Chapter 19:

- Attribute definitions
- Attribute blocks
- Editing fields and properties
- Attribute extraction
- Invisible attributes

REVIEW QUESTIONS

Answer the following based on what you learned in Chapter 18:

1. What is an attribute and how do you create one?
2. How do you edit attributes?
3. How do you extract data from attributes? Into what form?
4. How do you create an invisible attribute? Why can this be important?

EXERCISE

1A. Draw the following doors:

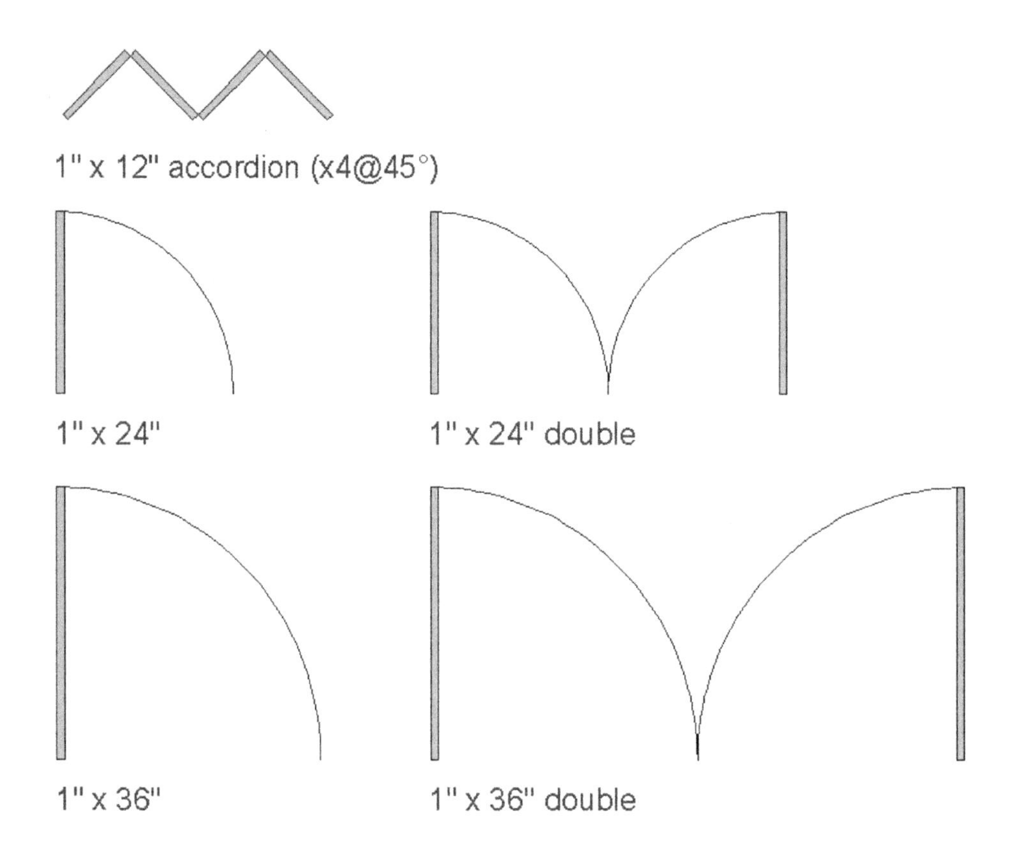

1B. Next, add text attributes to each, as shown by the sample that follows for the 1″ × 12″ accordion:

- *Type:* Accordion double.
- *Size:* 1″ × 12″.
- *Color:* Natural wood.
- *Mfg.:* XYZ Door Company.
- *Stock #:* 654-A-1 ×12.

1C. Create similar attribute data for the remaining four doors. Vary the colors and the stock numbers.

1D. Make a block out of each one and fill in the attribute information.

1E. Copy the entire set once for a total of 10 doors.

1F. Extract the data to Excel.

(Difficulty level: Intermediate; Time to completion: 30 minutes.)

Chapter 19

Advanced Output and Pen Settings

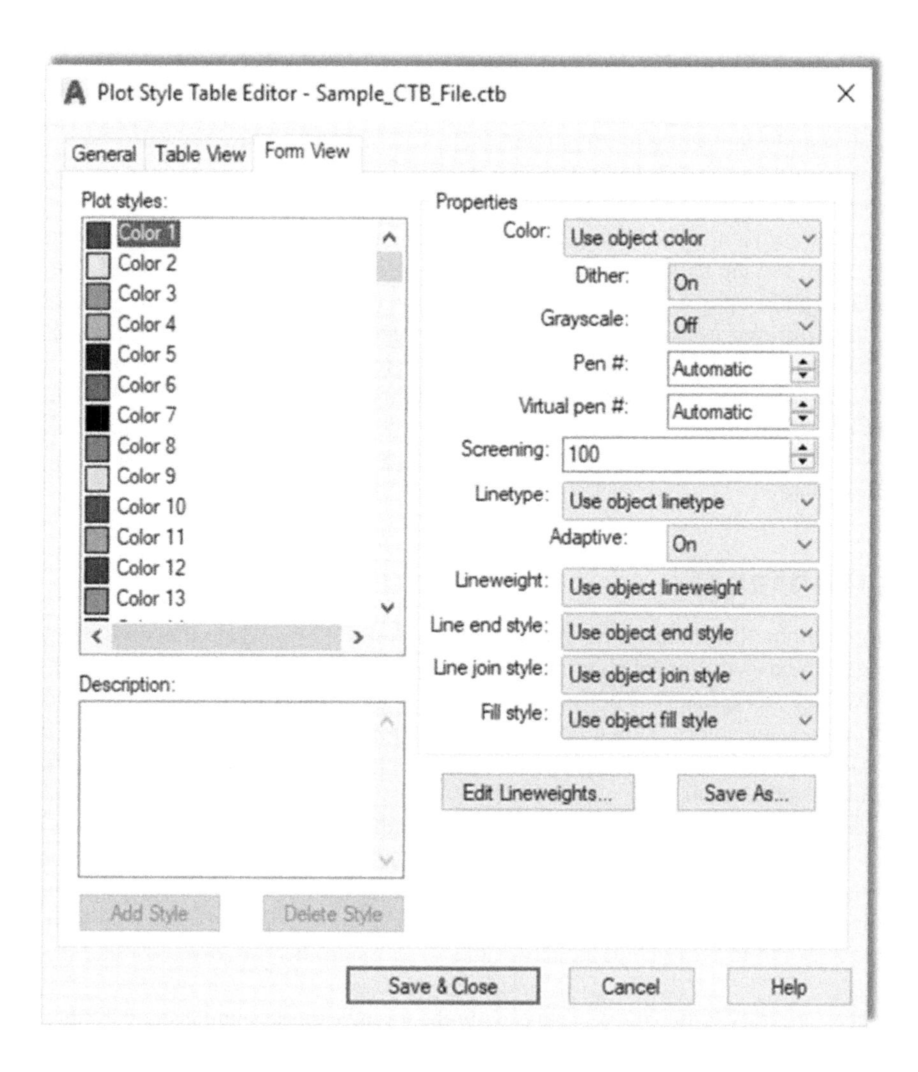

Learning Objectives

In this chapter, we introduce and thoroughly cover the concept of advanced output, which includes the concept of the CTB pen settings file. We specifically discuss

- CTB files and their purpose
- Editing lineweights
- Screening
- Lineweight settings

By the end of this chapter, you will be well versed in applying this critical concept to your design work.
Estimated time for completion of this chapter: 1 hour.

Up and Running with AutoCAD 2019. https://doi.org/10.1016/B978-0-12-816440-2.00019-7

19.1 INTRODUCTION TO ADVANCED OUTPUT AND PEN SETTINGS

This chapter is the last of three that talk about some aspect of printing and output. Although general and advanced printing was already explored in Chapters 9 and 10, a crucial piece of information was omitted. This information concerns pen settings and other advanced output techniques, essential knowledge for professional-looking output. The reasons we are doing this are outlined next.

Most designs are rather complex when everything has been added to the drawing. Lines cross other lines and text is everywhere. The eye needs to be able to quickly and easily focus on what is essential, e.g., the outlines of the walls of a floor plan or the outline of the mechanical part among the many other elements in the design.

This is analogous to a mix of a rock song. Not every instrument can be the loudest; something has to give, and a balance needs to be found so the nuances and color of the music can stand out. Usually, when the singer sings, other instruments have to be turned down a bit, but when a guitar solos, its volume gets turned up, and so on. Otherwise, it would be a loud, obnoxious mess with all sounds competing for attention.

Such is the situation with a pencil and paper drawing. All lines cannot be the same thickness and darkness. Such a monochrome mess would be confusing to look at. Draftspersons have known this for as long as the profession has been around and, as a result, have numerous pencils at their disposal, ranging from lightest (hard lead) to darkest (soft lead) and everything in between. They use the darkest for the primary design and the lighter leads for all else.

AutoCAD has a similar philosophy. Lines must have differences in thickness for some to stand out more than others. But, how do we show these differences now that we no longer use pencil leads? The answer is to use color. Colors are then assigned line thicknesses, and assuming of course a consistent and intelligent pattern of use on the part of the designer, differences in line thickness can easily be incorporated into the design by assigning certain colors to the primary geometry. As you can see, colors play a dual role. They allow you to tell geometries apart not only on the screen but also on paper. The challenge then is to set up a table that assigns colors to certain line thicknesses and, most important, stick to using these colors properly and consistently.

19.2 SETTING STANDARDS

The thing you need to determine before anything gets touched in AutoCAD is what colors will be used to represent the most important geometry. You actually did this back in Chapter 3 when layers were introduced. This, of course, is something done once by the CAD manager and usually is applied to all drawings. All we are really doing is establishing a "pecking order," so to speak, for colors. This order corresponds from the most important (hence, thickest) to the least important (the thinnest) pen setting. There are no hard rules here as far as what colors to use; it is more or less arbitrary, although convention favors some colors over others.

Let us say (based on common convention) that green is the A-Walls layer color in our architecture plan and is assigned the thickest lineweight. The A-Windows layer is typically red and the A-Doors layer yellow. Both those objects are almost as important as the walls but can be set a bit thinner. The A-Furniture and A-Appliances layer are magenta and thinner still. A-Text and A-Dims layers are cyan and thinner yet. Finally in our hypothetical example is hatch. Those dense patterns really need to be light, so as not to overwhelm the drawing, so we assign them a gray color and the thinnest lineweight of all.

Although this is just an arbitrary example, it is also exactly the thought process you need to go through to have a rough idea of how things will look. Once this is generally documented, you can move on to the actual settings.

19.3 THE CTB FILE

The CTB file is the file that needs to be modified to set the pen settings. It is where you assign pen thickness to a color and where we spend our time for the remainder of this chapter. We open the CTB file, give it a name, modify settings, and save. Then, as you set up plots, you assign the CTB file to those plots, which is a permanent, "do-once" action.

There are two ways to set up the CTB file: either by accessing it in the Page Setup Manager or as a separate action, which is what we do. Select Tools → Wizards → Add Plot Style Table… from the drop-down menu. The dialog box shown in Fig. 19.1 appears.

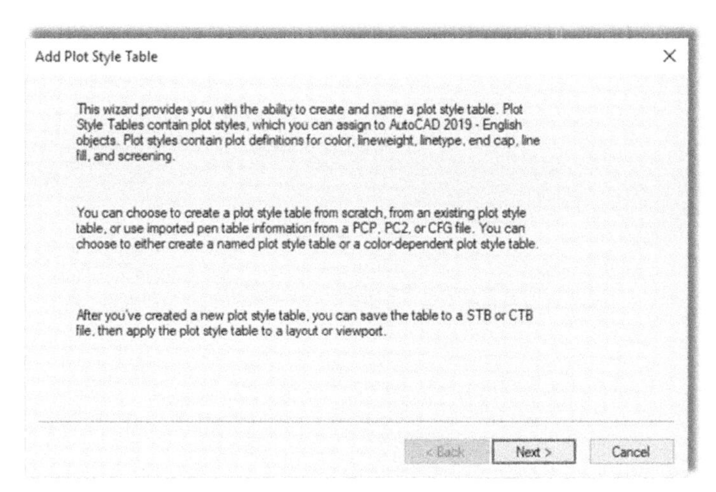

FIGURE 19.1 Add Plot Style Table.

Press Next> and you are taken to the next page of this dialog box (Fig. 19.2), where you select the first option, Start from scratch. This begins a new line setting procedure from the default base values.

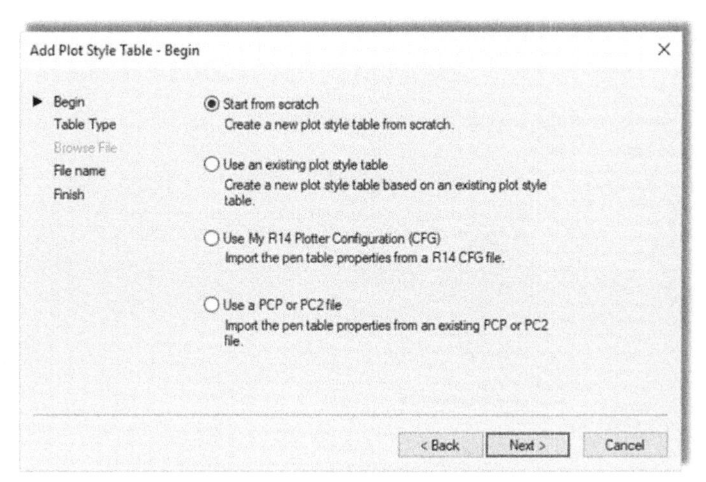

FIGURE 19.2 Add Plot Style Table—begin.

Press Next> and you are taken to the next page of this dialog box (Fig. 19.3); select the Color-Dependent Plot Style Table choice. This is the only choice we look at; the named plot styles are not covered and are rarely used.

Press Next> and you are taken to the next page of this dialog box (Fig. 19.4). Give the new CTB file a descriptive name, such as StandardCTB, ColorCTB, or anything else that will be recognized. Here, it is Sample_CTB_File.

Press Next> for the final step (Fig. 19.5). Here, we are interested in the Plot Style Table Editor… to actually set the lineweights. Press that button and you are taken to our final destination, the actual Plot Style Table Editor, shown in Fig. 19.6.

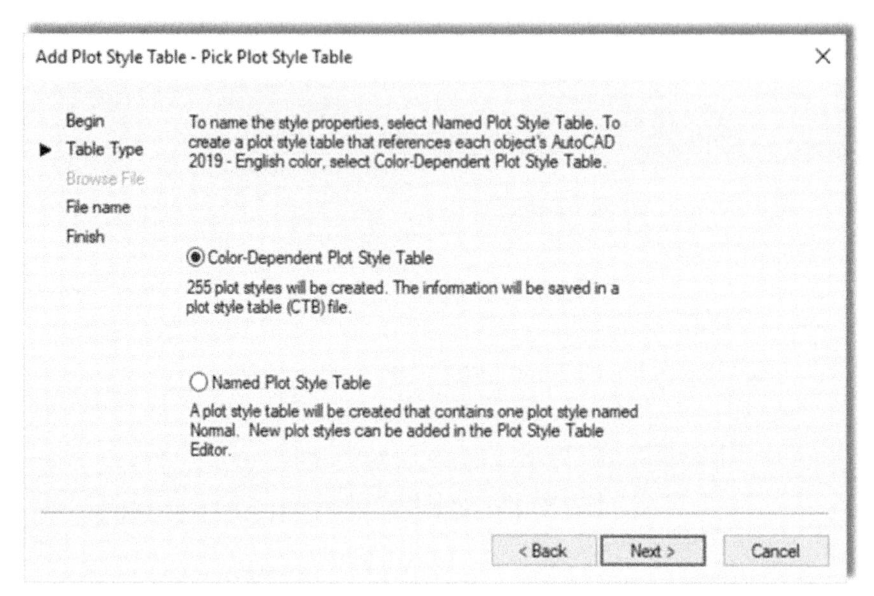

FIGURE 19.3 Add Plot Style Table—pick Plot Style Table.

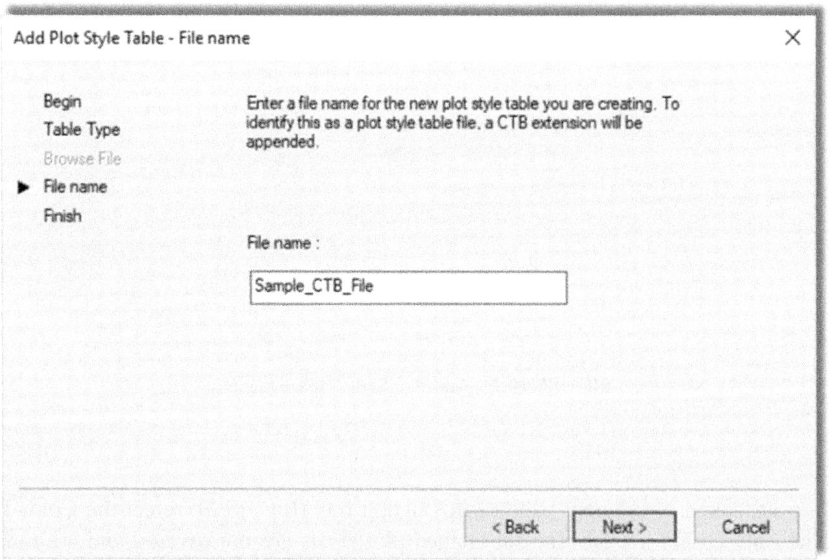

Add Plot Style Table - File name

Begin
Table Type
Browse File
File name
Finish

Enter a file name for the new plot style table you are creating. To identify this as a plot style table file, a CTB extension will be appended.

File name :

Sample_CTB_File

< Back Next > Cancel

FIGURE 19.4 Add Plot Style Table—file name.

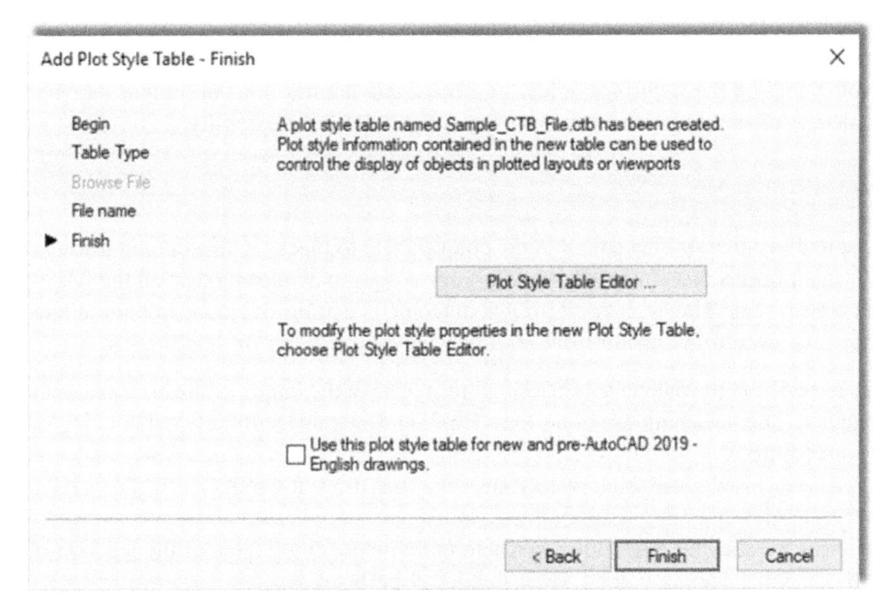

FIGURE 19.5 Add Plot Style Table—finish.

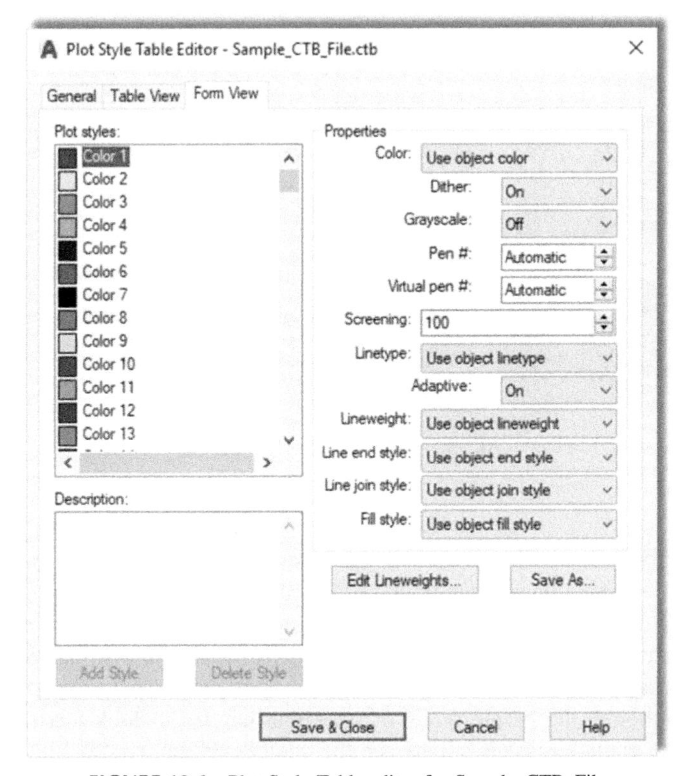

FIGURE 19.6 Plot Style Table editor for Sample_CTB_File.

Here is where we set the colors, weights, and a few other tricks. Stay in the default Form View (referring to the tabs on top), as we do not need the information in any other format. Familiarize yourself with this editor. On the left is the palette of the available 255 colors and on the right is a variety of effects and settings you can impart on whatever color or colors you select from the palette. All the tools are there; you just need to proceed carefully.

Step 1

The first thing you need to do is to set all the colors in the color palette to black. As the default settings are now, each color is interpreted literally, which means that if you are sending your design to a color plotter, all the linework will come out in color, exactly as seen on the screen. This is a very desirable outcome in certain situations but not in most (we talk about this more later on). For now, we need to monochrome everything so we can work on line thicknesses.

- Select all the colors (1–255) in the palette by clicking your mouse in the empty white area just to the right of the colors, holding down the button, and sweeping top to bottom. Holding down the Shift key and clicking the first and last entry accomplishes the same thing.
- Then, as all the colors are highlighted blue, select the very first drop-down choice in Properties, called *Color:*, and select Black (Fig. 19.7). All colors are assigned black ink.
- Clear the palette choices by clicking randomly anywhere in the white empty area of the left side of the palette and test out your changes by clicking random colors. They should all be black.

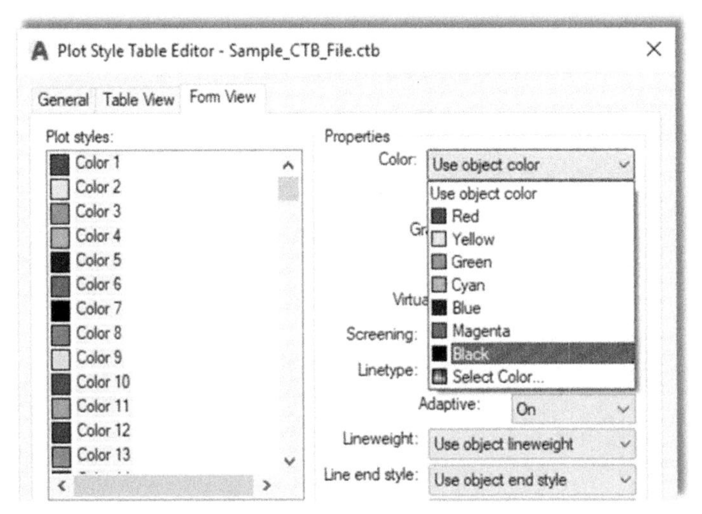

FIGURE 19.7 Selecting black for all colors.

Step 2

Next we need to set the actual thicknesses. This is done in the Lineweight: drop-down menu, about halfway down the list on the right. Currently, it says Use object lineweight, which simply means nothing is set specifically and all lineweights are the same (the default value is 0.1900 mm).

Here, we need to discuss a major point you must understand in working with this feature. The actual line sizes are of secondary importance to the relative differences between the sizes. This means that you need not worry too much about how big 0.2 or 0.15 mm is, although that is the range of sizes that happen to look just right on paper. What is more important is that there is a large enough difference between the choices (0.2 mm vs. 0.15 mm as opposed to 0.2 mm vs. 0.18 mm), so the eye can notice this difference. You can just as easily use 0.25 and 0.30 mm, but as you get up in the higher numbers, all lines end up being overly thick, so it is best to stick to the range of 0.25 mm on the upper end and 0.05 mm on the lower end. Let us put this all to use and set some values:

- Click on and select color 3 (green); set Lineweight to 0.2500 mm.
- Click on and select color 1 (red); set Lineweight to 0.1900 mm.
- Click on and select color 2 (yellow); set Lineweight to 0.1900 mm.
- Click on and select color 4 (cyan); set Lineweight to 0.1500 mm.

- Click on and select color 6 (magenta); set Lineweight to 0.1500 mm.
- Click on and select color 9 (gray); set Lineweight to 0.1000 mm.

Now, to put it all together, recall the Level 1 apartment drawing. The A-Walls layer is colored green and appears darkest (0.2500 mm), as required. The A-Doors and A-Windows layers were yellow and red, respectively, and are slightly lighter (0.1900 mm), followed by the A-Text and the A-Appliances (or A-Furniture) layers, colored cyan and magenta, respectively, lighter still (0.1500 mm). Last are the gray-colored hatch patterns coming in at (0.1000 mm).

So, will these particular lineweight choices look good? The drawing will certainly look more professional, and features will stand apart from each other. The exact values and colors are all determined by the architect, engineer, or designer, as this person inspects the results and tweaks the lineweights as needed. This is an extensive trial-and-error procedure, and not all individuals select the same settings; however, in any professional office, these techniques must be used regardless of exactly what specific settings are set. Nothing screams amateur more than a monochrome drawing with equal lineweights.

Step 3

To complete the procedure, press Save and Close and Finish in the Add Plot Style Table—Finish dialog box. Then, open a drawing, such as the Level 1 floor plan, if available. Type in plot and set the drawing up for plotting or printing as outlined in Chapter 10, being sure this time to select Sample.ctb instead of monochrome.ctb in the appropriate field.

19.4 ADDITIONAL CTB FILE FEATURES

We briefly cover two other useful features about which you should know. The first one was already alluded to earlier. When you selected all the colors and turned them to black, you could have left out a few. These colors would then have been plotted in color. So this action is certainly not an "all color" or "all black" deal. You can mix and match; and typically in a piping or electrical plan (for example), the entire floor or site plan is monochrome, but the pipes or cables and wiring are left in vivid red, blue, or green (yellow and cyan do not show up well on white paper). The intent of the design is much clearer. It is a great effect, although it can obviously be derailed by not having a color-capable printer or plotter.

The second feature has to do with the screening effect. This lightens any color of your choice so less ink is put on the paper during printing. The color is literally "faded out," and it is a very useful trick, as we see with an example. The color most often chosen to be screened is color 9 (gray), as it is a natural for this effect and already looks faded, although any color can be used.

To screen a color, select it in the editor as just outlined, and go to the Screening field (as seen in Fig. 19.8). Use the arrows or type in a value. Typically, the screening effect is most useful around 30%–40%. Then, proceed as before with saving and plotting.

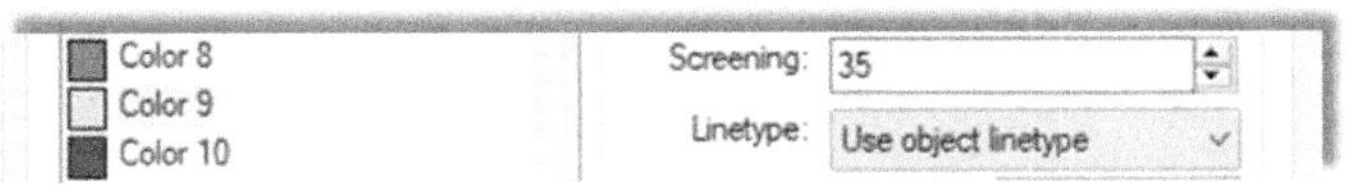

FIGURE 19.8 35% screening of Color 9.

So, what is the effect like? Fig. 19.9 shows two solid fills on the same solid white color: one with no screening (left), meaning it is 100%, and one with the 35% screening (right) as viewed through the plot preview window.

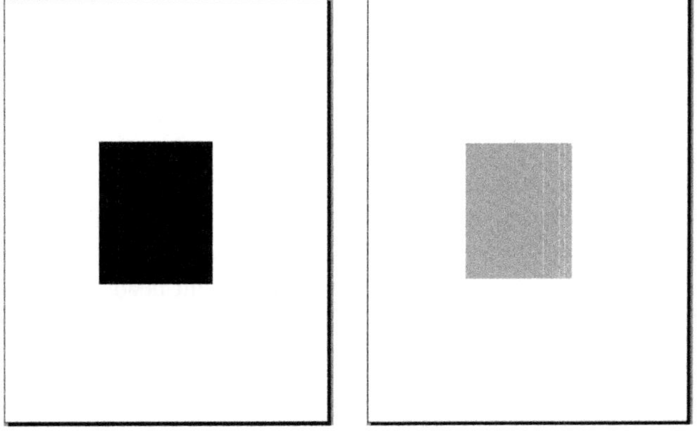

FIGURE 19.9 Plot preview without and with 35% screening.

- It is a great trick to use. Here are some ideas where you can do this:
- In a key plan, where the area of work is grayed out and screened, yet the underlying architecture is still visible.
- In an area of work on a demolition plan, where the area to be demolished for a redesign is grayed out, with all the features still visible.
- As a rendering tool, to create contrasting shading.
- As a rendering tool, to give a sense of depth or solidness to exterior elevations.

There are many more applications.

19.5 LINEWEIGHT SETTINGS

An additional Lineweight settings option is available to you. It is found all the way on the right among the bottom screen menu choices, as shown in Fig. 19.10. If it is not there already, you of course have to first bring it up via the main menu at the far right (the three horizontal bars). If the icon is blue, as shown in Fig. 19.10, the Lineweight feature (called LWT in older versions of AutoCAD) is active. If it is grayed out, then it is not.

FIGURE 19.10 Lineweight icon.

Right-clicking on this icon brings up the Lineweight settings... dialog box (Fig. 19.11).

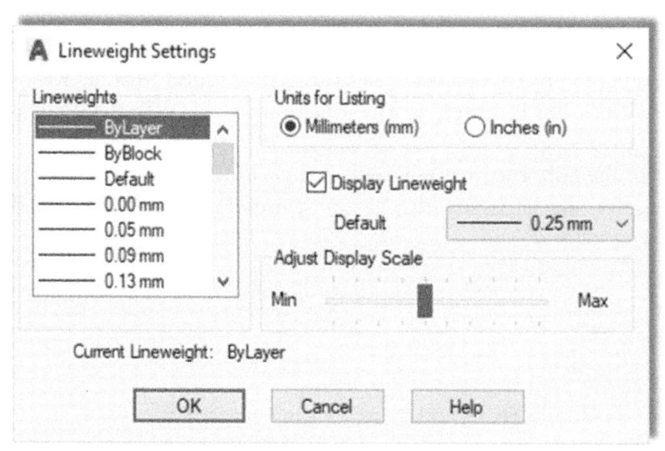

FIGURE 19.11 Lineweight settings.

This feature is not the same as covered earlier in this chapter. The CTB file and setting the lineweights in that manner gets you only plotted results. This means that, on screen, all the lines remain the same size, regardless of the settings. The results of the CTB settings can be seen only on paper.

In contrast, the Lineweight option allows you to set line thickness that is actually visible on screen. It is a rather interesting effect. Once set, the lines remain at that thickness regardless of how much you zoom in or out. Opinions on this effect vary among AutoCAD users. Some think it is very useful and accurately reflects what will be printed, whereas others see it as a distraction and are perfectly fine remembering what colors are set to what line thickness via the CTB method.

It is up to you to experiment and see where you stand. To set Lineweight globally and draw using that weight, use the dialog box shown in Fig. 19.11. Alternatively, you can set this layer by layer by going to the Layers dialog box, creating a new layer (usually with a color), and setting the value under the Lineweight header, as seen in Fig. 19.12 (with 0.50 mm used for clarity). The result is seen in Fig. 19.13.

After setting this layer as current, draw something. Notice that the line is thin, as if nothing happened. Now press in the Show/Hide Lineweight button to activate it; the line suddenly turns thicker, as seen in Fig. 19.14. This is exactly how it will print. Note, however, that CTB settings supersede the Lineweight settings, and you would leave the CTB settings as default if you are inclined to use this Lineweight feature instead.

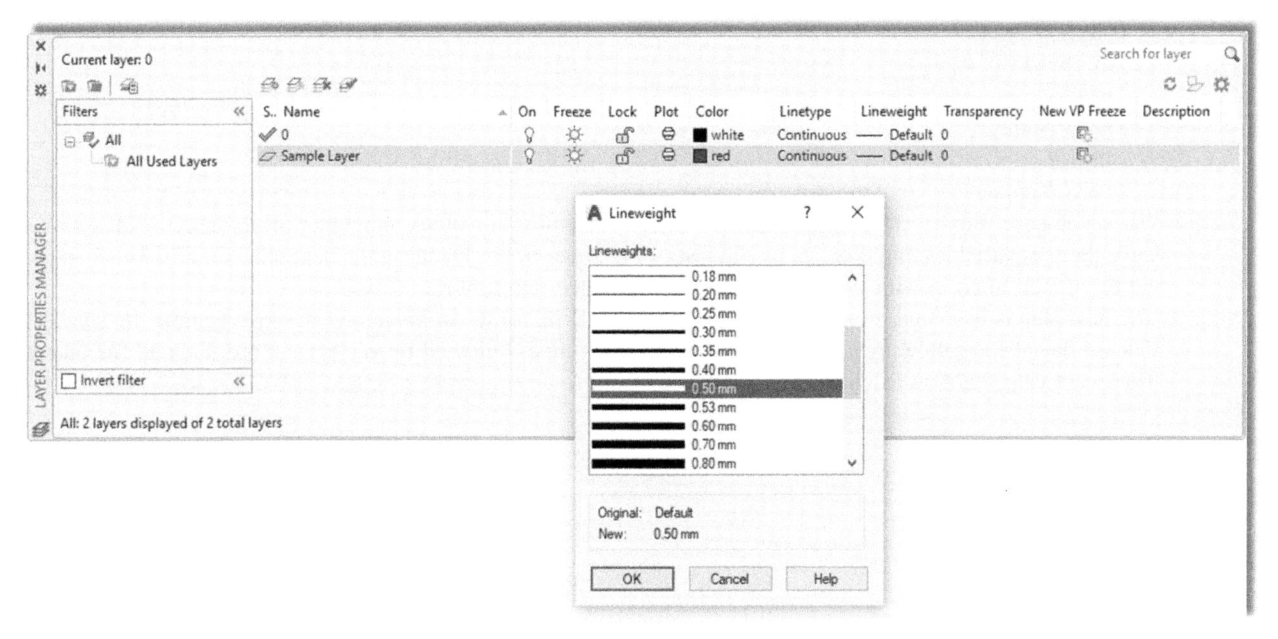

FIGURE 19.12 Lineweight settings.

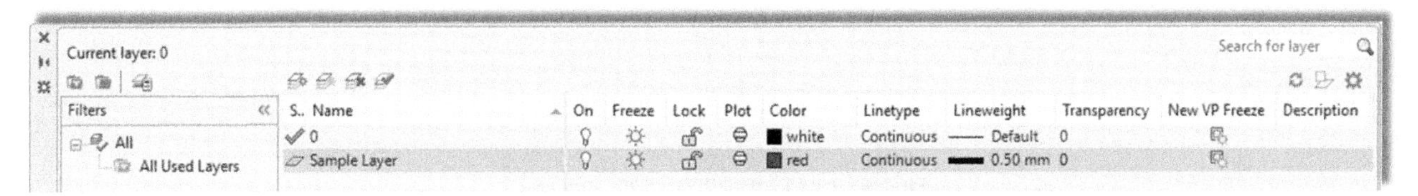

FIGURE 19.13 Lineweight set to 0.50 mm.

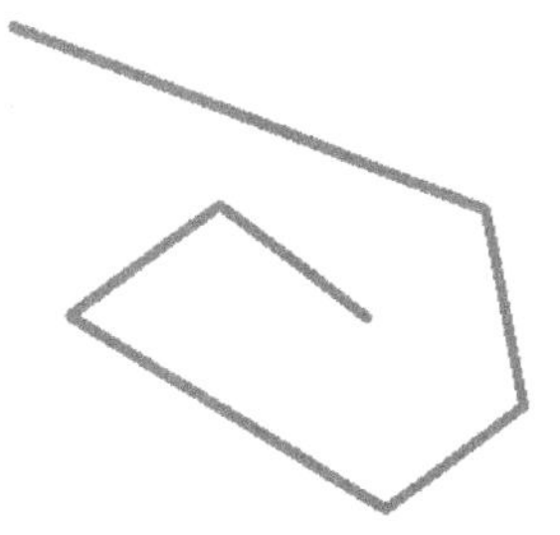

FIGURE 19.14 Line drawn with the 0.50 mm Lineweight setting.

19.6 LEVEL 2 DRAWING PROJECT (9 OF 10): ARCHITECTURAL FLOOR PLAN

For Part 9 of the drawing project, you need to do two things. You must first set up a master title block that contains attributes. It is reused numerous times when we set up full Paper Space layout tabs in Part 10, in the next chapter. Then, use what you learned in this chapter to create a useable CTB file.

Step 1. Apply attributes to create an intelligent title block. You can use a modified version of the title block first created in Chapter 10. Start out with that basic template, insert it into a Paper Space tab, and add some extra lines to the lower right part of the title block (see Fig. 19.15). Then, following what you learned in Chapter 18, add text for the following categories, followed by the attributes themselves. Make sure you are drawing on a title block layer (color: white).

- Drawn by:
- Checked by:
- Approved by:
- Date:
- DWG Name:
- Sheet #:
- DWG Number:

Step 2. When you have filled everything in, make a block of the entire title block and either fill in the Xs right away or leave them as variables until you set up the rest of the Paper Space layout in the next chapter. Fig. 19.16 shows the title block filled in with information after it was made into a block.

Step 3. The next step is to assign proper lineweights to the CTB file of the architectural design. Assume the standard that we have been following and make green the darkest line, followed by yellow and red, then all the others. The exact values are up to you.

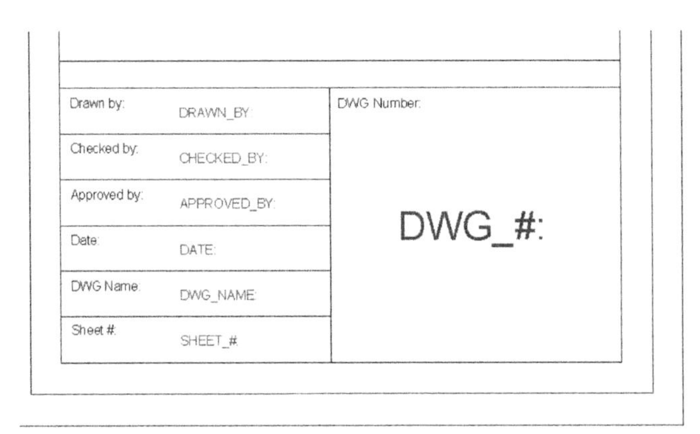

FIGURE 19.15 Title block with attributes.

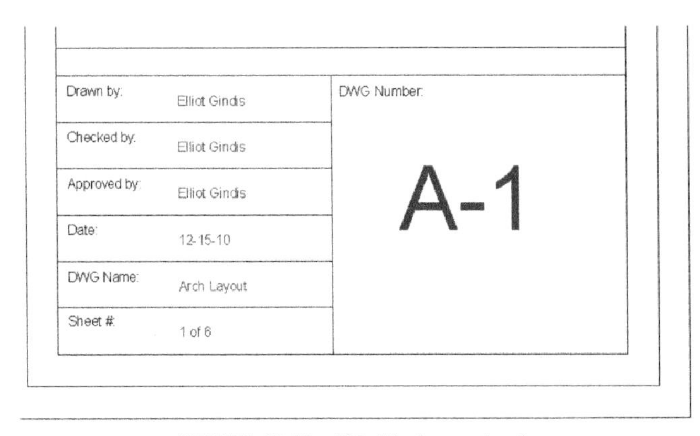

FIGURE 19.16 Title block completed.

SUMMARY

You should understand and know how to use the following concepts before moving on to Chapter 20:

- CTB file
- Editing lineweights
- Screening
- Lineweight settings

REVIEW QUESTIONS

Answer the following based on what you learned in this Chapter 19:

1. What is a CTB file?
2. Describe the process of setting up a CTB file.
3. What is screening?
4. What is the effect of Lineweight settings?

EXERCISE

1. For practice, create a brand new CTB file. Call it Ch_19_Sample_File. Monochrome all colors and assign the thicknesses to the listed colors as seen next:
   - Color 1, red, 0.2500
   - Color 2, yellow, 0.1900
   - Color 3, green, 0.1500
   - Color 4, cyan, 0.3000
   - Color 5, blue, use object lineweight
   - Color 6, magenta, 0.0900
   - Color 7, white, 0.1000
   - Color 9, gray, use object lineweight
   - Screen color 9%–30%.

   (Difficulty level: Easy; Time to completion: <10 minutes.)

Spotlight On: Drafting, CAD Management, Teaching, and Consulting

Overview

Our final Spotlight On focuses on drafting, CAD management, teaching, and consulting. The first two of these topics may seem like rather odd choices; after all, haven't you been studying them throughout the book? While it is true that you have been learning how to draft and may have even read the appendix dedicated to CAD management (see Appendix E), we really have not discussed what it is like to actually have a career working as a drafter or CAD manager. While all the engineering, architecture, and design professions we profiled earlier are excellent career paths, many students using this textbook began learning AutoCAD to be a drafter, and this is what we focus on. We also look beyond doing drafting to teaching drafting, either in a classroom or in a consultant capacity.

All these topics are very personal to me. While I am working as an engineer right now, I started off my career in the mid-1990s as an AutoCAD drafter and kept at it for 10 years full time (actually double time, you could say, considering the long days I worked). It was not until later in life I returned to engineering and away from pure drafting. Although today I still use AutoCAD regularly and still teach, write, and consult, I will never forget the decade of AutoCAD work, in the trenches of New York City. Drafting, CAD management, consulting, teaching, and corporate training was my life—14 h per day, 6 days per week. Although one could say that the "Golden Age" of AutoCAD has somewhat passed, with enough qualified candidates finally filling the unending industry demand, it still can be a rewarding and well-paid career.

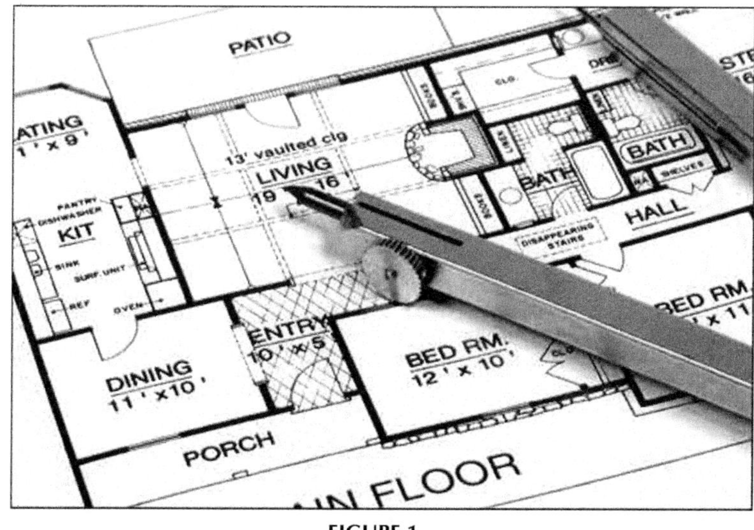

FIGURE 1

AutoCAD Training Requirements

The educational requirements for an AutoCAD drafting career are more rigorous now than at the peak of demand, when I got started. Although I had minimal experience back then, I was hired anyway, and the company trained me in-house over a grueling 6-month period; AutoCAD boot camp you could say. Now, 20 years later, this is much less common. You need to attend one of the many community college or tech school programs to get either a certificate or a full diploma in drafting. This can take anywhere from a few months (basic certificate) all the way up to 2 years (for an associate's degree). Either way, you graduate "ready to go" and can be productive at a company right away.

If necessary, take both of the exams offered by Autodesk: User and Professional (see Appendix K) for a good "foot in the door" to a company. However, I consider these exams inadequate in measuring the overall knowledge of an individual. They just are not comprehensive enough. The proof of your skill remains in the ability to draft quickly and efficiently, and many companies will make you sit in front of a CAD station to draft a given sample. I was on both sides of that fence, doing a few of those and later on devising tests for new hires. Being fast and efficient is still the best way to stand out from the pack. I cannot say it enough—practice your basics!

Architectural Drafting

So, what is it really like once you get hired as a draftsperson? This of course depends on where you go. You can draw a line between the world of architecture and engineering. They are different and have their own pros and cons. On your first day on the job, in either profession, you will be shown company standards and given some time to customize your CAD station. Be sure to get AutoCAD looking like you want it. Gone are the days when you shared workstations. You will be on your own computer, so be sure to customize anything that boosts your productivity (pgp file, toolbars, etc.).

Be sure to take note of the printer/plotter configurations, and what is available. Be sure to also identify the lead CAD designer or the one most knowledgeable in AutoCAD. Not only are you guaranteed to get stuck on some features of the software, but also you need to know exactly what standards are to be followed—not all of them are enforced—and there may be a reason why. You also need to learn all the network paths and where to retrieve and save drawings.

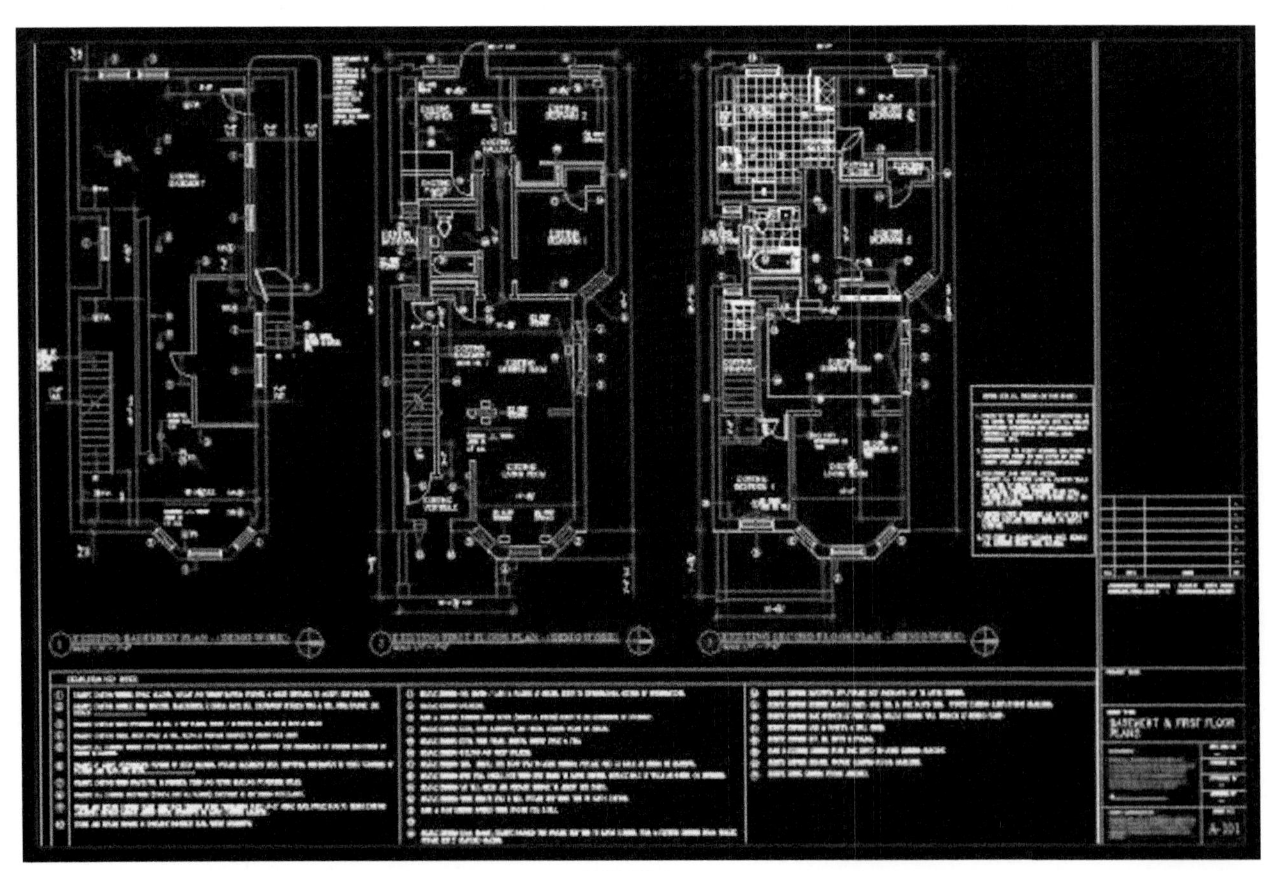

FIGURE 2

Day-to-day work at an architecture office revolves around whichever project is assigned to you. In my experience, it is more stressful in architecture, given the deadlines inherent in that profession. Engineering offices in larger companies (such as Siemens and IBM, two companies with which I spent significant time) operate on larger budgets and longer timeframes, giving you a chance to develop additional skills for the job. Architects bill per hour, and you need to be on your game from day one. Expect to see Paper Space and xrefs as a matter of routine. Expect also (in some offices anyway) an emphasis on 3D. More and more architects create computer renderings for clients, often using additional software such as Rhino, SketchUp, and Revit. By and large, however, regular 2D floor plans will be your bread and butter. Most architects I worked with used redline markups, basically red ink on a paper printout; and you will quickly learn the meaning of the phrase "he (or she) bled on it." You will take those changes and execute the actual drafting and then print a new sheet for more markups or a final check.

As you gain experience, the architect may allow you to lay out the design or make changes to the design, especially if they are routine, then just approve what you do. This is a great way to get to know the profession in case you do want to become an architect. Although I was never interested in an architecture career, somewhere out there exist entire floors in buildings that I designed. Expect to also be out in the field to do surveys or "as-builds." You show up with a camera and pen and paper and take down information, measurement, and pictures to bring back to the office. Sometimes you can bring AutoCAD with you on a laptop and do the drafting right on the spot. These surveys have gone high tech in recent years with widespread usage of hand-held laser measuring tools.

The money you can earn while drafting for architects is good but not great (except of course at the very top). Drafting time is submitted as part of the overall bid, but generally the allowed rate is not high by the time everything is negotiated. I left architecture after a few years to move to larger engineering companies, with a resulting boost in my pay. To that end, while I was working in architecture offices, I noticed the high rate of turnover in their drafting staff, with exceptions of course. One huge benefit of this type of drafting is the experience you get on AutoCAD. A sizeable architecture project, especially with a 3D component, uses virtually all of AutoCAD's capability. If you have a choice in the matter, start your career in architecture, and then move into engineering drafting. You will gain critical knowledge that will come in very handy if you want to be a CAD manager and for teaching or corporate training.

Engineering Drafting

Although this work is not fundamentally different from architecture—you are still using AutoCAD after all—as mentioned, some aspects of the job do change. In my experience, the projects are larger, more complex, and in many ways more diverse. At Siemens, where we worked on the New York City subway's signaling system, the work spanned the mechanical, electrical, and even structural fields. Although my use of AutoCAD was limited to mostly intermediate-level drafting and there was little in the way of 3D modeling, I did have to interact with varying CAD systems and other software. Most of Chapter 16 was written based on that experience. The money also was far better, as you would expect from a multimillion-dollar contract. Keep in mind of course that this was a major firm and engineering work done at a small office is not going to be the same.

Engineering drafting involves the same level of detail as architecture. You may encounter a great deal of electrical schematics, cross sections, and use of attributes to hold information. A lot of our layouts of complex communication equipment were simple from a drafting perspective, same lines, circles, and rectangles over and over again, but complex in its intricacy. When you have thousands of pieces coming together, you have to approach things differently. Use of blocks, nested blocks, and symbol libraries is routine. We had to be very careful in how we presented complex information, which brings us to another major point.

Especially in the engineering world, your job is not just drafting but designing a good layout. It is the next step up from just doing redline markups. You have to constantly take a step back from your drawing and ask if this will be easy to read by someone who is seeing it for the first time. Use wipeouts to hide what need not be shown (see Chapter 15) and use darker outlines for major pieces, reserving lighter lines for the secondary items. Be creative in positioning design elements on a crowded drawing and always ask yourself: Does this make sense? The purpose of drawings is to convey information. If it fails in that regard, you have not done your job as a draftsperson.

CAD Management

What is meant by CAD management is simply that you are the top draftsperson in the organization and it is your job to not only draft but also keep an eye on standards, resolve any AutoCAD questions and concerns, and perhaps supervise other draftspersons. No special training is needed, only experience. I spent several years as a CAD manager at Siemens, hiring and overseeing several individuals in that time, one of whom became my business partner in my company, Vertical Technologies, a few years later.

The key here of course is knowing AutoCAD inside and out and being able to explain to nontechnical managers what the issues are. You also have to be able to give reasonable estimates of how long something will take and any technical hurdles on which you need support. My ability to teach AutoCAD clearly to students was partially honed in my experience explaining issues to managers. You would be amazed how complex and mysterious this software and its abilities can seem to some. As a CAD manager, you will find yourself attending far more meetings than if you were just a draftsperson. Expect to also find yourself teaching others how to do things right.

The final piece of advice for CAD managers is this: Be careful hiring. As described in Appendix E, the best person to hire is the one that is fastest and most accurate on AutoCAD, as demonstrated by having the person do a simple drawing, such as the floor plan from Chapter 3. Experience has shown me that if someone does not know an advanced concept, that can be taught in a few minutes. But only extensive time spent drafting can instill speed and accuracy.

Teaching AutoCAD

Not everyone enjoys or wants to teach. Teaching is a calling more than a job, though teaching a small group of adult professionals is a far cry from a room full of fourth graders, so we will not pretend it is the same thing. However, it is still very rewarding and a lot of fun. You learn even more about AutoCAD through teaching because you really have to dissect it for students and troubleshoot some unique problems students get themselves into. I stepped into teaching by accident when, in 1999, a coworker at Liz Claiborne (where we did store planning and design) wanted to go to Europe for the summer after a bad breakup and asked me to fill in for her class at the prestigious Pratt Institute of Design in midtown Manhattan. I do not think she ever came back to the United States, and my temporary teaching gig stretched into many years. I also taught at New York Institute of Technology and other, private, training centers all over the New York City area.

If you think you may want to get into teaching, the best thing to do is to get as much diverse experience as possible. An instructor who has limited experience, even if he or she knows AutoCAD well, will not be able to speak about all the ways AutoCAD is used in the industry. If, however, you have worked across a wide array of specializations and have the patience to teach, this may be for you. Needless to say, you also have to know AutoCAD backward and forward. You have to display confidence in your knowledge and confidence while doing public speaking. You will be in front of anywhere from 5 to 20 students at a time.

To get a job teaching AutoCAD you really have to just watch for an opening and drop off your resume at potential schools. Most schools and educational centers want experience, so getting a foot in the door is not easy. It also matters a great deal what type of program you teach. A community college may offer an entire degree in drafting. Then, you have to be a true "educator," (such as Robert Kaebisch, the coauthor of this textbook), possibly with a degree in education, as you would also be responsible for teaching math, science, and maybe technical writing. I always taught in continuing education programs. This is far easier in terms of non-AutoCAD credentials. My students were all adults, and I had to teach only AutoCAD, usually for 8 weeks at a time. I preferred this approach, as I did not really wish to teach math and science. This may be the way to go for you as well, if you want to get into AutoCAD instruction, by seeking out these types of programs. They may be part of a university or college, such as Pratt or NYIT, or private companies such as Netcom, CDM, or RoboTECH, where I taught on and off.

One additional benefit is that the hourly pay is higher in these schools than in a full-time teaching position at a college. This is because you work on "contract" and do not get benefits, resulting in significantly higher pay. On the flip side, you may not work as many hours as a "full timer" would, and because of this may actually make less in the long run. Because I already had a full-time job in the daytime, and purchased my own benefits, this was perfect for me; and I taught in the evenings and often on weekends for many years.

AutoCAD Consulting

The final discussion here focuses on AutoCAD consulting. What is meant by *consulting* is any short- or long-term, and usually highly paid, specialized service that you offer a client. With AutoCAD that can be drafting, but generally you will rarely do that, as drafters can be hired cheaper through placement agencies. Most, if not all, of the consulting that I did refers to corporate training. These were short-term (2- to 5-day) training seminars, also referred to as *boot camps*. They are held at the company's offices and are intense 8- to 9-h per day classes, meant to educate the engineers or architects in proper use of AutoCAD, as well as problem resolution and optimizing AutoCAD to their specific needs.

This type of consulting is the pinnacle of AutoCAD use and AutoCAD training. You fly or drive to the corporate offices and immerse yourself in an intense, high-pressure environment, teaching a very demanding group of individuals, who paid a great deal of money to have you there, while at the same time giving them ideas on how to best optimize AutoCAD to their business and resolving thorny issues with their current approach. It is a tough environment, and it took me over 7 years to truly establish myself in this field, with the opening of my consulting firm in 2003. Prior to that, I was sent to these jobs through several private training centers with which I collaborated.

If you wish to get into this type of work, you need to overcome a few hurdles. You have to have absolute mastery of every aspect of AutoCAD, followed by a wide-ranging knowledge of many industries. You could be teaching structural engineers 1 week and electrical engineers the following week. I once even had the interesting experience of training a squad of FBI special agents assigned to counterterrorism. What did they need with AutoCAD, you ask? They were designing special non-fragmentation concrete barriers to secure a political convention and needed to learn 3D AutoCAD in a hurry. Why didn't they hire a company to create the barriers for them? The work was too sensitive in nature and they needed the product fast; too fast apparently for the lengthy bidding process to find a vendor to design these. The agents designed the barriers themselves (with input from structural engineers) and just had a company manufacture them.

You will also need a solid knowledge of computers, operating systems, and IT networks, as well as have experience on all versions of AutoCAD and LT. You just never know what you are going to get. You have to be willing to travel for a week at a time, often on short notice. The best way to get started is to first partner up with an established training center, then branch out on your own, when you get enough experience. A website advertising your services of course helps a great deal, as does reputation and performance. Many of my clients were repeat business.

Consulting work like this was very well paid for a long time. Typical fees for 1 day of work ranged from $1200 to $2200 or more. I typically charged on the lower end of that range (if I did not have to fly to the client) because I had no middleman to pay and wanted to be very competitive. Over the years, as more and more AutoCAD experts matured in the market, the prices I could charge dropped significantly. It is now a more a matter of convenience, the companies may have in-house experts, but just do not have the time to pull them off of projects to teach newbies or expand the skill sets of intermediate-level coworkers; hence the hiring of outside instructors. It is still a good business to pursue, and thought I do not do these gigs as often anymore, I still enjoy teaching, which almost makes up for the lower revenue per job. Well, almost…

Best of luck to you, in whatever you decide to do!

Chapter 20

Isometric Drawing

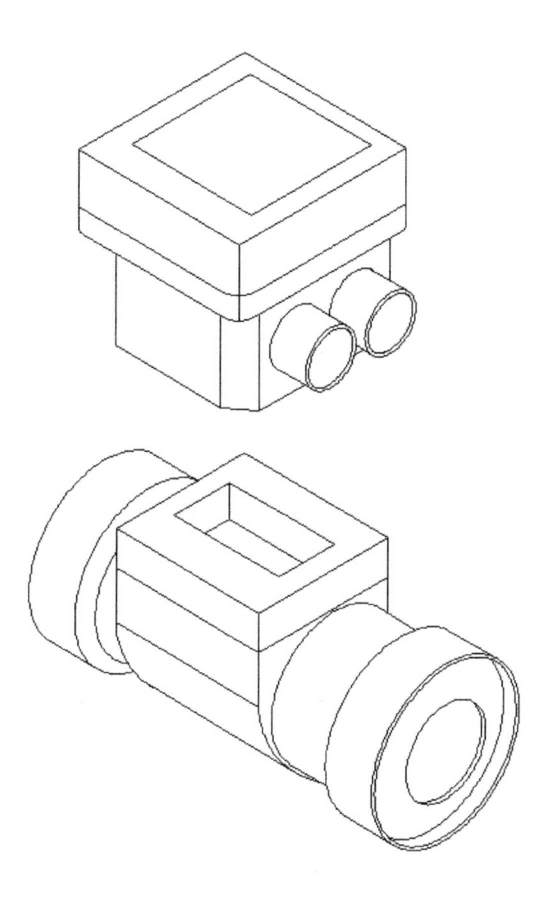

Learning Objectives

In this chapter, we introduce isometric drawing. We cover

- General types of graphical projection
- Isometric projection
- When to use it
- When not to use it
- Setting isometric projection
- Changing planes (F5)
- Ellipses in isometric
- Text in isometric

In the course of this chapter, you will create several isometric designs, including a computer desk and mechanical devices.

Estimated time for completion of this chapter: 1 h.

Up and Running with AutoCAD 2019. https://doi.org/10.1016/B978-0-12-816440-2.00020-3

20.1 INTRODUCTION TO GRAPHICAL PROJECTION

You really cannot talk about isometric projection without first discussing the general (and broader) topic of *graphical projection*. What we are trying to accomplish here is to project a 3D object onto a 2D plane. The various approaches for this are collectively referred to as graphical projection techniques. Isometric projection is just one of those techniques.

None of this is unique to AutoCAD or a new idea in any way. Technical drawing has been around since at least the industrial revolution, and engineers, designers, and architects have been seeking ways to show their designs in three dimensions on a piece of paper for generations. So, in essence (and as it applies to AutoCAD) it means we are drawing a 3D object without resorting to the complexities of 3D space; sort of "faking it" in 2D you could say. This creates the *illusion* of 3D and is often good enough to get across a design idea (or intent) to your customer or client.

Types of Projections.

There are essentially two broad categories of graphical projections—*parallel* and *perspective*. Each has numerous subtypes as described next:

- **Parallel Projection**—In this type of projection, the lines of sight from the object to the projection plane are parallel to each other. What this means in plain English is that the lines do not *converge* into the distance. This is quite the opposite of what your eyes see in reality; however, this approach is preferred for technical drawing. Examples of parallel projection include *orthographic*, *oblique*, and *axonometric*. Axonometric projection, which can reveal multiple sides, also includes several subtypes, including *isometric*, *diametric*, and *trimetric*. The most widely used of these is the isometric projection, which we will discuss in more detail.
- **Perspective Projection**—In this type of projection we are trying to approximate on a flat surface what is seen by the eye. This means usage of concepts such as *vanishing point convergence* and the appearance of *foreshortening*, which means that an object's dimensions along the line of sight are shorter than its dimensions across the line of sight. Perspective projection comes in several "flavors." A *one-point* perspective contains only one vanishing point. A typical example is a set of train tracks. If you stand on them and look into the distance, they converge to a single "vanishing point." A *two-point* perspective has two vanishing points, and a *three-point* has three vanishing points. Both of those perspectives can be easily seen with sharp-angled buildings, and these perspectives are most often represented in outdoor scenes. There is also a four-point and zero-point perspective, but we would not discuss those. AutoCAD does not make use of perspective projection in 2D but does in 3D via the DVIEW command. See the downloadable 3D chapters for more information.

20.2 ISOMETRIC PROJECTION

Using automated tools in 2D AutoCAD, you can quickly and easily draw in isometric projection. Using some additional manual methods, you can also draw in a type of *oblique* projection known as "cavalier," and even diametric and trimetric, but those last two cannot be automatically set, so they would be tedious to create.

So, after all the above definitions and descriptions, what do the main projections that we will be using look like in practice? Let us dive right in with Fig. 20.1 that represents isometric projection, created using the basic T-square and triangle. Note the various angles shown for this type:

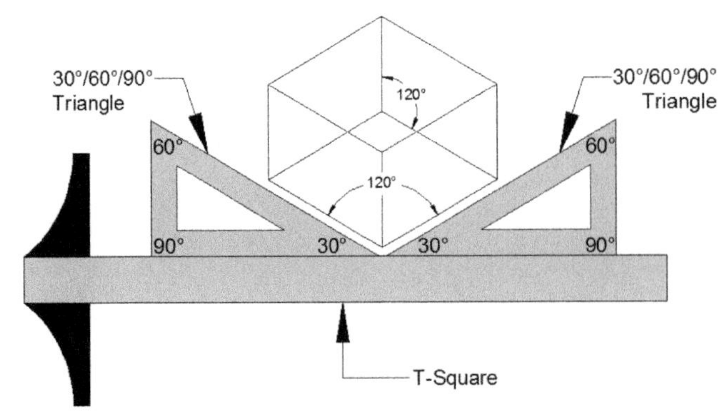

FIGURE 20.1 Axonometric (isometric) projection.

Next, take a look at Fig. 20.2, which represents the cavalier projection, a member of the oblique family. Note the various angles shown for this type:

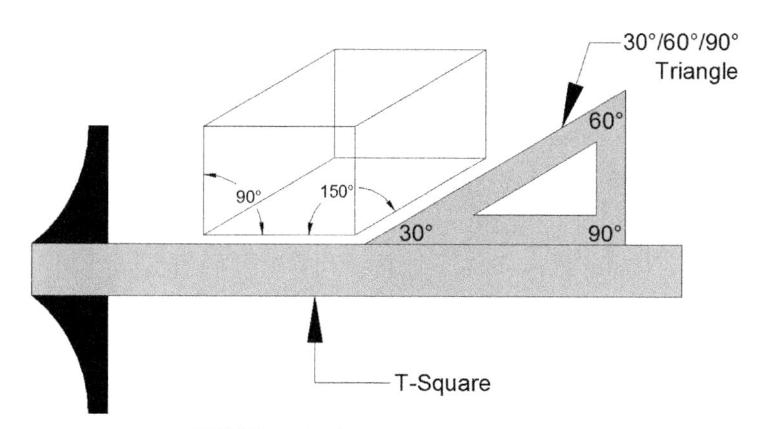

FIGURE 20.2 Oblique (cavalier) projection.

There is nothing inherently special about these two projections in regard to drawing them in AutoCAD. You can easily create both types manually by some simple application of the line and the rotate commands. In other words, draw a few lines at 30 degrees (or rotate some horizontal ones), then draw a few vertical ones, connect everything together neatly, and you will have the 3D-looking "box" shown in Fig. 20.1. However, the same is found in Fig. 20.2, just with more horizontal lines instead of rotated ones.

What we are going to do in AutoCAD is automate this process and create Fig. 20.1 faster and easier. Before we start, one last brief discussion regarding *when* you should use this technique and when you should not.

Why Use Isometric Projection Instead of 3D?

There are a few reasons. One is that it is much easier to learn isometric drawing than to develop proficiency in real 3D. So, for someone who needs only a quick 3D sketch occasionally, just to accent an otherwise excellent 2D presentation, this perfectly fits the bill. Another reason is that 3D takes up more time and computing resources, something that is not always available, and there is also some difficulty in combining 2D and 3D drawings on one sheet. Many students do not go on to take AutoCAD 3D right away (nor should they), and isometric fills that temporary void quite well. For those who do go on, isometric drawing serves as an excellent introduction to 3D and should be reviewed before starting the 3D instruction.

When *Not* to Use Isometric Perspective

Isometric drawing is inappropriate when precision 3D drawings are required for design, testing, and manufacture. Remember, isometric projection is only for visualization. These are *not* true 3D objects and have no real depth, only the illusion of depth. As a result, true measurements and the use of the offset command are not possible in the traditional sense.

20.3 BASIC TECHNIQUE

As just mentioned, nothing is inherently special about drawing an isometric design. A basic manual procedure was described earlier in just a few sentences. It is not hard, but it *is* tedious. There is a way to have AutoCAD preset the crosshairs so you are always drawing in isometric 30 degrees mode. To explain it, we need to introduce the idea of planes, so here is a brief introduction to the plane concept.

In isometric drawing, you will always draw on one of the three available planes: top, right, or left (see Fig. 20.3). Because you are using a 2D pointing device (a mouse), which exists only in a 2D world, you need to be able to easily move from one plane to the other to be able to draw on it. To toggle between planes, press the F5 key. Your crosshairs line up accordingly, and now you just draw using the line command. Here is an important rule: You almost always have Ortho on while in isometric mode; otherwise, the lines are not straight and you are no longer on any plane (though this *is* useful in rare cases). Let us put it all together and draw an isometric box. After you can do this, more advanced isometric drawings are surprisingly easy. One final piece of advice—set your crosshairs to 100% of the screen; it will help to visualize what you are doing better.

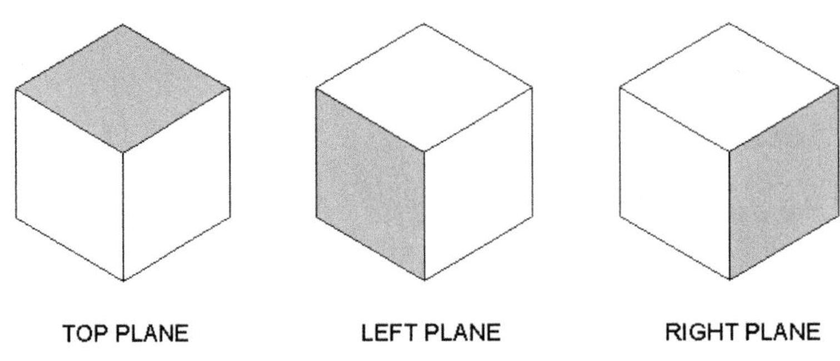

FIGURE 20.3 Isometric planes.

Step 1. Open a blank file.
Step 2. From the cascading menu select Tools → Drafting Settings..., and the dialog box seen in Fig. 20.4 appears.

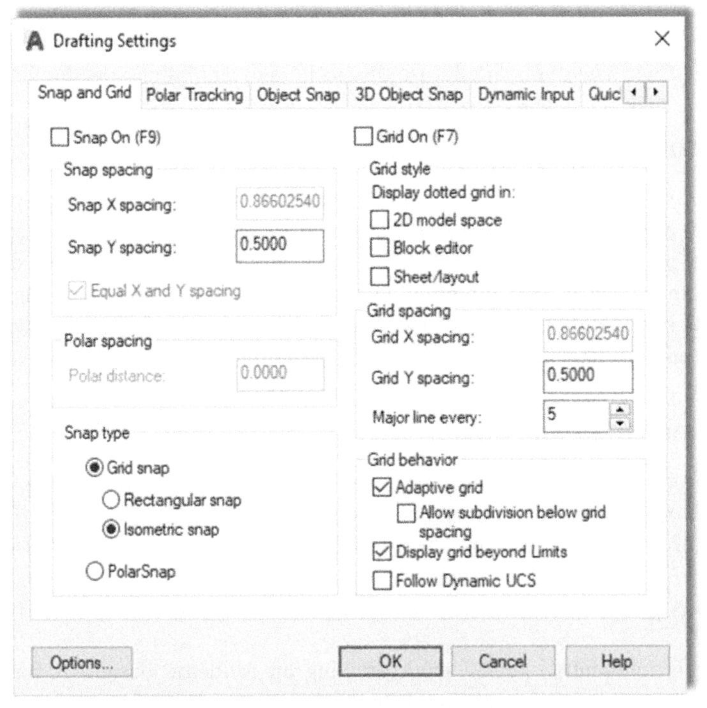

FIGURE 20.4 Drafting settings with Isometric snap selected.

Step 3. Pick the Snap and Grid tab, go to the Snap type category (lower left), pick Isometric snap, and press OK. Fig. 20.4 is a screen shot of the Drafting Settings dialog box with Isometric snap selected.
Step 4. Make sure Ortho is activated and the crosshairs are set at 100% for easier visualization of planes. Note also how the crosshairs are positioned now that you are in isometric mode.

Step 5. Draw straight lines (as they should be with Ortho) of any size and continue drawing until you have the front face of a box. Very important: Do not try to connect the final line to the first line; you will never get it exactly right. Instead overshoot and click after passing it by, as shown in Fig. 20.5.

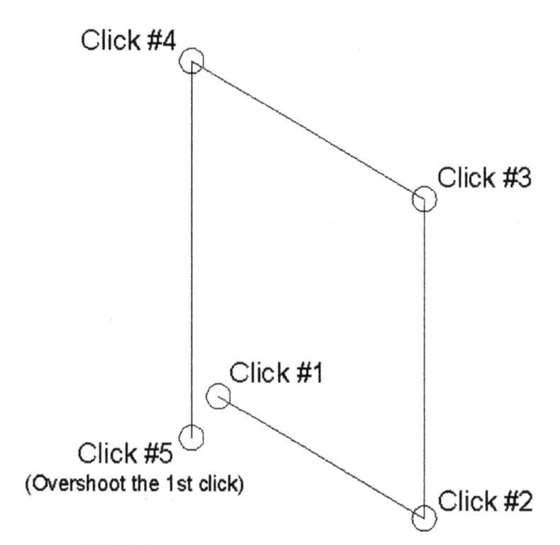

FIGURE 20.5 Isometric drawing, Step 5.

Step 6. Now, use the fillet command to fillet the line between Clicks 4 and 5 and Clicks 1 and 2. Fig. 20.6 shows the result.

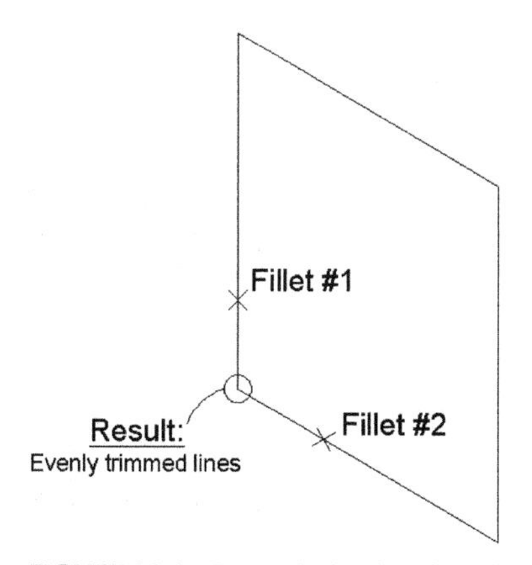

FIGURE 20.6 Isometric drawing, Step 6.

Step 7. Now, press F5. The crosshairs shift to another plane and you can draw another surface.

Step 8. You can press F5 again for the final surface or use the copy command. Be sure to use the ENDpoint OSNAP to cleanly connect lines. Fig. 20.7 shows what you get after a few lines.

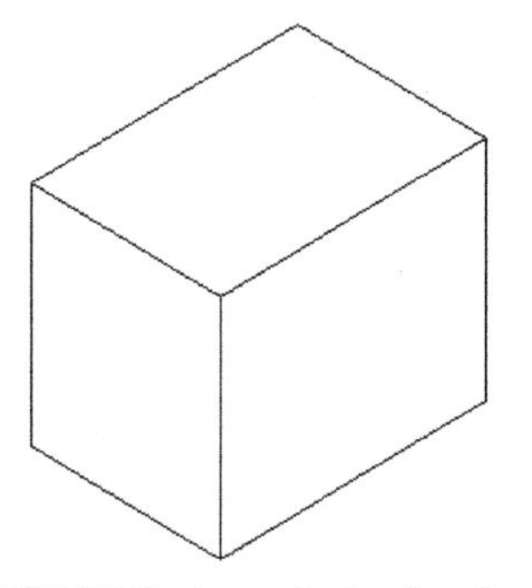

FIGURE 20.7 Isometric drawing, Step 8.

Step 9. And the result, after a few more lines, is seen in Fig. 20.8.

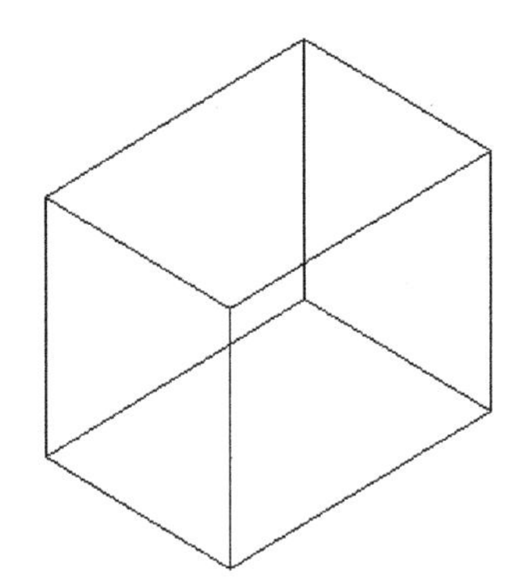

FIGURE 20.8 Isometric drawing, Step 9.

What makes isometric drawing easy is that the techniques do not change for more complex objects. There may be more lines, but it is the same over and over.

Here are some additional pointers:

- To draw circles in isometric perspective use ellipses. This command is introduced in Chapter 2 and used to make the bathtub in the apartment drawing at the end of Chapter 4. We say a few more words about using ellipses in isometric design next.
- Always stay in Ortho mode unless you need to deliberately angle a line at some odd angle (see Exercise 3 at the end of the chapter for an example of this.)
- Text can be angled up or down by 30 degrees to align with the Isoplanes. We will discuss this shortly.
- Have fun with this. Isometric drawing can be accurate and/or a semiartistic endeavor and not always as rigid as orthogonal drafting. Eyeball distances, but try to make the drawing look good. Do not be afraid to throw in some hatch patterns, as is done in some of the practice exercises in this chapter. Remember though, if you need more precision, you must learn 3D from Chapters 21–30 of the free downloadable 3D section of the book.

20.4 ELLIPSES IN ISOMETRIC DRAWING

A circle angled away from you is an ellipse. These shapes find wide application in isometric drawing, and you will need to use them often in the end-of-chapter exercises.

There are two ways to create an ellipse for isometric use—manually and using the Isocircle option of the Ellipse command. It's of course a bit more tedious to do it manually, but it is still a necessary skill if your isometric plane is not perfectly aligned to the expected values of 120 degrees between lines, as seen in Fig. 20.1.

Let us go ahead and first manually create and place an ellipse in an isometric plane. It may require some adjusting to get right and a bit of a creative eye to do so. Reproducing what was first shown in Chapter 2, we have Fig. 20.9.

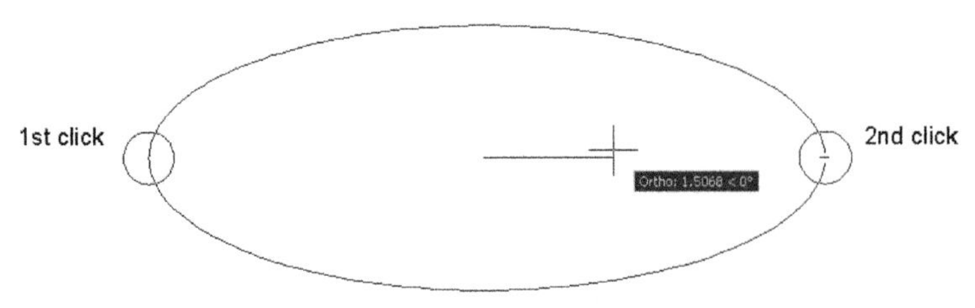

FIGURE 20.9 Basic ellipse.

Once the ellipse is positioned and is the right size, you can rotate it until it looks correct and in the same plane as the rest of the isometric geometry. Fig. 20.10 is an example of some ellipses in an isometric box.

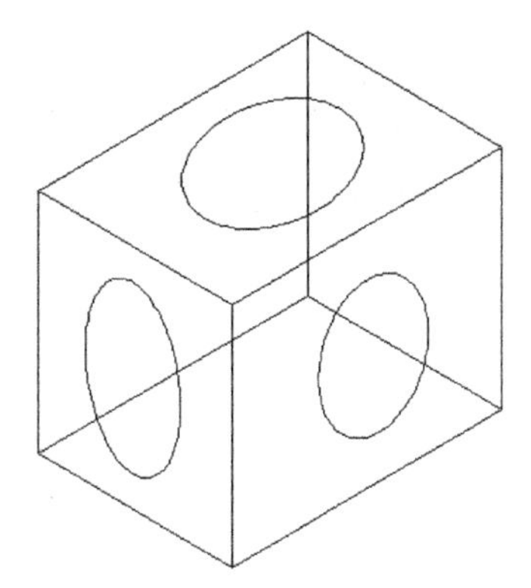

FIGURE 20.10 Isometric box with ellipse.

The more automated Isocircle option is of course easier. For this to work you need to be in one of the isoplanes via the isometric snap option as seen in Fig. 20.4. Cycle through the isoplanes via F5 to land on the one you want to have an ellipse in. Then begin the ellipse command and select the `Isocircle` option by pressing `i`. AutoCAD will say: `Specify center of isocircle:`, then as you draw the ellipse you will see: `Specify radius of isocircle or [Diameter]:`. When done, you will see an ellipse properly aligned to the plane. Simply press F5 and repeat for the other planes. You will again to get the design as shown in Fig. 20.10.

20.5 TEXT AND DIMENSIONS IN ISOMETRIC DRAWING

Text can easily be physically aligned with isoplanes by rotating it plus or minus 30 degrees, or other values depending on the plane, as shown in Fig. 20.11. There is, however, another, arguably better, a way to create isometric text for two of those three planes, by applying an oblique angle to it as described next.

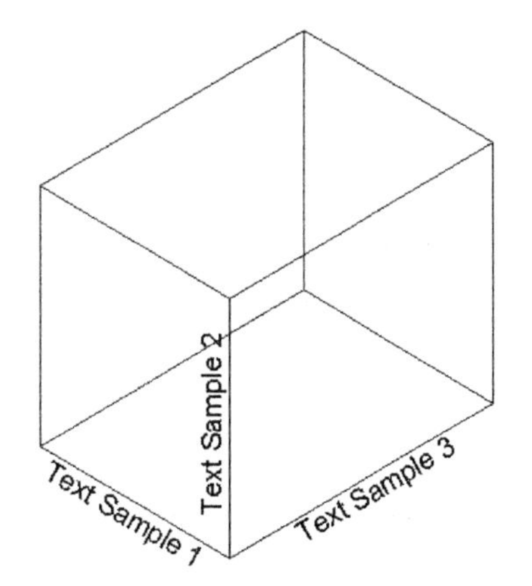

FIGURE 20.11 Text in isometric drawing—rotation.

Go to the Text Style dialog box and create a new style, called Oblique-Right, with an oblique angle of 30 degrees, as seen in Fig. 20.12. Size your text height as needed for your particular isometric design. When done, press Apply, Set Current, and Close.

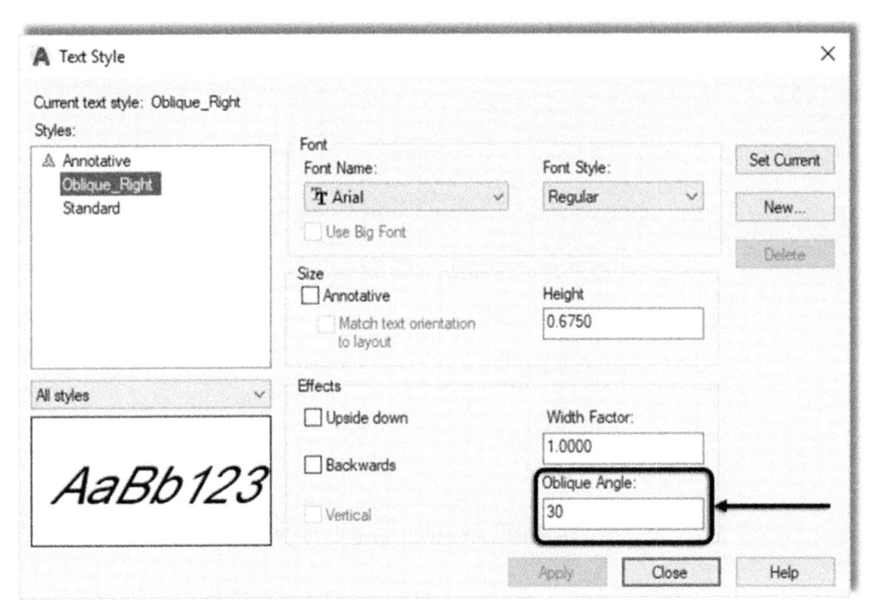

FIGURE 20.12 Setting the Oblique-Right text style.

Go ahead now and rewrite the same text at the bottom of the isometric box as seen in Fig. 20.13. Just as before, make sure it is rotated 30 degrees. Notice now how the new oblique text leans "in plane" with the box. A lot of designers prefer this approach as opposed to just pure rotation.

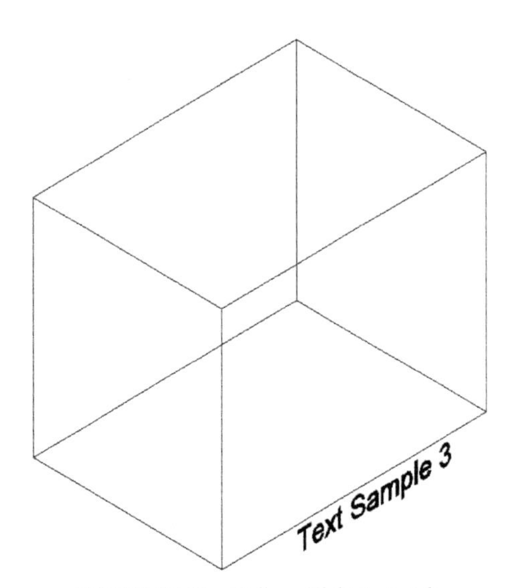

FIGURE 20.13 Oblique-Right text style.

You can also do this with the opposite plane by creating an Oblique-Left text font and leaning it negative 30 degrees. The result is seen in Fig. 20.14. For the straight vertical text (Text Sample 2 in Fig. 20.11), no obliqueness is needed of course, as it is pointing straight up and can be left as is.

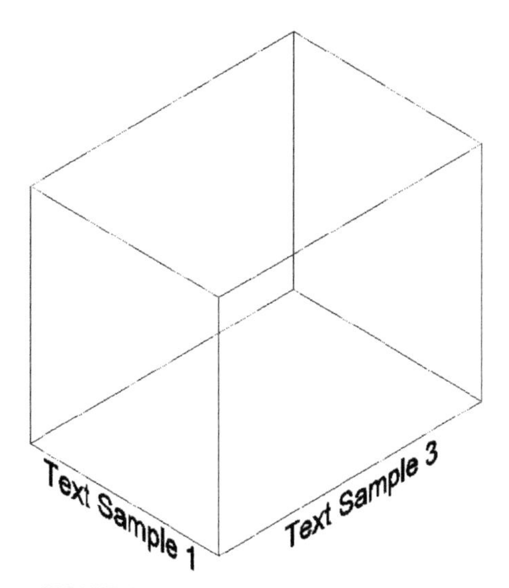

FIGURE 20.14 Oblique-Left text style added.

You can also easily dimension an isometric drawing using Aligned and Vertical dimensions, as seen in Fig. 20.15. However, just like with text, you can improve on this by using oblique angles. For dimensions this is a two-step process. First, if needed, set the dimension text to oblique (+30 degrees or −30 degrees) using the techniques just described for regular text. Next, set the dimensions themselves to an oblique angle. If everything is done correctly, the text and dimension's extension lines will match up and match the isoplane you are on.

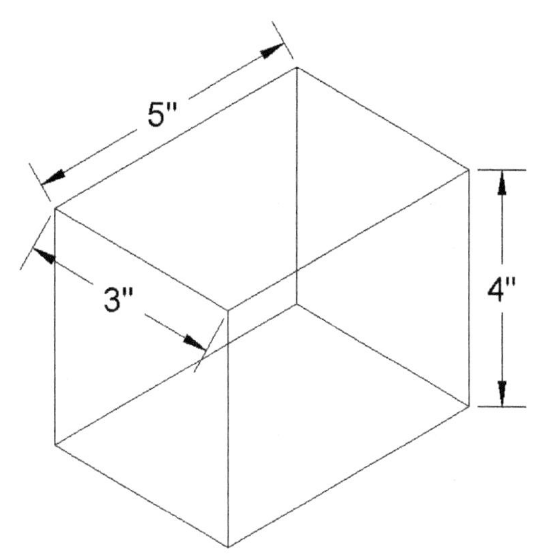

FIGURE 20.15 Dimensions in isometric drawing (no oblique).

To set the dimensions to oblique, first create and place them in position via Aligned or another approach, then type in `dimedit` or press the DIMEDIT button on the Ribbon's Annotate tab under the Dimensions drop down (see Fig. 20.16).

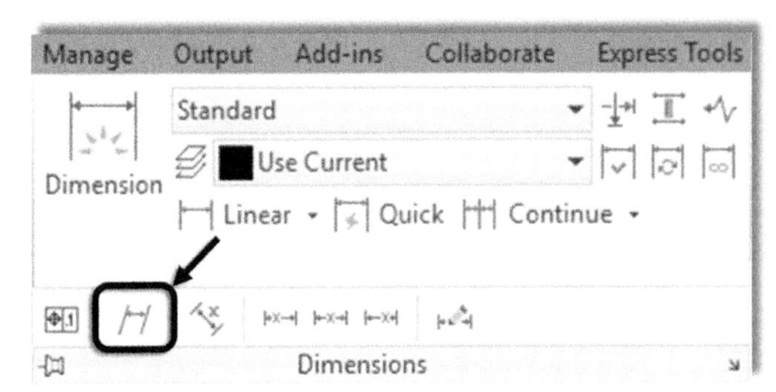

FIGURE 20.16 Dimedit for oblique dimensions.

AutoCAD will ask you to select the dimension and then`Enter obliquing angle (press ENTER for none):`
Go ahead and enter +30 degrees, −30 degrees or whatever value is appropriate, and the dimension will become oblique. If it does not look right, just undo and repeat with other values. Fig. 20.15 should now look like Fig. 20.17. Note that in this particular example there was no need to oblique the dimension's text, only the dimension itself.

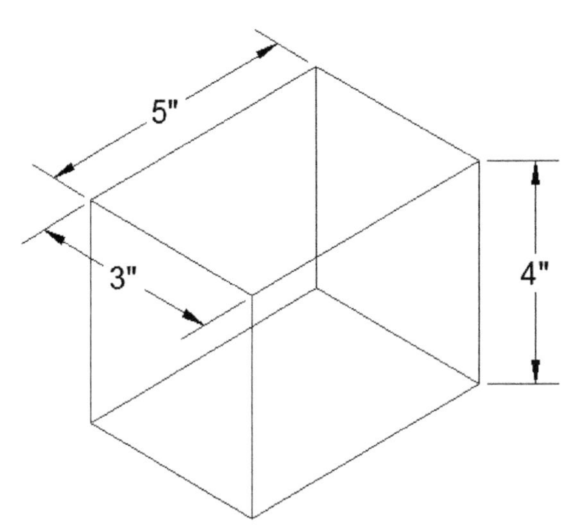

FIGURE 20.17 Dimensions in isometric drawing (oblique).

20.6 LEVEL 2 DRAWING PROJECT (10 OF 10): ARCHITECTURAL FLOOR PLAN

For Part 10 of the drawing project, you must finish the design by adding the Paper Space layouts. You need to put together a lot of techniques you learned in Levels 1 and 2.

Step 1. In Paper Space, create the following seven layouts:

- Architectural layout.
- Furniture layout.
- Carpeting layout.
- Electrical Layout.

- RCP_HVAC layout.
- Landscaping layout.
- Elevations.

Step 2. Insert a title block into each Paper Space layout. Label the title block as appropriate for that layout. Make sure all the attribute fields are filled in and page numbers are incremented.

Step 3. Create an mview window for each layout (VP layer, color: white), centering and zooming each layout to the proper scale using the techniques you learned in Chapter 10. The scale will be either ¼″, ⅜″, or not to scale. Freeze the VP layer when you are done with each of the layouts.

Step 4. In each layout, use advanced layer settings to freeze all layers that are not necessary for that particular view. You will likely need to create as many layer settings as there are Paper Space layouts, as each one shows something different.

Step 5. Add labels and any other title block information you may need.

Step 6. If your computer is connected to a printer or plotter, plot out all of the layouts on D-sized paper. Make sure all of the following are addressed:

- Paper Space layouts are correct.
- Proper scale is used in each mview window.
- Proper layers show in each mview window.
- The title block is correct for each layout.
- The CTB file is correct and there are varying lineweights.

All the layout plans are shown in Figs. 20.18–20.24.

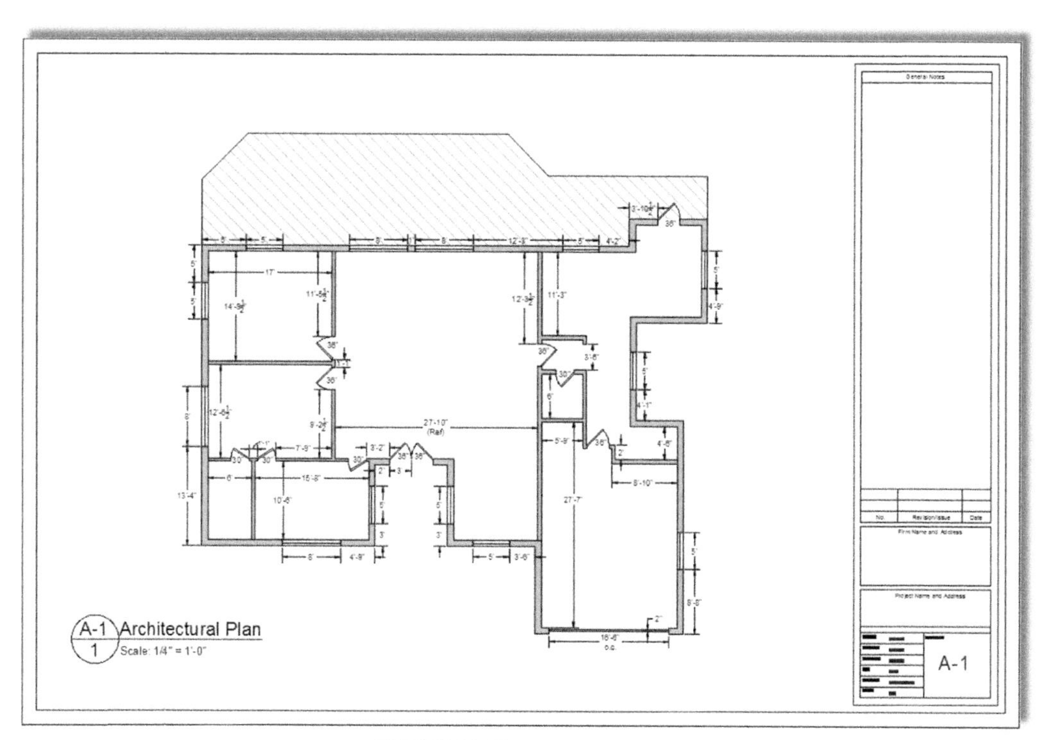

FIGURE 20.18 Architectural plan.

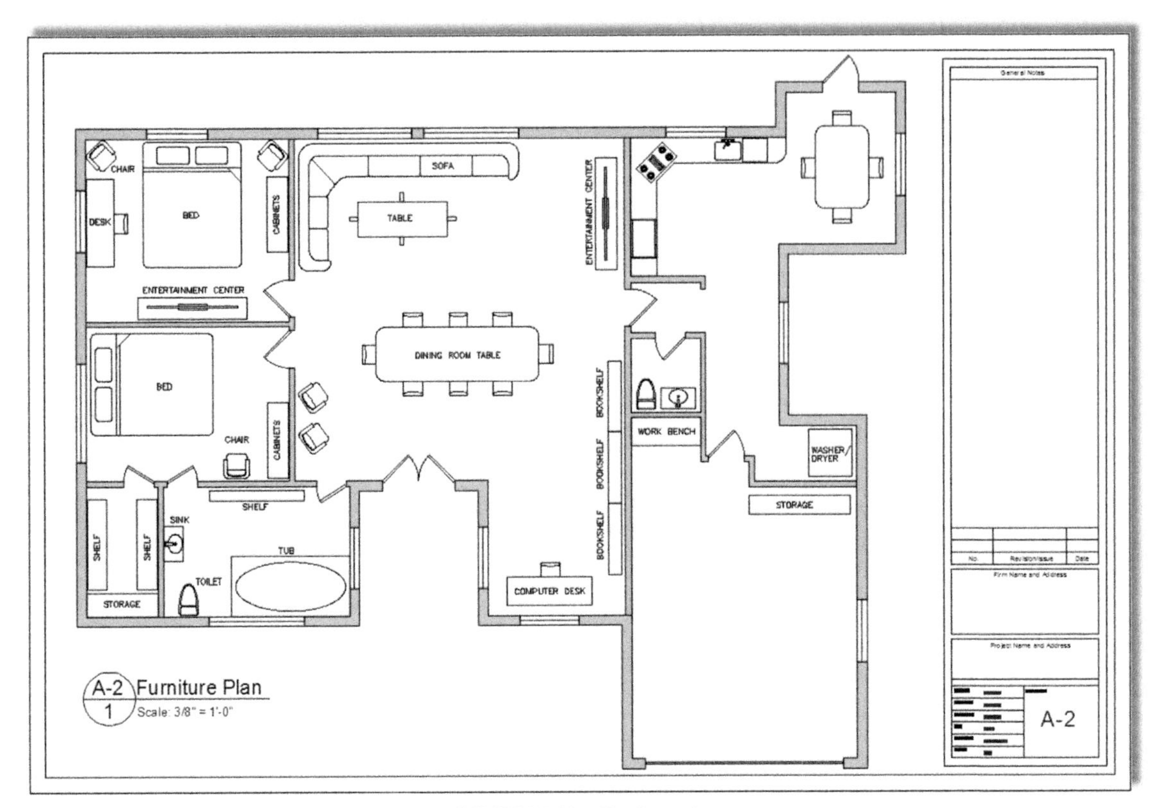

FIGURE 20.19 Furniture plan.

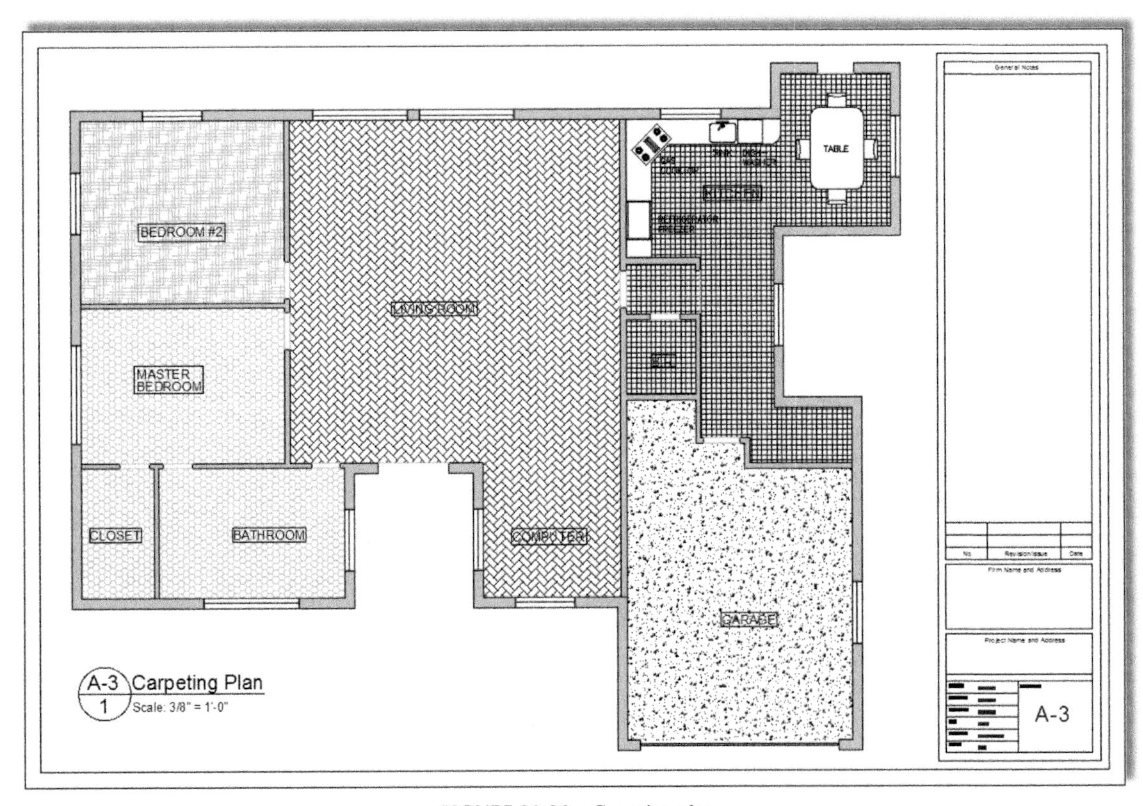

FIGURE 20.20 Carpeting plan.

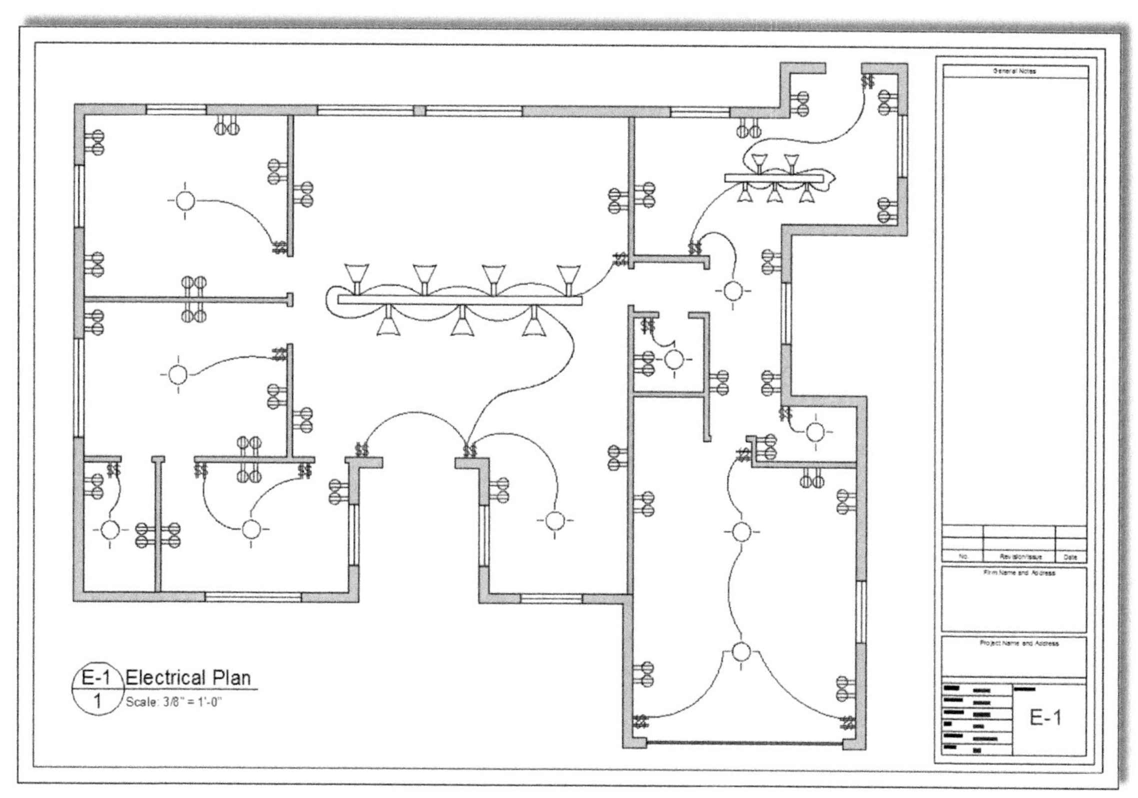

FIGURE 20.21 Electrical plan.

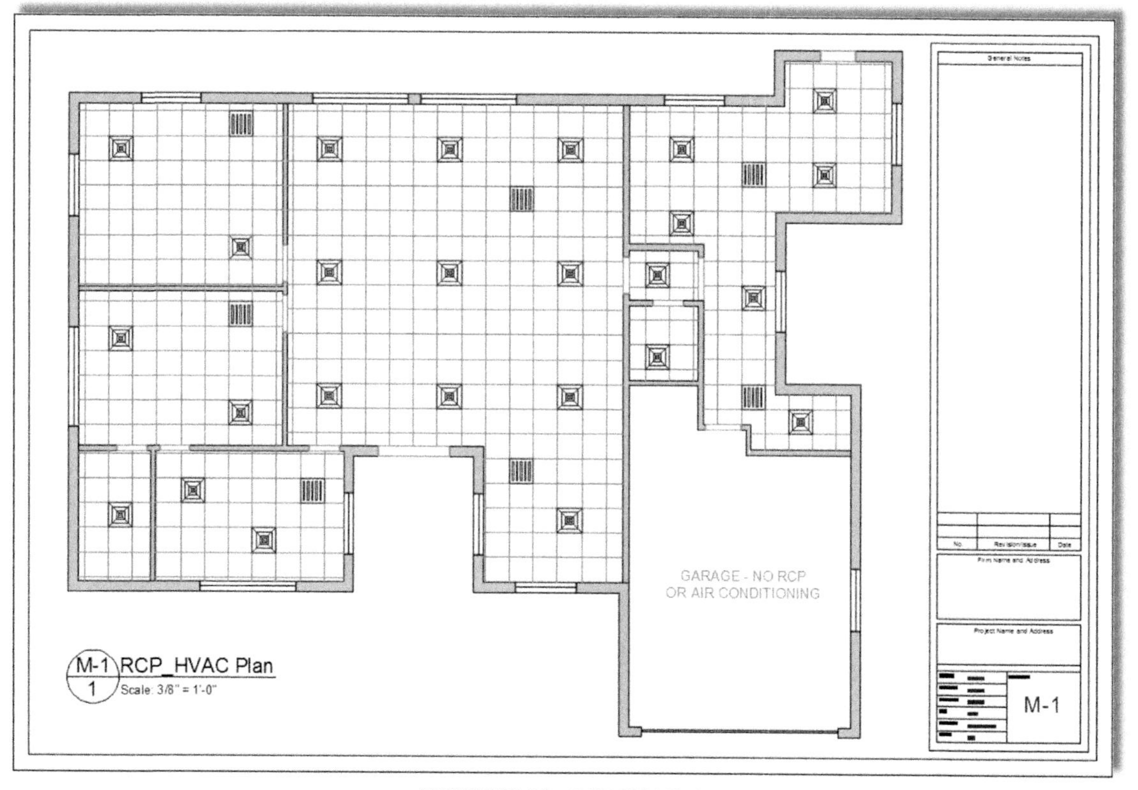

FIGURE 20.22 RCP_HVAC plan.

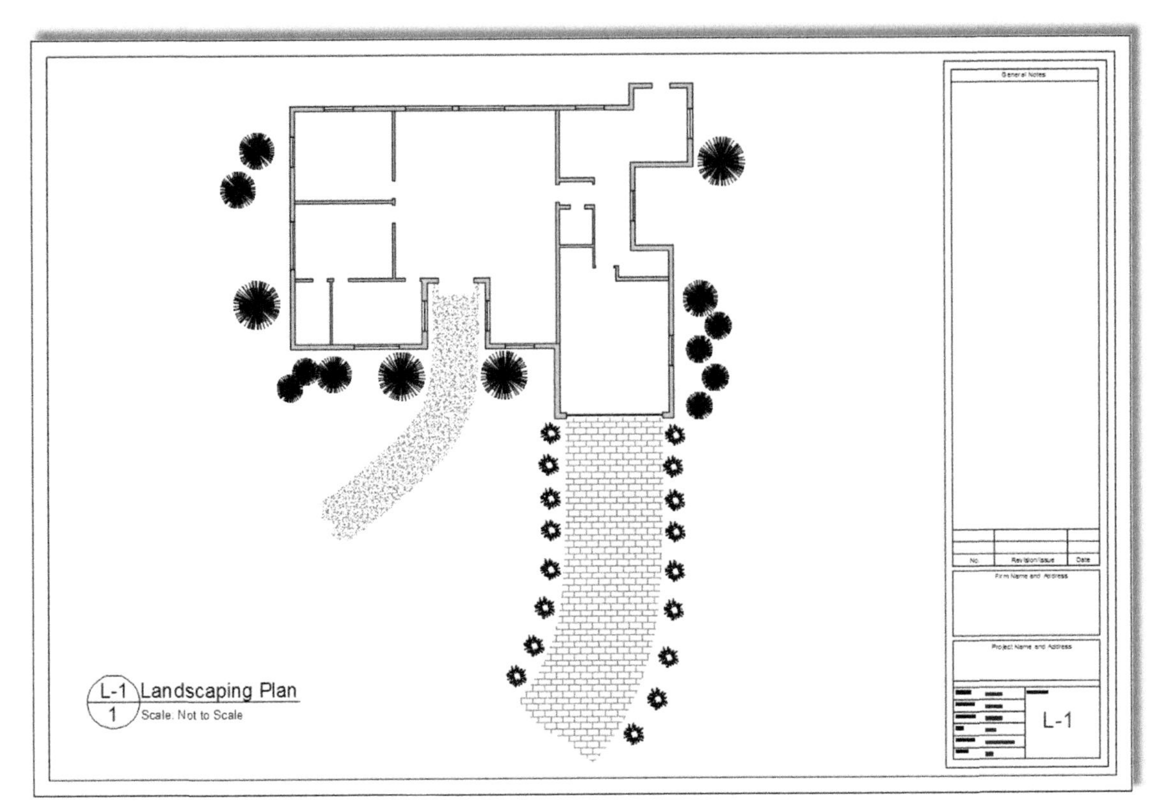

FIGURE 20.23 Landscaping plan.

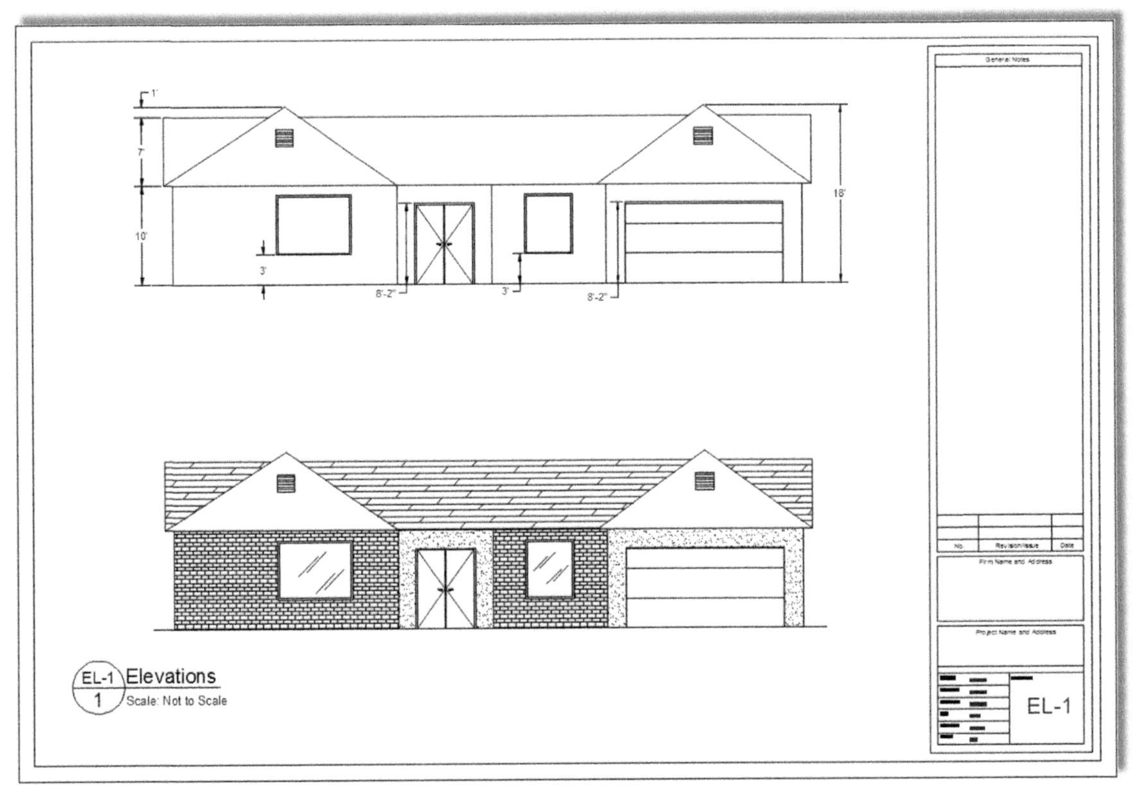

FIGURE 20.24 Elevations.

SUMMARY

You should understand and know how to use the following concepts at the conclusion of this chapter:

- General types of graphical projection
- Isometric projection
  - When to use it
  - When not to use it
- Setting isometric projection
- Planes and changing planes (F5)
- Ellipses in isometric drawing
- Text and dimensions in isometric drawing

REVIEW QUESTIONS

Answer the following based on what you learned in Chapter 20:

1. List some instances when isometric drawing is appropriate.
2. List some instances where 3D would be better than isometric drawing.
3. Name the three planes that exist in isometric perspective.
4. In what dialog box is Isometric snap found?
5. What F-key switches the cursor from plane to plane?
6. What is almost always on when drawing in isometric?
7. What is the equivalent of a circle in isometric drawing?

EXERCISES

1. Draw the following architectural detail. All sizes are approximate and may be estimated. You will need to cycle through all three Isoplanes for the main structure and use ellipses for the fasteners. (Difficulty level: Easy; Time to completion: 10–15 minutes.)

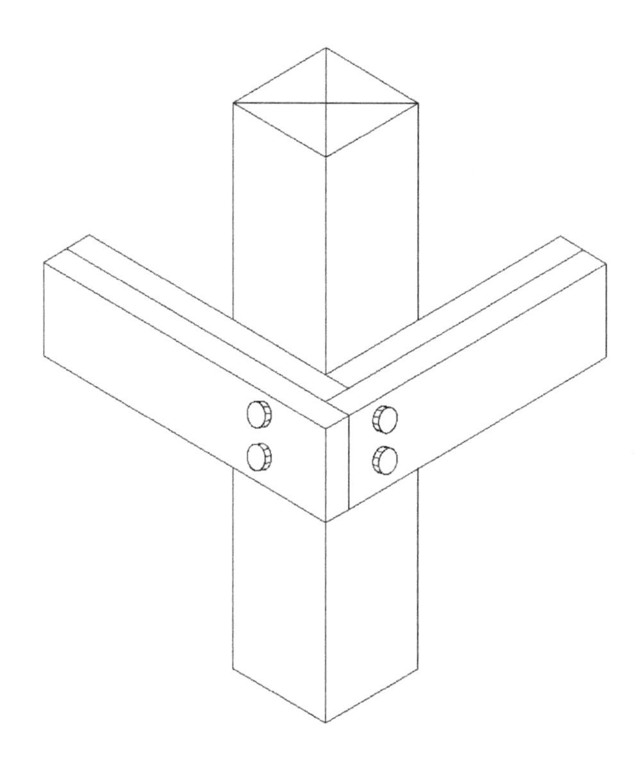

2. Draw the following architectural detail. All sizes are approximate and may be estimated. You will need to cycle through all three Isoplanes for the main structure, use ellipses for the fasteners and a spline for the cutaway. (Difficulty level: Easy; Time to completion: 10–15 minutes.)

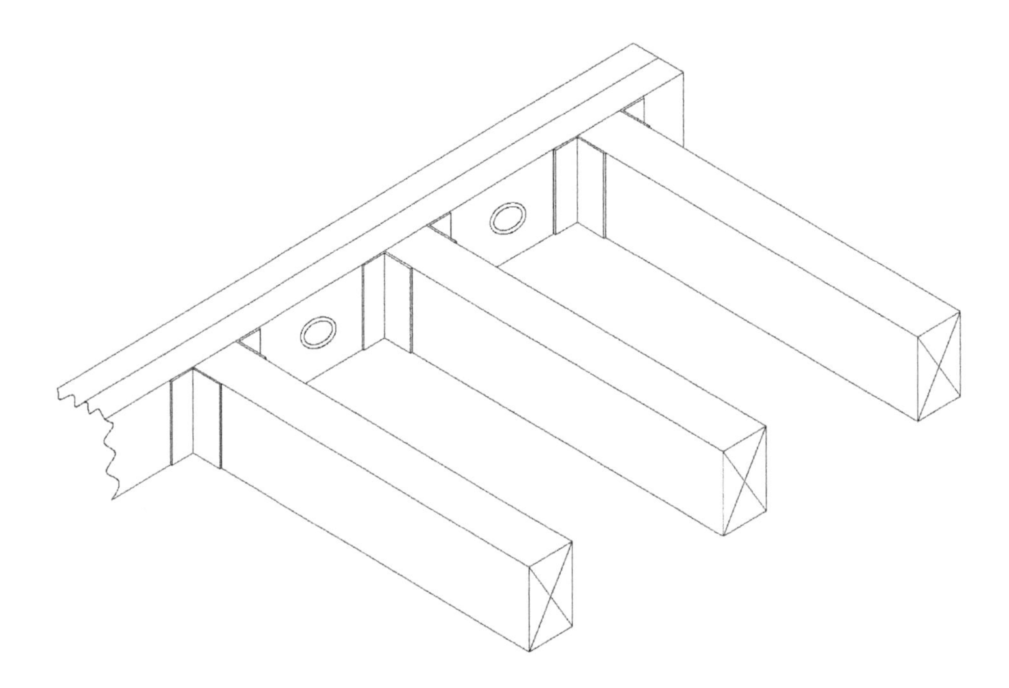

3. Draw the following detail of a concrete barrier. All sizes are approximate and may be estimated. You will need to cycle through all three Isoplanes for the main structure and use one ellipse. Be aware also that, for certain lines, you will break from the Isoplanes and draw freehand. As a last step, add solid hatching as shown. (Difficulty level: Easy; Time to completion: 10–15 minutes.)

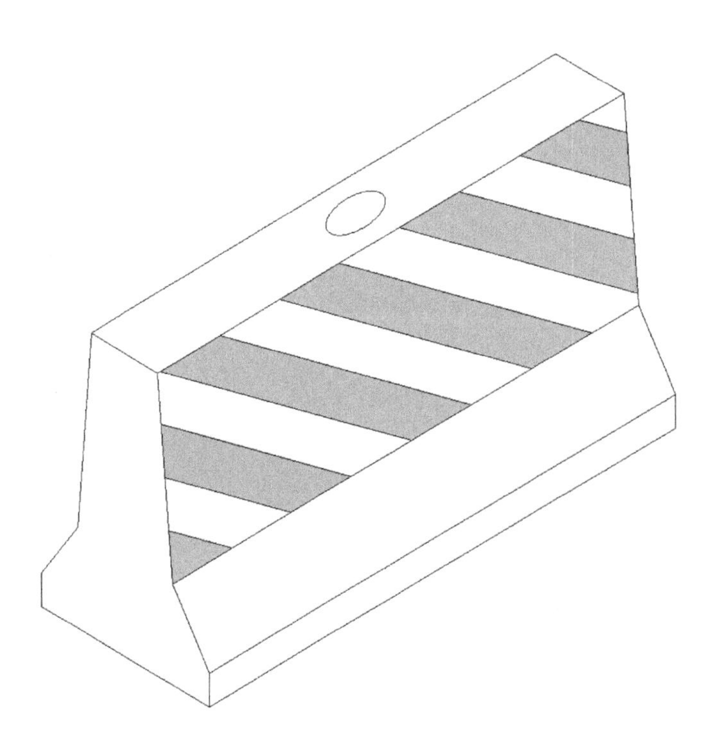

4. Draw the following detail of a nut and bolt. All sizes are approximate and may be estimated. There is more emphasis here on freehand drawing (including using arcs) after some basic Isoplanes linework is established. There is also more usage of ellipses and partial ellipses, which need to be trimmed as necessary. (Difficulty level: Easy; Time to completion: 10–15 minutes.)

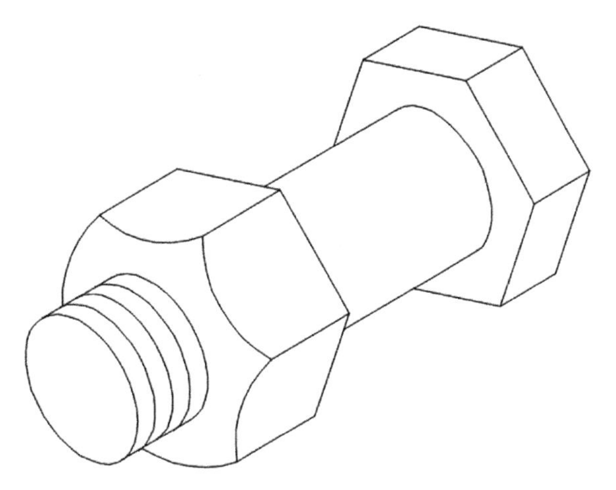

5. Draw the following valve/piping detail. All sizes are approximate and may be estimated. There is more usage of ellipses and partial ellipses, which need to be trimmed as necessary. (Difficulty level: Intermediate; Time to completion: 15–20 minutes.)

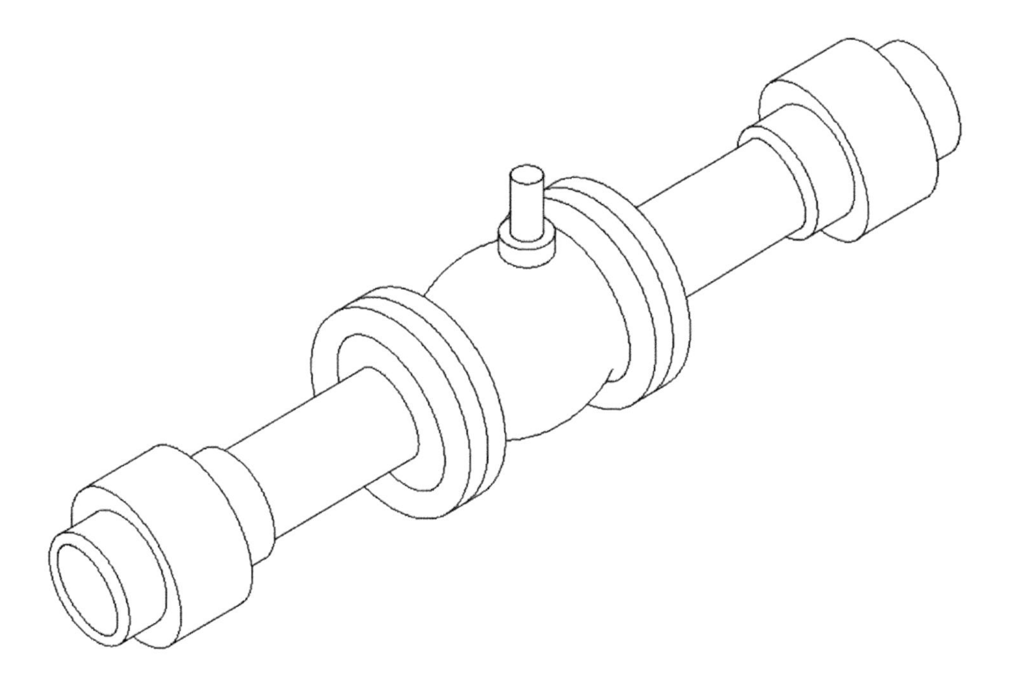

6. Draw the following piping detail. All sizes are approximate and may be estimated. This exercise is similar to the previous two in its usage of ellipses and partial ellipses, which need to be trimmed as necessary. (Difficulty level: Intermediate; Time to completion: 15–20 minutes.)

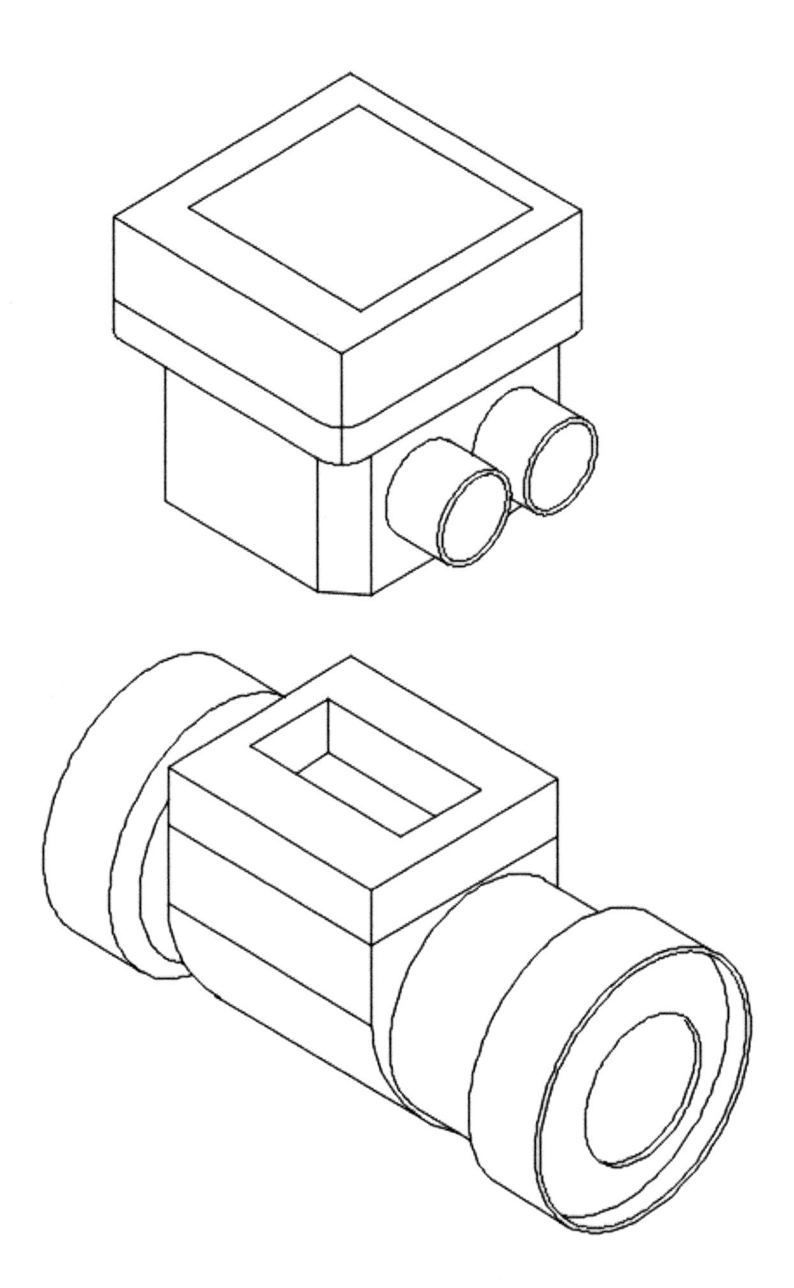

7. Draw the following computer desk. All sizes are approximate and may be estimated. You will need to cycle through all three Isoplanes for the main structure and use ellipses for a few elements. Add text and hatching as shown. (Difficulty level: Moderate; Time to completion: 20–25 minutes.)

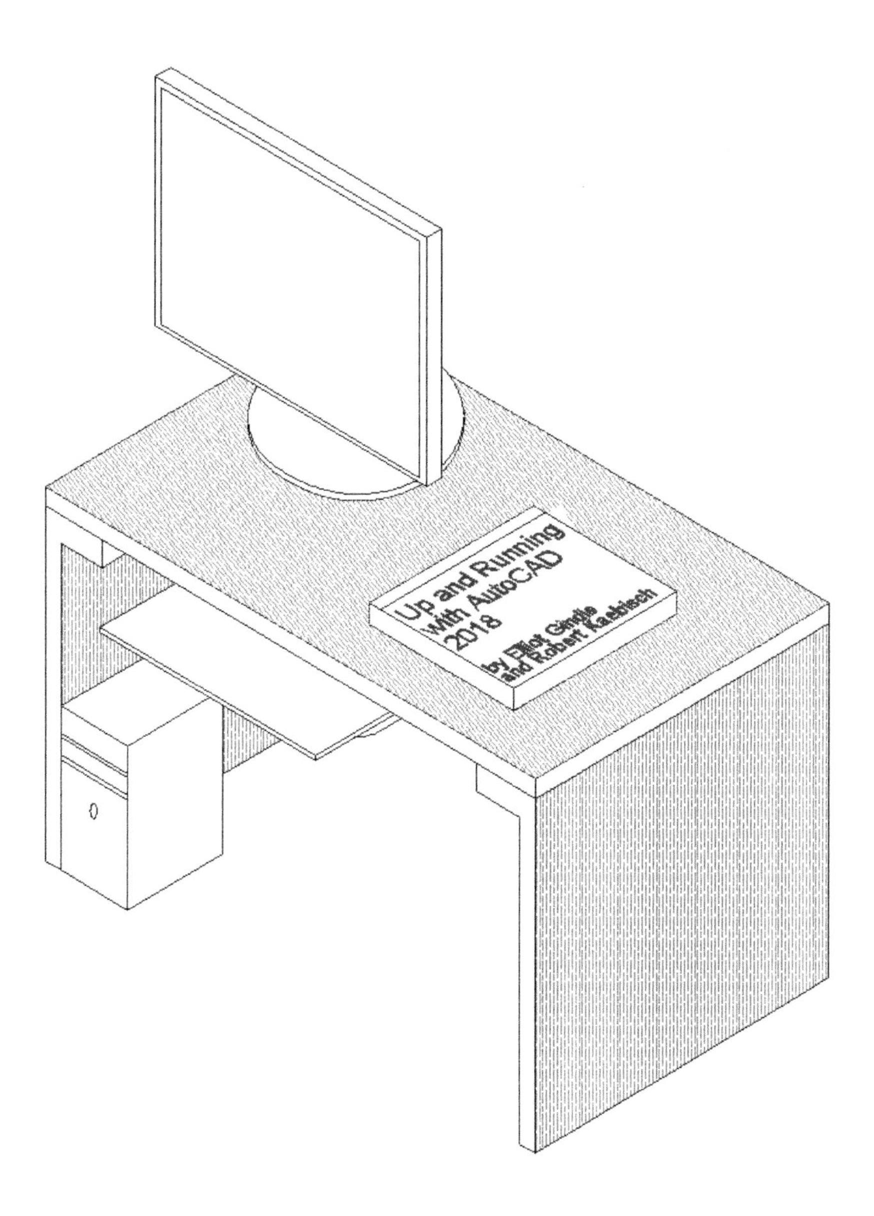

8. Draw the following architectural room detail. This exercise features not only isometric but also a perspective view, where the design converges to a "vanishing point" toward the back. Here, there is little Isoplanes work and quite a bit of improvising and freehand drawing, as isometric does not allow for perspective views. These types of layouts are not very common but still can be encountered in design work. They offer a very dynamic and interesting way to present an idea. A certain amount of artistic license is allowed, so do improvise. Note the various nuances in the design, such as slight rotation of the steel reinforcing bars in the side walls. Add in hatch and text as shown for the last step. (Difficulty level: Advanced; Time to completion: 60–90 minutes.)

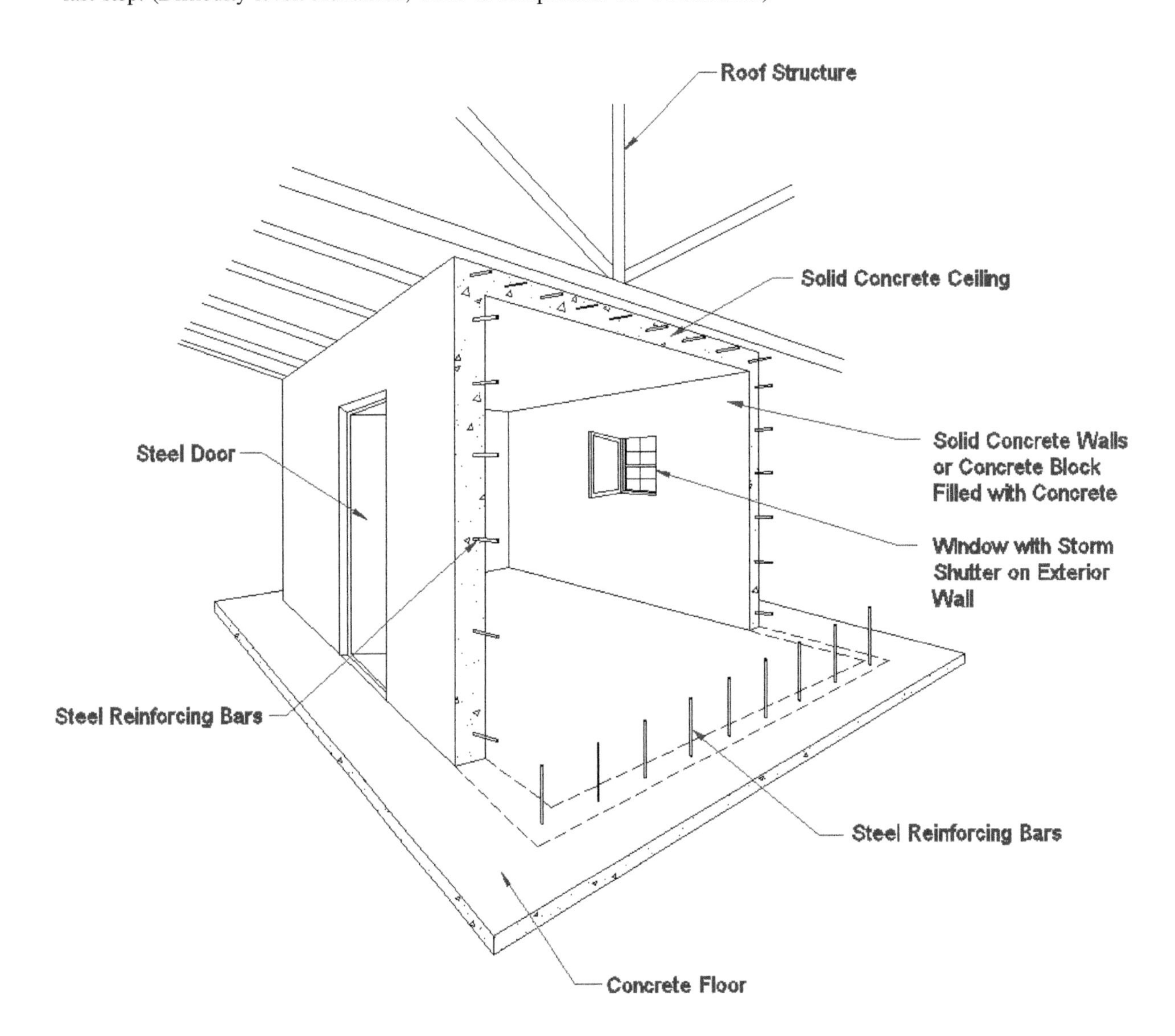

Level 2. Answers to Review Questions

Chapter 11 Review Questions

1. What are the two major properties a pline has and a line does not?
   Property 1: A pline's segments are joined together. Property 2: A pline can have thickness.
2. What is the key command to add thickness to a pline?
   Pedit.
3. What is the key command to create a pline out of regular lines?
   Pedit.
4. What option gives the pline curvature? What option undoes the curvature?
   Spline option. Decurve option.
5. What two major things happen when you explode a pline?
   It loses its thickness and is split into segments.
6. What other options for pline were mentioned?
   Making a pline with variable thickness, adding an arc, and various other vertex options.
7. What is the bpoly command?
   Bpoly creates a pline border on top of a group of ordinary lines forming a closed area. You need to select a point in the enclosed area to accomplish this.
8. Define an xline and specify a unique property it has.
   An xline is a continuous line that has no beginning or end.
9. How is a ray the same and how is it different from an xline?
   A ray has no end, just like an xline, but it does have a beginning, unlike an xline.
10. Describe a spline and list its advantages.
    A spline is a NURBS curve and is used to create organic curvature, something that cannot be done any other way.
11. List some uses for a spline in a drawing situation.
    Gradient lines, cables, electrical wires, pathways, landscaping, for example.
12. What is an mline? List some uses.
    An mline is a multiline. It can be used for everything from architectural walls to highway design.
13. What are some of the items likely to be changed using mlstyle?
    Number of lines making up the mline, distances between those lines, linetypes making up the lines, colors of the lines and fills, end caps, and other details.
14. What are some of the uses of mledit?
    Mainly to edit intersections that are made using mlines; also to create vertexes and make cuts in mlines.
15. Describe the sketch command. What is a record increment? What size is best?
    The sketch command allows you to draw freehand. A record increment is the smallest size piece of line that makes up the sketch pathway; 0.01 is an optimal size.
16. Describe the basic function of a Geometric Center OSNAP point
    The Geometric Center OSNAP allows you to snap from/to the center (centroid) of complex closed shapes.

Chapter 12 Review Questions

1. Describe what is meant by layer management. Why is it needed?
   Layer management describes a set of tools and concepts that, taken together, manage a large amount of layers, so they can be grouped together under some common theme and controlled as one unit (frozen, locked, etc.).
2. Describe the basic premise of a script.
   A script automates a series of input commands and saves the process under a name so it can be recalled and put to use automating a repetitive task.
3. What can a script file do for layer management?
   A script file can freeze and thaw a large amount of layers by simply executing the script. This was the original way to do layer management.
4. Describe what Layer State Manager does.
   It saves layer settings under a descriptive name, so this setting can be recalled to easily freeze and thaw (among other actions) large amounts of layers all at once.
5. What are the fundamental steps in using Layer State Manager?
   Once the layers are set up as desired, the setting is saved in the LSM.
6. What is layer filtering? What are the two types?
   Layer filtering sorts layers based on common traits (such as color or names), so they can all be worked with at once. Property filter and Group filter.

Chapter 13 Review Questions

1. Describe the important features found under the Lines tab.
   Nothing critical is found here, but the colors of the dimension and extension lines are changed.
2. Describe the important features found under the Symbols and Arrows tab.
   Here, the arrowheads are changed to an architectural tick.
3. Describe the important features found under the Text tab.
   Here, the text style and color are changed, with a box added around the text.
4. Describe the important features found under the Fit tab.
   Here, the most important feature is the setting of the overall scale.
5. Describe the important features found under the Primary Units tab.
   Here, everything under Linear dimensions, Zero suppression, and Angular dimensions is discussed.
6. Describe the important features found under the Alternate Units tab.
   Here, the approach to setting alternate units is discussed, including Zero suppression and placement.
7. Describe the important features found under the Tolerances tab.
   The types of tolerances are covered.
8. Describe the purpose of geometric constraints.
   To constrain the positioning of drawn geometry.
9. What seven geometric constraints are discussed?
   Perpendicular, parallel, horizontal, vertical, concentric, tangent, and coincident.
10. Describe the purpose of dimensional constraints.
    To constrain the sizing of drawn dimensions and the geometry they represent.
11. What six dimensional constraints are discussed?
    Horizontal, vertical, aligned, radius, diameter, and angle.

Chapter 14 Review Questions

1. What is important under the Files tab?
   It is recommended that nothing be changed under this tab.
2. What is important under the Display tab?
   Here, you can turn off some unnecessary screen tools such as scroll bars; change the background colors of the screen, command line, and other features; clean up the Paper Space layouts; and change the crosshair size.
3. What is important under the Open and Save tab?
   Here, you can save a drawing down to an older release of AutoCAD, delete Automatic save and backup files, and set security options.
4. What is important under the Plot and Publish tab?
   Here, you can set and modify plot stamp settings.

5. What is important under the System tab?
   It is recommended that nothing be changed under this tab.
6. What is important under the User Preferences tab?
   Here, you can set right-click customization.
7. What is important under the Drafting tab?
   Here, you can modify the look of the AutoSnap and Aperture boxes.
8. What is important under the Selection tab?
   Here, you can modify the pickbox and grip size as well as set the visual effects settings.
9. What is important under the Profiles tab?
   Here, you can save your settings and preferences under a name for future recall.
10. What is the pgp file? How do you access it?
    The pgp file holds AutoCAD shortcuts. It can be accessed via Tools → Customize → Edit Program Parameters (acad.pgp) in the drop-down menus.
11. Describe the overall purpose of the CUI.
    The Customize User Interface is used to modify the AutoCAD environment with changes to toolbars, palettes, shortcuts, macros, and much more.
12. Describe the purpose of the Design Center.
    The Design Center is useful for bringing entities such as layers and blocks into a drawing from another location as well as downloading symbol library blocks from the Autodesk Seek website.
13. Describe the important Express Tools that are covered.
    Some of the more useful Express Tools include Copy Nested Objects, Explode Attributes to Text, Convert Text to Mtext, Arc-Aligned Text, Enclose Text with Object, Delete Duplicate Objects, Flatten Objects, Break Line Symbol, Super Hatch, Convert PLT to DWG, and Layer Express Tools such as layfrz, layiso, and laywalk.

Chapter 15 Review Questions

1. What are the purposes of the audit and recover commands? What is the difference?
   Both commands are used to check up on the integrity of an AutoCAD drawing in cases of errors or a drawing not opening. Audit is the lower-level check for a drawing that is already opened but not running right. Recover is a higher-level check and may be used to try to open a corrupted drawing.
2. For join to work, lines have to be what?
   They have to be coplanar.
3. The Defpoints layer appears when you create what?
   Dimensions.
4. Does the divide command actually cut an object into pieces?
   No, it only inserts points to create sections, but the object is still intact.
5. What is the idea behind eTransmit?
   The eTransmit command allows you to send a drawing via email along with all associated fonts, linetypes, xrefs, and so on.
6. To what can you attach a hyperlink? What can you call up using a hyperlink? How do you activate a hyperlink?
   To an object or text. Either a website or a document. Hold down Ctrl and click.
7. What is the idea behind overkill?
   Overkill allows you to remove multiple stacked lines.
8. Can you make a revcloud out of a rectangle? A circle?
   Yes, via the Object option.
9. What three new selection methods are discussed?
   Window polygon, Crossing polygon, and Fence.
10. Do you use a Crossing or a Window with the stretch command?
    A Crossing.
11. Name a system variable mentioned as an example.
    Mirrtext.
12. Explain the importance of the wipeout command. What does it add to the drawing?
    The wipeout command hides objects by covering them up. It can be used to add depth to a drawing.

Chapter 16 Review Questions

1. What is the correct way to bring Word text into AutoCAD?
   Paste copied text from Word into mtext.
2. What is the procedure for bringing AutoCAD drawings into Word?
   For most applications, a Copy/Paste suffices. PrtScn can be used as well.
3. What is the correct way to bring Excel data into AutoCAD?
   Copy/Paste into an OLE.
4. What is the procedure for bringing AutoCAD drawings into PowerPoint?
   Copy/Paste or PrtScn works.
5. What is the procedure for importing JPGs and generating PDFs?
   The JPGs can be Copied/Pasted in, then adjusted. To generate a PDF, select it as the output printer or via the Ribbons output tab.

Chapter 17 Review Questions

1. What are several benefits of xref?
   Reduce file size by attaching a core drawing to multiple files, critical file security, and automated design change updates to core drawings.
2. How is an xref different from a block?
   An xref is attached to a drawing and is never truly a part of it. A block is inserted into a drawing and becomes part of it.
3. How do you attach an xref?
   Xref command, browse for the xref file, specify insertion point, and then OK.
4. How do you detach, reload, and bind an xref?
   By opening up a menu via a right-click on the drawing name in the xref palette.
5. What do layers in an xrefed drawing look like?
   They are preceded by the name of the xref itself.
6. What do layers look like once you bind the xref?
   They contain a 0 in the name.
7. What are some nesting options for xref?
   One xref attached to another then to another (etc.) and then to the main drawing or all attached to one drawing simultaneously.

Chapter 18 Review Questions

1. What is an attribute and how do you create one?
   An attribute is information inside of a block. To create one, you use the attdef command and fill in the required fields.
2. How do you edit attributes?
   Double-click on the attribute.
3. How do you extract data from attributes? Into what form?
   Data extraction is via the eattext command. It can be extracted into an Excel spreadsheet or an AutoCAD table.
4. How do you create an invisible attribute? Why can this be important?
   By checking off the invisible button during the initial creation of the attribute. This is important so that the information is not visible in the drawing but can still be used.

Chapter 19 Review Questions

1. What is a CTB file?
   This is the file that stores information on line thickness and other settings.
2. Describe the process of setting up a CTB file.
   You do this through a multipage Add Plot Style dialog box.
3. What is screening?
   Screening fades the intensity of any color.
4. What is the effect of Lineweight Settings?
   All lines that are made thicker via this option will appear as such on the screen.

Chapter 20 Review Questions

1. List some instances when isometric drawing is appropriate.
   If you do not yet have 3D skills or need only a quick and simple 3D-looking model. Also, if computing resources are limited or you do not want to spend time combining 2D and 3D.
2. List some instances where 3D would be better than isometric drawing.
   In any instances where a true 3D model is needed to convey the design intent, including for purposes of shading and rendering. Also when more accuracy is needed.
3. Name the three planes that exist in isometric perspective.
   Top, left, and right.
4. In what dialog box is Isometric snap found?
   ddosnap, Snap and Grid tab.
5. What F key switches the cursor from plane to plane?
   F5.
6. What is almost always on when drawing in isometric?
   Ortho.
7. What is the equivalent of a circle in isometric drawing?
   Ellipse.

Appendix A

Additional Information on AutoCAD

Appendix A outlines the additional information on AutoCAD, such as who makes it, what AutoCAD LT is, how AutoCAD is purchased and how much it costs, differences between releases, AutoCAD for the Mac, a history of AutoCAD, AutoCAD releases, other major Autodesk products, and AutoCAD-related websites.

A.1 WHO MAKES AUTOCAD?

As you may already know, AutoCAD is designed and marketed by Autodesk, Inc. (NASDAQ: ADSK), a software giant headquartered in San Rafael, California. Its website is www.autodesk.com. The company was founded in 1982 by programmers John Walker and Dan Drake along with several other associates and went public in 1985 (see the history). Autodesk today is headed by the president and CEO Carl Bass, has over 9000 employees worldwide, and posted revenues of approximately $2.51 billion in FY 2015. Although the company markets over 120 software applications, AutoCAD and AutoCAD LT remain their flagship products and biggest single source of revenue (about 44% of total).

It is difficult to pin down the exact size of AutoCAD's worldwide "installed base," meaning the number of copies of the software that have been purchased and used over the years. Some estimates from 2008 show this to be 2.8 million units of AutoCAD and 3.6 million units of AutoCAD LT. An Autodesk press release from the same year claimed 4.16 million seats but did not specify if that was only AutoCAD or LT as well. Industry analysts have estimated that Autodesk currently sells about 40,000 copies of AutoCAD per quarter or 160,000 seats per year. Due to piracy, the actual installed base is probably significantly higher. In addition, the number of seats can vary depending on how you count educational versions and those "add-on" vertical software purchases that need AutoCAD to run but are counted separately.

Any way you look at this, though, these numbers indicate that AutoCAD is indisputably the world's number one drafting software, with a market share often estimated to be somewhere in the 55%–60% range in the 2D world (some analysts say 85%) and 30%–35% among all CAD software (both 2D and 3D) worldwide. Even though AutoCAD has over a dozen major competitors and hundreds of smaller ones, this sort of market dominance is likely to assure AutoCAD's relevance for years to come.

A.2 WHAT IS AUTOCAD LT?

Contrary to popular belief, LT does not stand for "lite" (or as we joke in class, "AutoCAD with fewer calories"). It also does not mean "less trouble" or "learning and training" or even "limited," although Autodesk likes to throw that last term around during conferences. Instead, it is generally means for "laptop" and was originally meant for slower and weaker early laptop computers, a sort of portable AutoCAD, if you will. Today, laptops are just as powerful as desktops and LT exists to compete in the lower-end CAD market with similarly priced software. It is essentially a stripped-down AutoCAD. What is missing is detailed exhaustively on the Autodesk website, but it essentially boils down to no 3D ability, no advanced customization, and no network licensing. In addition to these, some other minor missing features are scattered throughout the software, but the differences are not that great; and aside from the lack of 3D, it takes a trained eye to spot LT. It really looks and acts like AutoCAD and with a good reason. It was created by taking regular AutoCAD and "commenting out" (thereby deactivating) portions of the C++code that Autodesk wanted to reserve for its pricier main product.

Having said this, however, AutoCAD LT is an excellent product (its estimated sales are said to outnumber AutoCAD) and is recommended for companies that do not have ambitious CAD requirements. The simple truth and something Autodesk does not always promote is that some users never need the full power of AutoCAD and do just fine with LT, while paying a fraction of the cost of full AutoCAD. Electrical engineers especially need nothing more than LT for basic linework and diagrams, as do architects and mechanical engineers who deal with schematics. The size of the files you anticipate working on, by the way, is of no consequence. LT lacks certain features, not the ability to work with multistory skyscraper layouts, if needed. It truly is a hidden secret in CAD. Autodesk must have been taken by surprise by the strong LT sales since 2002 and dramatically raised the price of the software from $700 to $1200 in 2008.

A.3 HOW IS AUTOCAD PURCHASED AND HOW MUCH DOES IT COST?

There was a major change to Autodesk's policy in regard to AutoCAD pricing and purchase options. It was hinted at for some time throughout 2015 and was finally implemented on January 31, 2016. You can no longer outright purchase AutoCAD for a one time price. Instead there exists a "subscription" model, where you pay as you go (monthly or annually) for usage and licensing of the software. This is a major departure from the last 30+ years and follows an overall trend among some software vendors to "rent" out their products as opposed to a one time sale. The specifics of this new policy are outlined next, along with the older approach presented for historical reasons and to put things into perspective.

For the most part, the way AutoCAD was sold, up until this change, was via a perpetual license or the "buy-it-once" method. It is the way we approach most purchases—you contact a seller, pay for the product, and it is yours to use forever, or until you want a newer version of that product or something else entirely.

To purchase a perpetual license of AutoCAD in the 1980s and 1990s you would have contacted an authorized dealer (or reseller). The sale was then usually arranged by phone, paid for with a credit card, and AutoCAD was shipped on floppy disks (then CDs and later DVDs) via registered mail. In more recent times the process moved online and since 2000 Autodesk's website was configured for direct purchase and the reseller was often passed over—although resellers still often had access to better pricing and offered tech support. Over the years, printed command references and manuals were phased out and even physical media was hard to find as customers were just downloading the software purchase and associated authorization codes. The full AutoCAD was never sold in physical or online stores, but its sibling AutoCAD LT was in fact found in such retailers as Staples and Office Depot as well as www.Amazon.com.

So how much did it all cost? Here are some recent historical data. In January 2015, AutoCAD 2015 was shipping or downloading for $4195. An upgrade from 2014 or 2013 was $2095. For its part, AutoCAD LT was a far more modest $1200. A maintenance subscription was also available as an option. For 2015, it was an extra $545 per license, bringing the cost of AutoCAD to $4740. For the extra cost you would get several layers of support and other benefits such as cloud access and discounts on future purchases, and of course free upgrades every year.

So where are we now with this new policy? Well, if you purchased AutoCAD before January 31, 2016, you were grandfathered into the perpetual license and could use the software for as long as you wished. All purchases since then, to include the new AutoCAD 2019 (to be released in late March 2018), you now have only the "rental" option. The cost for one license, as outlined on the Autodesk website, is either $185 per month or $1470 per year. There is also a 3-year ($4410) subscription available. As you probably noticed, with this model it takes just over 3 years to spend as much as you would outright purchasing AutoCAD. For AutoCAD LT, the prices are $50 monthly or $380 per year. LT also has a 3-year ($1140) subscription available. With this pricing model it takes almost 4 years to spend as much as you would outright purchasing AutoCAD LT. In essence, you have much smaller upfront costs and instead incur rental fees spread out over time.

So is this new pricing a good thing for AutoCAD customers? That depends on who you are and how you use and manage the software. It is certainly is a win for Autodesk when it comes to small firms, as these often do not upgrade for many years and may now have to. However, it should benefit some of the larger enterprise users who do upgrade annually and now can spread out their software expenses over a full year cycle. The monthly rental plan is also good for when you hire some drafting help for a few months and would not need the license once the overflow work is done.

For its part, Autodesk presents its yearly $1470 rental option as the best value for the customers and some basic math supports that—if you upgrade often anyway. For example, back before the change, to purchase one license and 3 years' worth of maintenance and upgrade subscriptions was $5830 ($4195 + $545 + $545 + $545), but in comparison, 3 years' worth of rentals is now $4,410, a better deal. It is also a better deal than purchasing an all-new license at full price every other year. However, with new releases featuring less and less critical new changes and upgrades, all this math ignores an important fact: many organizations simply do not upgrade that often. If you have a small engineering office running just two licenses of AutoCAD 2015, you can probably get by just fine until AutoCAD 2025 comes out! Then, you would have spent just $8390 to "own" those two licenses for 10 years. Now you will have to rent two copies for that same 10-year time frame that (assuming you can only "rent" for no more than 3 years at a time) works out to 3 × $4410 = $13,230 (for 9 years) + $14,700 (for the final 10th year). Of course, multiply that by two; so the final bill is $29,200 spread out from 2015 to 2025! "Hey, you get upgrades often," Autodesk may say. But do you need to?

Recent trends (e.g.,: AutoCAD 2010–18) point to the fact that AutoCAD functionality will continue to remain relatively similar, with only minor improvements, and no new game-changing "must-have" technology works its way into the software. It is possible that many vendors will resent the forced upgrades going forward. The bottom line is that whether you love or hate this new pricing depends on your (and your company's) exact situation. Any company has to consider its own policies of capital expenditures and how it allocates software funds. Sometimes, it makes more sense to pay it all up front, and sometimes it is better to spread it out, so consider all angles carefully.

A.4 ARE THERE SIGNIFICANT DIFFERENCES BETWEEN AUTOCAD RELEASES?

This is a common question at the start of training sessions, and the answer is hinted at in the last paragraph of the preceding software pricing discussion. Many schools are authorized training centers and, as such, have the latest version of AutoCAD installed. That may or may not be the case with the students and their respective companies. No issue seems to cause as much confusion and anxiety as this one. To put your fears to rest, if you are learning AutoCAD 2019 (this book) and have AutoCAD 2018, 2017, 2016, 2015, or even earlier versions at home or at work, do not worry. The differences, while real, are for the most part relatively minor and cosmetic (except perhaps for the introduction of the Ribbon in 2009). While AutoCAD has certainly been modernized, it has not fundamentally changed since it moved to Windows and away from DOS over 20 years ago and finally stabilized with R14 in 1997. If you were to open up that release, you would feel right at home with 90% of the features.

While Autodesk's marketing department may insist that the latest version of AutoCAD has some critically important new features or enhancement, the fact remains that the vast majority of what is truly important has already been added to AutoCAD, and new features are incremental improvements. Most of what you learn in one version is applicable to several versions backward and forward, especially at a beginner's level.

A.5 IS THERE AN AUTOCAD FOR THE MAC?

The short answer is "Yes." After a nearly 20 year absence, Autodesk announced in August 2010 that the Macintosh platform is once again supported, and AutoCAD for the Mac was shipped in November of that year. Why was AutoCAD offered for the Mac up until 1992, then removed from the market, only to be reintroduced in 2010? The reasons are purely financial ones. In the late 1980s and early 1990s, for those old enough to remember, the PC was not yet the dominant force at present. Mac and other manufacturers were all strong competitors and there were many choices on the market. Autodesk decided to go with the leading two, PC and Mac, and invested in both. As the PC became the de facto world computer standard and Apple went into a decline, resources had to be focused on the PC, as it takes a lot of effort to reprogram and support a major piece of software for another configuration.

We all know what happened to Apple Inc., of course. Under leadership from the late Steve Jobs, it roared back strong with a line of high-demand consumer products, such as the iPod, iPhone, and iPad. Its computers also regained some of its former market share with superb desktop and laptop offerings.

Autodesk took note of these developments. Such is the nature of the software business: Your fortunes are closely tied in with the ups and downs of hardware, operating systems, and chip makers. The Mac once again began to look like an attractive platform, and Autodesk invested in reprogramming AutoCAD to support it. When exactly this effort was started is not known outside of the company, but it was likely authorized a few years before the official 2010 launch, as it takes a lot of lead time to rework code of such complexity. This move should net Autodesk quite a few new sales of AutoCAD and spells major trouble for other Mac-based CAD software such as VectorWorks, which became popular precisely *because* AutoCAD was not available for the Mac and catered to those designers who refused to switch to a PC. Fig. A.1 is a screenshot of AutoCAD for the Mac from Autodesk's promotional literature.

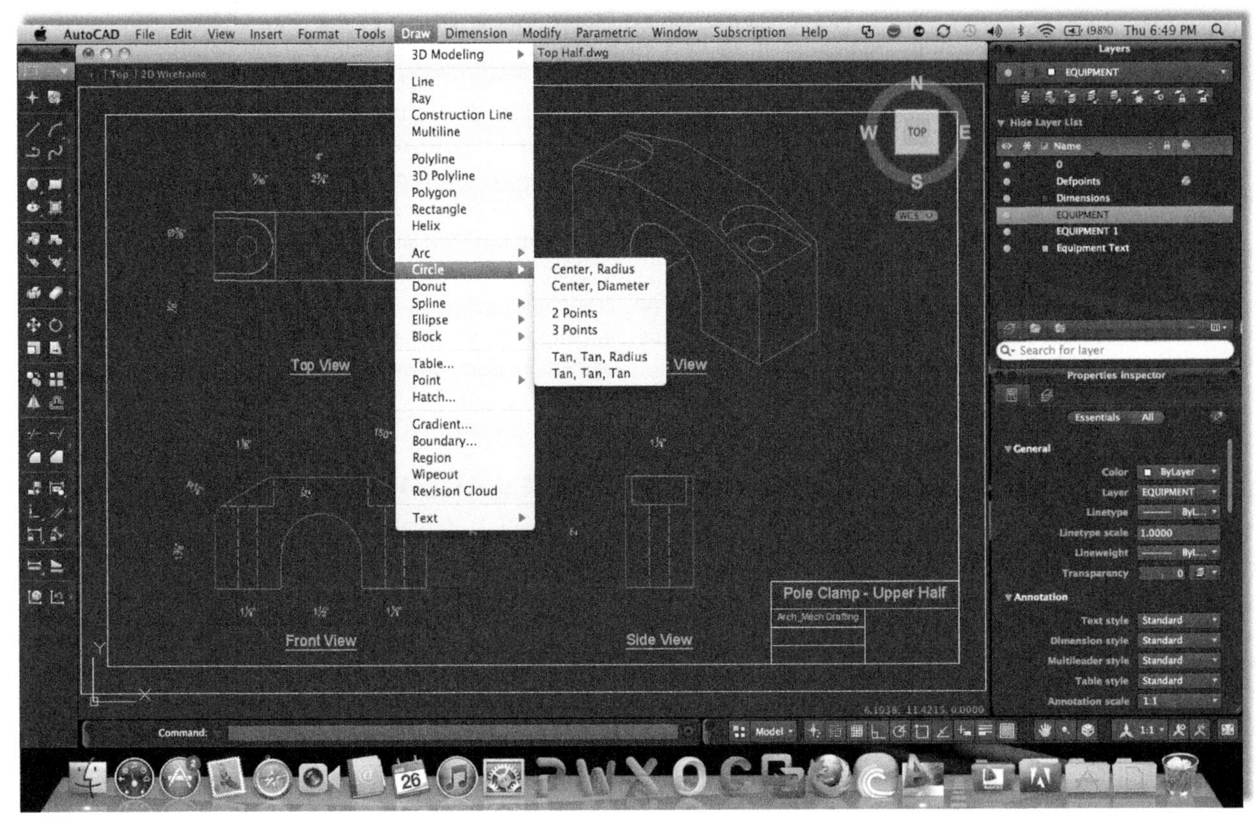

FIGURE A.1 AutoCAD for the Mac screenshot.

It remains to be seen how successful AutoCAD for the Mac will be in the long run, of course. The program has been rewritten from the ground up as a native Mac application, not simply as a port from Windows, so the software itself certainly seems to have a solid foundation. However, the original late-2010 release seemed, at the time, to be a bit of a work in progress, as evidenced by the fact that some standard AutoCAD features were not yet supported. Things have got better in the last few years, of course. The full (and shrinking) list of what is still not available can be found on the Autodesk website, but it includes the following. If you completed Levels 1, 2, and the 3D chapters of this textbook, you should recognize all these tools. Here is a partial list of AutoCAD for the Mac missing features (as of this writing):

- Quick Properties Palette
- Layer State Manager
- Tool Palettes
- Navigation Bar
- AutoCorrect Command Entry
- Hatch Preview
- Express Tools
- Material Mapping
- Advanced Rendering
- Camera Creation
- Walkthroughs
- DGN Underlay
- Hyperlinks
- dB Connect
- Visual LISP
- Action Recorder
- Drawing Password Protect
- CUI Import/Export

Some of these are reasonably important, some are more obscure; and it is likely that many of these issues will be resolved in the coming releases or even by the time this textbook goes to press. Other than that, AutoCAD for the Mac is not entirely different from the PC version; after all, it still has to behave like AutoCAD. Although the Ribbon is removed, the cascading menus and toolbars are still present, and the software can accept files from its PC-based cousins. AutoCAD for the Mac was priced the same as AutoCAD for the PC. Its subscription fees are currently the same $185 per month or $1470 per year as its PC-based cousin. There is an LT version for the Mac as well. It also mirrors the PC version in the past and present pricing.

A.6 A BRIEF HISTORY OF AUTODESK AND AUTOCAD

The father of Autodesk and AutoCAD is John Walker. A programmer by training, he formed Marinchip Systems in 1977 and, by 1982, seeing an opportunity with the emergence of the personal computer, decided to create a new "software-only" company with business partner Dan Drake. The plan was to market a variety of software, including a simple drafting application they acquired earlier. Based on code written in 1981 by Mike Riddle and called *MicroCAD* (later on changed briefly to *Interact*), the software would sell for under $1000 and would run on an IBM PC. It was just what they were looking for to add to their other products.

Curiously, the still unnamed company did not see anything special in the future of only MicroCAD and focused its energy on developing and improving other products as well. Among them were Optical (a spreadsheet), Window (a screen editor), and another application called *Autodesk* or *Automated Desktop*, which was an office automation and filing system meant to compete with a popular, at the time, application called *Visidex*. This Automated Desktop was seen as the most promising product; and when it came time to finally incorporate and select a name, Walker and company (after rejecting such gems as Coders of the Lost Spark) settled on *Autodesk Inc.* as the official and legal name.

Written originally in SPL, MicroCAD was ported by Walker onto an IBM machine and recoded in C. As it was practically a new product now, it was renamed *AutoCAD-80*, which stood for Automated Computer−Aided Design. AutoCAD-80 made its formal debut in November 1982 at the Las Vegas COMDEX show. As it was new and had little competition at the time, the new application drew much attention. The development team knew a potential hit was on their hands and a major corner was turned in AutoCAD's future. AutoCAD began to be shipped to customers, slowly at first, then increasing dramatically. A wish list of features given to the development team by customers was actively incorporated into each succeeding version—they were not yet called *releases*.

In June 1985, Autodesk went public and was valued at $70 million in 1989 it was at $500 million and in 1991 at $1.4 billion. With each release, milestones, large and small, such as introduction of 3D in Version 2.1 and porting to Windows with Releases 12 and 13, were achieved. In a story not unlike Microsoft, Autodesk, through persistence, innovation, firm belief in the future of personal computers, and a little bit of luck, came to dominate the CAD industry.

Fig. A.2 shows perhaps the most famous early AutoCAD drawing of all, a fire hose nozzle drawn in 1984 by customer Don Strimbu on a very early AutoCAD version. This was the most complex drawing on AutoCAD up to that point, used by the development team as a timing test for machines and featured on the covers of many publications, including *Scientific American* in September 1984.

FIGURE A.2 A nozzle drawn by Don Strimbu in 1984 on a very early version of AutoCAD.

A.7 AUTOCAD RELEASES

Autodesk first released AutoCAD in December 1982, with Version 1.0 (Release 1) on DOS, and continued a trend of releasing a new version sporadically every 1–3 years until R2004, when it switched to a more regular, once a year schedule. A major change occurred with R12 and R13, when AutoCAD was ported to Windows and support for UNIX, DOS, and Macintosh was dropped. R12 and R13 were actually transitional versions (R13 is better left forgotten), with both DOS and Windows platforms supported. Not until R14 did AutoCAD regain its footing, as sales dramatically increased. Fig. A.3 is AutoCAD's release timeline. A release history follows:

Version 1.0 (Release 1), December 1982.
Version 1.2 (Release 2), April 1983.
Version 1.3 (Release 3), August 1983.
Version 1.4 (Release 4), October 1983.
Version 2.0 (Release 5), October 1984.
Version 2.1 (Release 6), May 1985.
Version 2.5 (Release 7), June 1986.
Version 2.6 (Release 8), April 1987.
Release 9, September 1987.
Release 10, October 1988.
Release 11, October 1990.
Release 12, June 1992 (last release for Apple Macintosh until 2011).
Release 13, November 1994 (last release for UNIX, MS-DOS, and Windows 3.11).
Release 14, February 1997.
AutoCAD 2000 (R15.0), March 1999.
AutoCAD 2000i (R15.1), July 2000.
AutoCAD 2002 (R15.6), June 2001.
AutoCAD 2004 (R16.0), March 2003.
AutoCAD 2005 (R16.1), March 2004.
AutoCAD 2006 (R16.2), March 2005.
AutoCAD 2007 (R17.0), March 2006.
AutoCAD 2008 (R17.1), March 2007.
AutoCAD 2009 (R17.2), March 2008.
AutoCAD 2010 (R18.0), March 2009.
AutoCAD 2011 (R18.1), March 2010.
AutoCAD 2012 (R18.2), March 2011.
AutoCAD 2013 (R19.0), March 2012.
AutoCAD 2014 (R19.1), March 2013.
AutoCAD 2015 (R20.0), March 2014.
AutoCAD 2016 (R20.1), March 2015.
AutoCAD 2017 (R21.0), March 2016.
AutoCAD 2018 (R22.0), March 2017.
AutoCAD 2019 (R22.1), March 2018.
AutoCAD 2020 (R22.2), March 2019 (anticipated).

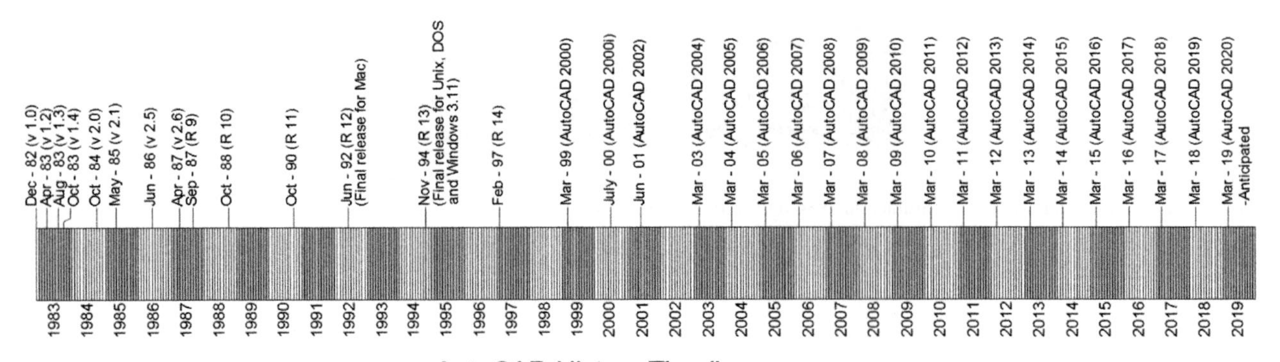

FIGURE A.3 AutoCAD timeline, 1982 through 2019 (projected).

A.8 MAJOR AUTODESK PRODUCTS

Autodesk currently markets just over 120 software products, some major and some obscure. Only one application sold today is the company's original product, which is AutoCAD. The rest have been developed in-house from scratch over the years, purchased outright from the original developers, or as is often the case, acquired in takeovers and mergers. Many products are designed to "hitch a ride" on the back of AutoCAD and extend its power to serve a niche market with additional functionality and menus not found in the base product. Those, often homegrown applications, are referred to as *verticals*. In other cases, Autodesk markets radically different software, such as one used for high-end animations and cartoons. Those have been acquired to extend the company's reach into new markets.

The following lists Autodesk's flagship products that you should know about, as you may come across them in the industry; also listed are the popular add-on products purchased with AutoCAD.

Flagship products:

- *AutoCAD*: Autodesk's flagship product for 2D and 3D computer-aided drafting and design.
- *AutoCAD LT*: A limited functionality version of AutoCAD for an entry-level 2D only market.
- *Alias*: Advanced software for creating complex surfaces for automotive and other industrial design.
- *3ds Max*: A powerful high-end rendering and animation software product.
- *Inventor*: A mid-to high-end parametric solid modeling and design package for engineering.
- *Revit*: A mid-to high-end parametric software for architectural and building design.
- *Maya*: A very high-end animation, effects, and rendering software.

Add-ons to AutoCAD (verticals):

- *AutoCAD Architecture*: Additional functionality for architecture design applications.
- *AutoCAD Civil 3D*: Additional functionality for civil engineering applications.
- *AutoCAD Electrical*: Additional functionality for electrical engineering applications.
- *AutoCAD MEP*: Additional functionality for building system design (mechanical, electrical, plumbing).
- *AutoCAD Mechanical*: Additional functionality for mechanical engineering.
- *AutoCAD P&ID*: Additional functionality for creating piping and instrumentation diagrams and designs.
- *AutoCAD Plant 3D*: Additional functionality for plant (factory) designs.

These are just some of the major software applications in Autodesk's inventory. A significant number of products, including all the less-known or obscure ones, have been omitted. A full and comprehensive list of all of them can be found on the Autodesk website (www.autodesk.com/products). You are encouraged to look them over to stay ahead of the curve and know what other design professionals may be using.

A special mention should be made regarding AutoCAD WS (recently renamed *AutoCAD 360*). WS was Autodesk's first mobile app for AutoCAD. It allows you to view cloud-based drawings and edit them in a limited manner on a mobile device, such as a phone or tablet. AutoCAD WS was free at its basic level but did feature several low-cost paid plans that in turn gave you greater access to tools, storage, and other goodies. Last year, WS was renamed *AutoCAD 360*, in reference to the Autodesk 360 cloud service. You can download this app from Google Play, Amazon, and other sources. It is a very convenient approach for managers and designers with limited drafting needs, who are often in the field or just need to quickly look at a drawing and approve it while on the go.

A.9 AUTOCAD-RELATED WEBSITES

Next is a listing of important and valuable websites for AutoCAD- and general CAD-related information. Online content has grown from occasionally useful to indispensable, and experienced AutoCAD users must have at least the following at their fingertips when the need arises. Note that link fidelity is not guaranteed, and except for the few major ones at the start of the list, the sites may and do change.

- *Autodesk Inc.* (www.autodesk.com): Mentioned previously, it is home base for the makers of AutoCAD. Here, you find locations of AutoCAD dealers and training centers as well as endless information on AutoCAD and other software products and services. It is well worth visiting occasionally.
- *Autodesk User Group International* (www.augi.com): The Holy Grail for all things related to AutoCAD. This site is the officially sanctioned user community for Autodesk and its products. AUGI puts out numerous publications and runs

conferences and training seminars all over the United States and worldwide. Its website is packed full of tips, tricks, articles, and forums where you can ask the community your hardest AutoCAD questions. Claiming over 100,000 members, AUGI has the distinction of being the only organization that can officially submit "wish lists" directly to Autodesk, meaning that, if you are a member and have an idea for a new feature or how to improve AutoCAD, submit it via the AUGI community, and Autodesk may just listen to you.

- *Cadalyst magazine* (www.cadalyst.com): A great magazine dedicated to all things CAD, including solid modeling and rendering but with a large amount of coverage dedicated specifically to AutoCAD. A must read for AutoCAD managers, the magazine features informative articles on all aspects of the job, including programming, industry trends, and advanced topics. Subscription was free up until 2010. You can still access free archived issues from as far back as 2004, but the magazine now charges a fee for the new issues.
- *Lynn Allen's blog* (http://lynn.blogs.com): You may have never heard of Lynn Allen if you are just starting out with AutoCAD, but hang around long enough and you probably will. She is what the AutoCAD community refers to as an Autodesk Technical Evangelist, an expert on all things AutoCAD. Although there are certainly many experts around, what sets Lynn apart is that she actually works for Autodesk and gets to learn AutoCAD directly from the developers. Her blog is the place to go to when a new release is about to hit the market. She will likely have already analyzed it and written a book by then. Lynn is also a prolific speaker, appearing at just about every AutoCAD conference, and has been a columnist for *Cadalyst* for many years as well. She writes and speaks well and employs humor in her approach to thorny technical issues. You can download a useful "tips and tricks" booklet on her website, so go ahead and check it out.
- *AutoCAD Lessons on the Web* (www.cadtutor.net): A well-organized and extensive site for learning AutoCAD that includes some tutorials on 3ds Max, VIZ, Photoshop, and web design.
- *The Business of CAD, Part I* (www.upfrontezine.com): The website describes itself thus: "Launched in 1985, the upFront.eZine e-newsletter is your prime independent source of weekly business news and opinion for the computer-aided design (CAD) industry." The site is a no-nonsense, cut-through-the-hype source for CAD (mostly AutoCAD but others, too) information, latest releases, and analysis of Autodesk business decisions; a great resource for a CAD manager.
- *The Business of CAD, Part II* (http://worldcadaccess.typepad.com): Edited by blogger Ralph Grabowski, this site is a spin-off blog from the preceding upFront.eZine site. It also addresses news in the media about all things CAD and how they will affect you, as well as general computer topics; it is recommended for the serious CAD manager or user.
- *John Walker's site* (www.fourmilab.ch): This site, while not essential AutoCAD reading, is still interesting, fun, and often quite informative. As cocreator of AutoCAD, former programmer John Walker has put together the definitive early history of how it all began. Fascinating reading from a fascinating man who is now "retired" in Switzerland and pursuing artificial intelligence research. His site is chock full of readings on numerous subjects, as you would expect from a restless and inquisitive mind; he is truly someone who changed computing history in a significant way.

As with many topics in this brief appendix, what is listed here only scratches the surface. Remember that you are not the only one out there using AutoCAD, so feel free to reach out and see what else is out there if you have questions or would just like to learn more. At a minimum, it is recommended to join AUGI and get a subscription to *Cadalyst*.

Appendix B

Other Computer-Aided Design Software, Design and Analysis Tools, and Concepts

While AutoCAD may be the market leader for drafting software, it certainly does not operate in a vacuum and is not the only game in town. The list that follows summarizes other major players in the drafting field. The reason you should know about them is because you may run into them in use and may have to exchange files if collaborating. Saying "I've never heard of that" is usually not a good way to start a technical conversation in your area of expertise, so take a note of the following products.

MICROSTATION

(www.Bentley.com)

A major, and really the only serious, competitor to AutoCAD, MicroStation is produced and marketed by Bentley Systems. For historical reasons, the software is widely used in government agencies and the civil engineering community. It comes in full and PowerDraft variations (akin to AutoCAD and LT). The current version is MicroStation CONNECT v10, and its file extension is .dgn. Relatively recently, AutoCAD began to open these files via the import command. MicroStation, on the other hand, has always been able to open AutoCAD's .dwg files. Of course, many techniques and even third-party software are available to quickly and easily convert one file format to another quickly as well.

MicroStation was designed from a different perspective than AutoCAD, and this shows mainly in the way the user interacts with this software. It is more icon and menu driven than AutoCAD, it allows little typing and had no Ribbon Interface until the most recent release. However, MicroStation executes the same exact commands, and you can use it to design anything that you can do in AutoCAD. One aspect that users may find frustrating when switching from AutoCAD is that MicroStation commands are for the most part named differently, so you need to learn a new vocabulary. Blocks, e.g., are called *cells*, layers are called *levels*, snap points are referred to as *tentative points*, and so on. Bentley designers have done their best to differentiate their product from AutoCAD, but the result can be like "using a different part of the brain" when working on it, as described by one former AutoCAD user in an online blog.

That said, however, MicroStation is very capable software and, in some areas, superior to AutoCAD, such as its ability to handle multiple Xrefs, 3D rendering, design simulation and review, change and design management, and digital file security, among other advanced features. Its main disadvantage (interface opinions aside) is that it is not an industry standard outside of some civil engineering communities, which creates occasional collaboration and file sharing issues as well as difficulty in finding qualified users and training classes. Bentley's pricing is also a bit cumbersome with various a-la-carte modules that can be purchased separately. Overall, the software costs about the same as AutoCAD.

MicroStation and AutoCAD users have historically divided themselves into somewhat fanatical partisan groups, each extolling the virtues of its respective software over the other. However, an objective view of both applications yields an opinion that, like any software, both have their pros and cons, and those who disparage one over the other usually do not know what they are talking about.

ARCHICAD

(www.graphisoft.com)

ArchiCAD is a very popular application among some architects. It is geared toward the Macintosh but is also available for Windows. Developed by a Hungarian firm, Graphisoft at around the same time as AutoCAD, ArchiCAD took a different approach, allowing for a parametric relationship between objects and easier transitions to 3D, as all objects are created inherently with depth. These features, common to BIM software, can also be found in Revit and other applications, and the direction of the industry seems to be headed as well (see , Appendix N), making ArchiCAD a bit ahead of its time. The software has a loyal installed user base of about 100,000 architects worldwide and can interoperate fully with AutoCAD files. The current release (from early 2017) is ArchiCAD21.

TURBOCAD

(www.turbocad.com)

TurboCAD is a midrange 2D/3D drafting software, first developed and marketed in South Africa in 1985 and brought to the United States the following year. It was marketed as an entry level basic drafting tool (it once sold for as low as $99), but with constant upgrading and development, TurboCAD has grown over the years to a $1600 software product for the architectural and mechanical engineering community. It is currently owned by IMSI/Design LLC and operates on both Windows and Macintosh. The current release is TurboCAD Pro Platinum 2017 (for Windows) or Mac Pro v10 (for Mac) and a midrange 2D/3D version TurboCAD Expert 2017 for $500. There is also a downgraded 2D-only TurboCAD Deluxe 2017 for $150 similar to AutoCAD LT. Some versions of the software have a subscription sales model as well, and many have specialized add-ons (civil, mechanical, etc.)

OTHER DESIGN SOFTWARE

Some other notable computer-aided design (CAD) applications include RealCAD, ViaCAD, SharkCAD, AllyCAD, PowerCADD, CorelCAD, DataCAD, VariCAD, VectorWorks (formerly MiniCAD), IntelliCAD, Trimble's Sketch-Up, Form Z, and Rhino, although the last two are often used only for modeling and visualization. This list is not exhaustive by any means, as hundreds of other CAD programs are available.

Many of my students are interested in other aspects of design, and questions often come up as to how AutoCAD fits into the "big picture" of a much larger engineering or design world. To address some of these questions, the following information, culled from a chapter written in 2007 for an undergraduate machine design theory workbook, is included. AutoCAD students, especially engineers, may find this informative. The subject matter is extensive and this brief overview only scratches the surface. Ideally, if the reader was not aware of these concepts and the engineering design and analysis software described, this serves as a good introduction and a jumping off point to further inquiry.

- **CAD:** This is a broad generic term for using a computer to design something. It can also be used in reference to a family of software that allows you to design, engineer, and test a product on the computer before manufacturing it. CAD software can be used for primarily 2D drafting (such as with AutoCAD) or full 3D parametric solid modeling. The latter is the software that aerospace, mechanical, automotive, and naval engineers commonly use. Examples include Computer-Aided Three-dimensional Interactive Application (CATIA), NX, Pro-Engineer (Creo), IronCAD, and SolidWorks, among others. Included under the broad category of CAD is testing and analysis software.
- **Finite element analysis (FEA):** This is used for structural and stress analysis, testing thermal properties, and much more as related to frames, beams, columns, and other structural components. Examples of FEA software include NASTRAN, ALGOR, and ANSYS.
- **Computational fluid dynamics (CFD):** This is software used to model fluid flow around and through objects to predict aerodynamic and hydrodynamic performance. Examples of CFD software include Fluent, ANSYS-CFX, and Flow 3-D.
- **Computer-aided manufacturing (CAM):** The basic idea is this: We designed and optimized our product, now what? How do we get it from the computer to the machine shop to build a prototype? The solution is to use a CAM software package that takes your design (or parts that make up your design) and converts all geometry to information useable by the manufacturing machines, so the drills, lathes, and routing tools can create the part out of a chunk of metal. Examples of dedicated CAM software include Mastercam and FastCAM.

- **Computer-aided engineering (CAE):** Strictly defined, CAE is the "use of information technology for supporting engineers in tasks such as analysis, simulation, design, manufacture, planning, diagnosis, and repair." We use this term to define an ability to do this all from one platform, as explained next.
- **Computer numerical code/G-Code (CNC):** This is the language (referred to as *G-Code*), created by the CAM software and read by the CNC machines. G-Code existed for decades. It was originally written by skilled machinists based on the geometry of the component to be manufactured (and sound manufacturing practices). This code can still be written by hand. It is not hard to understand. Most lines start with G, hence its name; and they all describe some function that the CNC machine does as related to the process. After some practice, the lines start to make sense and you can roughly visualize what is being manufactured, just by reading them.

As you read the CAD and CAM descriptions you may be wondering why CAD companies have not provided a complete solution to design, test, and manufacture a product using one software package. They have, but to varying degrees of success. For a number of years, an engineering team designing an aircraft, e.g., had to typically create the design (and subsystems) in a 3D program such as CATIA, then import the airframe for FEA testing into NASTRAN, followed by airflow testing in CFD software (and a wind tunnel), and finally send each component through CAM software such as Mastercam to manufacture it (a greatly simplified version of events, of course).

Obviously, data could be lost or corrupted in the export/import process, and a lot of effort was expended in making sure the design transitioned smoothly from one software application to another. While the final product was technically a result of CAE, software companies had something better in mind: "one-stop shopping." As a result, CAE today means one source for most design and manufacturing needs. CATIA, e.g., has "workbenches" for FEA analysis and CAM. Code generated by CATIA is imported into the CNC machines that (typically after some corrections and editing) produce virtually anything you can design, with some limitations, of course.

As you may imagine, not everything is rosy with this picture. As the specialized companies got better and better at their respective specialties (CAM, FEA, or CFD), it became harder and harder for the CAD companies to catch up and offer equally good solutions embedded in their respective products. Mastercam is still the undisputed leader in CAM worldwide, and NASTRAN is still the king of FEA. CATIA's version of these has got better but is still a notch behind. The situation is the same with Pro/E (Creo), NX, and other CAD software.

These high-end 3D packages are also referred to as *product life cycle management* (PLM) software, a term you will hear a lot in the engineering community. As described already, these software tools aim to follow a product from design and conception all the way through testing and manufacturing in one integrated package.

Here are some more specific details on the CAD software mentioned in the preceding section. You may run into many of these throughout your engineering career. It is well worth being familiar with these products. The information is up to date as of early 2018.

Computer-Aided Three-Dimensional Interactive Application

(www.3ds.com)

CATIA is the world's leading CAD/CAM/CAE software. Developed in France by Dassault for its Mirage fighter jet project (although originally based on Lockheed's CADAM software), it is now a well-known staple at Boeing and many automotive, aerospace, and naval design companies. It has even been famously adopted by architect Frank Geary for certain architectural projects. CATIA has grown into a full PLM solution and is considered the industry standard, although complex and with a not always user-friendly reputation. The latest version is CATIA v6 R2015/2016. Besides CATIA, Dassault Systems also markets other CAD software such as the more affordable and much lauded SolidWorks (discussed separately) and even an AutoCAD LT clone called DraftSight.

NX

(www.plm.automation.siemens.com)

Another major CAD/CAM/CAE PLM software solution that was developed by a company called UGS and was originally known as Unigraphics. NX is essentially a blend of this Unigraphics software and SDRC's popular I-DEAS software which UGS purchased and took over. A few years later, Siemens PLM Software took over UGS and renamed this blended product to NX. It is used by GM, Rolls Royce, Pratt & Whitney, and many others. The current version is NX 11.0. It runs on Windows and, surprisingly, the Mac OS. Similar to Dassault Systems, Siemens PLM Software makes an entire suite of CAD solutions, including a lower-end NX called SolidEdge (discussed separately).

PTC Creo Elements/Pro (Formerly Pro/ENGINEER)

(www.ptc.com)

The third major high-end CAD/CAM/CAE PLM software package, originally called Pro/ENGINEER, is historically significant. It caused a major change in the 3D CAD industry when first released in 1987 by introducing the concept of *parametric modeling*. Rather than models being constructed like a mound of clay with pieces added or removed to make changes, the user constructs the model as a list of features (or parameters), which are stored by the program and can be used to change the model by modifying, reordering, or removing them. The software also pioneered the use of feature trees and dimension-driven design in general. CATIA, NX SolidWorks, and many others work this way. The current software is no longer called Pro/ENGINEER; the name was changed to Creo Elements/Pro in the late 2010 (it was also called Wildfire), and then to just Creo Design Suite. The suite features 3D modeling software (Creo Parametric, Direct, Options Modeler) and 2D drafting software (Creo Sketch, Layout, Schematics) and others. The flagship product, and the direct descendent of Pro/ENGINEER, is the Parametric package. The current release is Creo Parametric 4.0.

SolidWorks

(www.solidworks.com)

SolidWorks is a midrange CAD/CAM/CAE software product, first introduced in 1995 by a Massachusetts company of the same name. Dassault Systems, the makers of CATIA, acquired the company in 1997. The SolidWorks installed base is estimated at over two million users, making the product a major player in the 3D CAD field. SolidWorks is marketed as a lower-cost competitor to major CAD packages, but the software still has extensive capabilities and caught on quickly in smaller design companies. It comes in three flavors: Premium, Professional, and Standard, along with many modules and options. The software has an enviable reputation of being powerful, yet easy-to-use, and many universities and colleges teach it. The purchase price is $3995 per license, with an annual subscription for 1 year costing you $1295. The current version is SolidWorks 2018.

Inventor

(www.autodesk.com)

Inventor is another midrange product, from the makers of AutoCAD, for solid modeling and design. Considered a low- to midrange product when first introduced, Inventor matured to become a major competitor to SolidWorks, SolidEdge, and Creo Parametric. The flagship versions are Inventor HSM Pro (original cost: just under $10,000) and Inventor HSM (original cost: $7500). There is also a downgraded Inventor LT (original cost: $1495). Like AutoCAD, Inventor moved to a subscription model last year, and the associated costs are $235 monthly and $1890 for 1 year, rising to $5105 for 3 years. The current version for all is Inventor 2017.

IronCAD

(www.ironcad.com)

IronCAD is another 3D CAD software package in the midrange market. Originally developed by Visionary Design Systems, the software gained ground in recent years due to a user-friendly reputation and a more intuitive approach to 3D design. It retails for about $1200. A downgraded lower-cost IronCAD Draft is also available for $600. The software recently added a new feature called *TraceParts*, which is a web-based content center (a library) with free 2D and 3D standard parts available from numerous manufacturers, making mechanical design easier. The current version is IronCAD 2018 SP1.

SolidEdge

(www.plm.automation.siemens.com)

SolidEdge is a midrange 3D CAD product from Siemens PLM Software, the same company that makes the high-end NX package. It is comparable to, and competes with, SolidWorks and Inventor in both cost and functionality. It comes in two flavors: Classic and Premium, and has many specialized modules for Design, Simulation, Manufacturing, etc. There is also a 2D version called SolidEdge 2D Drafting. The current release is SolidEdge ST10 2017.

Listed next are some of the major software products in FEA and CFD, as mentioned earlier.

NASTRAN

Originally developed for NASA as open-source code for the aerospace community to perform structural analysis, NAsaSTRucturalANalysis, or NASTRAN, was acquired by several corporations and marketed in numerous versions. NEiNastran is just one, commonly used, flavor of it. NASTRAN in its pure form is a solver for FEA and cannot create its own models or meshes. Developers, however, have added pre- and postprocessors to allow for this. The current release for NEiNastran is V9, though other companies have different designations for their products. NEiNastran was acquired by Autodesk in 2014, and renamed Autodesk NASTRAN. It remains to be seen what direction the software will take in the future.

ALGOR (Autodesk Simulation)

ALGOR is a powerful general-purpose FEA software package. Developed by ALGOR Inc. for the use by engineers in the aerospace industry, it has found a wide use in structural, thermal, CFD, acoustic, and electromagnetic simulations as well. The company was acquired by Autodesk in 2009 and renamed Autodesk Simulation. The old website, www.algor.com, redirects you to the Autodesk page for this software, with full descriptions and applications.

ANSYS

Ansys, Inc., a Pennsylvania-based company, is a major player in the engineering software industry and makes a range of products for FEA and CFD. One of its flagship offerings is ANSYS Fluent, the industry standard for CFD software, holding about 40% of the market. It imports geometry, creates meshes (using GAMBIT software) and boundary conditions, and solves for a variety of fluid flows, using the Navier—Stokes equations as theoretical underpinnings. The current release is Fluent 6.3. The entire suite of products is marketed as ANSYS 16.0 and includes not just Fluent but packages for structures, electronics, and systems.

Appendix C

File Extensions

Appendix C outlines a variety of AutoCAD and other commonly encountered software file extensions for student recognition and familiarity. File extensions are the bane of the computer user's existence. They number in the many hundreds (AutoCAD alone claims nearly 200), and even if you pare the list down to a useful few, it is still quite a collection. Here, we focus on those extensions that are important to an AutoCAD user, as chances are that you will encounter most, if not all, of them while learning the software top to bottom. They are listed alphabetically (with explanations and descriptions) but grouped in categories to bring some order to the chaos and set priorities for what you really need to know and what you can just glance over. Following AutoCAD's list is a further listing of extensions from other popular, often-used software, which you may find useful if you are not already familiar with most of them.

C.1 AUTOCAD PRIMARY EXTENSIONS

- **.bak** (backup file): A .bak is an AutoCAD backup file created and updated every time you save a drawing, provided that setting is turned on in the Options dialog box. The .bak file has the same name as the main .dwg file, generally sits right next to it, and can be easily renamed to a valid .dwg if the original .dwg is lost.
- **.dwf** (design web format file): The .dwf is a format for viewing drawings online or with a .dwf viewer. The idea here is to share files with others who do not have AutoCAD in a way that enables them to look but not modify. If this sounds like Adobe .pdf, you are correct; .dwf is a competing format.
- **.dwg** (drawing file format): The .dwg is AutoCAD's famous file extension for all drawing files. This format can be read by many of Autodesk's other software products and by some competitors. The format itself changes somewhat from release to release, but these are internal programming changes and invisible to most users.
- **.dwt** (drawing template): The .dwt is AutoCAD's template format and is essentially a .dwg file. The idea is to not repeat setup steps from project to project but use this template. It is equivalent to just saving a completed project as a new job and erasing the contents.
- **.dxf** (drawing exchange format): The .dxf is a universal file format developed to allow for smooth data exchange between competing CAD packages. If an AutoCAD drawing is saved to a .dxf file, it can be read by most other design software (in theory anyway). This, of course, is not always the case, and some CAD packages open .dwg files anyway without this step, but the .dxf remains an important tool for collaboration. The .dxf is similar in principle and intent to IGES and STEP for those readers who may have worked with solid modeling software and understand what those acronyms mean.

C.2 AUTOCAD SECONDARY EXTENSIONS

- **.ac$:** A temporary file generated if AutoCAD crashes; it can be deleted if AutoCAD is not running.
- **.ctb:** This color table file is created when pen settings and thicknesses are set (see Chapter 19).
- **.cui:** This customizable user interface file is for changes to menus and toolbars (see Chapter 14).
- **.err:** This error file is generated on an AutoCAD crash. You can delete it.
- **.las:** This layer states file can be imported and exported (see Chapter 12).
- **.lin:** This linetype definition file can be modified and expanded (see Appendix D).

- **.lsp:** This LISP file is a relatively simple language used to modify, customize, and automate AutoCAD to effectively expand its abilities.
- **.pat:** This hatch definition file can be modified and expanded (see Appendix D).
- **.pgp:** This program parameters file contains the customizable command shortcuts (see Chapter 14).
- **.scr:** This script file is used for automation (mentioned in Chapter 12).
- **.sv$:** This is another temporary file for the AutoSave feature. It appears if AutoSave is on and remains in the event of a crash. It can be then renamed to .dwg or deleted.

C.3 MISCELLANEOUS SOFTWARE EXTENSIONS

These extensions are for software and formats likely to be encountered daily or at least occasionally by AutoCAD users and are always good to know. Some extensions can be found in AutoCAD as export/import choices.

- **.ai:** Adobe Illustrator files.
- **.bmp:** Bitmap, a type of image file format that, such as jpg, uses pixels.
- **.dgn:** The MicroStation file extension. This competitor to AutoCAD can open native .dwg files. Appendix B, Other CAD Software, Design and Analysis Tools, and Concepts discusses this software.
- **.doc and .docx:** MS Word files.
- **.eps:** Encapsulated PostScript file.
- **.jpg:** The .jpeg (joint photographic experts group) is a commonly used method (algorithm) of compression for photographic images. It is also a file format (discussed in Chapter 16).
- **.pdf:** The famous Adobe Acrobat file extension. You can print any AutoCAD drawing to a .pdf with printer drivers installed with the AutoCAD program. It is one of the few extensions to become a verb, as in, "Can you PDF that invoice?"
- **.ppt and .pptx:** MS PowerPoint file extension.
- **.psd:** The famous Photoshop file extension. The software itself also became a verb, as in, "Can I Photoshop my ex-boyfriend/girlfriend out of this photo?"
- **.rfa:** Revit family files, similar yet very different than AutoCAd blocks)
- **.rvt:** Revit project files
- **.skp:** file extension used by SketchUP
- **.vsd:** MS Visio drawing file extension. Visio is a simple and popular "drag and drop style" CAD package for electrical engineering schematics and flowcharts, among other applications. It accepts AutoCAD files.
- **.xls and .xlsx:** MS Excel file extension.

This list is by no means complete, but it covers many of the applications discussed in Chapter 16.

Appendix D

Custom Linetypes and Hatch Patterns

This appendix outlines the basic procedures for creating custom linetypes and hatch patterns if you need something that is not found in AutoCAD itself or with any third-party vendor. All linetypes and hatch patterns were originally "coded" by hand, and there is nothing inherently special or magical about making them; they just take some time and patience to create. Though it is admittedly rare for a designer to need something that is not already available, creating these entities is a good skill to acquire. We just present the basics here, and you can certainly do additional research and come up with some fancy designs and patterns.

D.1 LINETYPE DEFINITIONS (BASIC)

All linetypes in AutoCAD (about 38 standard ones, to be precise) reside in the acad.lin or the acadiso.lin file. These are linetype definition files that AutoCAD accesses when you tell it to load linetypes into a drawing. In AutoCAD 2019, these files can be found in the Autodesk\AutoCAD_2019_English_Win_32 bit_dlm\x86\en-us\acad\Acad\Program Files\Root\UserDataCache\en-us\Support folder. Anything you do to these files, including adding to them, immediately shows up the next time you use linetypes. Our goal here is to open up the acad.lin file, analyze how linetypes are defined, then make our own.

Locate and open (usually with Notepad) the acad.lin file and take a look at it closely. Here is a reproduction of the first six linetype definitions: Border (Standard, 2, and X2) and Center (Standard, 2, and X2).

```
*BORDER,Border____ .____ .____ .____ .____.
A,.5,2.25,.5,2.25,0,2.25
*BORDER2,Border (.5x)__.__.__.__.__.__.__.__.__.__.__.
A,.25,-.125,.25,2.125,0,2.125
*BORDERX2,Border (2x)________ .________ .___
A,1.0,2.5,1.0,2.5,0,2.5
*CENTER,Center______________________________
A,1.25,-.25,.25,-.25
*CENTER2,Center (.5x)_______________________
A,.75,2.125,.125,2.125
*CENTERX2,Center (2x)_________________________
A,2.5,2.5,.5,2.5
```

Let us take just one linetype and look at it closer. Here is the standard size center line:

```
*CENTER,Center______________________________
A,1.25,2.25,.25,2.25
```

The two lines you see are the header line and the pattern line; both are needed to properly define a linetype. The header line consists of an asterisk, followed by the name of the linetype (CENTER), then a comma, and finally the description of the linetype (Center). In this case, they are the same.

The pattern line is where the linetype is actually described, and the cryptic-looking "numerical code" (A,1.25,2.25,.25,2.25) is the method used to describe it and needs to be discussed for it to make sense.

The A is the alignment field specification and is of little concern (AutoCAD accepts only this type so you always see A). Next are the linetype specification numbers, and they actually describe the linetype using dashes, dots, and spaces. Let us go over them, as really this is the key concept.

The basic elements are

- Dash (pen down, positive length specified, such as 0.5).
- Dot (pen down, length of 0).
- Space (pen up, negative length specified, such as −0.25).

Now, you just have to mix and match and arrange these in the order you want them to be to define a linetype you envisioned.

What would the following look like?

```
(A,.75,-.25,0,-.25,.75)
```

Well, you know that dashes are positive, so everywhere you see .75, there will be a dash of that length. Spaces are negative, so everywhere you see −2.25, there will be a space of that length, and finally a 0 is a dot. This is the result:

—— ·——

These are the essentials of creating basic linetypes. AutoCAD, however, also allows for two other types to be created, and they are part of the complex linetype family: string complex and shape complex.

D.2 LINETYPES (STRING COMPLEX AND SHAPE COMPLEX)

The problem with the previously mentioned linetypes is obvious; you have only dashes, dots, and spaces with which to work. While this is actually enough for many applications, you may occasionally (especially in civil engineering and architecture) need to draw boundary or utility lines that feature text (Fence line, Gas line, etc.) or you may want to put small symbols between the dashes. It is here that strings and shapes come in.String complex linetypes allow you to insert text into the line. Here is a gas line example:

```
*GAS_LINE,Gas line —GAS—GAS—GAS—GAS—GAS—GAS—
A,.5,-.2,["GAS",STANDARD,S=.1,R=0.0,X=20.1,Y=2.05],2.25
```

The header line is as discussed before, and the pattern line still starts with an A. After the A, we see a dash of length .5 followed by a space of length .2 so far it is familiar. Next is information inside a set of brackets, defined as follows:

```
["String", Text Style, Text Height (S), Rotation (R), X-Offset (X), Y-Offset (Y)].
```

Finally, there is a .25 space and the string repeats. The meaning of the information inside the brackets is as follows:

- String: This is the actual text, in quotation marks (GAS, in our example).
- Text Style: The predefined text style (STANDARD, in our example).
- Text Height: The actual text height (.1, in our example).
- Rotation: The rotation of the text relative to the line (0.0, in our example).
- X-Offset: The distance between the end of the dash and the beginning of the text as well as the end of the text and the beginning of the dash (a space of .1, in our example).
- Y-Offset: The distance between the bottom of the text and the line itself. This function centers the text on the line.

With these tools you can create any text string linetype. Shape complex linetypes are basically similar, aside from one difference: Instead of a text string, we have a shape definition inserted. Shape definitions are a separate topic in their own right and are not covered in detail in this textbook. Briefly, however, shape files are descriptions of geometric shapes (such as rectangles or circles but can be more complex) and are written using a Text Editor and compiled using the compile command. They are saved under the *.shp extension. AutoCAD accesses these when it reads the linetype definitions. If you are familiar with shapes, then you can proceed in defining shape complex linetypes as with string complex ones.

D.3 HATCH PATTERN DEFINITIONS (BASIC)

Knowing the basics of linetype definitions puts you on familiar ground when learning hatch pattern definitions. Even though they are more complex, the basics and styles of description are similar.

Hatch pattern definitions are found in the acad.pat and acadiso.pat files, and they reside in the same folder as the .lin files. Open the acad.pat file and take a look at a few definitions. Some of them can get long and involved due to the complexity of the pattern (Gravel, for example), but Fig. D.1 shows a simpler one to use as a first example: Honeycomb.

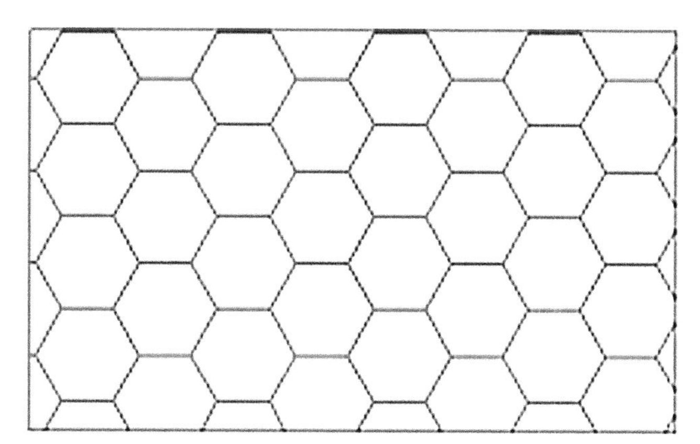

FIGURE D.1 Honeycomb pattern.

The pattern definition file that describes it looks like this:

```
*HONEY, Honeycomb pattern
0, 0,0, .1875,.108253175, .125,2.25
120, 0,0, .1875,.108253175, .125,2.25
60, 0,0, .1875,.108253175, 2.25,.125
```

The first line (header line) is as discussed previously with linetypes. The next three lines are the hatch descriptors, and they have somewhat different definitions as shown next:

Angle, X-Origin, Y-Origin, D1, D2, Dash Length, where (in this particular example)

- Angle: Angle of the pattern's first hatch line from the X axis (0 degree, in our example).
- X-Origin: X coordinate of the hatch line (0,0, in our example).
- Y-Origin: Y coordinate of the hatch line (.1875, in our example).
- D1: The displacement of the second line (.1083, in our example).
- D2: The distance between adjoining lines (.125, in our example).
- Length: Length of dashes or spaces.

The pattern is repeated two more times to form the overall shape and repeated many more times in the overall hatch pattern.

Appendix E

Principles of CAD Management

So, you have been hired on as a CAD manager or maybe forced into this due to your substantial AutoCAD skills—or was it because no one else wanted the job? Either way, you are now the top dog, head honcho, and you have to take the bull by the horns, well many bulls actually, coworkers who run around and draft whatever they please just to get the job out. Then again, maybe you are simply a student finishing your AutoCAD education. The bottom line is that you could use some ideas and a short discussion on this elusive science of CAD management. Well, you came to the right appendix.

All jokes aside, this is an important topic. You must take everything you learned in Levels 1 and 2 and apply it, while keeping others in line. You must be the keeper and the enforcer of standards and the technical expert who can resolve any issue. A true CAD manager who does nothing else is not all that common. You may have to multitask and do your own work while doing the management, especially in smaller companies.

Here are a few rules, ideas, and thoughts gathered from 20 years of experience as a CAD consultant, educator, and CAD manager. They outline some of the challenges and issues, and provide some possible answers, to those in charge of the software used to design a lot of what is around us.

E.1 KNOW THE EIGHT GOLDEN RULES OF AUTOCAD

1. Create a solid company template as a starting point for all projects.
2. Always draw one-to-one (1:1).
3. Never draw the same item twice.
4. Avoid drawing standard parts.
5. Use layers correctly.
6. Do not explode anything.
7. Use accuracy: Ortho and OSNAPs.
8. Save often.

Notice that most of these rules were discussed in some way in the first few chapters of Level 1. It is the basic stuff that people screw up, not some fancy new feature. Let us briefly go over these one by one; they bear repeating.

1. *Create a company template from which all projects are started:* At a minimum, everyone in a company should start new drawing files from the same template file (even a company of one person). The template should have the company standard layers (see below), dimension styles, text styles, leader styles, table styles, layout sheet(s) with title blocks, and units and limits set. If you have to start with a blank acad.dwt and start every project from scratch, it will lead to a standard-less environment. This also makes work more productive by giving someone a kit of tools to start with.
2. *Always draw one-to-one (1:1)*: This is the most important one. Always draw items to real-life sizes. Endless problems arise if users decide to scale items as they draw them. Leave scales and scaling to the pencil-and-paper folks. A drawing must pass a simple test. Measure anything that is known to be a certain value, for example, a door opening that is 3 feet wide, by using the list command. AutoCAD should tell you its value as 36 inches or 3 feet and nothing else. You must enforce this as a CAD manager.
3. *Never draw the same item twice*: In hand drafting, just about everything needed to be drawn one item at a time. Some inexperienced CAD users do the same in AutoCAD. That is not correct. AutoCAD and computer drafting in general

require a different mindset, philosophy, and approach. If you have already drawn something once, then copy it, do not create it again. While this one may be obvious, another overlooked tool is the block or wblock command. If you anticipate using the new item again (even if there is a small chance), then make a block out of it. As a CAD manager, you need to sometimes remind users of this. Remember: Time = $$.

4. *Avoid drawing standard parts*: This, like the preceding rule, aims to prevent duplicate effort. Before starting a project, as a CAD manager, you need to see to it that any standard parts (meaning ones that are not unique but constantly reused in the industry) are available for use and are not drawn and redrawn. Chances are good that you either already have these parts or can get the "cads," as they say, from the part supplier. Another resource is the web. Numerous sites have predrawn blocks available for use. Buy them, borrow them, or download them, but do not redraw items that someone already has done before. Again, Time = $$.
5. *Use layers correctly*: Layers are there to make life easier, and they are your friends. While overdoing it is always a danger (500+ layers is pushing it), you need to use layers and place items on the appropriate ones. As a CAD manager, you need to develop a layering standard (more on that later) and enforce it. Users in a hurry do not always appreciate what having all items organized in layers does ("it all comes out the same on paper!" they may say), so you must stand firm and enforce the standard. And do not use layer 0. It cannot be renamed with a descriptive name, so why use it?
6. *Do not explode anything*: How do you destroy an AutoCAD drawing and render weeks of work and thousands of dollars in work hours useless? Why, explode everything, of course. Short of deleting the files, this is the best way to ruin them. While this is generally a far-fetched scenario, remind users to not ruin a drawing in small ways, like once in a while exploding an mtext, dimension, or hatch pattern. They are liable to do so, if these items do not behave as they are supposed to and users explode them to manually force them into submission. Not a good idea; lack of skill is not an excuse. Remind them to ask you for help if a dimension just does not look right. There is always a way to fix something. Leave the exploding for the occasional block or polygon.
7. *Use accuracy: Ortho and OSNAPs*: Drawing without accuracy is not drafting, it is doodling. Most users learn the basics, such as how to draw straight lines and connect them, but you may still see the occasional project manager (who gets to draft on average only once a month) fix a designer's linework and forget to turn on OSNAPs. When zoomed out, the lines look OK and print just fine (reason enough to throw accuracy out the window for the occasional user), but a closer inspection reveals they are not joined together, causing a world of problems for a variety of functions, such as hatch, area, distance, and other commands. Besides, it is just plain sloppy. How accurate is AutoCAD? Enough to zoom from a scale model of the earth to a grapefruit (more on that later), so be sure to use all the available accuracy.
8. *Save often*: Let us state the obvious. There is no crash-proof software or operating system. Both AutoCAD and Windows can lock up and crash at some point. Remind users to save often, and do daily and weekly file backups. But you already knew this, right?

E.2 KNOW THE CAPABILITIES AND LIMITATIONS OF AUTOCAD

As a CAD manager, you have to know what you have to work with. All computer-aided design software, including AutoCAD, has an operating range for which the software is designed and optimized. Some tasks it excels at, you need to leverage that advantage, and some tasks are outside the bounds of expectation. AutoCAD cannot be all things to all people (although you would be hard-pressed to tell based on Autodesk's marketing). What follows is a short discussion on the capabilities and limitations and how you need to best take advantage of what you have.

Capabilities

Often, as a CAD manager, you are asked to provide an estimate on how long it will take to complete the drafting portion of a job or just recreate something in CAD. Sometimes, these requests come from individuals (senior management) who have had extensive experience in hand drafting, and they expect an answer in the form of a significantly smaller time frame. AutoCAD has an aura of magic about it, as if a few buttons just need to be pushed and out pops the design. Where *is* that "Finish Project" icon?

The surprising truth is that a professional hand drafter can give an average AutoCAD designer a run for the money if the request is a simple floor plan. After all, the hand drafter is not burdened with setting up layers, styles, dims, and other CAD prep work. However, this is not where AutoCAD's power lies. Once you begin to repeat patterns and objects, AutoCAD takes a commanding lead. A layout that took 1 hour to draft by hand or via computer can be copied in seconds on the computer but adds another hour to the hand drafter's task.

Do not be surprised by the seeming obviousness of this. I have met many users and students who do not always see where exactly CAD excels; they think work is always faster on a computer. It may not be and may lead to erroneous job estimates. However, in all cases with patterns, copying, and use of stored libraries, CAD is much quicker. So when estimating a job, look for

- Instances where the offset command can be used.
- Instances where there are identical objects.
- Minimal cases of unique, uneven, and nonrepetitive linework.
- Minimal cases of abstract curves.
- An abundance of standard items.

Knowing what to look for enables you to make more accurate estimates. Remember, the more repetition, the greater is the CAD advantage.

This is just one example, of course. Other advantages include accuracy (discussed next) and ease of collaboration. Another interesting advantage noticed by me is that nonartistic people or ones who would never have been able to draft with pencil, paper, and T-square often excel at CAD, and skills such as neatness and a steady hand become less critical.

In support of Golden Rule 7, a brief discussion on accuracy follows. This is one often brought up in class. Search the web for a drawing called SOLAR.DWG. Download and open the file. It is a drawing of the solar system (Fig. E.1) created in the early days of AutoCAD by founder John Walker and has become quite legendary over the years. You see, the drawing is to scale. This means that the orbits of all the planets are drawn to full size as well as correct relative to each other. Zoom in to find the Earth. Then, zoom in to see the moon around the Earth. Zoom in further and find the lunar landing module on the moon. Zoom in again and find a small plaque on one of the legs of the landing module. On it is a statement from humankind to anyone else who may read it. What does it say? And what is the date?

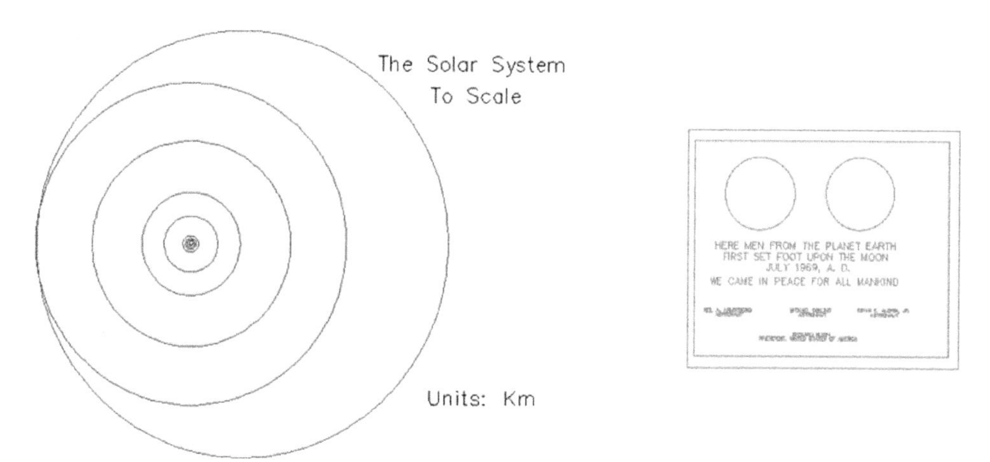

FIGURE E.1 The solar system and plaque to scale. *Source: John Walker.*

Think about what you just witnessed. You have zoomed in from the orbit of Pluto to look at a 5″ × 4″ plaque. That is quite an amazing magnification. To put things into perspective, this is the equivalent of looking at the Earth then at a grapefruit or looking at a grapefruit then an atom. Can you design microchips in AutoCAD? Safe to say, yes. Can you connect walls together in a floor plan? Absolutely.

LIMITATIONS

One of the chief limitations of AutoCAD is its 3D capabilities. Not that they are shabby, but sometimes they are pushed to do what they were not created for. The essence is that AutoCAD is not modeling and analysis software, rather it is visualization software when it comes to 3D. It was not designed to compete with CATIA, NX, Inventor, and SolidWorks. Parametric design, dimension-driven design, FEA, and CFD analysis are not what AutoCAD is about. These packages are dedicated engineering tools. AutoCAD was intended to be powerful 2D software with tools for presenting a design in 3D when necessary. As such, it is good at pretty pictures, to sell or present an idea.

Another limitation of the software, an obvious one about which a CAD manager needs to be aware, is that AutoCAD does not do the thinking for you. It contains little to no error checking, interference checking, or design fidelity tools, so garbage in definitely gives you garbage out. Be aware as a CAD manager that AutoCAD lets users do almost anything, and the word *anything* can swing in either a positive or a terribly negative direction. Finally, understand the learning curve of this software. It will take a few months for a novice user to get up to speed. Keep an eye on that person until then.

E.3 MAINTAIN AN OFFICE CAD STANDARD

A standard is a necessity for a smooth-running office, especially for one with more than a few AutoCAD users. A standard ensures uniformity and a professional look to the company's output, minimizes CAD errors, and makes it easy for new users to get up to speed. A manual needs to be created, published, and given to all users who may come in contact with AutoCAD. This manual should have, at a minimum,

- The exact location of all key files and the overall office computer file structure, indicating the naming convention for all jobs and the locations of libraries, templates, and standard symbol folders.
- A listing of the office layering convention. This may follow the AIA standard in an architecture office or just an accepted internal company standard.
- A listing of acceptable fonts, dimension styles, and hatch patterns.
- A listing of print and plot procedures, including *.ctb files.
- A listing of proper xref naming procedures.
- A description of how Paper Space is implemented. List title blocks, logos, and necessary title block information.
- A policy on purging files, deleting backup files, and AutoSave.

Keep the manual as short as possible while including all the necessary information. Do not attempt to cover every scenario that could possibly be encountered, as the longer the manual, the less likely it will be looked at cover to cover.

E.4 BE AN EFFECTIVE TEACHER AND HIRING MANAGER

Part of being a CAD manager is sometimes explaining the finer points of AutoCAD to new hires or updating the occasional user as well as promotion of better techniques and habits. If you have studied Levels 1 and 2 cover to cover, you may have noticed a concerted effort to simplify as much as possible. You should do the same with any potential students. While this is not meant to be a teaching guide, some basic points apply. If you are tasked with training new hires, then try to observe the following:

- Cover basic theory first—creating, editing, and viewing objects.
- Cover accuracy next, such as Ortho and OSNAP.
- Move on to the fundamental topics in the order of layers, text, hatch, blocks, arrays, dimensions, and print and plot, similar to Level 1.
- If advanced training is needed, cover advanced linework, followed by Xref, attributes, and Paper Space, in that order. The rest of Level 2 can wait, as some of it is geared mainly to CAD managers anyway.
- Stress to users the secrets to speed: the pgp file, right-click customization, and intense practice of the basics; remember the 95/5 rule from the second paragraph of , Chapter 1.
- Accurately assess the skill level of users and do not assign advanced tasks to those not yet ready, unless the drawings are not critical for a particular project. I have personally seen the subtle damage to a drawing that can be caused by an inexperienced drafter.

In addition, be ready to test potential hires with an AutoCAD test. This test does not have to be long; 45 minutes to an hour should be the absolute maximum, although truthfully, an experienced user can tell if another person knows AutoCAD within 3 minutes of watching him or her draft. The trick is to maximize every minute and squeeze the most amount of demonstrated skill out of a potential recruit, which means do not have that person engage in lengthy drawing of basic geometry but rather perform broad tasks.

For example, the floor plan from Level 1, , Chapter 3, is a good test for an architectural drafting recruit (the drawing can be simplified further if necessary by eliminating one of the closets or even a room). It should take an experienced user about 15 minutes total to set up a brand-new drawing (fonts, layers, etc.) and draft the basic floor layout (most students can do it in an hour after graduating Level 1—not too bad). Then, it should take another 15 minutes to add text, dims, and maybe some furniture, followed by another 5–10 minutes of hatch and touchups, followed by printing. So it is roughly a

45-minute test and runs the recruit through just about every needed drafting skill. When looking over the drawing, note the layers chosen, accuracy demonstrated, and how much was completed in what time frame.

Note the following. It is not critical if a new hire does not know xrefs or Paper Space. These can be explained in a few minutes and learned to perfection "on the job." However, nothing but hard time spent drafting teaches someone how to draft quickly and efficiently with minimal errors. This is a requirement, and do not hire anyone who does not have that, as it will take time climbing that learning curve. Exceptions should be granted, of course, for individuals hired for other skills, with AutoCAD being of only secondary importance.

The bottom line (in my hiring experience) is that a fast, accurate drafter is far more desirable than a more knowledgeable one with a sloppy, slow manner of drafting. Realize also that extensive knowledge is not always a good substitute for experience; and while many drafters are fine workers and can come in and be productive on an entry level, be sure you understand what you are getting. I have for a long time encouraged constant practice in class, even at the cost of occasionally not covering an advanced topic or two, knowing full well what these students will face in a saturated market full of experienced AutoCAD professionals.

As a final word, be sure the AutoCAD test is fair but challenging. Leave units in decimal form as the student needs to demonstrate a grasp of nuances and situational awareness and the ability to change units to architectural if the foot/inch input does not work. Include in the test

- Basic linework to be drawn.
- A handful of layers to set (colors, linetypes, etc.).
- Some text to write (set fonts, sizes, etc.).
- Some hatches to create.
- Some arrays to create.
- Some blocks to create.
- Some dimensions to put in.
- Plot of output.

E.5 STAY CURRENT AND COMPETENT

While we are on the subject of technical competency, you, as a recent advanced class graduate or CAD manager, are not off the hook for continuing to add to your skill set. It is much too easy to settle into a comfortable routine of drafting or designing and not learn anything outside the comfort zone (and not just with AutoCAD either). Some thoughts on this topic follow.

First of all, realize that it is virtually impossible for one individual to know everything about AutoCAD, although some claim to come close. It is truly an outrageous piece of software, developed and updated constantly by hundreds of software programmers and designers. This statement should make you realize that there is always more to learn (do not throw your hands up in despair, however), and you will not run out of new, easier, and creative ways to do a task. The main trait shared by all successful AutoCAD experts is that they are genuinely interested in the software and in the concept of computer-aided design in general.

To maintain your skills or advance further, look through the following list and determine what applies to you, then take the steps needed to acquire that knowledge. Some of this you will learn as part of studying all levels of the book; others you have to initiate on your own:

- *Learn the 3D features*. Many users do not know 3D well, and it is a good way to stand out from the crowd. It is fun to learn and is an important side of AutoCAD.
- *Learn basic AutoLISP*. This relatively simple programming language allows you to customize AutoCAD and write automation routines, among other uses. While its use dropped off somewhat in recent years, it is still a valuable skill.
- *Explore AutoCAD's advanced features that are sometimes overlooked*, such as sheet sets, dynamic blocks, eTransmit, security features, and CUI. Keep notes of key features and effects (a "cookbook" in software lingo).
- *Read up on the latest and greatest from AutoCAD*, new tips and tricks, and what the industry is talking about. Cadalyst.com and Augi.com are two good sites for this. You should also get a *Cadalyst* subscription. It is no longer free but still worth the cost. At the very least, check up on the latest in the AutoCAD world on all of the relevant websites.
- *Do not work in isolation*; talk to others in your field and see how they do things. In addition, be sure to take an AutoCAD update class if upgrading to a new release.

- *Be curious about other design software*, including AutoCAD add-on programs (verticals) such as AutoCAD Mechanical, Electrical, Civil, and MEP. Be knowledgeable about as much software as possible. In the world of CAD, Revit and SolidWorks are two software packages to keep an eye on in the future.
- *Enjoy what you are doing*. If you are not, you should seek other work duties in your profession. AutoCAD is very miserable to deal with if you hate it. Occasionally, you may think it is just one big software virus out to get you; just do not let this be a daily thought.

Appendix F

AutoLISP Basics and Advanced Customization Tools

You may have heard of AutoLISP, Visual LISP, VBA, .NET Framework, Active X, and ObjectARX. What exactly are they? These are all programming languages, environments, and tools used, among other things, to customize and automate AutoCAD beyond what can be done by simpler methods such as shortcuts, macros, and the CUI, all explored earlier in this book.

To properly discuss all these methods would take up quite a few volumes of text, and indeed much has been written concerning all of them. Our goal here is to introduce the very basics of AutoLISP and only gloss over the rest, so at least you have an idea of what they are. If advanced customization or software development work interests you, you can pursue it much further with other widely available publications.

F.1 OVERVIEW 1. AUTOLISP

This is really the very beginning of the discussion. AutoLISP is a built-in programming language that comes with AutoCAD and is itself a dialect of a family of historic programming languages, collectively referred to as LISP (list processing language). LISP was invented by an MIT student in 1958 and found wide acceptance in artificial intelligence among other research circles. The language and its developer, J. McCarthy, pioneered many expressions and concepts now found in modern programming.

AutoLISP is a somewhat scaled-back version of LISP uniquely geared toward AutoCAD. It allows for interaction between the user and the software, and indeed one of the primary uses of AutoLISP is to automate routines or complex processes. Once the code is written, it is saved and recalled as needed. The code is then executed as a script (no compiling needed) and performs its operation.

One advantage of AutoLISP over other methods, discussed shortly, is that it requires no special training, except some familiarity with syntax, and is thus quite well suited for nonprogrammers. While it is not a compiled, object-oriented language such as C++ or Java and therefore not suitable for writing actual software applications, it also does not need to be for its intended use.

F.2 OVERVIEW 2. VISUAL LISP

Over the years, AutoLISP has morphed into a significantly more enhanced version, called Visual LISP. More specifically, Visual LISP was purchased by Autodesk from another developer and became part of AutoCAD since release 2000. It is a major improvement and features a graphical interface and a debugger. Visual LISP also has Active X functionality (more on that later). However, Visual LISP still reads and uses regular AutoLISP, and that is still where a beginner needs to start before moving up. We will briefly touch on Visual LISP and its interface (the vlide command) later. The subject is an entire book all in itself.

F.3 OVERVIEW 3. VBA, .NET, ACTIVE X, AND OBJECTARX

There is evidence to suggest Autodesk is moving away from using Visual LISP toward other, more advanced tools for customization and programming.

VBA (Visual Basic for Applications) is a language based on Visual BASIC, a proprietary Microsoft product. VBA allows you to write routines that can run in a host application (AutoCAD) but not as stand-alone programs. It has some additional power and functionality over Visual LISP.

.NET Framework is another Microsoft software technology that was intended as a replacement for VBA. It features precoded solutions to common programming problems and a virtual machine to execute the programs.

Active X, also developed by Microsoft, is used to create software components that perform specific functions, many of which are found in Windows applications. Active X controls are small program building blocks and are often used on the Internet, such as for animations and unfortunately also for some viruses and spyware applications.

ObjectARX is the highest level of customization and programming you can do with AutoCAD, short of working on developing AutoCAD itself. Unlike the scripting-type languages and methods discussed so far, ObjectARX (ARX stands for AutoCAD Runtime eXtension) is an application programming interface (API). Using C++, programmers can create Windows DLLs (Data Link Libraries) that interact directly with AutoCAD itself.

Everything needed to do this is part of the ObjectARX Software Development Kit, freely distributed by Autodesk. This is not easy to master, as you would expect from professional-level software coding tools. Typical users are third party developers who need a way to write complex software that interacts seamlessly with AutoCAD. ObjectARX is specific to each release of AutoCAD and even requires the use of the same compiler Autodesk uses for AutoCAD itself.

Hopefully this discussion gives you some insight into what is out there for AutoCAD customization. This is only a cursory overview, of course, and you can pursue these topics much further. Any journey to learn advanced AutoCAD customization, however, begins with AutoLISP; and this is where we go next to introduce some basics of this programming language.

F.4 AUTOLISP FUNDAMENTALS

As mentioned before, despite the proliferation of new tools, AutoLISP is still very much in use, thousands of chunks of code are still being written annually, and many are easily found online for just about any desired function. The main reason for AutoLISP's past and present (to some extent) popularity remains the fact that AutoCAD designers are not programmers and AutoLISP is accessible enough for almost anyone to jump in and learn. The value it adds to your skill set is substantial. Let us cover some basics and get you writing some code. AutoLISP code can be implemented in two distinct ways:

- You can type and execute a few lines of code right on the command line of AutoCAD. This is good for a simple routine that is useful only in that drawing.
- You can load saved AutoLISP code and execute it, usually the case with more complex and often-used routines.

We generally use the second method of writing and saving our routines. At this stage, it is helpful to know how (and where) to save AutoLISP code and, by extension, how to load what is written.

All AutoLISP code is saved under the *.lsp extension. Once the code is written, it can theoretically be stored just about anywhere (AutoCAD Express folder contains a sizable number of built-in routines). To load the code, go to Tools → Load Application… or Tools → AutoLISP → Load Application… (or just type in `appload`) and the dialog box in Fig. F.1 appears.

Then, you just browse around, find the AutoLISP code you are interested in, click on it, and press Load. Once loaded, you can close the dialog box; the code is ready to execute with a command input. There is a way to automatically load all existing code on starting AutoCAD by adding it to the Startup Suite (lower right corner, the briefcase icon).

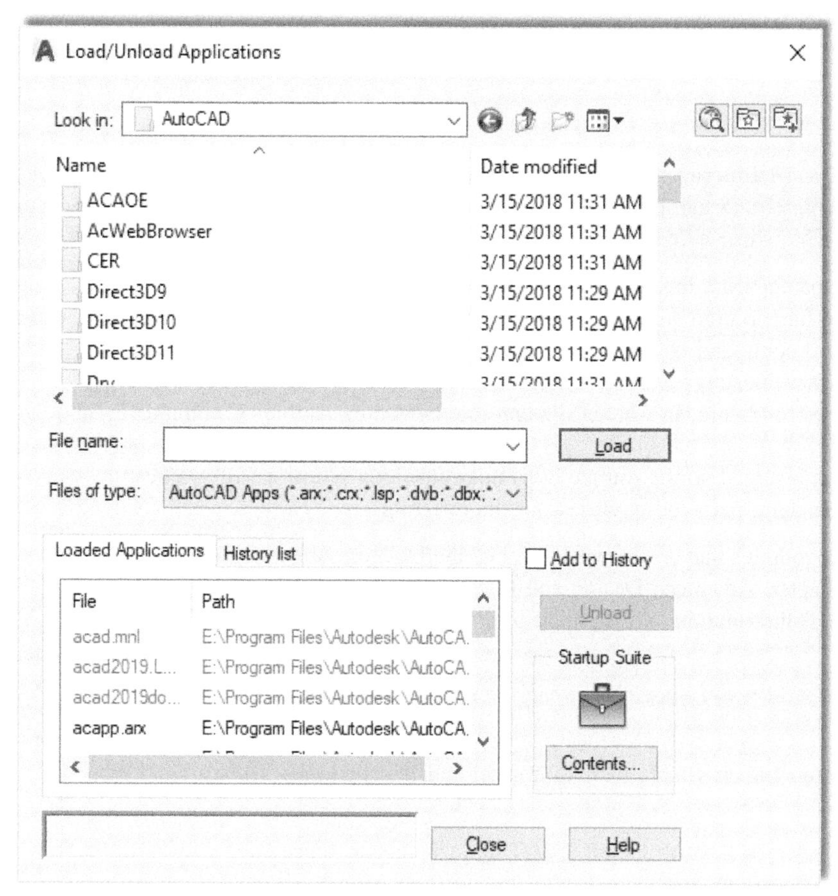

FIGURE F.1 Load/Unload Applications.

Now that you know how to save new code and where to find it, we should try writing a simple example that illustrates some AutoLISP basics. Here are two critical points that need to be understood before you begin:

- AutoLISP is closely tied to AutoCAD commands, so to write something useful, you need to know the commands well. If you do not know how to do it in AutoCAD manually, AutoLISP is of no use.
- Approach AutoLISP for an answer to a specific problem. State what you want to solve or automate, then write out the steps to do it manually. Only then, using proper syntax, write and test the AutoLISP code.

Following this philosophy, let us state a problem we would like to solve. We would like to modify the zoom to extents command by having it zoom to only 90% of the screen, not 100%. That way we can see everything and nothing gets lost at the edges of the screen. But, how do we do this by manually first?

You, of course, type in `zoom`, press Enter, select `e` for Extents, and press Enter again. The drawing is zoomed all the way. To bring it back a bit, you repeat the zoom command but type in `.90x` instead of `e`. The drawing then zooms in a bit to fill only 90% of the screen.

Let us now automate these steps so all you must do is type in `zz` (a random choice for a command name; you could pick anything) instead of the two-step process just described.

We need to:

- Open the Notepad editor (we use the built-in Visual LISP editor later).
- Save the blank file as ZoomSample.lsp in any folder you wish.
- Define the function zz by typing in exactly the following: `(defun C:zz ()`.
- List the relevant commands in order by typing in exactly the following: `(command "zoom" "e" "zoom" ".90x")`.
- Finally add `(princ)` and close the expression with another).

So, here is the result:

```
(defun C:zz ()
(command "zoom" "e" "zoom" ".90x")
(princ))
```

Save the file again and let us run it. Open any drawing, load the AutoLISP routine using appload or the drop-down menu, and run it by typing in `zz` and pressing Enter. The drawing should zoom out to 90% as expected.

This simple routine has some essential ingredients of all AutoLISP code:

- It starts out with a parenthesis, indicating that a function follows.
- The defun or define function (an AutoLISP expression) is next followed by C, indicating it is an AutoCAD command, and finally the actual command, zz.
- The set of opening and closing parentheses can hold arguments or variables.
- The next line features command, an AutoLISP expression meaning AutoCAD commands (in quotation marks, meaning it is a string) follow.
- The (princ) function is added so the routine does not constantly return a "nil" value.
- Finally, another, matching parenthesis closes out the program.

Here are two simple routines that can be used to rotate the UCS crosshairs to align them to some geometry (this general task was covered in chapter: Advanced Design and File Management Tools).

The first routine does the rotation and is called *perp*:

```
(defun C:perp ()
(command "snap" "ro" "" "per" pause "snap" "off")
(princ))
```

The second one brings the crosshairs back to normal and is called *flat*:

```
(defun C:flat ()
(command "snap" "ro" "" "0" pause "snap" "off")
(princ))
```

The perp and flat AutoLISP routines feature two new concepts. The first one is the pause function, which allows for user input. In this case, it is for the user to select the line to which the crosshairs are to be aligned. Then, the function resumes and turns off snap. The other concept is the two quotation marks back to back with nothing between them. They simply force an Enter.

Study these three examples closely. They are simple; many AutoLISP routines are, as that is all that is needed to get the job done.

Here is what we learned so far:

- *String*: Anything that appears in quotes (often an AutoCAD command).
- *Expression*: AutoLISP built-in commands, such as defun.
- *Functions*: These can be user-defined (perp, flat) or built-in (pause, princ, command).
- *Arguments*: Functions are sometimes followed by arguments in parentheses.

F.5 VARIABLES AND COMMENTS

Let us move on to other concepts. To take a step up in sophistication from what we did so far, we need to introduce program variables. Variables can be numbers, letters, or just about anything else that can change. The reason they are needed is because they allow for user interaction. Because the user input is not known when the code is first written, a variable is used. When that information is acquired, the program proceeds accordingly. Variables are a fundamental concept in all programming.

Variables in AutoLISP can be global (defined once for the entire routine) or local (defined as needed for one-time use). Local is usually how variables get defined in AutoLISP. The function that creates variables is *setq*. To set a variable equal to whatever the user inputs (such as the radius of a circle), you write setqRadCircle. The next logical step with variables is to pick up on whatever the user inputs. For this we need the getstring function, which returns anything the user types in.

Here is how these concepts look in practice, with the \n simply meaning that a new line is started.

```
(setqRadCircle (getstring "\nCircle Radius: "))
```

In addition to getstring are many other get functions, such as getfiled (asks user to select a file using a dialog box), getangle, and getdist.

As in any programming language, you are allowed to comment on your work, so you or someone else can understand your intent when examining the code at a later time. In AutoLISP, comments are preceded by a semicolon. You can add more than one semicolon to indicate sections or headers:

```
;This is a comment
;;So is this
```

F.6 ADVANCED FEATURES AND THE VLIDE COMMAND

We have only scratched the surface of AutoLISP, and if you enjoyed writing and running these simple routines, then consider taking a full class on this subject or reading specialized books. There is much more to learn. For example,

- *The IF → THEN statements*: This is a common feature in all of programming. If certain conditions are met, then the program does something; if not, it does something else.
- *The WHILE function*: This is another important concept in programming. This function continues to do something until a condition is met.

Finally, we come to the vlide command and the Visual LISP Editor, as shown in Fig. F.2. Everything you typed earlier into Notepad can be typed into the Visual LISP Editor (also referred to in programming as an Integrated Developers Environment), which has many additional useful features, such as debugging, error tracing, and checks for closed quotations and parentheses (formatting tools). Everything you type is also color coded for easy identification. Simply type in `vlide` and press Enter to call up this editor.

In-depth coverage of Visual LISP is beyond the scope of the book, but you are encouraged to pick up a specialized textbook and try to progress with this invaluable skill.

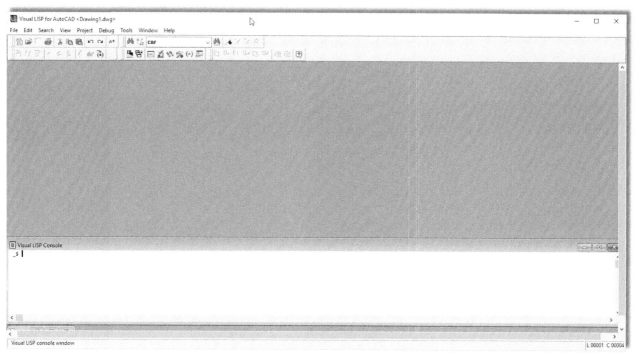

FIGURE F.2 Visual LISP Editor (Vlide command).

Appendix G

PC Hardware, Printers/Plotters, Networks, and the Cloud

This collection of topics can easily span volumes, and quite a bit has been written about all of them. We try to focus on hardware and networks from an AutoCAD user's point of view exclusively.

Software generally goes hand in hand with the hardware. In recent years, computers have become a lot like cars—very reliable and are usually packed full of standard features that were considered high-end not too long ago. With both you will likely get a solid product that will be right for most ordinary tasks. Remember though that CAD, in general, and AutoCAD specifically, is not ordinary software. It is a high-end, powerful application that places heavy demands on the associated hardware and does not run at its peak on a clunky PC. Much like the case where the owner of a sports car is more likely to be interested in (and look under the hood of) the car than the owner of a commuter economy car, so should AutoCAD designers express interest in what is "under the hood" of their computer. After all, they are manipulating and creating multimegabyte sophisticated drawings or 3D models, not merely typing up a memo on Word; and it is in their best interest to use a PC that meets or exceeds AutoCAD's demands.

Fortunately, as previously mentioned, it is easier than ever to have the right PC. Processor power and the downward spiraling costs of RAM and graphic cards have outpaced AutoCAD's requirements, and most folks, if they spend enough for a midrange system, will likely get everything they need. Furthermore, anyone who engages in gaming or the visual arts (such as digital film editing) will already have the right PC for the job—or Mac, if working with AutoCAD for the Mac. What is good for the goose (games and video editing) is also good for the gander (AutoCAD).

G.1 PC HARDWARE

By its nature, AutoCAD is not as fussy as some other CAD software out there (a few 3D solid modeling applications come to mind), but it does need a midrange PC to operate optimally. Here is a rundown of all the computer components:

- *Processor*: The "brain" of the computer, also referred to as the CPU, it is where all the calculations and operations are performed. Intel and AMD are the major players in the PC market. Which is better is a moot point; while Intel was once the only game in town, these days the two companies constantly one-up each other with performance benchmarks. At this point, both manufacturers produce high-quality products, and the Intel versus AMD comparison is essentially meaningless to the average user. Intel processors are organized by families such as i3, i5, i7, i9 (new), or Xeon. The numbers following that are the CPU's power and its intended placement (PC, mobile, laptop, etc.). If a letter follows that it is used for additional classification, such as K, meaning "unlocked for overclocking," or a power saving chip, and so forth. The top-of-the-line Intel processor for a PC (as of this writing) is the Intel Core i9−7900X processor, with a 3.3 GHz (4.5 Turbo) clock speed, 10-core, and with a 14 Mb cache. At a list price of over $1000, it is a bit much, and you can get close to the same performance out of the less expensive i7-8700 or the even cheaper i5-8600k for under $400. AMD has a roughly comparable top-of-the-line Ryzen 7 1800X, 8-core processor running at 3.6 GHz for $300, or the slightly tamer Ryzen 7 1700 for $230, as an alternative. All these CPUs feature multiple cores and something called *hyperthreading technology*. Multiple cores are independent processing units embedded on the main chip, and hyperthreading allows for multiple tasks to be done at once. To take advantage of this, your software must be written for it. As of this writing, AutoCAD is not optimized for these technologies.

- *Motherboard*: The motherboard is the main mounting and control unit for the CPU (and associated cooling hardware), the RAM, and the video card. The motherboard is also where you will find the connections for the hard drive, power, CD-ROM, and just about everything else. Most of the interconnection wiring also originates or terminates here. Motherboards are obviously very critical and expensive parts of the overall system. When selecting a motherboard, you have to match it to the CPU via the stated socket type and number.
- *RAM*: Random access memory is the next consideration, and the more the merrier. RAM has a direct effect on performance, as it is there that software resides while in operation. More available RAM allows for more functions to execute faster. Autodesk recommends 2 GB, but go as high as your motherboard supports. Older motherboards accept at least 4 GB and some of the newer ones, like the one from the ASUS P9X79 family, up to 64 GB of DDR3 memory. Some well-known RAM manufacturers are Corsair and Kingston.
- *Hard drive*: Also known as the storage disk, main drive, and the like, it needs to be at least 1.8 GB to install AutoCAD. Fortunately, this is rarely a problem, as most machines have drives on the order of 100–250 GB. It is not unusual to find a terabyte-sized (1 TB) and even larger drives. Hard drives are classified by the "old style" disk drive, which involve spinning magnetized platters. The newer type, solid state (SSD), features much faster data access, but for a significantly greater cost and currently smaller storage size than its spinning disk counterpart. A good combination is to use two drives, an SSD as the primary drive for blazing fast startups of Windows and various applications, with a secondary regular drive for data storage. Hard drives are hooked up to the motherboard via the older IDE hookup or the newer SATA. Some users opt for several drives stacked next to each other in what is known as a RAID configuration. This adds security in case of failure but, of course, increases the cost. Some well-known brands include Seagate, Toshiba, and Western Digital. Prices can range from $50 to $250 and up.
- *Video graphics card*: Perhaps the most overlooked part of the PC, the video or graphics card is similar to a processor but is specially designed to control the screen images or graphics. It mounts on the motherboard and needs to be of at least midrange level if you are planning on doing a lot of 3D work, including rendering. Autodesk recommends only a "workstation-class" graphics card, but get the best one possible under the allowed budget. High-quality graphic cards are not cheap; a high-end card can easily top out at over $1000 (and up). One in the $300–500 range should be just fine for AutoCAD applications. Some well-known video card manufacturers include ATI, NVidia, AMD, and EVGA.
- *OS*: Until the late 2010, operating systems that AutoCAD could run on numbered in the single digits (actually precisely one, Windows). In November of that year, after a 20-year break, Autodesk announced AutoCAD would once again be available for the Mac. AutoCAD still does not run on UNIX or Linux. Windows 7 SP1, Windows 8.1, and Windows 10 are currently the main Microsoft operating systems AutoCAD that is certified compatible for. Windows 8.1 was released in October of 2013. It has proved problematic in the computing world at large, but seems to be just fine for AutoCAD. Windows 10, released in 2015, will work just fine. As far as the older XP operating system, note the following warning from Autodesk: "AutoCAD 2014 is the final version of AutoCAD software that will be supported on the Microsoft Windows XP operating system." Keep this in mind if you are planning on installing AutoCAD 2015–19 and still have XP running.
- *Sound (audio) card*: This add-on is not technically required for AutoCAD, the way it may be for video editing and gaming, but it is good to have for a complete PC. Sound cards are not terribly expensive, and top-of-the-line ones can be had for under $200. Some major manufacturers include ASUS and Creative Sounds. The sound card enhances the audio output over the base motherboard and provides some outputs for speakers and microphones.
- *Optical drive (CD/DVD)*: An optical drive is needed if you plan to read from or burn to any disks such as CDs or DVDs. Most modern drives can handle all types of CDs, DVDs, and Blue Ray, both reading from and burning to. The current speeds are about 48x write, 24x rewrite, and 48x read, with prices in the $30–$60 range. Some machines come without CD drives now, relying on USB, internet downloads, and other methods for acquiring data, but I would keep the CD drive around for some time still—it is cheap and still very useful.
- *Monitor*: Get the best and largest flat screen monitor you can afford; your eyes will thank you many times over. Actually get two, and put them side by side with DVI connections. Once you have that sort of setup, there is no going back. Prices on monitors have also dropped in recent years and you should be able to find an excellent 24″ monitor for under $250.
- *Mouse and keyboard*: Get a laser precision mouse (forget roller ball, if you can even find one like that) and the best soft-touch keyboard you can find. Do not try to cut corners here, as you will be in physical contact and using these devices constantly while working on the PC, so get the good stuff. There are some ergonomic mice around that are formed for better hand positioning.

Whether building your own PC or ordering an assembled one, take the PC purchase seriously and try to set realistic budget expectations. CAD workstations are in the same family as gaming stations and you need to budget $1500 (or more) for a serious machine, not including the monitor. Some well-known vendors include Dell, but if you are inclined to assemble your own, check out online stores such as www.Newegg.com or www.PCPartsPicker.com and shop sales. You will get phenomenal performance for a fraction of the cost, as long as you know what you are doing when assembling the machine.

AutoCAD Requirements

Finally, as a conclusion to this section, here is what Autodesk recommends for AutoCAD 2018, as posted on its website. AutoCAD 2019 is expected to have the same requirements. As a reminder again, any release after AutoCAD 2014 does not run or install on Windows XP; you need at least Windows 7 to proceed. Note that, in all cases, AutoCAD comes in two versions: 32 bit and 64 bit. Generally, for both versions, you need, at a minimum,

OS:

- Windows 10 (64-bit only) (version 1607 and up recommended)
- Windows 8.1 with update KB2919355
- Windows 7 SP1 (32-bit and 64-bit)

Hardware:

- Intel Pentium or AMD multicore processors
  - 32-bit: 1 GHz or faster 32-bit (x86) processor
  - 64-bit: 1 GHz or faster 64-bit (x64) processor
- 2 GB RAM (8 GB recommended).
- 4 GB free disk space for installation.
- Conventional Display: 1360 × 768 (1920 × 1080 recommended) with True Color
- Windows Internet Explorer 11.0 or later web browser.
- Install from internet

As you can see these requirements are not astronomical by any means. They are the minimums, so aim to build or configure a machine that is above this.

G.2 PRINTERS AND PLOTTERS

There are essentially two ways to output a paper drawing from AutoCAD, a printer or a plotter. Printers are generally Inkjet or LaserJet and black and white or color capable. Prices can range from a $49 desktop Inkjet to a "sky's the limit" corporate high-speed color LaserJet. Which to get depends entirely on budget and desired level of quality for output. Once AutoCAD is installed, it automatically senses any printer attached to the PC (assuming it works and has drivers in the first place). Be aware that most small desktop printers cannot handle 11″ × 17″ paper. The 11″ × 17″ format is very important in AutoCAD production, as many jobs are handed in on this size paper or printed as half size check plots, so be sure your printer can handle it.

Plotters (Fig. G.1) are a familiar sight in architecture or engineering offices. Anything sized above an 11″ × 17″ sheet of paper needs to be routed to a plotter, all the way up to a 48″ × 36″, or larger in some cases. Plotters as a rule are more expensive than printers, and buying one should be a carefully researched decision. Most plotters are color capable and accept almost any type of paper, although bond or vellum is the norm. Hewlett–Packard is the premier manufacturer of plotters, with some models on sale well over $20,000, though their T120, T520, and T730 series plotters are much more affordable. Other companies, such as OCE, Canon, and Encad, round out the field. The typical average cost of a professional-grade plotter hovers around $5000–$6000. Do your research carefully when selecting one and be aware of recurring costs such as ink and paper.

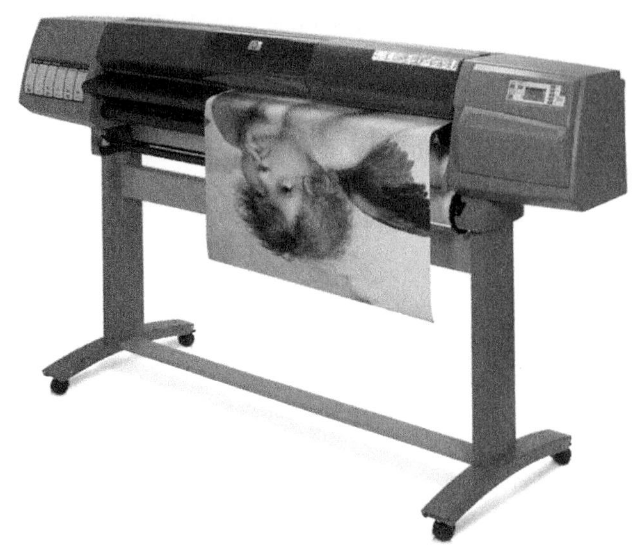

FIGURE G.1 A typical professional plotter. *Source: Hewlett–Packard.*

G.3 NETWORKS

This topic is an entire profession all to itself, of course. A computer network is defined as a group of interconnected computers. Networks can range from a local area network (LAN) to a city, country, or worldwide (wide area network) one. Networks allow for collaboration and sharing of information from computer to computer and have their own set of hardware familiar to any network engineer, such as routers, hubs, switches, repeaters, and other gear. Networks are operated via a protocol or set of instructions, such as the IEEE 802.3 protocol running on Ethernet technology for a typical LAN. The medium of data transfer between computers is usually a standard Cat5e cable but can be wireless or fiber optic for larger networks.

As an AutoCAD designer, you will likely encounter an LAN consisting of a server (host) with some other computers (clients) connected to it at your company. All the data are stored on the server, and once you log on to the network, your computer can access what it needs.

In regard to networks, AutoCAD can be installed on a per seat or network license method. Per seat is cost-effective to a certain limit (usually 10 seats). If your company requires more, a site license is purchased, which is cheaper and allows more computers to run the software, which now resides on the server only.

G.4 THE CLOUD

The *cloud* refers to two things here: Cloud computing and cloud storage. You are probably familiar with both, even if you did not specifically refer to them as such. Cloud computing is a general term for use of computing resources (hardware and software) that are delivered as a service over a network. You, the user, are the front end. The service is the back end, and you communicate with the service via the internet. A simple example is Gmail. You do not store the email messages and all the attachments on whatever computer you may be using, it is all on Google's servers. You simply access it remotely as needed. Cloud storage is a related concept. You store data, personal or company, on a remote server and upload and download as needed.

These explanations are, of course, highly simplified, and a great deal of technology and innovation lie "behind the scenes" of this broad high-tech industry. Cloud computing offers many great services, such as software, platform, a test environment, and data. Many companies and now individuals use this technology. It is not really new, with underlying principles dating back to the 1950s, but was held back mostly due to low-speed point-to-point connections up until the 1990s. After all, moving data and information is the cornerstone of cloud computing and storage, and until fast internet and other data transfer methods became the norm, a lot of computing and storage was local in nature. The term *cloud* and the cloud symbol first appeared in the early 1990s as well and was a popular metaphor to represent the complex (and unseen) internet computer architecture in network diagrams.

So what does this all have to do with AutoCAD? Well, AutoCAD's output—the drawings—are data. Such data need to be stored and sometimes shared as needed. Just like storing music in the cloud (such as with iTunes), you can store these drawings in a similar manner. This also allows for safe, remote archiving of a company's designs and easy collaboration between individuals that are far away.

Autodesk jumped in the cloud game a few years ago by offering something called *Autodesk 360.* This is a service that, according to Autodesk itself, "provides a broad set of features, cloud services and products that can help you improve the way you design, visualize, simulate, and share your work with others anytime, anywhere." The goal here is to create a collaborative work environment that is hosted completely by Autodesk on its servers. This is a new direction for the company, and no doubt a lucrative one. Users create an account, once they subscribe, and log in via secure, password protected means. They can then share designs and collaborate in a "digital workspace."

It is not known outside Autodesk how many companies signed on to the cloud service, but it is clear that this idea in general is the way things will be done in the future, not just with AutoCAD but in other areas of design. By outsourcing the physical and even the software component of a design firm to a cloud-based service, huge cost savings can potentially be realized. Keep an eye on this in the future, and if interested in learning more, research the topic online.

Appendix H

What Are Kernels?

If you have taken an active interest in 3D CAD software and are learning what is "under the hood," then you may have read or heard about kernels. What are they and what do farming and planting terms have to do with software? Well, as you may have already deduced, the word *kernel* is a metaphor for describing the heart of the software, a starting point from which everything else grows, the same as a kernel of corn. This analogy is quite accurate: A 3D modeling kernel is the computer code that is the "brains" of any 3D software package, such as CATIA, NX, SolidWorks, Inventor, and of course AutoCAD.

As it turns out, kernels are quite rare. This is because they are extremely difficult, expensive, and labor intensive to develop from scratch. As a result, while hundreds of 3D-capable CAD software applications are available, there are only a small handful of kernels. If you want to start a new CAD company and develop a competitor to any of the established products, then chances are good that you would simply purchase a kernel license and design your software around the kernel, not try to create your own. In many cases, this is indeed what happens in the industry.

Why would kernel developers license their kernels, as opposed to using them for just their own CAD programs? Well, for starters, they do use them in such a way, and most kernel developers have a popular 3D CAD program for sale. However, the licensing fee to "rent out" kernel code is substantial, so they do this as well to increase the revenue stream.

What customers get when they purchase the kernel is a large chunk of code made up of what are called *classes* and *components*—essentially mathematical functions that perform specific modeling tasks. The CAD company then writes its own code to interface with the kernel via numerous application programming interfaces and essentially creates the new software around the kernel.

This new CAD software is unique, according to the design philosophy and talent of the CAD company and may be different in how it operates compared with another company's software that uses the same kernel. In essence, however, most modelers are essentially the same, and one can switch relatively easily from, say, CATIA to NX or Creo Elements (Pro/ENGINEER). All three of these software applications feature parametric modeling, part trees, dimension-driven design, and other hallmarks of solid modeling. The differences are chiefly in the interfaces, ease of use, extra features offered with each package, and of course price.

One notable exception is AutoCAD. The kernel it uses, ACIS, is made by Spatial Technologies. Its modeling methods are somewhat different than the other kernels on the market, giving AutoCAD a unique (and not necessarily the best) approach to 3D. The essential difference has to do with the purpose and goal of the ACIS kernel, which is to simply represent objects in 3D, not to analyze and design with them. Models created by ACIS lean more toward "pretty pictures" than engineering tools. This is not surprising because that is exactly what is needed by AutoCAD design professionals—a representation of a 3D model to bring across an idea of a design. There is simply no need for highly sophisticated 3D operations found in the high-end solid modeling software and would be a waste of money to add it to AutoCAD, when most customers would never use it.

These other, more high-end, kernels include Parasolid (used by NX, SolidWorks, and SolidEdge), Granite (used by Creo Elements [Pro/ENGINEER]), and CAA (used by CATIA). Autodesk set out to develop its own kernel in recent years, and the fruit of that labor is in its Inventor software.

Appendix I

Lighting, Rendering, Effects, and Animation

The world of digital effects is a very expansive one, and under that umbrella falls everything from simple-shaded models (extensively used in the downloadable 3D chapters this book) all the way up to full-blown Hollywood-style effects. Architects and engineers use some of these digital effects and techniques, which are roughly in the middle of the pack as far as sophistication goes.

What architects look for includes visualization renderings that present a building or an interior scene in the most realistic way possible, so the clients see what they are getting and can offer input. They may also want some basic animation, such as a walk-through. Some high-end requests may even include photorealistic scenes that have weather effects (wind, precipitation, etc.) or realistic images of people interacting with the design.

Engineers have the same basic goals, to present a design to a client, although their renderings must often account for metals, plastics, and other materials, including material properties. Engineering animation needs are also more sophisticated. An architectural animation of a walk-through is essentially a collection of snapshots played back in succession. With engineering animations, complex part interaction must be accounted for and a full motion path must be shown. The entire animation must also be looped for constant playback. An example of this is a cutaway of an engine, with all the components (pistons, connecting rods, crankshaft, etc.) working in unison to demonstrate the design in operation.

As far as engineering applications go, AutoCAD plays a relatively minor role. From the beginning, the intent of AutoCAD was to be a superb 2D drafting tool, not a 3D solid modeling and testing software. Other software have evolved over the years (with some of these applications predating AutoCAD) for use in such cases. Examples, quoted elsewhere in this book, include CATIA, Creo Elements (Pro/ENGINEER), NX, SolidWorks, SolidEdge, IronCAD, Autodesk's own Inventor, and many others. These software applications often take care of the drafting, modeling, rendering, testing, presentation, animation, and manufacturing in one complete package. Therefore, AutoCAD rarely finds its way into truly high-end situations; it is simply not built for that.

In the world of architecture and related fields, however, AutoCAD 3D models often find themselves in the middle of the rendering world. These models are imported into a variety of software and are the "skeletons" on which advanced rendering and even animation techniques are applied. Studio VIZ was the affordable choice for many professionals until it was discontinued a few years back. Its replacement was the top of the line, more expensive 3ds Max Design. This software can create stunning photorealistic renderings that feature advanced use of lights, shadows, and materials. 3ds Max software can also do sophisticated animations, even to the point of full feature cartoon films, but it is not commonly used for that, as even more advanced software is available for that purpose. There are many other software solutions for creating renderings, of course. Two popular ones include Rhino and Form Z. Both software packages can generate their own shapes and solids but are often used as pure rendering engines with imported models.

Another excellent (and cost-effective) solution is to purchase add-on software that simply extends AutoCAD's built-in abilities. One well-known choice is AccuRender. It installs and integrates seamlessly with any AutoCAD, going back to AutoCAD 2002. The software creates sophisticated renderings and light effects and has a huge collection of materials, plants, and light fixtures to use. It also has a desirable feature called *Procedural Materials*. This allows materials to become part of the design, not just wrapped around it like wrapping paper. The material property penetrates through the part and it becomes that material. So a piece of wood actually has grains and jagged edges if viewed up close.

The world of animation is tied in with rendering, as the animator simply animates an already complete rendering and adjusts for changes in lights and shadows. A major player in that field is Maya, formerly of Alias Wavefront but now owned by Autodesk. Maya has extensive customizing capabilities, and studios take its core (the kernel) and write custom code around it, tuned to their specific production needs. Maya in various forms was one of the software applications behind movies such as *Jurassic Park*, *The Abyss*, and *Terminator 2*, although the amount of customization needed for such high-end productions practically turned Maya into whole new software.

The software listed thus far is just some of the high-end market products. Many other companies in the field produce software in many other price ranges. If you are interested in rendering and animation, it is a great field to get into, and many such students passed through my classroom, as they wanted to start with being able to model 3D designs (in plain old AutoCAD) and move on from there.

Appendix J

3D Printing

J.1 OVERVIEW AND HISTORY

3D printing is not really "printing" as most people use and understand that word; a more descriptive name is *additive manufacturing*. It is a process where material is precisely ejected from a specialized computer-controlled machine and added together, layer by layer, until a 3D object is produced. The data source used to create this object is a 3D CAD model or sometimes a 3D scan of an existing object; and the material used in the manufacture can be almost anything, as long as it is suitable for projecting out of a nozzle or something similar (a liquid, gel, or powder would work). It must then, of course, be able to solidify into a stable physical object that you can handle and use. That, in simplified terms, is the essence of 3D printing.

This technology has received a lot of publicity in recent years, but it is not as new as it seems. Evolving from the science of *stereolithography* (pioneered by Chuck Hall in 1984), early versions of additive manufacturing machines used soft polymers (plastics) as the source material. These polymers would be melted down before passing through a nozzle and solidify (or were cured) rapidly after being built up into layers, eventually forming an object. Polymers remain the most popular source material, but other materials (even edible food items) were used in the 1990s and 2000s, as the technology matured and dropped drastically in price. There are now home-based, relatively affordable 3D printers for hobbyists and entrepreneurs. Uses for 3D printing vary widely, with primary applications in product development, prototyping, biotech and dental devices, art, and education, among others. It is a rapidly growing field and one to keep an eye on.

There are four distinct parts of a 3D printing process, as described in the following.

J.2 PRINTING PROCESS—PART 1 (CREATING THE MODEL)

The first step in creating a 3D-printed model is to create a viable model of what you want to make. You can go about it in several different ways. The most direct is to design something yourself from scratch via any of the popular 3D solid modeling software packages, such as CATIA, SolidWorks (or AutoCAD). Alternatively, you can go on with one of the several 3D printing "marketplace" websites from companies such as Thingiverse, Shapeways, or Threeding and find, trade, or buy a premade model of something you want. Finally, if you want a 3D model of something you already have (in other words, you just want to replicate it), then you can also do a 3D scan of the object via a 3D scanner. These types of advanced scanners use lasers or soft touch probes to map the surface shape of an object, creating a resultant point cloud, then use a reconstruction process to build up a 3D model of it.

J.3 PRINTING PROCESS—PART 2 (EXPORTING AND PREPARING THE MODEL)

Once the model is created it needs to be exported to an STL (STereoLithography) file, which is the native language format for 3D printers and generally most computer-aided manufacturing and rapid prototyping software. Most solid modeling and drafting software can export STL, including AutoCAD.

There are a few ways to export your model. The most basic way is to just select File → Export, then choose the *.stl format from the drop-down menu at the bottom of the dialog box. You can accomplish the exact same thing by typing in `3dprintservice` and press Enter or select Publish → Send to 3D print Service on the Application menu (the AutoCAD symbol at the upper left).

However, if you use this approach, then exporting is not the end of the model preparation; you need to now "slice" your model. This converts it to a series of thin layers and generates the code needed to by the 3D printer. Several commercial and open-source slicers are available, such as Skeinforge, Cura, and Simplify3D.

Starting with AutoCAD 2017 there is another option—but only if you are running a 64-bit system; you are out of luck on a 32-bit one. This new approach is a separate application called *Print Studio*. It shipped with AutoCAD 2017 (should also come with AutoCAD 2019) but needs to be installed separately. If it is there, you will see the symbol for it in the Output tab of the Ribbon (in the 3D Modeling Workspace) or you can use the Application menu, Print → 3D Print. If it is not installed, you will be given the opportunity to install it. This application preps up your model and sends it to a 3D printer, if one is hooked up to your computer. If your AutoCAD has it ready to go, try it out!

J.4 PRINTING PROCESS—PART 3 (PRINTING THE MODEL)

The actual printing itself is almost fully automated and is the relatively easy part of the process, where after hours of designing and testing a model (if you did that vs. just purchasing one), you finally just get to watch it being built. There are a number of popular printing approaches of varying complexity and cost. Many of them continue to use historically proven polymer, powder, gel, or small beads as source material, heated up for easy passage through an applicator nozzle then rapidly cooled to form the solid object. The source material may be coiled up and unfurled into the nozzle mechanism (or be gravity fed from a hopper) at a steady pace.

Extrusion deposition uses this exact approach with polymers, which can be styrene, polycarbonate, polyethylene, or polystyrene. Other methods use liquids or, in the case of *bioprinting* (an active research field), even living tissue.

Regardless of the source material, the actual printing process is very similar. The movable nozzle passes over the base plate (on which the model rests), ejects material, and builds up the model one pass at a time, using the slices as a guide (Fig. J.1). The printer size varies widely, from smaller "desktop" units to larger commercial machines, including a very large unit called the *Big Area Additive Manufacturing machine*, used in several industries such as aerospace.

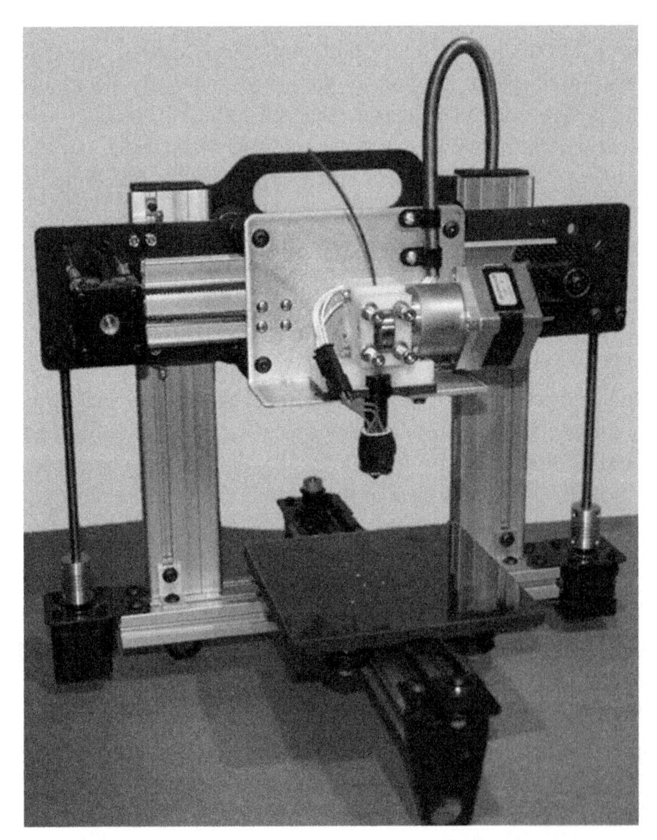

FIGURE J.1 A basic 3D printer. *Source: ORDbot Quantum.*

J.5 PRINTING PROCESS—PART 4 (FINISHING MODEL)

Once the model is completed, depending on the material, it may need to sit on the base plate for a while to cure fully. It then needs to be "finished," which can range from just a quick polish all the way to additional material removal. In latter case, material is removed because the model is printed slightly oversized. This is because a greater degree of accuracy and resolution can be achieved in the final product by removing some material in a secondary step, as opposed to trying to achieve this in the original manufacturing process. Finally, these 3D models can be coated, painted, or otherwise "postprocessed" for use or display (Fig. J.2).

FIGURE J.2 A 3D-printed architectural model. *Source: Rietveld Architects NY.*

J.6 APPLICATIONS, RELATED ISSUES, AND THE FUTURE OF 3D PRINTING

Needless to say, all of this just scratches the surface of this emerging technology. 3D printing has already found use in the manufacture of consumer apparel, automotive body panels, construction blocks, hand tools, medical implants, and of course extensive use in art, design, and for educational purposes (Fig. J.3).

The overall speed with which the model can be built is not all that fast, thereby limiting the usefulness of 3D printing in mass manufacturing. This, however, is likely to be overcome in the future; and already machines with multiple applicator heads are making use of faster curing materials. Other machines use multiple types of materials at one time.

The technology has had its share of controversy as well. With lower cost do-it-yourself 3D printers becoming widely available, issues of intellectual property, trademarks, and copyrights have arisen. There has been some concern as well of 3D home manufacturing of plastic guns that can evade detection at airports. Social commentators have written about a new approach to manufacturing, where almost anyone can be a producer of an in-demand product and a centralized factory no longer is employed.

As stated earlier—and it bears repeating—this is a rapidly expanding field and one to keep an eye on. This appendix will be updated from edition to edition as this technology matures and grows.

FIGURE J.3 A 3D-printed mechanical model. *Source: Stratasys.com.*

Appendix K

AutoCAD Certification Exams

It seems like there is an exam or certification for just about every profession or occupation. Although AutoCAD and drafting in general are not really part of the greater information technology world, where certification exams are the benchmarks of competence and almost as valuable as college degrees, there still exists a series of exams for AutoCAD designers to take if they wish to "formalize" their knowledge, hang up a certificate in their office and share a digital badge from Acclaim in their online profiles.

These exams have changed significantly over the years and varied in the number of questions, difficulty level, and the amount of time given to complete them. The typical method of delivery today is online, with users logging in and taking the exam in a timed fashion, although, in years past, there have been exams with no time limit. You are NOT able to use the Help files, manuals, or texts in the exam.

Are these exams important? This question does not have an easy answer. The belief among some established AutoCAD professionals is that the exams are quite limited and do not accurately assess an individual's true mastery of the software, which can really be tested only by having the person to draw a complex design. It is partially Autodesk's fault for not developing these exams into the powerhouse institutions that Microsoft, Cisco, Novell, and others have done with theirs. There is. however, a reason to believe that, at the very least, if you are able to answer the random set of questions on these exams, then you have a good grasp of a multitude of AutoCAD topics.

Do employers ask for these certifications? Historically, there was little demand for them in the past. As AutoCAD was exploding in popularity, anyone who was reasonably good was hired and allowed to mature on the job. In recent years, as the supply of knowledgeable AutoCAD designers finally caught up with the demand, employers wanted available experts and needed a way to separate the "wheat from the chaff." This would explain the significant growth of testing for AutoCAD competence and the newfound interest from students as to how to study for and pass these exams. So, yes, these exams may be important to get a foot in the door with certain companies, but having passed the exam does not necessarily guarantee a knowledgeable AutoCAD expert and should not be the sole reason to hire someone.

Let us go over the exams as they exist now for AutoCAD 2018; it is anticipated that the 2019 exams will be similar, as they are meant to be release-independent. Two levels (or tiers) are available:

- AutoCAD Certified User (ACU)
- AutoCAD Certified Professional (ACP)
- Other Autodesk Software Exams
  - Fusion 360 Certified User
  - Revit Architecture Certified User
  - Revit Architecture Certified Professional
  - AutoCAD Civil 3D Certified Professional
  - 3ds Max Certified User
  - 3ds Max Certified Professional
  - Inventor Certified User
  - Inventor Certified Professional
  - Maya Certified User
  - Maya Certified Professional

The *ACU exam* verifies entry-level skills in key Autodesk products and is designed for students and instructors who wish to demonstrate basic proficiency. It is recommended that students should have passed an AutoCAD course and have at least 50 hours of hands on experience with the software. The exam includes multiple-choice and performance-based questions. The following are some examples of topics covered in the exam. Note that these are subject to change without notice; use this list as a sample guide:

- Creating, organizing, annotating, and plotting drawings
- Manipulating, altering, and hatching objects
- Working with layouts
- Dimensioning
- Working with reusable content
- Creating additional drawing objects

The *ACP* validates more advanced skills, including complex workflow and design challenges, and is designed for students seeking a competitive advantage in a specific product area. It is recommended that students should have passed an AutoCAD course and have at least 400 hours of hands on experience with the software. The exam includes multiple-choice and performance-based questions. The following are some examples of topics covered in the exam. Note that these are subject to change without notice; use this list as a sample guide:

- Apply basic drawing skills
  - Create selection sets
  - Use coordinate systems
  - Use dynamic input, direct distance, and shortcut menus
  - Use inquiry commands
- Draw objects
  - Draw lines and rectangles
  - Draw circles, arcs, and polygons
  - Draw polylines
- Draw with accuracy
  - Work with grid and snap
  - Use object-snap tracking
  - Use coordinate systems
- Modify objects
  - Move and copy objects
  - Rotate and scale objects
  - Create and use arrays
  - Trim and extend objects
  - Offset objects
  - Mirror objects
  - Use grip editing
  - Fillet and chamfer objects
- Use additional drawing techniques
  - Draw and edit polylines
  - Apply hatches and gradients
- Organize objects
  - Change object properties
  - Alter layer assignments for objects
  - Control layer visibility
- Reuse existing content
  - Insert blocks
- Annotate drawings
  - Add and modify text
  - Use dimensions
- Layouts and printing
  - Set printing and plotting options

Both exams are timed and you must earn at least 80% to pass. Practice sample tests are available, and it is highly recommended to go over these at least to see what types of questions to expect. The cost for taking these exams is $75–$100 per exam, but that is subject to change, of course, so check the latest information on the Certiport website and with your local authorized testing center. Once you have passed the ACP, it is recommended that you maintain your certification by retaking the ACP exam once every 3 years.

Finally there is the *Autodesk Certified Instructor* (ACI) certification for those wishing to teach AutoCAD and have a credential to show those training centers that ask for it. The ACI requirements have changed dramatically starting in 2017. These requirements include obtaining access to the Autodesk Instructor Development Portal, completing an instruction methods course, completing 16 hours of software courses, and have a Certified Professional certification. There are also additional continuing education requirements to maintain your ACI status.

Information on purchasing and scheduling your exam can be found at https://www.autodesk.com/training-and-certification/certification. The exams themselves are administered by Certiport (certiport.com/autodesk).

Appendix L

AutoCAD Employment

AutoCAD students vary widely in their backgrounds. Some are already in midcareer and are registered architects or engineers looking to pick up new skills. Others are recent college graduates entering the job market with architecture or engineering degrees. These students need AutoCAD as a small part of their job, or maybe even the main part of their job, but are ultimately hired for their engineering or design knowledge and may draft in only the first few years of their career as they learn the ropes.

Another very large group of students, however, are attracted to CAD drafting in and of itself and attend one of the many programs at local community colleges or tech/vocational schools that ultimately lead to a certificate or even a 2-year associate's degree in architectural or engineering drafting. These students intend to enter the workforce as professional AutoCAD drafters, a career track in itself. This can lead to CAD management positions and serve as a springboard for a career in engineering or architecture, if the students decide to return to school to complete the bachelor's or master's degree. Even if they do not, a professional draftsperson is a well-paid professional who is a valuable asset and, as CAD software increases in sophistication, is needed more than ever. So the question is then: how do you get your "foot in the door" and begin an AutoCAD career?

Once you graduate and are ready to look for work, there are a number of things to consider. First of all, decide if you would like to concentrate on architectural, mechanical, electrical, or some other drafting field. Although I moved around a lot in my early (mid 1990s) days of consulting work, this approach does not work in today's saturated market. It is easier to succeed if you focus your interests instead of jumping around from company to company and field to field. What you ultimately decide to pursue should be something you are interested in (houses vs. mechanical devices, for example) and something you are good at. Generally, while learning AutoCAD in school, you are drawn to one type of work over another; go with your gut feeling. Then consider your options to actually find work. Two major sets of resources are presented in the following:

- *Staffing firms*: This is one of the best ways to find a CAD job, though sometimes staffing firms prefer experienced CAD designers, as those command higher salaries and therefore higher commissions and fees for the staffing agencies. These staffing firms can either place you outright with a client (collecting a onetime finder's fee) or place you on an hourly rate as one of their own workers "on loan" to a client (collecting a premium over your pay). It may or may not matter which route you go, although being hired on directly with benefits may be more desirable, even if the initial salary is lower. Search for staffing firms (also informally called *job shops*) on the internet in your area. Some well-known, large, nationwide firms include Kelly Staffing, Manpower Technical, and Aerotek.
- *Job boards*: These include Monster.com, Dice.com, and Craigslist.org, among others. Often these sites are populated with postings from staffing firms (see the preceding), but independent companies can be found as well. Simply enter AutoCAD and your city or town into the search engine and listings appear. Of special note is Craigslist.org. You may have already used it for everything from apartment hunting to selling something, and its job boards are excellent and very informal and direct. The only downside is that Craigslist.org is most utilized by residents and companies in large metropolitan centers, and if you live in a small town far from major cities, you may have slim pickings (even if the actual job scene is pretty good).

Other places to look include your school's employment office and of course social media. Sites such as Facebook and especially LinkedIn are a great way to get the word out that you are looking via a resume posting. And finally, never forget the value of word of mouth and personal contacts. Let everyone know you are looking for AutoCAD work and just maybe someone's cousin's friend knows of someone looking to hire.

Appendix M

AutoCAD Humor, Oddities, Quirks, and Easter Eggs

AutoCAD has its own brand of humor, with a lot of it being "inside jokes," best understood by those that have worked in the industry. Such is the nature of this first tongue-in-cheek listing. It's a bit outdated, but long-time users will nod in agreement, and if you are just getting started, you will soon see what the fuss is all about. Following this listing are some general engineer and architect jokes, followed by a few oddities and quirks. AutoCAD is definitely not serious all the time!

M.1 AUTOCAD, ARCHITECTURE, AND ENGINEERING HUMOR

Real AutoCAD users:

- Have been doing AutoCAD for six or more years.
- Never bother to keep .bak files.
- Have a graphics card that can support two monitors.
- Have two monitors.
- Have games on their system.
- Get bent out of shape when someone else uses their machine just to "look around."
- Know AutoLISP.
- Know how to edit an AutoLISP file.
- Wonder why Autodesk charges so much for an upgrade, and then AutoCAD gets pirated.
- Maintain at least three older versions of AutoCAD on their system.
- Never ask for the newest copy of AutoCAD—they already have it.
- Have a subscription to *Cadalyst*.
- Do not really bother to read *Cadalyst*.
- Could write an article that could go in *Cadalyst*.
- Do not run AutoCAD on a Mac.
- Do not have friends that run AutoCAD on a Mac.
- Despise Macs.
- Are not concerned with losing a drawing.
- Use the purge command about 60 times in a drawing.
- Use the spacebar, not return.
- Have drawn a picture of at least one of the following:
  - Their car.
  - Their house.
  - Their dream house.
  - Their dream house with their dream car parked in the drive.
- Like to waste time, such as like typing ZOOM, E twice in a very large drawing.
- Can talk on the phone and draw at the same time.

- Drink lots of liquid while drawing.
- Go to the bathroom a lot.
- Are not afraid to eat while drawing.
- Can leave the room while plotting.
- Do not care that the work system is on 24 hours a day; it is not their machine.
- Do not really like the system they run AutoCAD on; they are sure there is a better one.
- Would be damn lucky if they got the system they have at work, for their home.
- Always complain that AutoCAD on their system is too slow.
- Work in the dark with their AutoCAD screen color set as black.
- Get upset when someone turns on the lights.
- Listen to free downloaded MP3s while working.
- Are reading this while they should be working.

You know you are an architect if:

- The alarm clock tells you when to go to sleep.
- Coffee and cokes are tools, not treats.
- People get nauseous just by smelling your caffeine breath.
- You bought $50 magazines that you have not read yet.
- You fight with inanimate objects.
- You have used up your cameras entire SD card to photograph the sidewalk.
- You have more photographs of buildings than of actual people.
- You have taken your girlfriend/boyfriend on a date to a construction site.
- You can live without human contact, food, or daylight, but if you cannot make prints, it is chaos.
- You refer to great architects (dead or alive) by their first name, as if you knew them ... Frank, Corbu, Mies ...

You know you are an engineer if:

- You cannot write unless the paper has both horizontal and vertical lines.
- You are currently gathering the components to build your own nuclear reactor.
- You have "Dilbert" comics displayed anywhere in your work area.
- You spent more on your calculator than on your wedding ring.
- You think that when people around you yawn, it is because they did not get enough sleep.
- Your favorite character on Gilligan's Island was The Professor.
- Your IQ is higher than your weight.
- Your 3-year-old son asks why the sky is blue and you try to explain atmospheric absorption theory.
- You have no life—and can prove it mathematically.
- You have a pet named after a scientist.

Engineering rules:

1. Any design must contain at least one part that is obsolete, two parts that are unobtainable, and three parts that are still under development.
2. Nothing ever gets built on schedule or within budget.
3. A failure will not appear until a unit has passed final inspection.
4. If you cannot fix it, document it.
5. The primary function of the design engineer is to make things difficult for the fabricator and impossible for the serviceman.
6. Arguing with an engineer is a lot like wrestling in the mud with a pig. After a few minutes, you realize that he likes it.
7. To the optimist, the glass is half full. To the pessimist, the glass is half-empty. To the engineer, the glass is twice as big as it needs to be.
8. Normal people believe that if it ain't broke, don't fix it. Engineers believe that if it ain't broke, it doesn't have enough features yet.

A man was flying in a hot air balloon and realized he was lost. He reduced altitude and spotted a man below. He lowered the balloon further and shouted, "Excuse me, can you tell me where I am?"

The man below said, "Yes, you're in a hot air balloon, hovering about 30 feet above this field."

The balloonist said, with some irritation, "You must be an architect."

"I am," replied the man, surprised with his deduction. "But how did you know?"

"Because everything you have told me is technically correct, but it doesn't do me any good," said the balloonist.

The man on the ground laughed, "Well, then, you must be a building contractor."

"I am," replied the balloonist, "but how did you know that?"

"Well," said the Architect, "you don't know where you are, or where you are going, but you expect ME to be able to help you. Now, you're still in the same predicament you were in before we met, but now it's MY fault."

- A contractor, an engineer, and an architect were standing inside their recently completed building, looking out at the street. A very attractive woman walks by. The contractor whistles, the engineer says, "Did you see the legs on that woman?" The architect says, "Did I miss something, I was admiring my reflection."
- How many architects does it take to screw in a light bulb? Nobody knows for sure, it has never been witnessed.
- What is the worst name you can call an architect? An engineer (they call it the "EN" word).

M.2 ODDITIES AND QUIRKS

- AutoCAD has an "oops" command; try it to see what it does.
- AutoCAD had an "end" command. It used to shut down your drawing, even if you meant that as an OSNAP command and just forgot to type in `line` and press Enter first. End has been permanently discontinued and replaced with QUIT for obvious reasons.
- AutoCAD has a "battman" command. It has nothing to do with the caped crusader and everything to do with the Block Attribute Manager (similar to the Enhanced Attribute Manager of Chapter 18) Fig. M.1.

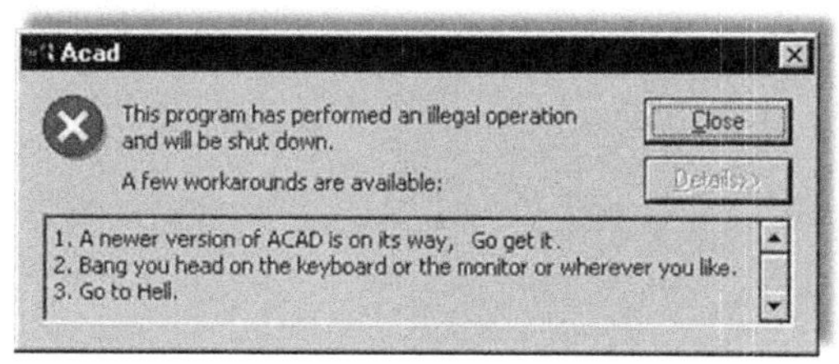

FIG. M.1

Appendix N

The Future of AutoCAD

It is a bit of a gamble to try to nail down future trends in technology, and the world of computer-aided drafting and design is no exception. One has to only look back a few years to find plenty of "experts" who were off in their predictions. New trends and innovations, appearing (and sometimes disappearing) rapidly, ensure that charting the long-term future is more guesswork than a hard science. Who would have predicted just a decade ago that AutoCAD would be again run on a Mac that we would be storing our data and even collaborating "in the cloud," and that you would be able to draft on a handheld tablet?

Still, however, it is possible to discuss where AutoCAD, as well as the CAD industry in general, may be going in the coming years by sticking to the relatively near future. In this compressed timeframe of a few years, we can look at technologies and trends that are already being implemented or will soon be here. Although the computer-aided drafting design industry as a whole can be considered relatively conservative, taking its time to implement new ideas (as opposed to the lightning fast telecom and entertainment industries), trends and directions do emerge within it, and some fundamental changes are on the way. Let us take a closer look by first revisiting a bit of history.

N.1 2D DESIGN, AUTOCAD, AND VERTICALS

Over its long history, AutoCAD has matured and grown into a stable, efficient, effective, and very popular 2D drafting application. By design, it was meant to be everything to everyone, and for many years it was. Spanning just about every design and manufacturing industry in existence, in its peak days (arguably 1994–2006), AutoCAD was found in the offices of nearly every small- to medium-size engineering and architectural firm. This was its greatest strength and, surprisingly, its greatest weakness. By being so "generic," it found a wide application in many industries, but on the flip side, it was not specialized software and could not be made such without reworking its fundamental core.

Autodesk, of course, did not want to reinvent AutoCAD and risk losing its main audience, but AutoCAD was clearly not specialized enough for many uses. Autodesk remedied this by adding "verticals" of all kinds. This was add-on software that ran on top of AutoCAD and extended its functionality for the mechanical, architectural, civil, electrical, and other markets. With Autodesk's marketing muscle, these verticals proved relatively successful, and an entire generation of CAD designers learned Mechanical Desktop, Architectural Desktop, and many others, whereas competitors closed in with specialized software of their own. A pattern emerged where just about any engineering or architectural subfield had a choice between AutoCAD plus a vertical add-on or another competitor or two to choose from.

N.2 3D DESIGN AND SOLID MODELING

Meanwhile, a parallel world was evolving and thriving on its own, quite apart from AutoCAD and most types of architecture. This was the high-end and somewhat exclusive world of 3D solid modeling. For a select group of companies with the need (and the resources), using some off-the-shelf 2D application, no matter how good, was never an option. These industries included the aerospace, automotive, naval, and rail design, among others. The products these companies were designing were simply too complex to invest in 2D drafting. When they left pencil and paper behind in the 1960s and 1970s, they moved right into the world of 3D by commissioning internal development (or purchase) of high-end solid modeling software. Though true 3D software did not appear until the early 1980s, such was the beginning of applications such as Unigraphics, CATIA, and their predecessor, the famous CADAM, a Lockheed staple.

These 3D modeling software applications, including later products such as Pro/Engineer (later renamed to Creo), NX (a merger of Unigraphics and I-Deas), SolidWorks, and others, had features and capabilities that were specialized to high-end engineering design, and they introduced new concepts along the way. For example, when designing a product with hundreds of parts (most of them interacting with each other), techniques such as part trees, parent/child relationships, and parametric design had to be invented to keep track of everything, including interferences, bills of materials, and much more. It made no sense to do anything in 2D, as parts had to be visualized and virtual models had to be built, analyzed, and tested. It was only at the very end of the process that 2D views of individual parts were produced and given to machinists to manufacture.

With the advent of computer numerical code even that was no longer necessary, information could be transmitted directly from a model to the manufacturing machine, with a machinist overseeing setup and quality control. All these techniques developed slowly and deliberately in response to a clear need in the industry. These software applications were complex, expensive, and virtually unrecognizable to an AutoCAD user. They also often required specialized training, and to this day, you generally do not sit down to learn CATIA or NX unless you are a working (or student) engineer. There really was no industry of pure drafters as there was with AutoCAD. You had to know what you were doing in your specialty and also how to interpret the results of your design or analysis.

N.3 MERGING OF INDUSTRIES

If you have downloaded and studied the 3D chapters of this textbook, you may already see where this narrative is going. It was inevitable that the folks from the 2D architecture, engineering, and construction world would take a look at their peers in the 3D world and ask, "Can we also do this?" The fundamental premise they saw was very simple: We all live in a 3D world, and this is the way all design should be done. Nowhere was this more evident than in the world of architecture. As early as the late 1980s, visionaries such as Frank Geary commissioned Dassault, the makers of CATIA, to modify their product for architecture/building design. Architects over the world still designed everything in 2D (some still via pencil and paper), and 3D modeling was the rare exception, not the rule. Because AutoCAD was the undisputed software of choice among architects and interest was building in a 3D approach, the ultimate clash of industries was bound to happen.

Autodesk's response was to borrow a number of techniques and incorporate them into the 2D and 3D capabilities of AutoCAD, a fact mentioned in a few chapters of this text. Dimension-driven design (see , Chapter 13) and 3D-to-2D layouts (see Chapter 27 of the 3D sections) are just some of the features that filtered their way down from solid modeling into plain AutoCAD. Clearly, however, AutoCAD itself was not fundamentally changing; and another approach was needed to fill the need of customers who wanted to take things into the 3D world, both in engineering and architecture. As they were doing this, competitors wasted no time in pushing their own solutions. In the world of solid modeling, numerous mid- to low-priced products for engineers emerged, such as IronCAD, IntelliCAD, SolidWorks, and Solid Edge. In the world of architecture and design, ArchiCAD and even Rhino touted their 3D capabilities. Specialized software was also available from MicroStation, and these companies captured a significant portion of the civil engineering market. Autodesk came back strong with Inventor, a successful solid modeling software of its own. But the biggest jump toward merging 2D and 3D design was yet to come.

N.4 REVIT AND BUILDING INFORMATION MODELING

If migrating to 3D was all there was to architecture and building design that would have been done a long time ago. Something else was missing from AutoCAD and similar 2D programs, a set of features that had been present in 3D solid modeling software for quite some time. The architecture world wanted those features as much, if not more, than the 3D capabilities alone. These features are best described as "relational databases" and "parametric change engines."

Simply put, solid modeling software had far more intelligence than drafting software. Objects in solid modeling were all part of a database that tracked their positions and attributes. Everything created was part of something bigger, and everything was linked together. If you created a bracket in 3D, the 2D views were already part of that bracket, the part was tracked in a bill of materials, and its physical properties were also cataloged. If you wanted a section view, e.g., it was just a click away. The bracket was not just a collection of lines, it was a dynamic entity.

To bring these features to the architecture world was the premise behind Revit, a "solid-modeling-for-architects" software acquired by Autodesk in 2002 (originally developed by Revit Technology Corporation). Technically, Revit was what is referred to as building information modeling (BIM) software. This type of software goes one huge step beyond 2D line drafting of a facility by digitally representing all of its physical and functional characteristics in a powerful relational database. This type of software is useful not just during the design phase but throughout the entire life cycle of the facility, including its management and operation. Revit (which is short for REVise InsTantly) is, of course, not the only such software on the market; and MicroStation, ArchiCAD, Vectorworks, ARCHIBUS, and CATIA's Digital Project are all viable competitors. Revit, however, is the leading software among its peers and is credited with bringing BIM to a new generation of users.

Revit, and other available BIM software, has revolutionized building and facility design. Solid building models are first constructed immediately in 3D. Then, section, details, and a bill of materials can easily be generated by "slicing" the model up into 2D views. All views are linked together, and any changes are instantly updated throughout the entire project. An architect/designer can now keep track of all entities (doors, windows, etc.) in the design and link cost and materials to them. Square footages and even energy usages can now be tracked in real time as the design progresses. Time and cost are also factored in (the fourth and fifth dimensions in BIM). Finally, Revit has advanced rendering tools and can produce excellent rendered images without third-party software.

All this was already incorporated into mechanical design for years, but only relatively recently it was widely available to architects. Considering the advantages, such as waste minimization, shorter turnaround times, and better error detection, it is surprising this was not available sooner. There are, of course, some disadvantages to BIM software such as Revit, with the chief among them being the significant retraining and upfront investment needed. It may also not be the right fit for very small projects, where basic AutoCAD is more than enough.

N.5 THE FUTURE OF AUTOCAD

So, after all this discussion of solid modeling, Revit, and BIM, what is the future of AutoCAD? It is probably better to split this into two questions: What is the future of AutoCAD itself, and what is the future of computer-aided design in general, in light of this overall trend toward 3D? Will AutoCAD stay the way it is? Will it morph into a different software altogether?

The overall consensus is a resounding "No!" AutoCAD enjoys a very special place in the overall scheme of things. It is a very well-established and successful 2D drafting package. There will always be a need for this software in its present form. Too many clients in the design world require nothing more than what AutoCAD already offers. By being so good at the simple and basic task of drafting, AutoCAD will remain relevant for the foreseeable future. It is unlikely Autodesk will significantly tinker with its formula for success and its main cash cow.

However, Autodesk is changing AutoCAD in small ways. It is, e.g., incorporating as much advanced 3D as possible without fundamentally changing the identity of the software. Autodesk is also doing its best to keep AutoCAD looking modern and cutting edge by incorporating such design techniques as "heads up" display for data entry. It is also trying its best to eliminate the archaic-looking command line—but for the sake of the legacy users, the command line is expected to remain available, if needed. AutoCAD has also followed basic software modernization trends over the years, such as incorporating the Microsoft Office–like Ribbon Interface design style and being offered for sale via download as opposed to physical DVDs or CDs.

The industry itself, however, is moving ever slowly, but surely, toward 3D solid modeling and BIM. A wave is spreading across the architectural profession to incorporate BIM into basic design from the ground up and not treat this approach as something exotic but rather as a new fundamental approach. Autodesk clearly recognizes this trend, and its Revit software is well positioned in the market. Expect to see this trend continue, and if you are inclined toward architectural drafting, you may be well advised to also learn Revit or another BIM-oriented software, such as ArchiCAD, Digital Project Designer, or Bentley Architecture.

In summary, it is doubtful AutoCAD itself will ever morph into another software under pressure from the industry trends; it is simply too profitable, and it will remain relevant "as is" for years to come. Autodesk will continue its yearly update and "sprucing up" of the interface and adding more and more 3D features up to the limit of its ACIS modeling kernel, but its core will remain the same. AutoCAD's market share, however, will likely shrink in the coming years, with some of that going to Revit and other BIM software. Some mechanical engineers will also switch to 3D, but since that type of modeling software was already available for a long time, the main drift away from AutoCAD will be in the architectural world, not engineering.

As with any predictions, it remains to be seen how things will actually turn out, but the overall consensus is that the just-discussed scenarios are the most likely for the immediate future.

Index

'*Note:* Page numbers followed by "f" indicate figures, "t" indicate tables.'

B

C

D

E

J

K

L

M

Q

R

Printed and bound by CPI Group (UK) Ltd, Croydon, CR0 4YY
08/06/2025
01896877-0003